ACCIDENT PREVENTION MANUAL
for Business & Industry

Environmental Management

Occupational Safety and Health Series

The National Safety Council's OCCUPATIONAL SAFETY AND HEALTH SERIES is composed of five volumes and two study guides written to help readers establish and maintain safety and health programs. The latest information on establishing priorities, collecting and analyzing data to help identify problems, and developing methods and procedures to reduce or eliminate illness and accidents, thus mitigating injury and minimizing economic loss resulting from accidents, is contained in all volumes in the series:

ACCIDENT PREVENTION MANUAL FOR BUSINESS & INDUSTRY
  (3-volume set)
  Administration & Programs
  Engineering & Technology
  Environmental Management
STUDY GUIDE: ACCIDENT PREVENTION MANUAL FOR BUSINESS &
  INDUSTRY: Administration & Programs, Engineering & Technology
OCCUPATIONAL HEALTH & SAFETY
FUNDAMENTALS OF INDUSTRIAL HYGIENE
STUDY GUIDE: FUNDAMENTALS OF INDUSTRIAL HYGIENE

Other safety and health references published by the Council include:
ACCIDENT FACTS (published annually)
LOCKOUT/TAGOUT: THE PROCESS OF CONTROLLING HAZARDOUS ENERGY
SUPERVISORS' SAFETY MANUAL
OUT IN FRONT: EFFECTIVE SUPERVISION IN THE WORKPLACE
PRODUCT SAFETY: MANAGEMENT GUIDELINES
OSHA BLOODBORNE PATHOGENS EXPOSURE CONTROL PLAN (National Safety
  Council/CRC-Lewis Publication)
COMPLETE CONFINED SPACES HANDBOOK (National Safety Council/CRC-LEWIS
  Publication)

ACCIDENT PREVENTION MANUAL
for Business & Industry

Environmental Management

Edited by Gary R. Krieger, MD, MPH, DABT

With a Foreword by William D. Ruckelshaus
Former Administrator, Environmental Protection Agency

National Safety Council

Itasca, Illinois

Editor-in-Chief: Gary R. Krieger
Project Editor: Patricia M. Laing
Associate Editor: Julie McIlvenny
Technical Adviser: Dale C. Haskin

DISCLAIMER

Although the information and recommendations contained in this publication have been compiled from sources believed to be reliable, the National Safety Council makes no guarantee as to, and assumes no responsibility for, the correctness, sufficiency, or completeness of such information or recommendations. Other or additional safety measures may be required under particular circumstances.

©1995 by the National Safety Council
All Rights Reserved
Printed in the United States of America
00 99 98 97 96   5 4 3 2 1

Library of Congress Cataloging-in-Publication Data
Accident prevention manual for business & industry : environmental
    management / edited by Gary R. Krieger ; with a foreword by William
    D. Ruckelshaus.
        p.   cm. — (Occupational safety and health series)
    Includes bibliographical references and index.
    ISBN 0-87912-170-X
    1. Industry—Environmental aspects.   2. Environmental protection.
    I. Krieger, Gary R.   II. Series: Occupational safety and health
    series (Chicago, Ill.)
    TD194.A27   1995
    658.4'08—dc20   94-19061

5M894   Product Number: 12156-0000

Contents

Foreword

"Humanity has the ability to make development sustainable—to ensure that it meets the needs of the present without compromising the ability of future generations to meet their own needs." *(Our Common Future, The Report of the World Commission on Environment and Development)*

As the World Commission on Environment and Development has pointed out, the environmental and economic issues of today must be managed so that the environmental, economic, and social needs of future generations can be met. Over the last 20 years, our understanding of business and industry's potential and actual impact on the global environment has significantly increased. Appropriately, this third volume of the National Safety Council's *Accident Prevention Manual* series *(APM)* is entirely dedicated to environmental management. It is a response to the need to understand and comply with national and international environmental regulations while simultaneously anticipating and planning for major trends in environmental science and policy.

To develop and promote a long-term strategy that emphasizes pollution prevention and sustainable development, an understanding of national and international regulatory and policy issues is essential. This volume provides detailed discussions of regulations, technical issues, and future policy directions so that economically viable and environmentally progressive policies can be developed and implemented within a national and global context.

Business leaders have discovered that a technically sound and forward-looking environmental program can provide a competitive advantage and new business opportunities while enhancing employee morale, investor support, and community acceptance. The National Safety Council's *APM: Environmental Management* is a valuable guide and resource—an important addition to your library.

WILLIAM D. RUCKELSHAUS
Former Administrator, Environmental
Protection Agency
(December 4, 1970, to April 30, 1973, and
May 18, 1983, to January 7, 1985)

Preface

This volume of the National Safety Council's *Accident Prevention Manual* is much like the mythical Greek figure Janus who simultaneously looked backward and forward. This dual perspective is appropriate because the future directions of the environmental movement will be driven by the past attitudes and opinions formed by the public and various worldwide legislative bodies. For an environmental professional or a newcomer to this field, it is critical to develop a broad understanding of the political, economic, ethical, scientific, and regulatory aspects of this broad and ever-expanding field.

The approach to this volume of the *Accident Prevention Manual* has been to present and discuss those topics and trends which we feel most impact the contemporary state of environmental affairs and are most likely to influence future directions. Our goal is to provide both new and experienced environmental professionals with enough basic information and analysis so they can successfully understand, initiate, and direct appropriate actions for their business or industry.

One goal of good management practice is the ability to anticipate trends and formulate timely and appropriate responses. To help facilitate this goal, we have organized the *APM: Environmental Management* into three general but sequential parts: (1) General Framework, (2) Hazardous Wastes, and (3) Special Concerns. Part 1 is divided into seven chapters covering the fundamental areas in the field of environmental affairs. Chapter 1 focuses on the political and historical roots of both the United States and international environmental movement. Chapter 2 discusses economic and ethical aspects of environmental decisions. Many observers see these two areas as mutually incompatible; however, both are essential for proactive and sound management of a company's environmental program. Chapters 3 and 4 cover the United States and international legal and legislative framework of environmental laws and policies, respectively.

Although this book is primarily directed toward U.S. issues, the United States is part of a global economy and many environmental issues cross national borders and require multinational approaches. Hence, the international aspects of environmental issues are emphasized in most chapters. Chapter 5 is a presentation of basic science concepts that govern the movement of potentially hazardous materials in air, water, and soil. This chapter is particularly important because many environmental managers are unfamiliar with the physical and chemical processes that move materials in various environmental media. Chapters 6 and 7 are directed toward management, strategy, and assessment and audit techniques. A fundamental paradigm of program management is the assess-plan-do-verify circle. Chapter 6 illustrates how to develop an environmental strategy and management approach, while Chapter 7 presents the fundamentals of environmental auditing techniques and concludes the first part.

Part 2, Hazardous Wastes, is directed toward an understanding of the complex laws and techniques used to manage the treatment, storage, and disposal of hazardous wastes. Chapter 8 is a detailed discussion of these laws and techniques and covers both chemical and radioactive wastes. Chapter 9 presents information on the training requirements surrounding the hazardous waste laws in the United States. Finally, Chapter 10 discusses both the growing importance of the pollution prevention and waste minimization movement and some of the emerging techniques used to treat waste materials.

Part 3, Special Concerns, is directed toward broader issues that impact both the external and internal environment in which we live. Chapters 11 and 12 deal with the external sphere of public health, epidemiology, risk communication, and risk assessment. Increasingly, regulatory and legal decisions are made on the basis of quantitative risk assessment techniques that examine both human and potentially exposed ecological systems. The indoor environment and the challenges of indoor air quality evaluations are discussed in Chapter 13. Finally, Chapter 14 analyzes the broader problems of sustainable development for the external environment, covering such issues as population growth, deforestation, soil depletion, acid rain, ozone depletion, and global warming so that environmental professionals can have a dispassionate and rational understanding of the scientific and policy approach. A series of illustrative case studies, additional references and sources of help, and glossaries of technical terms and acronyms are presented in the appendixes.

Overall, it is our attempt to present some of the key concepts and concerns that are driving the environmental field into the 21st century. Like Janus, we are understanding, reviewing, and evaluating the past so that we can look forward and appropriately prepare for and manage sustainable development in the future.

Acknowledgments

I would like to acknowledge the expert editorial advice and assistance given to me by the staff of the National Safety Council. This book would not have been developed and sheparded through the complex process of review, revision, and production without the unfailing efforts of Pat Laing and Jodey Schonfeld of the Council. In addition, I would like to thank the staff of the Dames & Moore Denver office. Particularly, Luann Langley, Sheree Robertson, Marci Balge, Nancy Huettl, Dave Hinrichs, and Peter Sinton, who provided technical reviews, logistical support, and encouragement during the writing and production phases of this book. Finally, no acknowledgment would be complete without extending my appreciation to my wife, Jeanne, and children, Lauren and Taylor, who put up with the phone calls, papers, books, and time commitment associated with the effort required to write a book. I take responsibility for any errors of commission or omission since, to quote the 1st century B.C. poet Horace, "A word, once sent abroad, flies irrevocably."

The National Safety Council and I would like to thank the following staff and volunteers who devoted their time and expertise to a technical review of the initial manuscript of this manual: J. David Amos, Peter D. Bowen, Corey Briggs, Lee M. Feldstein, William R. Gallagher, Michael Gaskill, Jay S. Goldstein, Jordan S. Harwood, Dale C. Haskin, Ronald J. Koziol, Richard S. Kraus, Robert D. Langtry, David Lee MacAllister, James P. McBain, John McHale, Pamela Murcell, Jill Niland, Fred Rine, Donald Robinson, Homer Sterner, and Bud Ward.

GARY R. KRIEGER, MD, MPH, DABT
Editor-in-Chief

Contributors

Deborah Anne Amidaneau, MS, PhD Candidate; environmental section manager/NEPA Compliance Officer, National Renewable Energy Laboratory. Ms. Amidaneau has experience in environmental sciences, geology, biology, engineering, and law gained through projects with the Forest Service; Bureau of Mines; private sector oil, gas, and mining industries; and public utilities. Her responsibilities include management of all environmental programs, inclusive of air, groundwater, surface water, drinking water monitoring, regulatory compliance, hazardous waste management and minimization, federal, state, and local permitting, remediation construction monitoring, NEPA compliance, and environmental restoration research. Compliance Officer responsibilities include evaluating proposed actions, including research projects, to identify the level of environmental regulation compliance required, evaluate proposed actions for compliance with the NEPA to identify the need for Environmental Impact Statements or Environmental Assessments. She is currently also an Adjunct Professor at the University of Denver and has conducted research in identification of new botanical species sensitive to metal concentrations due to contamination from old mining activities. Mailing address: Manager, Environmental Section, National Renewable Energy Laboratory, 1617 Cole Boulevard, Golden, CO 80401.

D. Ian Austin, PhD; consultant, Managing Principal-in-Charge, Dames & Moore, Tokyo. Dr. Austin has provided consultation for environmental assessments and environmental modeling for coastal projects and subsurface water quality studies, including those involving groundwater contamination, discharges into rivers and coastal waters, and oil spill risk. He has directed preliminary site assessments, audits, investigative site assessments and hazardous waste remediation projects in Japan and Korea. He has also been senior engineer and principal investigator in numerous projects in the United States, Abu Dhabi, Chile, and Peru. Mailing address: Dames & Moore, Asahi Building, 7th Floor, 38-3 Kamata 5-Chome, Ota-ku, Tokyo 144, Japan.

Stephen W. Bell, MBA, CSP, BSCE; manager, Safety, Health, Environmental and Claim Service, Hylant MacLean, Inc. Mr. Bell is Chairman of the National Safety Council's Occupational Health Hazards Committee and 1991 recipient of the Society of Safety Engineers' Safety Professional of the year award for Region V. He previously held positions of director, Safety and Loss Control, Waste Management of North America; and manager, Safety and Health, Quaker Oats Company. Mailing address: Manager, Safety, Health, Environmental and Claim Service, Hylant MacLean, Inc., 1505 Jefferson Avenue, PO Box 1687, Toledo, OH 43603-1687.

D. Jeff Burton, MS in Industrial and Environmental Health, PE, CSP, CIH; Adj. Associate Professor, Mechanical Engineering Department and Assistant Professor, College of Medicine, University of Utah. Mr. Burton is the author of workbooks and self-directed study programs in ventilation, IAQ, and industrial hygiene and has over 20 years experience in projects related to the maintenance of IAQ in occupational settings. Mailing address: IVE, Inc., 2974 South Oakwood Drive (2974 S, 900 E), Bountiful, UT 84010.

W. Douglas Costain, PhD; Senior Instructor, Political Science Department, University of Colorado/Boulder. Dr. Costain teaches American government and environmental politics. He has also held faculty positions at the University of Pittsburgh and Colorado State University. He has published articles in political science journals and book chapters on interest groups, social movements, and lobbying. Mailing address: Political Science Department, Campus Box 333, University of Colorado/Boulder, Boulder, CO 80309-0333.

John G. Danby, CIH; manager, Western Division Health and Safety, Dames & Moore. Mr. Danby has hazardous waste operations experience in both private sector environmental engineering firms and public sector environmental compliance agencies. He developed a guidance document for the preparation of hazardous waste site health and safety plans for the California Department of Toxic Substances Control. Mailing address: Dames & Moore, 8801 Folsom Boulevard, Suite 200, Sacramento, CA 95826.

Joan Gibb Engel, PhD; author and editor of essays reflecting cultural perceptions of nature, combining the fields of environment, culture, and natural history. Dr. Engel is co-author of the ethics section of Chapter 2, Economic and Ethical Issues and co-editor of *Ethics of Environment and Development* (Tucson: University of Arizona Press, 1990) and an annotated bibliography on environment, justice, and religion. Dr. Engel has taught at the University of

Illinois at Chicago and in public schools in the Chicagoland area. Mailing address: P.O. Box 717, Beverly Shores, IN 46301.

J. Ronald Engel, PhD; Professor of Social Ethics, Meadville/Lombard Theological School; Lecturer in Ethics and Society, Divinity School, University of Chicago. Professor Engel is co-author of the ethics section of Chapter 2, Economic and Ethical Issues; chair of the Ethics Working Group of IUCN (The World Conservation Union), headquartered in Gland, Switzerland; and a member of the executive committee of the International Development Ethics Association, the editorial board of the *American Journal of Theology and Philosophy*, and the advisory board of Beacon Press. He has served as consultant to the biosphere reserve project of UNESCO Man and Biosphere programme (MAB) and is currently doing research for the Libby Endowment on the spirit/nature split in Western thought. He is author of *Sacred Sands: The Struggle for Community in the Indiana Dunes* (Wesleyan University Press, 1983), numerous articles on religion, ethics, and the environment, and co-editor of *Ethics of Environment and Development* (Tucson: University of Arizona Press, 1990). Mailing address: Meadville/Lombard Theological School, 5701 South Woodlawn Ave., Chicago, IL 60637.

Susan Gesoff, CHMM; senior administrator-environmental, American Airlines. Ms. Gesoff has experience as a manufacturing engineer and environmental analyst. Ms. Gesoff joined American Airlines in 1990 as an Analyst in their IdeAAs Program and joined the Environmental Department following its creation in 1991 as an Environmental Analyst. She was a manufacturing engineer with the Boeing Helicopter Company and worked for Texas Instruments Defense and Electronics Group. Mailing address: Environmental Department, American Airlines, PO Box 619616 M/D 5425, Dallas/Ft. Worth, TX 75261-9616.

Joseph W. Gordon; Unit Leader, Government Risk Assessment Services group, Dames & Moore. Mr. Gordon has over 12 years of experience in analyzing human health, safety, and environmental risks using state-of-the-art transport, dispersion, and dosimetry models to evaluate the fate and risk of hazardous and radioactive contaminants in the biosphere. He has considerable experience in performing safety analyses, environmental impact assessments, and risk assessments related to facilities that process transuranic, low-level radioactive and/or hazardous materials. He has managed several large decontamination/decommissioning and WM projects at U.S. nuclear facilities and sites. He is a contributing author to many PSARs/FSARs, Public Health Risk Assessments, and NEPA documents at many DOE sites and is a project manager for several current Baseline Health Risk Assessments at the Rocky Flats Plant and Idaho National Environmental Laboratory. Mailing address: Dames & Moore, First Interstate Tower North, 633 Seventeenth Street, Suite 2500, Denver, CO 80202.

A. Roger Greenway, MS in Meteorology and Environmental Science, MBA, CCM; Principal of Dames & Moore, Cranford, NJ. He has over 25 years experience in environmental consulting, including more than 8 years in asbestos and indoor air quality. He is a Certified Consulting Meteorologist, licensed asbestos Project Designer, and a licensed Asbestos Safety Technician. Mr. Greenway's projects include air modeling and permitting for coal-fired electric generating stations, cogeneration projects, direction of air emission inventory development projects, design of control systems, and risk assessment studies. He has directed several hundred asbestos-related projects, including building surveys, development of management plans, preparation on removal specifications, and the monitoring of asbestos removal projects. He has published over 50 technical papers and presentations and has spoken at national conferences of the AWMA, AMS, National Academy of Sciences, HazMat, and other forums. He has organized and presented courses for government institutes and exec enterprises for over 5 years. He has testified before many agencies and boards, including the staff of the U.S. Senate and House of Representatives committees working on Clean Air Act revisions. Mailing address: Dames & Moore, 12 Commerce Drive, Cranford, NJ 07016-1101.

William L. Hall, PE; principal, Dames & Moore. Mr. Hall's experience includes roles as project manager and principal investigator for all phases of water and land resources planning and design, including harbor hydraulic studies, dredge disposal studies, hydraulic structure design, dam design, resource utilization, wastewater modeling studies and solid waste disposal. His responsibilities have included water resources and solid waste disposal facilities designs and studies in Africa, the Middle East, Europe, the Caribbean, and over 20 states. Mailing address: Dames & Moore, 6 Piedmont Center, Suite 500, 3525 Piedmont Road, Atlanta, GA 30305.

Holly A. Hattemer-Frey, MS; senior risk assessment scientist, Dames & Moore. Ms. Hattemer-Frey has more than eight years experience in the areas of human health and ecological risk analysis and exposure assessment, has authored multiple publications and technical reports. In addition, she co-edited *Health Effects of Municipal Waste Incineration* and edited a special edition of *Risk Analysis* on municipal waste incineration. While working as an Environmental Scientist with Oak Ridge National Laboratory's Office of Risk Analysis, she gained experience in assessing human exposure to organics through the food chain, evaluating human health risks associated with municipal solid waste incineration, evaluating the potential human impacts of genetically-altered organisms, and in using pharmacokinetics to improve the risk assessment process. Mailing address: 1100 Sanders Road, Knoxville, TN 37923.

Beth A. Hayes, manager, Waste Minimization, American Airlines. Ms. Hayes has experience as an industrial engineer and a speaker on recycling topics at the World Recycling Conference, national recycling Coalition Conference, and the National Safety Council. She is coordinating American Airlines' "Buy Recycled Campaign." Mailing address: Manager, Waste Minimization, American Airlines, PO Box 619616 M/D 5425, Dallas/Ft. Worth, TX 75261-9616.

Robert J. Hollingsworth, MBA, CSP, SPHR; Instructor, Engineering Technology Department, Western Washington University. Mr. Hollingsworth has 28 years experience in the primary metals industry with management assignments in safety, industrial hygiene, security, fire prevention, workers' compensation, medical, environmental affairs, and training. He is past president and is serving a fifth term on the State of Washington Governor's Industrial Safety and Health Board. Mailing address: 1877 Academy Road, Bellingham, WA 98226.

William J. Keane, PhD, CIH; president, Enviro Dynamics, Inc. Dr. Keane has over 20 years experience as an industrial hygienist, chemist, and manager. He is a pioneer in the development of occupational and environmental health computer systems. He has consulted for corporations, government and research agencies, and financial and academic institutions on projects including occupational health system needs analysis, risk assessment, expert testimony, crisis management, asbestos exposure evaluation and abatement procedures, facility design evaluation, engineering control recommendations, facility monitoring programs, indoor air quality evaluation, and employee training. Mailing address: Enviro Dynamics, Inc., 520 North Washington Street, Suite 300, Falls Church, VA 22046-3538.

Janet E. Kester, PhD; senior toxicologist, Dames & Moore, Dr. Kester works in the Toxicology, Risk Assessment, and Environmental Statistics Group at Dames & Moore. She has 10 years experience in toxicology, ecological, and human health exposure and risk assessment, and litigation support. She provides technical coordination and expert toxicology/risk assessment support for projects involving a variety of sites, chemicals, and receptors. Dr. Kester served as Adjunct Professor of Toxicology at the University of Rochester, where she developed and taught graduate courses in basic toxicology. She has published a number of research and symposium papers and book chapters on various aspects of toxicology and risk assessment. Mailing address: Dames & Moore, First Interstate Tower North, 633 Seventeenth Street, Suite 2500, Denver, CO 80202.

Brian J. King, JD; attorney, Holland & Hart. Mr. King has been practicing environmental, natural resource, and worker safety law throughout his 14-year legal career. He has represented clients in environmental and OSHA enforcement proceedings, environmental permit appeals, and Superfund litigation. He is also very active in the areas of environmental permitting, environmental auditing, and environmental compliance and works with clients to improve the quality of their environmental compliance programs. Prior to joining Holland & Hart, Mr. King served as associate general counsel for Boise Cascade Corporation. Mailing address: Holland & Hart, PO Box 2527, Boise, ID 83701-2527.

Gary R. Krieger, MD, MPH in Occupational Medicine/Toxicology, DABT; manager and Principal, Health Systems Group, Dames & Moore. Dr. Krieger holds simultaneous certification by the American Boards of Internal Medicine, Preventive (Occupational) Medicine, and Toxicology. He is an Assistant Professor of Toxicology at the University of Colorado, Graduate School and has served as a visiting professor at Johns Hopkins School of Hygiene and Public Health and the Mayo Clinic. He has extensive academic training in occupational medicine and toxicology as applied to hazardous waste investigations, 10 years full-time experience in the fields of public health/environmental toxicology and risk assessment; and more than seven years experience with RCRA facility investigations and CERCLA risk assessments. He has Public Forum experience as an expert witness on incinerators, cogeneration, and risk assessment of hazardous materials; has developed risk assessments for heavy metals, PCBs, dioxins, chlorinated and nonchlorinated solvents, asbestos, radioactive materials total petroleum hydrocarbons, and polyaromatic hydrocarbons; and has written many publications and presentations, including 8 chapters of *Hazardous Materials Toxicology: Clinical Principles of Environmental Health,* of which he is co-editor with J.B. Sullivan (Williams & Wilkens, 1991. Mailing address: Manager, Health Services Group, Dames & Moore, First Interstate Tower North, 633 Seventeenth Street, Suite 2500, Denver, CO 80202.

Mark J. Logsdon, MS; President and Principal Geochemist of Geochimica, Inc. He has 22 years of professional experience in applying geochemistry and hydrology to the environment and is a consultant in hydrogeochemistry, hydrology, geochemistry, and related earth sciences. He has worked with mining, energy, and industrial sector corporations, as well as native American tribes and federal, state, and local agencies on projects including hydrologic and geochemical studies of acid mine drainage, cyanide hydrogeochemistry, geochemical modeling programs, water-quality site investigations, studies of LNAPL and DNAPL contamination of soils and waters, radiological contamination investigations, and design and execution of remedial action programs. Mailing address: Geochimica, Inc., 1635 Downing Street, Suite 200, Denver, CO 80218.

Charles N. Lovinski, MS; executive director, Training and Consulting, National Safety Council. Mr. Lovinski has 20 years experience in safety and health positions ranging from Supervisor of Safety to Corporate Senior Manager of Safety and Hazardous Materials. He is a past chairman of the Air Transport

Association Dangerous Goods Board, member of the U.S. Office of Technology Assessment Committee for Hazardous Materials, and a Board Member of International Air Transportation Association. Mailing address: National Safety Council, 1121 Spring Lake Drive, Itasca, IL 60143-3201.

Robert J. Marecek, MALS; manager, Library, National Safety Council. Mr. Marecek received his master's degree in library and information science. He manages the most comprehensive safety and health collection in the United States, over 140,000 documents. His technical expertise is in reference/research work and online data base searching, and he has assisted Council staff and library users in extensive research involving printed sources, the Council's Library data base, and commercial data bases. Mailing address: National Safety Council, 1121 Spring Lake Drive, Itasca, IL 60143-3201.

John F. Montgomery, PhD, CSP, CHCM, CHMM; Corporate Manager, Environmental Department, AMR Corporation/American Airlines. Dr. Montgomery is a writer and speaker on industrial safety and health topics, Assistant Professor/Lecturer at Central Missouri State University, Texas A&M University, and other academic institutions, and has served in several safety positions with AMR Corporation. He has been elected to several *Who's Who* volumes, including *Who's Who in the Environment* and *Who's Who in the World*. Mailing address: Manager, Environmental Safety & Health, American Airlines, PO Box 619616 M/D 5425, Dallas/Ft. Worth, TX 75261-9616.

Joanna Moreno, consultant, Dames & Moore. Ms. Moreno has for 16 years provided mathematical modeling evaluations of contamination control and groundwater management for industry and government projects in North and South America, Europe, the Middle East, and Australia. Her experience relates to flow characterization and remedial planning at groundwater contamination sites, predicting and aiding to minimize environmental impacts associated with power plant, radioactive, and municipal solid waste disposal, mining and process liquid-waste disposal, and accidental spills and leaks. She has also taught workshops on the use of groundwater models to regulatory agencies, consulting firms, and industry.

Florence Munter, consultant, Dames & Moore. Senior environmental regulatory specialist with over 15 years experience on compliance issues and environmental impact studies. Her experience includes radioactive, air, water, and waste permitting as well as regulatory analysis on remediation projects. She frequently provides training seminars and talks to a variety of audiences. Her compliance expertise includes knowledge of various industrial processes, mineral development projects, and government-operated facilities. Her background includes working for the U.S. EPA, U.S. Department of Interior, Office of Surface Mining, and as an independent consultant. Since 1981, she has specialized in hazardous waste and hazardous substance cleanups under RCRA and CERCLA, RCRA permitting, compliance audits, and waste stream characterization. She has previously contributed to many published Environmental Assessments, EISs, and a book on the ecology of the Boston Harbor Islands. Mailing address: Dames & Moore, First Interstate Tower North, 633 Seventeenth Street, Suite 2500, Denver, CO 80202.

James M. Ohi, PhD; senior environmental analyst and manager, Environmental Analysis Group, Analytic Studies Division, National Renewable Energy Laboratory. The group analyzes the environmental aspects of renewable and conventional energy technologies, particularly those of the transportation and utility sectors. They are involved in analyzing the potential of renewable energy technologies to reduce greenhouse gas emissions; safety assessments of advanced propulsion systems, including fuel cells and batteries; environmental data base and modeling development; infrastructure needs for a "hydrogen economy"; and a total energy cycle assessment of electric vehicles. Mailing address: Environmental Analysis Group, Analytic Studies Division, National Renewable Energy Laboratory, 1617 Cole Boulevard, Golden, CO 80401.

Frank J. Priznar, vice president, Roy F. Weston, Inc. Mr. Priznar advises senior executives regarding the implications of environmental issues on their strategic goals and business plans. He is internationally recognized in the environmental auditing field, having directed or participated in over 100 audits since 1980. He is founder and past president of the Institute for Environmental Auditing, former editor of *Working Papers*, a newsletter on environmental auditing, technical chair of ASTM Committee E50.04.1 to develop a national standard practice for environmental regulatory compliance audits, author and speaker on environmental auditing and assessments, and co-leader of the White House environmental audit team. Mailing address: Roy F. Weston, Inc., 1395 Piccard Drive, Suite 200, Rockville, MD 20850.

Martha A. Rozelle, PhD; manager and Principal, Public Involvement Services Group, Dames & Moore. Dr. Rozelle has directed more than 200 public involvement/conflict resolution programs for the siting of transmission lines in urban and rural areas, transportation facilities, sanitary landfills, high- and low-level nuclear repositories, dams, and reservoirs, and other controversial facilities. She has worked with citizen groups to promote understanding of the levels of risk and the potential health effects of EMF and various chemicals. She has authored a chapter titled "Siting Issues and Public Acceptance" in the *Handbook of Incineration of Hazardous Waste* (CRC Press, 1991) and was appointed to the Arizona Comparative Environmental Risk Project. Mailing address: Dames & Moore, 7500 North Dreamy Draw Drive, Suite 145, Phoenix, AZ 85020.

William Doyle Ruckelshaus, JD; chairman and CEO, Browning-Ferris Industries. From 1960 through 1969, Mr. Ruckelshaus served in several appointed and elected positions in Indiana, including Deputy Attorney General, Chief Counsel of the Office of Attorney General, member of the Indiana House of Representatives, and its majority leader from 1967 to 1969. He was appointed by President Nixon as Assistant Attorney General in charge of the Civil Division, U.S. Department of Justice, 1969 and 1970. He became the first administrator of the U.S. Environmental Protection Agency, December 1970 through April 1973. He was appointed by President Reagan, and unanimously confirmed, as the fifth Administrator of the U.S. EPA from May 1983 to January 1985. In 1973 he served as acting Director of the Federal Bureau of Investigation and later as Deputy Attorney General. He now serves as a director of several corporations, including Cummins Engine Company, Monsanto Company, Nordstrom, Inc., Texas Commerce Bancshares, Inc., and Weyerhaeuser Company. He is a member of The Trilateral Commission, the President's Council on Sustainable Development, the Business Council for Sustainable Development, the National Commission on Superfund, and was the United States Representative to the United Nations World Commission on Environment and Development. Mailing address: Browning-Ferris Industries, PO Box 3151, Houston, TX 77253.

C. Ford Runge, PhD; professor, Department of Agricultural and Applied Economics, University of Minnesota. Dr. Runge wrote the economics section of Chapter 2, Economic and Ethical Issues. He has served on the staff of the House Committee on Agriculture, and as a Science and Diplomacy Fellow of the American Association for the Advancement of Science, working in U.S. AID on food aid and trade issues. In 1985, he served as Chairman of the Governor's Farm Crisis Commission, structuring recommendations on farm credit and land markets in Minnesota. In 1986, he was awarded an International Affairs Fellowship by the Council on Foreign Relations, and in 1987 was selected as a Bush Foundation Leadership Fellow and Ford Foundation Economist. He spent 1988 as a special assistant to the U.S. Ambassador to the General Agreement on Tariffs and Trade (GATT) in Geneva, Switzerland, working under Trade Representative and former Agriculture Secretary Clayton Yeutter. In 1988 he was named a member of the Council on Foreign Relations in New York, and in 1990 a Fulbright Scholar for study in Western Europe. From 1988–91, he served as the first director of the Center for International Food and Agricultural Policy at the University of Minnesota. He continues as subdirector in charge of Commodities and Trade Policy of the Center. His publications include several books, and a wide range of articles concentrating on agricultural trade and natural resources policy, with particular emphasis on the links of agriculture to environmental quality. His books include *Reforming Farm Policy: Toward a National Agenda*, co-authored with Willard W. Cochrane (Iowa State University Press, 1992) and *Freer Trade, Protected Environment* (1994). Mailing address: University of Minnesota, 332K Classroom Office Building, 1994 Buford Avenue, St. Paul, MN 55108.

Gary F. Vajda, MS, PE; managing principal, Midwest Region, Dames & Moore. Mr. Vajda directs environmental engineering services, including environmental audits, regulatory consultation and strategies, site investigation and remediations, and pollution prevention evaluations for a number of corporate and facility clients. Mr. Vajda has more than 20 years experience in environmental engineering. Previous responsibilities have included process and engineering design, project management, and regulatory liaison in the engineering/construction and manufacturing industries, and has worked in the United States, Great Britain, and Poland. Mailing address: Dames & Moore, 11701 Borman Drive, Suite 340, St. Louis, MO 63146.

Anthony Veltri, PhD, MS in Safety and Environmental Management Studies; Associate Professor of Safety and Environmental Management Studies, Oregon State University. Over 15 years experience and research in assessment models that evaluate a firm's organizational strategy and structure in safety and environmental management. He received Outstanding Teacher awards, 1979–1980 and 1982–1983 and an Outstanding Service Award from the Aerospace Section, Industrial Division, National Safety Council, 1992. Mailing address: Oregon State University, Safety Department, Corvallus, OR 97331.

Christopher P. Weis, PhD, DABT; regional toxicologist, U. S. Environmental Protection Agency, Region VIII. Dr. Weis has been involved with the development and oversight of over 40 risk assessments for hazardous waste sites throughout the intermountain west. His experience includes assessment of chemical munitions disposal, radiological waste assessment, and particularly metal toxicology and risk assessment. Additionally, he has participated as principle toxicologist on multiple emergency response actions in the six-state region. Dr. Weis received a doctorate in both medical physiology and toxicology from Michigan State University and completed a National Institutes of Health postdoctoral fellowship at the University of Virginia School of Medicine. He is a diplomate of the American Board of Toxicology and is presently conducting research in the pharmacokinetics of lead absorption from various environmental matrices. Mailing address: Toxicologist, U.S. EPA, Region VIII, 999 Eighteenth Street, Suite 500, MC HWM-SM, Denver, CO 80202-2405.

Matthew D. Ziff, MS; Assistant Professor, Department of Housing and Interior Design, University of North Carolina at Greensboro where he teaches Interior Environmental Systems, Structures and Materials and interior design studio courses. He is also a registered architect and NCIDQ certified interior designer.

Part 1

General Framework

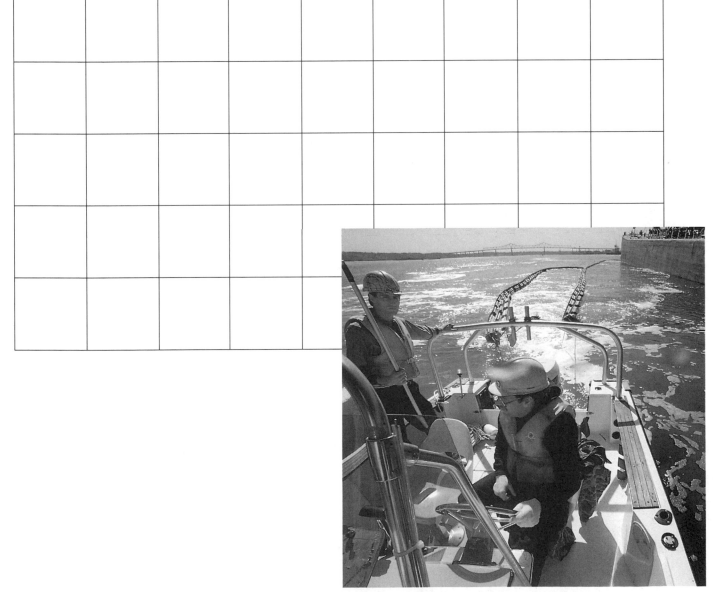

Oil containment boom used to control oil spill.

1

History and Development

W. Douglas Costain, PhD

Can the planet achieve sustainable development if world population continues to increase at its present rate? Can the power of the marketplace adequately protect the environment? Is sustainable development an achievable goal? Does the human species have an ethical imperative to protect and ensure global biodiversity of other living species? Do international accords and agreements improve the environment, or are they an ill-disguised form of trade protectionism? These questions lie at the center of the ongoing debate and evolution over worldwide environmental practices.

The United States has been a leader in the environmental movement during the last 25 years and has created a gigantic, multistatutory regulatory infrastructure. By 1990, the U.S. Environmental Protection Agency (EPA) administered 11 major statutes and more than 9,000 regulations. Direct and indirect annual costs are in the hundreds of billions of dollars (Abelson, 1993). Across the globe, many nations follow the lead of the United States and are similarly developing basic environmental regulatory infrastructures. The evolution of "bucks" to "greenbacks" is accelerating at a rapid pace. Part 1 of the *Accident Prevention Manual for Business & Industry: Environmental Management* explores and analyzes global environmental developments from a historical, ethical, economic, and legal/regulatory perspective. Because environmental issues frequently cross national boundaries, a major emphasis of Part 1 is the role of the United States in the ongoing international debate about the policies, philosophy, and economics of environmentalism.

The emergence of environmentalism as a social and political movement during the past 25 years has transformed human thought and behavior. The avant-garde concern for preserving and protecting nature in the 1960s is now embraced by nearly every part of the population and by every institution in society. Protecting the environment is also a big business, with U.S. sales of pollution control equipment estimated at $130 billion in 1991 and global sales exceeding $370 billion (Bryner, 1993). The costs to the economy of meeting the standards laid out in the 1990 Clean Air Act Amendments will exceed $20 billion *per year* by the end of the 1990s. This law is just one of the more than 30 major U.S. environmental laws enacted in the past 25 years. Moreover,

3

recent national polls show that more than 70% of Americans feel that environmental protection is so important that improvements must be made *regardless of cost* (Dunlap, 1992).

How did such a significant transformation come about? Where did environmental concerns spring from, and how did advocates for ecology come to have so much political clout? This chapter addresses these questions, first by briefly reviewing the origins of the contemporary environmental movement in the conservation and public health movements of the early 1900s. Next, the interaction between changes in U.S. public opinion and the activities of environmental groups in the past 25 years will be analyzed.

However, environmentalism is not just an American idiosyncrasy. It is a global phenomenon that spread from the wealthy nations of the Western Hemisphere throughout the world, so that there are now environmental protection laws and government agencies in almost every nation. Local citizens' groups are putting political pressure on their leaders as are worldwide nongovernmental organizations (NGOs). The local groups are frequently composed of indigenous peoples and have links to international networks of ecology organizations. Finally, in the latter part of this chapter, an overview of the globalization of environmental concerns will be discussed (see also Chapters 4, International Legal and Legislative Framework; and 14, Global Issues).

MODERN U.S. ENVIRONMENTALISM

The environmental movement that came to prominence in the late 1960s can best be viewed as a grassroots or "bottom-up" phenomenon in which the commitment to environmental protection arose out of deep-seated changes in public values about the relationship of humans to nature. By contrast, the conservation movement at the beginning of the 20th century was a "top-down" phenomenon in which technically trained experts and political leaders were entrusted with the responsibility and power to manage natural resources and to protect the environment on behalf of society.

The implications of this understanding of modern environmentalism are significant. The base of support for the current environmental movement consists of a larger sector of the public than that of the early 20th century conservation movement. Such broad support attracted the attention of politicians seeking to attach themselves to popular causes. Because strong programmatic national political parties

were in decline, temporary coalitions of opportunistic politicians advanced the issues of the movement. This transitory support became institutionalized in the laws passed by governments and in the regulations that were administered by new agencies like the U.S. Environmental Protection Agency (EPA). The media and the ambitious politicians did move on to newer topics, but the wide array of environmental groups and government agencies became firmly rooted in social networks and in national, state, and local communities. In this way, an upsurge in public concern became a permanent part of society.

For health and safety professionals concerned with environmental issues, such broad-based public support combined with active local and national environmental groups can be challenging. Politically sophisticated national NGO groups ranging from the National Wildlife Federation to the Sierra Club, Natural Resources Defense Council (NRDC), Friends of the Earth, and Greenpeace have a great deal of credibility and legitimacy with the public as the spokespersons for environmental issues. Their advocacy tactics range from the presentation of technical analysis in front of EPA scientific advisory panels to noisy demonstrations staged for television cameras during proceedings of major federal and state lawsuits. Many technically trained professionals are uncomfortable with the reduction of complex analyses to slogans and media sound bites, especially if their employer is the target of this attention.

Environmental groups have evolved with public evaluation of the severity of environmental problems and have grown as that evaluation has shifted. Since the 1800s, the U.S. public has moved from a primary concern over "natural resource issues" in the era of the conservation movement, to "environmental issues." Supplementing first-generation problems—public lands, irrigation and water rights, and park management—the public and organized groups added concern for second-generation problems—toxic waste, groundwater protection, and air pollution. Moreover, ordinary citizens and world leaders are now debating third-generation problems—global warming, thinning of the ozone layer, tropical deforestation, and the rate of species loss (see Chapter 14, Global Issues).

An ever-expanding environmental issues agenda advanced by experienced political groups that have broad public support embroils us all in a seemingly endless set of controversies and conflicts. In the United States, scientific specialists are no longer called on to resolve ecological disputes. Rather, specialists are used as a tool by each side to advance its arguments. In developing nations, indigenous peoples and their allies in environmental groups raise concerns about the ecological impacts of economic development projects in their impoverished nations. Democracy is an untidy process at best, and the worldwide expansion of democratic participation in environmental decisions seems irreversible. The people of the world are not going to return to a former era of quiet deference to established expert authorities. If health and safety professionals are going to participate in environmental controversies, they will have to recognize the need to convince either elected officials sensitive to public opinion or to win over concerned citizens directly.

Early Conservation Movement

At the turn of the 20th century, the Progressive movement, a broadscale tide of social and political reform, swept the United States. Although the agenda of the reformers was multifaceted, ranging from ending corruption in municipal government to advocating public health concerns such as clean drinking water, a common thread was the belief that public policy should be based on applied scientific knowledge. The reformers' disdain for the shortsighted greed and selfishness they observed in the rough-and-tumble business practices of that period led them to value honest, expert government administration for the public good, not private profit. Most histories of the conservation movement place it squarely within the wider Progressive reform movement, sharing a common worldview with other subsidiary movements, such as Prohibition and women's suffrage, as each developed over time.

However, one branch of the Progressive movement—urban public health and sanitation—has not been sufficiently linked to the conservationists. Modern environmentalism owes much of its public support and political influence to its public health content. Many environmental laws are concerned with public health as well as preserving nature. The Progressive period launched many reform impulses, and it was the reuniting of two of them in the 1960s that provides the durability and appeal of modern environmentalism.

"Nature" focus—conservation and preservation.
Although we can identify earlier attempts to protect the natural environment, there was not really much effort for environmental protection in the United States until after the Civil War. The years of resource exploitation by a rising industrial sector and subordination of natural resources to industrial development, homestead settlement, and the promotion of free enterprise raised concerns. Serious efforts to protect the nation's natural resources began in the late 1800s.

The essence of the conservation movement was rational planning *by government* to promote the efficient development and use of all natural resources (Hays, 1959). Gifford Pinchot and others were concerned about the shortsighted commercial exploitation of the nation's natural resources. Pinchot became an important link between President Theodore Roosevelt and the intellectual and scientific founders of the conservation movement. Indeed, the Theodore Roosevelt administration is as noteworthy for the drive and support it gave the conservation movement as it is for the initiation of new policies or legislative enactments (Hays, 1959). Under Pinchot's guidance, the Roosevelt administration greatly enlarged the area of the national forests from 41 to 159 national reserves, and Pinchot completely reorganized the Forest Service, infusing it with a new spirit of public responsibility.

Together, Pinchot and his colleagues formulated four basic doctrines that became the creed of the conservation movement:

- Conservation is not the locking up of resources; it is their development and wise use.
- Conservation is the greatest good for the greatest number, for the longest time.
- The federal public lands belong to all the people.
- Comprehensive, multiple-purpose river basin planning and development should be used with respect to the nation's water resources (Caulfield, 1989).

Nevertheless, the deepest significance of the conservation movement, according to historian Samuel Hays, "lay in its political implications: How should resource decisions be made and by whom?" Should conflicts be resolved through partisan politics, by compromise among various interest groups, or through the courts? To the conservationists, politics was an anathema. Instead, Hays notes that conservationists believed that scientific experts, using technical and scientific methods, should decide all matters of development and use of natural resources together with the allocation of public funds. The key "lay in a rational and scientific method of making basic technological decisions through a single, central authority" (Hays, 1959).

A second form of conservationism, preservationism, was similar to that of the conservation-efficiency advocates, except that preservationism was more concerned with *habitat* than rational use (Engelbert, 1961). Preservationism is associated with increased leisure and affluence, and the growth of outdoor recreation. It drew its support from the upper middle class and from working-class hunting and fishing groups. Although preservationists experienced internal conflicts about natural resource policy, they were largely confined to struggles between those who favored "multiple use" of public lands versus those who favored "pure preservation." The preservationists often had wealthy sponsors, like Laurance Rockefeller, who facilitated the preservation of major tracts of land surrounding the hotels that he built (Schnaiberg, 1980).

Another major difference between the two groups was that conservation-efficiency concerns resided with corporations and national and state agencies, whereas conservationist-preservationist concerns resided in local and especially national voluntary organizations, such as the Sierra Club, the National Wildlife Federation, and the Wilderness Society.

Urban focus—public health in cities. Although tracking the antecedents of modern environmentalism to the nature conservation movement is common, it is important to recognize that the urban-focused public health movement, part of the Progressive era reforms, shared many of the same values. The idea of a central role for government in promoting and preserving the health of individuals rather than relying on the private sector also emerged in the early years of the 20th century. Health was not simply a private decision. It was also a public concern because disease spread by inadequate sanitation, impure water, and poor sewer systems seemed to call for community-wide solutions. Just as foresters decried the loss of natural resources resulting from poor management of the land, so early public health advocates, including medical professionals, saw the need for government to assume responsibility for the prevention of communicable disease in crowded cities. As the germ theory of disease was accepted among health experts, it became clear that individuals alone could not protect their own health, because each of us was vulnerable to exposure to germs carried by others. Thus were mass vaccinations, childhood health screenings, nutrition education, purified drinking water, and municipal sewer systems justified. The idea that experts, working through governments, could identify and protect the interests of the community was common to the Progressive era. It is easy to see that many of these concerns are clearly within a modern understanding of environmentalism. The recognition that an individual's health can be placed at risk by the actions of others—whether by infection with a contagious disease (the early 1900s) or by toxic emissions from industrial smokestacks (the 1970s)—unites the old public health with the new environmental concerns.

Organizational links also join the older public health and the newer environmental issues. Throughout the 1950s and 1960s, the Health, Education, and Welfare Department administered the increasing amounts of federal government aid and advice to state and local governments for air and water pollution control. Upon its creation in 1970, the EPA was staffed by Public Health Service researchers and administrators. At the state level, it is common for modern environmental laws to be administered by state departments of health or newly created departments of the environment/ecology.

Evolution of Environmental Issues

Using data from the *New York Times Index* on the media attention to environmental topics through time, it is possible to track the changing pattern of environmental politics. By describing the evolution of the *content* of environmentalism during the period 1890–1990, with particular attention to the types of issues considered (using a categorization of the *New York Times Index*), it is also possible to map the changing focus of environmental concerns over the past century (Figure 1–1).

Most striking is the dramatic increase in the coverage of environmental issues in the past 30 years. For example, from 1890 to 1953, the *Index* space for a list of about 40 environmental topics and keywords reached an annual 100 column inches (Figure 1–2). In each year from 1970 to 1975 more than 1,000 column inches were given to these same subjects (Figure 1–3).

The transformation from the conservation issue, a low-visibility topic of concern only to an elite, to the environment, a highly visible subject involving mass participation, is significant. Yet, even within the less public era of conservation politics there are some notable patterns. The early peak in column inches of newsprint from 1906 to 1912 was followed by low coverage until the mid-1920s; the upsurge in news stories ushered in a higher plateau of media attention. Also worth noting is the steady increase in

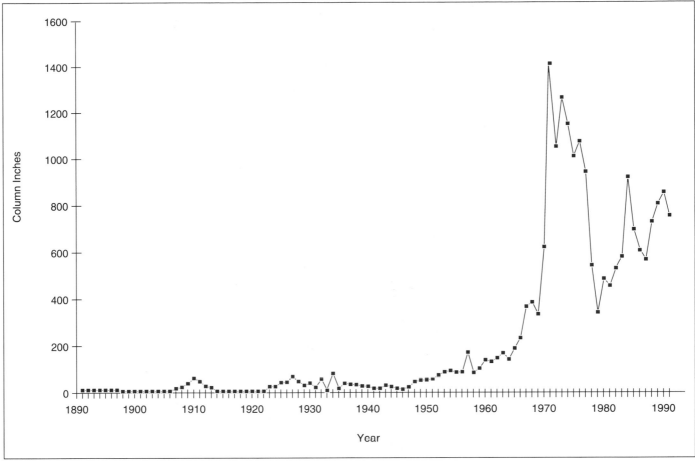

Figure 1–1. Coverage of environmental topics in the *New York Times Index,* 1890–1990.

environmental coverage from 1945 through the 1950s. At least in retrospect, the growth in space for environmental topics in the quiet 1950s seems a precursor to the dramatic upswing in attention during the 1960s.

In the 1960 to 1990 era, the 60-year peak of news coverage reached in the 1950s becomes the base for the dramatic surge in the late 1960s (Figure 1–3). Most visible, of course, is the 1970 to 1976 volume of environmental news stories: the *average* number of column inches was almost six times the *peak* amount of space in any year before 1963. The late 1970s drop-off in attention is almost as spectacular: The lowest coverage year, 1978, registers fewer than 400 column inches as compared to the 1970 high of more than 1,400 inches. Although the controversy over President Reagan's initial environmental policies is often credited with restoring the salience of ecological issues, the increase in news coverage began before the well-covered conflicts.

Similarly, the upswing in 1989 related to the Alaska oil spill came after two years of increasing

space for ecological news items. Taken together, the data suggest cycles of increased media attention to environmental topics followed by declines in news coverage. The amount of attention given to environmental news is higher since the mid-1960s, suggesting a different type of politics from earlier periods, one much more visible and controversial.

Thus far, the broad array of environmental issues has been aggregated into a single category. Is there a single environmental agenda, or do distinct components show how the agenda changes over time? The first pattern of *New York Times* coverage of stories is related to three "classic" conservation topics—public lands and forests, national parks and seashores, and fish and animals—as well as one "modern" issue—wilderness. (There were *no* entries for wilderness prior to 1959 in the *Index*.) These topics were combined to create the "conservation" measure used in Figure 1–4.

Within the combined conservation category, one of the most striking features in this data is the dominance of the "fish and animals" category in the

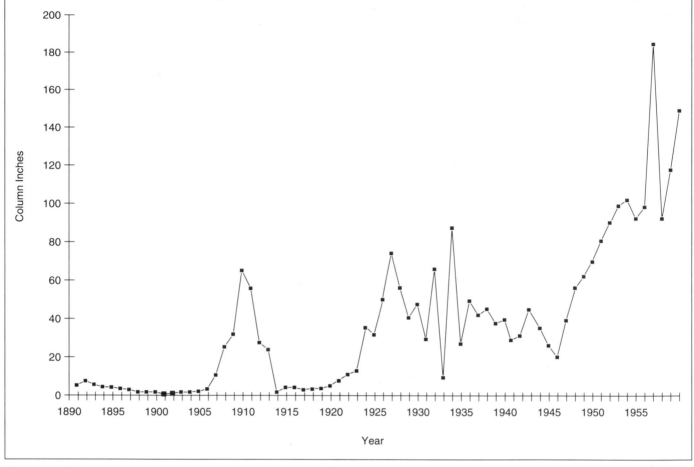

Figure 1–2. Swings in coverage of environmental topics in the *New York Times Index,* 1890–1959.

public agenda until the 1950s. "Public lands" issues surface in the years around 1910 and again in the mid- to late 1920s. "Parks" as an agenda item becomes visible in the late 1920s and briefly overshadows the other issues in the mid- to late 1950s. This older conservation agenda is much more visible than newer topics such as pollution in the decades prior to the 1960s. About twice as much *Index* space is given to stories from the conservation agenda (70 column inches maximum per year) as to newer issues (a maximum of 35 column inches).

Within the combined "environmental" topics category in Figure 1–4, there are several stories on air and water pollution in the years around 1910, whereas water pollution becomes quite visible in the late 1920s and early 1930s. Air pollution is overshadowed by water pollution until the late 1940s, when it becomes the more visible agenda item. The other two categories, "oceans" and "chemicals and toxics," have little space in the decades prior to the 1960s.

As Figure 1–5 shows, the older or conservation issues are much less visible in the past 30 years.

In an absolute sense, this agenda is given *more* attention than ever before. However, the dramatic transformation is in the pollution issues, which increase by a factor of 20 from pre-1960 to post-1960.

The striking increase in the overall attention given to all environmental issues during 1970 to 1974 is barely visible in the traditional conservation topics. The decline in media coverage in the late 1970s is apparent, as is the renewal of attention in the 1980s. Clearly, the environmental movement did not elevate the conservation agenda to prominence. But neither were the old issues neglected or ignored. The agenda broadened and newer topics were *added* to the existing agenda.

The increased attention is given to the newer environmental agenda. The dramatic increase in media space for environmental concerns in the early 1970s was largely for air and water pollution as well as for the multi-issue class. Most of the mid-1970s drop-off in attention is also visible in the space given to these issues. Interestingly, chemicals and hazardous and toxic materials become a relatively visible category only in 1985 and 1986.

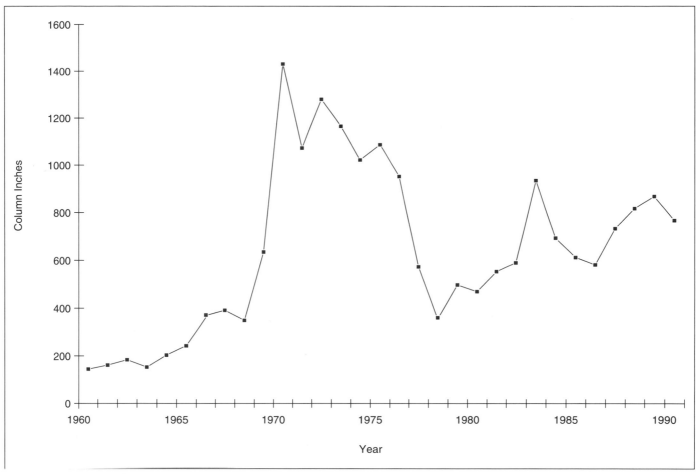

Figure 1–3. Dramatically increased coverage of environmental topics in the *New York Times Index,* 1960–1990.

In summary, the public agenda of the environmental movement, as revealed by the *New York Times Index,* built on the foundation of the issue agenda of the conservation movement. The conservation issues, such as public lands and forest management, national parks, and regulating hunting and fishing, continue to coexist with the newer concerns for air and water quality, toxic wastes, and wilderness preservation. However, the relative space given to most of these older topics is dwarfed by the media attention devoted to the newer dimensions of environmental concern.

Environmental Activist Groups—NGOs

Most environmental organizations, whether politically active or not, are less than 30 years old and are products of the modern environmental movement. Some of the largest, richest, most active groups are much older. These and other ecology groups were also revitalized in the 1960s and 1970s (Mitchell, 1979).

The founding date of environmental organizations provides some further evidence of the evolution of issues. In the early 1900s, both preservationist and hunting and fishing groups become visible participants in the policy debates. There is, however, a vast gulf between the two categories. The preservationist Audubon Society (founded in 1905) and Sierra Club (founded in 1892) were tiny in contrast to the mass-membership sportsmen's clubs, such as the American Game Protective and Propagation Association (founded in 1911). However, the members of preservationist organizations were unusually well-connected to the nation's political and economic elites, so they exercised an influence far out of proportion to their organizations' size.

By contrast, many of the organizations that were founded during the current environmental era, such as the Natural Resources Defense Council (founded in 1970) or the Environmental Defense Fund (founded in 1967), deliberately focused on using the political and legal systems to advance environmental protection. They were also membership groups,

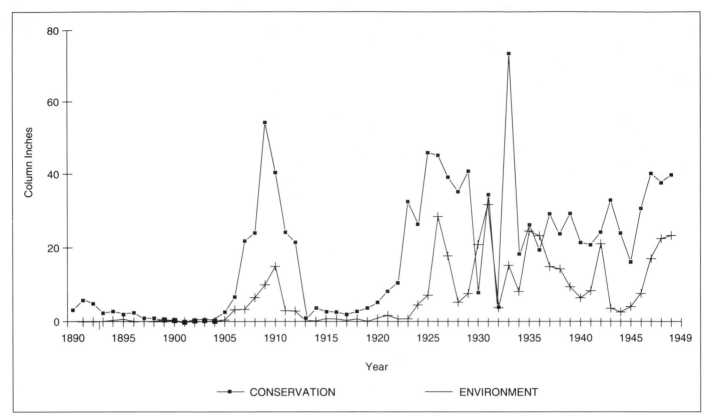

Figure 1–4. Conservation topics dominate environmental coverage in the *New York Times Index*, 1890–1949.

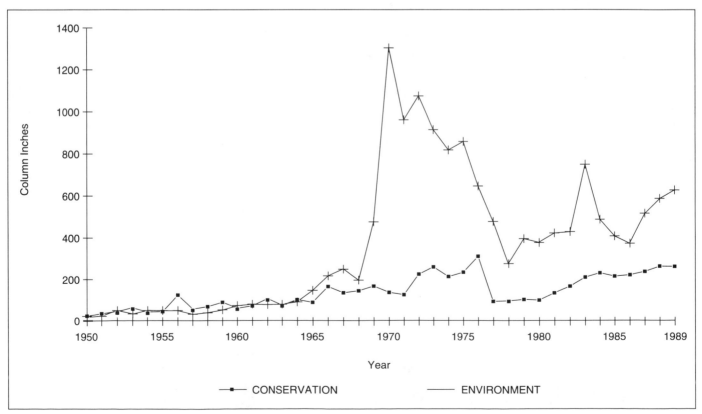

Figure 1–5. Pollution concerns overshadow conservation issues in the *New York Times Index*, 1950–1990.

relying on contributions from a broad base of members and donors to fund their advocacy, as well as soliciting larger grants from foundations. Environmental organizations now range from those that buy land for preservation, such as the Nature Conservancy, to those that use confrontational rhetoric and tactics, such as Greenpeace.

The sheer size and diversity of the groups mean that new participants often carve out new niches rather than directly compete with older groups. Moreover, this variety also suggests an expansion of the issues, with established groups and issues coexisting with emerging ones. In the late 1960s and early 1970s, old and new environmental groups were able to become full participants in policy making. This was also the time when there was an influx of new groups and a revitalization of existing organizations (Vig & Kraft, 1990).

Reunion of Nature Preservation and Public Health

The first tremors of the current public concern for the environment were recorded in public opinion polls taken in 1965: One-sixth to one-third of national samples named air and/or water pollution as a priority for government action. By the first Earth Day in April 1970, more than half of citizens surveyed shared this concern (Dunlap, 1992). The definition of "environmental concern" in the late 1960s was pollution, especially air and water pollution. Since the late 1960s, the environmental issues that embody threats to human health arouse almost universal public support for government action to protect the environment. Although philosophers of environmental ethics and ecology activists talk about the transformation of human values, a much more basic need is often being measured as environmentalism: a desire to protect oneself and especially one's children from threats to life and health.

Interestingly, the level of public attention to ecology topics declines after the federal government undertakes major new environmental policies. After the enactment of major laws addressing air and water pollution in the early 1970s and the creation of a highly visible EPA to carry them out, public interest declined. This downturn continued throughout President Carter's term, even as he appointed environmental advocates to major government positions (Mitchell, 1984; see also Figure 1–6).

In the 1980s, interest was revived by the intense controversies over President Reagan's environmental policies, which were widely seen as reducing the government's role and budgets (Bartlett, 1984). The public is alerted by visible disasters like oil spills or nuclear accidents, but the key to whether an incident will spark ongoing concern is whether the government seems to be dealing with the crisis. In the 1980s, the diminished advocacy for environmental issues by government leaders ironically served to heighten public concern. Part of this increased concern resulted from the publicity of environmental groups pressuring the government and recruiting new dues-paying members. It seems clear, however, that today Americans want and expect their elected leaders to "take care of" the environment. If the public is assured of their government's commitment, mass interest declines. This pattern will be tested during the highly visible environmental agenda of the Clinton presidency.

Environmental Political Activism: 1960–1990

In the early 1960s new intellectual and political forces emerged to redefine natural resource and conservation issues. The conservation movement of the late 19th and early 20th centuries had become institutionalized within a national policy community composed of economic users of natural resources as well as recreational interests (such as hunters) and a smaller set of nature lovers (on the earlier conservation movement, see Hays, 1987; on the emergence of environmentalism, see Andrews, 1980; Mitchell, 1985). Challenges to this limited network came initially from the scientific community as the concept of "ecology" broadened the scope of human-environment relationship studies. In addition, a new concern for the health impacts of pollution spurred government studies and the introduction of pollution-control bills into Congress (Schoenfeld et al, 1979; Hays, 1987).

Both public opinion and key political institutions early supported the central goals of environmentalism. The potential for ecological values to transform American society was little recognized, even by critics of the movement, until the mid-1970s. The environmental movement could, and would, refute charges that ecology was elitist with evidence of widespread public acceptance of its broad goals. The movement's key political problem in its formative period was how to sustain public interest while institutionalizing environmental protection within federal agencies and laws.

The movement began to routinize its structures and behavior in the early 1970s in response to growing opposition to the burdens of environmental

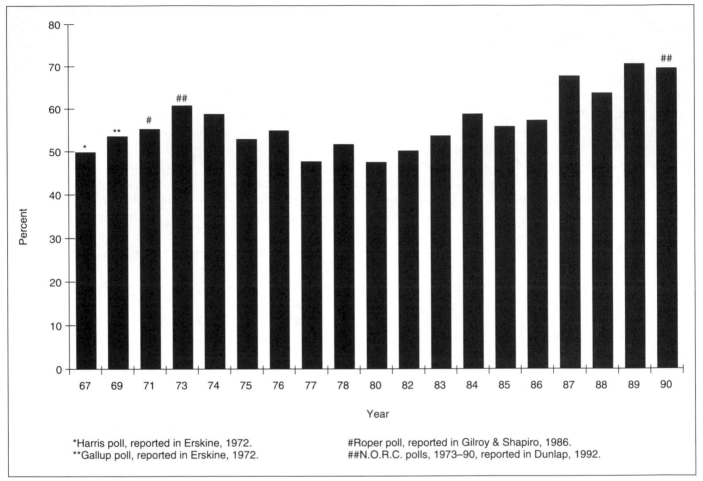

*Harris poll, reported in Erskine, 1972. #Roper poll, reported in Gilroy & Shapiro, 1986.
**Gallup poll, reported in Erskine, 1972. ##N.O.R.C. polls, 1973–90, reported in Dunlap, 1992.

Figure 1–6. Percent of the general public agreeing that "government is spending too little on the environment." Source: Dunlap, 1992; Erskine, 1972; Gillroy and Shapiro, 1986.

protection. Advancing and defending the cause of the environment against powerful economic interests required sophisticated political tactics and organization. Movement opponents raised concern about the costs of basic changes in the economic and social fabric of the nation. However, critics of environmentalism had limited success in arousing public opposition to ecological protection, especially because public health was a central focus of the laws. Environmentalists could portray their critics as selfish interests putting profit ahead of the general good (Costain & Costain, 1992).

During the mid-1970s the environmental movement faced a drop in the salience of its issues, a decline that would not be reversed until the end of the decade. President Carter incorporated both ecological personnel and values into his administration, moving the cause of environmental protection to a more privileged status in the political system. By contrast, President Reagan and his administration criticized the leaders and rejected many of the values of

the movement. This strategy actually served to mobilize supporters of environmentalism. Environmental issues returned to public salience in the early 1980s. Ecology groups worked to politicize this concern against the administration. The movement shifted its political efforts toward partisan politics, allying itself with the Democratic party, and succeeded in creating awareness of a significant "Green vote" in the elections of the mid-1980s (Mitchell, 1990).

The movement's organized groups had become institutionalized within the political system before the Reagan Revolution. The movement used traditional political tactics to defend itself in the 1980s. Its success in activating its supporters and gaining sympathy from the larger public seemed to influence a change of rhetoric (if not policies) in the Reagan administration. The movement demonstrated durability and a grasp of the techniques of working within the system and responding to dramatic shifts in the political climate. In the face of a powerful opponent, it used its skills to persevere and wait out adversity.

Although groups within the movement actively campaigned for the Democratic party in 1988 and 1992, their influence went beyond their ability to deliver a bloc of votes. As a candidate and as president, George Bush incorporated ecological concerns into his political agenda, only to be outdone by Bill Clinton's selection of an ardent environmental advocate as his running mate. Environmental groups found a new enthusiasm for their issues in the executive branch. Only a few politicians from either party are willing to oppose the expansion of federal environmental programs.

The environmental movement has regained the initiative for its agenda and has returned to the offensive after years on the defensive. Heightened media attention to pollution has increased the prominence of environmental issues. Political leaders are proposing new federal programs. Movement groups are taking advantage of this increased support to press for action on a broader range of issues, including protection of tropical rainforests and recycling.

As the major environmental groups were pursuing insider political tactics, a number of local groups engaged in civil disobedience and direct action aided by loosely structured national networks such as Earth First!. Overall, though, the environmental movement had become an institutionalized part of the mainstream of American political life through Environmental Impact Statements that changed bureaucratic planning, legislation that mandated federal leadership in improving environmental quality, and public attitudes that supported expenditures to clean up pollution.

Risk Perceptions and Policies

Two major reports commissioned by the EPA were published in the late 1980s and early 1990s: The 1990 report of the EPA's Science Advisory Board, *Reducing Risk: Setting Priorities and Strategies for Environmental Protection,* and the 1987, *Unfinished Business: A Comparative Assessment of Environmental Problems.* These reports attempt to set environmental priorities based on scientific estimates of the relative risk to human health and habitat. They seem to harken back to the early 20th century progressive reformers of public lands and public health who succeeded in placing those policies in the hands of experts rather than corrupt politicians or untrained amateurs. EPA administrator William Reilly noted that the "EPA's priorities appear more closely

aligned with public opinion than with estimated risk." Reilly's advocacy of risk assessment as a central instrument for making environmental decisions called for a better informed public "to make rational risk assessment a part of every citizen's common sense." He did not favor making experts more powerful and autonomous (Reilly, 1993).

By recognizing that democracy is at the center of modern environmental policy making and that experts must win their cases in the court of public opinion rather than simply convincing their scientific peers, the EPA administrator focused attention on the greatest frustration as well as the greatest strength of American environmental policy: its democratic character.

Although popular fears, such as water pollution or toxic waste contamination, dominated the 1970s (water pollution) and 1980s (toxic waste), the environmental issues of the 1990s appear to be on global initiatives that focus on "sustainable development" agendas (see Chapter 2, Economic and Ethical Issues; and Chapter 14, Global Issues). The "globalization" of the environmental movement represents a new political development that may well dominate the remainder of this century.

GLOBALIZATION OF ENVIRONMENTAL CONCERNS

Today, almost every nation has environmental protection laws and government agencies assigned to enforce the laws. Especially among industrialized countries, there is significant uniformity in the formal structure of "on the books" environmental protection. Air and water emissions from industrial sites and toxic waste disposal are subject to government controls and inspection. However, below the level of laws and agencies is the less visible and more problematic area of actual compliance and implementation of these laws. There is much more variance among nations in both the stringency of environmental regulations and the enforcement methods used. This diversity parallels national differences in other public policies. Moreover, examination of the implementation of national environmental protection measures may not measure the behavior of nongovernmental entities such as industries and state-owned enterprises. Finally, global generalization is complicated by the activity or inactivity of state, local, or provincial governments with responsibilities for environmental protection.

Thus, a great deal of variance *within* countries as well as among nations should be expected (see Chapter 4, International Legal and Legislative Framework). The level of citizen concern and activity to protect the environment also varies significantly among nations. Looking at the role of national and local environmental groups reveals no place matching the size, resources, and political clout of U.S. environmental groups. But the U.S. political system is uniquely hospitable to the formation and sustenance of interest groups and their integration into the structure of political power. In many other nations the political impact of environmental concern is visible in different types of organizations, especially the "Green" political parties in Europe. The potential for Green political parties depends on whether the political system is supportive or inhospitable to new parties as much as on the level of public concern for ecology.

Western Europe

During the past quarter century, Western Europe experienced two waves of public concern and government action to protect the environment, in the early 1970s and again in the late 1980s. In Great Britain, for example, memberships in national and local ecology groups soared in the early 1970s, and Parliament passed major new laws regulating toxic waste, and air and water pollution (Vogel, 1990). In the renewal of environmental concern in the late 1980s, Green political parties spread from their earlier visible roles in West Germany and Belgium to gain in representation in many countries including Sweden, France, and Italy. Throughout Europe, public opinion polls supported expansion of environmental protection and elevation of the priority given to ecology by governments (Williams, 1991). Even as European governments increased their environmental protection activities within their borders, they also acted collectively to strengthen continent-wide ecological standards (Pinders, 1991).

The European Community. The political popularity of environmental protection within the member states of the European Community (EC) has been reflected in the increasing role of the EC institutions, such as the Council, Commission, and Parliament in legislating and regulating EC-wide standards. (With the signing of the Maastricht Treaty on November 1, 1993, the European Community was renamed the European Union or EU.) During the 1980s, Green parties were more successful at gaining seats in the

EC Parliament than in winning in their national legislatures. They were vocal in advocating a stronger role for environmental protection in the activities of the EC (Axelrod, 1993).

Beginning in 1973, the EC embarked on a series of five-year plans designed to coordinate and harmonize the environmental policies of the 12 member states. The Environmental Action Programs began by emphasizing remediation and cleanup of pollution but have broadened significantly. By the end of the 1980s, the member states of the EC were required to prepare the equivalent of U.S. Environmental Impact Statements before firms or governments undertook major projects and to integrate environmental protection into other government programs, such as subsidies for agriculture and energy. Moreover, the 1987 Treaty uniting the continent, the Single European Act, provides a form of "constitutional" protection for the environment, because it specifies the central role of ecological values in the policies of the EC (Whitehead, 1992).

The combination of public support mobilized in Green parties as well as interest groups, and EC-wide institutions that can use their mandate to protect the environment as a means of centralizing power, has elevated the political importance of ecology in Europe. Although most of the environmental regulations imposed by the EC on its constituent states have been the result of consensus negotiated among the national governments, there is an emerging potential for EC rules to override the tradition of unanimous consent. For example, tougher auto exhaust standards were imposed by the European Parliament than had been agreed to by the national governments—an expansion of central control that was upheld by the European Court of Justice (Whitehead, 1992).

Interestingly, there is an explicit economic rationale for the imposition of EC-wide environmental standards that goes beyond the common notion of a conflict between protecting the environment and creating jobs or economic growth. The concern within the EC is to build a single economy without internal barriers to trade and investment. Low standards for environmental protection in a member state are seen as creating not only a public health risk to adjacent nations but also a threat to "the integrity of a single market" (Williams, 1991).

Non-EC countries. The political support for environmentalism extends throughout Europe, beyond the 12 EC nations. The Scandinavian countries, Sweden, Norway, and Finland, have at least as strong environmental protection policies as most of the continent. Switzerland and Austria are similarly not "pollution havens" despite their nonmember status, indicating the pervasiveness of ecological norms. There is, of course, more variance in the issues raised among these countries, because there is no common forum, like the European Parliament in the EC, to synchronize environmental political concerns.

Debates over natural resource policy and the processing of the raw materials have a higher profile in the Nordic states, reflecting the prominence of those economic sectors. Domestic political interests can also raise unique ecological issues as reflected in Norway's efforts to resume commercial whaling, which puts it at variance with the rest of Europe. The pharmaceutical and chemical industries in Switzerland have also lagged behind their German competitors in controlling pollution, although this difference is disappearing.

The end of the Cold War that had divided Europe into two competing military and economic blocs has removed the threat of nuclear or large-scale conventional war from the agendas of Europeans and their governments. In the absence of such an overwhelming threat that forced a high level of cooperation, Europeans are creating new economic, social, and environmental institutions to carry out shared concerns. Although some nations—notably Switzerland—will resist the integration into a continent-wide economic and political union, there is clearly less resistance to recognition of the need to protect their common environment.

Other Developed Industrial Democracies

The global scale of popular environmental concern is also apparent in developed nations outside of Europe. Environmental laws are in place and enforced by competent government agencies in Australia, New Zealand, and Canada. Environmental groups are well established and politically influential, and in the case of Canada, receive significant government subsidies. These countries parallel the United States in the range of public health, natural resources, and nature preservation issues encompassed by the environmental agenda. As with European nations, pollution regulations are usually enforced through private negotiations between government officials and the affected industry, in contrast to the highly public adversarial relations common in the United States.

The Japanese case is unique. Beginning in the late 1960s, pollution control laws centering on public health protection were enacted amidst significant public mobilization. By almost any standard, Japan has strict, well-implemented regulations for air, water, and hazardous waste pollution. Public concern has not returned to the high levels of the early 1970s, in contrast to the pattern in most other developed countries. Moreover, the agenda of Japanese environmentalism does not seem to include a major nature preservation component. David Vogel suggests that Japanese culture sees nature as a threat, rather than a solace, and has little reverence for animal life—as is evident in the Japanese whalers harvesting of marine mammals (Vogel, 1990). The Japanese government is, however, sensitive to foreign criticism, especially from the United States, so environmental groups have had some influence using this indirect channel.

Central Europe, Russia, and the Newly Independent States (NIS—Former USSR)

The collapse of communism as an economic and political system in Eastern Europe (now called Central Europe) and in the former Soviet Union has exposed a legacy of environmental devastation that will take decades to remediate. Air and water pollution, and toxic and nuclear waste contamination are so pervasive that literally millions of people are suffering the health consequences of their nations' economic and military competition with the West. Repairing the damage will also require hundreds of billions of dollars in pollution control and cleanup expenditures. On the one hand, this investment in water treatment facilities, power plant emissions controls, and the like competes with education and other infrastructure needs in nations that have limited funds. However, the impacts of past and current environmental hazards on the public health and on the investment climate are so significant as to warrant remediation as a prerequisite for economic development (Levy & Hertzman, 1992).

Until the last years of Communist rule in the late 1980s, individuals or local groups that raised environmental health concerns or pointed to ecological damage were suppressed for being anti-Communist or unpatriotic. Ironically, their counterparts in the West have sometimes been accused of being socialist or anticapitalist. Grass-roots groups protesting the extent of environmental damage flourished in the waning months of Communist rule, symptomatic of the relaxation of control called *glasnost* in the former USSR. The reconstruction of these societies is now underway—and with it the elevation of environmental protection to a status that it never enjoyed under communism (Jancar-Webster, 1993).

Communist environmental legacy. Although western economists frequently point to pollution as a consequence of the failure of markets to place a price on public goods such as air, the undervaluing of the environment was endemic also under Communist economics. Factory managers were allocated production quotas by economic planners in government ministries, with little attention paid to product quality, much less damage to the environment. Investment in pollution-control equipment was rare, and proper maintenance and operation of any such equipment was rarer still. Moreover, because factories were allocated energy and raw materials at subsidized prices, major industries were extremely inefficient, compared to those in free markets, at converting iron ore into steel or coal into electricity. In addition, environmental laws were seldom enforced, reinforcing the incentives to disregard ecological and public health concerns (Livermash, 1992). Without citizens' groups to pressure polluters or a free press to expose them, or government agencies to regulate them, polluters in Communist states held unchallenged power (Jancar-Webster, 1993).

Environmental restoration. The transition to market economics and to competitive political systems has already reduced the output of pollution in the former Communist states. Subsidies to inefficient, highly polluting state enterprises have been cut drastically in Central Europe and somewhat less in the former USSR, thus encouraging less wasteful production techniques. In some cases, notably the former East Germany (GDR), most of the worst pollution sources were simply closed. Local environmental groups, focusing particularly on public health, have continued to expose dangerous situations and to press for enforcement of environmental laws and the cleanup of the most significant health risks. Although environmental and public health officials now readily acknowledge the devastating consequences of the old regime, the new political and economic order has too few resources to do much more than to catalog the needs (French, 1991).

The creation of free societies in place of the old Communist systems challenges the victorious nations to replace military aid to allies with ecological cleanup aid to former adversaries. Interestingly, the West's technological edge in military hardware, often credited with forcing the collapse of communism, is exceeded by the superiority of western pollution control and environmental restoration technology. The Chernobyl nuclear plant accident is but one reminder of the potential for transborder ecological damage that can originate in former communist nations but imperil others.

Developing Nations

In the developing world, the political status of ecological concerns has been subservient to basic issues of poverty and economic growth. The leaders of many developing nations see environmental protection as a luxury that can only be sought *after* their people have obtained basic necessities such as adequate food, housing, and education. Rapid and unfettered development of their natural resources is often seen as the only way to access world trade and the foreign currencies to pay for needed imports and to repay foreign debt. Local environmental groups are usually tiny and politically marginal. Concerns for the impact of unregulated industries on public health are also limited to a few on the periphery of influence.

Many nations do have modern resource management and pollution-control laws on the books, but lack of money, trained personnel, and an abundance of corruption hampers these good intentions from implementation. Despite this bleak picture, there are some signs of change toward ecologically sound economic development practices and a higher priority for public health as a benefit to future prosperity (Okun, 1991).

Asia

The world's most populous nation, China, has reduced some of its threats to the environment. China's population control program has reduced the country's growth rate but has engendered controversy because of its coercive components. Unlike many other developing nations, China's government has the capability and experience to extend its laws to almost every part of the country. Outside of the heavy-industry sector, China has few vestiges remaining of its Communist economic policies, but its government system retains and may invoke totalitarian controls. Its Communist industrialization pattern emphasized heavy industry and neglected pollution control, but its current free-market-based rapid economic growth has yet to emphasize environmental protection. Rising standards of living in the 1980s and 1990s have continued to increase demand for electricity from coal burning power plants with their production of carbon dioxide (CO_2) and its impact on acid rain (Phillips, 1992) also increases.

China's centralized political bureaucracy does have some potential for enhancing its limited environmental protection efforts. For example, widespread access to education and basic services like clean water and a sophisticated public health system provide a basis on which to implement future clean air and water standards. Within the national and regional governments, environmental agencies have technical competence but only marginal impacts on important decisions. Citizens' groups are carefully controlled, so there is little visible grass-roots concern for environmental issues. However, neighborhood and workplace organizations are able to express their concern for the impacts of air and water pollution on public health.

On the Indian subcontinent, two societies coexist within each nation. In India and Pakistan, modern industrial urban sectors face the impact of pollution on public health. A large rural subsistence-agriculture sector is also living with the ecological impacts of rapid population growth amidst poverty. The pressure of fuelwood gathering and the expansion of lands for traditional agriculture is causing deforestation and species extinction in the rural environment. Although education is widespread, governments have limited capacity to provide basic public health services, such as clean water. Government environmental protection agencies are incapable of enforcing standards even if there were a political impulse to do so. Unlike the tightly controlled Chinese political scene, few constraints on organizing ecology groups exist. However, few effective large-scale environmental organizations exist, aside from neighborhood protests against local polluters.

The rapidly developing Asian nations of South Korea and Taiwan still have weak implementation of environmental laws. There are small environmental movements, largely focused on urban pollution issues, thus setting up a potential larger role for enforcement of public health standards. With affluence has come its impacts: urban air quality deterioration from the exhausts of increasing numbers of private autos. In Thailand, Malaysia, and Indonesia, these urban environmental issues are paralleled by controversial timber harvesting practices, as foreign (often Japanese) companies clear-cut large swatches of tropical forests (World Resources Institute, 1992). Environmental groups in developed nations have focused attention on these practices, often allying themselves with native peoples and villagers displaced by the lost forest. Under foreign scrutiny, new laws have been passed, but these countries are far from raising nature preservation to a national priority.

Latin America

The nations of Latin America can be seen as falling somewhere between the emerging potential of Asia and the bleaker ecological fate of Africa. Like much of Africa, many parts of Latin America find economic stagnation and ecological threats fueled by a growing population that outstrips economic and environmental capacity. Huge urban slums are growing, posing significant public health threats because of the lack of basic services, such as sewers and clean water. The once vast rainforests are shrinking, threatening the loss of much of the planet's biodiversity. Population pressures are bringing subsistence farmers with their slash-and-burn agriculture into the jungles. Governments are encouraging the rapid expansion of extractive industries to meet the interest payments from the high levels of foreign debt borrowed in the 1970s. However, the burden of these interest payments has provided opportunities for imaginative "debt for nature swaps" in which U.S. environmental groups buy heavily discounted Latin American government bonds and return this debt to a government in exchange for plans for ecological preservation. The governments themselves have called for the outright cancellation of the debts (Postel & Flavin, 1991).

U.S. environmental groups have participated at a significant level with their small Latin American counterparts, most notably alliances of some indigenous peoples. The traditional lives of the rainforest inhabitants are seen as having limited ecological impacts in contrast to the influx of peasant or commercial agriculture. Most of these governments are ineffective in enforcing the wide range of pollution control and nature preservation laws. In some cases, these laws are intended to be simply "window dressing" and are responses to pressure from outside interests, including international agencies. In other cases, the good intentions of domestic environmental interests are no match for the economic and political influence wielded by the opponents of pollution control and nature preservation. Latin America is often cited by critics of current international political and economic institutions as an example of the linkage of ecological peril with an inequitable international system. Although it is easy to point to the legacy of U.S. intervention and economic control in Latin America, it is less obvious that this history explains the current threats to the environment.

Africa

Africa is the focus of private, governmental, and international aid to protect and preserve its ecology, particularly its wildlife. Soaring population growth coupled with lagging economic growth threatens not only natural habitats but also the public's health. Moreover, a series of armed conflicts in many nations, such as recent civil wars in Sudan, Ethiopia, and Somalia, creates famine conditions and environmental devastation. There is a recent tradition of significant support from European ecology groups and governments to enhance the capacity of African states to enforce nature preservation laws. The world's media has given wide attention to Africa's large, photogenic animals such as elephants, lions, and zebras. Although a popular cause in the West, wildlife preservation projects are more difficult at home: The African central governments gain the most because of tourists' money, whereas local villagers lose their crops or cannot access former hunting grounds now inside game parks. Environmental and animal preservation groups are involved in new efforts to link economic benefits to local peoples with wildlife preservation (Pitman, 1990).

EMERGING INTERNATIONAL ENVIRONMENTAL STANDARDS

The impact of the elevation of environmental values is also apparent at the level of the world's political structures and processes. These changes include adding new international organizations, such as the United Nations Environmental Program, and expanding the responsibilities of existing entities to include environmental protection, such as occurred at the World Bank. In addition, international scientific bodies, like the International Union of Pure and Applied Chemistry, have added environmental topics to their global research agendas. Environmental concerns are also reshaping old ways of international decision making, because economic development loans administered by nations and international agencies now have new "green" criteria assessing the environmental impacts as well as simple economic feasibility. International treaties and conferences, such as The Protocol on Substances That Deplete the Ozone Layer, held in Montreal in 1987, create new institutions and pressures for further global actions (Caldwell, 1990).

Global Networks of Nongovernmental Organizations

Nongovernmental organizations (NGOs) are also playing more prominent roles in global environmental politics. In some cases the NGOs are supported and subsidized by governments, but most are truly nongovernmental. In developing nations, many environmental NGOs are supported by aid provided by environmental groups in the developed countries. Representatives of NGOs were very visible at the United Nations (UN) Conference on Environment and Development held in Rio in 1992, and the growing networks of NGOs permit significant global coordination of activities of the environmental movement. The organization is now established for worldwide mobilization on a range of issues, especially tropical deforestation and limits on trade in tropical timber as well as the protection of marine mammals. In other cases, the cooperation may take the form of tracking the international trade in hazardous waste, a focus of the national branches of Greenpeace, one of the few truly international groups. Environmental groups in the developed nations are frequently involved in the training and subsidization of ecology groups in developing nations, especially those involving indigenous peoples, such as tribes that live in threatened tropical rainforests (Livernash & Paden, 1992).

The environmental community is split between those groups who seek to make the existing world order "green" and those groups who wish to transform the international balance of power. A paradox is presented to environmental advocates. On the one hand, environmental groups are most influential in the wealthy Northern Hemisphere nations that fund—and for the most part, control—international political and economic organizations. These environmental groups have expanded their agendas to include global ecological issues and have the capacity to influence their national governments' activities in international forums. It is also clear that ecological protection mandates are being imposed by agencies such as the World Bank that limit aid to nations that do not follow environmental protection guidelines.

This "top down" imposition of green values by western governments, acting at the behest of domestic interests, constrains the autonomy of developing states and is described as "green imperialism." Yet, using the existing international system is a relatively fast route to implementing environmental concerns, and it can also be effective, because many developing nations depend on financial aid from the wealthy nations.

However, many environmental activists are also egalitarians who see the existing distribution of wealth and power in the world as the legacy of past colonialism and exploitation of the developing nations (the South) by the developed nations (the North). Adding ecological concerns to existing global structures does not alter the inequality between rich and poor nations and does little to correct what some see as the problem of overconsumption of the planet's resources by a small percentage of the world's people (Durning, 1991). The classic manifestation of this imbalance is the attention given by western ecology groups and governments to the value of tropical rainforests as the "lungs of the earth" and as a valuable "sink" for sequestering CO_2 in the form of plant carbon. The burden for protecting the earth's atmosphere is being placed on poor tropical countries by the world's industrial nations, who are benefiting from the energy and consumer goods that generate the CO_2. Egalitarians and representatives from poor nations argue that a more equitable sharing of the burden would follow the "polluter pays" principle and protect the atmosphere by reducing emissions as well as preserving rainforests.

Whatever the ultimate goals of the green groups, their alliances with ecology advocates in developing nations magnifies the political resources available to environmentalists of those nations and can often tip the balance of power. For example, a local environmental dispute between villagers and a subsidiary of a Japanese conglomerate logging in Malaysia can lead to demonstrations in front of an automobile dealer in the U.S. Midwest who sells cars made by another subsidiary of the conglomerate. Media coverage of the U.S. demonstrations serves to highlight threats to tropical rainforests by multinational corporations and increases calls for economic boycotts by U.S. consumers.

As another example, representatives of a Kenyan women's ecology group gain access to the media and a college speaking tour through the support of sympathizers in U.S. environmental organizations. Questions are raised in congressional hearings about whether U.S. aid to Kenya should be reexamined in light of the criticisms raised by the women's group. The bumper sticker slogan of "think globally, act locally" has become real in the internationalizing of developing-nation issues by ecology activists in the developed nations.

The "Greening" of International Organizations

The formal international system of organizations, treaties, and agreements has been transformed by the impact of ecological issues pressed by governments and NGOs. The World Meteorological Organization (WMO) can trace its origins to the exchange of weather information in the 19th century and is now on the cutting edge of data collection and analysis concerning the state of the earth's atmosphere. Hundreds of WMO stations in 78 nations now measure air pollutants, acid precipitation, and the health of the ozone layer as part of the Global Atmosphere Watch program (Koptyug, 1992). The International Labor Organization (ILO) began its efforts to monitor and disseminate information on pollution hazards in the workplace in the 1920s and now provides a global forum for environmental health and safety standards debates that have implications far from the factory floor. Similarly, the World Health Organization (WHO) has established guidelines for water quality and air pollution in addition to its older public health roles, such as standardizing vaccinations and fighting epidemics (Sand, 1991).

Landmark events like the 1972 Stockholm Conference on the Human Environment have the effect of revitalizing and reorienting existing bodies as well as spurring the creation of new ones, such as the United Nations Environment Program. Within the UN structures, the Food and Agriculture Organization (FAO) has a mandate from its birth in 1945 to promote the conservation of natural resources, and has become involved in the study of sustainable agriculture, using minimum amounts of chemicals, in contrast to its earlier advocacy of farm chemicals. The UN Development Program, long a venue for the concerns of poor nations, also has incorporated environmental issues as it grapples with the concept of "sustainable development," a possible route out of poverty that preserves ecosystems. The United Nations Educational, Scientific, and Cultural Organization (UNESCO) in addition to providing a common format for global exchanges of scientific information on the environment has a process for designating "biosphere reserves" under its 1976 Man and Biosphere Program (Abramovitz, 1992). Other UN offices have funded international conferences in developing nations on women and the environment, linking women's' educational and family-planning options to protecting the environment. In the 1970s the UN Development Program had come to recognize the need to explicitly incorporate women into plans for economic development, so this linkage to the environment was evolutionary (Livernash & Paden, 1992).

Although the UN system of organizations and conferences is usually seen as an arena for exchanges between diplomats representing national governments, over the past decade NGOs have become increasingly part of the activities of the world body. These NGOs act both as conduits for the carrying out of UN projects, such as family planning, and also as agents pressing for policy change in the system. As noted earlier, environmental groups from developed countries have begun to apply their sophisticated political strategies, honed in national politics, to influence planetary politics.

Other International Institutions

The world's financial and economic institutions, notably the International Monetary Fund (IMF) and the World Bank, have also become "greener," but much more recently and reluctantly than other organizations. For the World Bank, 1986 was the transition year when it formally accepted responsibility for evaluating the environmental impacts of its loans to economic development projects in developing nations. The bank has since enlarged its environmental assessment office and routinized the assessment of environmental impacts as criteria for the granting of loans. The Tropical Forestry Action Plan supported by the bank is, in part, a recognition that past loans had encouraged deforestation (Miller et al, 1991). The bank is now denying requests for major loans for multipurpose water projects in China and India based on ecological criteria. The IMF, which loans money to nations to help them balance their international trade and currency accounts, has enormous potential to add ecological criteria to its already long list of conditions for loans. Analysts note that the IMF has yet to exercise its powers to protect the environment and may, through some of its requirements for economic reforms as conditions for loans, actually encourage ecological damage (Mathews, 1991).

Trade and the Environment: GATT and NAFTA. Another pillar of the world's economic system, the 45-year-old General Agreement on Tariffs and Trade (GATT) has also become the center of controversy concerning its environmental implications. Headquartered in Switzerland, the GATT combines elements of a treaty with those of an international organization and is designed to eliminate trade barriers among nations (Charnovitz, 1992). The "trade versus environment" controversy became focused when

a 1991 GATT panel ruled that the United States could not restrict tuna imports based on U.S. dolphin protection laws. Unilateral ecology standards were seen as barriers to international trade. If the United States wished to protect dolphins from sloppy foreign tuna fishing, it would have to lead other nations into an international agreement and not take the shortcut of using trade restrictions to impose ecological standards. This case, although not a binding precedent, has aroused opposition from environmentalists—not only to the 1993 Uruguay Round of GATT negotiations but also to the 1993 approved North American Free Trade Agreement (NAFTA).

If, as it now seems, the United States is not going to be able to require that its trading partners/competitors uphold the same levels of environmental protection as do U.S. businesses, there are fears of a flight of industry and jobs out of this country and into nations with no environmental regulation costs (Mumme, 1993). The NAFTA adds Mexico to the existing Canada/U.S. free trade agreement, and most observers count Mexico among the nations that are unable to enforce their own environmental laws. Opponents of free trade, including trade unions fearing job losses and some businesses fearing new competition, found the environmental opposition to NAFTA convenient and joined coalitions with some of the environmentalist groups. Supporters of expanded trade point to the EC model, which allows some nations to impose *stricter* pollution control laws but forbids the creation of pollution havens within their free trade area (Cairncross, 1992). The popularity of environmentalism, especially when joined to the threat of job losses, imperiled the passage of NAFTA by Congress and forced President Clinton to add side agreements guaranteeing high environmental protection standards to the agreement.

Proponents of freer trade argued that the trade-induced increase in Mexico's economic growth rate would add revenues that can be used to improve private and public environmental protection. Conflict over whether environmental standards should be explicitly incorporated into international trade agreements promises to be a feature of environmental policy debate throughout the 1990s. In addition, the federal judiciary has entered the picture in a 1993 U.S. District Court ruling requiring an Environmental Impact Statement as part of the NAFTA deliberations. Although reversed on appeal, the challenging lawsuit was brought by several of the large

NGOs as a way of forcing political change via a judicial mechanism. This challenge illustrates the complexity and linkage of politics, trade, and the environment (see also Chapter 2, Economic and Ethical Issues).

International treaties and agreements. Agreements among nations to protect or manage aspects of the environment predate modern ecological concerns. Decades ago, the United States entered into agreements to preserve migratory birds with Canada and to regulate fur seal harvests in the Pacific Ocean. The leap from an agreement between two nations to manage a common river or among several nations to divide up fish catches in the North Atlantic to truly global efforts to protect the planet is a product of modern environmentalism. The 1972 United Nations Conference on the Human Environment held in Stockholm is seen as putting the environment on the world's diplomatic agenda (Caldwell, 1990). More recently, conferences have been held and agreements drafted to protect the atmosphere—truly a global resource.

Protection of wildlife, habitat, and oceans includes a series of conventions, beginning in 1959, on preserving the Antarctic, and a 1991 Protocol on Environmental Protection designating the continent a natural reserve and prohibiting mining for 50 years. Other agreements on waterfowl habitat and wetlands were signed in 1971, and a convention on international migratory species was effected in 1979. Among the best known animal preservation agreements is the ban on the international ivory trade promoted through The Convention on International Trade in Endangered Species (CITES), established in 1973. Several agreements from the 1970s (MARPOL 73/78) limit dumping waste and pollution from ships into the ocean, as well as the 1982 Law of the Sea Convention, which set up a legal framework for enforcement of marine pollution regulations. The status of these conventions and agreements varies. Some conventions, such as ones preserving the Antarctic, are ratified and in force; others, like the Law of the Sea, lack sufficient signatory nations to put them into effect (Soroos, 1993).

Perhaps the most far-reaching and controversial global environmental agreements are those involving the atmosphere (see also Chapter 14, Global Issues). The 1987 Protocol on Substances That Deplete the Ozone Layer was amended in 1990 to speed up the phase-out of chemicals such as chlorofluorocarbons (CFCs) linked to damage to the stratospheric ozone layer. Although some scientists continue to argue that the threat to the ozone layer has been overstated, the impetus to protect the ozone layer came from sophisticated international research cooperation (Benedick, 1991). Moreover, although the Reagan administration has been frequently blamed for neglecting international environmental issues such as acid rain, it is clear that the United States took a leading role in pushing other key nations to accept the agreement. The 1987 Protocol was also significant in trying to compensate developing nations for the loss of the benefits of CFCs as simple, inexpensive refrigerants. Initially, nations with low per capita use of CFCs were allowed to increase their use even as production in the developed countries decreased. In the 1990 amendments, a $240 million fund from the developed states was established to subsidize non-CFC refrigerants in poor nations, a recognition of the need to end CFC use worldwide with equity to those nations that had not benefited from decades of CFC use. Establishing a fund paid into by the countries that have gained from environmentally harmful practices in order to aid poor nations in *not* replicating this ecologically risky path is seen as a model for the move costly and complex global climate change policies. This debate occurred in the context of the "Earth Summit"/Rio Summit in 1992.

UN Conference on Environment and Development. The United Nations Conference on Environment and Development (UNCED), also called the Earth Summit, was held in Rio de Janeiro in June of 1992. Representatives of more than 150 nations, including more than 100 heads of state, attended along with 1,400 NGOs and 8,000 journalists, making the conference a global media event as well as a landmark in global environmental politics. The most visible results of the conference were the signings of two conventions, one on climate change, the other on biodiversity. However, these conventions are only part of what is likely to be the enduring legacy of Rio: the establishment of a process and institutions to place environmental protection and preservation at the forefront of the world's political agenda for years to come.

As part of the Climate Change Convention, governments issue reports on greenhouse gas emissions and pledge to hold CO_2 emissions to 1990 levels by the year 2000. This emissions limit is a nonbinding target, but the United States refused to sign the convention out of concern for the costs of the U.S. economy. One hundred and fifty countries did sign, including the EC, but the key will be whether their legislatures will ratify the convention. This step may entail much more controversy when the price tag for holding down the production of greenhouse gases is debated.

The biodiversity convention also gained 156 signatory nations, but the United States did not sign until June of 1993. The convention requires that nations promise to protect endangered species within their own territory, which is not controversial. Conflict arises over a global reimbursement system for part of the commercial value of genetic material from rainforests and other developing nations. This obligates countries to share the revenues with the nations where the original genetic materials were collected. This provision causes concern for biotechnology and pharmaceutical companies that create products based on developing-nation genes. The United States accepted the convention by adding explicit language on intellectual property rights issues and limiting royalty payments as part of its acceptance (U.S. State Department, 1993).

The institutional apparatus for continuing the work of the Earth Summit is in the hands of the Sustainable Development Commission, with a staff and organization originally scheduled to be established in New York in 1993 as part of the UN. The commission will be a focus for implementation of the massive Agenda 21 report hammered out in the months preceding the summit. The commission will coordinate international environmental research and act as a liaison with NGOs as well as receive reports from governments on national implementation of Agenda 21 (Haas et al, 1993). The biodiversity and Climate Change Conventions were part of this global environmental agenda.

Agenda 21, with its focus on the complex interconnections of economic development in developing nations as a prerequisite for environmental protection, faces uncertain financing and thus an uncertain future. In Agenda 21, the developed countries were willing only to reaffirm the goal of contributing 0.7% of gross domestic product (GDP) in official development assistance to developing countries. However, estimates are that $125 billion per year in Western aid is needed to implement Agenda 21. Currently, such aid runs about $60 billion per year. Although the magnitude is different, there is a precedent for a fund to aid environmental protection and economic development: the 1990 Global Environment Facility (GEF) loan fund. This fund, however, had only $1.3 billion but may be precursor to a future massive environment/development transfer from North to South. Given the current recession in most of the industrialized countries, it seems unlikely that these governments will be willing to allocate scarce resources to such a fund. However, the planned renewal of the GEF will occur in the 1993–1994 time frame. All countries with a per capita income of less than $4,000 per year are eligible for funds. GEF resources are allocated on a formula that gives 40% to 50% for projects designed to reduce global warming 30% to 40% to conserve biological diversity, and 10% to 20% to protect international waters.

The striking contrast between the focus of the Stockholm conference in 1972 on industrial pollution in developed nations and the Rio Summit in 1992 with its expansive Agenda 21 illustrates the transformation of global environmentalism. The World Resources Institute, for example, now lumps the nuclear test ban treaty—once a centerpiece in Cold War diplomacy—with conventions to end CFC production as efforts to protect the atmosphere. Similarly, biological and toxic weapons agreements, nuclear accident notification conventions, nuclear accident assistance protocols, and international hazardous waste movement (the Basel Convention) are categorized together (World Resources Institute, 1992). This blurring of once-crucial distinctions between civilian and military policy arenas exemplifies the emergence of the environment as a central concern in the post-Cold War world.

FUTURE ENVIRONMENTAL TRENDS

Within nations and globally, environmental protection and preservation concerns have expanded to include issues of equity that were once separate (see Environmental Ethics in Chapter 2, Economic and Ethical Issues). At home, the disproportionate siting of hazardous waste facilities in minority communities has lead to widespread concerns over "environmental racism," with the EPA devoting an issue of *The EPA Journal* (March/April 1992) to the topic of environmental equity. Internationally, the yet-to-be implemented Basel Convention on the Control of Transboundary Movements of Hazardous Waste and Their Disposal seeks to ensure that poor nations similarly do not become the toxic repositories for the wealthy, waste-producing countries. Along the same lines, environmental activists question whether developed nations should export to developing countries products such as agricultural chemicals that are banned for domestic use because of their ecological or health risks. The broader question—when we as individuals and a nation should accept responsibility for the global impact of our actions—is no longer simply a concern for seminars on ethics but is now an ongoing subject of corporate and government policy debate. An analogy from the U.S. Civil War may be relevant here: Young men of wealth at the time could avoid conscription by hiring a substitute to go to war in their place. The current possibility of wealthy communities and countries "hiring" poor neighborhoods or nations to bear the burden of environmental risks without commensurate benefits raises similar moral issues.

The implications the unequal distribution of environmental benefits is an emerging international issue that builds on existing debates over international economic inequality. The interaction between levels of economic development and levels of environmental preservation displays what is called the "paradox of development." Wealthy nations invest more in environmental protection, and, by most measures, have better environmental quality than do poor countries. However, the process of acquiring national wealth through industrialization places enormous burdens on the environment. Thus, the emphasis in Agenda 21 on harmonizing economic development with environmental preservation addresses exactly the key problem of environmental quality in the 21st century.

Only through increased wealth can environmental quality be provided; yet this increase imperils the health and the ecology we wish to preserve. Moreover, increases in individual as well as national wealth are associated with increased demands for resources, especially energy. Although Agenda 21 embraces the theme of "sustainable development," as an imperative for developing nations to move toward the most efficient use of resources and the least ecological damage, sustainability on a planetary scale will require those individuals and countries that consume the most to make proportionately greater contributions to sustainability. It remains to be seen how much citizens in the wealthiest countries are willing to pay for environmental protection in the developing nations, and how much they will curtail their freedom to consume for the sake of global sustainability.

One of the fundamental principles of ecology, that "everything is connected to everything else" also describes the relationship among local, national, and international environmental groups. Ecology groups are involved in an ever-widening range of social and economic issues. At the same time, existing community organizations, especially in minority neighborhoods, are becoming advocates for local ecology concerns. Ad hoc citizens' groups, often sparked by the lack of response from local government agencies to perceived threats to public health also are active. On one level, this burgeoning of local activity, whether in a middle-class suburb in the United States or a tribal village in a tropical rainforest, can be described as Not In My Back Yardism (NIMBYism). Critics of NIMBYism observe that neighborhood groups are willing to accept the benefits of, for example, municipal trash collection, so long as the landfill is located in someone else's back yard. Yet a more accurate understanding of the spread of grassroots environmentalism is that it reflects the extension of environmental concern throughout the consciousness of most of the inhabitants of the earth. At its root, environmentalism is a way of thinking about the relationship of humans to the planet. Organizing to implement this new viewpoint is changing the way the earth's inhabitants behave as well as the way they think.

SUMMARY

Environmentalism as a political movement has come full circle. National preservation and conservation movements of the 1900s have been transformed into 1990s sustainable development initiatives that act on a global stage. The environmentalism of the 1990s truly illustrates the phrase "think globally, act locally." Environmental managers in the 1990s will continue to be confronted with local problems that are small dramas in the larger global play of politics, economics, and ethics.

REFERENCES

Abelson PH. Pathological growth of regulations. *Science* 26:1854 (25 June 1993).

Abramovitz J. Wildlife and habitat. In World Resources Institute (ed), *World Resources 1992–93*. Oxford, UK: Oxford University Press, 1992.

Andrews RNL. Class politics or democratic reform, environmentalism, and American political institutions. *Nat Res J* 20:221–242, 1980.

Andrews RNL. Risk assessment: Regulation and beyond. In Vig N, Kraft M (eds), *Environmental Policy in the 1990s,* 2nd ed. Washington DC: CQ Press, 1990.

Axelrod RS. Environmental policy management in the European Community. In Vig NJ, Kraft M (eds), *Environmental Policy in the 1990s*, 2nd ed. Washington DC: CQ Press, 1993.

Bartlett R. The budgetary process and environmental policy. In Vig N, Kraft M (eds), *Environmental Policy in the 1980s*. Washington DC: CQ Press, 1984.

Benedick RE. Protecting the ozone layer: New directions in diplomacy. In Matthews JT (ed), *Preserving the Global Environment*. New York: Norton, 1991.

Bryner, GC. *Blue Skies, Green Politics: The Clean Air Act of 1990*. Washington DC: CQ Press, 1993.

Cairncross F. The environment: Whose world is it, anyway? Reprinted in Allens J (ed), *Environment 93/94*. Guilford, CT: Dushkin, 1992.

Caldwell LK. International environmental politics: America's response to global imperatives. In Vig N, Kraft N (eds), *Environmental Policy in the 1990s*, 2nd ed. Washington, DC: CQ Press, 1990.

Caulfield H. The conservation and environmental movements: An historical analysis. In James Lester (ed), *Environmental Politics and Policy*. Durham, NC: Duke University Press, 1989.

Charnovitz S. GATT and the environment: Examining the issues. *Int Environ Affairs* 4:203–233, 1992.

Costain WD, Costain AN. The political strategies of social movements: A comparison of the women's and environmental movements. *Congress & The Presidency* 19:1–28, 1992.

Dunlap RE. Trends in public opinion toward environmental issues: 1965–1990. In Dunlap RE, Mertig A (eds), *American Environmentalism: The U.S. Environmental Movement, 1970–1990*. Washington DC: Taylor & Francis, 1992.

Durning A. Asking how much is enough. In Brown L, et al (eds), *State of the World 1991*. New York: Norton, 1991.

Engelbert E. Political parties and natural resources policies: An historical evaluation. *Nat Res J* 1:224–256, 1961.

Erskine H. The polls: Pollution and its costs. *Public Opinion Quarterly* 36:120–135, 1972.

French H. Restoring the East European and Soviet environments. In Brown L, et al (eds), *State of the World 1991*. New York: Norton, 1991.

Gillroy JM, Shapiro RY. The polls: Environmental protection. *Public Opinion Quarterly* 50:270–279, 1986.

Haas PM, Levy MA, Parson EA. Appraising the earth summit: How should we judge UNCED's success? Reprinted in Goldfarb TD (ed), *Taking Sides: Clashing Views on Controversial Environmental Issues*. Guilford, CT: Dushkin, 1993.

Hays S. *Conservation and the Gospel of Efficiency: The Progressive Conservation Movement, 1890–1920*. New York: Atheneum, 1959.

Hays S. *Beauty, Health and Permanence: Environmental Politics in the United States, 1955–1985*. New York: Cambridge University Press, 1987.

Jancar-Webster B. *Environmental Action in Eastern Europe: Responses to Crisis*. Armonk, NY: ME Sharpe, 1993.

Koptyug V. International cooperation and some research needs to improve our understanding of the chemistry of the atmosphere. In Birks J, Calvert J, Sievers R (eds), *The Chemistry of the Atmosphere: Its Impact on Global Change CHEMRAWN VII Perspectives and Recommendations*. Washington DC: Agency for International Development, 1992.

Levy B, Hertzman C. Environment and health in Central Europe. In World Resources Institute (ed), *World Resources 1992–93*. Oxford, UK: Oxford University Press, 1992.

Livernash, R. Regional focus: Central Europe. In World Resources Institute (eds), *World Resources 1992–93*. Oxford, UK: Oxford University Press, 1992.

Livernash R, Paden M. Policies and institutions: Nongovernmental organizations: A growing force in the developing world. In World Resources Institute (ed), *World Resources 1992–93*. Oxford, UK: Oxford University Press, 1992.

Mathews JT. The Implications for US policy. In Mathews JT (ed), *Preserving the Global Environment*. New York: Norton, 1991.

Miller K, Reid W, Barber C. Deforestation and species loss: Responding to the crisis. In Mathews JT (ed), *Preserving the Global Environment*. New York: Norton, 1991.

Mitchell RC. National environmental lobbies and the apparent illogic of collective action. In Russell CS (ed), *Collective Decision Making*. Baltimore: Resources for the Future, Johns Hopkins University Press, 1979.

Mitchell RC. Public opinion and environmental policies. In Vig N, Kraft M (eds), *Environmental Policy in the 1980s*. Washington DC: CQ Press, 1984.

Mitchell RC. From conservation to environmental movement. Washington DC: Resources for the Future Discussion Paper QE 85–12, 1985.

Mitchell RC. Public opinion and the green lobby: Poised for the 1990s?. In Vig N, Kraft M (eds), *Environmental Policy in the 1990s,* 2nd ed. Washington DC: CQ Press, 1990.

Mumme S. Environmentalists, NAFTA and North American environmental management. *J of Env and Devel* 2:205–217, 1993.

Okun D. A water and sanitation strategy for the developing world. Reprinted in Allen J (ed), *Environment 93/94*. Guilford, CT: Dushkin, 1991.

Phillips R. Poor countries: Breaking the cycle of poverty, environmental degradation, and human deprivation. In World Resources Institute (ed), *World Resources 1992–93*. Oxford, UK: Oxford University Press, 1992.

Pinders J. *European Community: The Building of a Union*. Oxford, UK: Oxford University Press, 1991.

Pitman D. Wildlife as a crop. Reprinted in Allen J. (ed), *Environment 92/93*. Guilford, CT: Dushkin, 1990.

Postel S, Flavin C. Reshaping the global economy. In Brown L, et al (eds), *State of the World*. New York: Norton, 1991.

Reilly WK. Why I propose a national debate on risk. Reprinted in Goldfarb TD (ed), *Taking Sides: Clashing Views on Controversial Environmental Issues,* 5th ed. Guilford, CT: Dushkin, 1993.

Ridgeway J. Environmental devastation in the Soviet Union. Reprinted in Allen J (ed), *Environment 92/ 93*. Guilford, CT: Dushkin, 1990.

Sand P. International cooperation: The environmental experience. In Mathews JT (ed), *Preserving the Global Environment*. New York: Norton, 1991.

Schnaiberg A. *The Environment: From Surplus to Scarcity*. New York: Oxford University Press, 1980.

Schoenfeld AC, Meir RF, Griffin RJ. Constructing a social problem: The press and the environment, *Soc Prob* 27:38–61, 1979.

Soroos MS. From Stockholm to Rio: The evolution of global environmental governance. In Vig NJ, Kraft M (eds), *Environmental Policy in the 1990s,* 2nd ed. Washington DC: CQ Press, 1993.

U.S. EPA. *Unfinished Business: A Comparative Assessment of Environmental Problems*. Washington DC: EPA, 1987.

U.S. EPA. *Reducing Risk: Setting Priorities and Strategies for Environmental Protection*. Washington DC: EPA, 1990.

U.S. State Department. U.S. support for global commitment to sustainable development, by Vice President Gore. *Dispatch* 4(24). June 16, 1993.

Vig NJ, Kraft M. Environmental policy from the seventies to the nineties: Continuity and change. In Vig NJ, Kraft M (eds), *Environmental Policy in the 1990s,* 2nd ed. Washington DC: CQ Press, 1990.

Vogel D. Environmental policy in Europe and Japan. In Vig NJ, Kraft M (eds), *Environmental Policy in the 1990s,* 2nd ed. Washington DC: CQ Press, 1990.

Whitehead C. Introduction: Environment protection in the framework of the European Community legal system. In Commission of the European Communities, *European Community Environment Legislation,* Volume 7, Water. Luxembourg: Office for Official Publications of the European Communities, 1992.

Williams AM. *The European Community: The Contradictions of Integration*. Oxford, UK: Basil Blackwell, 1991.

World Resources Institute. *World Resources 1992–93*. Oxford, UK: Oxford University Press, 1992.

2

Economic and Ethical Issues

C. Ford Runge, PhD
Gary R. Krieger, MD, MPH, DABT
Joan Gibb Engel, PhD
J. Ronald Engel, PhD

In December 1992, then President-elect Clinton said, "Our future depends on maintaining a sustainable environment, and in doing that we can create economic opportunity." In this sentence, President Clinton linked two powerful forces: sustainability of the environment for future generations and increasing economic opportunity based on achieving some type of equilibrium with the natural environment. Can these two ideas be linked in a complementary fashion, or are they locked in a zero-sum game where there are only winners and losers? The environmental movement has been motivated by a perceived imbalance between the economic interests of individual nations or industries and the environmental interests of the general global community (see Globalization of Environmental Concerns in Chapter 1, History and Development; and Environmental Ethics, later in this chapter). Theoretically, economic analysis provides insight into the mechanisms for efficiently allocating resources so that pollution is minimized and environmental protection maximized in ways that avoid irreversible harm to a nation's economy.

This chapter will present and discuss some of the fundamental economic concepts that have been and are currently applied to environmental policies. The emphasis will be on general mechanisms that are applicable to both an individual country and the larger global economy.

During the last five years there has been increasing disenchantment in the United States with regulatory command-and-control mechanisms that mandate pollution control. Command-and-control mechanisms operate at two levels: (1) government specifically mandates the technology an industry must use for pollution control; and (2) an emission-rate cap that all pollution sources must meet is set regardless of economic costs (Alper, 1993). Since the 1970s, the United States has established a gigantic, multistatutory regulatory framework for environmental protection that involves hundreds of thousands of people and spends hundreds of billions of dollars (Abelson, 1993). A 1991 U.S. Senate environmental/economic study concluded that, "while command-and-control approaches can be effective in reducing pollution output, they tend to impose relatively high costs on society because some unnecessarily expensive means of controlling pollution will be used" (quoted in Alper, 1993). Clearly, these sentiments indicate that the United States may have or is beginning to reach the saturation point for command-and-control environmental policies. Increasingly, effort in the United States and worldwide has been directed toward the development of free-market environmentalism.

To many natural scientists, "free-market environmentalism" appears to be a contradiction in terms and is simply a political slogan. However, recent work by Repetto and other economists indicates that the marketplace is capable of sending appropriate economic signals if policymakers and economic models consider the economic value of natural resource stocks (Repetto, 1989 & 1993). Repetto recently was quoted, "If we can enact policies that adjust prices so that they more accurately reflect all the costs associated with producing a particular pollutant or using a particular resource, then society will make better decisions" (quoted in Alper, 1993).

Fundamental to Repetto's argument is the concept that the 50-year-old United Nations System of National Accounts (UN SNA) totally ignores the economic costs of natural resource degradation and the pressures on global life-support systems like climate and biodiversity (Repetto, 1992). The UN SNA is a standard method of measuring and reporting fundamental economic parameters of consumption, savings, and investment. While the UN SNA recognizes certain natural resources such as land, minerals, and timber as economic assets to be included in a nation's capital stock, the UN SNA income and product accounts do not (Repetto, 1992). Why does this difference matter and what impact does it have on economic/environmental policy? Consider this example: A country has substantial forest "assets." If this forest has been completely depleted, there is a substantial difference between the income and product accounts before and after resource utilization. Therefore, there should be a charge, or economic cost, for the asset depreciation between the two time intervals. In many developing nations, natural resource assets are in fact not depreciated, and the true cost of resource development is not accurately calculated. Repetto has bluntly stated,

> Codified in the UN SNA, the bias against natural resource assets gives false signs to policymakers. It reinforces the illusion that a dichotomy exists between the economy and the environment and so leads policymakers to ignore or destroy the latter in the name of economic growth. It confuses the depletion of valuable assets with the generation of income. The result can be illusory gains in income and permanent losses in wealth (Repetto, 1992).

Based on Repetto's argument, free-market environmentalism must consider and price the value of natural resource assets both in terms of their direct monetary and indirect societal value. Indirect values can include difficult-to-price attributes such as enjoyment, cultural relevance, and preservation of biodiversity. The utilization of natural resources is increasingly complicated by these types of problems. Economists have directed substantial effort toward determining the value of environmental protection. This type of economic valuation is extremely difficult and complex, because a large number of indirect costs, known as *externalities,* are associated with environmental valuation. As a general concept, externalities can include direct and indirect environmental impacts (such as loss of enjoyment) as well as other economic impacts on employment and human welfare.

As an example, consider the environmental costs associated with the pricing of energy resources such as electricity. Air emissions can equal between 50% and 100% of the conventional costs of typical electricity resources (Woolf, 1993). These costs can clearly have a major impact on the environmental planning associated with electricity production.

Increasingly, the ability to establish monetary values for environmental impacts is considered the most effective strategy for ensuring that the market sends accurate price signals to both producer and consumer.

In the United States, the ability to estimate monetary values of environmental externalities is a major consideration in natural resource damage actions. In general, there are two primary methods of cost estimation: (1) damage cost, and (2) control cost (Woolf, 1993). Damage costs are based on the identification of the amount and type of environmental damage and represent the dollar value of environmental protection. For example, historic mining activities may have impaired adjacent streams and lakes. The loss of this fishing resource would be associated with a calculated dollar amount. Alternatively, control costs represent the costs associated with avoiding the potential environmental impact.

Obviously, there is tremendous subjectivity associated with both the assessment of damage and the attachment of monetary value to the environmental resource. Despite this uncertainty, there are a number of approaches used to value nonmonetary goods (Woolf, 1993):

- *Direct market-based values.* Direct multiplication of the quantity of goods lost or damaged by the current market value can work for goods such as timber and fish, for which an existing market is in place.

- *Revealed preferences.* This is the ability to infer value by observing the behavioral preferences exhibited by consumers. For example, what costs are the public willing to pay for different types of recreational facilities (swimming, fishing, hiking) located at similar distances?
- *Hedonic pricing techniques.* What price is the public willing to pay to either obtain or avoid a given environmental impact and how are these reflected in property values? For example, would electric rate payers absorb the large cost increases associated with underground line burial in order to avoid electromagnetic field exposures?
- *Contingent valuation techniques.* A sample population is surveyed in order to determine the monetary valuation individuals will pay to avoid an untoward effect or how much compensation would be required if an adverse outcome occurs.

Based on these and other techniques, economists are providing a mechanism that allows companies, regulators, and the public the ability to "internalize" externalities. Tradable effects for air emissions, pollution fees, and energy taxes (such as taxes based on the carbon content of fuels) are examples of internalizing externalities so that potential control and damage costs are reflected in a given product market value.

In the United States, the 1990 Amendments to the Clean Air Act established a market for tradable emission allowances for sulfur dioxide (SO_2). This market rewards utilities that reduce pollution by investing in new technology, because the utility has the ability to recoup its investment and sell unused SO_2 emission credits to other utilities at market prices (see Chapter 14, Global Issues, for further discussion of SO_2 emissions and the effects of the 1990 Clean Air Act, Title IV Acid Rain Policy). The intent of this program is to reduce SO_2 emissions by more than 50% while saving utilities more than $1 billion in costs. If successful, this program will be a landmark in free-market environmentalism.

Compared to command-and-control policies, free-market environmentalism holds the promise of more effectively and efficiently allocating resources and distributing costs. However, the ability to establish market mechanisms within a given country's borders is different from internalizing externalities in a global economy.

U.S. ENVIRONMENTAL LEGISLATION IN AN INTERNATIONAL CONTEXT

Health and the environment are no longer purely matters of domestic policy. (The material in this section draws from Runge, 1990a; Runge et al, 1994.) In the 1980s, air pollution, acid rain, and global warming became major items on the international agenda. That shift reflects growing recognition of the global impact of economic development and the rising problem of international externalities as hazards spill over national borders and affect the oceans, air, and climate (see Chapter 14, Global Issues).

But just as health risks flow through the world's physical environment, they also flow through the world economy—and threaten to disrupt it. Risk tends to move to countries with the least regulation, in some cases because the advanced economies are directly exporting to the less advanced ones various products and production methods no longer considered safe at home. Ironically, the higher the advanced countries set their regulatory standards, the more they create incentives for a kind of "environmental arbitrage," that is, for making a profit by producing goods cheaply where regulation is lax and selling them dearly where regulation is strict. Some regulatory differences exist among countries at the same stage of development, but in the world as a whole the flow of environmental and health risks runs from the technologically advanced North to the developing nations of the South.

At the same time, growing consumer concerns about health in the advanced countries are prompting more attention to health hazards from imported products, particularly food. As a result, domestic agricultural interests and other producers seeking protection from foreign competition are finding a new source of support in the environmental and consumer movements. Import restrictions, when presented as a public health measure, gain a legitimacy that they might not otherwise enjoy.

These events suggest the new realities created by uneven environmental and health regulation. When nations exchange goods and services, they also trade environmental and health risks. These risks are the opposite of services—they are environmental and health *disservices* traded across national borders.

This trade in disservices is an emerging source of tension in the world. The United States and other signatories to the General Agreement on Tariffs and Trade (GATT) were committed to pursuing more open borders in the 1993 Uruguay Round of trade negotiations. But as national health, safety, and environmental regulations grow in importance, different national regulatory priorities pose several related problems for trade and development. First, groups in some countries will seek to gain competitive advantage over foreign producers by opposing strong environmental regulation at home. Alternatively, other groups may try to use health risks as an excuse to keep out imported products. Producers may also try to export products that threaten the health of consumers in foreign countries. All such actions could do unnecessary harm to *both* environmental quality and world trade, unless we devise new international arrangements to resolve the problems.

TRADE AND ENVIRONMENT: THE ISSUES

Three main issues dominate the debate over trade and environment. The first is the potential environmental impact of trade liberalization, both in the regional context of the North American Free Trade Agreement (NAFTA) between the United States, Canada, and Mexico, and in the global trade talks in the GATT. The second issue is the possible use of environmental measures as nontariff barriers to trade. The third is the relationship between trade agreements under NAFTA and GATT and the variety of international agreements affecting the environment, such as the Montreal Protocol agreement to protect atmospheric ozone.

Environmental Impacts of International Trade

The first and most politically charged issue is the concern that trade liberalization, whether in NAFTA or under GATT, will lead to increased levels of environmental damage. There are at least three different aspects of this concern, which has been expressed strongly by a variety of environmental groups and members of the U.S. Congress. At the most basic level, many of these environmental concerns derive from the "scale effects" of freer trade. *Scale effects* are the result of increases in the quantity of goods and services moving within countries and across borders. To the extent that increases in trade following liberalization lead to greater transportation needs, higher levels of manufacturing output, and general increases in the demand for raw and processed products, they can also impose greater wear and tear on natural ecosystems. Among these possible effects are increasing consumption of nonrenewable natural resources (including fossil fuels, minerals, and old-growth forests) and increasing levels of air and water

pollution. A particularly striking example of these impacts often cited by environmental groups is the pollution found in the rapidly growing *maquiladora* sector of Mexico (see also Chapter 11, Public Health Issues, for further information). It has also been suggested that differences in environmental standards, especially between North and South, will create "pollution havens" for firms and industries seeking less regulatory oversight. Finally, the proposed harmonization of environmental standards is argued to lead to a "lowest common denominator," in which higher levels of environmental protection are sacrificed in the name of competitiveness.

Trade Impacts of Environmental Legislation

In contrast to the environmental community's concern about the impacts of more liberal trade, those most directly involved in trade have tended to focus on a second main issue: the potential for protectionism disguised as environmental action. This can occur when a country or trading bloc protects internal markets in the name of environmental health or safety, such as the decision of the European Community (now known as the European Union or EU) to ban the import of beef from cattle treated with certain growth hormones. It can also occur when higher levels of environmental standards are used to bar market access to goods and services produced under lower levels of regulation, especially by developing countries. The fundamental issue concerns the ability to distinguish legitimate environmental measures, which may well distort trade, from those that are not only trade distorting but have little basis from an environmental standpoint. Developing such criteria involves complex legal, scientific, and institutional issues.

Interaction of Environmental and Trade Policies

The third issue involves the relationship between trade agreements and environmental agreements. In the last decade, a variety of new international agreements have been negotiated in response to global environmental challenges such as ozone depletion, species extinction, protection of Antarctica, and international management of the oceans. The Rio Conference on Environment and Development held in June 1992, resulted in a broad new mandate for environmental action, *Agenda 21,* together with the creation of a new UN Commission on Sustainable Development (see Chapter 4, International Legal and Legislative Framework).

Some of these agreements call on their signatories to refrain from trade in certain goods or processes. In the 1993 NAFTA negotiations, for example, a trinational commission was created with apparent authority over trade with damaging environmental effects. This commission has authority apart from the GATT and the existing GATT dispute resolution process. The question is: How are international environmental accords to be balanced with existing or new trade obligations under GATT? What body of international law and which international institutions should exercise authority over the intersection between multilateral environmental and trade policy?

Where different groups stand on these issues depends on their perspective. Some environmental and consumer groups have been highly vocal critics of more open trade under both NAFTA and GATT. Other environmental groups have taken a cautious but less openly critical approach. Overall, the *environmental* community generally sees risks in more open trade, whereas the *trade* community sees threats to economic growth and integration if environmental regulations are not harmonized, because these regulations can provide good cover for protectionism cloaked in a "green" disguise. Trade negotiators and environmentalists alike share concerns about how global environmental and trade agreements are to be linked, whether one should take precedence, and the methods by which conflicts should be resolved.

North/North Versus North/South

The increasingly competitive trade relations between the United States, EU, and Japan are one axis along which trade and environmental issues arise. In some respects, the high-income countries of the North are increasingly alike in placing relatively greater value on environmental quality. But these economies are also locked in a high-stakes game of competition for global markets, and their governments face domestic pressures to loosen regulatory oversight. Even given their similarities, differences exist in the North not only in scientific and environmental standards but in culture and social norms, which will continually confront efforts to harmonize environmental regulations. Challenges to these regulations as nontariff trade barriers, both within regional trading blocs such as NAFTA and among nations such as the United States, EU, and Canada, are likely to be recurrent themes.

The gap between the environmental regulations along the North/South axis is even wider, accentuating problems of harmonization and concerns about pollution havens and competitiveness. The NAFTA negotiations reflect these differences in microcosm, with Mexico attempting rapidly to upgrade its environmental regulations in order to address fears in the United States and Canada. These fears include lower costs of environmental compliance by competitors in the South, movement of firms and industries into these low-regulation areas, the import of goods (such as fruits and vegetables) tainted by treatments banned in the North for environmental reasons, and the use of production methods in the South (such as tuna fishing with nets that also kill dolphins) objectionable to environmental interests. Yet from the perspective of many in the South, the environmental regulations adopted in the North are unaffordable, even if desirable. In addition, many developing countries suspect the North of using its higher standards to discriminate against products and processes primarily for trade rather than environmental reasons.

ENVIRONMENTAL REGULATIONS AND INTERNATIONAL TRADE

How can the U.S. government regulate environmental impacts when these impacts extend beyond U.S. borders or arise from the actions of foreign nations, companies, or individuals? This is the challenge of environmental policy in an open world economy. Two main approaches to these problems exist (see also Runge, 1990b; Runge et al, 1994).

The first is a legal approach. When a nation develops an environmental standard such as a ban on products containing certain polluting chemicals, the action may place a burden on those importing such products from overseas. Is this burden justified by the environmental harm that it avoids? How are such justifications to be measured? What body or group should decide such questions? These are largely questions of law, both domestic and international.

The other main approach is an economic one. What are the environmental damages resulting from unregulated trade? If trade is restricted for environmental reasons, what harm is done to consumers and producers of goods and services at home and abroad? How can the environmental harm be reduced at least cost in terms of trade distortion, and how can trade be expanded at least cost to the environment? Comparing legal and economic approaches to trade and environment provides a basic understanding of the criteria that can guide future policy choices.

Both environmental and trade rules are legal instruments that regulate the actions of individuals and firms in the conduct of commerce and the economy. Environment and trade intersect when trade actions lead to claims of environmental harm, when environmental standards lead to claims of economic harm, or when international trade and environmental rules appear mutually inconsistent or ambiguous. Fortunately, there are many cases in which environmental and trade rules do not intersect at all and in which there is no claim of harm, inconsistency, or ambiguity. But the increasingly integrated global marketplace, combined with a rising awareness of environmental damages due to market activity, has led to a growing number of such claims.

Trade Measures and Environmental Damages

When trade measures lead to claims of environmental harm, it is often because expanded markets have created pressures on scarce natural resources or on the ability of governments to regulate environmental protection. As the EU seeks to reduce internal barriers to trade and creates additional flows of goods and services, many feel that new trade patterns will also increase the flows of "bads" and "disservices," such as pollution from higher levels of automobile and truck traffic. As NAFTA is implemented, similar fears exist about water and air pollution in the border region and the possible influx into the United States and Canada of products that fail to meet domestic health and environmental standards.

Potential environmental burdens arise not only from changes in trade rules that open access to markets but also from trade rule changes that close off access or guarantee differential access, such as special trading arrangements or commodity agreements. These arrangements or agreements may encourage resources to be allocated in ways that increase pollution or problems of hazardous waste, leading to environmentally unsustainable patterns of resource exploitation. One example is the export bans on tropical hardwood timber that have lowered domestic timber prices in countries such as Ecuador, discouraging conservation and encouraging unsustainable rates of cutting (Southgate, 1988; Barbier, 1989).

An important question in evaluating these claims is the degree to which the environmental burden is directly attributable to trade. Although trade may amplify and aggravate damages resulting from a lack of environmental regulation and enforcement, it is difficult to lay the whole burden on trade itself. If the benefits of trade are accompanied by enforced regulation that reduces or eliminates environmental damages, then its burdens have been internalized and its benefits remain. In some cases, damaging environmental consequences of trade and development are difficult to avoid, even with strict regulation. Export-oriented intensive agricultural production in the heartland of North America, for example, is responsible for a substantial amount of the nation's water pollution. Even if agricultural production were more strictly regulated, some damage to soil and water resources would continue. Yet, it is equally clear that this production is a highly competitive trade sector in the North American economy, consistently contributing surpluses to the balance of trade of both the United States and Canada. Hence, the burden imposed on the environment may at some point be argued to be justified by these trade benefits.

Although some might argue that no such burdens are justifiable, realistic policy making cannot treat either environmental or trade objectives as strictly dominant. Trade-offs must and will be made. Striking a balance between trade and the environment requires a careful assessment of environmental impacts and the cost of minimizing damages, with the possibility that additional environmental regulations may be necessary. Fundamentally, this is a matter of offsetting the damages associated with trade by "internalizing" them through some type of regulation.

A more complicated issue arises when trade appears to cause environmental damage, but the damage occurs outside the home market. Such was the case when the United States demanded that tuna fishing practices in foreign waters change because of dolphin kills. Regardless of whether the trade action has its primary environmental impact at home, abroad, or in the "global commons" (such as the atmospheric ozone layer), a politicoeconomic decision will be developed (see also Chapter 14, Global Issues). This process of decision is shown in Figure 2–1.

Once it has been determined that trade measures lead directly to some type of environmental damages, a variety of means are available to determine the level of damage that may result, and how best to respond. In the United States, the most often employed is the Environmental Impact Statement (EIS)

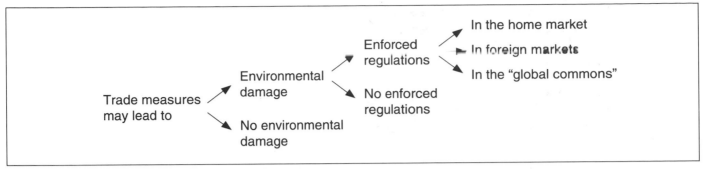

Figure 2–1. The impact of trade liberalization on the environmental decision process.

procedure that has evolved under the National Environmental Policy Act (NEPA) of 1969. Based on these detailed assessments, it is often determined that additional actions, including regulations, are required to protect the environment, although in many cases the preparation of the statement itself serves to reassure those who fear potentially adverse environmental consequences. If the damages occur outside the home market or as part of the global commons, the environmental response may involve a bilateral or multilateral agreement, which is simply a multinational regulation. Here, of course, issues of enforcement remain critical.

It is worth noting that if trade actions appear to cause no obvious environmental damages, it is wasteful to propose that all such trade be subjected to environmental assessments. If, for example, the United States were to increase textile quotas, allowing foreign shipments of sweaters or blue jeans into the domestic market, it would be hard to make a case for an EIS in the home (U.S.) market. Yet if a case can be made for linkage from expanded trade flows to environmental damages and the level of these damages and the appropriate response is in doubt, such an assessment can provide useful information.

The danger in calling for full-scale assessments of *all* trade actions is that the assessments will appear to (and in fact will) obstruct the expansion of trade, lending credence to the view that environmental claims are substituting for protectionism. It is also doubtful whether the environmental assessment process, which has been applied to specific projects (such as dams and other water diversions) and various U.S. regulations, is equally well-suited to trade agreements covering many different products and multiple sectors of the economy. Such a process would be unwieldy and probably so general in focus as to provide little meaningful detail.

For example, the environmental assessment undertaken by the Bush administration and released in August 1991, of the impacts of the then-proposed NAFTA on the United States-Mexico border was criticized by environmental groups because of its overly general and conjectural nature. However, as the EIS becomes more specifically focused, it provides a framework for developing legal and regulatory actions. Both NAFTA and European trade agreements have been subjected to detailed, proscriptive environmental/regulatory side agreements.

Unfortunately, if environmental damages seem clearly to result from a trade action, in "other" foreign markets or the global commons, bilateral or multilateral rules may be difficult to develop and enforce. As an example, the 1992 Rio Accords have been criticized as long on rhetoric and short on substantive specifics. Furthermore, as the Western community continues to suffer through slow or no economic growth, the ability to enforce potentially economically damaging environmental agreements is significantly lessened.

Environmental Measures and Trade Burdens

When domestic environmental measures lead to claims of *trade* damage, it is generally because a burden has been imposed on individuals or firms seeking to export or import goods or services in the name of domestic (and sometimes global) environmental protection. Now the shoe is on the other foot. The question is whether the environmental measure is justified primarily as a form of necessary environmental protection or is a disguised restriction to trade, in which harmful trade effects loom proportionately larger than beneficial environmental effects. Here again, a balanced judgment must be made concerning the costs (because of trade distortion) that should be borne in order to protect the environment.

The issue of whether a government environmental regulation is a nontariff trade barrier is a question faced domestically in the United States by the states under the commerce clause of the U.S. Constitution, and by the 12 member states in the European Community under the Treaty of Rome. As Robert Hudec and Daniel Farber argue, such questions typically break down into two parts:

- Does the measure create a burden on trade?
- Is the burden justified by the environmental benefits of the regulation (Hudec & Farber, 1992; see also Figure 2–2)?

(The discussion that follows is based largely on Hudec & Farber, 1992.) From a legal perspective the apparent burden imposed on trade is a "gateway concept." If a burden appears to be present, it opens the way to further inquiry as to its justification, in which its benefits for the environment are weighed against its damage to trade. If no burden is found, the trade effects of the regulation are not at issue. The decision problem is thus the converse of that shown in Figure 2–1.

Although nearly all environmental regulations impose some differential burdens on commercial transactions, to be trade-related this differential must exist between some foreign producers and their domestic competition. This differential may be relatively easy to see, as when foreign products are subjected to obviously different standards compared with domestic products. Under Section 337 of the 1930 Trade Act, for example, certain trade cases for foreign violators are heard before the International Trade Commission (ITC), whereas cases against U.S. firms charged with similar violations are sent to U.S. courts. In general, going before the ITC is regarded as more burdensome to the defendant. Not every differential rule clearly constitutes a burden, however, even though domestic and foreign products are treated differently. Auto safety glass inspected at U.S. auto manufacturers' factories is different from inspections of foreign vehicles' windshields at the border, but the border inspections do not appear to create a differential burden.

Less obvious are standards that appear neutral on their face but have a differential impact on foreign and domestic products. Provisions of the 1985 Farm Bill sought to apply sanitary processing and inspection standards to chickens from outside the United States that were "the same" as those standards used domestically ("the same" standards were substituted for previous language by members of Congress from Arkansas). The previous language had called for foreign standards "at least equal to" those used domestically. In a case brought before the federal court for the Southern District of Mississippi, the language calling for "the same" standard was upheld, despite warnings from the U.S. Department of Agriculture that "such a definitional finding would augur dire foreign trade implications . . ." (*International Trade Reporter*, 1992).

Overall, balancing legal and economic judgments must be made in order to extend the regulation of environmental risk into the international arena. When trade measures lead to environmental risks, these risks can be remedied through regulations. However, many such risks are not subject to regulations in the home market and may require negotiations with other countries, whether bilaterally or trilaterally as in NAFTA, or multilaterally, through international agreements. It is increasingly clear that unilateral U.S. action will not be tolerated by the global community.

When environmental actions lead to trade distortion, the burden on trade must be assessed in relation to the environmental benefits to determine whether the trade burden is justified. Again, the decision that a trade-distorting environmental regulation is justified cannot be unilateral. Consultation and agreement with other countries is likely to be necessary.

ENVIRONMENTAL IMPACT AND RISK ASSESSMENTS IN AN INTERNATIONAL CONTEXT

The U.S. regulatory approach to environmental issues depends heavily on the ability to estimate environmental impacts and environmental risks, often including technical cost/benefit studies and risk assessments. These methods are made especially complex by the issues just discussed.

Complexities of Impact Assessments

Environmental Impact Assessments (EIAs) depend on the capacity to trace the effects of actions, such as the construction of a dam, on the environment.

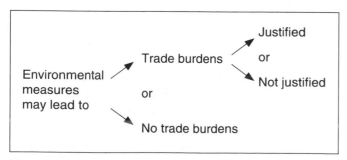

Figure 2–2. Environmental measures and trade damage.

However, if the impacts occur outside the home market or arise from foreign actions, the assessment is made that much more complex. Benefits and costs may be quite different if evaluated from the perspective of foreign nations, as may be the assignment of risks. Interestingly, EIAs are one of the first environmental measures enacted by newly industrializing countries (such as Taiwan, Singapore, Malaysia, Indonesia, South Korea).

In general, these differences are likely to cause assessments to depend more on qualitative information than on developing a single standard of quantitative evaluation. This is likely to be true whenever opinions over benefits and costs differ widely, whether domestically or across national borders.

Complexities Introduced by Risk

Quantifying risks is an extremely difficult job of estimation. However, assuming the probability of a risky event can be estimated, the problem of how much risk is *tolerable* presents itself (Runge, 1984).

Choosing an acceptable level of risk is fundamentally an ethical question (see Environmental Ethics later in this chapter). "Acceptable" health hazards defining a threshold of risk may still result in damage or even death if the standard chosen is low. Willingness to accept these hazards must result from an explicit or implicit trade-off between social benefits and human life and health. The assessment of risk is thus linked to the value attached to morbidity and mortality.

Treating risk in an international context requires consideration of the levels of risk generally deemed tolerable in different nations of the world. To decide how high the probability should be that a standard is within range of an unacceptable threshold of risk is to put a value on adverse effects to some individuals, which may vary especially from low- to high-income countries. A decision to accept high rather than low risks is a decision to treat different persons differently and different nations differently when the evaluation of risks varies. Hence, risk assessment is made even more complex by international comparisons. Currently, there is no worldwide accepted standard of risk. In the United States, one excess cancer per one million (1×10^{-6}) exposed individuals is considered the regulatory point of departure. (See Chapter 12, Risk Assessment.)

Complexities Introduced by International Trade

When risks *themselves* are traded, the problem is doubly compounded, because willingness to pay for risk reduction will vary country by country. What may appear in the United States as a prudent action of risk reduction may appear to importers (such as Chilean importers of grapes alleged to have been contaminated with traces of cyanide) as overt protectionism. Even if contamination can be shown to exist, different attitudes toward risk may lead to different conclusions about its tolerability. A one per thousand (1×10^{-3}) excess cancer or adverse reaction risk of a drug or product may be unacceptable in the United States but may be deemed an acceptable cost-benefit risk in another country.

The internationalization of the economy poses difficult challenges for environmental regulation, such as these:

- Health and environment can no longer be thought of purely as domestic policy issues. However, the tools of regulation remain largely domestic. Therefore, the result is a need for more coordinated policies with other countries through international environmental regulations.

- Trade liberalization may lead to environmental damages requiring such regulations. Developing a basis for determining when and how to "internalize" the damages that can accompany trade reforms will remain a priority.

- Environmental regulations may themselves pose burdens to trade. Criteria must be developed to determine when such burdens are justified by the environmental harm they avoid.

- Assessing environmental impacts and risks in an international context is made much more complex by the foreign origins of these impacts and risks, and the different weights attached to them, especially by low- versus high-income nations.

It is clear that these issues will continue to confront both the environmental and trade communities for many years to come.

Because the United States has become integrated into the global economy, it is difficult for the United States to take internal environmental regulatory actions that impose serious adverse economic consequences on domestic business and industry without potentially significant impacts on job creation and investment. Therefore, the economic environmental argument comes full circle and becomes a political and ethical debate on the balance between sustainable development and economic opportunity.

A provocative lead editorial in the influential journal *Nature* was entitled, "Environmental Protection or Imperialism?" This editorial clearly expresses the difficulty of imposing Western environmental standards on newly industrializing or developing countries:

> The truth is that where people are relatively prosperous, they spend increasing larger proportions of the resources available on the avoidance of personal disaster (through insurance) than on the avoidance of environmental risks. . . . Environmental protection is thus often a public purchase on behalf of a whole community. . . . But elsewhere, especially in the poor countries, the cost of environmental protection can be, and usually is, a drag (*Nature*, June 24, 1993).

The 1990s will be the decade that debates sustainable development versus economic opportunity. President Clinton's administration has begun to position the United States as the leader of the sustainable development *and* economic opportunity philosophy. However, whether other countries are willing and able to follow the United States lead is unknown. The major ethical issues involved in this debate are discussed next.

ENVIRONMENTAL ETHICS

As the issues relating to sustainable development and economic opportunity are debated, a number of ethical questions arise. This section discusses those ethical questions related to power, law, rights, justice, and responsibility. Further, it discusses the nature of ethics and its relationship to business, science, religion, and politics; the nature of global environmental ethics; changing worldviews; current political, religious, and social views; and corporate responsibility in relation to environmental policy. Many of these issues are illustrated by an industrial accident that occurred in 1984.

An Accident and Its Issues

Shortly after midnight on the morning of December 3, 1984, a gas leak at a small pesticides plant in central India spread gas over the capital city of Bhopal. In the shantytowns that lay to the south and southeast of the plant, many people died in their sleep. Others awoke feeling short of breath and rushed outside. Employees ran from the plant, eyes and throats stinging, but a deadly gas cloud covered their path of escape.

Estimates of the number of dead vary widely. At the lowest estimate, some 2,500 persons were killed and 150,000 more were injured. Lungs and eyes were most affected, but also immune systems, so that many survivors later fell victim to a variety of secondary infections. Some survivors became extremely sensitive to airborne pollutants, becoming unable, for instance, to be near smoke from cooking stoves without suffering an asthma attack. Many whose lungs were injured found themselves unable to do the lifting or walking that their jobs required. The rate of spontaneous abortions, stillbirths, and deformities increased, and women developed various gynecological diseases. Victims suffered depression, anxiety, impotence, loss of appetite, and nightmares. Large numbers of cattle, dogs, cats, and birds died, and vegetable crops in the outlying areas were destroyed (Morehouse & Subramaniam, 1986).

In the immediate aftermath of this tragedy, one of the worst industrial accidents that has ever happened, attention was focused on what went wrong and who was at fault. It became obvious early on that a runaway reaction occurred when water was accidentally introduced into a methyl isocyanate storage tank. Plant design, operating procedures, safety systems, and warning systems were scrutinized. The pesticide plant was an India-owned subsidiary of an American company. In the United States the question was raised, Can it happen here? The accident underscored the importance of accident prevention procedures and revealed the need throughout the industrial world for safer designs, better trained employees, and better planning for emergencies.

This question of the adequacy of industrial accident prevention measures, though important, is only one of many issues raised by the Bhopal leak, and what happened at Bhopal is itself merely a consequence, a monumentally tragic one to be sure, of decisions and attitudes that seem at first glance to be of little relevance for understanding how water got into a methyl isocyanate storage tank.

Power. Take, for instance, the question of "fault." Who is at fault in an accident is an important issue, involving enormous sums of money. Our society believes that, for justice to be done, the guilty party must accept blame and make reparations. But the question of fault is really a question of power, of who makes the decisions that can so destroy life and the environment. Having settled the question of who has such power, a larger question emerges—Who *should* have such power?

Law. During the investigation of any industrial accident, the focus is on what rules or laws were broken. Equally important is the question of whether the existing rules and laws are adequate to protect citizens from danger, especially given the environmental and social problems that face the industrial world today. Since Bhopal, the world has experienced Chernobyl, the nuclear power accident near Kiev that released radioactive material over Eastern Europe, Scandinavia, and Western Europe. The consequence of these and other industrial disasters is a new flurry of activity by international bodies to tighten and extend international law to protect human communities and environments around the globe.

Rights. Throughout the world, despite, or perhaps because of, their frequent abuses, the concept of human rights has weight and the sanction of law. Yet, A Universal Declaration of Human Rights (adopted by the United Nations General Assembly in Paris, December 10, 1948) says nearly nothing about the environmental bases of human welfare. Questions remain, such as:

- Should the neighbors of a chemical plant know its inherent dangers? What if the plant provides essential employment to the region?
- Do all humans have a right to clean air and water? How clean?
- What if this right conflicts with industrial development?
- Do all humans have a right to at least a fringe of natural beauty? Or is beauty itself a fringe benefit of success?

And if human rights turn out to be controversial, there are even more sticky questions of plant and animal rights.

Justice. Increasingly, in a world dominated by multinational corporations, the parent company is from a rich nation, the subsidiary and victims from a poor one. The questions of justice that this situation spawn go from the specific—should the families of poor victims be paid the same amount of money as they would have received had the accident occurred in the parent country—to the very broad—should such inequalities of wealth exist?

Responsibility to the future. Elements of the accident at Bhopal highlight a situation that potentially could grow much more conflictive—earth's increasing human population and industry's increasing potential for harming the environment. In general, laws are designed to benefit the living, although the continuity of law over time extends its protection to future generations also. Is there an ethical responsibility to more specifically address the needs of future generations?

All of these larger questions are questions of ethics. The following discussion examines environmental ethics—what it is and why it is important.

The Nature of Ethics

Ethics refers to the disciplined reflection about moral problems. *Moral* refers to right and wrong, so ethics concerns judgments about what is right or good, and what is wrong or bad. Another way of stating this definition is that ethics concerns values—questions of what is valuable or desirable—to do, to be, to live and die for, and questions of what is not valuable—what is evil, what is merely frivolous, what is a crime, what is a tragedy. Ethics is a branch of philosophy; its hallmark is reason. An ethicist examines ideas, testing their reasonableness, aiming for clear, orderly thinking rather than emotionalism or dogmatism.

A close relationship exists between ethics and morals but they are not synonymous; morals refers to standards of ethical conduct or right action. This distinction is somewhat academic because of the links between what is good and what should be done to promote good. Indeed, the adjective moral is used both to denote the capacity of a person to make ethical judgments and to characterize a person whose behavior is consistent with some standard of ethical behavior—a "good" person.

Although ethics is a thinking activity, its essence is relationship. Both private ethics and public ethics deal with relationships to others and questions as to what these relationships should be. Ethical thinking involves not only questions of what is good, but questions of what structures—community relationships, political systems, corporate structures—enable the good to be realized.

Ethical dilemmas. Ethics implies limits on individual and group freedoms. Depending on what is considered of value in our relationships to others, some things should and others should not be done. Because more than one value is nearly always at play, making an ethical judgment about the right course of action in a specific case is not easy. The following are some examples:

- If human life is valuable, one should not kill humans. But what if by not killing, someone gets injured? Is killing justified in self-defense, in the defense of one's country, to save another life, to deter would-be killers, to eliminate suffering?

- Honesty is generally considered good. Should a politician be honest if honesty will mean that a far less capable person will govern?
- Aware of dishonest policies within her company, should the sole support of a family risk unemployment by blowing the whistle?
- Where do the commitments of individuals, of corporations, of nations, lie? Should a wealthy nation give its excess grain to countries where persons are starving or stockpile it for future emergencies at home? Both actions do good.
- Is it permissible to corner the market on a rare mineral or patent a life-saving drug to fulfill the commitments of a company to its shareholders?
- Is it ethical to treat fetuses as commodities when doing so will cure human disease?
- Is it ethical to treat animals as commodities when the end is more food for more human beings?

Such dilemmas are the very stuff of ethics and make ethical thinking a most difficult and challenging art. No one person can answer these ethical questions for everyone. Nor is it the job of some specially trained person to do so. A knowledge of ethics is only helpful as a source of knowledge about what others have thought and wrestled with. Over time, ethical systems have developed with differing approaches to the problems posed by values in conflict. One approach is to assert that there are certain values that one is ethically bound to honor, no matter what the situation. In practice, this approach runs into problems unless all parties to the conflict agree on the imperative values. Opposite to it is the view that there are no moral or immoral acts as such; that the rightness and wrongness of an act depends solely on its consequences. The greatest good, for instance, might be defined as whatever benefits the greatest number of human beings. Another approach establishes ethical criteria—certain conditions, such as justice or freedom, that should be met and can be used to judge situations with conflicting values. Any number of broad ethical goals might be established by which to decide the right path to take. The right path might be one that best promotes human life, or one that leads to the sustainability of the diversity of life.

If at one time ethics was popularly conceived of as a subject for college students or the exclusive concern of the clergy, today ethics is known to be a necessary activity engaged in by persons from all walks of life. This changed perception accompanies a new awareness of the role of values in human activity. It is also linked to the new globalism—the modern age's ability to observe the effects of cultural attitudes and actions on lands and peoples around the globe. As a consequence, the relationship of ethics to business, science, religion, and politics has undergone significant change.

Relationship to business. In the past, "business ethics" was perceived, in the main, to involve fair dealings with individual customers and other businesses. But, as the history of business/labor relationships in the United States and elsewhere, and as recent concerns about the movement of large corporations have shown, the relationship of ethics to business is both broader and deeper than market relationships. It involves questions of the relative values of individuals and corporations—questions of the employee's right to share in policy making and the employer's responsibility to employees after they have outlived their "productive" years. How is an employee to be understood? As a cog? As a family member? As a citizen of a society greater than the corporation? In contemporary society, the basic nature of corporations has been the subject of much ethical thought, revolving around questions such as:

- Is a corporation a free-wheeling entity? Should corporations be free to uproot their facilities in order to achieve greater profit and benefit their stockholders? Or do they have an obligation to the communities they have shaped?
- What responsibility has the corporation to conserve raw materials?

Relationship to science. Scientists and ethicists share a desire to know the truth and to examine actions in light of their consequences, yet ethics is sometimes dismissed on the grounds that science alone gives valid knowledge, that conclusions derived in any way other than by the testing of large numbers of subjects are contaminated by desire and prejudice.

Philosopher and mathematician Bertrand Russell gave voice to this viewpoint when he said, "Whatever knowledge is attainable, must be attained by scientific methods; and what science cannot discover, mankind cannot know" (Russell, 1935). Since Russell's statement, made in 1935, discoveries in physics have shown the impossibility of the completely neutral, value-free observer posited by the scientific method. As philosopher of science, Stephen Toulmin writes:

Instead of viewing the world of nature as onlookers from outside, we now have to understand how our own human life and activities operate as elements within the world of nature. . . . Nowadays, scientists have always to consider themselves as agents, not merely observers, and ask about the moral significance of the actions that comprise even the very doing of science (Toulmin, 1982).

Humanity's increasing dependence on the work of scientists, and the development of technologies with significant social and environmental impacts—new weapons systems, biomedical engineering, artificial intelligence, and the like—require more than ever that disciplined reflection be applied to scientific endeavor.

Relationship to religion. Ethics and religion are closely linked: Both concern values and right behavior. Over the course of history, the religions of the world have been largely responsible for shaping our thinking about ethical behavior. The Ten Commandments, the Sermon on the Mount, the teachings of Muhammad, and numerous other codifications of ethical behavior influence our thinking today. Religious ethics is grounded in divine authority; what is believed to be good is believed also to be the will of God. Although a great deal of agreement about the nature of good exists among the many religions of the world, the areas of disagreement have led to wars and enduring hatreds. In order to ease world conflict, there is need for a wider understanding and an openness to the evaluation of the moral codes woven into cultures—what they are, how they developed, how they function to enhance or distort the relationships of human beings to one another and to the earth. This evaluation has begun, brought on by an awareness that issues of social justice—the welfare of peoples around the globe—are linked to environmental issues—the use of land and natural resources, the treatment of plants and animals—and that these, in turn, are influenced by religious understandings of the nature of the created universe.

Relationship to politics. Pessimistic about the ability of nations to serve the good, theologian Reinhold Niebuhr declared that "civilization has become a device for delegating the vices of individuals to larger and larger communities." Therefore,

the relations between groups must . . . always be predominantly political, rather than ethical . . . determined by the proportion of power which each group possesses at least as much as by any rational and moral appraisal of the comparative needs and claims of each group (Niebuhr, 1932).

Was Niebuhr correct? History and the morning newspaper testify to the selfishness of nations. Nevertheless, certain recent events provide hope that Niebuhr's viewpoint is dated. One of these events took place in space. When an orbiting satellite beamed back a photograph of the earth from space, a new consciousness of limits resulted. For the first time in human history, the ultimate powerlessness of power on a fragile planet where all humanity and nature are connected and interdependent became obvious to many. In part because of this, environmental ethics has emerged as an international concern.

The Nature of Environmental Ethics

Environmental ethics can be defined as disciplined reflection about moral relationships to the natural world. The natural world—and therefore environmental ethics—includes human beings and human culture, as well as plants and animals and inanimate matter. Because human beings are part of the natural world, and human culture influences how the natural world is viewed and treated, environmental ethics concerns human cultural traditions, religious beliefs, and political and economic relationships.

One of the major contributions of environmental ethics to knowledge is the new awareness of the role of culture in shaping the environment. For example, the cultural perception that particular places are sacred has contributed to their preservation. Every region of the planet has such sacred places—the Nile, the Ganges, Mt. Olympus, Mt. Everest, Mount Zion, Popocatapetl, Mt. Shasta. Although the sacred places that remain today are perhaps most often associated with mountaintops (in part because of their inaccessibility throughout history to invading cultures with different conceptions of sacrality), places providing water have been considered sacred by many cultures. The nomadic Australian aboriginal peoples reverenced a network of streambeds and waterholes they believed were byways of the Rainbow Serpent. When the British came to Australia, they brought with them a different set of beliefs, including one that said that cultivation of the land was an edict of divine authority. The watering places were turned into homesites and places for sheep without regard for the aborigines—whom the settlers regarded as little more than wild beasts (Wright, 1981). Both patterns of settlement (aboriginal and European) and their consequences for the total environment—persons, land, and animals—were directed by cultural belief systems. Thus, environmental ethics lays the foundation for preserving cultures as well as habitats.

An evolving perception. Throughout history, humans have been, and still are, at the mercy of such overpowering phenomena as plagues, floods, and famine. So destructive are these forces, so seemingly random and undeserved, that men and women have found it difficult to comprehend the impact of their own actions on the environment. They have spent more time shaping rituals, laws, and treaties to deal with interpersonal violence than with violence to natural things. Nevertheless, contemporary interest in environmental ethics has brought to light stories, myths, and rituals from the past of many nations and cultures that support an environmental ethic of respect and care for the earth.

Examples include the belief held by ancient peoples of Hawaii that the land, sky, sea, and all species of nature that have ever lived are conscious ancestral beings who care for and protect humans and who deserve similar protection in return. This belief lives on and is popularly recognized in the practice of never asking a fisherman whether he is going fishing so that the fish won't hear and swim away (Dudley, 1993). The story of Noah is an example of ancient concern that species not become extinct. Traditional Jewish law requires farmers to rest after working for six days and extends this day of rest to the farmer's ox and ass (Clark, 1990).

Today, with nonhuman nature in retreat everywhere, the need for environmental ethics is much more apparent. The environmental consequences of human actions and attitudes are better known, and discussions of their morality or immorality are possible. Also, as parts of the natural environment are threatened with extinction, the value of what remains has increased. At the same time, population growth, exponential growth in energy and resource use, and a worldwide hankering for "the good life" pressure for decisions that increase environmental destruction. A backlash against environmental safeguards by those who wish to see the environment used to its fullest for immediate human ends only underscores the need for reasoned discourse.

Yet, while the call for environmental ethics has increased in volume, it is also true that a segment of the public is unconcerned about the natural environment. This, too, may be a new phenomenon in the world, linked to increasing urbanization. Television, for many, is the primary means of finding out about the nonurban world, and it may prove no substitute for the real thing when conservation policies are at issue. Jerry Mander argues, for instance, that television cannot convey the "very odd, soft, sticky, wet and smelly" nature of a marsh. "And without the understanding who can care about the marsh?" (Mander, 1978). Thus, environmental ethics must deal with the ways humanity learns about the natural world and the adequacy of that knowledge.

Grounds. Humanity depends on the natural world for survival. Environmental ethics, therefore, does not start in a valueless vacuum; behind its disciplined reflection lies the assumption that the natural world is a source of values, not the least of which is self-preservation. Are there other values that form the grounds for deciding that the environment should be preserved? Professor of religion and science Ian Barbour finds three grounds for environmental ethics: short-run human benefits from the environment, duties to future generations, and duties toward nonhuman beings (Barbour, 1980).

Human benefits. If human life is valuable, then preserving high-quality air, water, and soil is valuable, because humans cannot live except by breathing air, drinking water, and eating food. Polluting air and water and depleting the soil of its richness is harmful or fatal to humans, and therefore morally bad. The more humans learn about the delicate chemical balances of the human body, the more we learn about environmentally linked diseases, the firmer are the grounds for environmental preservation on the basis of safeguarding human life. Human health and longevity also argue for preserving plant species, even plants not yet known to benefit humans. Discoveries of disease-curing plants—such as the rosy periwinkle that now keeps three-quarters of the world's children who suffer from leukemia alive, or the mayapple, from which is derived a treatment for testicular cancer—depend on there being such plants around to discover. The same is true of animals; not long ago it was learned that the armadillo can be useful in leprosy research and the manatee in blood-clotting research (Myers, 1984). Plants and animals need to be preserved both for their contributions as food and because new species add resistance and hardiness to present food varieties. In fact, the more that is learned about the "web of life," the ecological interconnections of plant and animal species, the firmer are the grounds for preserving all life forms. The ocean's fish rely on microscopic plankton; trees are nourished by microscopic algae. In case after case it becomes apparent that so threaded together is life, to remove a single species or even transfer a species from one locale to another, is to tear the fabric and ultimately hurt humanity.

The environment also benefits humans because it provides employment. All manufactured goods originate as earth's raw materials. This basic point is often overlooked in "jobs versus the environment" controversies.

In addition to material benefits, the environment provides spiritual and psychological benefits to humans. Wilderness preservation may be advocated on the basis of its recreational and spiritual benefits. It is said that wilderness fosters self-reliance and courage, that it brings about peace of mind and release from anxiety, and also that it is a source of inspiration, evoking feelings of awe and reverence.

Duties to future generations: sustainable development. To maintain that humans should preserve the environment because we have a duty to future generations extends what matters morally to a class of beings not yet born. Barbour suggests that this duty derives from the idea of justice between generations—if we did not know what generation we would be born in, we would recommend policies that would keep the environment in equally good shape for each generation. Some ethicists feel that there is more than duty involved, that the unborn have the right to a supportive environment. Others who have considered this question claim that rights cannot be extended to an entity that doesn't exist. Although there is general agreement among ethicists that some duty is required, the question of how much duty is perplexing. Many persons would agree that the basic needs of future generations have priority over the luxuries of present generations and that wasting resources is immoral. But should the present generation sacrifice for future generations?

To fulfill our obligation to future generations, Barbour suggests, for starters, that we not use more than the maximum sustainable yield of a renewable resource. This means, for instance, that we not cut forests down faster than they can grow back, thus conserving our biological "capital." He further suggests that each generation leave the environment no more polluted than when it arrived, and that we keep the resource base intact by not allowing it to be depleted "more rapidly than it can safely be extended by technology" (Barbour, 1980).

The moral problems posed by the depletion of nonrenewable resources are many and difficult. For example, even if new discoveries and extraction techniques ease its depletion rate somewhat, oil is one resource that is clearly running out. Who has the right to what is left? The present owners? Whoever can afford to buy it? Everyone now living? All future persons? Philosophy professor Richard T. DeGeorge points out that if we push the generations far enough into the future, each person ends up with "a thimble full" of oil (DeGeorge, 1979). Yet DeGeorge and others feel that we have a responsibility to preserve the environment in such a way that it will provide the necessities of life for our children and grandchildren. In the case of oil, that's going to mean some conservation measures, raising issues such as:

- Who should conserve?
- Is the individual morally responsible for buying a fuel-efficient car or using public transportation even if others do not?
- Does a country that keeps its population within limits have to share with all the individuals born in countries where parents had no regard for whether the children could be supported?
- By what right do some countries use more of the world's resources per capita than others?

Duties to nonhuman beings: species preservation and biodiversity. There is less agreement about the ethical grounds for environmental preservation apart from human benefit. Also, most arguments for and against animal preservation are based on the degree of likeness of animals and humans.

Aristotle, Aquinas, Kant, and other Western philosophers who held that reason was of highest value argued that humans have no duties toward animals because animals cannot reason. However, a more recent philosophy indebted to evolutionary biology—process theology, developed by Alfred North Whitehead—sees no such radical break between humans and other creatures. In process philosophy, experience is good and every creature is capable of enjoying it and of contributing to the experience of others. This common participation in experience, believed also to be a contribution to God's experience, is reason for preserving other beings. But because more complex beings are capable of greater experience and of contributing in greater measure to the experience of others, so-called simple beings are less valuable and can be sacrificed to save complex ones. Thus, for example, mosquitoes can be killed to prevent malaria.

Peter Singer, author of *Animal Liberation* (1975), argues that we have duties toward animals because they, like us, are capable of suffering. Animals incapable of suffering—primitive beings such as oysters—may be eaten for food, as may eggs from free-ranging chickens, whereas an animal that suffers may not be made to suffer by being killed for food, nor may an animal be raised in such a way that it suffers. Most persons today probably do not know how the food they eat was raised, but exposés on factory farming methods such as the book *Animal Machines* (1964) by British animal welfare campaigner Ruth Harrison, have raised public consciousness, and the concept of suffering has proven a powerful concept for promoting the notion that humans have duties toward nonhuman beings.

The reasons just cited for preserving animals are all *hierarchical,* meaning that they place animals on a ladder of complexity with the higher ones more deserving than the lower ones, and to a large degree they are *anthropocentric,* meaning that they interpret reality in terms of human values and experience. A *biocentric* ethic avoids all hierarchical postures and acknowledges the tight ecological relationships, the web of life that connects all living and nonliving entities. Aldo Leopold's "land ethic" is an example of biocentrism: "A thing is right when it tends to preserve the integrity, stability, and beauty of the biotic community. It is wrong when it tends otherwise" (Leopold, 1949). When plants and animals and other environmental entities are preserved because of their usefulness to human beings, they are said to be preserved because of their *instrumental* value. When plants and animals and other natural entities are believed to be valuable in and of themselves and not because of what they may add to human experience, they are said to have *intrinsic* value. The "deep ecology platform" of the Norwegian ethicist Arne Naess is based on intrinsic value. Its first two points are:

- The flourishing of human and nonhuman living beings has value in itself. The value of nonhuman beings is independent of their usefulness to humans.
- Richness of kinds of living beings has value in itself (Naess, 1990).

The notion of intrinsic value is often present in religious reasons for preserving the environment. A few examples are Albert Schweitzer's "reverence for life," Eastern notions of the underlying unity of all beings, and the Hebrew declaration that "the earth is the Lord's and the fullness thereof."

Relationship to justice. Eduardo Gudynas (1990) set out the basic interconnection of environmental ethics and social justice:

> Nature and humanity will be liberated together or not at all, for every step in environmental destruction has the effect of increasing social injustice, and every act of social injustice has the effect of increasing environmental destruction.

These words also represent a focus in environmental thinking sometimes missing in debates pitting "jobs versus the environment," "the North versus the South," and "preservationists versus conservationists." In these debates, persons concerned with saving the environment are perceived as unconcerned with human welfare.

The charge of elitism. Unfortunately, environmentalists are often perceived to be rich and selfish—well-heeled travelers with the means and the leisure to visit world nature preserves, to live in places surrounded by natural beauty, and to care about saving exotic animals, but with little or no knowledge or concern for the poor, for minorities, for those less affluent than themselves. The concern of environmentalists for wilderness preservation is often labeled elitist—a preference for pretty parks located in places the majority of urban dwellers will never visit. Environmental scientists are often similarly depicted as primarily concerned with maintaining their university status by conducting research on the vagaries of animal behavior while pressing problems of human health and well-being go unattended. When North Americans and Western Europeans seek the preservation of Brazilian rainforests or African wildlife, the charge of elitism swells.

These charges stem from long-standing patterns of injustice. Environmental battles usually begin at home with concern over the degradation of a local area. Although this is only natural, the NIMBY (not-in-my-backyard) mentality, especially when "not" means "anywhere but," fuels the elitist charge. In the United States, environmental causes have not been high on the agenda of minorities still seeking justice on other fronts nor, until recently, have environmentalists made social concerns part of their agenda. When the globe's poorer nations—the South, or the developing nations, as they are designated—receive lectures on resource conservation from the affluent nations of the North, who can fault the developing nations' resistance? The United States, for example, consumes about 12.5 times as much paper and paperboard as all of Latin America; its oil use per person is double that of Western Europe, and 24 times that used in sub-Saharan Africa. Americans use more than 4 times as much steel and 23 times as much aluminum as their neighbors in Mexico. In developing countries, waste is a luxury; the United States landfills over 80% of its waste (Myers, 1984).

A false dichotomy. But as the quote from Gudynas indicates, the apparent dichotomy between environmental and social concern is ultimately false. The needs of humans and nature are linked, as is their suffering. The poorest humans bear the brunt of environmental destruction. It is to poor countries that rich nations ship their garbage, and it is poor countries that sell their natural resources and the labor of their people for a fraction of their worth. The damage from oil and gas exploration and coal mining is experienced in the exporting country. It was the lure of the export market that encouraged the people of Costa Rica to reduce their rainforest so that they could grow beef cattle to satisfy the demand for hamburgers elsewhere.

Poverty breeds environmental destruction as well as attracts it. In some places the numbers of the poor and their requirements, such as for firewood for cooking, have magnified the problems of erosion and desertification. Military expenditures and welfare payments drain the economy of resources that could be used for environmental restoration. Polluted cities become more so. Even in small matters, poverty works against environmental preservation. In cities, bushes and trees are cut down or not planted because they provide hiding places for persons intent on crime. Environmental ethics concerns human inequalities and the means to their rectification.

Ecofeminism. The issue of women's rights, one of the serious contemporary issues of social justice, also figures importantly in environmental ethics. Around the globe, the caring functions traditionally assumed by women, which contribute to earth's well-being, are often downplayed. Women's work is often not recognized as work—not included in the gross national product (GNP) and not recompensed. Worldwide, women own less than 1% of the world's property (Sallah, 1990). In many countries, women and children produce most of the food, yet they are usually not part of the decision-making relating to land use or to the economics of food distribution.

This situation is changing. Identified with nature throughout the history of Western thought, often to their and nature's denigration, women have begun to assert themselves on nature's behalf. One well-known example, the Chipko movement, is representative of the new trend. In Uttar Pradesh, India, in March 1973, women formed a human chain and hugged earmarked trees to keep them from being felled for a nearby factory producing sports equipment. They recognized not only the beauty of trees but also the importance of trees to prevent the flooding of their villages. This same protest has been repeated in other communities and has led to a greater respect for the voices of women in matters of environmental importance.

Global Environmental Ethics

Our Common Future, the 1987 report of the World Commission on Environment and Development (chaired by then Prime Minister Gro Harlem Bruntland of Norway, and commonly referred to as the "Brundtland Report") has profound meaning for environmental ethics. Just as there are no longer social matters without an environmental component or environmental matters without a social side, so there are no longer local or regional environmental problems without a global dimension (see also Chapter 14, Global Issues).

The northern spotted owl controversy of the U.S. Pacific Northwest illustrates the need for global environmental ethics. When the owl's protection by the U.S. Endangered Species Act forced the abatement of timber harvesting in the old-growth forests where it lives, the popular press characterized the problem as one of an elusive bird versus the livelihood of Northwest timber cutters and their families. But the elusive bird was primarily a symbol—a symbol of the forest itself—of the remnant of a forest that once covered the continent, a symbol of all the plants and animals and trees and the mystery and the stories attached to the forest. And the timber cutters were themselves a symbol of the ruggedness of a proud occupation, the remnant of an occupation associated with the settlement of a nation and cherished in the tall tales of Paul Bunyan. Overlayering these deeply held values is a history of nonsustainable logging aided by government subsidies, of an industry where levels of employment have steadily fallen because of increased mechanization and raw log exports. And overlayering these are the many facets of the global dimensions of wood harvesting and wood use, so many facets that they can be only briefly discussed.

Among these facets are the following:

- The worldwide disappearance of primary forest, not only in the United States where, exclusive of Alaska, only 5% of the forest is intact
- The change from wood exporter to wood importer by countries, such as Thailand and the Philippines, forcing U.S. timber companies to look elsewhere, to the world's few remaining virgin areas such as

the untouched forests of the Soviet Far East, to set up subsidiaries (Postel & Ryan, 1991)

- The twin discoveries of global atmospheric warming and the forests' capacity to slow warming by acting as "sinks" or "banks" that absorb and hold atmospheric carbon
- The changes wrought by the timber industry in response to the disappearance of primary forests combined with an increased demand for wood: clear-cutting (causing soil erosion, nutrient loss, and water runoff) and experimentation with selective cutting techniques
- The establishment of industrial tree plantations with a consequent elimination of plant and animal diversity and a narrowed genetic tree base, making them vulnerable to climatic change, pollutants, pests, and pathogens
- The uses to which wood is put: the increased price of wood and what this means for new-home construction, the fact that 50% of U.S. paper production goes into packaging, that paper and paperboard use has risen dramatically in industrial nations during the past few decades
- The search for wood substitutes
- The use of less wood in building
- The increased recycling of paper.

Therefore, ultimately, the northern spotted owl controversy is a global matter, touching, to some extent, all of earth's citizens. Each segment of the problem involves values—a concern for their preservation and a fear for their loss. A global environmental ethics is needed to evaluate the total picture, to begin to find ways to prioritize its pieces, ways that help to heal divisions rather than inflate passions.

Almost any environmental or sociopolitical problem on the globe today has far-reaching consequences. For example, when an Ethiopian villager is unable to find firewood to use as cooking fuel and substitutes animal dung, the soil is deprived of nutrients and essential organic matter. It becomes more compact, soil erosion accelerates, and a process leading to desertification, hunger, human death, and species loss sets in. Further, the loss of homeland by indigenous peoples is linked to deforestation and global warming and may ultimately result in coastal loss. Even garbage disposal is a global problem as well as a matter of local pickup.

The world has relied in the past on military power to solve its thorny problems. As a result, the environmental costs of international conflict are a staggering burden affecting combatants and noncombatants alike. In fact, "far from being the savior of the biosphere," according to Worldwatch Institute staff writer Michael Renner (1991), "the world's armed forces are quite likely the single largest polluter in the world."

Population issues. The most pressing, and at the same time the most difficult, problem that environmental ethics must address is the issue of increasing population. So highly do we value human life that we fear even the mention of a possible need for curbing it. Yet, here too, reasoned discourse is a duty. It is an area in which environmental ethics encounters its greatest challenges. Ethicist Holmes Rolston, III, provides insight into some of the factors with a case involving 240 animals, the total population of the mountain gorilla, *Gorilla gorilla beringei*. This animal survives in only one place in the world—the Parc National des Volcans, a 30,000 acre national park in a small country with the highest population density in Africa. The park has lost 40% of its acreage to land cultivation; if it were eliminated entirely, 36,000 persons could live on the property at subsistence level. As Rolston says, "a few gorillas for the price of tens of thousands of impoverished humans." However, he goes on to add:

But an ethic that seeks to sustain life by taking a global view can reverse the presumptions and reply that the lop-sided score in fact favours the gorilla minority, a relict population and the last of an endangered species, against the human majority, of whom there are 3 billion in the world. Further, these 36,000 humans do not now own any rights to this land; the gorillas live there. Indeed, only some of these humans are alive now as subsistence farmers elsewhere; in most cases these humans are as yet unborn. Although humans are individually valuable after they exist, like an additional child in a family, an additional human in a society is not always an appropriate development. Development pits 240 existing gorillas, the last of their kind, against 36,000 largely potential humans, an excess of their kind. [This country] already has too many humans, and to sacrifice a species to make place for a quarter of one year's population growth is only to postpone a problem. Such development is not sustainable, and even if technology could make it so, it would not be appropriate because it would not sustain life in the biotic sense. The gorillas would be a casualty of human inability to control cultural development. The global community would be poorer (Rolston, 1990).

International initiatives. It is nearly always less painful to deal with problems affecting a far-off country than to deal with those in one's own backyard. In order to develop environmental ethics acceptable to the global community, there must be global participation and exchange, and an agreement that economic and social order should be based on principles of justice and human rights. The founding of the United Nations in 1945 was based on these principles, and through the years respected world figures have supported its founding ideals. Among the most hopeful signs of global cooperation during the last decades of the 20th century have been a series of international environmental ethics initiatives (see also Chapter 4, International Legal and Legislative Framework). Exclusive of initiatives by the world's religions, the most significant of these are:

- *The United Nations Conference on the Human Environment at Stockholm in 1972.* At this conference and its associated meetings, developing nations placed economic justice on a par with the concern of many industrialized nations for environmental protection.
- *The launching of the World Conservation Strategy (WCS) in 1980 by the International Union for the Conservation of Nature and Natural Resources (IUCN) in cooperation with UNEP, FAO, UNESCO and the World Wildlife Fund (WWF).* This marked the first official notice that ethics was a matter of explicit concern to the international conservation movement. This historic document declared that "ultimately the behavior of entire societies towards the biosphere must be transformed if the achievement of conservation objectives is to be assured." It called for "a new ethic, embracing plants and animals as well as people" so that human societies might "live in harmony with the natural world on which they depend for survival and well-being."
- *The adoption, by the United Nations General Assembly, of the World Charter for Nature, sponsored by Zaire and 32 other developing nations, in 1982.* Among the remarkable statements of this document was the declaration that "every form of life is unique, warranting respect regardless of its worth to man."
- *The meeting of the International Conference on Conservation and Development in Ottawa in 1986 to review the progress of the World Conservation Strategy and to map its revision.* Moral concerns were prominent in the deliberations of the conference, as seen by the title chosen for its proceedings, *Conservation and Equity,* the inclusion of a workshop on Ethics, Culture, and Sustainable Development sponsored by the Ethics Working Group

of IUCN, and the vote of the conference to include a new section on ethics in the revised WCS.

- *The 1987 report, Our Common Future, by the United Nations Commission on Environment and Development, chaired by Norwegian Prime Minister Gro Harlem Brundtland.* This report concluded that "human survival and well-being could depend on success in elevating sustainable development to a global ethic."
- Caring for the Earth: A Strategy for Sustainable Living, *the successor document to the World Conservation Strategy, in preparation under the guidance of UNEP, IUCN, and WWF from 1987 to 1991, officially launched in October 1991.* This document set out the elements of a world ethic for living sustainably, and included practical guidelines for their implementation. *Caring for the Earth* provides a framework of guiding principles that correlate with and can serve to undergird Agenda 21.
- *The United Nations' Conference on Environment and Development (UNCED), meeting in Rio de Janeiro in June 1992.* UNCED, attended by some 110 heads of state or government, generated an unprecedented degree of government involvement and media interest. Products of UNCED include "The Rio Declaration on Environment and Development," which outlines the rights and obligations of states and individuals in respect to sustainable development, and Agenda 21, a comprehensive set of programs and actions to promote sustainable development.
- *Executive Order 12898 signed by President Clinton February 1994: Federal Actions to Address Environmental Justice in Minority Populations and Low Income Populations (59 FR 7629).*

Human environmental rights. *Our Common Future* declares that "All human beings have the fundamental right to an environment adequate for their health and well-being." *Caring for the Earth* states that every human being has rights "within the limits of the earth, to the resources needed for a decent standard of living." The Rio Declaration on Environment and Development has as its third principle the right to development and equity in meeting human needs.

Discussion of human environmental rights blossoms in a way never seen in the world before. Still, there is no agreement that human rights should embrace environmental matters, some persons feeling that the focus should be on responsibilities rather than rights. James Nash, Executive Director of the Churches' Center for Theology and Public Policy, argues that human environmental rights are imperatives for our time:

> Since environmental health is essential for human survival and creativity, human environmental rights are no less important than social, political, and economic rights. In fact, the possibility of realizing nearly every other human right depends on the realization of environmental rights. Ecological integrity, then, is a precondition of justice (Nash, 1992).

Nash proposes a Bill of Environmental Rights that includes the right to:

- public disclosure of environmental hazards
- participation in public decisions affecting environmental quality
- freedom from the types of risks that could lead to catastrophe
- adequate ecological education
- protection of soils, air, waters, and atmosphere from pollution exceeding absorptive capacities of ecological processes
- just standards for the distribution of burdens in the disposal of trash and toxic materials
- birth control information, education, services, and means
- preservation of biodiversity and ecosystems
- government management to assure just, frugal, and sustainable use of all natural resources
- global cooperation by all governments to meet transboundary ecological problems
- equitable distributions within and among all nations of the essential natural resources to satisfy basic needs.

Changing Worldviews

A *worldview* is a set of assumptions about what is real, what is true, and who can be trusted to know the facts about reality and truth. All persons living at any one time do not have the same notions about what constitutes the real, but at any one time there is a prevailing worldview among persons with a similar culture. Constituting this view is a set of standards, beliefs, values, and habits sometimes referred to as the *dominant social paradigm*. Events are always challenging a worldview, and when the current view is no longer perceived to explain events, it is replaced by whatever is thought to be more appropriate, and its spokespersons are replaced by those thought to have a firmer grasp on reality. Thus, the modern age gave scientists an authority once reserved for kings and priests.

The "Western" worldview. Key figures are usually associated with a particular worldview, although many changes—economic, social, ecological, and political—combine to bring about such a general change in perception. Along with Bacon and Newton, the philosopher Descartes is associated with what is often termed the "Western" worldview, or the "scientific revolution" of the 17th and 18th centuries. Descartes developed a logic that separated mind and body, subject and object, value and fact, spirit and matter, that set human beings apart from and over nature, and that depicted value judgments as too subjective and unreliable to constitute proper knowledge. Cartesian reality approximates a machine, with everything working according to simple, discoverable laws.

This worldview is largely responsible for technological progress. It has tended to favor looking at problems by isolating them and taking them apart. As a consequence, certain areas of knowledge—areas such as molecular physics and genetics—have made great strides, whereas complex problems involving overlapping fields, social problems, and problems involving both human and natural factors have been neglected.

The ecological worldview. Today, the Western worldview, though still dominant, is being challenged on many fronts. As we have seen, developments in physics questioned the notion that scientists are independent of reality, that their discoveries are not influenced by their own worldviews. More importantly, the dualistic Cartesian worldview has been much criticized for not providing answers to—and worse, for helping to create—the most pressing global problems faced today. By focusing on parts rather than systems, the whole was allowed to deteriorate. By separating fact and value, the Cartesian worldview permitted technology to proceed without regard for its human and environmental consequences. To reverse this trend, many persons see the need for an ecological worldview, a "holistic" or whole-including view in which fact and value are related. Elements of this view would include:

- seeing people and nature as inseparable
- seeing nature as made up of interrelated wholes greater than the sum of their parts
- concern for quality and well-being
- an emphasis on cooperation rather than competition

- an integration of all systems, social, cultural, and economic (Sterling, 1990).

Technology is not unimportant to the ecological worldview, but it functions *with* earth's systems, *for* the realization of earth's goals, rather than as a separate end. This approach keeps the larger picture in the forefront of every issue. It is an approach exemplified by Rachel Carson in *Silent Spring* (1962), a book that alerted the world to the dangers of the indiscriminate application of pesticides by explaining the role played by microbes and worms in the maintenance of healthy soil and in the chain of life that includes fish and birds and human beings. Such awareness of the problems associated with narrowly conceived technological "fixes" in no way minimizes the roles of scientists; in fact, their jobs become more challenging as they search for solutions that promote the values of larger communities.

The Ethic of Sustainable Development

Change is difficult, and the ecological worldview has yet to dominate world thinking. No place is this more apparent than in the controversies surrounding growth and development. In today's world, the expectations of unlimited material progress and ever-growing consumption still dominate economic thought and business practice. Every country aims to export more than it imports. The GNP remains the measure of a nation's health, despite its exclusion of environmental costs and despite resource depletion, pollution, rising health costs, and the threat of climatic upheaval. Yet side by side with the concepts of unlimited growth and bigger is better, a new concept—that of sustainable development—is taking hold. Increasingly written about and discussed, sustainable development is for many today an ethically superior alternative to modern global development.

What does sustainable development mean? To some, the term itself is an oxymoron, embracing irreconcilable opposites. They fear it is essentially a camouflage for maintaining business-as-usual, for the destructive limits-denying economic practices that have created our present world crisis. Those who embrace the term point out that *sustainable* means nourishing, as the earth is nourishing to life, and *development* embraces the concept of fulfillment. Thus, *sustainable development* may be defined as the kind of human activity that nourishes and fulfills the whole community of life on earth. (See also Chapter 14, Global Issues, for additional discussion of sustainable development.)

Caring for the Earth defines this term as an ethic for "improving the quality of human life while living within the carrying capacity of supporting ecosystems" and suggests nine actions that individuals, communities, and nations must take to create a sustainable society:

- Respect and care for the community of life.
- Improve the quality of human life.
- Conserve the earth's vitality and diversity. This requires each responsible group to conserve life-support systems, conserve biodiversity, and ensure that uses of renewable resources are sustainable.
- Minimize the depletion of nonrenewable resources.
- Keep within the earth's carrying capacity.
- Change personal attitudes and practices. This requires that humans reexamine their values and alter their behavior, that society promote those values that support the ethic of sustainable development and discourage those that do not, and that information be disseminated through both formal and informal educational systems.
- Enable communities to care for their own environments.
- Provide a national framework for integrating development and conservation.
- Create a global alliance.

ENVIRONMENTAL POLICY AND CORPORATE RESPONSIBILITY

The Nobel Prize-winning economist, Milton Friedman, states that, ". . . the overriding obligation of corporate managers is to maximize corporate profits within the constraints imposed by the laws and customary moral rules governing business activities" (quoted in Rodewald, 1987). This perspective places the corporate manager as an agent of capital (that is, the manager has both a legal obligation and a moral responsibility to protect and promote the interests of corporate investors) (Boehlje, 1992; Rodewald, 1987).

This attitude is not, however, universally held. A contrary viewpoint argued by Stone is that beyond "profit-making obligations, corporate managers should have social responsibilities to avoid significant human and environmental harm and help solve some of our persistent societal problems" (quoted in Rodewald, 1987). Thus, this perspective places the manager in a position as an agent of society, responsible to use resources and produce products in a manner that does not harm either individuals, society, or the environment, regardless of potential loss of profitability (Boehlje, 1992).

The agent of capital versus the agent of society debate highlights the ethical dilemma that senior corporate managers face. Interestingly, the center of this debate has clearly shifted toward the agent of society position. A 1992 survey of business executives (Winsemius & Guntram, 1992) found that:

- Fifty-six percent agreed that consumers will increasingly ask "How green is your company?" before buying a product.
- Sixty-eight percent agreed that organizations with a poor environmental record will find it increasingly difficult to recruit and retain high-caliber staff.
- Seventy-two percent agreed that environmental legislation has become a decision factor in plant location.
- Forty-nine percent agreed that government regulations supporting a 50% reduction in environmental pollution are likely by the year 2000.

Corporate response to these trends can be found in several major corporate social and environmental responsibility documents:

- Valdez Principles by the Coalition for Environmentally Responsible Economics (Figure 2–3).
- Charter for Sustainable Development at the Second World Industry Conference on Environmental Management.
- Responsible Care Program developed by the U.S. Chemical Manufacturers Association. This program includes 10 guiding principles and 6 management practices, including an agenda for preventing accidents, guidelines for informing the public and responding to its concerns and a program of pollution prevention via waste minimization.
- Environmental policy statements by major corporations such as Du Pont, Dow, Shell, 3M, Nissan, and Sony.
- 1991 "Business Charter for Sustainable Development." Established by the International Chamber of Commerce, this document begins with the statement, "There is widespread recognition today that environmental protection must be among the highest priorities of every business" and sets out 16 principles that companies should follow.

- Business community participation in the 1992 Rio UN Conference on Environment and Development. Business interests were represented by the Business Council for Sustainable Development, a group of 48 chief executives of major worldwide corporations. The viewpoint of this group is expressed in Schemidheiny's *Changing Course* (1992). Schemidheiny argues for greater reliance on economic instruments (pollution taxes, reduced subsidies for coal and electricity, and so on) as opposed to direct government regulation.

These industry efforts represent a significant paradigm change from the agent of capital viewpoint; however, there remains considerable skepticism from many environmental organizations. This watchful waiting attitude is reflected in the Chemical Manufacturers Association responsible care slogan, "Don't Trust Us; Track Us." The evolution from agent of capital to agent of society will be difficult and has yet to meet its ultimate test: What level of profitability can or will be sacrificed for environmental protection? The ethical and economic dialogue will continue and is far from concluded.

Politics Today

Because it is basic to environmental ethics that freedoms of thought, conscience, religion, enquiry, expression, association, participation in government, and access to education be available to all the peoples of the earth—in some parts of the world environmental ethics awaits stable and inclusive forms of government. Where such has been achieved, there is a call today for the environmentally concerned to become more involved in governmental decision making on local, national, and international levels. To an extent, this is being achieved, and the "greening" of politics is a real phenomenon (see Chapter 1, History and Development). In some European nations, the Green party, with a platform committed to environmentally responsible politics, has had an influence beyond its numbers. Yet, the world is also experiencing an "environmental backlash," a resistance to the phasing-out of environmentally destructive industries and practices, and the curtailment of personal freedoms that must accompany a move to sustainability. Many who believe in energy conservation, for instance, cannot fathom life without the family car. Addiction to luxury is the common lot of the inhabitants of the world's wealthier nations. For

The Valdez Principles

1. **Protection of the biosphere**
   We will minimize and strive to eliminate the release of any pollutant that may cause environmental damage to the air, water, or earth or its inhabitants. We will safeguard habitats in rivers, lakes, wetlands, coastal zones, and oceans, and will minimize contributing to the greenhouse effect, depletion of the ozone layer, acid rain, or smog.

2. **Sustainable use of natural resources**
   We will make sustainable use of renewable natural resources, such as water, soils and forests. We will conserve nonrenewable natural resources through efficient use and careful planning. We will protect wildlife habitat, open spaces, and wilderness, while preserving biodiversity.

3. **Reduction and disposal of waste**
   We will minimize the creation of waste, especially hazardous waste, and whenever possible, recycle materials. We will dispose of all wastes through safe and responsible methods.

4. **Wise use of energy**
   We will make every effort to use environmentally safe and sustainable energy sources to meet our needs. We will invest in improved energy efficiency and conservation in our operations. We will maximize the energy efficiency of products we produce and sell.

5. **Risk reduction**
   We will minimize the environmental, health, and safety risks to our employees and the communities in which we operate by employing safe technologies and operating procedures, and by being constantly prepared for emergencies.

6. **Marketing of safe products and services**
   We will sell products or services that minimize adverse environmental impacts and that are safe as consumers commonly use them. We will inform consumers of the environmental impacts of our products or services.

7. **Damage compensation**
   We will take responsibility for any harm we cause to the environment by making every effort to fully restore the environment and to compensate those persons who are adversely affected.

8. **Disclosure**
   We will disclose to our employees and to the public incidents relating to our operations that cause environmental harm or pose health or safety hazards. We will disclose potential environmental, health, or safety hazards posed by our operations, and will not take any action against employees who report any condition that creates a danger to the environment or poses health and safety hazards.

9. **Environmental directors and managers**
   We will commit management resources to implement the Valdez Principles, to monitor and report upon our implementation efforts, and to sustain a process to ensure that the Board of Directors and Chief Executive Officer are kept informed of and are fully responsible for all environmental matters. We will establish a Committee of the Board of Directors with responsibility for environmental affairs. At least one member of the Board of Directors will be a person qualified to represent environmental interests to come before the company.

10. **Assessment and annual audit**
    We will conduct and make public an annual self-evaluation of our progress in implementing these Principles and in complying with all applicable laws and regulations throughout our worldwide operations. We will work toward the timely creation of independent environmental audit procedures, which we will complete annually and make available to the public.

Figure 2–3. The Valdez Principles. Source: Sanyal RN, Neves JS. The Valdez Principles: Implications for corporate social responsibility. *J Bus Ethics* 10:13, 1991. Reprinted by permission of Kluwer Academic Publishers.

this reason, many who are concerned for the environment look to the world's religions for help in instilling and underscoring values that will benefit the environment.

Environmental Ethics and World Religions

Within the world's religions, interest in environmental ethics is certainly not new; humanity's inability to live in harmony with the environment has always been an important subject for religion. It is often conceived as a turning away from the true ground of existence, as is seen in the biblical myth of the fall of Adam and Eve in the Garden of Eden. Today, however, within all the religions there is new emphasis on discovering and highlighting those texts and inherent beliefs with environmental consequences. In part this is a response to criticism of religion's shortcomings; in part an effort to stem increasing materialism, the separation of spirit and matter that has been the world's lot since the time of Descartes; in part a response to the severity of the global environmental crisis.

Shortcomings. How adequately the world's religions address the environmental crises of contemporary civilization has been the subject of much probing debate during the closing decades of the 20th century. In 1967, in a widely read essay, "The Historical Roots of Our Ecological Crisis," medieval historian Lynn White, Jr., argued that the notion of "human domination" over nature, the notion that the world was created and planned by a transcendent God for the exclusive purpose of serving humanity, links Christianity and Judaism to the present environmental crisis. The essay, the source of much controversy, was not easily dismissed; other Christian theologians agreed with White's analysis. Subsequently, White's article has prompted the examination of other religious faiths that have also been found to have shortcomings in their relationships to the environment. Classic Hinduism, for example, has been found to be world-denying; Buddhism, ethically passive and ecologically unrealistic.

New emphases. Today the tide has turned from finding ways in which the world's religions are anti-environmental to seeking ways in which they can be sources of environmentally positive teachings. Representatives of the world's religions, including His Holiness the Dalai Lama, Pope John Paul II, and Chief Oren Lyons, have spoken on behalf of the environment. New fields of religious interest, variously called ecotheology, ecojustice, and creation spirituality, have developed. The teachings within each religion that are helpful as guides to a better relationship with the environment are more widely disseminated today. What can be mentioned here are only a few examples from this immense body of concern and scholarship.

In the Jewish tradition, a strong sense of belonging to a people that will continue provides grounds for the preservation of nature (Barbour, 1980). Within the Jewish and Christian traditions, the biblical ideas of stewardship and covenant are foundations for environmental ethics. Stewardship provides a religious justification for respect toward other creatures and the earth itself. God's covenant with Noah (Genesis 9:12) includes nonhuman life and posterity. God is said to value creation, the basis for creation spirituality, according to the writings of theologian Matthew Fox. Creation spirituality holds up the vision of human beings acting as co-creators with the cosmic Christ to bring the cosmos to harmonious fulfillment (Fox, 1988).

New models of the relationship of God to the world have been presented by feminist theologians such as Sally McFague and Rosemary Radford Ruether. These views balance the masculine aspects of divinity—God as lord, king, and patriarch—with the feminine aspects of birthing, healing, and caring for creation.

Islam permits the use of the natural environment but not its destruction, because the earth is a source of blessedness and a place for the worship of God. Earth's preservation is a duty to be undertaken, not for personal benefit but because it is right to do so.

In illustration, the Prophet Muhammed said, "When doomsday comes, if someone has a palm shoot in his hand he should plant it."

Within Hinduism, the doctrine of *ahimsa*—nonviolence against animals and humans—is an environmental ethic. Also important to the treatment of animal life is the belief that the Supreme Being was incarnated in the form of various species, including fish, tortoise, boar, and man-lion. The Hindu worship of trees and plants occasioned the heroic defense of trees—an inspiration for the later (20th century) Chipko movement—by the villagers of the Bishnois community in Rajasthan, India, in the 15th century. To prevent their trees from being cut down by soldiers, the villagers hugged the trees with their bodies. Three hundred sixty-three persons were thus killed defending trees.

In Buddhism, human beings are not the center of the universe, but like all other beings, dependently co-arising parts of it (Callicott & Ames, 1989). Contemporary Mahayana Buddhists stress that a Buddha-nature awaits realization in all humans and many living things, and its essence is a pure, disinterested love for all being. Zen Buddhists practice arduous spiritual exercises to achieve a vivid esthetic experience of nature, called *satori*. Theirs is an active concern for harmonizing human dwelling with nature on the model of Japanese landscape architecture.

Today, there is renewed interest in traditional African, Amerindian, and other indigenous peoples who developed over time patterns of sustainable relationships with the land and its creatures upheld by religious beliefs. For instance, traditional African societies honored both "ownership rights" and "possession rights." Property was owned by the community, including all ancestors and future generations, who had been given it by God. It was possessed only temporarily by an individual who had the right to farm and settle on it. This arrangement meant that everyone in the community had a responsibility to see that land was well cared for (Omari, 1990). Aspects of Amerindian faith most often cited as environmentally sustainable include:

- belief in the familial relationship of all life. Black Elk speaks of "us two-leggeds" who "with the four-leggeds and the wings of the air and of all green things . . . are children of one mother and their father is one Spirit" (Neihardt, 1979)
- belief that all beings share in a universal Spirit
- a religious attachment to specific places

- the use of ritual to restore harmony with nature
- the belief that land belongs to all.

The Ethics of Home

This discussion of environmental ethics began with an accident that, in addition to its other dire consequences, left many thousands of persons homeless. Homelessness as a result of environmental degradation is now commonplace. Indeed, environmental refugees have become a large class of displaced persons in the world (Jacobson, 1989). Environmental refugees are refugees of industrial accidents, such as Bhopal, and of other accidents, such as the 1983 one in Times Beach, Missouri, when oil laced with dioxin was sprayed on city streets. They are refugees of nuclear reactor accidents, of which Chernobyl in 1986 is the most horrendous example, causing the evacuation of more than 100,000 persons from what remains uninhabitable land. They are refugees of pollution, such as the former residents of the community of ironically-named Love Canal, New York, whose homes were built on a chemical waste dump. They are, most of all, refugees of land degradation and other untenable changes in their habitat.

In Ethiopia they flee from desertification; in Haiti, from political repression combined with poverty and the environmental devastation brought on by deforestation and topsoil loss; in the Philippines from the ravages of tropical storms magnified by mining and lumbering operations that radically changed patterns of water runoff. If the threat of a global warming sufficient to melt earth's icecaps were to become a reality, millions more would flee the inundation of earth's coastal areas.

Ethical considerations are one of the guiding forces behind the sustainable development movement that has become prominent in the 1990s. At the 1992 Earth Summit, Norway's Prime Minister defined sustainable development as "the development that meets the needs of the present without compromising the ability of future generations to meet their needs."

Thus, environmental ethics is most basically an ethics of home. It is, at the least, an ethics of material maintenance, of seeing to it that present and future generations have the bodily comforts of home, have trees to roof them, food and pure water to drink, clean surroundings, a place of rest, and safety. Environmental ethics must also be an ethics of spiritual restoration, of seeing to it that present and future generations have freedom, education, and exposure to natural and other values. Environmental ethics is about both maintenance and restoration; both material and spiritual welfare. Fundamental to the concept of environmental ethics is the notion that the spiritual and the material are inseparable, that they nourish each other, that earth, our only home, needs both.

SUMMARY

Environmental economic and ethical considerations are intertwined. Although there appears to be an occasional dichotomy between these two forces, the worldwide business community has increasingly discovered that "the degree to which a company is viewed as being a positive or negative participant in solving sustainability issues will determine, to a very great degree, their long-term business viability" (Woodhouse, 1992). Environmental economics can be the leading force that moves business activity from a frequent situation of inefficient consumption and pollution toward pollution prevention and more efficient allocation of resources. However, it is clear that this movement will only achieve significant success when a thorough airing of ethical considerations is undertaken.

REFERENCES

Abelson PH. Pathological growth of regulations. *Science* 26:1854 (25 June 1993).

Alper J. Protecting the environment with the power of the market. *Science* 26:1884–85 (25 June 1993).

Barbier EB. Cash crops, food crops, and sustainability: The case of Indonesia. *World Devel* 17(6): 879–895, 1989.

Barbour IG. *Technology, Environment, and Human Values.* New York: Praeger, 1980.

Boehlje M. Environmental Policy and Corporate Responsibility (or Liability!). Proceedings of the 3rd Annual Conference on Agricultural Policy and the Environment. Center for International Food and Agricultural Policy. Dept. of Agriculture and Applied Economics, University of Minnesota, St. Paul, June 1992.

Business Charter for Sustainable Development, The. Paris: International Chamber of Commerce, 1991.

Callicott JB, Ames RT. *Nature in Asian Traditions of Thought.* Albany: State University of New York Press, 1989.

Caring for the Earth. Gland, Switzerland: IUCN, UNEP, and WWF, 1991.

Carson R. *Silent Spring.* Boston: Houghton Mifflin, 1962.

Clark B. "The range of the mountains is his pasture"—environmental ethics in Israel. In Engel JR, Engel JG (eds), *Ethics of Environment and Development.* Tucson: University of Arizona Press, pp. 183–188, 1990.

De George RT. The Environment, Rights, and Future Generations." *Ethics and Problems of the 21st Century.* Goodpaster KE, Sayre KM. Notre Dame, IN: University of Notre Dame Press, 1979.

Dudley MK. Traditional native Hawaiian environmental philosophy. In Hamilton LM (ed), *Ethics, Religion and Biodiversity.* Cambridge, UK: White Horse Press, pp. 176–182, 1993.

Ethics of Environment and Development. Engel JR, Engel JG (eds). Tucson: University of Arizona Press, 1990.

Fox M. *The Coming of the Cosmic Christ.* San Francisco: Harper & Row, 1988.

Friedman M, quoted in Rodewald RA. The corporate social responsibility debate: Unanswered questions about the consequences of moral reform. *Am Bus Law* 525(3):443–466, Fall 1987.

Gudynas E. The search for an ethic of sustainable development in Latin America. In Engel JR, Engel JG (eds), *Ethics of Environment and Development.* Tucson: University of Arizona Press, pp. 139–149, 1990.

Harrison R. *Animal Machines.* London: Vincent Stuart, 1964.

Hudec R, Farber D. Distinguishing environmental measures from trade barriers. University of Minnesota, Workshop on International Economic Policy. November 17, 1992.

International Trade Reporter. May 27, 1992. (*Mississippi Poultry Association Inc. v. Madigan,* No. J91–0086(W), DC SMiss 4/23/92).

Jacobson JL. Abandoning homelands. In *State of the World 1989.* New York: WW Norton, pp. 59–76, 1989.

Leopold A. *A Sand County Almanac and Sketches Here and There.* London: Oxford University Press, 1949.

Mander J. *Four Arguments for the Elimination of Television.* New York: Morrow Quill, 1978.

Morehouse W, Subramaniam MA. *The Bhopal Tragedy.* New York: Council on International and Public Affairs, 1986.

Myers N. *Gaia.* Garden City, NY: Anchor Press/Doubleday, 1984.

Naess A. Sustainable development and deep ecology. In Engel JR, Engel JC (eds), *Ethics of Environment and Development.* Tucson: University of Arizona Press, pp. 87–96, 1990.

Nash JA. Human rights and the environment: New challenge for ethics. *Theol and Pub Pol* 4:42–57, 1992.

Nature 363(6431):657–58, June 24, 1993.

Neihardt JG. *Black Elk Speaks.* Lincoln: University of Nebraska Press, 1979.

Niebuhr R. *Moral Man and Immoral Society.* 1932.

Omari CK. Traditional African land ethics. In Engel JC, Engel JG (eds), *Ethics of Environment and Development.* Tucson: University of Arizona Press, pp. 168–175, 1990.

Postel S, Ryan JC. Reforming forestry. In Brown L (ed), *State of the World, 1991.* New York: WW Norton, pp. 74–92, 1991.

Renner M. Assessing the military's war on the environment. In Brown L (ed), *State of the World 1991.* New York: WW Norton, 132–152, 1991.

Repetto R. Accounting for environmental assets. *Sci Am* 266(6):94–100, June 1992.

Repetto R. Wasting assets: Natural resources in the national income accounts. World Resources Institute, 1989.

Repetto R, Dower RC. Green fees. *EST* 27(2), 1993.

Rodewald RA. The corporate social responsibility debate: Unanswered questions about the consequences of moral reform. *Am Bus Law* 525(3):443–466, Fall 1987.

Rolston H. Science-based versus traditional ethics. In Engel JR, Engel JG (eds), *Ethics of Environment and Development.* Tucson: University of Arizona Press, pp. 63–76, 1990.

Runge CF. Economic criteria and net social risk in the analysis of environmental regulation. In Smith

VK (ed), *Environmental Policy Under Reagan's Executive Order: The Role of Benefit Cost Analysis.* Chapel Hill, NC: University of North Carolina Press, 1984.

Runge CF et al. *Freer Trade—Protected Environment: Balancing Trade Liberalization and Environmental Interests.* New York: Council on Foreign Relations, 1994.

Runge CF. Trade protectionism and environmental regulations: The new nontariff barriers. *NW J of Int Law and Bus* 11:1, 47–61, Spring 1990a.

Runge CF. Environmental risk and the world economy. *Am Prospect.* 1:114–118, Spring 1990b.

Russell B. *Religion and Science.* London: Oxford University Press, 1935.

Sallah A. Living with nature: Reciprocity or control? In Engel JR, Engel JG (eds), *Ethics of Environment and Development.* Tucson: University of Arizona Press, pp. 245–253, 1990.

Schemidheiny S. *Changing Course.* Cambridge, MA: MIT Press, 1992.

Singer P. *Animal Liberation.* New York: Avon Books, 1975.

Southgate D. The economics of land degradation in the third world. Working Paper No. 2. Environment Department. World Bank, 1988.

Sterling S. Towards an ecological world view. In Engel JR, Engel JG (eds), *Ethics of Environment and Development.* Tucson: University of Arizona Press, pp. 77–86, 1990.

Stone, quoted in Rodewald RA. The corporate social responsibility debate: Unanswered questions about the consequences of moral reform. *Am Bus Law* 525(3):443–466, Fall 1987.

Toulmin S. *The Return to Cosmology.* Berkeley: University of California Press, 1982.

White L, Jr. The historical roots of our ecological crisis. *Science* 155:1203–7 (10 March 1967).

Winsemius P, Guntram U. Responding to the environmental challenge. *Bus Horizons*, March-April 1992.

Woodhouse B, quoted in Schmidheiny S. *Changing Course.* Cambridge, MA: MIT Press, 1992.

Woolf T. It is time to account for the environmental costs of energy resources. *Energy & the Environment.* 1993.

World Commission on Environment and Development. *Our Common Future.* Oxford, UK: Oxford University Press, 1987.

Wright J. *The Cry for the Dead.* Melbourne: Oxford University Press, 1981.

3

United States Legal and Legislative Framework

Brian J. King, JD
Deborah Anne Amidaneau, MS

During the last 25 years, the United States has developed an extensive web of environmental laws, regulations, and compliance procedures. Many of the laws are complex and proscriptive (i.e., they detail precise compliance criteria). However, other laws are performance driven and specify endpoints rather than procedures. This chapter is divided into two sections: (1) a legal overview of the main federal laws that affect business and industry, and (2) the regulatory basis for environmental compliance at federal facilities. (For additional information, see Chapter 8, Hazardous Wastes.)

FEDERAL ENVIRONMENTAL LAWS

One of the most complicated problems in environmental law is defining the respective roles of the state and federal government in the administration of environmental law. Most federal environmental statutes provide for federal primacy in enforcement and implementation while creating a significant role for the states. The system is complicated by the fact that the enacting legislation often fails to create clear boundaries among the respective authorities of the federal government and the states.

Generally speaking, a state may administer its own environmental program as long as the program requirements are as stringent as federal requirements. Many states have environmental programs that are more stringent than the corresponding federal program.

National Environmental Policy Act

The National Environmental Policy Act (NEPA) of 1969 provides the basic framework for protection of the environment. The National Environmental Policy Act establishes policy, sets goals, and provides a means for implementing the policy.

Section 101 of NEPA. The National Environmental Policy Act establishes a "public trust" concept for future generations. The federal government shall

use all practicable means and measures to create and maintain conditions under which man and nature can exist in productive harmony and fulfill the social, economic, and other requirements of present and future generations of Americans.

Section 102 of NEPA: NEPA's "Action-Forcing Provision." Section 102 of NEPA directs all federal agencies, to the fullest extent possible, to meet the following requirements:

- Use a systematic interdisciplinary approach in planning and decision making that may affect the environment.
- Include an Environmental Impact Statement (EIS) in every recommendation or report on proposals for legislation and other *major federal actions significantly affecting the quality of the human environment.*

It is important to remember that NEPA does not apply directly to private entities. Rather, an EIS is required only for "major federal actions" that significantly affect the environment. The requirements of NEPA can become important to private industry when a federal permit, approval, or funds are involved in a proposal. The federal permit or approval can necessitate the preparation of an EIS, or, at a minimum, the preparation of an environmental assessment prior to the decision to grant a permit or approval.

The Council on Environmental Quality (CEQ) defines major federal actions as "actions with effects that may be major and which are potentially subject to federal control and responsibility." Here *major federal action* includes adoption of rules, regulations, and policy; adoption of formal plans; and approval of specific projects.

- Study, develop, and describe appropriate alternatives to recommended courses of action.

Section 202 of NEPA: Creation of the CEQ. The CEQ's role was to advise the president on federal programs and policies affecting the environment and issue regulations governing federal agency implementation of NEPA. The role of the CEQ has been diminishing during the years as the U.S. EPA assumed the advisory role to the president.

State programs. Many states have developed environmental impact legislation similar to NEPA.

Clean Air Act

The Clean Air Act (CAA) is a comprehensive federal statute designed to protect public health and regulate air emissions from stationary and mobile sources. Major federal clean air legislation includes the Air Quality Act of 1967, the Clean Air Act Amendments of 1970, the Clean Air Act Amendments of 1977, and, most recently, the Clean Air Act Amendments of 1990. Table 3–A lists the CAA Amendments of 1990 and presents a synopsis of each amendment.

Ambient air quality standards. Ambient air quality standards are national in scope. State implementation plans (SIPs) set air quality goals to be achieved by each state. The standards are stated in terms of annual concentration levels or annual mean measurements for the air.

Based on the criteria established for each pollutant under Section 108(a) of the CAA and "allowing an adequate margin of safety," the EPA is directed to promulgate national primary and secondary ambient air quality standards for each pollutant.

Primary air quality standards are designed to protect the public health. Secondary standards are intended to protect the public welfare from any known or anticipated adverse effects associated with the presence of such pollutants in the air.

Primary and secondary standards have been established by EPA for seven pollutants. The major criteria pollutants are:

- carbon monoxide
- nitrogen oxides
- sulfur dioxide
- particulates and particulate matter (dust)
- ozone
- hydrocarbons
- lead.

State implementation plans (SIPs). National air quality regulations are applied to individual sources through state implementation plans (SIPs). Under Section 110 of the CAA, a SIP provides for implementation, maintenance, and enforcement of national standards by the state in each air quality region within its boundaries. States must meet the national ambient air quality standards through the SIP, which imposes specific emission limitations on individual sources located within the state. If a state fails to adopt an acceptable SIP, the EPA is directed to formulate and enforce one.

Prevention of significant deterioration. The prevention of significant deterioration (PSD) provisions were added in 1977 to the requirements of the CAA. The program is designed to prevent significant deterioration of air quality in regions where the air is already cleaner than ambient standards.

In order to implement the regulatory program for PSD, all air quality control regions that are cleaner than ambient standards are designated under one of three classes. Regions are designated Class I, II, or III areas for the purpose of specifying the amount or "increment" of air pollution that can be permitted in each area.

Class I increments permit only minor air quality deterioration; Class II increments permit moderate deterioration; and Class III increments permit

Table 3–A. The U.S. Clean Air Act Amendments of 1990

Title I—Provisions for Attainment and Maintenance of National Ambient Air Quality Standards

Section 101   General planning requirements.

Section 102   General provisions for nonattainment areas.

Section 103   Additional provisions for ozone nonattainment areas.

Section 104   Additional provisions for carbon monoxide nonattainment areas.

Section 105   Additional provisions for particulate matter (PM_{10}) nonattainment areas.

Section 106   Additional provisions for areas designated nonattainment for sulfur oxides, nitrogen dioxide, and lead.

Section 107   Provisions related to Indian tribes.

Section 108   Miscellaneous provisions.

Section 109   Interstate pollution.

Section 110   Conforming amendments.

Section 111   Transportation system impact on clean air.

The impetus for development of this title was the fact that many areas of the United States have not achieved attainment with the national ambient air quality standard for ozone. Much of this title pertains to the establishment of new deadlines for meeting ozone standards and other requirements for ozone nonattainment areas. Sections on nonattainment areas for carbon monoxide and particulate matter less than 10 micrometers in diameter (PM_{10}), and other issues are also included.

Title II—Provisions Relating to Mobile Sources
Part A—Amendments to Title II of Clean Air Act

Section 201   Heavy-duty trucks.

Section 202   Control of vehicle refueling emissions.

Section 203   Emission standards for conventional motor vehicles.

Section 204   Carbon monoxide emissions at cold temperatures.

Section 205   Evaporative emissions.

Section 206   Mobile source-related air toxics.

Section 207   Emissions control diagnostics systems.

Section 208   Motor vehicle testing and certification.

Section 209   Auto warranties.

Section 210   In-use compliance—recall.

Section 211   Information collection.

Section 212   Nonroad fuels.

Section 213   State fuel regulation.

Section 214   Fuel waivers.

Section 215   Misfueling.

Section 216   Fuel volatility.

Section 217   Diesel fuel sulfur content.

Section 218   Lead substitute gasoline additives.

Section 219   Reformulated gasoline and oxygenated gasoline.

Section 220   Lead phasedown.

Section 221   Fuel and fuel additive importers.

Section 222   Nonroad engines and vehicles.

Section 223   New Title II definitions.

Section 224   High altitude testing.

Section 225   Compliance program fees.

Section 226   Prohibition on production of engines requiring leaded gasoline.

Section 227   Urban buses.

Section 228   Enforcement.

Section 229   Clean-fuel vehicles.

Section 230   Technical amendments.

Part B—Other Provisions

Section 231   Ethanol substitute for diesel.

Section 232   Adoption by other states of California standards.

Section 233   States authority to regulate.

Section 234   Fugitive dust.

Section 235   Federal compliance.

As its name implies, this title pertains to air pollutant emissions related to the operation of vehicles. The amendments in this title not only establish tailpipe emission standards, but also impose new requirements for vehicle fuels, including fuel volatility requirements.

Title III—Hazardous Air Pollutants

Section 301   Hazardous air pollutants.

Section 302   Conforming amendment.

Section 303   Risk assessment and management commission.

Section 304   Chemical process safety management.

Section 305   Solid waste combustion.

Section 306   Ash management and disposal.

Perhaps the most far-reaching of all the changes in the Clean Air Act Amendments of 1990, this title greatly expands federal requirements for control of hazardous air pollutants at industrial facilities. Three components are at the heart of this title. First, the list of substances governed by hazardous air pollutant policies expands from the few regulated prior to the adoption of the Clean Air Act Amendments of 1990 to an initial list of 189 pollutants (see Table 3–C for list). Second, designated industrial emission sources will have to achieve maximum achievable control technology standards to limit emissions of hazardous air pollutants to the atmosphere. Third, new and existing facilities will have to comply with the provisions of a new accidental release program. This program will require preparation of a risk management plan that includes an assessment to identify potential impacts of accidental releases on the surrounding community.

(continued)

Table 3–A. *(Continued)*

Title IV—Acid Deposition Control

Section 401 Acid deposition control.

Section 402 Fossil fuel use.

Section 403 Repeal of percent reduction.

Section 404 Acid deposition standards.

Section 405 National acid lakes registry.

Section 406 Industrial SO_2 emissions.

Section 407 Sense of the Congress on emission reductions costs.

Section 408 Monitor acid rain program in Canada.

Section 409 Report on clean coal technologies export programs.

Section 410 Acid deposition research by the U.S. Fish and Wildlife Service.

Section 411 Study of buffering and neutralizing agents.

Section 412 Conforming amendment.

Section 413 Special clean coal technology project.

After years of research and debate, the Bush administration and the Congress decided that specific controls were needed to reduce the emissions of pollutants that produce acid deposition, more commonly referred to as acid rain. The chief target of this title is the release of sulfur dioxide from the combustion of fossil fuels by power plants. The goal of the legislation is to reduce sulfur dioxide emissions by approximately 10 million tons annually compared to 1980 levels. This goal is to be achieved by the creation of "allowances" that can be used, traded, or sold as needed to meet overall goals, thereby allowing market forces to play an important role in environmental improvement. Emissions of nitrogen oxides will also be controlled, but not through an allowance program.

Title V—Permits

Section 501 Permits

Like the hazardous air pollutant provisions in Title III, Title V also establishes far-reaching requirements. This title creates a new national operating permit program that will be modeled after the national pollutant discharge elimination system program of the Clean Water Act. Owners of affected facilities will have to obtain an operating permit that will require renewal at least every 5 years. Associated with the permit program is the additional provision that owners of emission sources requiring an operating permit must pay an annual fee.

Title VI—Stratospheric Ozone Protection

Section 601 Part B repeal.

Section 602 Stratospheric ozone protection.

Section 603 Methane studies.

This title focuses on the control of substances that contribute to depletion of ozone concentrations in the protective ozone layers of the stratosphere. The most important part of this title is a schedule for phasing out the production and consumption of ozone-depleting substances. Chemicals affected by this title will include, at a minimum, listed chlorofluorocarbons, listed hydrochlorofluorocarbons, listed halons, carbontetrachloride, and methyl chloroform. Methane studies are also required by this title.

Title VII—Provisions Relating to Enforcement

Section 701 Section 113 enforcement.

Section 702 Compliance certification.

Section 703 Administrative enforcement subpoenas.

Section 704 Emergency orders.

Section 705 Contractor listings.

Section 706 Judicial review pending reconsideration of regulation.

Section 707 Citizen suits.

Section 708 Enhanced implementation and enforcement of new source review requirements.

Section 709 Movable stationary sources.

Section 710 Enforcement of new titles of the act.

Section 711 Savings provisions and effective dates.

Tougher enforcement provisions are now part of the Clean Air Act. Among the more important of these provisions are (1) criminal violations are now felonies rather than misdemeanors; (2) EPA can assess civil penalties administratively; (3) a field citation program is established; (4) stiffer imprisonment sentences can be imposed (up to 15 years for certain types of violations); and (5) citizens can sue for penalties.

Title VIII—Miscellaneous Provisions

Section 801 OCS [outer continental shelf] air pollution.

Section 802 Grants for support of air pollution planning and control programs.

Section 803 Annual report repeal.

Section 804 Emission factors.

Section 805 Land use authority.

Section 806 Virgin Islands.

Section 807 Hydrogen fuel cell vehicle study and test program.

Section 808 Renewable energy and energy conservation incentives.

Section 809 Clean air study of southwestern New Mexico.

Section 810 Impact on small communities.

Section 811 Equivalent air quality controls among trading nations.

Section 812 Analyses of costs and benefits.

Section 813 Combustion of contaminated used oil in ships.

Section 814 American-made products.

Section 815 Establishment of program to monitor and improve air quality in regions along the border between the United States and Mexico.

(continued)

Table 3–A. *(Concluded)*

Section 816 Visibility.

Section 817 Role of secondary standards.

Section 818 International border areas.

Section 819 Exemptions for stripper wells.

Section 820 EPA report of magnetic levitation.

Section 821 Information gathering on greenhouse gases contributing to global climate changes.

Section 822 Authorization.

The variety of provisions contained within this title can best be seen by referring to the section headings for Title VIII. These provisions range from requirements for control of outer continental shelf air pollution emission sources to information gathering on "greenhouse" gases.

Title IX—Clean Air Research

Section 901 Clean air research.

The commencement or continuation of various research programs are directed by this title. These programs include air pollutant monitoring and modeling research, environmental health effects research, ecosystem research, and continued research on acid precipitation.

Title X—Disadvantaged Business Concerns

Section 1001 Disadvantaged business concerns.

Section 1002 Use of quotas prohibited.

This title specifies that 10% of any EPA funds used for research required by the Clean Air Act Amendments of 1990 be made available to disadvantaged business concerns.

Title XI—Clean Air Employment Transition Assistance

Section 1101 Clean air employment transition assistance.

The intent of this title is to provide assistance to individuals who may lose their jobs as a consequence of compliance with the Clean Air Act.

Source: Dames & Moore, Clean Air Act Amendments of 1990. Used with permission.

---

deterioration up to the secondary ambient air quality standard.

The increment of deterioration that EPA will allow for a particular pollutant is the maximum allowable increase from all sources above a certain "baseline concentration." Baseline concentration is the ambient concentration levels that exist in a certain area as of August 7, 1977.

The PSD regulations usually apply to the construction or modification of a major source of any pollutant located within an area designated attainment or unclassifiable for any criteria pollutant. To obtain a PSD permit, the source must:

- submit monitoring data for one year
- apply best available control technology (BACT) to each pollutant that the source emits in more than the minimum amounts
- model the projected emissions to demonstrate the increment of ambient air quality consumed by the proposal and demonstrate future compliance with ambient air quality standards
- undergo a review and public hearing.

New source performance standards (NSPSs). Performance standards for new sources are designed to allow industrial growth without undermining the national program for achieving air quality goals. NSPSs are established at the national level in order to prevent states from becoming "pollution havens" and attracting industry by using lenient emission standards.

NSPS differ from ambient standards in both purpose and form: NSPS are oriented to particular sources of pollutants and types of technology rather than to air quality in general. For example, special NSPS are designed for kraft pulp mills.

Nonattainment areas. Nonattainment areas are air quality control regions that do not meet the ambient air quality standards. EPA allows new major sources in nonattainment areas only if stringent conditions are met, including a greater than one-for-one offset of emissions from existing sources in the area.

Although all of the major criteria pollutants are affected by the new nonattainment area provisions, three pollutants (ozone, PM_{10}, and carbon monoxide) are not affected because (1) there are more ozone nonattainment areas than there are nonattainment areas for other pollutants and (2) attaining the ozone standard in large metropolitan areas has proven to be a difficult task. The amendments establish nonattainment area classifications and compliance deadlines for ozone, PM_{10}, and carbon monoxide. PM_{10} represents the respirable fraction of airborne particulate matter (total suspended particulates or TSP). Usually 60% of TSP is considered as PM_{10}. Table 3–B lists the ozone nonattainment area classifications and associated requirements.

The emissions offset policy was adopted as part of the 1977 Amendments to the Clean Air Act to allow industrial growth in areas currently violating the ambient air quality standards and to ensure continued progress towards attainment of the standards.

Table 3–B. Ozone Nonattainment Area Classifications and Association Requirements

Nonattainment Area Classification	1-Hour Ozone Concentration Design Value (ppm)	Attainment Date	Major Source Threshold Level (tons VOCs/yr)	Offset Ratio for New/Modified Sources
Marginal	0.121–0.138	Nov. 15, 1993	100	1.1 to 1
Moderate	0.138–0.160	Nov. 15, 1996	100	1.15 to 1
Serious	0.160–0.180	Nov. 15, 1999	50	1.2 to 1
Severe	0.180–0.190	Nov. 15, 2005	25	1.3 to 1
	0.190–0.280	Nov. 15, 2007	25	1.3 to 1
Extreme	0.280 and up	Nov. 15, 2010	10	1.5 to 1

Source: Public Law 101–549, Title I.

Nonattainment Areas—PM$_{10}$
Important provisions for PM$_{10}$ nonattainment areas include:

- *Nonattainment Area Classification*—Initially, PM$_{10}$ nonattainment areas are those classified by EPA as Group I areas prior to November 15, 1990, or areas which had a monitored violation of the NAAQS before January 1, 1989. All such areas are initially classified as "moderate" nonattainment areas. Areas can also receive a "serious" nonattainment area classification. If a moderate nonattainment area has not been determined to be attainment within six months after the attainment deadline for that area, it will be reclassified as a serious nonattainment area.
- *Deadlines for Compliance*—The deadline for achieving attainment in moderate nonattainment areas is 6 years after an area has been designated as nonattainment, or, in the case of areas that were initially designated as nonattainment areas on the effective date of the amendments, the deadline is December 31, 1994. The attainment deadline for serious nonattainment areas is 10 years after an area has been designated as nonattainment, or December 31, 2001, in the case of areas that were initially designated as nonattainment areas on the effective date of the amendments.
- *Major Source Definition*—Major sources in moderate nonattainment areas are those that emit 100 tons per year or more of particulate matter. Major sources in serious nonattainment areas are those that emit 70 tons per year or more.
- *Control Requirements*—A permit program and adoption of reasonably available control measures (RACM) are required for PM$_{10}$ nonattainment areas. In serious nonattainment areas, provisions for adoption of best available control measures (BACM) are required. EPA is required to issue guidance on RACM and BACM no later than May 15, 1992.

Nonattainment Areas
Nonattainment areas for carbon monoxide are classified as either moderate or serious. Moderate nonattainment areas are those with a 1-hour concentration design value of 9.1 to 16.4 ppm, and serious nonattainment areas are those with a 1-hour concentration design value of 16.5 ppm and above. The attainment deadline for moderate nonattainment areas is December 31, 1995. The attainment deadline for serious nonattainment areas is December 31, 2000. If EPA determines that stationary sources contribute significantly to carbon monoxide levels in serious nonattainment areas, major stationary sources will be defined as sources with a potential to emit 50 tons per year of more of carbon monoxide.

Sanctions
Two primary types of sanctions can be applied by EPA in areas where attainment deadlines are not met. First, with the approval of the U.S. secretary of transportation, EPA can prohibit federal highway projects and grants. Second, EPA can require that emission offsets for the permitting of new and modified sources be increased to a ratio of at least 2 to 1. Construction bans in existence as of November 15, 1990, will continue in effect.

It has become part of EPA's "controlled trading approach" that allows controlled interfacility transfers of emission reduction requirements.

EPA's present policy also allows development of a procedure to permit the "banking" (saving) of offsets. Excess emission reductions, beyond the point needed for immediate trade, can be banked for use in future offset trades. Banked emission reductions become assets of the owner company.

The offset policy means that a proposed source wishing to locate in a nonattainment area must negotiate enforceable agreements from existing sources for emission reductions greater than the amount of its expected emissions. The proposed source must also meet emission limitations based upon the "lowest achievable emission rate" (LAER) for such source and must certify that all other sources owned by it are in compliance with SIP requirements.

EPA's bubble concept. The general definition of the bubble concept provides that all emissions from stack, fugitive, and area sources within a given facility site (the bubble) are to be considered as coming from one source and that the emissions from that "bubble" as a whole may be reduced to a required level by reductions of emissions from the individual sources within the bubble. This policy allows a site to meet air quality standards by reducing the overall emissions from a plant rather than requiring it to reduce emissions from individual sources within the plant.

Hazardous air pollutants (HAPs). Section 112 of the Clean Air Act is EPA's mandate to control hazardous pollutants discharged into the nation's air (Table 3–C). The agency is authorized under this section to promulgate National Emission Standards for Hazardous Air Pollutants (NESHAPs) for both new

Table 3–C. Hazardous Air Pollutants Regulated by the Clean Air Act Amendments of 1990

CAS* Number	Chemical Name	CAS* Number	Chemical Name
75070	Acetaldehyde	64675	Diethyl sulfate
60355	Acetamide	119904	3,3-Dimethoxybenzidine
75058	Acetonitrile	60117	Dimethyl aminoazobenzene
98862	Acetophenone	119937	3,3'-Dimethyl benzidine
53963	2-Acetylaminofluorene	79447	Dimethyl carbamoyl chloride
107028	Acrolein	68122	Dimethyl formamide
79061	Acrylamide	57147	1,1-Dimethyl hydrazine
79107	Acrylic acid	131113	Dimethyl phthalate
107131	Acrylonitrile	77781	Dimethyl sulfate
107051	Allyl chloride	534521	4,6-Dinitro-o-cresol, and salts
92671	4-Aminobiphenyl	51285	2,4-Dinitrophenol
62533	Aniline	121142	2,4-Dinitrotoluene
90040	o-Anisidine	123911	1,4-Dioxane
1332214	Asbestos		(1,4-Diethyleneoxide)
71432	Benzene (including benzene from gasoline)	122667	1,2-Diphenylhydrazine
92875	Benzidine	106898	Epichlorohydrin
98077	Benzotrichloride		(1-Chloro-2,3-epoxypropane)
100447	Benzyl chloride	106887	1,2-Epoxybutane
92524	Biphenyl	140885	Ethyl acrylate
117817	Bis(2-ethylhexyl)phthalate (DEHP)	100414	Ethyl benzene
542881	Bis(chloromethyl)ether	51796	Ethyl carbamate (Urethane)
75252	Bromoform	75003	Ethyl chloride (Chloroethane)
106990	1,3-Butadiene	106934	Ethylene dibromide (Dibromethane)
156627	Calcium cyanamide	107062	Ethylene dichloride (1,2-Dichloroethane)
105602	Caprolactam	107211	Ethylene glycol
133062	Captan	151564	Ethylene imine (Aziridine)
63252	Carbaryl	75218	Ethylene oxide
75150	Carbon disulfide	96457	Ethylene thiourea
56235	Carbon tetrachloride	75343	Ethylidene dichloride (1,1-Dichloroethane)
463581	Caronyl sulfide	50000	Formaldehyde
120809	Catechol	76448	Heptachlor
133904	Chloramben	118741	Hexachlorobenzene
57749	Chlordane	87683	Hexachlorobutadiene
7782505	Chlorine	77474	Hexachlorocyclopentadiene
79118	Chloroacetic acid	67721	Hexachloroethane
532274	2-Chloroacetophenone	822060	Hexamethylene-1,6-diisocyanate
108907	Chlorobenzene	680319	Hexamethylphosphoramide
510156	Chlorobenzilate	110543	Hexane
67663	Chloroform	302012	Hydrazine
107302	Chloromethyl methyl ether	7647010	Hydrochloric acid
126998	Chloroprene	7664393	Hydrogen fluoride (Hydrofluoric acid)
1319773	Cresols/Cresylic acid (isomers and mixture)	123319	Hydroquinone
95487	o-Cresol	78591	Isophorone
108394	m-Cresol	58899	Lindane (all isomers)
106445	p-Cresol	108316	Maleic anhydride
98828	Cumene	67561	Methanol
94757	2,4-D, salts and esters	72435	Methoxychlor
3547044	DDE	74839	Methyl bromide (Bromomethane)
334883	Diazomethane	74873	Methyl chloride (Chloromethane)
132649	Dibenzofurans	71556	Methyl chloroform (1,1,1-Trichloroethane)
96128	1,2-Dibromo-3-chloropropane	78933	Methyl ethyl ketone (2-Butanone)
84742	Dibutylphthalate	60344	Methyl hydrazine
106467	1,4-Dichlorobenzene(p)	74884	Methyl iodide (Iodomethane)
91941	3,3-Dichlorobenzidene	108101	Methyl isobutyl ketone (Hexone)
111444	Dichloroethyl ether (Bis(2-chloroethyl)ether)	624839	Methyl isocyanate
542756	1,3-Dichloropropene	80626	Methyl methacrylate
62737	Dichlorvos	1634044	Methyl tert butyl ether
111422	Diethanolamine	101144	4,4-Methylene
121697	N,N-Diethylaniline (N,N-Dimethylaniline)		bis(2-chloroaniline)

(continued)

62

Table 3–C. *(Continued)*

CAS* Number	Chemical Name	CAS* Number	Chemical Name
75092	Methylene chloride (Dichloromethane)	95807	2,4-Toluene diamine
101688	Methylene diphenyl diisocyanate (MDI)	584849	2,4-Toluene diisocyanate
101779	4,4'-Methylenedianiline	95534	o-Toluidine
91203	Napthalene	8001352	Toxaphene (chlorinated camphene)
98953	Nitrobenzene	120821	1,2,4-Trichlorobenzene
92933	4-Nitrobiphenyl	79005	1,1,2-Trichloroethane
100027	4-Nitrophenol	79016	Trichloroethylene
79469	2-Nitropropane	95954	2,4,5-Trichlorophenol
684935	N-Nitroso-N-methylurea	88062	2,4,6-Trichlorophenol
62759	N-Nitrosodimethylamine	121448	Triethylamine
59892	N-Nitrosomorpholine	1582098	Trifluralin
56382	Parathion	540841	2,2,4-Trimethylpentane
82688	Pentachloronitrobenzene (Quintobenzene)	108054	Vinyl acetate
87865	Pentachlorophenol	593602	Vinyl bromide
108952	Phenol	75014	Vinyl chloride
106503	p-Phenylenediamine	75354	Vinylidene chloride (1,1-Dichloroethylene)
75445	Phosgene	1330207	Xylenes (isomers and mixture)
7803512	Phosphine	95476	o-Xylenes
7723140	Phosphorus	108383	m-Xylenes
85449	Phthalic anhydride	106423	p-Xylenes
1336363	Polychlorinated biphenyls (Aroclors)	0	Antimony Compounds
1120714	1,3-Propane sultone	0	Arsenic Compounds (inorganic including arsine)
57578	beta-Propiolactone	0	Beryllium Compounds
123386	Propionaldehyde	0	Cadmium Compounds
114261	Propoxur (Baygon)	0	Chromium Compounds
78875	Propylene dichloride (1,2-Dichloropropane)	0	Cobalt Compounds
75569	Propylene oxide	0	Coke Oven Emissions
75558	1,2-Propylenimine (2-Methyl aziridine)	0	Cyanide Compounds [a]
91225	Quinoline	0	Glycol ethers [b]
106514	Quinone	0	Lead Compounds
100425	Styrene	0	Manganese Compounds
96093	Styrene oxide	0	Mercury Compounds
1746016	2,3,7,8-Tetrachlorodibenzo-p-dioxin	0	Fine mineral fibers [c]
79345	1,1,2,2-Tetrachloroethane	0	Nickel Compounds
127184	Tetrachloroethylene (Perchloroethylene)	0	Polycylic Organic Matter [d]
7550450	Titanium tetrachloride	0	Radionuclides (including radon) [e]
108883	Toluene	0	Selenium Compounds

Note: For all listings that contain the word "compounds" and for glycol ethers, the following applies: Unless otherwise specified, these listings are defined as including any unique chemical substance that contains the named chemical (that is, antimony, arsenic, and so forth) as part of that chemical's infrastructure.
* Chemical Abstract Service.
[a] X'CN where X = H' or any other group where a formal dissociation may occur. For example, KCN or Ca(CN)$_2$.
[b] Includes mono- and di-ethers of ethylene glycol, diethylene glycol, and triethylene glycol R-(OCH$_2$CH$_2$)n-OR' where

      n = 1, 2, or 3
      R = alkyl or aryl groups
      R' = R, H, or groups which, when removed, yield glycol ethers with the structure:
        R-(OCH$_2$CH)n-OH. Polymers are excluded from the glycol category.

[c] Includes mineral fiber emissions from facilities manufacturing or processing glass, rock, or slag fibers (or other mineral-derived fibers) of an average diameter of 1 micrometer or less.
[d] Includes organic compounds with more than one benzene ring and which have a boiling point greater than or equal to 100 C.
[e] A type of atom that spontaneously undergoes radioactive decay.

Source: Public Law 101–549, Title II.

and existing sources. EPA's past activities in promulgating NESHAPs for hazardous air pollutants such as asbestos were widely perceived as a major failure. Consequently, the 1990 Amendments established an initial list of 189 HAPs and gave EPA the authority to periodically review and add HAPs to the list.

EPA also is required under the amendments to promulgate technology-based limitations for industrial source categories and issue standards for each category. It is estimated that as many as 250 categories will be established with major sources subject to maximum achievable control technology (MACT) limitations.

EPA is also required to develop a strategy to reduce human cancer risks from area source emissions by at least 75%. By 1996, EPA must report to Congress on any health risks remaining ("residual risk") from major source HAP emissions after the implementation of the technology-based limitations. Either Congress or EPA must impose additional controls, as needed, to protect the public health or prevent significant, widespread environmental harm and provide an "ample margin of safety." Further, if the lifetime residual risk of cancer from the remaining HAPs is not reduced to less than 1 in 1 million, provisions to control known carcinogens will be automatically triggered.

In addition to a new HAP program, a comprehensive program for accidental hazardous substance release prevention, reporting, and investigation has been established. The owners/operators of stationary sources have a "general duty" to identify the hazards of accidental releases, take steps necessary to prevent releases, and minimize the consequences of accidental releases. Table 3–D lists the implementation schedule for the HAP regulations.

Visibility protection for federal Class-I areas. Section 169A of the Clean Air Act requires visibility protection for mandatory Class-I federal areas where it has been determined that visibility is an important value. Mandatory Class-I federal areas are all international parks and certain national parks, national monuments, and wilderness areas that are identified in Section 162 of the Clean Air Act.

Acid rain control program. Under the Clean Air Act Amendments of 1990, a new program has been established that requires a nationwide reduction of sulfur dioxide (SO_2) and nitrogen oxide (NO_x) emissions from fossil fuel-fired combustion devices that produce electricity for sale. A complex system of "allowances," which, in essence, can be bought and sold, will be used to regulate SO_2 emissions. Electric utilities will be directly affected by these statutory changes. New source performance standards for industrial power boilers also will be promulgated. (Chapter 14, Global Issues, discusses the acid rain program and presents the deposition control compliance requirements.)

Enforcement. As a result of the Clean Air Act Amendments of 1990, all requirements of the act will be subject to tougher enforcement by EPA and private citizens. The 1990 Amendments have established citizen awards for providing information that leads to a criminal conviction or civil penalty for air quality violations. Further, citizen suits previously limited to injunctive relief only may now request civil penalties.

Clean Water Act

The Clean Water Act (CWA) makes it unlawful for any person to discharge any pollutant from a point source into navigable waters, unless a permit is obtained under the act. The basic concept of the CWA is to force compliance with both uniform technology-based effluent limitations, regardless of the quality of receiving waters, and with more stringent limitations, if necessary, to meet state-established water quality standards.

Major federal clean water legislation includes the Federal Water Pollution Control Act of 1948, the Water Quality Act of 1965, the Federal Water Pollution Control Act Amendments of 1972, the Clean Water Act of 1977, and the Water Quality Act of 1987.

National Pollutant Discharge Elimination System (NPDES). Section 402 of the 1972 Amendments to the Clean Water Act established the National Pollutant Discharge Elimination System (NPDES). NPDES permits include specific discharge limitations for pollutants discharged by a facility. An NPDES permit can be required by the local authority for discharges into a publicly owned treatment works (POTW). The NPDES permit program is delegated to state or local authorities whose programs meet certain conditions and are approved by EPA.

Federal effluent limitations. In developing effluent limitations, EPA has divided the universe of point source discharges into industrial categories and subcategories. EPA then established effluent limitations for specific pollutants in the various categories and subcategories. In the absence of effluent limitations, permitted authorities may establish limitations on a case-by-case basis through the exercise of best engineering judgment (BEJ).

The Clean Water Act provides that existing sources must achieve effluent limitations based on:

1. Best Practicable Control Technology Currently Available (BPT) by July 1, 1977
2. Best Conventional Pollution Control Technology for certain pollutants (BCT) by July 1, 1984

Table 3–D. Implementation Schedule for Hazardous Air Pollutant Regulations

Deadline	Task
Nov. 15, 1991	Listing of categories/subcategories of major and area sources of hazardous air pollutants (HAPs) subject to maximum achievable control technology (MACT).
Nov. 15, 1991	Promulgation of new source performance standards (NSPS) for incineration units combusting >250 tons per day of municipal waste.
May 15, 1992	Guidance regarding relative hazards to human health from the initial list of 189 HAPs.
Nov. 15, 1992	MACT standards for first 40 categories/subcategories of HAPs.
	Schedule for regulation of source categories required to implement MACT standards.
	Initial list of 100 extremely hazardous substances subject to accidental release provisions.
	Promulgation of NSPS for incineration units combusting <250 tons per day of municipal waste and units combusting hospital waste, medical waste, and infectious waste.
	Report to Congress on hydrogen sulfide and hydrofluoric acid.
Dec. 31, 1992	MACT standards for coke oven batteries.
Nov. 15, 1993	Regulations for prevention, detection, and response to accidental releases.
	Proposal of NSPS for incineration units combusting commercial or industrial waste.
	Report to Congress on hazards to public health from utility HAP emissions.
Nov. 15, 1994	MACT standards for 25% of all categories/subcategories of HAP sources.
	Promulgation of NSPS for incineration units combusting commercial or industrial waste.
	Report to Congress on mercury emissions.
Nov. 15, 1995	Compliance deadline for achieving MACT standards for existing sources in first 40 categories/subcategories.*†
	Listing of categories/subcategories for area sources.
	MACT standards for publicly owned treatment works (POTWs).
Nov. 15, 1996	Report to Congress on health risks remaining after full compliance with MACT standards is achieved.
	Compliance deadline for achieving regulations for prevention, detection, and response to accidental releases.
	Final report to Congress on coke oven production emission control technologies.
Nov. 15, 1997	MACT standards for 50% of all categories/subcategories of HAPs.
	Compliance deadline for achieving MACT standards for 25% of all categories/subcategories of HAPs.*†
Nov. 15, 2000	MACT standards for 100% of all categories/subcategories of HAPs.
	Compliance deadline for achieving MACT standards for 50% of all categories/subcategories of HAPs.
Nov. 15, 2001	Residual risk standards for first 40 categories/subcategories of HAPs.
Nov. 15, 2002	Residual risk standards for 25% of all categories/subcategories of HAPs.
Nov. 15, 2003	Compliance deadline for achieving MACT standards for 100% of all categories/subcategories of HAPs.
Nov. 15, 2005	Residual risk standards for 50% of all categories/subcategories of HAPs.
Nov. 15, 2008	*Residual risk standards for 100% of all categories/subcategories of HAPs.

* One-year extensions are available.
† Six-year extensions are available for any existing source that voluntarily reducesH 90% below 1987 levels.
Source: Dames & Moore. Used with permission.

3. Best Available Technology Economically Achievable (BAT) for 65 listed toxic pollutants by July 1, 1984.

New sources must achieve effluent limitations based on Best Available Demonstrated Control Technology. Deadlines for complying with the effluent limitations on toxic, conventional, and nonconventional pollutants were extended as a part of amendments to the Clean Water Act through the Water Quality Act of 1987. Compliance must be achieved as expeditiously as practicable, but in no case later than March 31, 1989.

Major revisions were also made to the CWA concerning effluent limitations and the establishment of individual control strategies for toxic pollutants through the 1987 Water Quality Act. The enforcement of additional limitations will occur when state water quality standards and/or technology-based effluent limitations have not reduced toxic pollutant concentrations to acceptable levels. Each state must identify waters—after applying technology-based effluent limitations, NSPSs, and categorical pretreatment standards—that cannot attain or maintain either state water quality standards or a level of water quality that will:

> assure protection of public health, public water supplies, agricultural and industrial uses, and the protection and propagation of a balanced population of shellfish, fish and wildlife, and allow recreational activities in and on the water.

Individual control strategies must be developed for each identified source. The EPA is then required to establish effluent limitations for waters in which individual control strategies have been developed. Clearly these "area-specific" effluent limitations will create another layer of permit requirements for dischargers located in areas where water quality standards have not been achieved despite compliance with already stringent permit requirements.

Obligations of indirect dischargers. An indirect discharger refers to a source introducing pollutants into a POTW. The EPA is authorized to establish pretreatment standards applicable to certain indirect discharges for controlling pollutants determined not to be susceptible to treatment by a POTW or that would interfere with the operation of the treatment works. Indirect dischargers subject to pretreatment standards may not discharge effluent into a POTW unless they comply with such standards.

Water quality standards. A water quality standard for a particular body of water consists of a designated use (such as public water supply, recreation, or agriculture) and criteria for various pollutants, expressed in numerical concentration limits, necessary to support that use. Water quality standards may serve as the basis not only for imposing effluent limitations on point source dischargers but also for establishing controls for nonpoint sources under water quality management plans.

Stormwater discharges. Section 402 of the Clean Water Act was amended by the Water Quality Act to add requirements and time limits for permitting municipal and industrial stormwater discharges into waters of the United States. Under these amendments, the EPA is required to establish a stormwater discharge permit program. The stormwater permitting strategy developed by EPA is complex to allow flexibility. Facilities subject to the stormwater requirements may apply for an individual, group, or general permit, depending on the facility's specific type and circumstances.

Under EPA's approach, specific industry categories are targeted for individual or industry-specific general permits. These permits enable EPA and state agencies to focus attention and resources on industries of particular concern. EPA's general permits address the vast majority of industrial sources. The permits require industrial facilities to develop stormwater control plans and pollution prevention practices. Group and individual permit applications also have detailed application requirements.

Industries that are indirect dischargers into POTWs do not have to apply for a permit but are required to notify POTWs of their discharge.

Discharge of oil and hazardous substances. Section 311 of the Clean Water Act prohibits the discharge of oil or hazardous substances into navigable waters in quantities that may be harmful. Preparation of a Spill Prevention Control and Countermeasure (SPCC) plan is required when discharges to such waters are possible (see Chapter 8, Hazardous Wastes). Immediate notice must be given of any such discharge.

Discharge of dredged or fill material. Under Section 404 of the Clean Water Act, discharges of dredged or fill material into navigable waters are prohibited except in compliance with a permit issued by the U.S. Army Corps of Engineers and EPA. In many cases, a permit under the River and Harbor Act of 1899 will also be required for placement of structures in activities involving discharge of dredged or fill material.

Resource Conservation and Recovery Act

The Resource Conservation and Recovery Act (RCRA) of 1976 provides a comprehensive federal program for the regulation of solid wastes with a special emphasis on the regulation of hazardous waste. Major changes were made to RCRA in 1984 through the Hazardous and Solid Waste Amendments of 1984. Under these amendments, Congress directed EPA to tighten restrictions on waste recycling; land disposal operations; and contaminant releases from old, abandoned facilities not covered by the original RCRA. The 1984 amendments also established changes to the small quantity hazardous generator rule and established a new program to address the problems created by leaking underground storage tanks.

Hazardous waste. The major regulatory program under RCRA is contained in Subtitle C and covers the management of solid wastes that are defined as hazardous. In accordance with the directives of RCRA, EPA has developed a cradle-to-grave regulatory scheme for controlling hazardous wastes and their ultimate disposal. EPA's RCRA program prescribes standards and regulations applicable to hazardous waste generators, transporters, and facility owners and operators, as well as permit requirements for owners and operators of facilities that store, treat, or dispose of hazardous waste. (See Chapter 8, Hazardous Wastes, for more information.)

Pursuant to the 1984 Solid and Hazardous Waste Amendments, EPA issued regulations in 1985, revising its regulatory definition of the term *solid waste* and its rules governing the recycling and reuse of hazardous wastes. A hazardous waste under RCRA is a material that by definition is a *solid waste*—a solid, liquid, or gas that is a discarded material and is abandoned, recycled, or otherwise "inherently waste-like." EPA is considering a number of methods to define solid waste and hazardous waste and will be redefining these terms.

Identification and listing of hazardous wastes. Once it is established that a material is a "solid waste," there are two ways in which it is designated a hazardous waste: It is specifically listed, or it exhibits one of four characteristics. The second type of waste exhibits one or more of the EPA-defined characteristics of ignitability, corrosivity, reactivity, or toxicity.

Listed hazardous wastes. The listed wastes include hazardous wastes from nonspecific sources, hazardous wastes from specific sources, and chemical product wastes that are hazardous.

Hazardous wastes from nonspecific sources are generic wastes that may be generated by any number of industrial processes. An example of a hazardous waste from a nonspecific source is dewatered air pollution control scrubber sludges from coke ovens and blast furnaces.

Hazardous wastes from specific sources are wastes generated by processes in a particular industry. An example of a hazardous waste from a specific source is wastewater treatment sludges from the manufacturing, formulation, and loading of lead-based initiating compounds.

This list also contains a number of wastes identified either by trade name or chemical name that EPA considers hazardous. An example of a hazardous waste from this list is pentachlorophenol.

Hazardous characteristics. In addition to the foregoing listed wastes, solid waste is hazardous if a "representative sample of the waste" has any of the following four characteristics: (1) ignitability, (2) corrosivity, (3) reactivity, or (4) extraction procedure toxicity.

In March 1990, the EPA issued final regulations changing the test used for the toxicity characteristic. Additional changes in 1990 also included an addition of 25 organic chemicals, with specified concentration levels, to EPA's existing list of chemicals that are subject to the hazardous waste toxicity characteristic. The solid waste analysis procedure changed from the EP toxicity-leaching procedure to the toxicity characteristic leaching procedure (TCLP). Expansion of the chemicals considered toxic hazardous waste affects a significant volume of additional industrial wastewaters, solid waste, and sludge that may, for the first time, be subject to EPA's hazardous waste regulations. For example, chloroform, one of the new characteristic organic chemicals, must now be tested for the regulatory concentration level in industrial wastewaters to determine whether the rule applies.

Standards applicable to generators of hazardous wastes. The hazardous waste regulations define *generator* as any person, by site, whose act or process produces hazardous waste. Under the regulations, generators of hazardous wastes must obtain an EPA identification number. The generator must also prepare a manifest for each shipment of hazardous wastes for off-site transport, treatment, storage, or disposal at a licensed facility. The manifest system is designed to track hazardous waste from the generator through the transporter to the disposal facility by having the disposal facility return to the generator a signed copy of the manifest. The existence of the manifest system is central to the legal issues surrounding generator liability.

In addition to these requirements, generators are now subject to EPA's Land Disposal Restriction (LDR) regulations that were issued by EPA pursuant to the 1984 Hazardous and Solid Waste Amendments. Restricted wastes subject to the LDR requirements have been phased in over the last few years pursuant to several different rulemakings.

All restricted wastes, whether treated and disposed of on-site or sent off-site to a RCRA treatment, storage, or disposal facility, are subject to verification and record-keeping requirements. Generator responsibilities include (1) determining whether the waste is subject to the LDR rules; (2) determining what treatment standards, technology, or numeric standards apply; and (3) certifying that the waste sent to a disposal facility meets the applicable treatment standard and that the certified information is true, accurate, and complete.

Hazardous waste generator categories. There are three categories of hazardous waste generators defined by EPA. These categories and the applicable requirements are as follows:

- *Conditionally exempt small-quantity generator.* If one does not generate more than about 220 lb or 25 gal (100 kg) of hazardous waste and no more than 2.2 lb (1 kg) of acute waste in a month, one is considered a "conditionally exempt" small-quantity generator and is not subject to the full spectrum of hazardous waste management and permitting requirements.
- *220–2,205 lb/month (100–1,000 kg/month) small-quantity generator.* If one generates more than 220 lb (100 kg) and less than 2,205 lb (1,000 kg) (or about 25 to under 300 gal) of hazardous waste in a month, one must comply with the hazardous waste management regulations. Under this category, hazardous waste may be accumulated on-site for up to 180 days (or up to 270 days if the waste has to be transported across a distance of 200 miles [322 km] or more) without a permit, provided the waste accumulated on-site never exceeds 13,230 lb (6,000 kg).

  These requirements do not apply to small-quantity generators of 100–1,000 kg per month when the waste is reclaimed under a contractual agreement, and the generator maintains a copy of the reclamation agreement for at least three years after termination or expiration of the agreement. An example of this exclusion for recycling hazardous waste is the use of a national solvent recycler that is used by facilities to recycle their solvents and waste oil.
- *2,205 lb/month (1,000 kg/month or more) generator.* If one generates 2,200 lb or approximately 300 gal (1,000 kg) or more of hazardous waste in one month or more than 2.2 lb (1 kg) of acute waste, one must comply with the hazardous waste management requirements in managing a permitted unit. Generators of hazardous waste in this category are required to obtain a hazardous waste

storage permit if they store their waste for longer than 90 days.

Standards applicable to transporters of hazardous wastes. Generally speaking, the EPA hazardous waste transportation regulations apply to any off-site transportation of hazardous waste to a licensed treatment, storage, or disposal facility. The primary obligation of the hazardous waste transporter is to deliver the specified waste to the designated facility (see also Chapter 8, Hazardous Wastes).

Standards for hazardous waste treatment, storage, and disposal facilities. In order to operate a facility that treats, stores, or disposes of hazardous wastes, the owner of the facility must obtain a RCRA Part B permit and comply with certain design and operating standards under Part 264 of the hazardous waste regulations (see also Chapter 8, Hazardous Wastes).

Facilities that treat, store, or dispose of hazardous wastes are required to comply with certain regulations including personnel training, emergency response programs, inspection and reporting responsibilities, and other standards designed to protect the public health. The regulations also require the owner of a treatment, storage, or disposal facility to provide appropriate financial assurance for closure and post-closure costs of the facility.

A treatment, storage, or disposal facility is also subject to the LDR. These facilities must comply with the record-keeping requirements and document that all applicable restricted wastes are properly treated and disposed of in accordance with the RCRA requirements.

A generator who stores hazardous waste may avoid the burden of meeting the hazardous waste storage and permit requirements by never accumulating more than 2,205 lb (1,000 kg) per month [including never exceeding 2.2 lb (1 kg) of acute waste] and never storing the hazardous waste for more than 180 days or 270 days, depending on the distance the generator must have the waste hauled for proper treatment and disposal. Specific local requirements should be consulted to verify that they are not more stringent within a state or county.

Underground storage tanks. In the Hazardous and Solid Waste Amendments of 1984, Congress established a new program under Subtitle I of RCRA governing underground storage tanks. EPA has promulgated regulations covering all facets of tank-system management from installation to final closure, including technical requirements for leak detection, corrosion protection, and closure of existing tanks (see Underground Storage Tanks in Chapter 8, Hazardous Wastes). The program is to be implemented at the state and local level with the possibility of state and local level regulations being more stringent. There also are separate financial requirements for tank owners and operators to ensure proper corrective actions for tank releases and to provide funds to respond to accidental releases.

- *Technical standards.* The technical regulations are directly applicable to tank systems used to contain "regulated substances" in which 10% or more of the volume of the tank and associated pipes is located underground. There are exemptions from the regulatory requirements for certain tanks such as farm and residential tanks, septic tanks, and tanks used for heating oil for consumptive purposes on the premises (such as heating oil for generating heat and electricity). The "regulated substances" include a wide range of Superfund hazardous substances (except hazardous waste) and petroleum products. The regulations provide that all new underground storage tank systems must be protected from corrosion, equipped with spill and overfill prevention devices, and provided with leak detection. By December 1998, underground tanks that were installed before December 1988 must have corrosion protection for steel tanks and piping, and devices that prevent spills and overfills. Leak-detection requirements are being phased in for existing tanks, depending on their age. For example, the leak-detection requirements for tanks older than 25 years took effect in December 1989. The regulatory program also established requirements for responding to leaking underground tanks and permanently or temporarily closing underground tanks.
- *Financial assurance requirements.* Separate regulations were issued by the EPA for financial responsibility of tank owners and operators. Financial assurance—available through various mechanisms such as liability insurance, self-insurance tests, guarantees, trust funds, and letters of credit—must be established for a per-occurrence liability coverage of $500,000 or $1 million, depending on the amount of petroleum handled at a facility per month. An annual aggregate coverage of $1 or $2 million must also be provided, depending on a company's total number of tanks that are subject to the regulations.

Compliance deadlines were established according to the number of tanks owned by a company and the company's tangible net worth. The earliest compliance date was January 24, 1989, for all petroleum-marketing firms owning 1,000 or more tanks and all other tank owners that report a tangible net worth of $20 million or more to the SEC, Dun and Bradstreet, the Energy Information Administration, or the Rural Electrification Administration.

The Comprehensive Environmental Response Compensation and Liability Act of 1980

In December 1980, Congress passed the Comprehensive Environmental Response Compensation and Liability Act of 1980. Commonly known as CERCLA or "Superfund," the act provides authority and funding for cleaning up (1) spills and discharges of hazardous substances into the environment and (2) inactive hazardous waste dump sites.

Spills and discharges of hazardous substances. Section 103(a) of Superfund provided that any person in charge of a site where hazardous materials are located must notify the EPA as soon as he/she is aware of an unpermitted release, discharge, or spill of hazardous material. This requirement is designed to allow quick cleanup of any hazardous substance, release, or spill.

Inactive sites. Section 103(c) of Superfund required all businesses to notify the EPA by June 9, 1981, of any site or area where a hazardous substance had been deposited or disposed of. The legislation required that this notification include the amount and type of hazardous substance to be found at the site and any "known, suspected, or likely discharge of hazardous substances from that location."

Remedial action. Under Section 106 of Superfund, EPA is empowered to order potentially responsible parties (PRPs) to clean up or contain any release of hazardous substances or pollutants into the environment. Section 107 identifies four types of parties potentially liable for cleanup costs: (1) the current owner and operator of the site; (2) the owner and operator of the site at the time of disposal of the hazardous substance; (3) any party that transported hazardous material to a disposal or treatment site of its own choosing; and (4) any party that arranged for the disposal or treatment of hazardous material at the cleanup site or for the transportation of such material to the site.

If the responsible parties fail to respond, the government can expend money from the Superfund to do the job. There is a strong incentive for private parties to obey EPA's order; if they refuse to clean up and later are found liable, the government can sue for reimbursement plus punitive damages up to triple the cost of cleanup. Although language in the original Superfund legislation calling for "strict, joint and several liability" was dropped, through the years the courts have imposed joint and several liability on PRPs.

Superfund Amendments and Reauthorization Act (SARA) of 1986. The Superfund Amendments of 1986 added numerous provisions to the original Superfund legislation. The major changes require an acceleration of cleanup activities, expanded public participation, and expanded review of remedial action cleanup criteria. Cleanup standards require more permanent treatment techniques that in turn, lead to higher remedial costs for PRPs.

New provisions also were added to the Superfund law to encourage voluntary settlements. EPA now has authority to finance remedial actions with a mixture of funding from private parties and from Superfund. To further expedite settlement and relieve small contributors from the burdens of lengthy negotiations and litigation, the SARA amendments authorize EPA to enter into *de minimis* settlements with PRPs if the amount of the hazardous substances contributed by the PRP is minimal.

An "innocent landowner" provision has also been added to the Superfund law that under certain limited circumstances will protect an innocent landowner of property from Superfund liability. Subsequent court cases have supported different opinions on the innocent landowner provision. The amendments further provide that parties held liable for, or who contribute to, a cleanup may bring suit in federal court to obtain contribution from other responsible parties. Finally, the Superfund law now authorizes EPA to release settling parties from future liability under certain circumstances.

Emergency Planning and Community Right-to-Know Act

The Community Right-to-Know Act (EPCRA) was established as a free-standing provision of law contained in Title III of the Superfund Amendments and Reauthorization Act of 1986 (SARA). It represents an effort to ensure that (1) a mechanism exists through which citizens can be made aware of the existence, quantities, and releases of various hazardous materials in their communities; and (2) state and local governments undertake planning measures to respond to emergencies involving these hazardous substances.

Under EPCRA, the governor of each state was required to establish a State Emergency Response Commission (SERC). The SERC for each state was required to designate Local Emergency Planning Committees (LEPC). Each LEPC must develop and maintain an emergency response plan for its community. The major provisions of the law are summarized as follows.

Emergency planning (Sections 301–303). A facility must notify the State Emergency Response Commission within 60 days if an "extremely hazardous substance" is present at a facility in an amount that exceeds the applicable threshold planning quantity established for each listed substance. A list (EHS list) of approximately 340 extremely hazardous substances was issued by EPA. Examples of extremely hazardous substances include chlorine, ammonia, and hydrogen sulfide. Any facility subject to the EPCRA was also required to designate and notify the Local Emergency Planning Committee of its emergency coordinator who participates in the local emergency planning process.

Emergency release notification (Section 304). Emergency release notification is required by any facility where there is a release of a reportable quantity of any "extremely hazardous substance" or a reportable quantity of a "CERCLA hazardous substance" that results in exposure to persons outside the boundaries of the facility. A new executive order signed in August 1993 requires reporting of releases from federal facilities.

The emergency notification must be made to the Local Emergency Planning Committee, State Emergency Response Commission, and the National Response Center if the released substance is both on the EHS list and the CERCLA hazardous substance list. A written follow-up notice must also be submitted as soon as practicable after the release.

Material Safety Data Sheet (MSDS) reporting requirements (section 311). The owner or operator of any facility that is required to have available an MSDS for a hazardous chemical under OSHA must submit the MSDS to the appropriate local emergency planning committee, state emergency response commission, and the local fire department if the chemical is present at the facility in amounts greater than designated threshold levels.

Under EPA regulations, MSDSs must be submitted for "hazardous chemicals" that are present at a facility in an amount greater than 10,000 lb (4,536 kg) at any one time or "extremely hazardous substances" that are present in amounts greater than 500 lb (227 kg) at any one time or the applicable threshold planning quantity, whichever amount is less.

The federal OSHA requirement for MSDSs does not apply to any consumer product for which the employer can demonstrate it is used in the workplace in the same manner as normal consumer use and that use results in a duration and frequency of exposure that is not greater than exposure experienced by consumers.

Hazardous chemical inventory reporting (Section 312). Beginning March 1, 1988, and annually thereafter, facilities subject to the MSDS requirements under Section 311 of the EPCRA are also required to submit hazardous chemical inventory forms to the local emergency planning committee, state emergency response commission, and the local fire department. Inventory reporting forms (Tier I and Tier II forms) have been issued by EPA or the state. These forms basically request information from a facility on the quantity, location, and manner of storage of hazardous chemicals.

Toxic chemical release reporting or Form R (Section 313). The toxic chemical release reporting requirement is to inform government officials and the public of annual releases of toxic chemicals into the environment. Congress adopted a list consisting of approximately 300 chemicals as the Section 313 list. Examples of listed toxic chemicals include ammonia, chlorine, and friable asbestos. Section 313 requires facilities with 10 or more employees who manufacture, process, or use a "toxic chemical" in excess of the statutorily prescribed quantity to submit annual information on the chemical and releases of the chemical into the environment. This information must be filed with EPA and the State Emergency Response Commission by July 1, 1988, and annually thereafter. Each report is to cover estimated releases occurring during the prior calendar year.

The Section 313 (Form R) report requires the following information: (1) whether the toxic chemical is manufactured, processed, or used at a facility; (2) an estimate of the total accumulated quantity present during the preceding year; (3) a description of waste treatment and disposal methods used by the facility; and (4) the annual quantity of the toxic chemical entering any environmental medium.

The thresholds that trigger the need to file a toxic chemical release form are as follows:

- If a toxic chemical is otherwise used at a facility, 10,000 lb (4,536 kg) must have been present through the period of one year.
- If a toxic chemical is manufactured or processed at a facility, 75,000 lb (34,020 kg) per year must have been used during the year's time on or before July 1, 1988, 50,000 lb (22,680 kg) per year on or before July 1, 1989, and 25,000 lb (11,340 kg) per year on or before July 1, 1990.

Supplier notification. Effective January 1, 1989, facilities that are identified under SIC Codes 20–39, manufacture or process a toxic chemical, and sell or otherwise distribute a mixture or trade name product containing a toxic chemical, are required to provide written notification to their customers regarding Section 313 toxic chemicals. The primary purpose of the Section 313 notification requirement is to provide information to users of products regarding the presence and composition of listed toxic chemicals for purposes of complying with the annual release reporting requirements of Section 313.

Safe Drinking Water Act of 1974

The Safe Drinking Water Act (SDWA) of 1974 imposes federal drinking water standards on virtually all public water systems. The act delegates primary enforcement authority to the states. The EPA was required under the act to establish primary and secondary drinking water standards for public water systems. In order for a state program to conform with the federal requirements, the state must adopt primary drinking water regulations that are at least as stringent as the federal requirements. The act requires the establishment of drinking water standards for maximum contaminant levels for organic and inorganic chemicals, turbidity, coliform bacteria, and various measures of radioactivity.

In 1986, Congress passed amendments to the Safe Drinking Water Act because of growing public concern over contamination of public drinking water supplies and a lack of adequate federal standards. Over the years, drinking water monitoring revealed the existence of chemical contaminants for which there were no federal standards. In order to correct this problem, the statutory amendments required EPA to establish maximum contaminant levels for approximately 83 chemicals. An additional 25 contaminants are to be regulated every three years.

The amendments also required the promulgation of regulations for lead notification by public water systems concerning lead contamination in the drinking water. The lead notification had to be issued by owners and operators of community and nontransient, noncommunity water systems by June 1988. Certain facilities are subject to the Safe Drinking Water Act as "noncommunity" public water systems because the facility provides drinking water on a regular basis to at least 25 persons over a six-month period.

Toxic Substances Control Act

The Toxic Substances Control Act (TSCA) of 1976 governs the manufacture and use of chemical products. In implementing TSCA, EPA has promulgated detailed regulations regarding the use, management, storage, and disposal of polychlorinated biphenyl (PCB) materials. These regulations establish a complex inspection, record-keeping, and reporting program for PCBs. TSCA also establishes a variety of record-keeping, reporting, and testing requirements for manufacturers, distributors of chemicals, and processors of chemicals other than PCBs.

Importing and exporting of chemicals are regulated by EPA under TSCA. The lists of chemicals required to be reported as imported or exported chemicals is extensive and should be consulted prior to shipments across U.S. boundaries. Chemicals present in hazardous wastes or materials need to be known in order to assess an owner/operator's responsibility to notify EPA. The lists are published by EPA and are periodically updated with new chemicals. The EPA regulations on the subject of TSCA are extensive and complex, and may require regulatory expertise to verify that activities in which chemicals are processed at a facility and shipped elsewhere are in compliance with the various provisions of TSCA.

ENVIRONMENTAL COMPLIANCE AT FEDERAL FACILITIES

The federal perspective, as related to the laws previously discussed in this chapter, follows two general courses: enforcement and compliance. It is generally recognized that aggressive compliance is the most effective protection against aggressive enforcement and other efforts to assess liability. (A detailed discussion of the ethics involved in environmental compliance is provided in Chapter 2, Economic and Ethical Issues.)

One of the greatest enforcement efforts is from the EPA with responsibility for ensuring compliance with federal statutes under its control. Civil and criminal enforcement actions, as authorized by the federal statutes, are referred to the U.S. Department of Justice (DOJ). Although the EPA lacks a statutory charter, its general purpose is to protect and enhance our environment to the fullest extent possible under the existing laws. The eight major statutes directly enforced by the EPA include:

- The Federal Insecticide, Fungicide, and Rodenticide Act (FIFRA)
- The Marine Protection, Research, and Sanctuaries Act (MPRSA)
- The Safe Drinking Water Act (SDWA)
- The Solid Waste Disposal Act; amended by the Resource Conservation and Recovery Act (RCRA)
- The Clean Air Act (CAA)
- The Comprehensive Environmental Response, Compensation, and Liability Act (CERCLA or Superfund; includes EPCRA, or Emergency Planning and Community Right to Know Act, as Title III of the Superfund Amendments and Reauthorization Act of 1986)
- The Federal Water Pollution Control Act; also known as the Clean Water Act (CWA)
- The Toxic Substances Control Act (TSCA).

In addition, the EPA also has some authority under the National Environmental Policy Act (NEPA) and limited duties under the Surface Mining Control and Reclamation Act, the Noise Regulation Act, the Atomic Energy Act, and the Uranium Mill Tailings Radiation Control Act.

Most environmental enforcement activities in the past have been media specific; their focus was on water *or* air *or* waste *or* land in specific locations. The current trend, however, appears to be toward multimedia enforcement, leading to enforcement actions being filed against a single facility under several different statutes. Table 3–E illustrates administrative enforcement actions taken under various statutes from 1972 through 1991.

Compliance activities on the federal level have historically not been enforced in the same way as in the private sector. The DOJ has been reluctant to prosecute a "sister" agency, and many federal facilities, particularly those operated by the U.S. Department of Energy (DOE) and U.S. Department of Defense (DOD), operated under restrictions supported by the executive branch on the basis of national security. This type of operation was necessary because of the sensitive political situation that existed in the world during the 1970s and 1980s.

Table 3–E. Administrative Enforcement Actions Taken Under Statutes, 1972–1991

Fiscal Year	CAA	CWA/SDWA	RCRA	CERCLA	FIFRA	TSCA	EPCRA	Total
1972	0	0	—	—	880	—	—	880
1973	0	0	—	—	1,274	—	—	1,274
1974	0	0	—	—	1,387	—	—	1,387
1975	0	738	—	—	1,614	—	—	2,352
1976	210	915	0	—	2,488	0	—	3,613
1977	297	1,128	0	—	1,219	0	—	2,644
1978	129	730	0	—	762	1	—	1,622
1979	404	506	0	—	253	22	—	1,185
1980	86	569	0	0	176	70	—	901
1981	112	562	159	0	154	120	—	1,107
1982	21	329	237	0	176	101	—	864
1983	41	781	436	0	296	294	—	1,848
1984	141	1,644	554	137	272	376	—	3,124
1985	122	1,031	327	160	236	733	—	2,609
1986	143	990	235	139	338	781	0	2,626
1987	191	1,214	243	135	360	1,051	0	3,194
1988	224	1,345	309	224	376	607	0	3,085
1989	336	2,146	453	220	443	538	0	4,136
1990	249	1,780	366	270	402	531	206	3,804
1991	* 137	** 1,745	364	269	300	422	179	3,416

* Does not include asbestos data.
** Does not include SDWA data.

Source: U.S. Environmental Protection Agency, FY 1991, *Enforcement Accomplishments Report* (Washington, DC: EPA, 1991).

Today's managers of federal facilities, however, are not only trying to find ways to clean up their polluted sites but are also working aggressively on environmental protection and pollution prevention. In 1991, the Superfund National Priorities List (NPL) included 116 federal sites. These federal sites have been determined to be among the worst hazardous substances sites that warrant emergency cleanup. The majority of these sites (94) are attributable to the DOD's military installations, but the DOE is responsible for remediation of sites used in conjunction with the nuclear weapons program. Both DOD and DOE have committed significant funding and resources toward remediation of their contaminated sites.

Federal statutes are used not only for pollution abatement and environmental protection from hazardous substances but also as a mechanism to develop new technologies that would ultimately result in improved environmental quality. The following are some examples:

- RCRA directed the EPA to implement guidelines for federal procurement of recycled materials, including paper products. The Federal Energy Management Improvement Act of 1988 requires all federal agencies to reduce energy consumption by 10% by 1995 based on 1985 energy consumption.

- The DOE has set mandatory energy design standards for all new building construction, as codified in 10 CFR Part 435.
- The Solar Energy Research, Development, and Demonstration Act of 1974 authorized a federal program aimed at developing solar energy as a viable source of the nation's future energy needs. This program was implemented through the creation of the Solar Energy Research Institute (SERI). In 1991, SERI became one of the DOE's national laboratories (National Renewable Energy Laboratory, or NREL), responsible for taking the lead in conducting research in renewable energy technologies.

The following discussions will specifically address federal compliance requirements under several major statutes.

Major Federal Requirements

The four so-called media statutes—the Resource Conservation and Recovery Act (RCRA), the Clean Air Act (CAA), the Clean Water Act (CWA), and the Safe Drinking Water Act (SDWA)—all have requirements that obligate federal facilities to comply with environmental laws at the federal, state, and local levels. The following sections will primarily describe the federal compliance requirements under these and other selected statutes. See Federal Environmental Laws earlier in this chapter for detailed information on the general requirements for each of these statutes. See also Chapters 8, Hazardous Wastes, and 10, Pollution Prevention Approaches and Technologies.

Resource Conservation and Recovery Act. RCRA, an amendment to the Solid Waste Disposal Act, enacted in October 1976, establishes requirements for hazardous waste management from generation through final disposal. This is commonly referred to as "cradle-to-grave" control of hazardous waste, and this statute is enforced to provide safe deposition of hazardous and solid wastes. Subtitle F of RCRA describes federal responsibilities and states that all federal agencies "shall be subject to, and comply with, all Federal, state, interstate, and local requirements, both substantive and procedural . . . in the same manner, and to the same extent as any person is subject to such requirements" (42 USC § 6961). The president may "exempt any solid waste management facility of any department, agency, or instrumentality in the executive branch from compliance . . . if determined to be in the paramount interest of the United States to do so" (42 USC § 6961). It appears clear from these two statements that federal facilities are required to comply but can be exempted from compliance by the president.

RCRA also has requirements for cooperation with the EPA, stating that "all federal agencies . . . shall promptly make available all requested information concerning past or present agency waste management practices . . ." (42 USC § 6963). Despite this requirement, and the enforcement authority granted to the EPA under Section 3008 of RCRA (42 USC § 6928), the authority to fine other federal agencies has not been clear. Litigation in various federal district and circuit courts has resulted in vastly different interpretations of the law. Although several of the courts have decided that agencies of the federal government are immune from such penalties, others have determined that the waiver of sovereign immunity under RCRA allows the EPA to levy fines against violators. Passage of the Federal Facility Compliance Act (FFCA) of 1992, which will be discussed later in this chapter, eliminates the uncertainty with regard to EPA's authority. Federal agencies are now subject to monetary penalties to the same degree as the private sector for violations of federal, state, and local solid and hazardous waste laws. Compliance efforts at the federal level with RCRA will continue to increase as a result of the passage of the FFCA.

Clean Air Act. The CAA, as amended in 1990, currently contains the broadest compliance demands noted in the media statutes. This law is viewed as the most lengthy and complex environmental legislation ever enacted by Congress. It has grown from modest beginnings in 1967 to hundreds of pages of statutory requirements that are implemented through thousands of pages of regulations. The intent of the CAA is to improve ambient air quality, thereby reducing damage to the ozone layer, reducing acid rain, and improving visibility in areas with "brown clouds" from excessive air pollution.

The CAA specifically addresses control of pollution from federal facilities. Section 118 states that

> each department, agency, and instrumentality of the executive, legislative, and judicial branches of the Federal Government . . . shall be subject to, and comply with, all Federal, state, interstate, and local requirements, administrative authority, and process and sanctions respecting the control and abatement of air pollution in the same manner, and to the same extent as any nongovernmental entity (42 USC § 7418).

The CAA requires this to include all substantive and procedural requirements, "to the exercise of any federal, state, or local administrative authority, and to any process or sanction whether enforced in Federal, state, or local courts, or in any other manner" (42 USC § 7418). This act expands the compliance language noted in RCRA by requiring that recordkeeping, permitting, reporting, and fee payment obligations be met by the federal agencies.

Although Section 118(a) of the act states that the requirements specified in that subsection should apply in spite of "any immunity of such agencies, officers, agents, or employees under any law or rule of law," presidential exemption is still provided for in Section 118(b)(42 USC § 7418). Any emission source of any department, agency, or instrumentality in the executive branch may be exempted from compliance if determined to be in the interest of the United States. This language is similar to that provided in RCRA and again allows for sovereign immunity, but there are two notable differences. First, the CAA qualifies this exemption authority by stating that no exemption may be granted from the requirements of Section 111, which pertains to standards of performance for new stationary sources (42 USC § 7411). Second, exemptions from Section 112, National Emission Standards for Hazardous Air Pollutants, may only be granted if the technology to implement the standard is not available and the exemption is in the national security interests of the United States (42 USC § 7412). Further, this exemption is valid for no longer than two years.

Clean Water Act. The CWA is predominantly a technology-based statutory program that sets a national goal for elimination of the discharge of pollutants into the waters of the United States. In addition, the CWA provides for protection of wetlands through the Section 404 "dredge and fill" permitting program (33 USC § 1344) and the Section 402 discharge program. Statutory language regarding federal compliance with the CWA is similar to that found in the CAA and requires that

> each department, agency, or instrumentality of the executive, legislative, and judicial branches of the Federal government shall be subject to, and comply with, all Federal, state, interstate, and local requirements, administrative authority, and process and sanctions respecting the control and abatement of water pollution (33 USC § 1323).

It requires that the control and abatement requirements apply in the same manner, and to the same extent, as any nongovernmental entity. This includes payment of service charges, permitting, record-keeping, and reporting and "any other requirement whatsoever" (33 USC § 1323). The president, however, again has the authority to exempt any effluent source from compliance in the "paramount interest of the United States" (33 USC § 1323). As seen in the CAA, there are limits to the president's use of sovereign immunity under the CWA, and these include prohibiting exemptions of Sections 306 and 307 of the act.

Section 306 sets national standards of performance, based on various source categories, for "control of the discharge of pollution which reflects the greatest degree of effluent reduction . . . determined to be achievable through application of best demonstrated control technologies" (33 USC § 1316). It is probable that although presidential exemption does not apply to this section of the act, the statutory wording relating to "best demonstrated control technology" is vague enough to allow for wide variations in interpretation.

Section 307 provides requirements for toxic pollutants and pretreatment effluent standards. Listed toxic pollutants are subject to effluent limitations resulting from the application of "best available technology which is economically achievable for the applicable category or class of point sources" (33 USC § 1317). Again, the phrase *economically achievable* allows for a liberal interpretation of the requirement.

Section 404 of the CWA contains specific language relating to dredge and fill activities by federal agencies. The discharge of dredged or fill material, as part of the construction of a federal project specifically authorized by Congress, is not subject to the requirements of Section 404, provided the information relating to the impacts of the dredge and fill activities is contained in an approved NEPA document. An NEPA document can be an Environmental Assessment (EA) or Environmental Impact Statement (EIS) (33 USC § 1344). However, language included near the end of the section indicates that state and interstate agencies have the right to "control the discharge of dredged or fill material in any portion of the waters within the jurisdiction of that state. . . including any activity of any Federal agency . . ." (33 USC § 1344). This would appear to give the states control over federal agencies operating within their jurisdiction, but the use of sovereign immunity—if applied through presidential exemption—could preclude states' taking action against federal violators.

Safe Drinking Water Act. The SDWA provides for protection of *public water supplies,* defined as those that serve at least 25 people for at least 60 days per year. Primary drinking water standards (maximum contaminant levels, or MCLs) are established for 84 chemicals. Secondary standards are also set for aesthetic values, to regulate the appearance and taste of the water. In addition to regulating drinking water quality, the SDWA provides for protection of underground drinking water from injection of pollutants into wells and for protection of wellheads from surface activities that could affect water quality.

Federal agencies again

> shall be subject to, and comply with, all Federal, state, and local requirements, administrative authorities, and process and sanctions respecting the provision of safe drinking water and respecting any underground injection program in the same manner, and to the same extent, as any nongovernmental entity (42 USC § 300 j-6).

Record-keeping, reporting, and permitting-requirement compliance is also specified.

The compliance waiver seen in this act is significantly different from those previously discussed. In the SDWA, the administrator of the EPA can waive federal compliance on "request of the Secretary of Defense and upon a determination by the President that the requested waiver is necessary in the interest of national security" (42 USC § 300 j-6). The administrator is further required to maintain a written record of why the waiver was granted and make this record available when relevant for judicial review. A notice that the waiver was granted on the basis of national security is also required to be published in the *Federal Register,* unless the secretary of defense requests nonpublication because the publication itself would be contrary to the interest of national security. If the record cannot be published, the administrator submits notice of the waiver to the Armed Services Committees of the Senate and House of Representatives.

National Environmental Policy Act. Many of the functions that affect society have been delegated to federal agency establishments, including management of forest and water resources, management of public and national park lands, interstate highways, and nuclear power development. The National Environmental Policy Act of 1969 (NEPA) is a short, generalized statute that provides a declaration of general environmental policy and requires consideration of environmental impacts by federal agencies (42 USC § 4321 et seq.). This is the first statute that directly targets the actions of federal agencies, and its importance results from the large body of case law that has developed. NEPA's ultimate importance and success may be difficult to measure, as there is no way to measure the countless numbers of federal projects canceled because of the threat of NEPA scrutiny.

In addition to the requirements for federal agencies to ensure that impacts are evaluated, NEPA also resulted in the creation of the Council on Environmental Quality (CEQ), which oversees NEPA implementation by the federal agencies. Under the Clinton administration the authority of the CEQ may be diminished, and some oversight responsibilities may come under the control of another, perhaps newly formed, organization. As this law now enters its third decade, the role NEPA plays in federal decision making continues to be significant.

Numerous "action-forcing" provisions require compliance on the part of the federal agencies. The most important, and probably most familiar to the public, is Section 102(2)(C), which requires all agencies of the federal government to include a detailed environmental analysis in every recommendation or report on proposals for legislation and other major federal actions significantly affecting the quality of the human environment. This analysis must include impacts of the proposed action, alternatives, and adverse impacts that cannot be avoided (42 USC § 4332). To document this analysis, agencies generally

prepare either an EIS or EA. Certain classes of actions, however, defined by the individual federal agencies and approved by the CEQ are "categorically excluded" from the preparation of environmental analysis documentation. It should be noted that the emphasis on incorporating NEPA early in the planning process is designed to ensure that the negative impacts of actions are avoided or mitigated.

Other provisions require federal agencies to (1) use a systematic interdisciplinary approach to planning and decision making through use of natural and social sciences and environmental design arts, (2) develop alternatives in any proposal that involves unresolved conflicts related to use of available resources, (3) initiate and use ecological information in the planning of resource-oriented projects, and (4) recognize the worldwide and long-range character of environmental problems.

Federal compliance with NEPA is largely due to the effectiveness of numerous court decisions. There are hundreds of federal agency projects and proposals for legislation, however, that legally require NEPA review and are not in compliance because no one sues the agency. Although virtually all federal agencies have now adopted regulations for compliance with NEPA, the pattern of compliance varies because of the difficulty of some agencies' incorporating environmental values into their other immediate, mission-oriented goals. Table 3–F illustrates litigative actions from 1980 through 1990.

In 1991, federal agencies filed a total of 456 EISs, continuing a gradual downward trend that began early in the 1980s. This decrease may be due to the decrease in development of nuclear power plants and new military installations, as well as an increase in the quality of the agencies' NEPA implementation procedures.

Federal Liability

The President and Congress sent the mandatory compliance message to federal facilities in Executive Order 12088, signed on October 13, 1978. Although this message has been repeatedly emphasized in the federal compliance section of the environmental statutes, it has often been ignored. Many federal facilities have continued to operate as if they made their own law, with a general disregard for the federal, state, and local requirements to which they are statutorily obligated to comply.

The compliance message of the 1990s, however, is now being drafted and enforced by the facilities themselves. The Federal Facility Compliance Act of 1992 was, in part, drafted by representatives of the

Table 3–F. NEPA Litigation Survey, 1980–1990

Year	Cases Filed	Injunctions	Most Frequent Defendant	Most Common Complaint	Second Most Common Complaint
1980	140	17	DOT, DOI, DOD, HUD, EPA	No EIS	Inadequate EIS
1981	114	12	DOD, DOT, DOI, USDA, HUD, NRC	No EIS	Inadequate EIS
1982	17	19	DOI, COE, DOT, USDA, HUD, EPA	Inadequate EIS	No EIS
1983	146	21	DOI, DOT, USDA, FERC, NRC	No EIS	Inadequate EIS
1984	89	14	USDA, DOI, DOT	Inadequate EIS	No EIS
1985	77	8	DOT, DOI, COE, FERC	No EIS	Inadequate EIS
1986	71	16	DOT, DOI, USDA	No EIS	Inadequate EIS
1987	69	3	DOI, DOT, USDA	No EIS	Inadequate EA and EIS
1988	91	7	COE, DOI, USDA, DOT	No EIS	Inadequate EIS
1989	57	5	USDA, DOT	No EIS	Inadequate EIS
1990	85	11	DOT, COE, EPA	No EIS	Inadequate EIS

Abbreviations for Federal Agencies

COE	DOD Department of the Army, Corps of Engineers
DOD	Department of Defense
DOI	Department of the Interior
DOT	Department of Transportation
EPA	Environmental Protection Agency
FER	Federal Energy Regulatory Commission
HUD	Department of Housing and Urban Development
NRC	Nuclear Regulatory Commission
USDA	Department of Agriculture

Source: CEQ, 1991.

DOE and DOD, and the effects that this new act will have on federal compliance will be discussed in the following section. The message that environmental law compliance is required, not optional, is further emphasized by the fact that violators at all levels are made liable through the administrative and permitting process, through CERCLA bounty-hunter provisions, in contempt and injunction proceedings, citizen suits, tort liability, and civil or criminal penalties.

In *U.S. v. Curtis,* the Ninth Circuit U.S. Court of Appeals ruled that federal employees can be held criminally liable under the CWA (CA 9, No. 92–30235, 3/8/93). Curtis was indicted on five counts of knowingly discharging jet fuel, a known pollutant, into surface waters while acting as fuels division director at a naval air station in 1988 and 1989.

In March 1992, Rockwell International agreed to plead guilty to 10 separate violations of RCRA and the CWA and to pay a total of $18.5 million in criminal fines to the United States and to the State of Colorado (*United States v. Rockwell International Corporation*, No. 92-CR-107, June 1, 1992). These violations occurred while Rockwell was the contractor/operator of DOE's Rocky Flats Plant, located northwest of Denver, Colorado. This fine is currently the largest RCRA penalty ever assessed and the second largest environmental fine ever imposed. The acceptance of this plea agreement by District Court Judge Sherman G. Finesilver on June 1, 1992, ended a four-year environment, safety, and health investigation at the Rocky Flats Plant by the Department of Justice, EPA, FBI, Colorado Department of Health, and DOE's Office of Inspector General. Judge Finesilver stated that the punishment was just and fair and in the public interest, and that the amount reflected the severity of the offenses and should act as a deterrent to potential violators in the future. What is interesting to note, however, is Rockwell's explanation of why the violations occurred, citing DOE for the "inconsistencies in the regulatory position taken by DOE, inadequate funding, arbitrary personnel ceilings, pressure to put production first, and a refusal to fund needed or required environmental projects." The impacts of these two incidents, as well as several other recent court decisions, reflect a new trend in judicial attitudes toward federal compliance.

State Enforcement

Federal Facility Compliance Act of 1992. The Federal Facility Compliance Act (FFCA) of 1992 was enacted as an amendment to the RCRA. The primary purpose of this act was to ensure that a complete waiver of sovereign immunity was mandated, allowing for imposition of administrative fines under RCRA Section 6001 (42 USC § 6961). The first substantial provision of the FFCA is Section 102, which provides for a more substantive description of compliance obligations, designed to clarify that federal agencies are required to comply with state waste "management" laws as well as disposal regulations. This section also provides for a complete waiver of sovereign immunity and requires the agencies of the United States to comply with all state, interstate, and local requirements, including all civil and administrative fines that may be imposed for either isolated, intermittent, or continuing violations. See the Effects section below for a discussion of a recent application of this portion of the act.

Section 102(a)(4) provides for limitation of civil liability when federal employees are acting within the scope of their official duties. This statutory language is similar to that found in the federal facilities provisions of the CWA and CAA. However, there is now increased potential for criminal liability, as is evident in the addition of this language: "An agent, employee, or officer of the United States shall be subject to any criminal sanction (including, but not limited to, any fine or imprisonment) under any Federal or state solid or hazardous waste law."

Additional sections have been added to RCRA Section 6001 relating to EPA enforcement. The EPA may initiate administrative enforcement actions against any department, agency, or instrumentality of the executive, legislative, or judicial branch of the federal government. These actions are initiated in the same manner and under the same circumstances as the action taken against any other person. Here again is the explicit waiver of sovereign immunity.

Section 104 of the FFCA now requires the EPA to perform an annual inspection of federal facilities and provides for reimbursement to the EPA for the cost of the inspection by the agency being inspected. The first inspection of each facility must also include a review of its groundwater protection policy. In addition, states are given the right, although not an obligation, to inspect the facilities as well. The implications of these environmental audits will be further discussed in Chapter 7, Environmental Audits and Site Assessments.

The FFCA contains specific new requirements for the handling of mixed waste (waste streams with both a hazardous and a radioactive component), including requirements for DOE to develop comprehensive plans to treat mixed wastes generated at all DOE facilities, to prepare extensive planning and reporting documents, and to develop a national inventory of all mixed waste on a state-by-state basis. Additional information on the specific requirements for handling of RCRA hazardous mixed wastes is provided in Chapter 8, Hazardous Wastes.

New requirements specifically applicable to DOD operations are also covered in this act, including handling of excess munitions as waste and new standards of pretreatment for DOD sewage treatment plants. Clarification about the applicability of RCRA to "public" vessels, such as Navy ships, is provided by stating that waste generated on these vessels is not subject to RCRA's storage, manifest, inspection, or record-keeping requirements. The waste only becomes regulated when it comes ashore. Several interesting questions may arise here, such as what would prevent shipboard personnel from dumping the waste at sea and therefore avoiding the RCRA regulations altogether? Perhaps the best deterrent for this type of activity is the court decision and public disclosure of *U.S. v. Curtis*, previously discussed, in which Curtis was indicted on five counts of knowingly discharging pollutants into surface waters (CA 9, No. 92–30235, 3/8/93).

Effects. On March 11, 1993, the Washington State Department of Ecology announced that it had assessed fines against the DOE and Westinghouse for hazardous waste violations at the Hanford facility located near Richland, Washington. In addition to the fines, an enforcement order was issued for shipment of the waste to a disposal facility and for development of a plan for the testing and safe storage of the waste prior to disposal. The U.S. Department of Ecology issued the fines and order jointly against DOE, the owner; and Westinghouse, the operator of the reactor. The owner of the facility is responsible for oversight, and the operator has an obligation to comply with the federal and state laws.

This action is a reflection of what may begin to occur with increasing frequency as a result of the passage of the FFCA of 1992. Although penalties against an agency for noncompliance may not be a motivator, increased accountability on the part of federal employees should ensure increased diligence to the environmental regulations. Those people with direct line responsibility for RCRA compliance are increasingly aware that noncompliance may result in not only inconvenience in answering to auditors and enforcement officials but also ultimately in personal penalties such as fines and jail time, loss of stature in the professional community, and loss of employment.

Federal Emergency Exemptions from Compliance

Previous examples have shown that it is not advisable to avoid or circumvent environmental regulations from either an ethical or a legal stance, but it is often tempting to do so when faced with performance requirements from superiors to finish projects quickly or to operate under restricted funding. This is particularly true for federal agencies that must consider their actions under NEPA. Although proper planning should allow for adequate time to assess the environmental consequences of a proposed action, there are occasionally circumstances where use of NEPA analysis could result in the delay of an action required by other equally critical issues. It is for this reason that there are exemptions provided from NEPA compliance, and some examples of these will be provided in the following paragraphs.

Although rarely used, 40 CFR, 1506.11 in the CEQ regulations for implementation of NEPA does contain an emergency provision:

> Where emergency circumstances make it necessary to take an action with significant environmental impact without observing the provisions of these regulations, the Federal agency taking the action should consult with the Council about alternative arrangements. Agencies and the Council will limit such arrangements to actions necessary to control the immediate impacts of the emergency. Other actions remain subject to NEPA review.

Use of this emergency exemption is rare and generally involves consultation with the council over the phone. Concurrence with the emergency exemption is made the subject of a memorandum at a later date.

An example of such an exemption would be the emergency application of pesticides to combat an outbreak of encephalitis. Several exemptions have been granted for protection of fish and wildlife, including permission to destroy a generation of Chinook salmon due to the presence of a viral infection that threatened other fish, and the capture of the last five endangered California condors for placement in a captive breeding program. Following the invasion of Kuwait by Iraq in 1991, DOD requested and was granted an exemption for Operation Desert Shield and Operation Desert Storm based on the immediate requirement for protection of national security interests in the Middle East.

Explicit exemptions have also been granted by Congress, where requirements for NEPA compliance have been either limited or eliminated completely in a number of statutory programs. Most notable are the specified agency actions that are not "deemed [to be] a major Federal action significantly affecting the quality of the human environment." This statutory language, which is a paraphrase of NEPA Section 102(2)(C), has been used, for example, in sections of the CAA, CWA, Strategic Petroleum Reserve Plan, Defense Production Act, and the Disaster Relief Act.

In addition, there have been exemptions recognized by the courts that have found the NEPA requirements to be implicitly overruled by conflicting requirements in other statutes, exemptions based on the "functional equivalence" between NEPA and procedures required under other environmental statutes, and exemptions specifically granted by the president and executive branch. Further exemptions may be noted in that NEPA was not intended to be applied to federal agencies whose function is to protect the environment. Endangered species, for example, must be classified as such regardless of the environmental impacts of being classified as endangered.

SUMMARY

Federal agencies have commonly been known for their enforcement role in environmental statute compliance. The end of the Cold War and the increased public awareness of the environmental price paid in the name of national security has resulted in an increased role by federal facilities in the compliance area as well. This trend toward increased compliance and acceptance of their role as stewards of the environment should result in continued and expanded compliance by facilities and environmental restoration activities by federal agencies well into the 21st century.

4

International Legal and Legislative Framework

Gary R. Krieger, MD, MPH, DABT
Ian Austin, PhD

International (non–United States) environmental policies are extremely variable and are clearly in a transitional phase as individual countries or trading units (1) develop "transfrontier" perspectives on environmental rules and regulations, and (2) deal with both the positive and negative economic impacts of environmental laws. This chapter focuses on the broad policy outline of international environmental accords and general rules and regulations in four global areas: (1) Europe, including Russia and the Newly Independent States (NIS); (2) North and South America; (3) Asian/Pacific countries; and (4) developing nations worldwide. Because of the rapid regulatory and legal changes that can and will occur within an individual country, the reader should research and reconfirm specific, current regulatory requirements in a given country.

Since 1972, the scope of international environmental agreements has widened dramatically. Hence, the effects of environmental control policies on a given country's overall investment and resource allocation can be significant. In the early 1970s, research and cleanup by major western democracies on transborder pollution was directed toward internalizing both the costs of pollution and the costs of abatement (Nicolaisen et al, 1991). One result of this research was the development of a "mutual compensation principle." This principle is directed toward the efficient distribution of abatement and pollution costs among countries. Specifically, the polluter would pay for the transborder pollution damage, whereas the "recipient" country would pay for abatement. If put into practice, the mutual compensation principle would cause polluter countries to account for the external costs (that is, transfrontier damages) and the recipient country would have an incentive only to accept or "absorb" pollution as long as this strategy was cost-effective.

The mutual compensation principle has not been applied directly to a major pollution issue (e.g., acid rain, air pollution, large-scale releases of pollutants into surface water sources). To date, the international community has directed its efforts toward broad-based international agreements and conventions or general policy goals and not toward detailed compensation principles (Nicolaisen et al, 1991). During the last 25 years, the United Nations Environmental Program (UNEP) has served as the main forum for these agreements and accords. Table 4–A illustrates some of the major UNEP agreements of the last 25 years. Significantly, these agreements and conventions do not direct *specific* policy actions;

Table 4–A. Major International Agreements and Conventions on Environmental Protection[1]

Conventions to Protect the Marine Environment

- The 1972 London Agreement on Prevention of Marine Pollution from Dumping of Waste and Other Materials
- The 1974 Paris Convention for the Prevention of Marine Pollution from Land-Based Sources
- International Convention for the Prevention of Pollution from Ships (MARPOL)—1973, modified 1978
- Conventions regulating the use of regional seas (UNEP regional seas program), such as the 1976 Barcelona Convention in which the 16 signatory countries agreed to the Mediterranean Action Plan, the 1974 Helsinki Convention on the Protection of the Marine Environment of the Baltic Sea Area (7 countries), and the 1987 London Ministerial Declaration on the protection of the North Sea (8 countries).

  Other international agreements on resource preservation include the Law of the Sea (1982; 160 signatories), agreements on whaling, wetlands, migratory birds, Antarctica and the trade in tropical timber, and endangered species. UNEP lists altogether 140 various international agreements, conventions, and protocols as of 1989.

Economic Conventions and Treaties for Europe

- International Whaling Convention
- International Tropical Timber Agreement
- Convention on International Trade in Endangered Species (CITES)
- The ECE Convention on Long-Range Transboundary Air Pollution: Signed by 34 countries in November 1979, and entered into force in March 1983, having been ratified by 24 countries; to this convention was added:
  a. The Helsinki Protocol, which was signed by 20 countries and entered into force in September 1987. In broad terms, the Helsinki Protocol states that the signatories reduce their national annual sulfur emissions or their transboundary fluxes by at least 30% by 1993 at the latest (using 1980 emission levels as the base year);
  b. A protocol on nitrogen oxides (NO$_x$), which was signed by 25 countries in October 1988 in Sofia. This protocol will enter into force when ratified by 16 signatory countries. As a first step, the signatory countries agreed to take measures against further increases of NO$_x$ emissions so that national NO$_x$ emissions do not increase beyond 1987 levels after 1994.
- UNEP 1985—Vienna: Protection of the Ozone Layer
  a. 1987 Montreal Protocol, signatory countries (46) agreed to halve their production of five chlorofluorocarbons (CFCs) and three halons by 2000; phase-out of CFCs agreed in London by about 100 countries.
- UNEP 1989—Basel Treaty: Control International Trade in Hazardous Waste
  a. March 1989 (34 countries and the European Union, EU); signatories agree in principle to ban and establish notification procedures for all trade in hazardous waste.

  b. January 1991 (enters into force on May 5, 1993); Organization for Economic Cooperation and Development (OECD) Council Decision—Recommendation on the Reduction of Transfrontier Movement of Wastes; applicable to all wastes subject to controls under the Basel 1989 Treaty; OECD member countries (includes United States, Japan, Canada, Australia, EU countries, among others) must take steps to keep their wastes "at home" for those (wastes) destined for final disposal; and OECD must collect harmonized data on waste imports and exports.
- 1990
  a. Establishment of the Global Environmental Facility (GEF) administered by the World Bank, UNEP, and UN Development Program. GEF was established to help poor countries comply with environmental agreements.
- Rio—June 1992
  a. Agenda 21—Blueprint for sustainable development in the 21st century.
  b. Chapter 38 of Agenda 21—International institutional arrangements; confirms UNEP's mandate and its "need for an enhanced and strengthened role"; endorses and expands UNEP's role building on the Stockholm 1972 conference.
  c. Strengthens Earthwatch, the environmental early-warning system (established 1972 Stockholm conference).
  d. Emphasis on environmental emergencies, natural resource accounting/economics, environmental impact assessments, and international law, decertification, forestry, marine resources.
- Biological Diversity Convention[2]—(1) Aims to save species of animals and plants and their habitats from extinction or destruction; (2) stipulates that industrialized countries must help developing nations technically and financially over and above current levels of assistance; (3) rural communities and indigenous people must be the first beneficiaries of profit based on products from wild plants and animals.
- Framework Convention on Climate Change—Calls for (1) stabilization of greenhouse gas emissions at levels that would not harm the climate system; (2) levels to be reached early enough to allow ecosystems to adapt naturally to climate change. This convention does not specify timetables and targets for limiting greenhouse gas emissions.
- December 1992
  a. UN General Assembly established high-level Commission on Sustainable Development to oversee implementation of Agenda 21.
- 1993 and beyond
  a. Refunding GEF with emphasis on climate change, pollution of international waters, destruction of biodiversity, and depletion of stratospheric ozone.
  b. Negotiations on an international treaty to combat spread of deserts.
  c. Cleanup and protection of polar (especially Arctic) region.

[1]A "convention" is an agreement in which a general framework is established but must be supplemented in the future by "protocols" that list specific procedural details. A "treaty" is an agreement in which all the details are specified and expressed in the signatory document.
[2]Signed by the United States in June 1993.

hence, their overall effectiveness is difficult to accurately measure. In addition, when specific quantitative goals are stated (e.g., sulphur dioxide emissions), the published "compliance" deadlines are in the middle and late years of the 1990s. Nevertheless, the UNEP agreements have served as a touchstone for the global environmental movement and have clearly elevated global environmental practices to a center stage issue.

ENVIRONMENTAL POLICY IN THE EUROPEAN UNION (COMMUNITY), RUSSIA, AND THE NEWLY INDEPENDENT STATES (NIS)

European Union (Community)

Prior to the enactment of the Single European Act on July 1, 1987, the European Economic Community (EEC), established by the Treaty of Rome on January 1, 1958, did not focus on environmental issues. The original objective of the EEC Treaty was to establish a common market to facilitate the free movement of goods, services, persons, and capital. No explicit powers were granted to either the European Council or Commission for environmental policies. [EEC and Economic Community (EC) became the Economic Union (EU) as of the November 1, 1993 Maastricht Treaty.]

Movement toward unified environmental planning and policy coordination began with a 1967 directive concerning dangerous substances (Jans, 1993; Commission of the European Communities, 1967). The intent of this directive was to allow the free movement of goods, notwithstanding characterization of the material as a "dangerous substance." The directive was based on Article 100 of the EEC treaty, which states that the Council has the power to issue directives geared toward the harmonization of national legislation that has a direct impact on the operations of the common market. The free movement of goods and/or the relative competitive position of companies can be affected by national environmental legislation, so the Council can directly intervene based on Article 100. By July 1987, the Council had issued a substantial number of environmental directives affecting air and water pollution, waste products, waste oil, environmental impact statements, and ecological protection (Jans, 1993).

The Single European Act of July 1, 1987, explicitly specified community objectives in Article 130R:

- preserve, protect, and improve the quality of the environment

- contribute toward protecting human health
- ensure a prudent and rational use of natural resources.

In general, three overall environmental principles were articulated by the 1987 Act:

- Preventive action (minimization) should be taken.
- Source control and mitigation should be emphasized.
- The polluter should pay.

Furthermore, under the Single European Act, environmental protection requirements were emphasized, with the implication that environmental protection should be a general component of the EC's policies affecting trade, transportation, foreign aid, and so on.

Clearly, this type of broad-brush language allows considerable latitude in what environmental measures or directives can be issued by EU governing bodies. According to Article 189 of the EEC treaty, a directive is binding about the results that must be achieved; however, each Member State has discretion over the choice of form and methods. As analyzed by Jans (1993), "If an environmental directive regulates a certain matter exhaustively, Member States cannot in principle introduce or enforce more stringent national environmental legislation by invoking the need to protect the environment. On the other hand, they are also unable to ignore the level of environmental protection required by the directive, by reason of economic or financial factors." Therefore, if an environmental directive does not regulate a given situation, the Member States are able to develop and apply their own standards. Thus, Member States can and do develop specific attitudes, approaches, and standards for home country protection of air, water, and soil. As a result, there exists within the EU tremendous variability regarding in-country standards for hazardous materials rules and regulations.

Hazardous wastes. Where community rules and directives do have a major impact is on the interstate waste trade. Specifically, the Organization for Economic Co-operation and Development (OECD) estimates that 10% of all hazardous waste generated in Europe will pass at least one national border before final disposal (OECD, 1985) (see Table 4–B). Total waste production in the European Community is estimated at $2,755 \times 10^6$ short tons ($2,500 \times 10^6$ tonnes) per year, of which 1% is defined as hazardous (Junger & LeFevre, 1988). Therefore, it is

Table 4–B. Organization for Economic Co-operation and Development (OECD)

Milestone	Description
Establishment	September 30, 1961
Dedication of organization	—achieving highest sustainable growth and employment in member countries while maintaining financial stability —contributing to sound economic expansion in member and nonmember countries —contributing to expansion of world trade
Membership in early 1994	24 Nations: Australia, Austria, Belgium*, Canada, Denmark*, Finland, France*, Germany*, Greece*, Iceland, Ireland*, Italy*, Japan, Luxembourg*, The Netherlands*, New Zealand, Norway, Portugal*, Spain*, Sweden, Switzerland, Turkey, United Kingdom*, United States

*One of the 12 members of the European Union.

obvious that large quantities of hazardous and non-hazardous materials are crossing national frontiers. Whether the EU goals of free movement of goods and services apply to hazardous waste exports is unclear. This is particularly important given the tremendous political changes in Eastern Europe, because a significant proportion of the hazardous waste in Eastern European disposal sites is of Western origin (Brand, 1993).

Significant differences exist in Member State regulatory frameworks and disposal infrastructures, so only 50% of the total amount of hazardous waste generated in the EU can be legally disposed of and still comply with EU directives (Brand, 1993). Table 4–C illustrates the disposal infrastructure for selected EU Member States based on a study by the *Agence Nationale pour la Récupération et l'Elimination des Déchets* (ANRED, 1989). Several observations can be made from Table 4–C:

- The Netherlands (NL) has substantial incineration capacity.
- The United Kingdom (UK) has clearly concentrated its efforts on the use of controlled landfills for hazardous waste disposal.
- Denmark has a central publicly operated and owned disposal facility, whereas the UK has a large number of small privately owned disposal companies (Brand, 1993).

Table 4–D lists the hazardous waste generation of Member States as of 1989 (ANRED, 1989). This table predates the formation of a United Germany;

hence, West Germany (FRG) is listed. Obviously, a United Germany will have substantially increased quantities of hazardous materials.

Since the 1987 Single European Act, the EU has begun the arduous process of revising and clarifying its hazardous waste policies. Specifically, the EU has developed a classification scheme for defining hazardous wastes (Yakowitz, 1988). Hazardous wastes, as defined by the 1989 Basel Convention, are classified as either "amber" or "red." These wastes may not be shipped across borders unless all the pass-through countries give their consent. In Europe, over 154×10^6 short tons (140×10^6 tonnes) of wastes cross borders. The overall value of this activity exceeds $20 billion.

However, the ability to uniformly classify materials as hazardous does not solve the essential problem of creating uniform treatment and disposal standards. Without such standards, there will clearly be a cross-frontier differential in costs based on the varying national standards of Member States.

Thus, the fear of "garbage tourism" is real to those countries that have low disposal standards (Brand, 1993). If EU-wide treatment and disposal standards were in place, the economic incentive to export waste within the community would cease to exist; however, the theoretical possibility of exporting waste or hazardous materials to non-Member States in either Eastern Europe or Africa would remain. Because the 1988 classification scheme for hazardous wastes would produce a significant (up to 50%) increase in the quantity of hazardous waste— approximately 36×10^6 short tons (33×10^6 tonnes) per year versus 22×10^6 short tons (20×10^6 tonnes) per year—the export of hazardous waste is a significant potential problem (Yakowitz, 1989).

Currently, despite the continued movement toward EU harmonization, there are clear differences between Member States and the EU's administrative drive toward development of an internal market for hazardous waste. Thus, considerable uncertainty is likely to persist for the foreseeable future, and individual Member States will continue to develop and implement their own standards.

Although formal EU action has not occurred, the OECD has established a control system for moving recyclable wastes across frontiers. The EU's Member States are part of the OECD, which also includes the United States, Japan, Canada, and Australia, among other nations (Table 4–B). Under the OECD system, nonhazardous wastes that meet a specified criterion

Table 4–C. Disposal Infrastructure of Selected Member States of the European Community as of 1989

Facilities and Capacities	Member State					
	B	**DK**	**FR**	**NL**	**FRG**	**UK**
Number of Collective Treatment Facilities[a]	6	1	40	20	200[b]	
Incineration	1	1	25	7	17	4
Capacity (1,000 Tonnes)	40	90	600	160	620	80
Phys-Chem Treatment	1	1	10	6	23	13
Capacity (1,000 Tonnes)	24	20	300	30	280	?
Landfill	4	1	12	?	22	1,200
Capacity (1,000 Tonnes)	?	?	500	20	2,250	2,290

[a] Because a treatment facility can use several methods of disposal, the number of treatment facilities does not necessarily equal the number of separate disposal units.
[b] Both collective and private facilities.

Source: Agence Nationale pour la Récupération et l'Elimination des Déchets, 1989.

Table 4–D. Hazardous Waste Generation in the Member States of the European Community, 1989

Member State	Hazardous Waste Generated (× 1,000 Tonnes)
Belgium	1,000
Denmark	67
FRG	4,500
France	4,300
Greece	—
Ireland	76
Italy	2,000–5,000
Luxembourg	—
The Netherlands	1,000
Portugal	—
United Kingdom	3,750
European Total	19,943

Source: Agence Nationale pour la Récupération et l'Elimination des Déchets.

are labeled "green" and may cross national boundaries under existing trade rules.

Transfrontier pollution legislation. Recently, the EU has considered significant new laws on civil liability for environmental damage (Rose, 1993). The EU has proposed a system of strict civil liability for all Member States. Under the proposed system, three significant changes would occur: (1) enforcement agencies could deal with plant operations in any EU Member State; (2) the burden of proof would change so that it would be necessary to prove that the polluter was deliberately negligent; and (3) fees would be placed on polluting industries (to be retained in a fund for cleanup and liability costs). These changes would parallel the current United States Superfund system and would have substantial impact on the Member States. The event triggering these changes was the 1986 Sandoz Chemicals fire in Switzerland. Although Switzerland is currently not a member of the EC, the environmental impacts of the fire were felt in several Member States.

In order to further address the problem of transfrontier pollution, especially involving non-Member States, the Europeans have negotiated a special Convention on Civil Liability for Damage Resulting from Activities Dangerous to the Environment. This convention, effective June 1993, is the product of a five-year negotiation led by the 26-member Council of Europe. The convention will have multiple far-reaching effects:

- It defines environmental damage and who will pay for remediation.
- It allows environmental groups to bring action against alleged polluters.
- It allows plaintiffs the ability to file claims in either the country where the alleged damage occurred or where the hazardous activity originated.

The proposed legislation is in the review and comment process; the final form is unknown. However, substantial changes in both the EU and non-EU countries' attitudes toward environmental rules, regulations, and costs are occurring.

European countries—general and industry-related environmental policies. Individual governments are constantly faced with the dilemma of how to balance environmental protection with maintenance of international industrial competitiveness. To date, pure market solutions have rarely achieved environmental objectives; hence, direct government intervention is a frequent policy choice. Industry-related government policies include mandating emission standards or use of best available control technology.

According to a recent OECD analysis, "increasing emphasis has been put on the application of economic instruments," i.e., regulatory measures that try to exploit the efficiency of price signals (OECD, 1990). Examples of this strategy include emission tracking procedures, charges, or taxes.

The United States Environmental Protection Agency (EPA) has recently instituted trading emission offset procedures for air pollutants, and the Clinton administration considered initiating a variety of energy-related taxes that were rejected by Congress and not part of the 1993 budget. In general, economic instruments are preferred by industry over direct command and control policies such as emission standards, because a certain degree of individual industrial decision making is preserved. Companies can decide whether to pay or reduce pollution versus being subject to direct regulation (OECD, 1992).

In Europe, two policy approaches have evolved:

- policies directed toward reducing pollution burdens.
- integration of environmental concerns into other related aspects of industrial policy, e.g., energy.

Two examples of environmental policies that illustrate these two approaches are (1) global problems—climate change, ozone layer depletion, and deforestation; and (2) efforts directed toward reduction of the amount of hazardous waste generated by industry and private households. (See also Chapter 14, Global Issues, and Chapter 10; Pollution Prevention Approaches and Technologies.) Table 4–E provides a brief overview of significant environmental rules and regulations in a variety of European countries (Schlickman et al, 1992). OECD provides a yearly update of environmental rules or initiations that impact the industrial and energy sectors (OECD, 1990, 1991, and 1992). Table 4–F provides a brief synopsis of important industry-related environmental and energy policy initiatives for several European countries.

As Table 4–F illustrates, the European countries are managing their industrial and energy sectors with both economic incentives (e.g., CO_2 emission or carbon taxes) and direct regulatory initiatives. These economic and regulatory decisions are having a variety of repercussions throughout the private industrial sector. In general, based on their individual emphasis, European countries are developing focused commercial skills in several environmental technologies: (1) Germany and Sweden—equipment to treat water and effluents; (2) The Netherlands—development of advanced technologies for soil remediation;

and (3) United Kingdom—environmental consulting services. Significantly, the United States holds major positions in Europe with regard to waste management and environmental services. The Japanese have typically dominated the air and noise pollution control markets; however, Germany has emphasized these areas, and several German firms rank in the top five based on financial benchmarks (OECD, 1992). The combination of both economic and regulatory pressures are expected to drive a substantial increase in the commercial European environmental services sector.

Sweden

Like the United States, Sweden is frequently a bellwether country for environmental initiatives. Therefore, it is worthwhile to separately review the Swedish system of environmental protection. Environmental policy in Sweden is under the Ministry of Environment. Since 1990, the Swedish Environmental Protection Agency (SNV) has formulated national environmental goals. For example, ground and surface water goals are directed so that "Pollution does not limit surface water to fishing, recreation, or water supply [i.e., restrict use of surface water for fishing, and so on]. Ground water should have a quality that it can be consumed without purification" (SNV, 1990). Obviously, this is a broad-scoped goal that lacks specific language directing the manner and means of implementation.

In Sweden, the environmental goals are an important reference for legal decision making regarding water protection, limitation of emissions, and where to locate polluting activities. The environmental goals are translated into a form acceptable to the regional units by 24 county administrations. Regional action programs are then developed in order to meet the national goals. Finally, regional plans are implemented by municipalities and local administrations.

In 1991, the national government introduced a comprehensive government bill on environmental issues. This bill was based on five SNV-developed action programs: (1) "Air 90," (2) "Water 90," (3) "Nature 90," (4) "Sea 90," and (5) "Chemicals 90." These national action programs are scheduled to be integrated into the various country-wide industrial sectors (energy, transport, agriculture, forest, and so on).

Waste management and site remediation programs. The SNV has initiated a national program for the investigation and possible remediation of

---

Table 4–E. Country-Specific Environmental Rules and Regulations

---

German Environmental Legal Structure

Basic Legal Principles

- Principle of Providence (Varsovge Prinzip)—Environmental authorities have the right to intervene in circumstances when environment risk is very low (i.e., activities may be regulated even though they have not yet been proven harmful).
- Polluter Pays Principle (Verursacher Prinzip).
- Cooperation Principle (Kooperations Prinzip)—Requires the exchange of information (e.g., public hearings) and showing of environmental responsibility among all parties (e.g., state, society, individuals).

Environmental Liability Act, January 1991

- Public environmental law(s); prohibit certain activities, regulate issuance of licenses for hazardous installations, products, activities; provide authority for environmental taxes.
- Imposes comprehensive strict liability on the operators of certain installations for damages resulting from the environmental impacts of these installations.
- Environmental Impact Assessment (effective February 12, 1990) requires an assessment of the effects that certain public and private projects may have on the environment (e.g., power stations, airports, waste management facilities, etc.).
- Federal Emission Control Act, Water Act, Waste Avoidance and Waste Management Act, Biotechnology Act (i.e., genetically modified micro-organisms licensing requirement).
- Federal Chemicals Act on the Protection Against Dangerous Substances; requires compulsory testing and notification for hazardous substances, compulsory classification, labeling and packaging of hazardous substances.

Special Liability Rules for Environmental Cleanups in the New Länder (East Germany)

- Based on previous waste, water or police laws.
- Thuringia and Saxony have enacted new State Waste Acts.
- Art. 1, § 4, par. 3 of the General Environmental Act provides complex provisions and *possible* exemptions for purchasers of old plants.
- State Agency for Privatisation of State-owned Industry in the New Länder (Treuhandanstalt) *may* assume environmental liability for pre-July 1, 1990 contamination.

French Environmental Rules and Regulations

- 1971 Creation of the Ministry of the Environment.
- Strong influence of EC directives on overall policy.
- General environmental statutes: Law No. 76–663 of July 19, 1976 decrees Nos. 77–1133 and 77–1134 of September 21, 1977.
  a. Laws protect all components of the environment (i.e., neighborhoods, public health and security, agriculture) and nature against any kind of danger of nuisance created by a public or private facility known as "installation classées."
  b. Installation classées are divided into two types: (1) those that require prior authorization before an activity can begin ("installations classée sourmise à autorisation") and (2) those that simply require notification to the government ("à déclaration").
- General regulations covering air, water, noise, and waste.
- Regulations covering specific industries such as chemical, agricultural, and biotechnology.

United Kingdom Environmental Rules and Regulations

- Environmental Protection Act of 1990
- Her Majesty's Industrial Pollution (HMIP) is a major regulatory agency.
- Cross-media regulatory system for certain processes that have significant potential for pollution: *Integrated Pollution Control (IPC)*. Less polluting processes subject to an Air Pollution Control (APC) system administered by local authorities.
- IPC and APC systems in effect as of April 1991. IPC is administered by HMIP.
- 1989 National Rivers Authority (NRA) established for water pollution issues, discharge consents.
- Proposed 1994: sustainable development: The UK Strategy; sustainable forestry: The UK Programme; biodiversity: The UK Action Plan; climate change: The UK Programme.

Italian Environmental Rules and Regulations

- Creation of Ministry for the Environment—1986.
- National legislation supplemented by statutes enacted by regional parliaments. Primary enforcement arm is the Unita Sanitaria Locale (USL). USLs are local health units.
- General environmental laws covering water, solid waste disposal, air pollution, chemical substances, and noise.
- Use of risk assessment in order to quantify the cost of an activity related to potentially toxic chemicals and its future effects on exposed individuals.

---

contaminated sites. The program is directed toward five major types of sites:

- mine waste disposal
- municipal waste disposal
- inorganic chemical industrial production facilities
- specialty industries (e.g., wood preservation, coal gasification, leather tanning, and surface plating)
- fibrous sediments from the pulp and paper industries.

At present, there is no national site registry (like the U.S. EPA's National Priority List) or reporting system. However, best estimates indicate that there are approximately 4,000 municipal landfills, 80 major mine waste sites, 300 wood preservation plants, 2,700 industrial facilities, and 300 sites with contaminated sediments that require investigation. Investigations are supervised by local agencies with case-by-case guidance from SNV.

Table 4–F. European Industry-Related Environmental Initiatives

Austria

- Imported tropical wood and products eco-labeled.
- Clean air legislation directed toward stratospheric ozone protection, greenhouse gas reduction, and possibly a climate protection tax.
- Packaging material decree requiring 80% return quota of all packaging to be reused or recycled.
- Establishment of an environmental information law which will require enterprise to give authorities all environmentally relevant information unless constitutionally protected.
- General wastewater emission decrees.
- Changes in long-term environmental liability provisions with industries potentially responsible for damages allegedly due to long-term emissions.
- Law demanding obligatory test of environmental compatibility of both private and publicly owned enterprises.

Germany

- Highest priority is cleanup of eastern states, especially contaminated soils.
- Accelerate the licensing process for industrial facilities.
- Legislative approaches are generally regulatory (command and control) instruments; however, use of fees and taxes (economic controls) is increasing.
- Increased levies on water pollution, new regulations on the discharge of sewer water.
- Further implementation of the ordinance on large combustion plants and clean air instructions, emphasis on reducing greenhouse gas emissions.
- Obligations for the distributor/producer to take back wrapping material for recycling. Private companies may be required to take back products such as automobiles, electronic goods, batteries, and printed materials. Duales System Deutschland (DSD), general recycling laws for German householders.
- Adoption of environmental liability stipulations directed toward preventing damage in advance.
- Planned introduction of special waste and CO_2 levies.
- Planned transformation of the motor vehicle tax into a pollutant-oriented tax.
- More stringent rules for underground waste storage.
- Introduction of volume-related and pollutant-related economic incentives to avoid waste.
- Restrictions on highly volatile halogenated hydrocarbons, prohibition on the use of CFCs, and halons for large-scale users.
- Extension of the Act on Detergents and Cleansing Agents to the New Länder.
- Review of current energy policies in terms of the new, limited Germany.

Italy

- Application of the polluter pays principle versus government subsidies for cleanups.
- Comprehensive legislation on packaging and recycling will be introduced.
- Implementation of EC directives on "green" fuels and genetically modified organisms.

United Kingdom

- 1989 Environmental Programme directed toward business management: (1) waste minimization emphasis; (2) best

available control technology development; and (3) development of pollution control technology and service.
- Development of an Advisory Committee on Business and the Environment (May 1991) with emphasis on recycling and waste minimization:
  — July 1991 guidance to local authorities for recycling plans under the Environmental Protection Act.
  — "Green levy" to help finance the recycling or safe disposal of used tires.
  — Increase the proportion of recycled fibre in newsprint to at least 40% by the year 2000.
  — Development and emphasis on environmentally benign technology.
  — Development of a voluntary "eco-labeling" scheme for consumer products.
- Canceled a planned registry of contaminated land that would have included all sites which had been used in the past by a suspect industry.

The Netherlands

- National Environment Plan (1989); sets pollution targets for the next two decades. Costs, 1989 2% GDP, projected 3.5% GDP after the year 2000.
- Planned agreement with industry to establish environmental management systems for 10,000–12,000 firms by 1995. Systems include drafting of an environmental programme and registration of substances.
- Voluntary agreements between government and industry aimed at agreed environmental targets.
- Introduction of one integrated environmental license covering these acts: air, chemical waste, noise, and waste substances.
- Ecological taxation covering waste and groundwater under consideration.

Scandinavia

Finland

- Incremental duty on fossil-fuels.
- Cessation of CFC use by 1995.
- Guidelines on nitrogen oxide emissions.
- Requirements to use recycled paper in paper production.

Norway

- Industry emission standards.
- Environmental levies on gasoline, automotive diesel, and fuel oils.
- CO_2 emission taxes extended to coal and coke used for heating purposes.
- Reconsideration and restriction of emission permits especially directed toward pulp and paper industry.

Sweden

- CO_2 tax introduced (1990–1991), with a tax on oil, coal, natural gas, LPG, petrol, motor fuel, and domestic air traffic.
- Fee on nitrogen oxide emissions.
- Development of an environmental industry and service sector.

Denmark

- January 1992 Environmental Act has particular emphasis on ecoaudits, environmental labeling schemes.

(continued)

Table 4–F. *(Concluded)*

• CO_2 emissions tax on certain products used for the production of energy.

• Law modification to Law 42 of 1975 to promote return and recycling of glass, metal, cardboard, plastic, and cork packaging.

Spain

• Proposed National Hydrological Plan (PHN) involving movement of water to low moisture areas; water treatment systems upgrade costs 30 billion (United States) over 20 years.

Portugal

• Overhaul of environmental administrative structures with emphasis on water resources.
• Completion of implementing mechanisms for the basic environment framework law.

Sites are divided into four general categories (Table 4–G) based on the degree of contamination as compared to background, reference, or no-observed-effect levels. A major site consideration is both the total and time-dependent potential for leaching into groundwater. In general, the Swedish ranking system shares many characteristics with the EPA's Hazard Ranking System.

After site investigation, remedial alternatives are evaluated based on these factors:

• permanence
• reduction of toxicity and other hazardous properties
• limitation of volume, area affected, and mobility.

In general, the Swedish land-use goal is directed toward "multifunctionality" (land should be remediated so that contamination is reduced to levels that are acceptable for either residential or agricultural usage). The multifunctional land-use strategy is similar to the approach advocated by the Dutch government. The Dutch "A, B, C" system is similar to the Swedish site classification system presented in Table 4–G. Generally, countries like Sweden and The Netherlands that advocate multifunctional land use pursue cleanup goals that reduce contamination to background or reference levels (Dutch A Level). Other countries (e.g., United States and United Kingdom) allow site-specific cleanup based on multimedia (air, water, soil, and sediment) human health and ecological risk assessments. However, proposed changes in the U.S. Superfund law would move toward a system of national reference standards (see Chapter 8, Hazardous Wastes). The Swedish system is directed toward precalculated cleanup levels; a formal multimedia risk assessment is considered part of the site assessment and would influence the selection of remedial alternatives.

Liability issues. Like many other countries, the Swedish Environmental Protection Act adheres to

Table 4–G. Site Classification

Category*	Remedial Requirements
Very Large Risk	Remedial or protective measures are required.
Large Risk	Remedial or protective measures can be required.
Moderate Risk	Only simple measures are required.
Low Risk	Remedial and protective measures are not deemed necessary.

* From the Swedish Environment Protection Act.

the principle that the polluter should bear primary financial responsibility for cleanup. In addition, if there are multiple source contributors, cleanup or reimbursement responsibility is shared, based on the degree of contribution. If legally responsible parties are unavailable or insufficient for the remediation financial liability, then municipalities may have to assume primary liability under the Health Protection Act.

There is no statute of limitations for environmental damage; however, liability is typically limited to activities that occurred after the 1969 Environmental Protection Act. This liability restriction is valid even if the activities causing the damage were legal or permitted, as long as the effects could not be foreseen (SNV, 1992). Finally, liability for third-party damages (adjacent property) is the responsibility of the operator; however, if there is not a financially viable responsible party, then costs can be reimbursed by the national environmental damage insurance.

Eastern Europe, Russia, and the Newly Independent States (NIS) of the Former Soviet Union

The sudden collapse of the centrally planned economies of Eastern Europe and the Soviet Union has opened these countries to worldwide review and scrutiny regarding their previous environmental practices. The transition of a country from a Marxist centrally planned economy to a free market democracy is obviously painful and difficult. From a public health perspective, the legacy of apparent environmental degradation is economically and financially devastating.

The origins of the former Warsaw countries' environmental problems lie in politics and economics. Based on conventional Marxist theory, the state owned the entire means of production and was directed to ensure the welfare of its citizens. Therefore, in theory, the state was responsible for not allowing the welfare of its citizens to suffer because of environmental degradation arising from the production of goods and services. Thus, as the western democracies developed environmental policies in the 1970s and 1980s, the Warsaw Pact countries also passed environmental rules and regulations.

Unfortunately, as many countries have discovered, it is difficult to simultaneously be the regulator and the regulatee. A paper published by the OECD (Juhasz & Ragno, 1993) described how poor the water and air quality are in the former Warsaw Pact countries because of the low priority given to environmental quality compared to that given to production objectives. In addition, preliminary data from Upper Silesia in Poland indicates that residents may have been severely affected by environmental pollution. Compared to the Polish national average, Upper Silesia residents have:

- 155% more circulatory illnesses
- 30% more cancer
- 47% more respiratory illnesses (Juhasz & Ragno, 1993).

Similar health impacts have also been reported for the former Soviet Union (Green, 1993). The origins of environmental failure invariably lie in politics and economics. In the previous analysis of the Western European countries, it is clear that governments are attempting to balance economic or market incentives with regulatory demands. In most socialist (Marxist) countries, there are no economic feedback signals from the marketplace. Thus, environmental policies are invariably of the command and control style. Review of the environmental policies of the former Warsaw Pact countries indicates that:

- Policies were in place.
- Administrative frameworks were established.
- Laws and regulations were in place.

- Reporting and monitoring framework existed.
- Enforcement mechanisms complete with charges and penalties were available.

However, Juhasz and Ragno (1993) have analyzed the former Eastern Bloc countries and attribute their environmental failure to three main causes:

1. Government failed to coordinate environmental and economic policies. Specifically, production quotas were always more important than environmental costs. Under a centrally planned system, the artificially low subsidized prices for raw materials, energy, and agricultural inputs overwhelmed any environmental costs.
   - The industrial sector was significantly weighted toward heavy industries (metallurgy, cement, bulk chemicals) that consume high quantities of energy and materials. Typically, the industrial sector in OECD countries is 30–40% of gross domestic product (GDP), 55% for services, and less than 10% for agriculture. In the Warsaw Pact countries, the industrial sector was 35–65% of GDP, 20–50% for services, and 5–35% for agriculture.
   - Most factories were based on inefficient, energy-intensive, and antiquated production methods. For example, industrial energy intensity is five times higher in Poland than in the United States (Perkins, 1991).
   - Many countries relied on indigenous coal, which was frequently of low-quality grades with high sulfur content.
   - Emphasis on heavy industry was frequently at the expense of public environmental infrastructure (e.g., sewage and waste treatment plants).
2. Government did not enforce their policies. Large state enterprises dominated individual regions and were often the only employer. Thus, these facilities were impervious to pollution fines or charges and had no incentive to install pollution control devices.
3. The public was voiceless and environmental data was considered a state secret. Environmental movements frequently play powerful roles in the Western democracies and stimulate environmental initiatives (see Chapter 1, History and Development).

Ironically, with the fall of Communist rule in these countries, environmental issues are temporarily submerged in the chaotic transition to market economies. As real GDP falls and unemployment rises, environmental issues have become secondary.

However, this is probably a temporary condition, because environmental issues can be efficiently integrated into industrial restructuring (e.g., new facility construction with pollution abatement equipment).

Internationally, both Western Europe and the United States have a significant interest in environmental improvements in the former Soviet Union and its satellites, because:

- Transfrontier sources of pollution (e.g., acid rain, air pollution, greenhouse gases, and industrial waste water) affect Europe.
- The successful economic and political integration of the former Warsaw Pact countries into the worldwide global market economy is in everyone's interest.

The opportunities and problems are obvious; however, there is no clear roadmap indicating the correct path.

ENVIRONMENTAL POLICY IN THE AMERICAS: UNITED STATES, CANADA, MEXICO, AND SOUTH AMERICA

United States

The United States is one of the major innovators and contributors to global environmental affairs. Specific United States regulations and practices are discussed in Chapters 3, United States Legal and Legislative Framework; 8, Hazardous Wastes; and 9, Health and Safety Training for Hazardous Waste Activities. This discussion is directed toward the United States' role in general international environmental affairs. The United States' role in the North American Free Trade Agreement (NAFTA) and the General Agreements on Trade and Tariffs (GATT) is discussed in Chapter 2, Economic and Ethical Issues.

As the world's largest industrial power, the United States occupies a unique and powerful role in international environmental affairs. However, this role is highly dependent on the political climate prevailing in United States domestic affairs. The Clinton administration has assumed a more prominent role in international forums.

For example, under the Clinton administration, in June 1993 the United States signed the UNEP Biological Diversity Convention negotiated at the 1992 Earth Summit in Rio. The Bush administration had declined to sign this accord, in part because of opposition from the biotechnology industry over complex issues involving licensing and patent protection.

The Clinton administration is faced with a variety of international environmental issues that fall into the following categories:

- Implementation of Agenda 21—sustainable development.
- Implementation of the UN Framework Convention on Climate Change and the possible institution of domestic tax increases on energy sources (i.e., carbon or BTU (British thermal unit, a measure of heat content) taxes versus an increase in the federal gasoline tax. In the 1993 budget, carbon taxes were rejected in favor of a modest increase in gasoline taxes.
- NAFTA and GATT issues—see Chapter 2, Economic and Ethical Issues.
- Arctic protection agreements—specifically directed toward radioactivity from sunk nuclear submarines, waste, and plutonium extraction.
- Regulations that tighten controls on ozone-depleting chemicals, specifically, possible phaseout of methyl bromide, a widely used fumigant. The EPA has a proposed rule that would phase out production of this chemical by the year 2000. Opposition from the U.S. Department of Agriculture and farm lobbyists is expected to be substantial.
- Renewed funding and reform of the Global Environment Facility (GEF)—GEF is tasked to financially aid developing countries with compliance to international environmental agreements. The United States is a major financial contributor and has withheld renewed funding pending management review and reform of GEF activities.
- Revamping of the role of the United States Agency for International Development (U.S. AID). Specific planned actions include the following:
  1. Transfer of "green" technology from United States companies to foreign countries.
  2. Mandatory review of potential environmental impacts for all foreign aid projects.
  3. Closer ties between human rights issues and environmental issues. Chapter 2, Economic and Ethical Issues, gives a complete discussion of this topic.

Although the relative strength of the domestic U.S. economy will substantially influence the magnitude of U.S. efforts, it is clear that a major policy shift has begun that will substantially impact global environmental policies for the remainder of the 20th century.

Canada

Canada is a parliamentary democracy with division of powers between federal and provincial (state) governments. Canada comprises 10 provinces and 2 territories. The federal government has clear jurisdiction over transboundary air and water, fisheries, coastal waters, and interprovincial and international matters. The provincial governments have jurisdiction over land and resources; therefore, site remediation (cleanup) levels are established by provincial governments.

Organizationally, Canadians devised a dual system in which provincial governments have established environmental departments in addition to the federal department, Environment Canada. To develop regulatory consistency, each provincial environmental minister is a member of the Canadian Council of Ministers of the Environment (CCME). The CCME develops consensus documents and guidelines; however, in some areas, the federal government can impose baseline national standards, providing federal decisions are not ruled constitutionally invalid by the Supreme Court of Canada.

Federal rules. The transportation of hazardous wastes in Canada is governed by the Transportation of Dangerous Goods (TDG) Act and Regulations (TDGR, 1985) (see also Chapter 8, Hazardous Wastes). TDG regulations define *waste* as "a product or substance intended for disposal." Hazardous waste is classified as any of the following:

- any discarded goods listed in a schedule of dangerous goods
- any industrial waste listed in a schedule
- any waste mixtures or solutions, not fully specified, that have hazardous properties described by specified criteria.

Table 4–H presents the nine hazardous classes of materials. Further amendments to the TDGR include the addition of U.S. EPA-based Leachate Extraction Procedures. In general, the maximum concentrations in waste extract (in units of mg/l) are 100 times the corresponding Canadian drinking water guidelines (Health and Welfare Canada, 1989). Although all provinces use the TDGR regulations, several provinces (Alberta, Manitoba, Ontario, and Quebec) have additional waste categories covering (1) severely toxic wastes, (2) pathologic wastes, (3) acute hazardous waste chemicals, and (4) wastes containing PCBs greater than 50 parts per million (ppm).

In 1988, the Canadian Environmental Protection Act (CEPA) brought under a central umbrella four previously freestanding major environmental acts: (1) Canada Water Act, (2) Clean Air Act, (3) Environmental Contaminants Act, and (4) Ocean Dumping Control Act.

Additional regulations developed under CEPA control the import and export of hazardous waste. For example, under CEPA rules, all overseas exports of PCBs were banned as of August 1, 1980. Two subsequent policy directives initiated under CEPA were the Domestic Substances List (DSL) and the Priority Substances List (PSL).

The DSL attempts to catalog all substances currently in use in Canada. Any substance not on the DSL requires formal notification of Environment Canada for importation or manufacture. Thus, the burden of information is clearly placed on the importer or manufacturer to adequately describe and characterize any non-DSL substances.

The PSL describes substances that must be reviewed both for environmental toxicity and the possible need for federal regulation. For example, both PCBs and polychlorinated dibenzofarane are on the initial PSL of 44 substances.

Treatment, disposal, and cleanup criteria. Currently, Canada has limited capacity for off-site treatment or volume reduction of hazardous wastes. Therefore, extensive and prescriptive regulations governing the treatment and disposal of hazardous waste are not in effect. The existing individual facilities are operating under permits or licenses granted by specific provincial government.

Not surprisingly, a variety of soil cleanup criteria have been discussed and proposed by several provincial governments. These soil criteria are similar to the previously described Dutch A, B, C system. Recently, more generic soil criteria approaches have been developed so that site cleanup does not proceed on a completely ad hoc basis (CCME, 1991a). The CCME approach, in general, favors site-specific criteria derived from risk assessment techniques unless the generic criteria are clearly applicable (CCME, 1991).

Overall, the Canadian system is a two-pronged approach based on broad federal leadership and specific provincial government direction. The overall approaches are based on federal provincial consensus guidelines developed by the CCME. Soil cleanup criteria have initially been based on the Dutch A, B, C criteria; however, cleanups are moving toward site-specific risk-based approaches (see Chapter 12, Risk Assessment). Additional details of the Canadian system and approaches can be found in a variety of

Table 4–H. List of Toxic Substances—Schedule I of the Canadian Environmental Protection Act

Description	Type of Regulation Applicable
Chlorobiphenyls that have the molecular formula $C_{12}H_{10-n}Cl_n$ in which n is greater than 2	Prohibited commercial, manufacturing, or processing uses. Maximum concentrations in products. Maximum quantities and concentrations that may be released into the environment.
Dodecachloropentacyclodecane	Prohibited commercial, manufacturing, or processing uses.
Polybrominated biphenyls that have the molecular formula $C_{12}H_{10-n}Br_n$ in which n is greater than 2	Prohibited commercial, manufacturing, or processing uses.
Chlorofluorocarbon: totally halogenated chlorofluorocarbons that have the molecular formula $C_nCl_xF_{(2n+2-x)}$	Prohibited commercial, manufacturing, or processing uses.
Polychlorinated terphenyls that have a molecular formula $C_{18}H_{14-n}Br_n$ in which n is greater than 2	Prohibited commercial, manufacturing, or processing uses.
Asbestos	Limited atmospheric releases from mines and mills.
Lead	Limited atmospheric releases from secondary lead smelters.
Mercury	Limited atmospheric releases from chloralkali mercury plants.
Vinyl chloride	Limited atmospheric releases from vinyl chloride and polyvinyl chloride plants.

CCME publications (CCME, 1991a and 1991b) and in an excellent review article by Hrudey (1992).

Mexico

The debate about the North American Free Trade Agreement (NAFTA) illustrates the relationship between environmental practices and economic/trade issues. Cognizant of this relationship, the Mexican government has undertaken a variety of environmental initiatives. In addition, there is growing public environmental activism in Mexico that will continue to push for stronger environmental regulations. In the mid-1980s, former President Miguel de la Madrid elevated the head of the environment agency to cabinet rank in a new ministry, the Secretariat of Urban Development and Ecology (SEDUE), renamed in June 1992 as Secretariat for Social Development (SEDESOL).

In addition, de la Madrid improved environmental management efficiency, revamped existing environmental law, and raised public awareness of environmental issues (Mumme & Sanchez, 1992). De la Madrid's elected successor, Carlos Salinas de Gortari, has continued to stress environmental issues and has clearly articulated this as a basic priority of his government. Salinas has stressed a number of pressing environmental issues: (1) Mexico City's general environmental predicament, particularly air pollution; (2) deterioration of the Lacandon Forests; and (3) degradation of the Coatzacoalcos and Lerma River basins. In addition, Salinas has initiated active public participation and cooperation of the private sector in financing solutions (Mumme & Sanchez, 1992).

Mexican environmental management. Since March 1, 1988, environmental law and regulatory rules and requirements in Mexico have undergone significant revision:

- Detailed administrative and regulatory initiatives have been established and enforced by the National Ecological Institute and the attorney general's Office for Environmental Protection.
- SEDESOL's role in enforcement and compliance issues has increased, e.g., requiring an operating license when a fixed emission source is present on site.
- Environmental impact assessments are required for all federal public works, mining and tourist developments, and sanitary facilities.
- Registration (a manifest) is required if hazardous wastes are generated.
- Development of state and local ordinances to reinforce and supplement federal legislation has been encouraged.
- Registration of all waste water discharges is required.
- SEDESOL has been directed to issue technical guidelines and regulatory edicts.

- Emphasis is on prevention and control of atmospheric contamination and hazardous waste with required registration and permitting system. Risk analysis may be required.
- Public participation is encouraged.

In June 1992, two new environmental agencies were made responsible for developing and enforcing Mexican environmental legislation: (1) National Ecology Institute (INE—Instituto Nacional de la Ecologia) and (2) the Environmental Protection Enforcement Agency (PFPA). The government has two main policy initiatives: (1) increased emphasis on directed regulations and (2) program prioritization.

Mexican environmental reforms are still highly dependent on continued economic recovery, divestiture of state industries, liberal foreign aid, and technical assistance. The passage of NAFTA will spur continued Mexican environmental development.

As American and Canadian companies develop Mexican plants and facilities, the pressure on these businesses will be to conform to their home country's environmental standards. Just as European countries have acted to prevent transfrontier environmental imbalances regarding hazardous wastes, it is quite probable that NAFTA side environmental agreements will require similar standards. Public scrutiny over existing *maquiladora* industries and their potential health impacts has already occurred and will continue (Krieger, 1992 and 1993). Mexican environmental policy is in a formative state but will be significantly influenced by its two large northern neighbors.

South America

In general, South and Latin American environmental laws, regulations, and enforcement have significantly lagged behind North American and European efforts. As urbanization and industrialization have occurred, substantial air, water, and waste disposal issues have been raised by both international aid agencies and national level political organizations. Increasing pressure on logging, mining, and oil/gas development projects has increased, especially with the UNEP's Agenda 21 and emphasis on species preservation/biodiversity agreements. Environmental impact statements and sensitivity to indigenous (Indian) populations are now the norm rather than the exception. Table 4–I presents a brief synopsis of environmental activities in several South American countries.

ENVIRONMENTAL POLICY IN ASIAN/PACIFIC COUNTRIES

The environmental policies of Asian/Pacific countries reflect the full spectrum of possibility and opportunity:

- presence of an economic superpower (Japan) that is beginning to develop an in-house regulatory environmental infrastructure while aggressively marketing world-class global pollution abatement and control technologies
- transition from centrally planned Marxist to market-based economics (China)
- emergence of newly industrialized countries (e.g., Singapore, Taiwan, Malaysia, Indonesia, Thailand, and South Korea)
- established democracies of Australia and New Zealand that have environmental infrastructure that is similar to Canadian and United States models.

Japan

To ameliorate the damage to human health caused by its rush toward economic growth in the 1950s and 1960s, Japan has begun to institute a comprehensive environmental legislative system. Because the legislation is focused on protecting the health of the workers necessary to power growth, there are notable gaps and weaknesses in the legislation, in particular with regard to the natural environment and public involvement.

Administrative structure. The Japan Environmental Agency (JEA) is the principle national-level executive body responsible for environmental administration. It is part of the prime minister's office and comprises four bureaus and two departments (Figure 4–1). Advisory committees, chaired by the prime minister, including the Central Council for Environmental Pollution Control, are composed of the heads of other administrative bodies and assist in developing environmental policy. The Health Damage Compensation Grievance Board, which is intended to settle environmental disputes, and a number of issue-specific councils and institutes complete the environmental administration (Figure 4–1).

Many other ministries and agencies, 19 in total, have jurisdiction over specific aspects of environmental legislation. These ministries often have agendas different from the JEA's, and coordination is difficult. The JEA is perceived as politically weak and has often seen its efforts stymied by the economic

Table 4–I. Environmental Rules and Regulations—South America

Columbia

- Inderena is lead environmental agency with possibility of establishing formal Ministry of the Environment.
- Binational Agreement with Venezuela for enforcement of the Montreal Protocol on Substances that deplete the ozone layer.
- Establishment of high level committee on oil for granting licenses and conducting environmental impact statements.
- Committee on coal established with similar mission of oil committee.
- Increased international and United States funding with environmental budget to exceed 70 million (United States) in 1993.
- Ratification of UN Framework Conventions on climate change and biodiversity.

Chile

General Environmental law is presently being considered by the government. Current relevant legislation includes:

- State Political Constitution, Article 19, No. 8, that assures the right of every person to live in an environment free of pollution.
- Law No. 3.133, dealing with the neutralization of wastes from industrial plants.
- Supreme Decree No. 185 of the Ministry of Mining, 1991; Supreme Decree No. 4 of the Ministry of Health, 1992; No. 211 of the Ministry of Transportation and Telecommunications, 1991; and Supreme Decree No. 240, 1990, of the Ministry of National Property.
- Supreme Decree No. 349, 1990, of the Ministry of the Interior, under which the Special Commission for the Decontamination of the Metropolitan Region was created.

Environmental Agencies/Ministries:

- CONAMA—(Comission Nacional de Medio Ambient)—Coordinating Environmental Agency
- Ministry of Health—Human Health, Sanitation, and Environmental Hygiene
- Ministry of Agriculture—Terrestrial Resources and Land Use
- Ministry of Economy—Fisheries
- Ministry of Public Works—Water Supply and Discharge
- Ministry of Transportation—Pollution from Mobile Sources
- Ministry of Mining—Impacts from Mining Operations

Brazil

Environmental Laws:

- Governing Environmental Law (#6938) enacted on August 31, 1981, establishes national policy on environmental matters.
- Laws 7804 and 8028 of July 18, 1989, and April 12, 1990, respectively, amend Law 6938.

Environmental Agencies:

Federal

- SEMA—Secretaria Especial do Meio Ambiente
- IBAMA—Instituto Brasileiro do Meio Ambiente e Recursos Renovaveis
- CONAMA—Conselho Nacional do Meio Ambiente

State

- CETESB—Companhia de Tecnologia de Saneamento Ambiental (São Paulo)
- FFFMA—(Rio de Janeiro)
- FATMA—(Santa Catarina)

agendas of the powerful Ministries of Transportation and Construction, International Trade and Industry, and the Economic Planning Agency.

Japan has a three-tier legislative system, with the national government overseeing prefectural and local governments. The prefectures may introduce legislation equal to or stricter than the national legislation. The local governments often have responsibility for enforcement of prefectural and national legislation.

Environmental legislation. The Basic Law for Environmental Pollution Control was enacted in 1967 to "promote environmental policies, thereby ensuring the protection of the peoples health and the conservation of their living environment." The law establishes the basic policies and assigns responsibilities for legislation. Based on this law, a series of laws related to air pollution, water pollution, soil pollution, noise, vibration, land subsidence, and odors was enacted. The framework of the environmental legislation as established in the 1970s is shown in

Figure 4–2. The laws have been updated while maintaining the original policy of human health protection.

Standards for environmental regulations are established by the national and prefectural government based on local conditions. Standards for pollutant emissions have been set considering existing land use and historical conditions. Compliance with environmental legislation is achieved through restrictions on discharges, and enforcement is usually achieved through the use of "administrative guidance," or if necessary, civil or criminal actions.

Notable differences between Japan's environmental law and that of current environmental thinking in the United States and Europe include: (1) lack of legal requirements for environmental impact assessments, (2) lack of protection for the natural environment except where human health is involved, and (3) lack of public involvement in policy making. Japan has no current legislation incorporating the principle that polluters must pay. In addition, the public

94

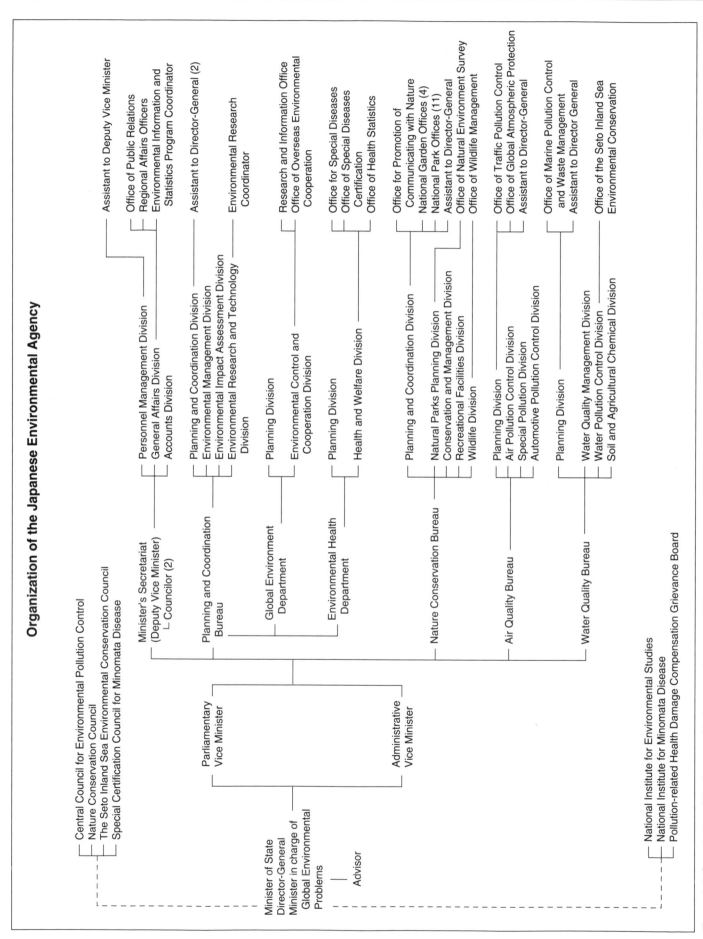

Figure 4–1. Organization of the Japanese Environmental Agency (JEA).

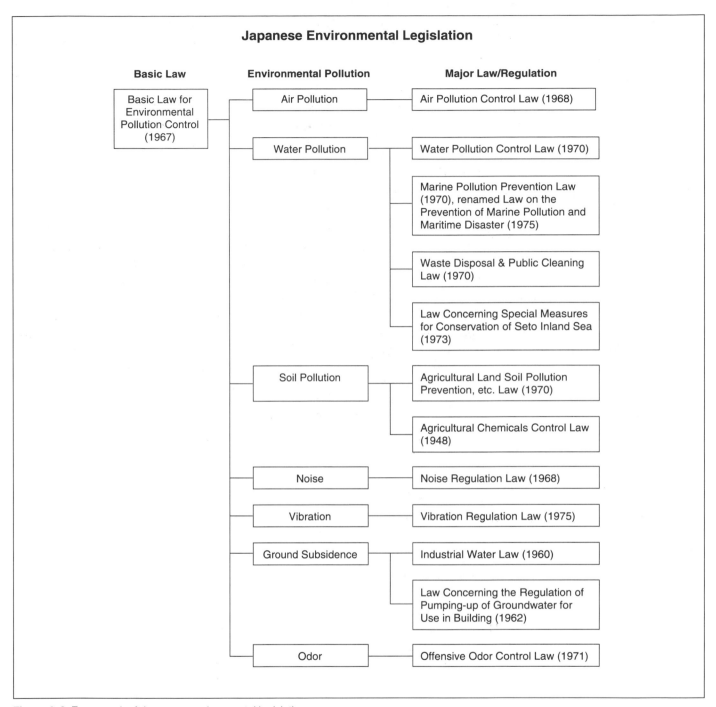

Figure 4–2. Framework of Japanese environmental legislation.

does not have right-to-know privileges regarding industrial activities. These policies are probably due to close government and industry cooperation but are aided by Japan's preference for nonlitigation dispute solving.

Legislative trends. The focus of U.S. and European environmental legislation evolved from the "protect" concept of the 1970s, through cleanup programs in the 1980s, to the present focus on further prevention of environmental damage. However,

Japanese bureaucratic battles appear to have prevented the Japanese environmental philosophy from progressing beyond the 1970 health protection focus.

In an attempt to catch up with the 20 years of philosophic progress elsewhere, a new Basic Law on the Environment is currently being debated in the Japanese Diet. Contentious issues in the proposed Basic Law are the JEA plans for environmental impact statements and audits to monitor industries' environmental performance. Previous attempts in 1984 by the JEA to introduce environmental impact statements were unsuccessful; in addition, the current pessimistic Japanese business climate may also decrease the likelihood of new regulations that could adversely affect home business/industry.

Significantly, the Basic Law reflects a shift in policy away from direct regulation; instead, economic instruments are used for environmental control. The JEA favors a form of environmental tax to protect the environment and encourage energy efficiency. However, the extremely influential Ministry of International Trade and Industry (MITI) favors its own system. MITI has launched a series of research and development projects to develop new environmentally oriented industrial technologies as part of the Japanese program known as "The New Earth 21." The New Earth 21 is a global strategy to combat the accumulation of CO_2 and other greenhouse gases. Table 4–J lists the essential features of this program.

As outlined, this program could significantly position Japan as a dominant supplier and source of environmentally oriented industrial technologies. Global markets for pollution abatement goods and services are estimated at $200 billion (U.S. dollars) and are expected to rise during the remainder of this century (Stevens, 1992). "End-of-pipe" equipment accounts for three-quarters of this $200 billion total. The world market for environmental equipment and services is forecast to increase at an average annual rate of 5–6% per year to $300 billion by the year 2000 (OECD, 1992). Obviously, both commercial and political forces come into play in MITI's interest in environmental issues.

Because Japan has expressed its intention to be a global environmental leader at the Rio Earth Summit, there will be internal and external political pressure on the government to fulfill its environmental commitment. Consistent with past practice, it should also be expected that Japan will strongly develop, market, and sell commercial applications for pollution abatement equipment and waste minimization process control. Whether these technologies will be applied domestically is uncertain; however, in other industrial sectors, the Japanese have clearly demonstrated the ability to catch up and subsequently establish a dominant leadership role.

China

From 1955 to 1980, the Chinese five-year economic development plans were void of provisions for environmental protection. Similar to the former Soviet Union, China followed a doctrinaire course of rigid central planning that did not provide incentives for efficient use of energy, resources, or new industrial technologies. In 1978, a provision for environmental protection was added to the Chinese Constitution. In a 1989 extension of this law, additional environmental and ecosystem degradation protections were incorporated into economic and social planning. Pollution control and environmental management were also added during the 1980s (Perlack & Russell, 1991):

- Marine Protection: 1982
- Water Pollution Control Act: 1984
- Air Pollution Prevention and Control Act: 1987
- Noise Pollution Prevention and Control: 1989
- Solid Waste Control Act/Hazardous Waste Management Regulation Act: 1992
- Regulations on the Management of Nature Reserves: 1993.

Although the Chinese have the administrative and policy framework in place, it is crucial to ask whether these various environmental directives are

---

Table 4–J. Japanese Action Programme for the 21st Century: "The New Earth 21"

Goals
- Promotion of energy efficiency, the conservation of energy and other resources;
- Development and incentives for the use of clean energy sources;
- Development of future-generation energy technologies;
- Enhancement of greenhouse gas sinks; and
- Development and transfer of environmentally oriented technologies.

MITI has established a new research and development center, The Research Institute of Innovative Technology for the Earth (RITE), to provide a strong institutional base for developing the new environmentally oriented industrial technologies. RITE will work closely with other major research and development institutions, both domestic and foreign.

Research Projects
- Development of full carbon cycle technology (e.g., CO_2 fixation and reutilization);
- Development of environmentally benign materials (e.g., substitution of CFCs, biodegradable plastics); and
- Development of environmentally benign production processes (e.g., bio-reactor systems, biological hydrogen production technology, separation, and reutilization technology for metal scraps).

enforced in a systematic fashion. Environmental conditions in the former Soviet Union illustrate the problem of administrative framework uncoupled with true economic incentives.

At this time, the true Chinese situation is difficult to accurately assess. China has clearly made dramatic economic changes during the last 10 years and now has a mixed economy of both market-responsive industries and old, money-losing, energy-inefficient (polluting) heavy industries. China has embraced the "polluter pay" principle and has developed an extensive network of 5,000 stations for monitoring atmospheric emissions, surface water, acid rain, radiation, and noise pollution (Perlack & Russell, 1991). Theoretically, data are available for enforcement; however, the actual effectiveness of the database for environmental management is unclear. Overall, a clear indication of Chinese environmental intentions will be the aggressiveness with which energy price reform is undertaken.

Worldwide experience demonstrates that freeing energy prices has a dramatic impact on energy consumption and subsequent source emissions. Currently, China has a large inefficient government-owned industrial sector that uses vast quantities of coal. In addition, the urban household sector relics heavily on energy sources derived from direct coal burning. When China diversifies and modernizes its primary energy structure, this will clearly reduce pollution and create significant energy savings. Chinese environmental rules and directives are present and available. The last five years have demonstrated that the political will is present to make the transition from a Marxist, centrally planned economy to a free-market system. Environmental changes and improvements are likely and will undoubtedly accompany the economic and political changes that are occurring in the world's most populous country.

Newly Industrialized Countries (NICs)

The region composing the Association of South East Asian Nations (ASEAN) is one of the fastest growing regions in the world. Extensive manufacturing facilities exist throughout the region. Although ASEAN nations have welcomed the investment and job formation, the serious environmental consequences of rapid development are becoming apparent. Four of the six ASEAN members' environmental structures will be reviewed: Singapore, Malaysia, Indonesia, and Thailand. Brunei and the Philippines are excluded. In addition, a brief summary of South Korean policies will be discussed.

Singapore. Singapore is a city-state with a strong active government. In 1983, the Ministry of the Environment (MOE) received administrative control over all environmental affairs. Recently, the government announced a "Singapore Green Plan" formalizing its commitment to environmental protection with a comprehensive suite of liquid waste, air emissions, noise, and toxic and hazardous waste legislation (Austin & Koontz, 1992). Singapore's plan is to establish environmental regulations that are on par with those of the United States and European Community. All indications are that Singapore will be successful in this goal and will become the Southeast Asian leader in environmental affairs among the NICs.

Taiwan. In 1986, Taiwan established an Environmental Protection Administration (EPA). Since then, major waste-related environmental laws have been passed:

- Toxic Substances Control Act (1988)
- Waste Disposal Act (1988)
- Industrial Waste Storage and Treatment Method and Facility Control Act (1989)
- a variety of air and water pollution control laws.

Like Japan, Taiwan is directed toward "end-of-pipe" solutions and applications rather than enforcement mechanisms. Although government funding for environmental projects is significant, a slowdown in private industrial investment is causing decreased enforcement pressure. Overall, Taiwan is far more likely to stress economic behavioral incentives rather than directed regulatory enforcement actions. The development of an environmental services sector, with emphasis on pollution abatement equipment, is clearly a government goal. Currently, solid waste management is a critical problem, and 21 incinerators are planned or under construction island-wide.

Malaysia. Environmental regulations are promulgated at the federal level through the Department of the Environment (DOE). The DOE is part of the Ministry of Science, Technology, and the Environment. Regional governmental offices are responsible for enforcement of the regulations.

Regulations are based on the Environmental Quality Act of 1974 and its various amendments and revisions. In June 1991, an Environmental Law Review Committee was established to identify existing gaps in present regulations and propose improvements. The major government focus is directed toward economic growth, and environmental protection will probably remain a secondary concern.

Indonesia. In late 1990, the Environmental Management Agency (BAPEDAL) was established to monitor, coordinate, and enforce environmental legislation. This agency is responsible for the 1989 Clean Streams Program, air pollution programs, domestic wastes, hazardous wastes, and coastal pollution programs. Environmental impact assessments are required for all existing or planned activities that have the potential for adverse effects. Despite the national direction of the BAPEDAL, provincial and local governments have the primary responsibility for pollution enforcement and management. Indonesia is highly dependent on bilateral and multilateral assistance programs for technical expertise.

Thailand. Thailand, like many of its Southeast Asian neighbors, has experienced significant unrestricted industrial growth coupled with severe environmental degradation. Although the Thai government enacted a number of significant environmental laws in 1992, there is a severe mismatch between the number of industrial plants (51,000) and the number of government employees (356) dedicated to monitoring enforcement (Austin & Koontz, 1992). Currently, the focus of government spending is on water quality and wastewater treatment systems, forest preservation, and solid waste disposal systems and upgrades.

South Korea. South Korea is similar to Japan in the structure of government-industry relationships and in its drive toward economic growth. In August 1990, a comprehensive set of six new environmental laws were enacted. Concomitantly, the Environmental Administration was upgraded to the Ministry of Environment (MOE). In February 1991, the Basic Law for National Environmental Policy came into effect. In addition, laws covering air and water quality, noise and vibration, hazardous chemical management, and environmental dispute resolution were passed. As discussed with the other Southeast Asian countries, lack of trained manpower and sophisticated equipment continue to restrict the effectiveness of the Korean program.

Australia

Australia is a federation comprising six states. In general, commonwealth powers are concurrent with state powers; however, in the absence of national regulation, each state can regulate activities within its jurisdiction. Australian environmental law is statutory and operates on two levels: (1) licenses must be obtained for activities that are likely to cause pollution, and (2) strict liability standards apply for certain conditions (e.g., contaminated sites). In general, specific environmental statutes fall into three categories: (1) environmental planning and protection; (2) legislation intended to protect the natural, built, or cultural environment; and (3) legislation controlling the development of natural resources. A more recent trend is the introduction of industry-specific controls for categories such as forestry and sandmining.

In many respects, the Australian environmental framework is quite similar to the Canadian system (i.e., a national framework with specific state-by-state regulatory discretion and enforcement). An extensive series of environmental rules and regulations has been passed. Particular interest in Australia is focused on contaminated sites. In general, Australia has leaned toward development of the multifunctional land use approach. Site-specific risk assessments are permitted and are increasingly common. Australia is a signator to the major UNEP accords and treaties, and local environmental movements play an active role in national and state policy formation.

ENVIRONMENTAL POLICY IN DEVELOPING NATIONS: PROGRESS AND PERIL

At the 1992 Earth Summit, Maurice Strong, secretary-general of the UNCED said, "No place on the planet can remain an island of affluence in a sea of misery. We're either going to save the whole world or no one will be saved. We must from here on all go down the same path." Despite this clarion call, the developing nations have frequently become an unwitting laboratory where the triad of uncontrolled population growth, unequal resource access, and decreased quality and quantity of renewable resources has been mixed with devastating potential. A variety of global initiatives to address these problems have been introduced by both the United Nations and other international lending agencies; however, the solutions are long-term while the current potential social instability is immediate (see also Chapter 14, Global Issues).

The growing scarcities of renewable resources can lead and substantially contribute to both social instability and civil strife (Homer-Dixon et al, 1993). In most developing countries (nations with a per capita GNP equivalent to less than $4,000 U.S. per year), population growth has far exceeded economic growth, and substantial falls in per capita income have occurred. The free fall in income has been combined with a significant financial outflow for foreign debt service. These debts totaled $1.35 trillion in early 1992; in addition, over one-quarter of the cash outflow was directed toward payment for past energy projects (Lenssen, 1993).

Because most developing countries have natural resource-based economies (e.g., soils, forests, species, parks), the unfortunate trend has been toward rapid short-term resource use at the expense of long-term sustainable development (Table 4–K). In addition, severe shortages of water resources are projected. Table 4–L illustrates the current available and future projected water resources for a variety of developing nations. Similar bleak figures are available for forestry loss (MacNeil, 1989; Gupta, 1993):

- Ethiopia   Forest cover, 1950: 30%
             Forest cover, 1990: 1%

Table 4–K. Resource Dependence in the Developing Nations

Country	Exports of Primary Products as a Percentage to Total Exports (1986 Figure)
India	36
Sri Lanka	59
Ethiopia	99
Ghana	98
Kenya	84
Senegal	71
Tanzania	83

Source: MacNeil, 1989, and World Bank, 1988.

Table 4–L. Water Resources (Cubic Meters per Person per Year)

Country	Year 1990	2025*
Algeria	750	350
Ethiopia	2,550	950
Haiti	1,600	950
Kenya	600	200
Nigeria	2,650	1000
Somalia	1,500	600
South Africa	1,400	800
Tanzania	2,800	900
Tunisia	550	350

1000 cubic meter/person/year is considered the mininimum amount necessary for an industrialized native.

* Projected available cubic meters/person/year based on projected population growth and utilization patterns.

Source: Based on data from Homer-Dixon et al, 1993.

- India   Past forest cover: 20–25%
           Forest cover, 1990: less than 7%.

The crushing level of economic deprivation led African ecologist Mohamed Suliman to remark, "The North is now mainly concerned about the environment. Developing countries are concerned about development, less about the environment" (quoted in *BNA International Environment Daily,* February 22, 1993). This emphasis on rapid development via direct cash infusion is illustrated by the trade in hazardous wastes from industrialized nations to developing nations.

Greenpeace estimates that more than 3 million tons of wastes were shipped to the developing nations between 1986 and 1988 (Vallelte, 1989). The cost differential between the "exporting" and "importing" countries can be substantial: an average of $250 per metric ton, long-haul transportation included (Asante-Duah, 1992). Several African countries have reportedly negotiated multimillion dollar industrial waste deals. Because of public outcry and widespread press reporting, there has been considerable pressure to eliminate the transboundary trade of hazardous materials to the developing nations.

The Organization of African Unity (OAU) in 1988 adopted a resolution barring any waste shipments from the industrialized countries. Finally, the 1988 Banko Convention on the Ban of Imports into Africa and the Control of Transboundary Movement and Management of Hazardous Wastes specified strict controls and policies:

- a ban on import of hazardous and radioactive wastes
- hazardous waste generation audits
- broad liability on hazardous waste generators
- a ban on the import of hazardous substances that have been banned, cancelled, or refused registration, or voluntarily withdrawn in the country of manufacture for either human or environmental reasons
- close supervision of any transfer of polluting technologies to Africa.

Similar conventions (Lomé IV Convention) have also involved African, Caribbean, and Pacific countries.

In March 1994, European Union countries, Canada, Japan, and Australia agreed to ban all exports of toxic wastes to developing countries, including exports intended for recycling. These agreements offer hope that developing nations will not continue to be a dumping ground for hazardous waste. However, the basic problems of moving from short-term natural resource use to long-term sustainable productivity and replenishment remain. Without fundamental political and economic reform, the environmental impacts of population growth and rapid nonsustainable use of natural resource capital will be severe. Western aid is essential, but less-developed nations must find the political and economic leadership required to avert future social and environmental disasters.

SUMMARY

International rules and regulations are in a constant state of flux and development. As discussed in Chapter 2, Economic and Ethical Issues, and Chapter 14, Global Issues, there is an ongoing tension between the current emphasis on sustainable development and the need for environmental policies that do not undercut economic development. During the 1970s and 1980s, there was substantial debate over the emergence of a North-South Split (a highly developed North—United States, Canada, Western Europe, Japan) and the less-developed South (Latin America, South America, Subsahara Africa). A similar split, described with environmental terminology, appears to be possible over sustainable development issues affecting rainforests, biodiversity, and timber and other natural resources that are now under intensive development in the "Southern" countries. How this issue can be resolved is unknown.

Long-term international environmental trends and goals are clearly driven by the 1992 Earth Summit in Rio, as countries come to grips with the implications of Agenda 21 and other conventions affecting biodiversity, greenhouse gases, and ozone. The European battle over command and control regulations versus economic incentives will clearly face its sternest test as the EU Member States struggle with balancing the political pressure for environmental initiatives against the long-term economic implications of these measures. The current European policy initiatives for recycling packaging materials and other consumer products are an excellent illustration of this tension between business and government.

REFERENCES

ANRED. Agence Nationale Pour la Récupération et l'Elimination des Déschets. Data published in 1989 by ANRED and cited by Brand in his 1993 paper.

Asante-Duah DK, Saccomanno FF, Shortreed FH. The hazardous waste trade: Can it be controlled? *Environ Sci Tech* 26(9):1684–1693, 1992.

Austin I, Koontz J. Asian countries balance growth with environmental protection. *Pacific Press*:35–38, November 1992.

Brand EC. Hazardous waste management in the European Community: Implications of 1992. *Sci of the Total Environ* 129:241–251, 1993.

CCME. Interim Canadian environmental quality criteria for contaminated sites. *Report CCME EPC-CS34*. August 1991a.

CCME. The national contaminated sites remediation program. *1990–1991 Annual Report,* 1991b.

Commission of the European Communities. Official Journal of the European Communities. No. L. 196, 1967.

Green E. Poisoned legacy: Environmental quality in the Newly Independent States. *Environ Sci Tech* 27(4):590–595, 1993.

Gupta BN. The Rio Conference: A view from India. *Environ Sci Tech* 27(2):212–218, 1993.

Health and Welfare Canada. *Guidelines for Canadian Drinking Water Quality,* 4th ed. Ministry of Supply and Services Catalogue, No. H48–10/1989E, 1989.

Homer-Dixon TF, Boutwell JR, Rathjeng GW. Environmental change and violent conflict. *Sci Am* 268(2):38–47, 1993.

Hrudey, SE. Hazardous waste management approaches in Canada. *Water Sci Tech* 26(1–2):1–10, 1992.

Jans JH. Legal grounds of European environmental policy. *Sci of the Tot Environ* 129:7–17, 1993.

Juhasz F, Ragno A. The environment in Eastern Europe: From red to green? *OECD Observer* 181:33–36, April/May 1993.

Junger JM, LeFevre B. Waste management at the European Community level. *Ind Environ* 11:15–17, 1988.

Krieger GR. The investigation of a cluster of neural tube defects in Cameron County, Texas. *OEM Report* 6(12): 89–92, 1992, and editorial response, March 1993, pp. 23–24.

Lenssen N. The new energy equation for developing countries. *Environ Sci Tech* 27(2):220–221, 1993.

MacNeil J. Strategies for sustainable economic development. *Sci Am* 26(3):154–165, 1989.

Mumme SP, Sanchez RA. Profile: New directions in Mexican environmental policy. *Environ Man* 16(4): 465–474, 1992.

Nicolaisen J, Dean A, Hoeller P. Economics and the environment: A survey of issues and policy options. *OECD Economic Studies* 16, Spring 1991.

OECD (Organization for Economic Cooperation and Development). *Industrial policy in OECD countries. Annual Review* 1985, 1990, 1991, and 1992.

Perkins S. Energy policies for Poland. *The OECD Observer* 170:30–34, June/July 1991.

Perlack RD, Russell M. Energy and environmental policy in China. *Ann Rev Energy and the Environ* 16:205–33, 1991.

Rose, J. Civil liability. *Environ Sci Tech* 27(5):784, 1993.

Schlickman JA, McMahon TM, Van Riel N et al. *International Environmental Law & Regulation.* Austin, TX: Batterworth Legal Publishers. 1992.

SNV (Statens Natarvards Verk). The Swedish Government Bill 1990/91:90. A Living Environment—Main Proposals. Solna, Sweden: National Environmental Protection Board, published in English 1991.

SNV. The Swedish Remediation Program. Swedish EPA, 1992.

Stevens C. The environment industry. *The OECD Observer* 177:26–28, August/September 1992.

TDGR. Regulations respecting the handling offering for transport and transporting of dangerous goods and subsequent amendments. Canada Gazette, Part H: 6, Vol. 119, 1985.

Vallelte J. *The International Trade in Wastes—A Greenpeace Inventory,* 4th ed. Washington DC: Greenpeace USA, 1989.

Yakowitz H. Identifying, classifying, and describing hazardous wastes. *Ind Environ* 11:3–10, 1988.

Yakowitz H. Possibilities and Constraints in Harmonizing Definitions of Hazardous Wastes. Presented at the SOTA Workshop. Brussels, 25–26 April, 1989.

Basic Principles of Environmental Science

Gary R. Krieger, MD, MPH, DABT
Mark J. Logsdon, MS
Christopher P. Weis, PhD, DABT
Joanna Moreno

Environmental assessments involving hazardous materials frequently depend on both predictions of how materials move through soil, water, and air and the underlying toxicity of the material. The process of describing the movement of hazardous materials in the environment is known as *fate and transport characterization* or *environmental chemodynamics*. Environmental chemodynamics is the general title given to the study of the transport of chemicals in the environment in terms of: (1) physical-chemical properties that influence transport, (2) factors that influence persistence in the biosphere, (3) partioning or relative ability to move into biota, and (4) the ability to make toxicological and public health predictions based on a material's physical-chemical properties (Thibodeaux, 1979).

Environmental chemodynamics is a complex field and draws extensively on a broad array of multidisciplinary concepts that can be found in the chemistry, physics, hydrology, geology, and engineering fields. The focus of this chapter is on the fundamental concepts that move chemicals between soil, water, and air. The ability to make accurate chemodynamic predictions is a critical skill for environmental managers, because many environmental problems are analyses of the unintended movement of a potential hazardous chemical across the three major interfaces: air-soil, soil-water, and water-air (Thibodeaux, 1979).

The approach of this chapter is to first introduce some of the key physical-chemical terms and concepts that affect intermedia (e.g., soil to air and water to air) transfer of chemicals. In the context of this chapter, *chemical* is used to generically refer to any natural or synthetic compound or single element that is undergoing movement or transformation in the environment. The second section focuses on the three major interfaces involving air, water, and soil. Finally, the third section presents basic concepts of environmental toxicology.

Because many of the terms and concepts in this chapter may be unfamiliar to many readers, a series of vocabulary tables will be presented within each section. The definition of many terms is highly medium-specific (i.e., water versus soil versus air), and there are subtle but important differences when the same general term is used in the context of a specific medium. Whenever possible, however, the general definition will be used unless otherwise noted. Finally, a series of illustrative examples will be presented in each section so that the importance and utility of environmental chemodynamics is better understood.

Overall, these concepts are used to define and predict the concentrations of pollutants at specified exposure points. Exposure points are locations where a biologic receptor contacts a pollutant (Figure 5–1). Exposure point concentrations are used in a wide variety of regulatory risk assessments and compliance procedures; hence, knowledge of the concepts of both the fate-transport process and the toxicology of potentially hazardous materials is critical (Table 5–A). (See also Chapter 12, Risk Assessment.)

PHYSICAL/CHEMICAL PROPERTIES— GENERAL CONCEPTS

To understand the importance of the physical and chemical properties of potentially hazardous materials, consider these common situations: (1) A small or midsized company has both above- and below-ground storage tanks on its property. After a major storm, the owner/operators discover that several of the tanks have leaked both bulk chemicals (various solvents) and petroleum hydrocarbons (gasoline, diesel fuel). (2) A major tanker spill has released thousands of gallons of crude oil into the ocean or other large water body. Aside from the legal reporting requirements (see Chapter 8, Hazardous Wastes), what preliminary chemodynamic predictions can be made? Table 5–B summarizes some of the important questions that will be asked.

Reasonable predictions can be made if four broad categories of information are known:

- physical/chemical properties of the spilled or leaked material
- bioaccumulative properties of the starting material(s)
- physical geography of the original setting, i.e., the soil types on the facility; depth to the water table; proximity of rivers, streams, lakes, or other surface water bodies; general climate conditions such as temperature, wind direction and speed, and general atmospheric stability during and immediately after the release
- volume and type of release, i.e., sudden and accidental or long-term slow release from an old underground storage tank(s).

Each of these categories supplies a piece to the overall puzzle of determining the likely impacts to air, water, and soil.

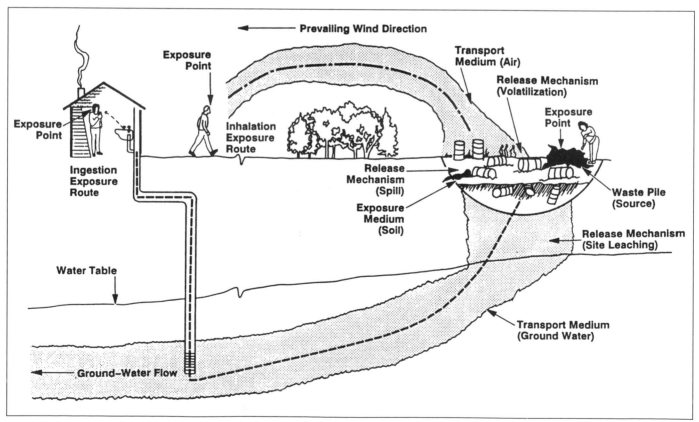

Figure 5–1. Exposure pathways. Source: U.S. EPA, 1984.

Table 5–A. Common Chemical Release Sources at Sites in the Absence of Remedial Action

Receiving Medium	Release Mechanism	Release Source
Air	Volatilization	Surface Wastes—Lagoons, Ponds, Pits, Spills Contaminated Surface Water Contaminted Soils Contaminated Wetlands Leaking Drums
	Fugitive Dust Generation	Contaminated Soils Waste Piles
Surface Water	Surface Runoff Episodic Overland Flow	Contaminated Soils Lagoon Overflow Spills, Leaking Containers
	Groundwater Seepage	Contaminated Groundwater
Groundwater	Leaching	Surface or Buried Wastes Contaminated Soils
Soil	Leaching Surface Runoff Episodic Overland Flow	Surface or Buried Wastes Contaminated Soils Lagoon Overflow Spills, Leaking Containers
	Fugitive Dust Generation/Deposition	Contaminated Soils Waste Piles
	Tracking	Contaminated Soils
Sediment	Surface Runoff, Episodic Overland Flow	Surface Wastes—Lagoons, Ponds, Pits, Spills Contaminated Soils
	Groundwater Seepage Leaching	Contaminated Groundwater Surface or Buried Wastes Contaminated Soils
Biota	Uptake (Direct Contact, Ingestion, Inhalation)	Contaminated Soil, Surface Water, Sediment, Groundwater, or Air Other Biota

Source: U.S. EPA, 1989.

Table 5–B. Important Considerations for Determining the Environmental Fate and Transport

- What are the principal mechanisms for change or removal in each of the environmental media?
- How does the chemical behave in air, water, soil, and biological media? Does it bioaccumulate or biodegrade? Is it absorbed or taken up by plants?
- Does the agent react with other compounds in the environment?
- Is there intermedia transfer? What are the mechanisms for intermedia transfer? What are the rates of the intermedia transfer or reaction mechanism?
- How long might the chemical remain in each environmental medium? How does its concentration change with time in each medium?
- What are the products into which the agent might degrade or change in the environment? Are these products potentially of concern?
- Is a steady-state concentration distribution in the environment or in specific segments of the environment achieved?

Source: U.S. EPA, 1989.

Some of the most important physical properties of a chemical are melting and boiling points, vapor pressure, and water solubility. The melting and boiling points are useful to determine the physical state of the material at the ambient temperature during the release. A sudden chemical release attributable to a severe winter freeze will have different fate and transport characteristics than a chemical spill during a severe summer thunderstorm. The physical state of a chemical can also be predicted based on a knowledge of the ambient temperature conditions and the boiling and melting points of the chemicals. For example, a chemical that is a solid at room temperature but a liquid at 100 F (37.8 C) will behave quite differently if the release occurs during below-freezing conditions or on a hot summer day.

Vapor pressure and water solubility are physical characteristics that can be described under the general category known as *partioning properties*.

Partitioning is the separation of a chemical between two mutually, totally, or nearly immiscible phases in both of which it is soluble (Thibodeaux, 1979). The chemical distributes itself between the two phases (e.g., air-water, water-soil) in fixed proportions that are independent of the quantity of the chemical. The *partition coefficient* is the ratio of the concentrations in the two phases at a constant temperature. The partition coefficient is constant and equals the ratio of the values of the solubility of the chemical in each of the two phases (Thibodeaux, 1979). This relationship can be expressed in general terms as: $C_1/C_2 = K_{1-2}$ where C_1 is the equilibrium concentration in phase 1; C_2 is the equilibrium concentration in phase 2, and K_{1-2} is the partition coefficient between phases 1 and 2. Depending upon the phase 1 and 2 medium, there are several partitioning terms and properties that are critical for the most important and common interfaces. Table 5–C lists and briefly describes some of the important physical properties and partition coefficients in the most important interfaces: gas-liquid, gas-solid (soil), liquid-liquid, liquid-solid (soil), and biological. A sixth potential interface, solid-solid, is considered unimportant for most environmental problems (Thibodeaux, 1979).

In general, if a chemical is present in the environment on one side of an interface, it will eventually move, to some degree, into the other phases. The speed and ultimate concentrations across the interface boundary can be affected by a variety of processes; however, in the absence of any external forces, a spontaneous equilibrium will be achieved.

This physical/chemical "reality" explains why it is possible to find small pockets of pollution that persist many years after the original release event. For example, underground pipe leaks of hydrocarbons can sometimes produce persistent trapped layers of the original material. This would be an example of a liquid-soil interface equilibrium. For environmental practitioners, interface transfer situations, including those across a biological interface, are extremely common; therefore, the major physical interfaces will be presented and discussed.

Air-Water or Gas-Liquid Interfaces

The interaction across a gas-liquid interface is, simplistically, a function of three general conditions: (1) the solubility of the gas in the receiving liquid; (2) the nature of the receiving liquid, e.g., there are significant differences between gas solubility in freshwater and seawater; and (3) a pure liquid-air interface in a closed space such as a partially filled storage tank.

The solubility of the gas is typically measured in both freshwater and seawater. For example, two common solvents, benzene and toluene, are significantly (22%) more soluble in freshwater than seawater. This has important planning implications for spill response teams that must respond to both ocean and freshwater spills of crude oils. Assuming constant pressure and temperature, the expected air concentration over an ocean spill would, in general, be more than a similar lake or river spill since more of the solvents will be dissolved in the freshwater.

Similarly, when a pure liquid-air interface in a closed space is present, the *vapor pressure* of the chemical (liquid) is the dominant predictor of air space chemical concentration. For example, benzene and anthracite are two common chemicals that are frequent environmental contaminants. Anthracite is a member of the polyaromatic hydrocarbon (PAH) family and is a by-product of incomplete combustion. The solubility of benzene is vastly different than

---

Table 5–C. Physical Properties

- Melting point and boiling point describe the physical state at ambient temperature and pressure and provide data relevant for changes in the physical state.
- Vapor pressure is crucial for assessing partitioning to air; the higher the vapor pressure, the more likely a chemical is to exist in a gaseous state. Vapor pressure changes as temperature and pressure change.
- Water solubility is crucial for assessing partitioning to water and is temperature dependent.
- Partitioning describes whether substances are water-seeking (hydrophilic) or water-avoiding (hydrophobic). Partitioning also characterizes the ability of a substance to bioaccumulate, transport through skin, and sorb to soils. Octanol is used as a surrogate for lipids (fats) and can predict bioconcentration in aquatic organisms. Kow is the measure of the octanol-water partition coefficient.
- Henry's Law Constant (H) is the air-water partition coefficient and the rate a substance will evaporate from water. The higher H is, the more likely a chemical will move from water to air. H equals the vapor pressure of a substance divided by the aqueous solubility.
- Adsorption coefficients are the distribution between water and a solid [e.g., soils, sediments (Kd or Koc)]. Koc is the most commonly used abbreviation and is the micrograms adsorbed per microgram of organic carbon (soil solid phase) divided by the microgram per milliliter of solution. Kd is unadjusted for dependence upon organic carbon. To adjust for the fraction of organic carbon (Foc), Kd = Koc times Foc.
- Bioconcentration factor (BCF) is the measure of the extent of chemical partitioning at equilibrium between a biological medium (e.g., fish tissue, plant tissue) and water. The higher the BCF, the greater the accumulation.

the PAHs, 1780 g/m³ versus 0.075 g/m³ for anthracite. Not surprisingly, the vapor pressures are also quite different, i.e., 0.125 atm for benzene and 5.04 × 10⁻⁵ atm for anthracite. Therefore, it would not be surprising to find substantial benzene vapors present in a partially filled underground storage tank while it is unlikely that a "solution" of PAHs would produce the same level of vapor emissions. As the ambient temperature increases, the rate of vaporization also increases. Thus, the temperature of the liquid surface at the interface will have a substantial impact on the vaporization and the measured vapor pressure. It would be expected, based on this discussion, that a crude oil spill in cold waters would produce different impacts than a similar warm-water release.

Another common way that air-water (liquid) interfaces are expressed is by using Henry's Law Constant (H). This constant describes the rate at which a substance will evaporate from water and can be calculated by dividing the vapor pressure of a chemical by its aqueous solubility. As H increases, it becomes more likely that a chemical will move from water to air. For example, the H for benzene is approximately five times greater than the value for anthracite and four thousand times greater than the most toxicologically significant PAH, benzo(a)pyrene. Clearly, the knowledge of vapor pressure, solubility, or Henry's Law Constant provides a powerful predictor of air-water interface behavior.

Air-Solid (Soil) and Liquid-Solid (Soil) Interfaces

Knowledge of the air-soil and liquid-soil interfaces is extremely important for environmental practitioners because direct spills and leaks to the soil are extremely common. The subsequent behavior of the released material is partially a function of a soil-specific property known as adsorption.

Adsorption is the process of chemicals attaching to the surfaces of minerals and organic matter in soil and rock. Adsorption is generally considered a reversible process, although in some circumstances, a material will so tightly adhere to the soil that very little is released to either the air or water. Mathematically, adsorption can be defined by the simple equation:

$$Kd = Cs/Cw \qquad \text{(Equation 5-1)}$$

where Kd is the water-soil partition coefficient; Cs is the mass of the chemical adsorbed on the soil per unit of bulk dry mass; and Cw is the mass of the chemical dissolved in a unit of water that is in contact with the soil. Kd is unadjusted for dependence on organic carbon. This is important, because the amount of adsorption is highly dependent on the organic content.

The organic content in soil is reflected in another commonly used term known as *Koc*. Koc is the micrograms adsorbed per microgram of organic carbon divided by the micrograms per milliliter of solution. Typically, soil organic contents less than 1% are considered low, whereas levels of 3%–5% are more representative of most soils. Koc and Kd can be connected by the formula:

$$Kd = Koc \times Foc \qquad \text{(Equation 5-2)}$$

where Foc is the fraction of organic carbon.

Knowledge of either Kd or Koc is very useful since a variety of fate-transport predictions can be made based on the value of these terms. For example, for chemicals with a Kd of less than 1.0, it is quite likely that the chemical will easily move in a water system. Similarly, a Kd greater than 10 implies that a chemical is preferentially adsorbed to the soil phase. For Koc, chemicals with values over 10,000 are highly adsorbed to soils, whereas a Koc of less than 1,000 is indicative of a more mobile substance. A comparison of the Koc values of benzene and anthracene illustrates this observation: benzene Koc = 83 ml/g, anthracene Koc = 14,000 ml/g. Not surprisingly, solvents like benzene are highly mobile in soil and frequently migrate through the soil column and impact groundwater. This is in contrast to poorly mobile chemicals typified by PAHs (e.g., anthracene), which tend to adhere tightly to the soil phase and are less likely to impact groundwater at depth.

In this situation, Kd is directly related to another descriptive term, the retardation factor. (See the definition in the section about Earth Sciences Principles and Nomenclature.) In general, as chemical movement is retarded in the soil column, it is potentially subjected to a variety of degradation or transformation processes (e.g., microbial, digestive, hydrolytic) that can convert the starting molecule into a more or less toxic by-product. For example, 1,1,2-trichloroethylene (TCE) is converted under anaerobic soil conditions to vinyl chloride. Similarly, under certain soil conditions, perchloroethylene (PCE), a common dry-cleaning and degreasing solvent, can be converted to TCE. Hence, it is possible to begin with one chemical at the soil surface and find different compounds at the groundwater level.

The soil adsorption of a chemical also impacts the predicted impact on the air phase. The adsorption of chemicals onto soil significantly affects the vaporization rate of the chemical. For example, pesticides generally are well adsorbed to soil; therefore, the vapor pressure of pesticides adsorbed to soil is quite low and the expected air concentration at the interface boundary would be commensurately diminished. Other factors that affect the air-soil and liquid-soil interface include the water content of the soil and the water solubility of the chemical.

While air-soil and liquid-soil interfaces are quite complex, they are extremely important because most spills and leaks occur into the soil medium. However, the final major physical interface, liquid-liquid, is similarly complex and has attracted substantial scientific interest because of the potential widespread ecological impacts associated with this type of release.

Liquid-Liquid Interfaces

Liquid-liquid interface situations are best typified by sudden and dramatic releases of hydrocarbons (crude oil) into saltwater or freshwater bodies. In these problems, there is an environmental interface between the underside of, for example, an oil slick and the water phase. In this situation, the density of the hydrocarbon phase is usually less than the water phase so the oil floats on the water surface and creates a slick. If the density of the released material is greater than water, the material will sink and create a different type of liquid-liquid interface problem.

When a water surface slick is created, there are a variety of processes that substantially impact the nature of the interface: evaporation, photochemical and oxidative reactions, dissolution and emulsification, and biological (microorganisms) activity (Thibodeaux, 1979). Many of these processes are influenced by the ambient air and water temperatures; in addition, rough sea surface conditions can significantly alter the interface boundary conditions.

Oil slick situations illustrate the interaction between a liquid-liquid and an air-liquid interface. The lower boiling-point components of the oil slick evaporate or dissolve within a period of several hours. In general, the liquid to air transfer accounts for the majority of the transfer, assuming stable sea conditions. There is a mass transfer from the hydrocarbon layer to the water phase; however, evaporation is the dominant mechanism of release. Evaporation to the air phase is rapid and the transfer rate decreases by as much as two orders of magnitude within eight days (Thibodeaux, 1979).

As the lighter hydrocarbon fractions evaporate, the residual oil becomes more viscous and affects the kinetic parameters of air transfer. At this stage in the natural history of an oil slick, a film forms on the surface of the unmixed oil and prevents further evaporation of the lighter fractions unless chemical dispersants are utilized. The other physical, mechanical (wave action), chemical, and biological processes become dominant. The heavy fractions can coalesce and form tar balls which initially float but eventually sink due to the accumulation of sand and shell material. This process results in the creation of a liquid-liquid interface below the water surface. Thus, a crude oil or other type of hydrocarbon spill illustrates the intricate interaction of the different types of interface problems. Ultimately, the role of biological organisms becomes critical both for their positive ability to impact a spill and for the potential negative effects of the chemical(s). The next section presents some of the partitioning factors that are useful for predicting the potential impacts of chemicals on biological organisms.

Biological Interfaces

Assessing the chemodynamics of chemicals in biological organisms is largely determined by the extent to which chemicals accumulate in living organisms. This ability to accumulate is known as *bioaccumulation*. Bioaccumulation is important because living organisms can concentrate chemicals in their tissues at levels significantly greater than the concentration in water or food. This phenomena occurs because organisms tend to accumulate chemicals in the lipid (fat) portions of their tissues. This process of migration into the lipid compartment is a function of the lipophilicity of the absorbed chemical(s).

Although it is difficult to directly measure lipophilicity, there is an adequate surrogate measure: octanol-water partition coefficient (Kow). Octanol is a reasonably good surrogate for lipid, so that the ability of a chemical to preferentially partition into octanol (versus water) is considered a relatively accurate predictor of bioaccumulation. Kow is formally defined as the ratio of the concentration of a chemical in the octanol phase to its concentration in the aqueous phase. Kow is directly related to the tendency to bioconcentrate and inversely correlated with water solubility. Highly water soluble chemicals are very mobile and less likely to volatilize and persist; hence, they are less able to bioconcentrate. Chemicals with large Kow values (e.g., PCBs, DDT, and dioxins), tend to accumulate in soil, sediment, and biota but not in water. Conversely, chemicals such as solvents (e.g., benzene, toluene) with small Kow values tend to preferentially partition to air and water.

The Kow numbers are frequently presented as logarithms or log Kow. For example, the log Kow of benzene is 2.12, while the log Kow for DDT is 6.19. This difference does not appear significant until the log values are transformed into their "natural" values: benzene 131.8 and DDT 1,548,817. The difference is readily apparent and it is easy to see why DDT is so persistent in living organisms.

The partition factors provide a sound predictive set of concepts for analyzing the movement of chemicals in the environment. The next subsections describe the physical framework (earth, water, and air) through which chemicals are transported in the environment.

Earth Science Principles and Nomenclature

The physical framework for waste management in the environment is the solid phases through which water (and any contaminants it may be carrying) moves. This section describes and discusses some of the major concepts and terms of physical geology and soil science. Table 5–D is a list of some of the common terms and definitions in physical geology and Figure 5–2 illustrates some of the soil/rock configurations discussed in the subsequent sections. Because many environmental problems are associated with surface contamination, the focus of this section will be directed toward a better understanding of the properties and problems affecting soils.

The physical properties of soils are critical and directly affect how chemicals move and react within the soil column. In this context, the term *soil column* refers to the entire distance between the surface and bedrock. The major physical properties include: (1) size of primary and secondary (aggregates) soil particles; and (2) size, distribution, and quantity of pores. These properties, along with color and texture, affect the absorption and radiation of energy in the soil column; in addition, the conductivity of water, heat, and gases are also altered by size and pore variability (Gardner, 1993).

Because the soil column is composed of a variety of minerals and rocks, it is not surprising that sci-

Table 5–D. Important Terms in Physical Geology

- *Mineral:* A naturally occurring solid with a defined crystal structure and a limited range of composition.
- *Rock:* A natural aggregate of one or more minerals. Three major classes of rocks are recognized:
  1. *Sedimentary rock:* Rock formed by accumulating sediments. Sedimentary rocks are characteristically stratified. They include chemical sedimentary rocks—such as limestone, gypsum, and salt deposits—as well as physical sedimentary rocks, such as sandstone and shale.
  2. *Igneous rock:* Rock formed by cooling of a magna or melt. Common igneous rocks include granite, basalt, and tuff.
  3. *Metamorphic rock:* Rock formed from a preexisting rock or sediment by subjection to heat and/or pressure without complete melting. Metamorphic rocks include gneiss, schist, and marble.
- *Geologic structures:* Features produced by the deformation of rocks. Geologic structures may occur at scales ranging from microscopic to continental (or oceanic). Significant types of geologic structures include
  1. *Fold:* Bending or crumpling produced in stratified rocks (and sediments) by compressive stress.
  2. *Fracture:* Break in rocks.
  3. *Fault:* Fracture or fracture zone along which there has been displacement of the opposite sides relative to one another.
- *Soil:* Earth material that has been so acted upon by natural physical, chemical, and biological agents that it can support rooted plants. Whether a soil does, in fact, support plants depends on factors such as moisture content and chemical composition, as well as on its degree of weathering.
- *Sediment:* Solid material—mineral or organic—that has been moved from its site of origin by air, water, or ice and has come to rest on the earth's surface. Sediments may be classified by grain size as well as by composition. Grain sizes for sediments, from coarse to fine, are gravel, sand, silt, and clay.

entists have classified soils by particle size. In this context, size refers to the diameter of a spherical particle that would fall in a viscous fluid at the same rate as the irregularly shaped soil particle (Gardner, 1993). A simpler definition is based on the size of the opening in a wire mesh screen: (1) clay—less than 0.002 mm, (2) silt—0.002 to 0.02 mm, and (3) gravel—greater than 2 mm in diameter. The various combinations of clays, silts, and gravels are known as soil textures. In the United States, soils are also categorized by horizons. Horizons are layers that are approximately parallel to the surface and show distinct differences by virtue of color, texture, porosity, and consistency. This layering of different horizons forms a soil profile that can be established by either visual or formal laboratory methods.

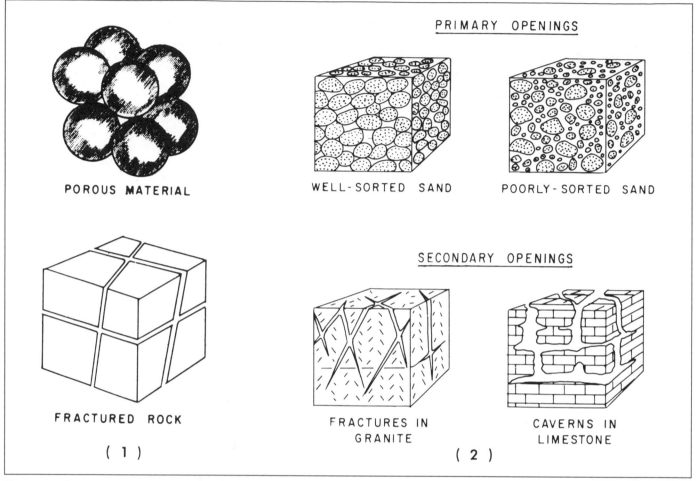

PRIMARY OPENINGS

POROUS MATERIAL

WELL-SORTED SAND

POORLY-SORTED SAND

SECONDARY OPENINGS

FRACTURED ROCK

(1)

FRACTURES IN GRANITE

CAVERNS IN LIMESTONE

(2)

Figure 5–2. Soil and rock formations have voids through which groundwater moves. Source: Heath, 1989.

In practice, the three most important horizons are:

1. *A horizon*—layer of maximum biological activity and of removal of dissolved or suspended materials in water.
2. *B horizon*—layers of suspended materials, including residual oxides, silicate clays, or other transformed materials.
3. *C horizon*—weathered mineral material that is unaffected by soil-forming processes except for the accumulation of salts or oxides of varying solubilities (Smith, 1993).

The concepts of soil horizons and profiles are important because they are standard terms of reference during environmental investigations of potentially hazardous materials. For example, when background soil samples are required, it is common practice to sample the B horizon of the reference area. This is particularly important when naturally occurring chemicals such as lead, arsenic, and cadmium are under investigation at the study site. There is considerable natural variation in metals concentration in soils; therefore, it is often important to differentiate between naturally occurring levels of metals and man-made or anthropogenic contributions. (See also Chapter 12, Risk Assessment, for a further discussion of the problems of differentiating between natural and anthropogenic affected concentrations.)

Within the different soil horizons, there is important variability due to the presence of clays, silts, and sands. The clay fraction is considered "active" because it has an extremely large surface area per unit mass of particles and is capable of forming colloids (suspensions of finely divided particles in a continuous medium). In addition, clays have another critical attribute, i.e., the ability to swell and retain surface ions due to adsorption. This ability affects the movement of both water and any chemicals that have entered the soil column. In contrast to clays, silts generally have smaller surface areas, no swelling properties, and poor capacity to absorb ions (Gardner, 1993). Finally, the presence of decomposed organic matter (humus) and inorganic compounds such as iron and aluminum oxides is also a critical component of soil because these materials can bind soil particles together and form aggregates that impact the pore size and physical behavior of a particular soil layer (Gardner, 1993).

While the various types and sizes of soil material form the physical framework of the soil column, several other important chemical and physical processes occur. Ion exchange, chemical speciation/precipitation and complex formation, and retardation are defined as follows:

- *Ion exchange*—An ion exchange process occurs when a dissolved chemical in the soil column substitutes itself for another chemical that is already adsorbed onto a mineral or soil surface. The primary ions affected by this process are cations (+ charge) and the measured factor in soil is known as the *cation exchange capacity* or *CEC*. The CEC of a soil will vary directly as a function of the amount and distribution of clay and organic matter present. The CEC has a significant impact on the uptake of chemicals into plants and can have potentially serious consequences for the home vegetable garden impacted by contaminated soils.
- *Chemical speciation/precipitation and complex formation*—This term generally refers to the important trace metals such as iron (Fe), copper (Cu), zinc (Zn), arsenic (As), lead (Pb), selenium (Se), nickel (Ni), cobalt (Co), and cadmium (Cd). The trace metal cations in a soil solution form complexes and precipitates as a function of the content and type of organic matter, the presence of other oxide minerals, soil pH (measure of acidity or alkalinity), soil Eh (oxidation-reduction potential), and the CEC. The bioavailability of trace metal and organic (synthetic) chemical uptake is dramatically affected by the factors that influence complex formation. These effects have serious impacts when soil treatment options are considered due to chemical contamination (see Chapter 10, Pollution Prevention Approaches and Technologies).
- *Retardation*—The retardation factor is a measure of the lack of movement of a chemical in the soil due to chemical reactions like ion exchange and adsorptive processes. The retardation factor, Rf, is a commonly used term and is related to the distribution coefficient Kd: Rf = 1 + Kd × the bulk density of the solid (Ps) ÷ the total porosity of the soil (N). This equation is formally:

$$Rf = 1 + Kd \times Ps/N \qquad \text{(Equation 5–3)}$$

This factor is significant when problems of groundwater flow and transport velocities of chemicals are addressed. This issue will be discussed in the next section on movement of chemicals through soil and water.

The geology of a site provides the physical framework; however, the hydrogeology is the engine that drives the system and produces the movement of materials from one area to another. This movement, or fate and transport, is the critical factor that determines whether a spill or leak will ultimately affect the underground water supply.

FATE AND TRANSPORT THROUGH SOIL AND WATER

Hydrology

In the United States, more than half of the population uses groundwater as its primary source of water. In rural areas, more than 95% of the daily-use water is derived from deep subsurface sources. Hence, it is easy to see why anything that impacts either the groundwater supply or purity is a major concern. In fact, although only approximately 1% to 2% of the total U.S. groundwater supply is contaminated, 33% of urban center groundwater supplies are adversely impacted. Some of the major sources of this groundwater contamination are:

- 75,000 industrial on-site hazardous waste landfills
- 100,000 sanitary landfills and dumps
- 5 million underground storage tanks (up to 25% may leak)
- 5,000 RCRA hazardous waste treatment storage and disposal facilities
- 1,200 uncontrolled hazardous waste sites on the National Priorities List (NPL)
- 27,000 potential Superfund NPL sites
- 20 million domestic septic systems (33% are defective)
- 200 commercial hazardous waste landfills
- 180,000 surface impoundments
- 1,000 hazardous waste surface impoundments
- 300,000 barrels of hazardous materials relevant from pipeline each year
- 60,000 oil and gas operations
- 15,000 mining operations
- 1 billion lb (1×10^9 lb or 4.5×10^8 kg) of pesticides applied over 280 million acres (McLane, 1992; U.S. EPA, 1987).

Figure 5–3 is a schematic representation of these same sources.

Knowledge of several of the key hydrology terms and concepts is essential in order to understand how groundwater can become impacted by potentially hazardous materials. Table 5–E presents these terms, and Figures 5–4 through 5–11 illustrate many of the concepts.

The terms and concepts in Table 5–E form a core vocabulary that is required in order to understand and discuss groundwater problems and impacts. Because the investigation and remediation of groundwater contamination is so complex and expensive, it is worthwhile for any environmental manager to gain some insight into the science of hydrology.

The fundamental concept of hydrology is the hydrologic cycle. The *hydrologic cycle* is the endless circulation of water among the ocean, atmosphere, and land. Components of the cycle include precipitation, storage, runoff, and evaporation. The total amount of water is constant; however, its form (e.g., solid, liquid, gas) may change. Figure 5–4 is a simplified version of the hydrologic cycle.

The portion of the hydrologic cycle below the land surface is known as the *subsurface flow system*. Precipitation (e.g., rainfall, snowmelt) recharges the subsurface flow system. Outflow or discharge occurs as evapotranspiration or evaporation and as flow to surface water bodies (U.S. EPA, 1988). Both artificial recharge and discharge are possible via injection or pumping.

In general, as water infiltrates the land surface, it moves vertically downward toward the saturated zone. Saturated water moves laterally from areas of greater hydrostatic pressure to areas of lesser hydrostatic pressure. The prime driving force is gravity.

The movement, storage, and occurrence of groundwater are affected by a wide variety of geologic and physical factors: porosity, lithology, hydraulic conductivity, climate, surface water, and gradient. As the earlier definitions explained, an aquifer

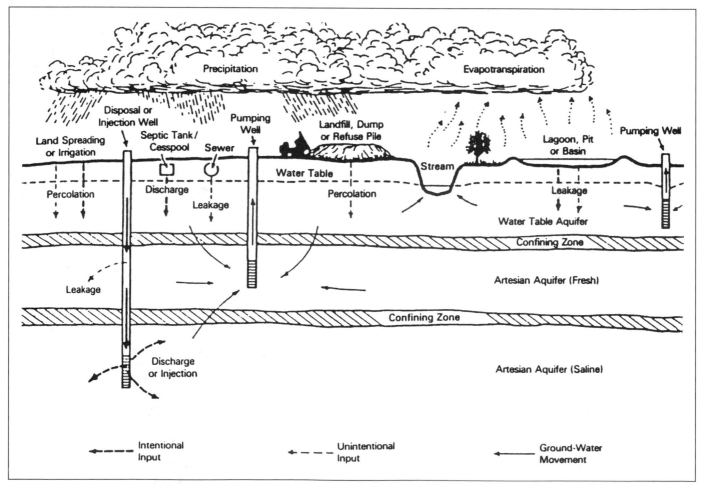

Figure 5–3. Sources of groundwater contamination. Source: *Ground Water Handbook,* U.S. EPA, 1987, p. 7, (625/6-87/016 Robert S. Kerr Environmental Research Lab, Ada, OK) from Geraghty & Miller, *Fundamentals of Groundwater Contamination: Short Course Notes,* Geraghty & Miller, Inc.: Syossett, NY, 1985.

Table 5–E. Basic Terminology for Hydrology

- *Advection:* Process of dissolved chemicals within an aquifer unit moving at the average velocity of the groundwater.
- *Aquifer:* A saturated, permeable geologic unit that can transmit significant or usable quantities of water under ordinary hydraulic gradients. There are many alternative definitions for aquifer, including that of a water-bearing rock. It is imperative for an industrial manager concerned with groundwater issues to understand how local and federal authorities define the term.
- *Aquitard:* Geologic unit that inhibits groundwater flow due to its low permeability. Aquitards frequently separate aquifer units, but flow across an acquitard is possible. If little flow occurs, the unit is termed on *aquiclude.*
- *Capillary fringe:* Transition zone from the partially saturated vadose zone to the water table (phreatic) surface.
- *Confined aquifer:* An aquifer that has an overlying layer that does not allow direct contact of the aquifer with the atmosphere. Water in a confined aquifer is under pressure and wells penetrating into the aquifer will have a water level that reflects the pressure in the aquifer at the point of penetration. Also called an *artesian aquifer.* Compare with *Perched aquifer* and *Unconfined aquifer.*
- *Darcy's Law:* Volumetric flow rate (Q) through a porous medium is directly proportional to the hydraulic conductivity (K), the hydraulic gradient (i), and the cross-sectional area of flow (A): $Q = -KiA$. The minus signifies that flow is from areas of high hydraulic head to areas of low hydraulic head.
- *Degradation:* The transformation of a contaminant to other forms through decay, biodegradation, or other process.
- *Discharge:* The quantity or process of water being lost from the saturated zone. Compare with *Recharge.*
- *Dispersion:* Spreading of a contaminant source as it flows through an aquifer.
- *Evapotranspiration:* Water that is returned to the atmosphere by evaporation from the surface and by transpiration from plants.
- *Hydraulic conductivity:* A coefficient of proportionality between specific discharge and hydraulic gradient. The hydraulic conductivity is related to the permeability of an earth material for a given fluid at a given degree of saturation. Unless otherwise stated, it is assumed that the hydraulic conductivity is stated for a geologic material that is fully saturated with water. Determined from well-pumping tests. Compare with *Hydraulic gradient.*
- *Hydraulic gradient:* The change in hydraulic head between two points divided by the distance between the points. Determined from a water level map. Compare with *Hydraulic conductivity.*
- *Hydraulic head:* The product of fluid potential and the acceleration due to gravity of groundwater in an aquifer. Head is the sum of two components—the elevation head (elevation of the measuring point) and the pressure head (the incremental elevation due to the pressure exerted by the water at the point of measurement).

- *Infiltration:* Percolation of water through earth materials to the saturated zone of the groundwater system.
- *Perched aquifer:* Beds of clay, silt, or other materials of limited areal extent that present a restriction to flow of downward-moving water in the vadose zone may cause local areas of saturation above the regional water table. An unsaturated zone is present between the bottom of the perching bed and the water table. Compare with *Confined aquifer* and *Unconfined aquifer.*
- *Permeability:* The ability of an earth material to transmit a fluid (usually taken to be water). Compare with *Porosity.*
- *Porosity:* The percentage of an earth material that is open space. Compare with *Permeability.*
- *Potentiometric surface:* The surface to which water in an aquifer would rise under hydrostatic pressure. For an unconfined aquifer, the potentiometric surface is the water table surface. For a confined aquifer, the potentiometric surface lies above the elevation of the aquifer in which the water is found.
- *Recharge:* The quantity or process of water being added to the saturated zone. Compare with *Discharge.*
- *Saturated zone:* The zone in earth materials in which the pore space is completely filled with liquid water. In the saturated zone, the pressure head is greater than atmospheric. Sometimes called the *phreatic zone.* Compare with *Soil zone* and *Vadose zone.*
- *Soil zone:* Area in which evaporation and transpiration of water occurs. Compare with *Saturated zone* and *Vadose zone.*
- *Sorption:* Transfer process in which dissolved chemicals in the groundwater become attached to sedimentary materials and/or organic matter. This process is described using the concept of partitioning.
- *Specific discharge:* Flow rate divided by the cross-sectional area across which the flow occurs.
- *Storage:* The quantity of water that is held in the pore space of an aquifer but that may be released (and begin to flow) when a hydraulic stress is applied to the aquifer.
- *Unconfined aquifer:* An aquifer in which the top of the saturated zone (water table) is in direct contact with the atmosphere through the open pores of the earth material above. Also called a *water table aquifer.* Compare with *Confined aquifer* and *Perched aquifer.*
- *Vadose zone:* The zone in earth materials in which the pore space is not completely filled with liquid water, although there ordinarily is moisture present. Sometimes called the *unsaturated zone.* The pressure head is less in the vadose zone than in the atmosphere. Compare with *Saturated zone* and *Soil zone.*
- *Water table:* The surface at which the water pressure in the pores of the porous geologic materials is exactly atmospheric. The elevation of the water table is identified by determining the elevation of water in a well that penetrates the vadose zone and is open to the top of the saturated zone.

is a geologic unit that contains sufficient permeable material to yield significant pumpable quantities of water (Figure 5–5). The two major parameters defining an aquifer are hydraulic conductivity (Figure 5–6) and porosity (Figure 5–7). There are three main types of aquifers:

- confined (artesian)
- perched
- unconfined (water table).

Four additional terms relating to aquifers describe flow in a groundwater system: hydraulic conductivity, porosity, transmissivity, and storage coefficient.

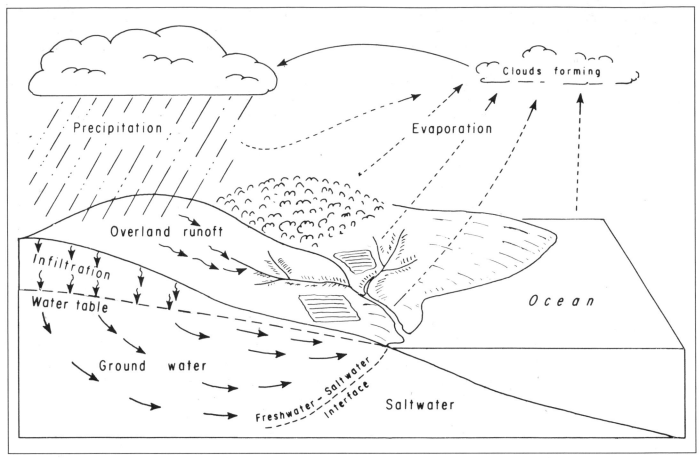

Figure 5–4. The hydrologic cycle. Source: Heath, 1989, p. 5.

The first three of these terms appeared in Table 5–E. *Storage coefficient* for an unconfined aquifer is the specific yield, that is, the amount of water that drains from a soil due to the force of gravity. Storage coefficient for a confined aquifer expresses changes in storage due to elasticity of the soil and compressibility of water.

In 1856, Henry Darcy reported that for a given soil in a column of constant diameter flow between two points in the soil column is directly proportional to the difference in potential (energy) between the points and inversely proportional to the distance between the points. *Darcy's Law* is a measure of the average or bulk velocity through a given cross section of a porous medium and is valid for steady flow with constant flux. If turbulence is present, Darcy's Law is not valid. Fortunately, for most groundwater systems, laminar or streamlike, smooth flow rather than turbulent flow is present. Thus, Darcy's Law, combined with the conservation of mass, is the basic groundwater flow equation and along with contaminant transport processes [i.e., advection, dispersion, sorption or adsorption (retardation), and

degradation] can be used to quantify the nature and extent of potential contamination.

Further detailed information is available from a wide variety of sources (Freeze & Cherry, 1979; Miller, 1984; Mercer & Faust, 1986; Heath, 1989; and U.S. EPA, 1987 and 1988).

The following generalizations apply to contaminant transport in near-surface, granular aquifers (common conditions). The primary transport process is *advection*, which transports most solutes at the velocity of the groundwater flow. This process is augmented by *dispersion*, which is spreading of contaminants due to the tortuous flow paths through the soil grains. Dispersion dictates the shape of contaminant plumes but is usually not the primary transport process. Contaminant transport is impeded by *adsorption*, or transfer of contaminants from mobile (in groundwater) to immobile (attached to soil) state, which is usually a reversible process resulting in delayed contamination. Degradation results in transformation of the contaminants to other forms, which may be either less or more toxic.

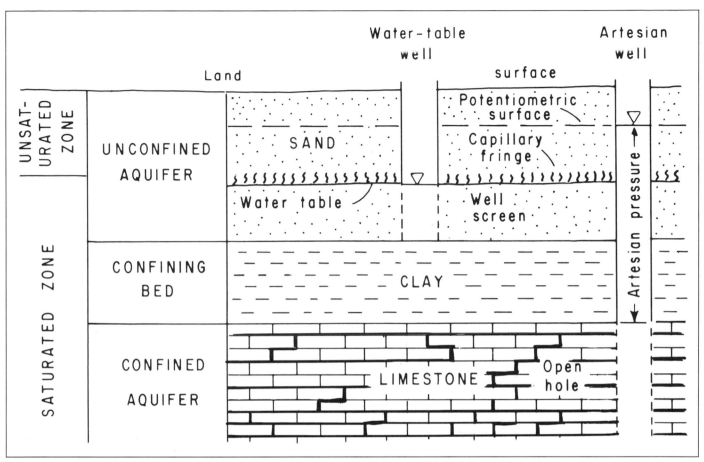

Figure 5–5. Aquifers and confining beds. Source: Heath, 1989, p. 6.

An example of a study in which these considerations arose involved mathematical simulation of chemical seepage from power plant solid waste in support of a risk evaluation of power plant integrated control configurations. The purposes of the study were to (1) predict the likely concentrations of nine priority pollutants in groundwater, as a result of the 30-year operation of hypothetical landfills at three sites, and (2) to assess the sensitivity of the predicted results to uncertain parameters (namely; adsorption distribution coefficient, moisture content of the waste pile, hydraulic conductivity of the waste and liner, and presence or absence of an impermeable cap). The predicted results were presented in terms of dilution factors for contaminants ranging from very mobile (antimony and nickel) to fairly immobile (arsenic, lead, and selenium).

Predicted concentrations at the water table were found to be sensitive to the groundwater recharge rate and waste characteristics. Predicted concentrations down gradient of the landfill were found to be sensitive primarily to adsorption and, to a lesser degree, to dispersion. Figures 5–8 and 5–9 illustrate

some of the common groundwater pollution situations.

Surface Water

The analysis of pollutant fate and transport in surface water bodies depends on several hydrologic transport processes. Some of these processes were covered in the groundwater discussion; however, the terms and definitions can be altered to account for the difference in medium (i.e., surface versus groundwater). Table 5–F presents some of the key surface water terms and definitions.

Obviously, many of these concepts are relevant to the other major media (groundwater, air, and soil) and have previously been discussed. Therefore, this discussion merely explores the ability to predict and calculate movement of chemicals in surface waters.

The predictive accuracy of fate-transport models is complex and controversial. The complexities arise from the multiple scales of motion, turbulence flow, boundary-layer effects, and a multiplicity of transport mechanisms. Consequently, predictions of contaminant migration in surface water are uncertain

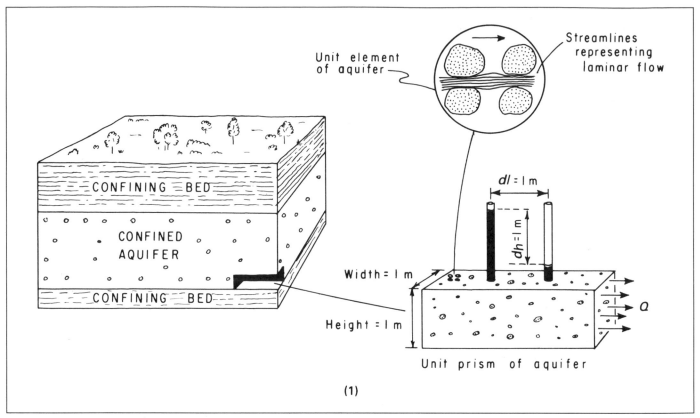

Figure 5–6. The hydraulic conductivity of an aquifer can be calculated by Darcy's Law: Q = KA(dh/dl), where Q is the quantity of water per unit of time; K is the hydraulic conductivity; A is the cross-sectional area, at a right angle to the flow direction, through which the flow occurs; and dh/dl is the hydraulic gradient. Source: Heath, 1989, p. 12.

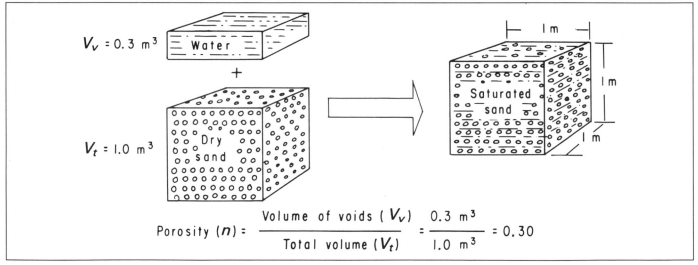

Figure 5–7. Features of porosity. Source: Heath, 1989, p. 7.

and open to debate. The environmental fate of a pollutant entering a surface water body is highly dependent on the type of water body: (1) rivers and streams, (2) impounds, and (3) estuaries and oceans. Rivers and streams exhibit turbulent pipe-flow; impounds offer a near-stagnant environment; and estuaries and oceans involve consideration of tidal and salt/freshwater effects.

Like groundwater, surface water has its own unique vocabulary and fundamental controlling mathematical equations. The equation of continuity is the fundamental concept for most surface water problems. This conservation of mass equation is: the time rate of change of mass within the volume plus the net rate of mass flow in and out of the volume equals that produced by sources and reduced by

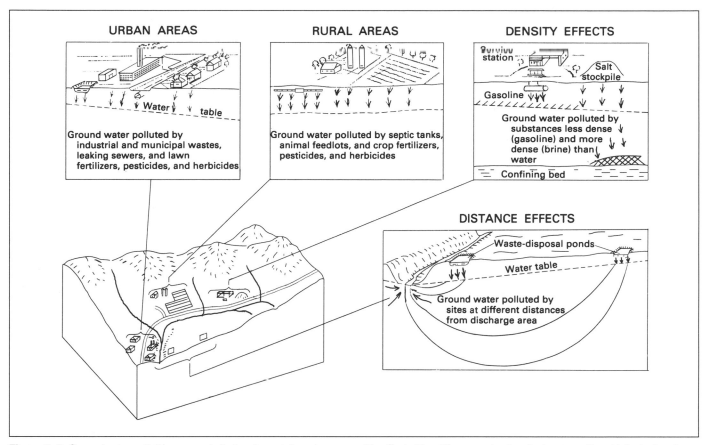

Figure 5–8. Groundwater pollution occurs in both urban and rural areas and is affected by differences in chemical composition, biological and chemical reactions, density, and distance from discharge areas. Source: Heath, 1989, p. 66.

sinks. This equation describes the relationship of mass transport through a volume of water and the sources and sinks of mass within it. For example, a sink may refer to (1) pipe discharges, (2) runoff along the length of a river (distributed nonpoint source), and (3) distributed source along the boundaries of the water body due to a previous impact (U.S. EPA, 1987).

These sources and sinks are critical to surface water calculations and must be considered so that conservation of mass occurs. A case in which the behavior of the source governed design considerations involved mathematical modeling of the cooling-water system for a nuclear power plant in the Middle East. The cooling-water system used the discharge of heated water into, and intake of cool water from, the Persian Gulf. The purpose of the model was to help to design the discharge structure and location, and to predict whether recirculation between the cooling-water discharge and intake was likely or possible.

A range of physical processes—entrainment, turbulence, bottom-surface interaction, buoyancy effects, wind effects, surface heat-exchange, and ocean currents—was of necessity embodied in the model.

First, the calculation procedure was validated against a physical model tested in the laboratory, and then—on application to the site conditions—the temperature excesses above ambient for various tidal conditions and cooling system designs were predicted. The model predictions were then used to finalize design of the offshore cooling system, which has subsequently been operational for almost 20 years.

In Chapter 8, Hazardous Wastes, the rules and regulations governing surface water and discharges are discussed in detail. These laws are based on many of the scientific concepts and terms discussed here.

ATMOSPHERIC FATE AND TRANSPORT

The effects of atmospheric releases of hazardous or toxic substances can be evaluated by analyzing several categories of information (U.S. EPA, 1993):

- source characteristics
- meteorological conditions
- geographic scale
- topography
- contaminant properties.

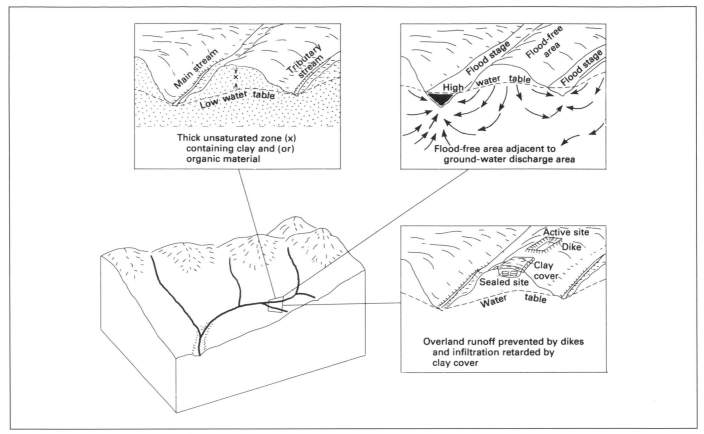

Figure 5–9. Selection of waste-disposal sites involves consideration of the unsaturated zone, flood danger, groundwater discharge, overland runoff, and infiltration. Source: Heath, 1989, p. 67.

Source Characteristics

Source characteristics refer to (1) type of source, (2) rate of pollutant release, (3) the *plume height,* or height of the release source, and (4) a variety of other site-specific parameters. Sources are defined as point, nonpoint, or multiple release points. The differences between source types are important and have a substantial effect on the ultimate contaminant concentration of a human exposure point (see Human and Health Risk Assessment in Chapter 12, Risk Assessment).

Point sources refer to a specific emission point, such as an industrial stack associated with a refinery or other industrial process. This source can be defined in terms of height above ground, stack diameter, gas velocity, and temperature at the stack opening. *Nonpoint* or *area sources* are several sources distributed over a homogeneous surface area.

Usually, the individual sources within an area are small and are dominated by a point source. If there are multiple overlapping sources along a specified surface area, they are known as a line source. An example of a line source is urban traffic within a defined geographic area.

Multiple source refers to the combination of point and nonpoint sources, such as ones typically found in urban or industrial settings. For example, a major highway (line source) may be adjacent to a large refinery composed of both point and nonpoint sources.

The characterization of source is important because the fate-transport calculation depends on what type of question is posed. For example, if on-site worker exposures from a near ground-level source(s) are calculated, emissions are generally modeled as an area source. However, if the population adjacent to the refinery is under study, then the emissions could be calculated as a single point source.

For each source type, the rate of release or emission rate must be defined. For point sources, emissions are expressed as mass per unit of time (grams/second), or parts per million (ppm) per second. Area sources must have both mass per unit of time and spatial distribution [i.e., micrograms (μg) per square meter per second (μg/m^2/sec)]. Line sources (urban traffic) are expressed as grams per second per kilometer. Multiple source emission rates must incorporate time and spatial distribution of emissions from all important sources (U.S. EPA, 1993).

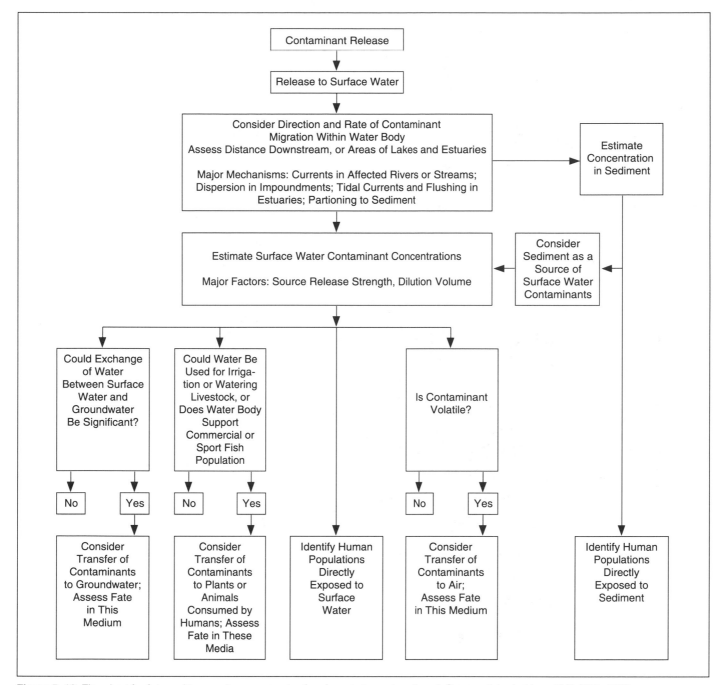

Figure 5–10. Flowchart for fate and transport assessments of surface water and sediment. Source: Adapted from U.S. EPA, 1989.

Plume height is the physical height of the stack adjusted by factors that can either raise (buoyancy or momentum) or lower (downwash or deflection) the plume. Plume height affects the time required for released contaminants to reach ground level. This is important because peak ground-level exposure concentrations are typically not located close to a point source with a high plume height (100 meters). For example, elevated ground-level concentrations from smelter stacks are usually found a significant distance (0.5 to 3 km) from the source. Other factors that affect plume height include (1) contaminant release temperature (e.g., greater buoyancy and greater rise), (2) stack-tip downwash (i.e., velocity of the stack contaminant emitted is low relative to ambient wind speed, thus effective plume height is lower, and (3) downwash due to adjacent structures (again, plume height is lowered).

Site-specific factors that affect the source include (1) operational variations in emissions and/or meteorologic conditions over time, (2) use of emission controls such as scrubbers, and (3) fugitive emissions

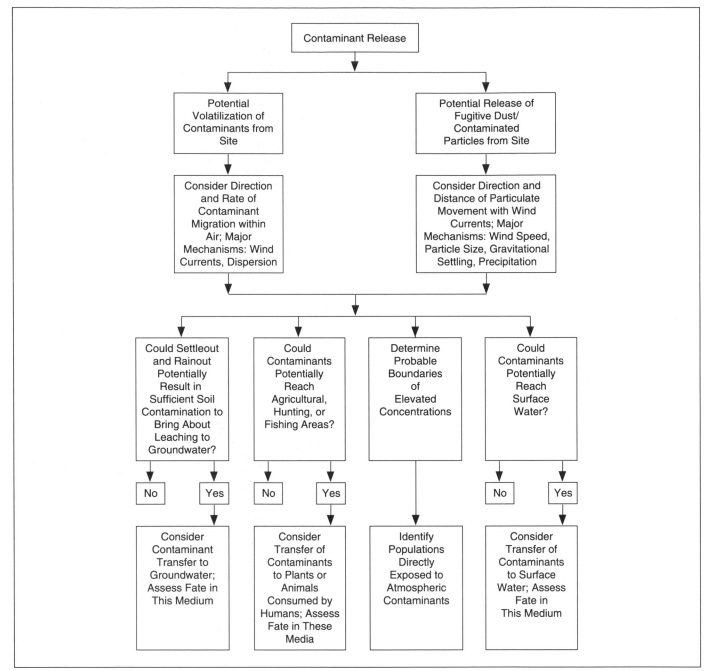

Figure 5–11. Flowchart for fate and transport assessments of the atmosphere. Source: Adapted from U.S. EPA, 1989.

from other ground-level sources such as storage facilities, piping, flanges, valves, and fittings.

An example of air emissions from hazardous waste sites is the remediation of a site by consolidation of soils contaminated with heavy metals. The site of a former steel mill was being developed for office building use. The soil on the site had been found to be contaminated with chrome and other heavy metals. After excavating several hot spot areas that had been identified, the rest of the soil needed to be consolidated (because of its expansive nature) so that construction could be undertaken. The consolidation process was accomplished by dropping a heavy weight from a crane repeatedly on the soil. Each time the weight impacted the soil, clouds of dust were raised 30 to 50 ft (9.15 to 15.25 m) from the impact site.

The concern was that this dust might contain chrome and other heavy metals. In order to ascertain that heavy metal concentrations leaving the site were

Table 5–F. Basic Terminology for Surface Water Transport Processes

- *Advection:* The transport of water flowing in a particular direction (more or less horizontally), such as water flowing because of the current in a stream or river.
- *Biodegradation:* Biodegradation transformations are reactions due to the metabolic activity of aquatic microbes, primarily bacteria. Depending on the specific chemical, the transformations may be very fast due to the presence of enzymes; for other compounds, the process may be very slow. For chemicals where the transformation is fast, biodegradation is often the most important transformation process in the aquatic environment.
- *Convection:* The transport of water because of density gradients. In this form of transport, the driving forces of the currents are density gradients resulting from temperature differences in deep lakes, and temperature and salinity differences in estuaries.
- *Dispersion:* The scattering of particles because of the combined effect of shear and diffusion (molecular and turbulent). Usually, the combined effect of shear and transverse diffusion, represented as an effective dispersion, is orders of magnitude greater than other diffusive mechanisms acting in the direction of flow in rivers and estuaries.
- *Hydrolysis:* The breaking of bonds in a molecule due to reaction with water. Compounds are altered in a hydrolytic reaction by the replacement of some chemical group with a hydroxyl group. These reactions are commonly catalyzed by the presence of hydrogen or hydroxide ions and, hence, the reaction rate is strongly dependent on the pH of the system. Generally, the new compound is usually less toxic than the original compound.
- *Ionization:* The fate of toxic organics that are either acids or bases can be strongly affected by the concentration of hydrogen ions in a water body. An organic acid or base that is extensively ionized could be markedly different from the corresponding neutral molecule in solubility, adsorption, bioconcentration, and toxic characteristics. For example, the ionized species of an organic acid is generally absorbed by sediments to a much lesser degree than is the neutral form. The solubility of an ionic form of an organic chemical will likely be greater than for the neutral species. Therefore, as a chemical is ionized under environmental conditions, the change in physical properties as well as the chemical reactivity will change with pH. The pH values found in most aquatic systems range from approximately pH 4 to 9, with extreme values down to pH 2 and up to pH 11 (see Chapter 14, Global Issues, for discussion of acidic pHs associated with acid rain).
- *Molecular diffusion:* The scattering of particles by random molecular motion, commonly characterized by Fick's law of diffusion.
- *Oxidation-reduction:* The transfer of electrons from the reduced species to the oxidized species. The oxidation-reduction potential is an important process in that it can control the oxidation number of the metals present in solution and may also change the oxidation state and structure of organics. In addition, redox reactions have been observed to be important for mercury, toxaphene, and DDT.
- *Particle deposition:* The settling of particles from the water body to the underlying bed.
- *Particle entrainment:* The picking up or lifting or particles from the underlying bed of a water body by turbulent motion over the bed.
- *Particle settling:* The sinking of particles having densities greater than the fluid of the water body, such as sediments or suspended solids.
- *Photolysis:* The degradation process whereby radiant energy in the form of photons breaks the chemical bonds of a molecule. Direct photolysis involves direct absorption of photons by the molecule. Indirect photolysis involves the absorption of energy by a molecule from another molecule that has absorbed the photons.
- *Shear:* Mixing due to variations in the fluid velocity at different positions in the water body. One example of this could occur in a lake where a significant decrease in temperature occurs with depth, thereby causing a thermal resistance (resistance of colder and, therefore, denser and lower-lying water to be displaced by warmer, lighter, and higher-lying water). A shear plane divides the surface current that follows the wind from the return currents that run counter to the wind (Fair et al, 1968).
- *Sorption:* Sorption is a transfer process whereby dissolved chemicals in the eater become attached to sedimentary materials. Some problems associated with field application of this concept include: (1) rapid movement of water and sediments in rivers and estuaries may not satisfy the assumption of system equilibrium; (2) some chemicals may exhibit non-reversible sorption characteristics; hence, desorption from sediments to the water column may not be correctly represented; (3) different particle sizes (sand, clay, and silts) exhibit different properties and should be accounted for separately; and (4) salinity in estuarine systems may affect the sorption process. Some of the compounds that may be strongly affected by sorption include heavy metals and many hydrophobic nonpolar compounds.
- *Turbulent diffusion:* Scattering of particles by random turbulent motion (advective transport via turbulent motion in the form of eddies).
- *Volatilization:* Volatilization is a physical transfer process where a chemical is transferred between the water body and the atmosphere at the water-air interface. Uncertainty in rates for natural systems is complicated by variations in velocity, depth, stratification, salinity, wind speed, and diurnal atmospheric conditions.

Source: U.S. EPA, 1987.

not a problem, an air quality monitoring program was undertaken to collect dust samples at the site boundary. Dust samples collected by high-volume samplers were analyzed to determine the total suspended particulate concentration and were fur- ther chemically analyzed to identify the chrome content of the dust leaving the site. Concurrent meteorological data consisting of wind speed and wind direction were also obtained. Results of this monitoring showed that both the dust levels and the heavy metal levels escaping the site were within acceptable standards, and so site remediation proceeded. The site is now an office park.

Meteorological Conditions

A variety of meteorological conditions or parameters can drastically affect the transport of pollutants in air. These parameters include:

- wind speed and direction, causing increased dilution and dispersion
- precipitation that can increase ground deposition near the source
- turbulence produced by kinetic and thermal energy transfers between air and the terrain; dispersion is enhanced at lower atmospheric levels, but the reverse may occur at higher atmospheric levels if conditions are stable and wind speed is high.

Geographic Scale

Dispersion of contaminants is categorized as a function of distance from the source: (1) near-field [< 30 mi (50 km) from the source] and (2) far-field [> 30 mi (50 km)]. In general, near-field scale exposures are not affected by atmospheric chemical reactions and the plume is assumed to spread both laterally and vertically in a Gaussian (bell-shaped) statistical distribution under steady-state atmospheric conditions. The EPA has a variety of Gaussian plume models, such as the Industrial Source Complex (ISC) short- and long-term models that are used for exposure point concentration determinations. These models typically assume constant wind speed and flat terrain and are accurate within a factor of two versus actual measured values (U.S. EPA, 1993). Far-field calculations and models are more complex and are typically used for radiological assessments.

Topography

By definition, the area surrounding a source point is either flat or complex. Flat terrain is characterized by land and building elevations below the stack height of the source and where steady-state atmospheric conditions are in effect. Conversely, in a complex terrain, either land or building elevations are greater than stack height or where terrain deflects or alters air movement (e.g., mountains, valleys, large bodies of water). For regulatory purposes, most atmospheric fate-transport calculations assume flat terrain.

Contaminant Properties

As discussed earlier for ground and surface water, the ultimate exposure point concentration can be significantly affected by both chemical and physical processes that affect a given pollutant. Transformational processes such as photolysis and oxidation may alter the initial pollutant concentration and produce reactive secondary contaminants (e.g., photochemical smog produced by nitrogen oxides, reactive hydrocarbons, and sunlight). Photochemical smog produces a secondary series of contaminants (e.g., ozone) with greater toxicity than the initial reactants. Removal processes such as dissolution, adsorption, settling, and precipitation also significantly impact the ultimate exposure concentration. As previously mentioned, many near-field atmospheric dispersion models do not account for either removal or transformation effects and assume simple exponential decay. Therefore, these models tend to overestimate exposure point concentrations. Table 5–G defines some of the key terms used for the characterization of atmospheric fate and transport processes (U.S. EPA, 1993).

These basic principles of pollutant fate and transport in air, water, and soil are used throughout the process of evaluating whether a substance will cause a problem in the environment. Summary flowcharts from EPA's *Risk Assessment Guidance for Superfund* (U.S. EPA, 1989) for the major environmental media are presented in Figures 5–10 through 5–12. These flowcharts systematically organize the approach to fate-transport evaluation. As regulatory assessment and compliance issues are presented in subsequent chapters, it is useful to reexamine these charts since they frequently guide EPA oversight actions and provide a framework for general fate-transport assessments.

The principles of environmental chemodynamics initially appear quite complex and difficult. However, it is important for environmental professionals to have a grasp of fate-transport principles and methods. The ability to accurately and appropriately evaluate and treat chemical contamination in the environment is substantially based on an understanding and appreciation of chemodynamics.

The next section presents a brief overview of the fundamentals of environmental toxicology. The previous discussion of fate-transport mechanisms is critical, because many potentially hazardous materials are either transformed to less toxic substances (e.g., microbial degradation in soil) or are diluted to levels that do not pose significant threats to humans

---

Table 5–G. Basic Terminology in Atmospheric Fate and Transport

- *Advection:* The movement of contaminants with an air mass in a predominantly horizontal direction. The process is dependent on wind speed and direction.
- *Area source:* Numerous, usually small, sources of contaminant emissions that are distributed across a specified surface area.
- *Atmospheric stability:* A meteorological condition that affects dispersion of airborne contaminants. The atmosphere is said to be stable when there is little or no vertical movement of air masses. With little or no vertical mixing of the contaminant with the air, contaminant concentrations accumulate at ground level. Unstable atmospheric conditions result in vertical mixing of the air masses.
- *Box model:* A model in which the entire modeling region is contained in a single cell (box); it is used to obtain estimates of area source concentrations.
- *Deposition:* The process by which particulates and reactive gases are deposited on the earth's surface from the atmosphere. Both wet and dry deposition occurs. Wet deposition occurs sporadically during specific rain or snow events; dry deposition occurs continuously under dry atmospheric conditions.
- *Diffusion:* The dispersion of a contaminant relative to its advective movement. Diffusion reduces the central concentrations in a contaminant mass (plume), increases the concentration at the periphery of the mass, and expands the periphery.
- *Effective plume height:* The physical height of the stack adjusted by factors that raise the plume (as a result of buoyancy or momentum) or lower it (as a result of downwash or deflection).
- *Gaussian plume model:* A model most commonly used to represent plume dispersion in the near-field range. Simple

expressions (Gaussian functions) are assumed to represent the dependence of pollutant concentration on lateral or vertical distance from the plume centerline (i.e., the advective path).
- *Gravitational settling:* An intermedia transfer mechanism in which particulate contaminants or contaminants adsorbed onto suspended particulates settle to surface media via gravitational attraction. The particulates are normally more than 20 µm in diameter (U.S. EPA, 1988a).
- *Pasquill-Gifford dispersion coefficients:* Numerical values of the standard deviations of atmospheric displacements about any point moving with the mean wind. The values are defined as a continuous, empirical function of downwind travel distance (or pollutant from a source) for each discrete stability class.
- *Pasquill-Turner atmospheric stability classification:* The most often used inferential method of estimating the turbulent state of the atmosphere, in discrete classes, using solar intensity and wind speed as surrogate indicators.
- *Precipitation:* An intermedia transfer mechanism that removes particulate and aerosol matter from the atmosphere. Particulate and aerosol matter serve as a nucleus for the condensation of raindrops. Raindrops generally remove particulates and aerosols greater than 1.0 µm in diameter from the air (U.S. EPA, 1988).
- *STAR (stability array) data:* Summaries of meteorological data, including seasonal or annual joint frequencies for each stability class, wind direction, and wind speed category. STAR data is available from the National Climatic Center (NCC), Asheville, North Carolina, for all National Weather Service (NWS) locations in the United States. STAR data from the NWS station most representative of the site should be used for modeling purposes.

Source: U.S. EPA, 1993.

---

or the environment (e.g., air dispersion). Hence, the interaction between fate-transport mechanisms and toxicity is crucial. Further discussion of the practical application of this interaction is presented in Chapter 11, Public Health Issues.

FUNDAMENTALS OF ENVIRONMENTAL TOXICOLOGY

Environmental toxicology is the science of poisons as they relate to earth's biosphere. As such, it is multidisciplinary and draws from the diversity of life, biophysical, and environmental sciences. The environmental toxicologist must be skilled at physical and chemical sciences relating to the fate and transport of potentially harmful materials through the environment. Equally important are the biological sciences of biochemistry, physiology, ecology, and pharmacology. Integration of principles used in these life and physical sciences to realistically characterize

human and environmental health problems is the challenge that the toxicologist faces. This section provides a brief overview of the fundamentals of environmental toxicology. For a more comprehensive review of the science of toxicology, refer to a variety of alternate texts (Amdur et al, 1991; Hayes, 1989; Williams & Burson, 1985).

Dose-Response Relationship

Characterization of the relationship between the amount of a substance absorbed into the body and the physiological response to that dose is a fundamental objective in toxicology. The 16th century Greek physician Paracelsus recognized the importance of the dose-response relationship when he wrote "the dose makes the poison." Implicit in this statement is the notion that *any* substance may be toxic at a high enough dose. Some commonly encountered substances and their approximate lethal doses are presented in Table 5–H.

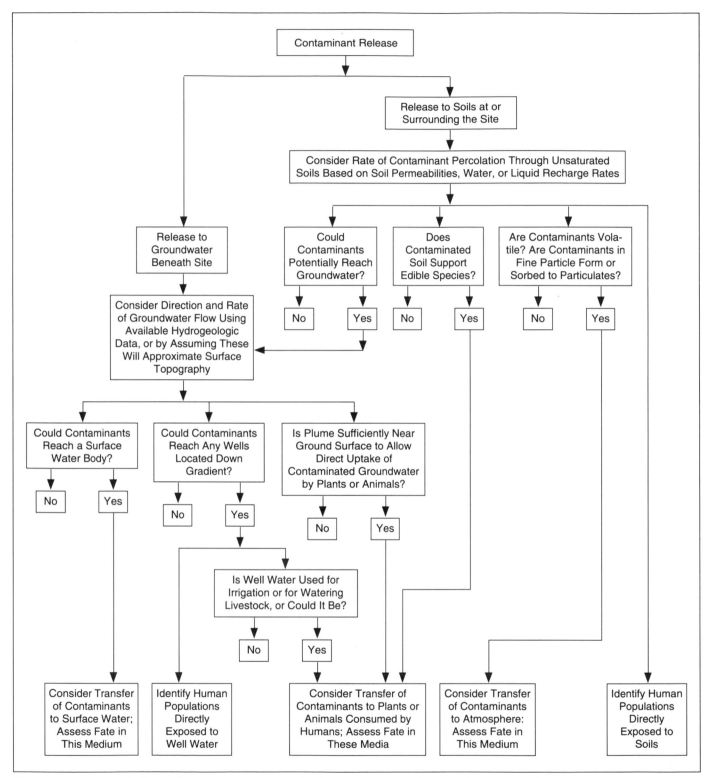

Figure 5–12. Flowchart for fate and transport assessments of soils and groundwater. Source: Adapted from U.S. EPA, 1989.

Lethality is, however, a crude measure of toxicity. Modern toxicologists study a variety of more subtle toxic effects, such as immune system disturbances, nervous system deficits, genotoxicity, effects on the developing fetus, and effects on both the male and female reproductive systems.

Equally important to understanding a substance's toxicity is the relationship between the dose-rate and the response. The toxicologist may be interested in measuring an effect and understanding the dose-response relationship following short-term (acute) exposure. Alternatively, the toxicologist may study the

Table 5–H. Approximate Lethal Doses in Mammals of Some Common Substances

Toxicity	Compound	LD$_{50}$ (mg/kg)
Moderate	Ethanol	10,000
	Sodium chloride	4,000
Extreme	Pentachlorophenol	17
	Nicotine	1

dose response following longer-term [subchronic (less than 90 days) or chronic (greater than 1 year)] exposure. The target organ(s) for toxicity, and therefore the toxic effect, may be quite different following acute versus chronic exposure, even with the same chemical. Certain toxic responses are not expressed immediately, such as a substance's effects on the female or male reproductive systems that remain unnoticed until pregnancy and certain cancers—such effects may not be expressed for more than 30 years following exposure. An effect may be well understood and quite specific to a particular chemical's inherent chemistry, such as the interference of oxygen transport by carbon monoxide, or it may be poorly understood and general to a class of chemicals. An example of the latter is general nausea and central nervous system effects associated with excessive exposure to hydrocarbons such as those in gasoline.

Typically, to set reference limits for the public, the toxicologist focuses on the lowest dose at which adverse effects appear for the specific chemical. The measured endpoint may be frank, as is the case with studies focusing on the LD$_{50}$ (the lethal dose of a chemical for 50% of the population). Or the endpoint may be as subtle as a small histological (cellular) change in a target organ such as the liver, kidney, or brain. The endpoints measured in any toxicological study may be based on the judgment of the toxicologist directing the study. In cases such as industrial compliance with federal government regulations in the Toxic Substances Control Act (TSCA, 1976) or Federal Insecticide, Fungicide, and Rodenticide Act (FIFRA), the details of toxicological study design and endpoints to observe are outlined by the U.S. EPA or other federal regulatory agencies.

Measures of exposure or dose. Figure 5–13 shows a typical dose-response curve. *Dose* is the mass of toxicant per mass receptor (e.g., mg toxicant/kg animal). Often the investigator is interested in the dose rate and then the dimension of time is added (e.g.,

mg/kg/day). The U.S. EPA uses an alternate approach to measuring exposure to airborne toxicants. In the case of inhalation exposure, the EPA uses units of milligrams toxicant per cubic meter of air (mg/m^3). This allows direct comparison between units of measure collected by field air sampling and potentially harmful levels. Careful consideration of comparative anatomy and physiology of toxicant absorption are conducted in the presentation of inhalation exposure units (Jarabek et al, 1989).

Any assessment of exposure should include clearly defined terms. To communicate units of dose, either *delivered dose* (the amount of material fed to an experimental animal) or *absorbed dose* (the amount of material absorbed into the systemic circulation) should be specified. Similarly, individuals may define *exposure* in different ways. Contact with a chemical or contaminated media is typically called exposure. *Intake* of that material can occur through inhalation or oral ingestion of the substance or contaminated media. *Uptake* (absorption) refers to the movement of the contaminant across a biological boundary such as the intestinal mucosa, alveoli of the lung, or epidermis of the skin. A more detailed discussion of exposure assessment may be found in the *Guidelines for Exposure Assessment* (U.S. EPA, 1992).

The route by which a chemical enters the body may greatly influence its toxic expression. Alteration of a chemical by gastrointestinal pH or flora can inactivate or activate that chemical. Inhalation of the same chemical may result in a different expression of toxicity. As an example, although exposure to cadmium via either the oral or inhalation routes may result in damage to the proximal tubule of the kidney, in humans the expression of lung cancer due to cadmium exposure is seen only through exposure by inhalation. By contrast, exposure to the nonspecific herbicide paraquat can cause extensive and often fatal damage to lung tissue whatever the route of exposure. Extrapolation of data derived from one exposure route to another should be conducted, if at all, only by a qualified toxicologist.

Examples of dose response. Dose-response relationships display a variety of characteristics that depend on the chemistry of the specific toxicant being studied, the endpoint being observed by the investigator(s) and, the timeframe during which the observations are made. The health, age, nutritional state, and physiology of the receptor can influence the dose-response relationship. Some examples of dose-response relationships are presented in Figure 5–14a,

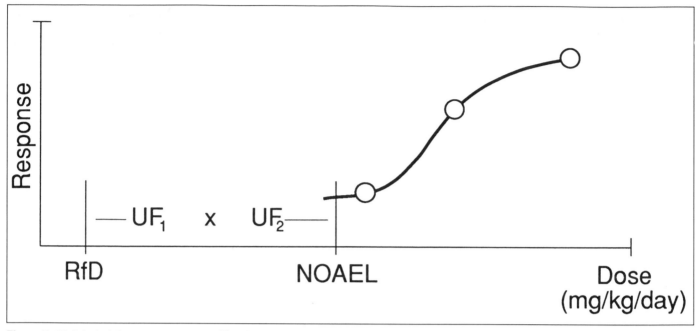

Figure 5–13. A typical dose-response curve. The dose may be presented as the log transformation of those used in the study. Plotting the x-axis logarithmically allows all doses to appear on the same graph. Uncertainty factors (UF) are applied to derive safe human exposure levels.

b and c. The dose is often presented as the log transformation of the quantity of chemical delivered in the experiment. Any particular dose-response investigation may have as its objective the characterization of a range of responses to a chemical within a population. Here the investigator studies the range of doses required to produce the same endpoint in all members of the population (Figure 5–14a).

Alternatively, the objective may be to establish an individual's graded response to a chemical. The severity of the endpoint(s) of interest is likely to increase with increasing dose. Different endpoints of concern may appear at different doses (or times) and may increase in a graded fashion across the range of doses applied in the study (Figure 5–14b).

The shape of the dose-response relationship is filled with information for the toxicologist. The form of the relationship between dose and response for a study employing a large population and many doses is often a classical sigmoid (Figure 5–13). The sigmoidal dose-response relationship shows the normal distribution of biological variance commonly found in living systems. The variance in response shows that members of the population have different susceptibilities to the effects of the toxicant.

The slope of the relationship is an important indicator of toxicant behavior (Figure 5–14c). A steep slope suggests that toxicity of the chemical may become severe with only small increases in dose. By contrast, a shallow dose-response slope indicates

that larger increases in dose are necessary to cause a toxic effect and that signs or symptoms that warn of impending toxicity may gradually increase over the range of doses employed.

Another important concept to investigative and regulatory toxicologists is that of the threshold. A fundamental concept for most systemic (noncancer-causing) toxicants is that there is some dose below which the body's innate repair mechanisms can prevent the onset of adverse effects. Establishing this threshold dose is one principal objective of most toxicological studies that use experimental animals. The threshold concept has important implications for the regulatory toxicologist and public health officials and is discussed in greater detail in the next section.

Interpretation of the dose-response for the purpose of regulation. Often, an important assumption in the interpretation of the dose-response relationship is that there is some dose below which no adverse effect is observable. This is the *threshold dose*; it is of particular interest to the regulatory toxicologist (Figure 5–13). This threshold dose for the most sensitive toxicological endpoint is used as a point of departure for setting regulatory limits on exposure to environmental toxicants. Threshold doses, called the *NOAEL (no observable adverse effect level)* or the next highest dose, the *LOAEL (lowest observable adverse effect level)*, are used as benchmarks for establishing acceptable contaminant levels in drinking water, soil, foods, and air. The NOAEL or

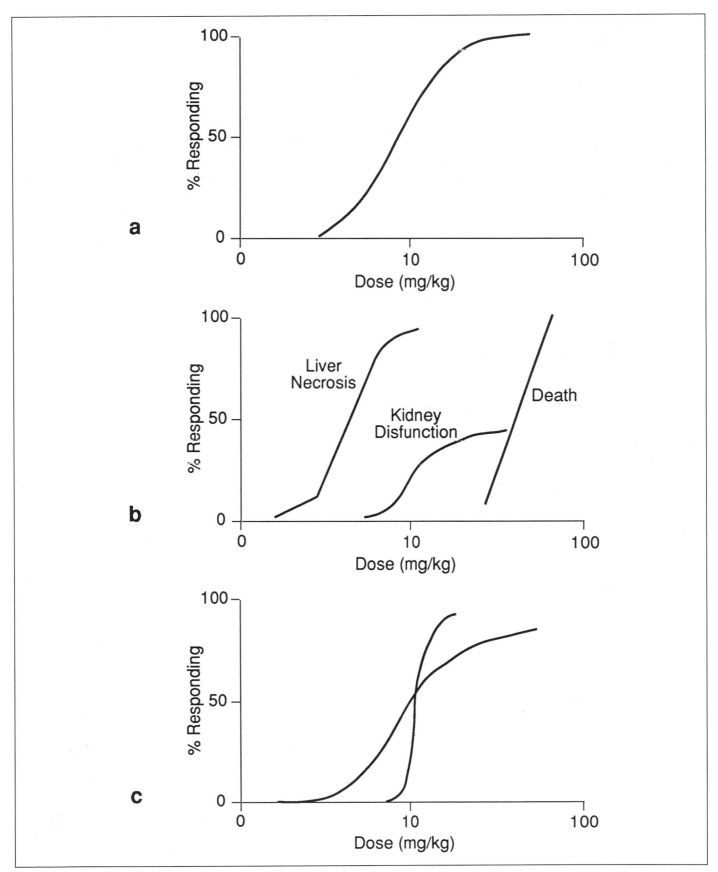

Figure 5–14. A typical dose-response curve, displaying the range of doses at which members of the population respond *(a)*. Natural variability in the population is expected. Different toxicological endpoints may arise at different doses *(b)*. The slope of the dose-response curve *(c)* can provide insight into a chemical's safety.

LOAEL is commonly divided by one or more uncertainty factors to account for possible interspecies and interindividual variability in the toxicity of a chemical (Table 5–I). Further uncertainty factors are employed if the confidence in the database is compromised.

Unfortunately, this approach ignores sometimes valuable information found in the shape of the dose-response curve. Specific procedures used for interpreting experimentally established NOAELs or LOAELs to human health-based exposure limits are presented in greater detail by Barnes and Dourson (1988).

By setting the acceptable exposure limit through the use of uncertainty factors and modifying factors based on the characteristics of the available information, both health professionals and the public can be reasonably assured that no toxicity will occur at the exposure limit set. Such exposure limits for non-cancer endpoints through the oral route of exposure are called *reference doses (RfDs)* by the U.S. EPA and are presented in units of dose (mg/kg/day). The mathematical and qualitative definitions of the reference dose are:

> An estimate (with uncertainty spanning perhaps and order of magnitude or greater) of a daily exposure level for the human population, including sensitive subpopulations, that is likely to be without an appreciable risk of deleterious effects during a lifetime (Barnes & Dourson, 1988, 480).

Corresponding values derived for inhalation exposure are called *reference concentrations (RfCs)* and are presented in units of milligrams per cubic meter (mg/m^3) (Jarabek et al, 1989). In contrast to the dose-response relationship for most noncancer-causing toxicants that are curvilinear and display a threshold or NOAEL, some toxicants such as lead and most cancer-causing chemicals are thought to display linear nonthreshold dose-response relationships. The assumption of linearity in the dose-response for carcinogenic chemicals has been controversial and is discussed in more detail later in the chapter.

Types of Adverse Effects

Cancer effects in humans and experimental models. Taken together, the incidence of the many diseases categorized as "cancer" may truly be viewed as a national epidemic. Approximately one American in three will suffer the effects of some type of cancer during his or her lifetime. Understanding the etiology of the many diseases referred to collectively as cancer is far from complete, and even a cursory review of current understanding is beyond the scope of this section. Contribution to this cancer incidence is thought to be attributed mainly to genetic, lifestyle, and dietary habits. A relatively small percentage of the overall cancer incidence in the U.S. can be attributed to exposure to synthetic chemicals.

Several natural and synthetic chemical carcinogens *have* been shown to cause cancer in laboratory animals after having been suspected of such in human populations. The discovery in the 18th century by the English physician Purcival Pott of testicular cancer in chimney sweeps following exposure to coal tar is, perhaps, a classic example. Experimental animal studies conducted after Dr. Pott's observations confirmed the causative agents as dibenz[a,h]anthracene and benzo[a]pyrene. Specific examples of synthetic environmental chemicals producing human cancers are relatively rare. Approximately 20 such chemicals have been identified to date.

Understanding of the mechanism(s) of cancer induction is presently undergoing significant refinement. Advances in the area of molecular carcinogenesis and molecular epidemiology are likely to decrease the uncertainty in human risk assessment in the future. Recent work has shown that tumor expression is suppressed by cellular proteins. Mutation of these suppressor proteins can lead to cancer. Nearly half of all cancer cases result in mutation of a suppressor gene called p53 (Harris, 1993). Different environmental carcinogens may lead to characteristic mutations at specific locations on the p53 gene. This allows identification of specific causative agents and may improve our ability to identify individuals at increased risk of cancer.

Before recent advances in understanding, the interaction of an etiological agent with the genetic

Table 5–I. Uncertainty Factors Used by the U.S. EPA to Develop Reference Doses for Lifetime Human Exposure

Uncertainty Factor	Rationale
10 ×	Intraspecies variability
10 ×	Interspecies variability
10 ×	Extrapolation of LOAEL to a NOAEL
10 ×	Less-than-lifetime to lifetime exposure
5 ×	Modifying factor

material of a cell was thought to be a requirement for the expression of cancer. Examples of such primary, direct-acting carcinogens include: (1) benzo[a]pyrene, an active component of coal tar; (2) dimethylnitrosamine (NDMA), a constituent of cigarette smoke, synthetic plastics, and breakdown product of rocket fuel; and (3) bis(2-chloroethyl)amine, mustard gas. Compounds that interact with and alter the structural/functional characteristics of the genetic material (DNA) resulting in the conversion of the cell to the preneoplastic stage (cancer precursor cells) are called *genotoxic carcinogens.* Naturally occurring viruses are suspected of playing a genotoxic role in some types of cancer. Preneoplastic cells may be kept from full expression of cancer for long periods of time by natural defense mechanisms or chemical antagonists within the body, making the identification of a causative agent extremely difficult.

Evidence has been discovered suggesting that some agents associated with the progression of preneoplastic cells to cancer may do so by causing cell division and proliferation without direct interaction with the genetic material. Such agents are called *epigenetic carcinogens,* because they can stimulate (or promote) the growth of preneoplastic cells within the target tissue without evidence of interaction with the genetic material. Epigenetic carcinogens may act by a variety of mechanisms, including suppression of the body's immune system, as happens with azathioprine, or by enhancing the carcinogenicity of another agent, as with ethanol. Some direct-acting or *complete carcinogens* may interact with the genetic material *and* stimulate cell proliferation through one or more of the mechanisms indicated above.

Curiously, the body's own metabolic mechanisms may play a role in the development of cancer. The metabolism of relatively innocuous compounds to reactive metabolites within the body occurs in both the cytoplasm of the cell and in association with a cellular organelle called the endoplasmic reticulum (Figure 5–15).

Cellular components collectively called the *mixed function oxidases (MFO)* may alter *xenobiotics* (foreign compounds) to speed their excretion from the body. In doing so, these enzymes may produce metabolites that are more reactive than the parent compound. Such cellular toxification or detoxification of xenobiotics by the MFO occurs primarily in the liver and kidney cortex but may also occur in the lining of the gut and epidermal layers of the skin or other tissues.

Alternatively, a toxicant may be metabolized at the target site to a more or less reactive compound. This is the case with the activation of the hormone diethylstilbestrol in the reproductive system. Thus, the route of exposure, the metabolism, and the route and time-course of excretion may play a significant role in the development of cancer or precancerous lesions.

Epidemiological evidence. The discovery of a causal relationship between coal tar exposure and testicular cancer by Purcival Pott is among the first uses of epidemiology in cancer assessment. Since then, cancer registries have become common throughout the United States and much of the developed world. These registries are used to estimate incidence of specific cancers and to provide a database for research into the etiology of the many diseases called cancer. It has long been recognized that many cancers, such as colon, rectal, and breast cancer, have strong environmental as well as genetic components. Recent advances in the tools available to geneticists are extremely promising and have resulted in the identification of specific genes associated with colon cancer, Huntington's disease, and amyotrophic lateral sclerosis.

The discovery in 1971 that the hormonal analogue diethylstilbestrol (DES) was associated with the development of clear cell adenocarcinomas has led researchers to aggressively investigate the role of synthetic and natural estrogen-like compounds in cancer progression (Herbst et al, 1971). In experimental animals, the carcinogenic activity of DES has been shown to be enhanced by the concurrent exposure to another primary carcinogen 7,12-dimethylbenzanthracene (DMBA) (Rustia & Shubik, 1979). Synergism is at work here: The cumulative effect of these two compounds is greater than the sum of the effects were the compounds given separately. Opposite or antagonistic interactions are also of consequence.

Interestingly, disturbances in natural hormonal balance, such as that associated with ovariectomy, which occur concurrent with DES exposure may produce a similar synergistic effect as exposure to DMBA (Rustia & Shubik, 1979). Because certain organochlorine pesticides such as DDT and its metabolite DDE and symmetrical chlorinated hydrocarbons such as the PCB, dioxins, and furans display hormone-like activity, some concern has arisen for the role of chlorinated organics in the etiology of human breast cancer (Wolff et al, 1993).

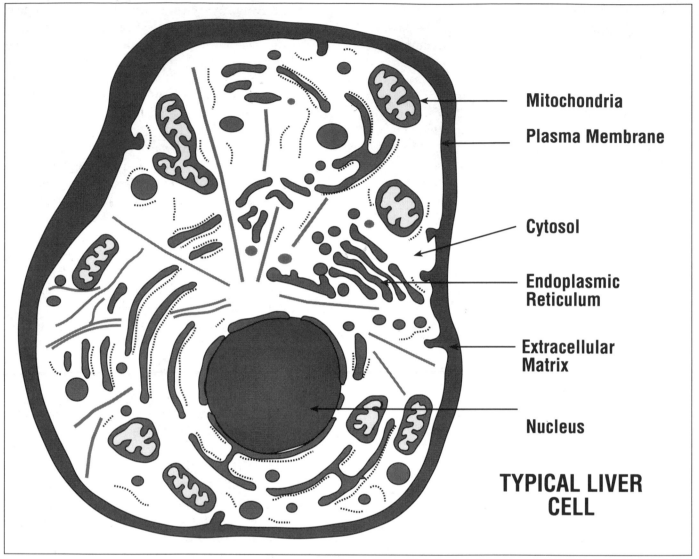

Mitochondria

Plasma Membrane

Cytosol

Endoplasmic
Reticulum

Extracellular
Matrix

Nucleus

**TYPICAL LIVER
CELL**

Figure 5–15. Cellular metabolism may activate or inactivate environmental toxicants or may enhance their elimination from the body. The mixed function oxidase system (MFO) is enriched in the cellular organelle known as the endoplasmic reticulum.

It is often presumed that human data available from epidemiological studies are preferable for the purpose of assessing risk to people. However, epidemiological evidence may be severely limited in its usefulness for risk assessment when quantitative information about exposure is missing or other uncertainties fog the interpretation of the results. Workplace exposure to a chemical or metal of interest often occurs concurrently with exposure to other potentially active agents. This complicates the estimate of dose for the compound of specific concern and limits the usefulness of the study. Differences in diet, lifestyle, and social class can also make study results difficult to interpret. Detailed information giving clues about exposure; the physical-chemical nature of the contaminant of interest; its particle size, oxidative state, and environmental disposition should be carefully noted by those responsible for quantifying exposure. Unfortunately, this information is rare in epidemiological studies.

Extrapolation from animals to humans. Assessing potential risk of cancer following exposure to environmental toxicants has been dominated by the notion that the growth and development of a cancerous lesion is probabilistic. Further, the probability of developing cancer from a nominal dose of any particular carcinogen is commonly thought of as being directly and linearly proportional to the dose. The U.S. EPA usually employs linear or near-linear extrapolation of the data available from animal studies to exposures found in the workplace or residential environments (1986 Cancer Guidelines). Criticism of this approach has focused on questions regarding the relevance of high-dose experimental studies conducted in rats or mice to human exposures that are likely to be thousands of times lower (Figure 5–16). Researchers typically determine the highest dose that experimental animals can tolerate over a lifetime to begin cancer testing. A cancer bioassay conducted using rats or mice may include this Maximum Tolerated Dose (MTD) and only one or two lower doses.

Critics of presently employed methods of cancer risk assessment argue that at the high doses delivered during an experiment, cellular toxicity and the resulting cellular proliferation are the cause of the cancer not the chemical being tested *per se*. Additional questions regarding the ability of this model to reflect researchers' current understanding of the carcinogenic process in humans have been raised. Beyond criticism of the use of a maximum tolerated dose in cancer bioassays, concern has arisen over predominant use of linear extrapolation from the high doses

used to the low doses of concern to the human population.

A commonly employed algorithm for extrapolation of experimental animal data to low dose is the linearized multistage model (Armatage & Doll, 1961). This mathematical model assumes that cancer develops by multiple steps and that these steps may be additive (Crump et al, 1976):

$$RfD = \frac{NOAEL\ (LOAEL)}{Uncertainty\ Factor(s) \times Modifying\ Factor}$$

The use of the upper confidence levels of the data representing α_1 is common and causes this first term to dominate the expression. U.S. EPA refers to this value as the q* and it is scaled for regulatory use by accounting for the difference in surface area between the animal used in the critical study and humans. This protective approach assures that the actual probability of encountering cancer under the experimental conditions is likely to be below the estimate obtained and may be as low as zero. The result obtained from this approach is functionally equivalent to the one hit model, and dose-response linearity is

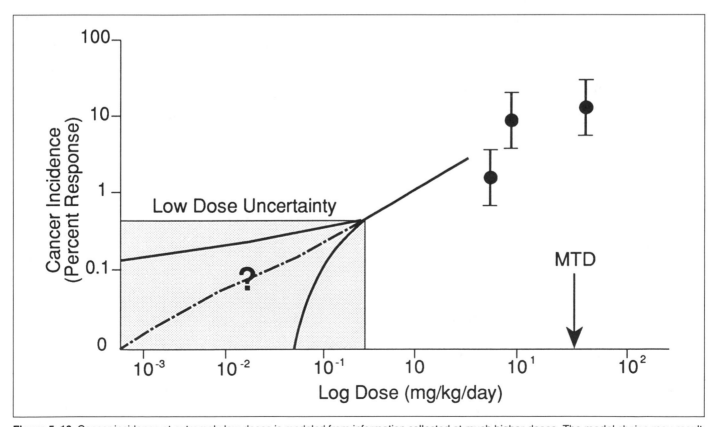

Figure 5–16. Cancer incidence at extremely low doses is modeled from information collected at much higher doses. The model choice may result in widely differing cancer risk estimates at the low doses commonly encountered in human exposure assessment.

assured. Implicit in the linear models for cancer risk assessment is the notion that there is no threshold for the induction of cancer and that any dose—no matter how small—has some finite probability of producing an effect. Several alternative algorithms that predict cancer risk at low doses are available in the literature. The reader is referred to more comprehensive texts for in depth discussion of these issues (Gad & Weil, 1991; Hayes, 1989).

Noncancer effects in humans. Significant advancement in the science of toxicology has led to a more broad, in-depth understanding of many toxic mechanisms. Researchers have come to appreciate that noncancer endpoints may be both important and widely diverse. An explosion of research in the area of immunotoxicology, in part associated with the interest in the AIDS epidemic, has led to advances in the understanding of toxic effects of environmental agents on the immune system. Researchers have learned, in part through tragedies such as the inappropriate use of diethylstilbestrol and thalidomide as therapeutic agents during pregnancy, that potential effects on the development of the fetus cannot be ignored. However, given the large number of new chemicals developed each year, thorough toxicological investigation of each new product is not feasible.

Toxicologists are working to develop predictive tools to understand the potential toxic activity of a chemical based on its structure. Computerized simulation of the physiological fate of a chemical is a rapidly expanding field of toxicology. Such physiologically based pharmacokinetic (PBPK) modeling is likely to play a more important role in the future of regulatory toxicology and risk assessment.

Most toxicants display inherent properties that may be quite specific. The chemical or physical nature of the substance causes a particular effect on a particular target site. For example, ethylene glycol (commonly used as antifreeze) is converted by metabolic activity in the body to oxalate. Ingestion and metabolism of ethylene glycol causes the formation of crystals of oxalate in the kidney where the cellular damage is greatest and can quickly become lethal. Interestingly, the same metabolic activity involved in converting ethylene glycol to oxalate is involved with the biotransformation of other alcohols. Ethanol—the alcohol found in liquor—interferes with ethylene glycol metabolism and, therefore, is commonly employed as an antidote for ethylene glycol poisoning.

Other toxicant mechanisms, although studied extensively, are not as well understood. For example,

organic hydrocarbons are suspected of causing toxicity by disturbing the integrity of cellular membranes in the nervous system, but little specific information is available regarding the exact mechanism of toxicity of these compounds.

Epidemiological studies. Human data that are of sufficient quality for determining the health effects of chemicals is relatively rare. To be most useful, such epidemiological data must:

- be conducted on a large enough population to allow meaningful statistics
- have adequate measures of exposure or dose
- have sufficient quality control such that the study can be scientifically interpreted or repeated by independent investigators
- have a comparable control population that has not been exposed to the toxicant of concern.

When epidemiological studies are conducted well, they are extremely important for learning the effects of exposure to environmental contaminants. Recent advances in scientists' understanding of health effects related to childhood lead exposure are testaments to the usefulness of sound investigations into human exposures to environmental toxicants. Promising work on the health effects of urban air pollution is likely to improve researchers' understanding of health effects related to ozone, particulate, and other noxious airborne contaminants (see Ozone in Chapter 14, Global Issues). Still, further epidemiological work is needed on exposure-related contaminants affecting the nervous system, reproductive system, and development of the fetus.

Extrapolation from animals to humans. In the absence of viable human data on the effects of environmental contaminants, animal data are used for setting safe exposure limits. By far, most information used for the regulation of exposure to contaminants is derived from animals. Such data are useful if carefully interpreted by trained toxicologists. Differences in physiology and biochemistry of experimental species when compared to humans can be important and should always be considered. Scaling for weight, surface area, or metabolic rate differences between humans and the experimental species of interest is a minimal requirement for extrapolating from animals. Differences in absorption, gastrointestinal transit time, biochemical metabolism, and excretion rates are only a few of the important characteristics to consider in comparing animal data to that for humans. Such studies have resulted in advancing modern medical science to its present state and remain the cornerstones of modern biomedical science and the subdiscipline of toxicology.

Interpretation of animal studies for the purpose of environmental regulation is conducted by a variety of federal and state agencies. U.S. EPA interprets animal data for the purpose of setting limits for most nonfood exposures and for food exposures under the Federal Insecticide, Fungicide, and Rodenticide Act (FIFRA). The process of setting exposure recommendations for noncarcinogens is conducted by EPA's Reference Dose Workgroup under procedures outlined earlier. (Chapter 11, Public Health Issues, presents detailed discussion of the use of basic toxicology principles for characterizing risk in an environmental setting.)

Toxicological information services. The task of interpreting scientific data and using this information for sound regulatory decision making is borne by local, state, and federal regulatory agencies. The U.S. EPA, through activities that support and maintain the Integrated Risk Information System (IRIS) database interprets experimental and epidemiological information for cancer (and noncancer) risk assessment. Many other toxicological information services are available through a variety of data bases accessible to the environmental health professional at nominal cost:

- the peer-reviewed Hazardous Substances Databank (HSDB)
- Registry of Toxic Effects of Chemical Substances (RTECS), which is maintained by the National Institute for Occupational Safety and Health (NIOSH)
- the Chemical Carcinogenesis Research Information System (CCRIS) sponsored by the National Cancer Institute
- Gene-Tox, which is supported by U.S. EPA and contains mutagenicity information on over 3,000 chemicals
- the Toxic Release Inventory, mandated by the Superfund Amendments and Reauthorization Act of 1986 (SARA) and also maintained by the U.S. EPA
- the Developmental and Reproductive Toxicology Database (DART), which is maintained jointly by U.S. EPA and the National Institutes for Environmental Health Sciences (NIEHS).

TOXNET, a computerized system of files on toxicology, integrates access to DART files and is available through the National Library of Medicine.

SUMMARY

As initially discussed, the dose is critical to a rational determination of potential toxicity. In an environmental setting, the dose presents the result of complex fate-transport activities. Thus, the ability to accurately predict the fate-transport of a material in soil, water, and air provides essential information for toxicity assessments. The fundamental concepts presented in this chapter are the building blocks for the discussions presented in the Waste Management (Part 2) and Special Concerns (Part 3) sections of this book.

REFERENCES

Amdur MO, Doull J, Klassen CD (eds). *Casarett and Doull's Toxicology: The Basic Science of Poisons,* 4th ed. New York: Pergamon Press, 1991.

Armitage P, Doll R. Stochastic models for carcinogenesis from the Berkeley symposium on mathematical statistics and probability. Berkeley, CA: University of California Press, pp. 19–38, 1961.

Barnes DG, Dourson M. Reference dose (RfD): Description and use in health risk assessment. *Reg Tox and Pharm* 8:471–486, 1988.

Crump KS, Hoel DG, Langley CH et al. Fundamental carcinogenic processes and their implication for low dose risk assessment. *Cancer Res.* 36:2973, 1976.

FR 51, 185, 1986 Cancer Guidelines. Washington DC: U.S. Government Printing Office, September 24, 1986.

Freeze RA, Cherry JA. *Groundwater.* Englewood Cliffs, NJ: Prentice Hall, 1979.

Gad S, Weil CS. *Statistics and Experimental Design for Toxicologists,* 2nd ed. Boca Raton, FL: CRC Press, 1991.

Gardner WH. Physical properties (of soils). In Parker SP, Corbitt RA (eds), *Encyclopedia of Environmental Science and Engineering,* 3rd ed. New York: McGraw-Hill, 1993.

Harris CC. At the crossroads of molecular carcinogenesis and risk assessment. *Science* 262:1980–1981, 1993.

Hayes WA (ed). *Principles and Methods of Toxicology,* 2nd ed. Raven Press, New York, 1989.

Heath RC. *Basic Ground-Water Hydrology.* United States Geological Survey Water-Supply Paper 2220.

North Carolina Department of Natural Resources and Community Development, 1989.

Herbst AL, Ulfelder H, Poskanzer DC. Adenocarcinoma of the vagina: Association of maternal stilbesterol therapy with tumor appearance in young women. *N Engl J Med* 284:878, 1971.

Jarabek AM, Menache MG, Overton Jr. JH et al. Inhalation reference dose (RfD$_i$): An application of interspecies dosimetry for risk assessment of insoluble particles. *Health Physics* 57:177–183, 1989.

McLane CF. The science of ground water hydrology. In *Practical Environmental Science Course.* Arlington, VA: Government Institutes, September 1992.

Mercer JW, Faust CR. *Ground-Water Modeling.* National Water Well Association, 1986.

Miller DW. Chemical contamination of groundwater. In Ward CH, Giber W, McCarty PL (eds), *Ground Water Quality.* New York: John Wiley & Sons, pp. 39–52, 1985.

Miller DW. *Groundwater Contamination.* Plainview, NY: Water Information Center, Inc., 1984.

Rustia M, Shubik P. Effects of transplacental exposure to diethylstilbestrol on carcinogenic susceptibility during postnatal life in hamster progeny. *Cancer Res* 39: 4636, 1979.

Smith, GD. Soil. In Parker SP, Corbitt RA (eds). *Encyclopedia of Environmental Science and Engineering,* 3rd ed. New York: McGraw-Hill, 1993.

Thibodeaux LJ. *Chemodynamics: Environmental Movement of Chemicals in Air, Water, & Soil.* New York: John Wiley & Sons, 1979.

U.S. EPA. Guidelines for exposure assessment. *Federal Register* 57(104):22888–22932, 1992.

U.S. EPA. Selection criteria for mathematical models used in exposure assessments:
— Atmospheric Dispersion Models (EPA/600/8–91/038, March 1993).
— Surface Water Models (EPA/600/8–87/042, July 1987).
— Ground Water Models (EPA/600/8–88/075, May 1988).

U.S. EPA. *Human Health Evaluation Manual, Part A. Risk Assessment Guidance for Superfund.* December 1989.

Williams PL, Burson JL (eds). *Industrial Toxicology.* New York: Van Nostrand Reinhold, 1985.

Wolff MS, Toniolo PG, Lee EW et al. Blood levels of organochlorine residues and risk of breast cancer. *J Nat Cancer Inst* 85:648–657, 1993.

6

Managing Environmental Resources

Anthony Veltri, PhD

During the past two decades, businesses and industries in the United States have been challenged by environmental regulatory authorities to formulate a reliable management strategy and organization structure for sustaining environmental resources. In addition, international initiatives to incorporate environmental considerations in strategic planning efforts of multinational firms have increased. Some firms are confronting this challenge by formulating a management strategy and organization structure that not only complies with environmental regulations but also places the same pressure on environmental performance that shareholders place on economic performance. These firms see a relationship between environmental performance and economic performance. As a result, their environmental management functions are solidly funded and considered a critical operating function that significantly contributes to the firm's long-term competitiveness.

Environmental issues have become so dominant that leading corporations are finding ways to integrate environmental planning into their business strategies (Barnes & Ferry, 1992). Organizations such as the International Chamber of Commerce (ICC) and the Business Council for Sustainable Development (BCSD) are having a major impact on global environmental management performance. Their mission is to enhance the environmental performance of business and industry so that sustainable development is achieved. A 16-point program for environmental management has been developed by the ICC:

1. *Corporate priority.* To recognize environmental management as among the highest corporate priorities and as a key determinant to sustainable development; and to establish policies, programs, and practices for conducting operations in an environmentally sound manner.
2. *Integrated management.* To integrate these policies, programs, and practices fully into each business as an essential element of management in all its functions.
3. *Process of improvement.* To continue to improve corporate policies, programs, and environmental performance, taking into account technical developments, scientific understanding, consumer needs, and community expectations, with legal regulations as a starting point; and to apply the same environmental criteria internationally.

135

4. *Employee education.* To educate, train, and motivate employees to conduct their activities in an environmentally responsible manner.

5. *Prior assessment.* To assess environmental impact before starting a new activity or project and before decommissioning a facility or leaving a site.

6. *Products and services.* To develop and provide products or services that have no undue environmental impact and are safe in their intended use, that are efficient in their consumption of energy and natural resources, and that can be recycled, reused, or disposed of safely.

7. *Customer advice.* To advise and, where relevant, educate customers, distributors, and the public in the safe use, transportation, storage, and disposal of products provided; and to apply similar considerations to the provision of services.

8. *Facilities and operations.* To develop, design, and operate facilities and conduct activities taking into consideration the efficient use of energy and materials, the sustainable use of renewable resources, the minimization of adverse environmental impact and waste generation, and the safe and responsible disposal of residual wastes.

9. *Research.* To conduct or support research on the environmental impacts of raw materials, products, processes, emissions, and wastes associated with the enterprise and on the means of minimizing such adverse impacts.

10. *Precautionary approach.* To modify the manufacture, marketing, or use of products or services or the conduct of activities, consistent with scientific and technical understanding, to prevent serious or irreversible environmental degradation.

11. *Contractors and suppliers.* To promote the adoption of these principles by contractors acting on behalf of the enterprise, encouraging, and, where appropriate, requiring improvements in their practices to make them consistent with those of the enterprise; and to encourage the wider adoption of these principles by suppliers.

12. *Emergency preparedness.* To develop and maintain, where significant hazards exist, emergency preparedness plans in conjunction with the emergency services, relevant authorities and the local community, recognizing potential transboundary impacts.

13. *Transfer of technology.* To contribute to the transfer of environmentally sound technology and management methods throughout the industrial and public sectors.

14. *Contributing to the common effort.* To contribute to the development of public policy and to business, governmental, and intergovernmental programs and educational initiatives that will enhance environmental awareness and protection.

15. *Openness to concerns.* To foster openness and dialogue with employees and the public, anticipating and responding to their concerns about the potential hazards and impact of operations, products, wastes, or services, including those of transboundary or global significance.

16. *Compliance and reporting.* To measure environmental performance; to conduct regular environmental audits and assessment of compliance with company requirements, legal requirements, and these principles; and periodically to provide appropriate information to the board of directors, shareholders, employees, the authorities, and the public.

Likewise, many national and international firms—such as the Southern California Edison Company's electromagnetic field research, 3M's Pollution Prevention Pays 3P Program, Dow Chemical's Work Reduction Always Pays Program, and McDonald's Environmental Packages Program—are constantly building distinctive competencies and capabilities for creating environmentally sustainable corporations.

In 1989, the Conference Board performed an extensive survey of corporate environmental policy issues. The survey included 1,200 U.S. companies, 1,100 European Economic Community (EEC, renamed Economic Union, EU) companies, and 535 Canadian companies. The Conference Board is a large multinational business roundtable organization founded in 1916 that provides a forum for senior executives to explore and exchange ideas on business policy and practices. At the Board's September 1989 International Industrial Conference of world business leaders, environmental issues were the subject most often mentioned at Boardroom sessions. Thus, it is clear that environmental issues occupy the center stage in the minds of senior corporate leaders. The 1989 Conference Board survey was published in 1991 (Conference Board Report Number 961, 1991). The Conference Board found that 70% of the surveyed firms had a formal system in place for identifying key environmental issues (Conference Board, 1991).

Many firms are integrating environmental management into the corporate strategy and scanning the environmental horizons for business opportunities that will provide a new source of competitive advantage (Winsemius & Guntram, 1992). However, other firms are reluctantly accepting this challenge, claiming that the implementation of a reliable environmental management strategy and organization structure would significantly impair their ability to compete. These reactive firms view environmental management as an unnecessary cost center and see no economic value associated with environmental management performance. They tend to automatically oppose all such calls for environmental control and improvement. Because of the expense involved, the reactive firms seek to delay the installation of problem-preventing equipment and respond to the calls for such items only after bitterly contested laws and regulations force their reluctant compliance. These firms would much rather budget for environmental problems as they occur. As a result of this mindset, reactive firms approach environmental problems with short-term financial solutions. Invariably, this thinking leads to costly after-the-incident response and recovery efforts that subsequently impact competitive performance.

Because environmentally reactive firms tend to be more interested in short-term solutions than in long-term environmental sustainment, they fail to see the benefits of a reliable environmental management strategy and organization structure. But modern environmental management strategy and organization structure is not about the short term. It is about staging for long-term environmental resource sustainment. If not solidly funded and strategically managed, lack of an environmental management strategy may leave a company at a distinct disadvantage relative to competitors with greater foresight and seriously affect the long-term viability of the firm.

The conflict between environmental resource sustainment and economic competitiveness is a false dichotomy. This dichotomy stems from a narrow view of the sources of a firm's economic competitiveness and a belief that the environmental management movement will stagnate (Porter, 1989). As this dichotomy becomes increasingly debated, the need for a reliable environmental management strategy and an organizational structure that enhances compliance with regulatory requirements while contributing to sustainable development and competitive

performance is becoming more apparent for business and industry.

The purpose of this chapter is twofold. (1) to present to organizations that do not yet have an environmental program or that wish to audit or upgrade their existing program a scheme for managing the environmental challenge and (2) to provide specific examples of existing programs developed by major multinational companies. The scheme is presented in three stages or dimensions. The first dimension presents a strategy, a structure, and a preferred positioning-location arrangement within the firm's overall organizational chart. The second dimension focuses on an implementation strategy that translates and converts environmental strategy and structure into actions that get completed. Finally, the third dimension concerns evaluating performance, as Table 6–A shows.

Much of the strategy and structure formulated for the environmental management function has been directed toward large firms, essentially ignoring small firms. Environmental management strategy and structure for large firms is not wholly generalizable and transferable to small ones. Small firms perform under conditions that distinguish them from large ones; they are not simply smaller versions of large organizations. Therefore, the size of a firm and its specific environmental regulatory requirements and range of environmental risk and liability appear to be relevant variables to consider in formulating an environmental management strategy and structure.

Nevertheless, this chapter is intended to guide environmental management decision-making capabilities for both large and small firms in terms of the three-dimensional scheme. Therefore, the chapter makes no basic distinction among and between environmental managers. The firms in which managers serve are different, the scope of authority held may

Table 6-A. Three-Dimensional Environmental Management Scheme

Dimension		
1	*2*	*3*
Management strategy Organizational structure Positioning arrangement	Implementing strategy	Evaluating strategic performance

vary, and the type of environmental compliance and magnitude of liability may be considerably different. However, environmental managers, all of whom are focused on a mission intended to prepare, protect, and preserve environmental resources, can employ the three-dimensional scheme to help guide the overall management of their environmental functions.

DIMENSION 1—MANAGEMENT STRATEGY, ORGANIZATIONAL STRUCTURE, AND POSITIONING ARRANGEMENT

Much of the strategy formulated for the environmental management function tends to focus on factors external to the firm (i.e., maintaining and enhancing compliance with regulatory requirements). Although justifiable as far as it goes, this external focus has tended to distract the environmental management function from focusing on internal factors that are likely to contribute to the firm's environmental sustainability and competitive performance. To some extent this imbalance of strategic focus has resulted in senior-level executives becoming disillusioned with the environmental management function. The matter has become so serious that senior-level executives may exclude the environmental management function from the competitive performance strategy process of the firm. Such a decision should not be a viable alternative.

To correct this circumstance and revitalize senior-level executives, organizations need a whole new conceptual model for formulating strategy, organizing structure, and positioning the environmental management function within the overall organization chart of the firm. This conceptual model should go beyond maintaining regulatory compliance to one that contributes to the firm's environmental sustainability and competitive performance.

Strategy Formulation

Strategy formulation is the process of determining and describing the long-term manner in which the environmental management function plans to contribute to the regulatory compliance, environmental sustainability, and competitive performance standards of the firm. This discussion outlines the application of typical fundamentals of strategy formulation within the context of the environmental management function.

Advancements in strategy formulation have been (refer again to Table 6–A):

- the envisioning of a "strategic intent" that focuses on building distinctive environmental management competencies and capabilities
- the development of a mission statement that serves as a guideline for strategic management and technical decision making and operating actions
- the construction and use of achievement-level statements (objectives, operational strategies, and tactical maneuvers) that are intended to be implemented and evaluated in terms of their efficacy
- the identification of performance and/or reform initiatives for transforming environmental management from a reactionary-centered strategy to one that focuses on contributing to the firm's regulatory compliance, environmental sustainability, and competitive performance.

Thus, fundamental advancements in strategic formulation are:

- envisioning a strategic intent
- developing a mission statement
- using achievement-level statements
- identifying initiatives.

Envisioning a strategic intent. Any modern attempt to formulate strategy and organization structure for the environmental management function needs to begin with a strategic intent that exceeds the organization's grasp and existing resources. Then, conditions must be arranged to close the gap by setting strategic and tactical challenges that focus efforts upon a mission statement. *Strategic intent* basically means the envisioning of building and sustaining distinctive organizational competencies and capabilities for becoming recognized as world class (i.e., being better than every other company in the industry in at least one important aspect of environmental performance). Strategic intent captures the essence of winning by (1) motivating personnel to tap their creativity, (2) stretching their resources, (3) leveraging their competencies and capabilities, and (4) communicating the value of the world-class challenge to all levels in the organization.

Companies that have risen to global leadership during the past 20 years invariably began with work ambitions that were out of all proportion to their resources and capabilities (Hamel & Prahalead, 1989). Similarly, if any environmental management function is to become recognized as a "world-class" function within its industry type, it too will need to establish and manage ambitious strategic and tactical work intentions, stretch its resource base, and leverage its competencies and capabilities.

"To become recognized as and remain the best in the industry in at least one important aspect of environmental management performance" is an example of a strategic intention statement that has broad appeal. Examples of distinctive competencies and capabilities that the environmental management function could envision include these:

- to lead the way into formulating new strategic and technical aspects for preparing, protecting, and preserving environmental resources within its industry
- to sustain low-cost, high-quality, and short developmental and implementation cycles compared to similar industries
- to maintain the lowest hazardous material disposal costs in the industry
- to become the industry leader in the minimization of hazardous waste generation, release, and emissions
- to achieve the best rate of reduced capital costs for "end-of-pipe" control equipment
- to be the quickest to adapt imaginatively and effectively to environmental protection compliance changes using multidisciplinary approaches and multilevel teams
- to maintain a staff so competent and capable that other companies in the industry are continually seeking to recruit them
- to become recognized in the industry as the leader in developing new, lower-cost pollution reduction technologies.

These strategic intention examples give organizations the only achievement level that is worthy of total commitment, that is, to become the best in their industry type.

Development of a mission statement. A *mission statement* describes the ambitious long-term strategic purpose of the function and is intended to guide decision making and operating actions. The statement of mission is the cornerstone on which the organization structure is developed and strategies and tactics implemented and evaluated. The basic question that should be answered in order to formulate an environmental management mission statement is "What is our strategic purpose?"

"Optimal competitive performance within the enterprise derived from cost effective use of resources" is recognized as being applicable to any business and industrial enterprise. Such a statement has an overarching relationship to the various strategic business units and operating functions, including the environmental management function. This example of a firm's competitive performance strategy serves as a guideline statement for the environmental management function to develop a statement of mission that supports the organization's competitive performance strategy. One way of supporting the competitive performance strategy is to ensure that the environmental management mission statement is congruent with the intent of the firm's competitive performance strategy.

"Optimal preparation, protection, and preservation of the firm's environmental resources" is an example of an environmental management mission statement that describes the congruency between the function's mission and the firm's competitive performance strategy. The rationale and analytical model shown in Figure 6–1 illustrate this congruency.

A firm's ability to fulfill its competitive performance strategy depends on three factors: first, the attractiveness of the industry in which it is located; second, its establishment of competitive advantage over competing industries; and third, appropriate use of assets and resources. Because assets and resources are used to accomplish the firm's competitive performance strategy, its resources possess unique value and worth of sufficient importance that *risks* (potential harm), *liabilities* (vulnerability to sources of harm), and *losses* (actual harm sustained) to which they might be subjected must be effectively and efficiently counteracted. The environmental resources of the firm play a pivotal role in the competitive performance strategy that a firm pursues.

An environmental management function within any given enterprise should possess a statement of mission on which those strategies and tactics are based to counteract risk, liability, and loss. As history has of course proven repeatedly, preparing, protecting, and preserving environmental resources is much cheaper than responding and recovering from environmental incidents. It is not good business practice to subject the organization's environmental resources to risks, liabilities, and losses. Therefore, strategic and tactical decision making and operating actions must address ways to effectively and efficiently prepare for risk(s), protect against danger(s), and prevent loss(es) of the firm's environmental resources.

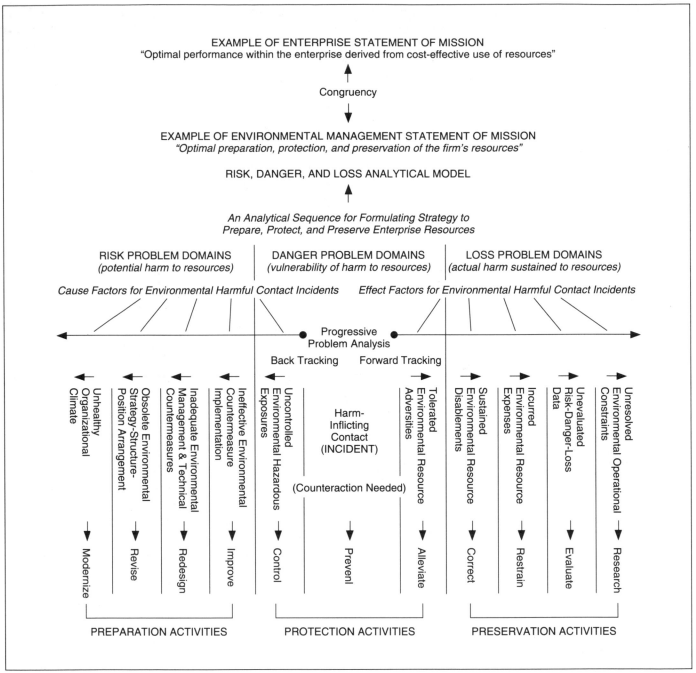

Figure 6–1. Example of enterprise statement of mission. (Source: C. Everett Marcum, Professor of Safety Studies, West Virginia University; Anthony Veltri, Associate Professor, Safety Studies, Oregon State University.)

Verifying the agreement between the environmental mission statement and the firm's competitive performance strategy and communicating this agreement to senior-level executives is an important step in staging for any long-term commitment. Verification is vital for building stability and continuity in the environmental management function and for resolving any elusive relationship that could exist between environmental sustainment and competitive performance.

Using achievement levels. *Achievement levels* are a hierarchy of objective statements that describe the intended focus and benchmarks for the environmental management function. This hierarchy of achievement levels in which the strategic intent and statement of environmental mission sets the stage for strategy formulation tends to ensure balance and consistency within the overall environmental management plan. Figure 6–2 illustrates a typical achievement-level statement and its intended focus.

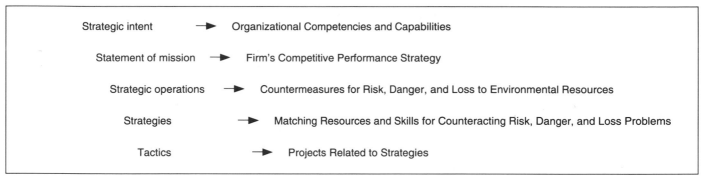

Strategic intent   →   Organizational Competencies and Capabilities

Statement of mission   →   Firm's Competitive Performance Strategy

Strategic operations   →   Countermeasures for Risk, Danger, and Loss to Environmental Resources

Strategies   →   Matching Resources and Skills for Counteracting Risk, Danger, and Loss Problems

Tactics   →   Projects Related to Strategies

Figure 6–2. Environmental management achievement level diagram.

Different names are given to the different levels of achievement, primarily to promote, identify, understand, and communicate. Additionally, each of these kinds of statements must be planned, regardless of the separate achievement level being addressed, as a uniquely formulated objective guideline that is intended to serve two basic purposes, namely:

- to ensure that a "blueprint" exists to remain focused on, and later
- to evaluate later how effective and efficient the various levels of achievement seem to be operating.

Examples of the kinds of achievement-level statements include the strategic intention statement, mission statement, strategic operations, and strategies and tactics. Each achievement level guideline should be considered as a kind of uniquely constructed objective. Definitions for each achievement level are:

1. *Strategic intent:* The envisioning of building distinctive organizational capabilities and becoming recognized as world-class (i.e., being better than every other company in the industry in at least one important aspect of environmental management performance).
2. *Mission:* The highest level of achievement that is expected by the environmental management function that describes the ambitious, long-term strategic purpose of the function and is intended to guide decision-making and operating actions. The basic question that must be answered in order to determine an environmental management statement of mission is, "What is our business purpose?"
3. *Strategic operations:* The major efforts necessary to fulfill major purposes or resolve major problems, thus improving the potential for accomplishing the environmental management mission. Strategic operations (sometimes called *programs*) are countermeasure efforts that are designed and

intended to prepare for risk(s), protect against danger(s), and preserve against environmental loss(es).

4. *Strategies:* The statements that represent the major portions of the strategic operations intended to guide decision-making and operating actions necessary to achieve the strategic operation goal. Strategies can also mean the actions needed to obtain and match available resources, under advantageous arrangements, and to improve the potential for achievement.
5. *Tactics:* The statements that represent the most basic level of achievement and that are intended for use in implementing strategies. Tactics relate to small-scale actions that serve the strategy and include information that emphasizes the degree of attainment expected in terms of measurable quality, quantity, order, time, money, sequence, object, machine device, human factor, or behavior demanded for successful completion. Tactics can also mean the skillful and effective use of resources to solve a portion of some particular problem and contribute toward achievement of strategy.

Usually, technically prepared specialists are best suited to perform most tactical activities, and environmental strategists must be willing to use their expertise to good advantage. Tactical maneuvers might include:

- designing devices intended to provide environmental protection
- analyzing exposures to environmental hazards
- determining use of specialized environmental instrumentation and monitoring equipment
- developing new lower-cost pollution reduction procedures or technologies
- providing data necessary for setting realistic emission-reduction targets and environmental impact considerations

- performing environmental audits and permit proposals; and interpreting data pertaining to environmental risk, danger, and loss resource problems.

Other tactical activities do not require the high level of technical expertise implied by the previous examples in the definition and may, in fact, be performed by persons who possess only technician-grade competencies. Some of these examples include:

- maintaining material safety data sheets
- monitoring data on location
- monitoring quantity of hazardous substance used
- keeping certain environmental records
- writing reports.

Relevant education and training seem to be all that is necessary to prepare individuals to fulfill such responsibilities. Personnel throughout all business and operating units may be needed to provide advice and perform these kinds of tactical activities.

Tactical activities that focus on counteracting environmental hazardous exposures and related compliance should not be disproportionately emphasized within the environmental management strategic plan. Attempting to formulate any strategic plan on a diversified array of tactical activities will usually become a disorganized experience and one that becomes extremely difficult to measure, evaluate, and justify to senior-level executives. Yet, one major reason for unsuccessful environmental management efforts involves the attempts to bypass the formulation of necessary strategies and the continuing overemphasis on the employment of tactical activities.

Strategies should become the basis for selection and implementation of tactical activities rather than vice versa. Tactical activities are the ones that require the use of strategically procured and assembled resources. The strategic intent and the statement of environmental mission should be used consistently to secure resource allocations proposed to senior-level executives rather than simply relying on tactical maneuvers for complying with mandates from regulatory sources. Senior-level executives tend to budget strategy in terms of its potential for gaining a sustainable competitive advantage rather than on its compliance value.

The concept of using achievement levels grows from a variety of strategic management techniques. The evolving idea is a product of the current thinking about management as well as a product of the history of management. Regrettably, for many firms the idea of designing environmental management performance achievement levels creates fear of structure and of being held accountable for any preestablished strategy and structure. There is no denying the descriptive validity of this concern, yet the personnel and financial resources that must be applied to the firm's environmental risk, danger, and loss problems are limited. Therefore, the use of achievement-level statements provides a model for managing activities pertaining to this important organizational function.

Reform initiatives for transforming environmental management strategy. Transforming the environmental management strategy from the traditional to one that contributes to the competitive performance strategy of the firm requires perceiving and structuring a set of new reform initiatives and mobilizing commitment to those initiatives. As discussed earlier in the chapter, the strategic intent of the environmental management function is to build distinctive organizational capabilities along one or more dimensions of environmental management performance.

But if one intends to develop distinctive competencies and capabilities, one's efforts must be directed not entirely toward regulatory compliance but rather toward the development of specific initiatives that contribute to competitive performance. Once it has been decided what kinds of distinctive competencies and capabilities the environmental management function and the firm is going to seek, the function has to transform itself in such a way that it can achieve and continually enhance them. This requires making a series of coordinated decisions that are transforming in nature. The following reform initiatives refer to such "bricks and mortar" decisions.

- *Initiative I.* Rethink how environmental management strategy is formulated.
- *Initiative II.* Reconstruct the organization structure of the environmental management function.
- *Initiative III.* Rethink the organizational positioning-location arrangement of the environmental management function within the organization chart.
- *Initiative IV.* Modify the financial tools in determining (1) financial and economic impact of environmental incidents, (2) the fiscal allocations needed for establishing countermeasures, and (3) economic performance ratios needed for tracing financial performance.
- *Initiative V.* Improve the process for preparing senior-level executives, managerial, supervisory,

and line-level employees to make solid decisions about environmental resource sustainment.

- *Initiative VI.* Revise the strategies for complying with environmental legislation.
- *Initiative VII.* Revise strategies for minimizing environmental liability exposure to the firm.
- *Initiative VIII.* Modify the methods for establishing performance criteria for measuring, evaluating, and benchmarking performance.
- *Initiative IX.* Develop and install management information systems to improve decision-making and operating actions' capabilities for setting priorities concerning environmental performance.
- *Initiative X.* Establish the means for designing and conducting environmental management operations research projects.

These initiatives are designed and intended to create a sense of direction that shapes the decisions and operating actions the firm will take and determines the character of the environmental management function. These initiatives also underpin the architecture for becoming a world-class company.

Organizational Structure

The purpose of this discussion is to provide a conceptual framework for structuring environmental management strategy. Specifically, this discussion (1) reviews approaches and arrangements, (2) addresses the concept of structuring a strategy that fits, and (3) presents two contrasting approaches for structuring environmental management strategy.

Approaches and arrangements to structuring strategy. There is no boilerplate approach and arrangement that works best for structuring environmental management strategy. In general, firms have structured their environmental management strategy in two ways:

- decentralized arrangements in which strategic business units and operating functions perform all strategic and tactical decision making
- centralized arrangements in which strategic and tactical decision making is done by senior-level executives and strategic business units and the operating functions carry out strategic and tactical assignments.

Recent efforts in revamping organization structure have been designed to draw on the information and beliefs of the employees. It seems reasonable that at least some employees—based on their first-hand experiences, their second-hand information, and

their own creative perception and insight—possess certain well-thought-out strategies and tactics for counteracting risk(s), liability(ies), and loss(es) to the firm's environmental resources. The challenge in implementing this approach to structure has been to effectively tap this store of creativity, knowledge, and judgment from the different organizational levels.

Currently, small, multilevel, interdisciplinary employee teams are used. These teams possess complementary competencies and capabilities. They have a common strategic intent and mission, a set of strategic and technical performance benchmarks, and an implementation approach that includes an evaluation phase. The challenge is to identify the environmental problem, performance expectations, and parameters for the team and then to allow team members to craft their strategy, structure, implementation, and evaluation plan for the environmental problem identified.

The purpose of organization structure is to weld the environmental management strategy and structure together to help the various strategic business units and operating functions within the firm to focus their attention and expertise on specific environmental risk, liability, and loss problems. Usually, environmental management structure tends to fall into one of three developmental levels, characterized as follows.

Level 1. Reactive. A structure for organizing strategy does not clearly exist. Environmental management responsibility tends to be diversely shared among various staff members with no fixed arrangement, approach, or accountability. The environmental management function takes its directions from committees that are organized by the company for creating and maintaining a commitment to environmental protection. The organization structure tends to be regulatory compliance driven.

Level 2. Ordinary. At this level, the arrangement and approach to structuring strategy is to integrate (downward) in the organization and to be near the operating levels. Primary focus is on solving regulatory compliance and environmental exposure problems affecting operating levels. The organization structure tends to be technically driven with emphasis on inspection/enforcement/training/engineering practices.

Level 3. Extraordinary. The arrangement and approach to structuring strategy at this level is to integrate (vertically and laterally) within the overall company organization chart and provide strategic management and technology transfer advisory services. Environmental management functions at this level view the organization structure as a means to remain constantly tuned into environmental risk, liability, and loss problems, needs, expectations, and requirements of each strategic business unit and operating function user. The organization structure tends to be a well-balanced mix of strategies and tactics that are congruent with the competitive performance strategy of the company. The function takes its direction from the findings obtained from special assessments, audits, and studies conducted on environmental risk, liability, and loss problems.

Concept of structural fit. Any modern attempt at structuring environmental management strategy that desires to contribute to the company's competitive performance should begin with a strategic intent that stretches its resource base and leverages, its competencies, and its capabilities. Next, conditions must be arranged to close the gap by setting strategic and tactical challenges that focus efforts on a mission statement. Then, an organizational chart (pictorial representation) can be drawn that configures and welds the function together in a manner to achieve structural fit. When these structural fit concepts are mismatched—when the wrong strategy is put together—the environmental structure will tend to be ineffective and inefficient. Approaches to structuring environmental strategy should be primarily focused on achieving *structural fit* (i.e., ensuring congruency between the strategic intent of the environmental management function, mission, strategy, and structure).

Constructing the organization structure after formulating the strategy tends to *enhance* structural fit. Some of the worst mistakes in designing for achieving structural fit have been made by imposing a concept of an ideal structure without attending initially to the strategy formulation process.

Although there is no universally agreed-on framework for structuring environmental management strategy, the following guidelines should be used to improve decision making for structuring environmental management strategy (Veltri, 1985):

- Methods of organizing that emphasize structuring of the environmental management function *after formulating its strategy* tend to arrange conditions for implementation practices to be executed.
- Methods of organizing that emphasize structuring the environmental function *in terms of low specialization of labor, heterogeneous workgroups,* *wide spans of control, and decentralized authority* tend to permit flexible and adaptive operations in changing and diversified organizations.
- Methods of organizing that emphasize arranging conditions for *environmental managers and supervisors to have a small number of subordinates under their immediate jurisdiction* tend to foster a layered organizational structure that establishes the basis for an increase in supervisory control.
- Methods of organizing that emphasize arranging conditions for *environmental managers and supervisors to have a large number of subordinates under their immediate jurisdiction* tend to foster a flat organizational structure that encourages more employee initiatives.
- Methods of organizing that emphasize arranging conditions for *decision making about operating action to be made at the organizational levels where the information is available* tend to influence responsible organizational decision making and improve the upward and downward flow of communication.
- Methods of organizing that emphasize arranging conditions for *specialized work assigned to individuals within limits of human ability and skill* tend to influence the potential for effective and efficient work performance.
- Methods of organizing that emphasize *downward delegation of authority* tend to influence organizational personnel to exercise more autonomy and motivate participation in problem solving that leads to improved creativity.

Now, not all firms need all of these guidelines. Some use few and are simple; others combine all of them in rather complex ways. Nevertheless, to effectively and efficiently structure strategy intended to prepare, protect, and preserve environmental resources, one must start with strategy that assures a structural fit. Structural fit, therefore, is the glue for bonding structure and strategy.

Peter Drucker, world-renowned organizational theorist, writes that:

. . . designing the organizational parts is the "engineering phase" of organization design. It provides the basic materials. And like all materials, these building blocks have their specific characteristics. They belong in different places and fit together in different ways. We have also learned that "structure follows strategy." We realize that structure is a means for attaining the mission of a function. And if a structure is to be effective and sound, we must start with strategy. This is perhaps the most fruitful new insight we have in the field of organization. Strategy that is the answer to the question: What it will be? determines the purpose of structure (Drucker, 1977).

One can conclude from Drucker's finding that the central purpose for organizing the environmental management function is to structure environmental management strategy toward a strategic intent and statement of mission. How that coordination is achieved dictates what the environmental management organization structure will look like.

Two contrasting approaches for structuring environment management strategy. Since the inception of the National Environmental Protection Act (NEPA), two contrasting approaches to structuring environmental management strategy have emerged. One approach, *compliance structure* (upon which most environmental strategy is organized), focuses on a strategic intent and a mission of maintaining compliance with mandates from environmental regulatory sources, as the left-hand column in Table 6–B indicates.

This approach to structuring strategy is to integrate downward in the organization and to solve regulatory compliance problems affecting line-level operations by using a command-and-control model. The operating staff is minimized, consisting of technical specialists engaged in environmental hazard recognition, control, and evaluation and regulatory compliance strategies and tactical maneuvers. Minimal use is made of the other employees in the producing and servicing units who could serve in a collateral duty capacity.

The other approach, *competitive performance structure,* focuses on a strategic intent of building distinctive organizational competencies and capabilities and on a mission of preparing, protecting, and

preserving the firm's environmental resources (Table 6–B, right-hand column). This approach to structure justifies its long-term budgetary existence by economic considerations as well as regulatory compliance considerations. The operating staff consists of specialists engaged in strategic management advising, technology transfer, and financial economic and evaluation services. The support staff is focused on the implementation of environmental management strategies and tactics within strategic business units and operating functions in which they are housed and serve. As is apparent from the table, the two structures are not mutually exclusive, but they do represent a significant difference in emphasis that deeply affects how strategy is formulated and how structure is approached.

Similarities and differences. Both structures recognize the problem of working in diverse, complex, and ever-changing firms with expanding regulations and usually with limited budget and resources. But the strategy in the first approach (compliance structure) is to match available resources and competencies with regulatory compliance demands. The strategy in the second approach (competitive performance) is to stretch its resource base and leverage competencies and capabilities in ways that enhance competitive performance.

Both structures recognize that reducing environmental incidents influences profitability motives. The first approach seeks to reduce financial-economic loss by maintaining compliance with standards. The second approach seeks to reduce financial-economic loss by managing a well-balanced and sustained attack on the risks, liabilities, and losses to the environmental resources of the company.

Both structures recognize the need for consistency in strategy and structure within the function. In the first approach, consistency is largely a matter of conforming to regulatory compliance demands by designing and implementing a wide array of tactical maneuvers that emphasize enforcement (command-and-control) and engineering controls usually aimed at line-level business and operating units. In the second approach, consistency comes from allegiance to a strategic intent and mission that is designed to contribute to the firm's competitive performance by formulating a wide array of strategies and tactics. This is accomplished by integrating vertically and horizontally within the various organizational and line levels, and providing strategic management advising and technology transfer services.

Table 6–B. The Two Approaches for Structuring Environmental Management Strategy

Compliance Structure	Vehicle	Competitive Performance Structure
Enhance compliance	Strategic intent	Build distinctive competencies and capabilities
Maintain compliance environmental regulation	Statement of mission	Optimal preparation, protection, and preservation of environmental resources

Both structures recognize the need for arranging conditions for environmental sustainment learning (i.e., the process by which an organization obtains and uses new information for detecting and correcting operational errors that could lead to environmental incidents). The first approach emphasizes arranging environmental learning situations that are regulatory-compliance oriented by ensuring conformance to regulatory mandates, standard operating procedures, and accepted industry practices and standards. In the compliance structure, when the error is detected and corrected in ways that permit the organization to carry on its present environmental management policies, strategies, and tactics, the error detection and correction process is maintenance learning. In environmental maintenance learning, performance is compared only with minimal standards, not with what might have been excellent performance or what is yet to be.

The second approach emphasizes environmental learning situations that search for advantages that are inherently sustainable and build distinctive environmental capabilities. In the competitive performance structure, environmental learning occurs when error is detected and corrected in ways that involve the modification of environmental management strategies. This error detection and correction process can be called *innovative environmental learning*. In innovative environmental learning, performance is assessed according to the amount of value provided to the firm's competitive performance. Innovative environmental learning focuses on anticipating and preparing to engage the firm in new environmental risk, liability, and loss situations. Because there are no familiar environmental contexts within which innovative environmental learning can take place, the construction of new environmental learning contexts is one of the aims of this approach. These new environmental learning contexts may be unique, so that there is no opportunity to learn by trial and error or by following a set of already existing standards.

From an organizational structure standpoint, environmental learning primarily concerns the ways in which members of the organization learn to detect risks to environmental resources before they result in incidents and losses. This is the ultimate goal of environmental management.

Both structures recognize the role that research and development play. In the compliance structure, research and development endeavors are focused on tactical maneuvers for improving methods, techniques, and approaches for complying with environmental regulations. In the competitive performance structure, research and development endeavors are focused on the development of innovative models for improving decision-making and operating action capabilities among the various organizational and line levels of the firm.

Both structures recognize the need for information systems to improve decision-making and operating action capabilities. In the compliance structure, information systems are concerned with collecting and reporting the effects of environmental incidents and losses. In the competitive performance structure, information systems are concerned with developing a database system for interrelating the smallest number and variety of environmental risk factors that appear to be connected to the largest number and variety of environmental incident and loss factors.

Almost all environmental management functions began their lives as compliance structures. The heyday of the compliance structure probably occurred shortly after the period of the passage of the NEPA. Even today the compliance structure remains widespread and a necessary structure for building up most new environmental management organizations. However, the competitive performance structure has been receiving an increasing amount of attention among environmental strategists who must justify their existence economically to senior-level executives and contribute to the firm's ability to compete.

Matching the environmental structure with performance. As previously mentioned, the purpose of organization structure is to weld the environmental management strategy and structure to help the various strategic business units and operating functions to focus their attention and expertise on specific strategies and tactics. There are various ways in which an environmental function can choose to structure itself; each approach will be influenced by its strategy. The approach also will be affected by how individuals and teams communicate with one another and integrate their activities.

A firm's environmental structure should reflect its concern for both compliance and competitive performance. But environmental organization structures tend to evolve in unpredictable ways and sometimes thwart the original strategy. Unless these dynamics are understood, harmless input will absorb all the energies of the participants without producing any significant output.

Positioning Arrangement

A dilemma commonly encountered by senior-level executives is determining the optimal organization location for positioning the environmental management function within the overall organizational chart of the firm. In fact, this issue has traditionally affected the occupational safety and industrial hygiene, disaster and emergency preparedness, corporate security protection, and fire prevention functions. This task is particularly difficult when one discovers that (1) there is no organizational location that is universally accepted and sufficient for positioning the environmental management function, and (2) there are no research studies or conceptual models to review that could facilitate decision making in this area.

In general, there are four positioning principles to guide organizational location efforts:

- The organizational location and positioning arrangement should be influenced by its formulated strategy and structure.
- The organizational location and positioning arrangement should be congruent (i.e., strategically fit) with the strategic focus of the company.
- The organizational location and positioning arrangement ought to accommodate the need for integrating work activities both vertically and laterally within the overall organizational structure of the firm.
- The organizational location and positioning arrangement ought to accommodate strategic opportunities (i.e., the ability of the function to be focused on the long-term and remaining flexible enough to solve day-to-day problems and recognize new opportunities).

These positioning principles are essential to deciding where to locate the environmental management function. However, they do not uniquely identify a particular solution. Based on the concept that structure follows strategy and location-positioning-charting of the function follows structure, Figure 6–3 shows a proposed organizational location and position for the environmental management function.

Figure 6–3 is an example of a location and positioning arrangement that should be favorable to accomplishment of the environmental management mission, i.e., optimal preparation, protection, and preservation of the firm's environmental resources. Solid lines indicate the "direct line" relationship in which authority-responsibility is usually delegated downward and compliance-accountability proceed upward to higher levels of superiors. The dotted lines indicate the "functional staff" relationships in which environmental managers have the needed authority regarding specialized mission-related activities of the environmental management function throughout the enterprise. Obviously, each "functional" aspect would include limited management powers. "Staff" aspects include the advisory services expected to be requested in terms of needs, expectations, and requirements. The proposed structure has three principle advantages:

- It provides centralized control of environmental policies, strategies, and tactical maneuvers.
- It requires a unique set of specialized competencies and capabilities of each member of the environmental management staff.
- If budgeted, staffed, and managed effectively, environmental management problems become more recognized within the company as problems that need to be solved.

The principal disadvantages are the following:

- Issues of coordination often occur.
- Overspecialization may result in tight control.
- Tight control may stifle innovation and creativity.

As an environmental management function develops from a compliance structure to a more competitive performance structure, the charting for the function often requires structural shifts that are more innovative and team oriented.

Although this dimension provides insight into the enormously complex and important issue of how to structure environmental management strategy, it is merely a first step. The propositions presented here are somewhat narrowly focused. They do not cover the full range of strategic, structural, and technical issues environmental managers face. Nor have their implications been formally tested within an environmental management context. The propositions are intended to stimulate further conceptual and empirical investigations into this important area of inquiry.

DIMENSION 2—IMPLEMENTING STRATEGY

After formulating the strategy, structure, and positioning-location arrangement, the environmental management function is now ready to consider the second dimension for managing the environmental sustainment challenge. This dimension is concerned

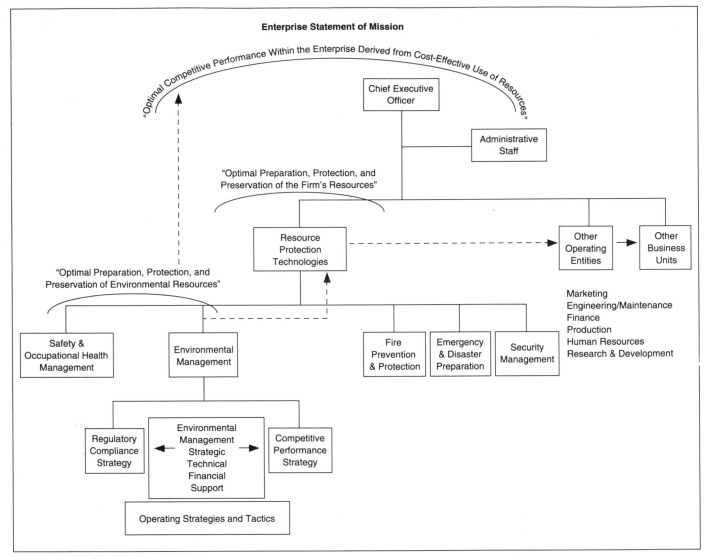

Figure 6–3. An environmental management organization favorable to achieving the environmental management mission.

with implementing (leading and controlling) the already formulated strategy and structure. Specifically, this dimension focuses on (1) an approach for implementing environmental strategy and structure and (2) the leading and controlling competencies and capabilities that are needed. Throughout the second dimension, keep in mind the strategy, structure, and positioning arrangement that were formulated in dimension 1.

An Approach to Implementation

Usually at this point, senior-level executives have reviewed and assessed the proposed environmental management strategy, organization structure, and positioning arrangement. With the environmental management staff, they identified the strategies and tactics that need to be both implemented and sustained for gaining a strategic advantage within their

industry. In determining an approach to implementation, the following can serve as useful guidelines:

- Implement environmental management strategies and tactics that create the best fit with the competitive performance strategy of the firm. For example, obtain a crystal clear picture of the business, regulatory, and innovative performance standards of the firm and then implement strategies and tactics to achieve those performance standards.

- Implement environmental management strategies and tactics that create the highest economic impact. For senior-level executives, the objectives for economic impact will typically be efficiency (maximizing net benefits) or cost-effectiveness (choosing the least costly method for gaining on economic impact).

- Implement environmental management strategies and tactics that are best equipped to succeed and that are likely to become visible throughout all levels of the firm. For example, a successfully implemented environmental management strategy or tactic provides a demonstration effect. If one strategic business unit or operating function sees another increase, its ability to create and sustain environmental protection or reduce its environmental risk/liability, it is far easier to convince other strategic business units to adopt whatever change was tried in the pilot study attempt.
- Implement environmental management strategies and tactics first where there are clearly identifiable supporters. For example, the environmental management staff should attempt to implement strategies and tactics with those strategic business units that already understand and accept the environmental sustainment challenge. In such a situation, the desire for help has already been extended, and the implementation effort can be initiated and conducted far more easily.
- Implement environmental management strategies and tactics that tend to be the least systemic. This means that an environmental management strategy or tactic that affects a strategic business unit or operating function of the organization is preferable to one that affects the entire firm. For example, try piloting a strategy or tactic before integrating it throughout the organization.

Implementation Skills Needed

Implementing environmental management strategy and structure requires the use of leading and controlling competencies and capabilities. The ability to lead and control tends to determine the success or failure of a strategy or tactic. Just being competent in strategy formulation and organization structure and positioning will not be enough. The environmental management staff will have to possess the leading and controlling competencies to convert strategy and structure into actions that get completed.

What leadership skills should the environmental management staff possess? Of course, as one moves away from the strategy formulation process to implementation, the specifics and dynamics of the particular situation become more important. With that in mind, the following principles are listed as guiding rules for action during the leading of the particular environmental management strategy or tactic:

- Techniques of leading that emphasize the *initiation and use of achievement-level type guidelines* tend to promote high degrees of organizational-level efficiency and effectiveness.
- Techniques of leading that emphasize *consideration* (i.e., showing regard for the comfort, well-being, status, and contributions of followers) tend to promote greater "follower" initiatives and positive results from initiatives.
- Techniques of leading that emphasize the *promotion and integration of a given environmental management philosophy and mission* throughout the entire organization tend to ensure that managerial decision-making and operating actions will be consistent and cohesive.
- Techniques of leading that emphasize the *development of team approaches* (i.e., arranging conditions for "followers" to participate as a group in the formulation of environmental management strategy and tactics) tend to be positively related to accomplishment of achievement-level guidelines.
- Techniques of leading that emphasize *promoting "follower" autonomy* (i.e., independence to make decision for influencing environmental management and/or counteracting constraints) tend to improve the quality of work performance and organizational fitness of the environmental management function.
- Techniques of leading that emphasize *person-oriented behavior qualities* (i.e., democratic, permissive, follower oriented, participative, considerate) as bases for sustained motivation tend to be related positively to "follower satisfaction."
- Techniques of leading that emphasize *satisfying the needs of the organization, protecting group members from the organization when necessary, and arranging conditions for members to gain satisfaction that would otherwise be outside their reach* tend to determine the degree and extent that a leadership position can be maintained.
- Techniques of leading that emphasize the *use of expert or referent power* (i.e., existence of a close interpersonal relationship or knowledge and expertise in an area) tend to promote favorable "follower" response in the design, implementation, and evaluation of objective-type guidelines.
- Techniques of leading that emphasize *satisfying the expectations of "superiors" as well as "followers" who depend upon the former as a mutually beneficial arrangement* tend to determine the degree to which the leader can count on "follower" support of persons who occupy the "superior" positions.

Also important is the ability to control environmental management strategies and tactics in such a manner that desired results are achieved. There are basically six different approaches to control and a number of principles. The approaches can be categorized as follows:

- *Leadership control* based on past practices, precedence, and intuition and perception of the strategy or tactic leader.
- *Strategic and tactical control* that involve anticipating and minimizing potential deviations from intended outcomes.
- *Adaptive control* that focuses on determining the most appropriate way in which to respond to changes.
- *Real-time control* that deals with information systems that provide current information about the status of strategies and tactics.
- *Opportunistic control* that focuses on the long-term strategy and structure of the environmental management function while remaining flexible enough to solve day-to-day problems and recognize new opportunities.

What controlling skills should the environmental management staff possess? Determining how best to control the implementation of certain strategies and tactics requires the use of certain controlling principles. The following principles are listed as guiding rules for action during the controlling phase of environmental management implementation:

- Measures of controlling that emphasize the *establishment of performance standards* tend to be more effective when the measures are interrelated to achievement-level guidelines designed by personnel who work in the area in which the activity is being initiated.
- Measures of controlling that emphasize *monitoring of performance standards* tend to confirm the value and worth of performance standards and provide information for improvement activities.
- Measures of controlling that emphasize *correction of deviations from performance standards* tend to accomplish achievement-level guidelines.
- Measures of controlling that emphasize the *analysis of differences between actual objective-type guideline results and intended achievement-level guidelines* tend to facilitate managerial actions to correct deviations.
- Measures of controlling that emphasize that *achievement-level guidelines reflect the mission of*

the organization tend to confirm the philosophical bases employed during planning, organizing, leading, controlling, and evaluating activities.
- Measures of controlling that emphasize *tracing the effects of failures back to the related level of manager responsibility* tend to discover the basic causes of the failures and establish ways to correct ineffective management practices.
- Measures of controlling that emphasize *tracing the effects of failures back to their most remote causes* tend to reveal the specific patterns followed by managerial personnel during employment of erroneous and impertinent management practices and principles.
- Measures of controlling that emphasize that *controls be exercised, primarily, by personnel at the level of performance execution* tend to verify deviations in performance effectiveness, to hasten corrective action, and to make controlling more effective.

For the environmental management function to help the firm to become environmentally sustainable, the function must both "implement strategies and tactics right" (be efficient), and "implement the right strategies and tactics" (be effective). Effectively and efficiently *finishing* each strategy and tactic must be the central theme of any strategy implementation process.

DIMENSION 3—EVALUATING STRATEGIC PERFORMANCE

Evaluating strategic performance is an important management practice. What does it take, then, to effectively evaluate environmental management performance? This dimension focuses on:

- a definition and purpose of environmental management evaluation
- preconditions necessary for conducting an evaluation
- the types and approaches to environmental management evaluation
- guidelines for planning an evaluation.

The need for a model to evaluate the efficacy (i.e., intended effectiveness and efficiency) of strategic performance continues to build among a distinct group of environmental managers who are required to justify their continued budgetary existence in a strict economical sense. Interest in evaluating environmental management performance is a relatively recent development, and although various efforts date back a couple of decades, widespread concern for evaluating can be considered essentially a post-EPA phenomenon. Although the literature related to environmental management evaluation continues to grow and increasingly sophisticated methods have been proposed, there is a need for more models to be published in professional journals.

The search for models to evaluate environmental management performance is part of a widespread effort for improving communication with senior-level executives. As a result, environmental management evaluation is moving from narrow inquiries about the ability to comply with regulatory mandates from government agencies to a broader concern for contributing to the competitive performance of the firm. The impetus for this approach is coming from senior-level executives who are beginning to hold the environmental management function accountable for its budgetary existence. In these instances strategic evaluation is expected to provide data on economic performance to supplement the traditional approach of evaluating regulatory compliance performance.

Definition and Purpose

Environmental management evaluation is the management's practice of determining the efficacy of formulated strategy compared against the objectives and expected effects (benchmarks) it set out to achieve. This type of evaluation aims at improving subsequent environmental management strategy. Within this definition are three key parts:

- *determining* refers to the methodology that is employed for studying and appraising the efficacy of environmental management strategy
- *comparison* stresses the use of indicators and/or explicit criteria for measuring objectives and expected effects (benchmarks)
- *contributing* to the improvement of subsequent strategy formulation denotes the purpose of evaluation.

Conducting evaluation solely for the purpose of determining the current status of environmental management strategy is not an advisable practice; improvement of subsequent environmental management strategy is a better reason, particularly if one desires to create and sustain long-term excellence. The environmental management evaluation process is a circular one, stemming from and returning to the formulation of environmental management strategy. The steps in the process are outlined in Figure 6–4.

Preconditions Necessary for Conducting an Evaluation

Evaluating the effectiveness of the environmental management strategy assumes certain preconditions:

- a clearly formulated strategy and structure
- clearly specified objectives and expected effects
- a rationale for linking strategy to the objectives and expected effects.

When the preconditions are missing or inadequately developed, the evaluation team must collaborate with the environmental manager and articulate the preconditions before initiating the evaluation. This discussion provides a framework and a model specifying the preconditions that are necessary for conducting an evaluation of the environmental management function (shown in Figure 6–5).

Presence of a formulated strategy and structure. It is essential for the environmental management function to possess a sound strategy and organization structure before it makes any attempt to evaluate. *Strategy* is the process of determining and describing the long-term manner in which the environmental management function plans to contribute to the competitive performance of the firm, while determining ways to comply with regulatory demands. The essence of environmental management strategy is to *prepare, protect,* and *preserve* the environmental resources of the firm. This demands analyzing the current and past operating risk-, liability-, and loss-producing patterns and forecasting expected risk-, liability-, and loss-operating patterns.

The organizational structure for the environmental management function is the operating arrangement and approach to environmental management strategy. The essence of structure is to organize environmental management strategy toward objectives and expected effects, while allowing various strategic business units and operating functions to focus their attention and expertise on specific environmental risk, liability, and loss countermeasures. As mentioned in the first dimension of this chapter, when formulating the strategy and organization structure of the environmental management function, it is important that the strategy and structure be congruent with the competitive performance strategy of the enterprise.

The emphasis on articulating environmental management strategy and structure reflects a way of evaluation that is called *congruency evaluation* (Conrad & Roberts-Gray, 1988; Conrad & Miller, 1987). The purpose of congruency evaluation is to show the

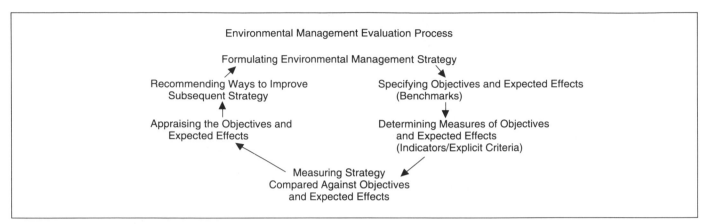

Figure 6–4. Environmental management evaluation process.

accord and harmony between environmental management strategy and structure and the competitive performance strategy of the organization.

Presence of objectives and expected effects (benchmarks). Objectives and expected effects (benchmarks) serve as a principal basis for determining environmental management success, so they must be clearly specified in order to develop indicators and explicit criteria for measurement. Senior-level executives like the process of objective and expected effect specification, because this process provides a basis for holding the environmental management function accountable for its strategy and structure. *Accountability* is the extent to which the environmental management function staff, first-line supervisors, and midlevel managers are expected to be answerable to senior-level executives for proper execution of preplanned strategies and tactics. This does not mean that the environmental management function is expected to be held liable for environmental risks, dangers, and losses incurred under the jurisdiction of other managers and supervisors. For environmental managers, the process of objective and expected effect specification provides assurances that the crucial aspects for justifying continual budgetary existence of the environmental management function will be included in the evaluation.

These comments reflect an objective and expected-effect approach to environmental management evaluation. Other viewpoints are held by those who prefer a maintenance-reactionary approach to environmental management evaluation (e.g., maintaining compliance with standards). However, this does not necessarily mean that objective and expected effect specification for the purpose of evaluation should be abandoned. Rather, a balance should be maintained among specified objectives, expected

effects that contribute to the firm's competitive performance, and the latent effects of dealing with reactionary demands. The latter effects can be included in the results section of the evaluation and added as supplemental information.

Presence of a rationale for linking effects to strategy. After the objectives and expected effects are specified, it is necessary to identify the linkage between environmental management strategy and its objectives and expected effects. The questions for this precondition are:

- Is the formulated environmental management strategy directed at the objectives and expected effects?
- What is the basis for stating that the strategy can potentially accomplish the objective and expected effects? That is, are the casual assumptions of the environmental management strategy linked to the objective and expected effects?

Lacking a strategy or pursuit of the objectives and expected effects, then why should the environmental management function be evaluated or even be held accountable for its performance? Yet, many environmental management functions state objectives and expected effects even though the function has no strategy or does not appear to pursue them. If an evaluation still measures the environmental management function in terms of such objectives and expected effects, negative results are predictable. When this happens, it becomes extremely difficult for environmental managers to communicate with senior-level executives for an improved budget.

Once the strategy directed at the objectives and expected effects are identified, the following question must be answered: What is the basis for stating that the strategy can potentially accomplish the objective

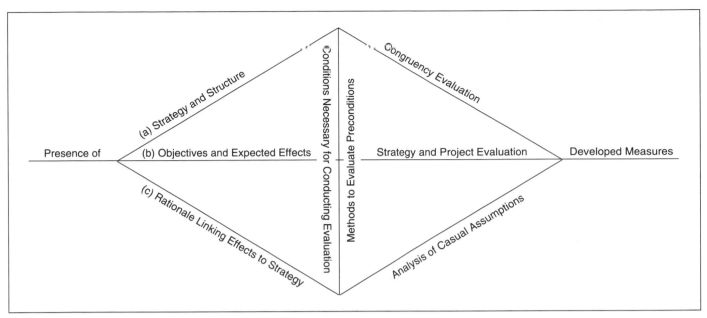

Figure 6–5. Model of preconditions for conducting an environmental management evaluation.

and expected effects? Raising this question is useful because evaluations of the environmental management function often have weak strategic and tactical interventions yet are expected to accomplish ambitious objectives and expected effects. The answer lies in analyzing the various projects that have been designed, making sure that they are strategic and tactical enough to contribute to the firm's competitive performance while complying with regulatory demands.

Types of Environmental Management Evaluation

Before selecting the best methods for conducting an environmental management function evaluation, one must know why and what to evaluate. One must also determine the kind, amount, and quality of data required to interpret results of the evaluation study. Evaluations for the environmental management function are conducted for various reasons. Descriptions of several types of environmental management function evaluations appear here. These types are defined both by the purpose of the evaluation effort and by the kinds of methods that are stressed and data required to measure. Some of the categories are associated more with some environmental management function contexts and settings than with others, and, as a consequence, the work of the environmental management evaluation team is likely to fall more in one type than another.

Front-end feasibility analysis, needs assessment evaluation. Front-end environmental management evaluation estimates need or ascertains operational and/or financial feasibility. This environmental management evaluation takes place prior to the installation of a particular environmental management strategy or tactic. Front-end analysis bridges recognizing a need and deciding what is feasible to do about it. It is also a way to look before leaping, improving its chance of survival by finding out what it requires financially and operationally to succeed. Participants have no way to do other, more complete environmental management evaluations without understanding the need and what is financially and operationally feasible. Traditionally, this is the difference between what is and what ought to be, between the target state and the present state.

Impact evaluation. Impact evaluation corresponds to one of the most common definitions of evaluation: finding out how well environmental management strategy or a tactic worked. The results of impact evaluations are intended to provide information useful in making major decisions about the continuation, expansion, or reduction of particular environmental management strategies and projects. The challenges for the environmental management evaluation team are to devise appropriate indicators of economic and regulatory impact and to become able to attribute types and amounts of economic and regulatory impact to formulated strategies and tactics rather than to other influences. Some knowledge or estimate of the status before the interventions were applied or of status conditions in the absence of strategic and tactical interventions is usually required.

Process evaluation. Environmental management process evaluation includes testing or appraising the procedures or processes of an ongoing environmental management project in order to make modifications and improvements. Activities can include the evaluation of work tasks, including the interactions among personnel, property, processes, and environmental aspects. In some cases, process evaluation means field testing (piloting) an environmental management project on a small scale before installing it.

Project monitoring. Project monitoring is the least acknowledged but probably most practiced type of environmental management evaluation. Substantial information can be gained from monitoring the strategies and tactics of the environmental management function that have already been structured into the operations of the enterprise. The methods involved in monitoring ongoing environmental management strategies and tactics vary widely, ranging from periodic checks of compliance with standards, to relatively straightforward tracking of technology transfer and training and education services delivered, to checking on how environmental hazardous exposures are controlled. Monitoring can include serious reexamination of whether the needs of the environmental management strategy and related projects as originally intended still exist, or it may suggest modification, updating, or revitalization.

The preceding descriptions of the types of environmental management function evaluations make it clear that a broad range of meaning can be attached to the statement that an environmental management function is being evaluated.

Guidelines for Planning Environmental Management Evaluation

Increasing interest in environmental management evaluation is related to the contribution it is expected to make toward improved planning, better management, and greater financial and regulatory accountability. The following guidelines and the model in Figure 6–6 are identified to provide an overview of the considerations in planning an environmental management evaluation. This discussion assumes that the environmental management function *can* be evaluated—a clearly formulated strategy and structure exists, objectives and expected effects (benchmarks) are specified, and a way exists to link objectives and expected effects to the formulated strategy.

Formulating evaluation strategy. After determining the evaluability of the environmental management function, the evaluation team should formulate the strategy for conducting the evaluation. The team should understand what is to be done, how it is to be done, why it is to be done, and the possible constraints. Potential users of the evaluation results, such as senior-level executives and other managerial and supervisor personnel, should be identified and their information needs and expectations spelled out. The most appropriate type of evaluation (front-end analysis, process, impact) should be identified and the method, measurement criteria, data required, and range of activities to be undertaken should be specified. An estimate of the cost and time schedule of the proposed evaluation should be provided. At the outset, all concerned should agree that the evaluation can produce information valuable enough to justify its cost. Any restrictions on access to the data and results from an evaluation should be agreed on between the environmental management evaluation team and the client at the onset. Finally, accountability for the management, financial, and technical aspects of the evaluation should be clearly defined.

Determining operating structure. The operating structure for any type of environmental management evaluation is influenced by logistical, ethical, political, organizational, and fiscal concerns. These issues as well as methodological requirements must be taken into account. Operating structure varies in vigor, and not all evaluation methods and measurements are equally objective. Despite broad variations, the following guidelines apply. For example, the structure of an impact evaluation is as subject to specifications as is the structure of a front-end evaluation. The reliability of judgments is as much an issue as the reliability of measures. Regardless of the type of environmental management evaluation—front-end analysis, process, impact—the evaluation's structure should be specified and justified as appropriate to the type of environmental management evaluation conducted; to the type of measurements, methods, and instruments used; and to the type of results drawn.

Data collection and preparation. A data collection and preparation scheme should be developed before the data are collected. The environmental management evaluation team should be selected, trained, and supervised to ensure completeness and consistency in the data collection process. All data collection activities should respect and protect the welfare

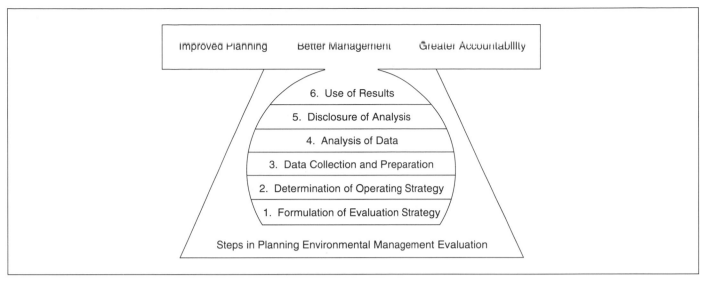

Figure 6–6. Steps in environmental management evaluation and their expected outcomes.

of all organizational levels. The validity and reliability of data collection instruments and procedures should be verified against the reasons of their use. Data collection activities should be conducted with minimum disruption to the organization and handled and stored so that release is limited to those with a critical need to know.

Analysis of data. The choice of analytic procedures, like the choice of data collection methods, is dictated by the type of environmental management evaluation. During the data analysis stage, the environmental management evaluation team no longer is free to change the strategy and operating structure, and is required to analyze according to the data collected. The analytic procedures should be matched to the type of environmental management evaluation, its strategy, operating structure, and data collection scheme. All analytic procedures used should be described, and the reasons for choosing the procedures should be explained. During the interpretation phase, data should be reported in a manner that distinguishes among objective findings, opinions, judgments, and speculation.

Disclosure of analysis. Most environmental management function evaluation efforts produce certain formal reports, intermediate and final, written and oral, and there are guidelines these reports should follow. Findings from any type of environmental management evaluation should be organized and stated in terms understandable and appropriate to the needs of the users. Recommendations should clearly relate to the findings. Findings and recommendations should be presented in a framework that indicates their relative importance. A complete description of how findings were derived should be provided, along with any limitations caused by constraints on time, resources, and data availability.

Use of results. The reasons for conducting an environmental management evaluation are functional: to find ways to improve operations, determine feasibility, and assess impact. The use of evaluation results cannot be guaranteed, but they are more likely to be used if careful attention is given to the information needs of the eventual users of the results. Evaluation results should be made available to appropriate users before relevant decisions are made. The environmental management evaluation team should bring to the attention of users any suspected side effects of the evaluation process. Evaluators should distinguish between the findings of the evaluation and recommendations based on the findings. In making recommendations about countermeasure actions, the environmental management evaluation team should carefully consider and indicate what is known about the probable effectiveness and costs of the recommended countermeasure.

CURRENT CORPORATE PRACTICES

Large and multinational corporations and industry trade organizations have made significant progress in developing environmental policies and strategies. These business responses are due to three converging forces: (1) regulatory pressure; (2) external constituent pressure from nongovernmental organizations, shareholders, and the media; and (3) pressure within the firms' own management teams (Conference Board, 1991). For example, as reported by the Conference Board, these large trade groups and businesses have publicly announced a wide variety of programs:

- American Petroleum Institute: an 11-point set of guiding environmental principles. Adherence is mandatory for corporate members.
- Conoco: a nine-point environmental program.
- Phillips Petroleum: new mission statement emphasizing environmental protection and safety.
- Disney: creation of corporate vice-presidency for environmental policy.
- Chevron: Save Money and Reduce Toxins (SMART) program.
- ICI America: mandatory business unit environmental report.
- Dow Chemical: Product Stewardship System, which directs employees not to sell a product if he or she believes it will be used or stored in an unsafe manner.
- Kodak: extensive community involvement and awareness programs.
- Chemical Manufacturers' Association: Responsible Care program for all members; Community Awareness and Emergency Response Program providing an opportunity for dialogue between plant managers and community leaders.
- 3M Corporation: Pollution Prevention Pays program.

This list represents only a small fraction of the major corporate environmental initiatives begun since the mid-1980s. Although these programs cover a wide variety of topics, the Conference Board found consistency in the overall strategic directions of their surveyed members. Specifically, four strategies tended to dominate corporate thinking and resource allocation:

- Installation of new pollution control devices
- Control of pollution through product and package redesign
- Improvement of existing pollution abatement
- Changes in raw materials to reduce pollution (source reduction).

Interestingly, in the United States, companies placed greater emphasis on source reduction than did their Canadian or European counterparts (see Chapter 10, Pollution Prevention Approaches and Technologies). The Conference Board felt this difference reflected the anticipation of the market incentives built into the U.S. Clean Air Act of 1990. The effects of different economic incentives on environmental practices was presented in Chapters 2, Economic and Ethical Issues; and 4, International Legal and Legislative Framework.

Overall, major corporate thinking has clearly undergone a paradigm shift since the 1970s. As Paul Huard of the National Association of Manufacturers stated, "... The debate is no longer clean versus dirty, it's warm versus warmer" (quoted in Conference Board, 1991).

SUMMARY

The dimensions of environmental management have now been covered. The conceptual thinking presented was that the firm's environmental resources need to be prepared, protected, and preserved in order to make contributions to the competitive performance standards of the firm. By employing the three-dimensional classification scheme, one can dramatically improve the management of the environmental function and secure an understanding of the strategy formulation, structure, implementation, and evaluation process. The future of environmental management presents opportunities for those who have the competencies, capabilities, and wherewithal to contribute to the competitive performance requirements of the firm. A wide variety of corporations and industry trade groups have clearly embraced the need to formulate and integrate environmental strategies in their daily business practices.

This paradigm shift represents a major development in corporate thinking and behavior. Although many large and medium-sized companies have embraced this new strategic approach, the critical test is still whether words can be translated into concrete actions. All parties are converging on agreement of the ultimate goal; however, many paths and speeds can be used to reach the same result. The approach presented in this chapter presents one vision of how to develop environmental strategies and manage environmental resources.

REFERENCES

Barnes AJ, Ferry JK. Creating a niche for the environment in the business school curriculum. *Bus Horizons*, 3–8, March-April 1992.

Conference Board. *Managing Environmental Affairs: Corporate Practices in the U.S., Canada and Europe.* Report Number 961. 1991.

Conrad KJ, Miller TS. Measuring and testing program philosophy. *New Dir for Program Eval* 3:19–42, 1987.

Conrad KJ, Roberts-Gray C. Checking the congruency between a program and its organizational environment. *New Dir for Program Eval* 40:63–82, 1988.

Drucker P. *People and Performance.* New York: Harper and Row, 1977.

Hamel G, Prahalead, CK. Strategic intent. *Harvard Bus Rev,* May–June 1989.

Hunt C, Auster, E. Proactive environmental management: Avoiding the toxic trap. *Sloan Man Rev* 31(2):7–18, Winter 1990.

Porter M. Commentary. *Sci Am,* September 1989.

Veltri A. *Expected Use of Management Principles for Safety Function Management,* Doctoral Dissertation, West Virginia University, 1985.

Winsemius P, Guntram U. Responding to the environmental challenge. *Bus Horizons,* 12–20, March–April 1992.

7

Environmental Audits and Site Assessments

Frank J. Priznar

The terms *environmental audits* and *environmental site assessments* are familiar to health and safety professionals. However, the terms historically were interpreted as something outside the health and safety professional's area of expertise—"environmental" and "health and safety" activities were traditionally separate. Environmental concerns and health and safety concerns were previously perceived to be so different that they required significantly different skills, knowledge, and expertise. Typically, staff responsible for those areas would report to different functions.

The implementation of both environmental audits and assessments, however, has more than casual similarities to the *safety inspection* process. In all cases, a professional is seeking to compare actual physical conditions against expectations and to apply judgment about the quality of the conditions. Consequently, environmental audits and assessments are environmental activities about which health and safety professionals should feel comfortable.

This chapter orients the reader to the fundamental issues underlying environmental audits and assessments. It clarifies the terms that are in use in this area today, offers practical information on current practices, and explores some of the most significant developments likely to have an impact on the future. For those who wish to learn more, references to professional associations and standards-setting organizations are included. Newcomers to environmental auditing and assessing should be able to use this chapter as a one-stop primer. For those who have had past involvement in these practices, this chapter will serve as an update on contemporary practices.

As companies expand internationally, the need for environmental audits and assessments has grown dramatically. Multinational companies are systematically investigating business opportunities in new growth areas, such as former communist countries or the newly industrialized countries in the Pacific Rim or Latin America. Because of past practices in these locations, many acquisitions carry significant financial liabilities. (See Chapter 4, International Legal and Legislative Framework.) The documentation developed in the assessment or audit of these liabilities has become a critical function of the acquisition process.

In summary, this chapter provides a current, detailed, and practical overview of both environmental audits and assessments.

ENVIRONMENTAL AUDITS VERSUS SITE ASSESSMENTS

Audits differ from assessments in the degree of judgment that is applied. In an audit, an auditor is seeking to verify expectations. More specifically, the auditor is seeking to confirm or deny a specific condition. Typically, in an environmental context, the auditor seeks to understand whether regulatory or policy requirements are being met (compliance audit). The answers to an auditor's questions are limited to "yes," "no," and "do not know."

Greater judgment is involved in an assessment, which is similar to an appraisal. An assessor seeks to estimate, or judge, environmentally important factors that affect value or character. Answers to an assessor's questions are typically judgmental and often require lengthy descriptions of conditions. Environmental site assessments usually include judgments on the likelihood of groundwater contamination, estimated costs for remediation, and similar characterization items. An assessment does not measure expectations of conditions documented in standards against actual conditions, whereas audits should. To make these distinctions clear, consider these definitions:

- **Environmental audit:** A process that seeks to verify documented expectations, typically regulations and policies, by conducting interviews, reviewing records, and making first-hand observations; also called a *compliance audit.*
- **Environmental site assessment:** A process that seeks to characterize a physical property or operation from an environmental view with an overall objective of understanding site-specific conditions. Information is collected through interviews, record reviews, and first-hand observations, and may also involve testing environmental media (e.g., air, water, soil) and facility characteristics (e.g., wastewater, insulation, paint, airflow).

These terms are not used consistently throughout the profession; hence, the reader is cautioned to clarify their meaning in discussions with other parties. Also, other organizations have issued definitions of these terms (U.S. EPA, 1986; ASTM, 1993) that are more rigorous and should be reviewed for further information on the subject.

Purpose

To avoid confusion, the environmental/health/safety professional should clarify whether the facility requires an audit or assessment. This is especially important if an audit of the operation or facility is being requested because, if the audit is being done on a competitive contract basis, often the low bidder really is offering an assessment with a narrower scope than that of a compliance audit. Whether the audit is internal or external, the activity, purpose, and scope must be precisely defined.

Environmental audits are performed for a number of reasons; however, typically, requestors want assurance that their organization will not be surprised by fines, negative publicity, and related distractions if they are caught in noncompliance by regulatory agencies. Audits also serve to demonstrate to internal staff and external entities the organization's good faith with regard to environmental management.

Environmental site assessments are also completed for a number of reasons, typically that the seller or buyer of a property wishes to learn whether there are any financial consequences of environmental issues associated with the property. In this case, the buyer commissions an assessment to help understand the property's value. The seller, on the other hand, may not reveal that he or she has completed an assessment but knows what environmental issues may be present. The knowledge gained from these assessments can affect a business deal, because the cost of fixing problems associated with some environmental issues can be high and exceed the value of the property.

Some proactive commercial property stakeholders complete a periodic assessment of their property to determine whether environmental issues affect the value of their investments as they are managed. This is most common when there is an opportunity for contamination to occur, such as a site with a current or previous light manufacturing, gas station, or dry cleaning operation. A lending institution might have an assessment performed if property is collateral for a loan.

In general, audits and assessments are performed to collect information so that management can make appropriate business decisions. Consequently, the procedures taken must be consistent and meet accepted industry standards.

Procedure

Procedures for completing environmental audits and environmental site assessments differ, although there are some similarities. A typical procedure for completing an environmental audit is called a *protocol.* The protocol outlines what is to be done, by whom, and in what sequence. Because environmental audits have been easier to standardize, a protocol for a typical audit can be generalized, as presented in Figure 7–1. Generally, the procedure outlined in Figure 7–1 occurs with a team of two or three auditors spending a total of about 80 to 160 hours to complete the audit. More information on general expectations for an environmental audit are detailed in *Draft Standard Practice for Environmental Compliance Auditing* (ASTM, 1994).

Procedures for completing an environmental site assessment are less standardized, but generally four steps occur:

1. *Records review* of facility, environmental, and regulatory files
2. *Site reconnaissance* consisting of visual and physical inspection of the property (Figure 7–2); samples may be taken for testing
3. *Interviews* of facility staff and regulatory agencies
4. *Judgment and report* on the probability of deleterious environmental issues and characterization of financial implications.

Further information on what constitutes appropriate procedure for environmental site assessments can be found in ASTM's *Draft Standard Practice for Environmental Compliance Auditing* (1994).

Trends

Because both audits and assessments are relatively new environmental activities, they are subject to the effects of rapid evolution. Auditing first surfaced around 1980, and environmental site assessments became popular in the mid-1980s (Priznar, 1993a). In this short period, auditing and assessment activity has evolved to become an important business tool. Several trends have developed to improve the auditing and assessment activities:

- *Standardization:* Many auditing and assessment organizations have formed to create and promote standard practices for these two activities, such as the Institute for Environmental Auditing (1740 Ivy Oak Square, Reston, VA 22090). IEA members have participated in a number of standards-setting activities for audits and assessments since 1986. (See Standards-Setting Organizations later in this chapter for more detail.)
- *Professionalization:* No one has been able to satisfactorily answer "Who is qualified to conduct an audit or assessment?" A number of certifying organizations exist that have no professional standing and add little to the value of an auditor's or

assessor's qualifications. One exception is the California-sanctioned Registered Environmental Assessor program. The program, managed by the California Environmental Protection Agency in Sacramento, actually verifies credentials of applicants and sets minimum standards for registration. Although there is no equivalent program for auditors, the IEA can provide the latest information on other certification programs.

- *Automation:* Computers are used in a number of ways for both audits and assessments. For audits, a computer can store all the environmental regulations needed for developing a checklist, provide a database for tracking the audit findings, and create a report from a template. Today, character-recognition electronic touch pads are being used to record an auditor's notes in the field, making it possible to create reports while the team is at the facility being audited. Similar information products are now available for site assessments, as Table 7–A lists.

The result of automation will be higher productivity, greater consistency, and improved quality. These areas seem to dominate the interests of auditors and assessors. They will probably not change in the short term, because they have been among the most important issues debated in the past few years and there seems to be no agreement on standards and certification. Also, technology improvements will change the way teams do auditing and assessing.

ENVIRONMENTAL AUDIT LEGAL ISSUES

Because an environmental audit focuses on the compliance status of legally enforceable regulations, it cannot be completed without understanding the legal issues that can apply to a company's activities. It would be irresponsible to gather information on noncompliance issues without knowing the organization's obligations or the consequences of disclosure. Thus, liability and confidentiality are critical legal issues. This section discusses some of the legal issues surrounding environmental audits, along with some common practices to address them.

Liability

Three types of legal liability are associated with environmental auditing (Hall & Case, 1987): statutory liability, common law liability, and site-specific obligations. Each of these is summarized with a few examples that highlight important points. This

---

Environmental Audit Protocol

Previsit Activities

At least four weeks before a planned facility visit, the audit team will undertake these activities:

- *Establish contact* with facility manager to
  — Confirm audit scope
  — Understand management commitment to the audit
  — Verify schedule
  — Confirm that health and safety issues for the auditors are not expected
  — Inform facility of audit team needs for duration of audit, including
    + Work space for audit team with working telephones and 110 V outlets, desks, chairs, and adequate lighting
    + Access to copy machines and computer printers
    + Staff to be interviewed and proposed schedule
    + Technical escorts
    + Hours of work (typically 8:30 AM to 6:00 PM) while on site
    + Records that are to be made available in advance and while on site
- *Prepare and send previsit questionnaire*
- *Update audit protocols* for state-specific or operation-specific issues
- *Establish reporting relationship* and internal points of contact for visit and postvisit activities
- *Confirm commitment* of audit team and individual assignments.

In general, the audit team will seek to provide sufficient detail, verbally and in writing via letter and previsit questionnaires, to ensure that the facility manager is fully prepared for audit team needs and is not surprised by last minute requests.

Visit Activities

During the visit the following activities typically occur:

- *Entrance briefing* with facility manager and selected staff to discuss plans, schedule, and expectations. This usually takes an hour.
- *Facility tour* to acquaint the audit team with the scope of activities and locations to be audited. For most audits, this can be accomplished in less than two hours.
- *Review previsit information* to seek newly recognized information needs and verify expected conditions.
- *Review team audit plan*, if necessary, adjusting the approach for completing the audit based on preliminary overview information.
- *Record reviews* of the documents that were requested in advance. These records must be used by the auditors for the duration of their visit, but most record reviews are completed in 4 to 16 hours, depending on their magnitude, complexity, and orderliness.
- *Interviews* of at least the plant manager, plant engineer, environmental coordinator, and usually staff in maintenance, health and safety, purchasing, security, shipping and receiving, and production. These individuals are interviewed once or twice for 20 to 40 minutes each. Notes are taken during these interviews, but they are not tape recorded.
- *Inspections* of specific operations or practices are made to gather information related to compliance or management systems. The audit team takes photographs (up to 36 per facility) to document observed conditions. Typically, the audit team is escorted by facility staff while in the plant for safety and access reasons. As some facilities are large (more than 1 million square feet), more than one escort may be needed. Inspection time varies from two to eight hours.
- *Findings* will be reviewed each evening by the audit team at their hotel in their rooms (so as not to be overheard in restaurants or other public places) to evaluate progress and plan the next day's agenda. The draft audit report is started on the first evening and then redrafted each evening. The audit team will have portable computers available to do this. Facility staff are welcome to participate in these meetings to help clarify issues and learn more about the audit process. Alternatively, the audit team will provide to the facility a daily summary of findings to ensure open communication on issues observed.
- *Exit briefing* will be conducted on the last day of the scheduled visit with the facility manager and whomever the manager feels is appropriate. Frequently, there is some corporate participation by telephone, but this is at facility's discretion. At this time, the audit team will summarize its findings and recommendations and answer questions. The audit team will also present its written audit report. The audit team will make its best efforts to include photographs in this report. Exit briefings average one to two hours.

Postvisit Activities

Postvisit activities can be the least intensive in this approach if a report is provided to the facility before leaving the site. If this is not possible, postvisit activities relating to the report production task take several weeks. The best, most accurate, and meaningful reports are prepared while at the site, so postvisit activities will consist of report improvement activities:

- Following up and resolving outstanding and unresolved questions and issues.
- Providing a revised report that includes photographs if they cannot be included in the original.
- Providing a final audit report after the facility has reviewed the draft. The audit team does not expect that substantial revisions will be needed.

The audit team will destroy and dispose of all working papers associated with each audit on assignment completion, unless instructed otherwise by the facility manager.

Figure 7–1. Generalized environmental audit protocol.

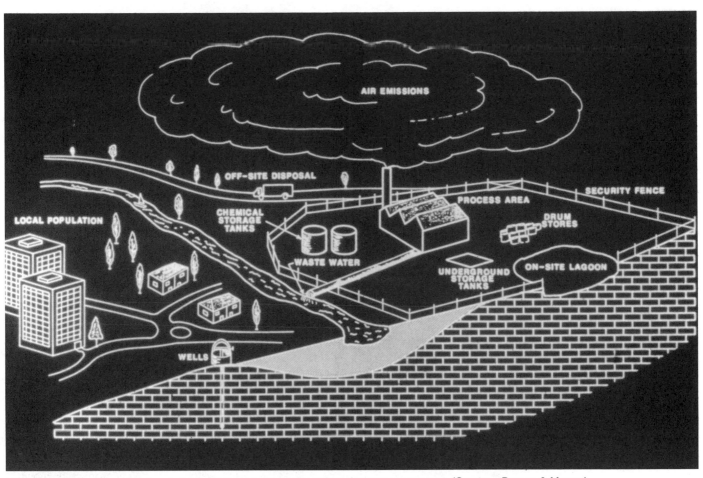

Figure 7–2. Some general factors to be considered during a site reconnaissance assessment. (Courtesy Dames & Moore.)

summary is not a comprehensive survey of relevant law but a description of the nature and scope of these laws. The auditor should study in greater detail those laws that apply to the facilities or operations that are to be audited. Additionally, in planning the audit, legal counsel should assist in identifying and interpreting all applicable laws and in finding ways to maintain audit confidentiality. (For more detail on environmental law, see Chapter 3, United States Legal and Legislative Framework; Chapter 8, Hazardous Wastes; and Arbuckle et al, 1991.)

Statutory liability. The environmental statutes described in other chapters of this manual generally impose reporting and record-keeping requirements. There are varying but significant penalties for failure to report, the submission of false information, and/or the destruction of required records (Cahill & Kane, 1989). Criminal penalties are imposed for knowingly and willfully giving false statements. For example:

- The Clean Water Act requires the submission of Discharge Monitoring Reports. It also imposes

criminal penalties for knowingly submitting false information in those reports.
- The Toxic Substances Control Act requires manufacturers to submit reports and retain certain information regarding the processing and/or manufacture of chemical substances. Civil penalties of up to $25,000 per violation can be imposed for noncompliance.
- The Resource Conservation and Recovery Act requires the submission of extensive information in permit applications. Violations of this act are subject to civil penalties of up to $25,000 per violation. Each day of any violation constitutes a separate violation. Criminal fines may be up to $50,000 for each day of the violation, and imprisonment up to three years is also possible.
- The Comprehensive Environmental Response, Compensation, and Liability Act of 1980 (Superfund) and subsequent amendments require the reporting of an on-site spill of a reportable quantity of a hazardous substance to the National Response Center. Failure to do so may result in fines and

Table 7–A. Automation Options for Environmental Auditing

Function	Medium	Sample Vendors	Perception of Features	
			Benefits	**Liabilities**
Field data collection	Pen-based table PCs	Grid Systems	Easy to use, possible interface with other automated products	New application, no proven track record compared to paper, legal requirements sometimes mandate paper evidence
Regulatory scanning	Software	ERM Computer	Easy updates, state coverage, ease of CD-ROM use	CD-ROM speed, hardware costs
		Regulation Scanning	Ease of use, speed, modular design and purchase	Hard drive utilization, scope, lack of cross-module searching
		Virtual Media	Speed, scope, operating features, environmental and safety coverage	Difficult to learn, lacks built-in, cross-module search
		IHS Regulatory Protocols	Networking capability, can display pictures and diagrams in regulations	CD-ROM speed, hardware requirements
		Washington Post	Search capabilities, minimal equipment requirements	Inconvenience of on-line system utilization
Complete automated auditing information program software	Software	Utilicom	Easy to use, can be customized, all-in-one, lots of installations	Pricey subscriptions, no states included, not compatible with other systems
		Audit PC	Integrated system, easy to use	New product, unknown performance, no states
Complete hybrid auditing information	PC	Do-it-yourself system with current office technology and readily available references	No learning curve, fully compatible with information organization, low cost, exists and ready to go, can always go to another product and enter old data	Customization time, sense of missing something or suboptimal operations

imprisonment of up to three years, or five years for subsequent violations.

Many federal laws, including the Clean Air Act, the Clean Water Act, and the Resource Conservation and Recovery Act, require or allow adoption of corresponding state laws. Many states have voluntarily gone beyond federal laws to adopt parallel or more stringent state-level requirements. There are also state and local zoning requirements, wetland protection laws, and other environmental controls often administered and enforced by state agencies. All of these potentially affect the operations of any company and must be considered in planning and conducting an audit.

Violations of legal responsibilities to report and record are likely to be discovered during the course of an audit. Consequently, auditors must understand these statutory requirements in order to assess past and present violations and assist in implementing procedures to promote future compliance.

Common law liability. Under common law (used in the United States and the United Kingdom but not universally in other major countries, such as France), one who injures another through environmental pollution can be held liable for damages and can be subject to more disruptive legal actions, if appropriate. The risks imposed by common law are minimized by insurance and good business practices. Environmental audits can be used to set priority to risks and therefore develop recommendations for appropriate actions.

The principal legal theories under which polluters have been held liable under common law are (1) negligence, (2) trespass, (3) nuisance, and (4) strict liability. There are also special common law rights and obligations related to the use of water. Because common law doctrines have evolved almost entirely as a matter of state law, the scope of liability varies from state to state. Each of the four is briefly summarized here:

- *Negligence:* conduct that falls below the standard established by law for the protection of others against unreasonable risk of harm. In the context of environmental law, if a person discharges pollutants negligently so as to cause injury to another person's property, the discharger is liable for the damage that results.
- *Trespass:* wrongful interference with another person's property rights. One whose pollution causes physical damage to another's property is liable for resulting damages in a trespass action. Thus, a neighboring landowner can sue if a company causes pollution to the land, groundwater, or air.
- *Nuisance:* can be either private or public. A private nuisance is an unreasonable interference with another's use and enjoyment of his or her land or related personal or property interests. A public nuisance is one that involves interference with a general public right.
- *Strict liability:* arises because some courts have decided that the handling of hazardous wastes is similar to the law of strict product liability, which applies to the manufacturers and distributors of defective products. Thus, there may be strict liability on those who generate or handle hazardous wastes for injuries or damages caused as a result of such activity. Also, the use of land to manage hazardous wastes may be considered an abnormally dangerous activity that justifies strict liability for any resulting harm.

Site-specific obligations. The environmental auditor will need to understand all of the environmental laws just discussed to properly evaluate the practical aspects of these laws on the site or process that is being audited. This is a complex task. One recommended way of doing it is to use a checklist to review the applicability of all environmental requirements to a facility. The following five areas should be part of the checklist:

- *Permits:* required by most of the environmental statutes. These permits must be identified in the course of the audit. Their terms must be reviewed to determine whether the permitted facilities are in compliance. If the facilities are not or compliance cannot be determined, the auditor should record this observation as a finding. In addition, the auditor must seek situations for which permits are required, but either have not been applied for or have not been issued.
- *Compliance schedules:* specific timetables by which an organization must bring itself into compliance with an applicable requirement. These schedules may be set forth in regulations, special judicial orders issued by a court, administrative orders issued by the U.S. Environmental Protection Agency (EPA), or less formal correspondence from a regulatory agency at any level. These compliance schedules should also be reviewed for compliance.
- *Judicial or administrative orders:* result from litigation and typically direct that certain actions be taken. If the company has settled a case with the enforcement agency, the results are often embodied in a consent decree. These orders and consent decrees should be identified and reviewed for compliance, because they impose legally binding obligations.
- *Statutory and regulatory requirements:* directly imposed obligations upon facilities apart form permits, orders, and schedules just mentioned. Items such as air emissions, water discharges, solid waste, and transport vehicles have specific requirements that must be identified and reviewed for applicability for legal obligations.
- *Records:* required to be kept in connection with each statutory program and each source of environmental release just mentioned. In planning the environmental audit, the operations of each facility should be considered in order to prepare a list of applicable reporting and record-keeping obligations.

Most environmental audits conducted today focus on statutory requirements. Enforcement issues and more complex environmental legal issues covered by common law are left to attorneys. Attorneys have the training and expertise needed to properly understand the implications of the situation, and there are certain protective measures that attorneys can invoke to benefit their client's position. Some of these are covered in the next section.

Confidentiality

Environmental audit results may contain sensitive information that, if revealed, could be damaging to the audited organization. The audit may inadvertently identify materials, products, or processes that are closely guarded trade secrets. Alternatively, the audit may turn up evidence of previously undetected violations that could lead to liability to the organization and its officers and employees. A decision to disclose sensitive information should be made by management and not left to chance. This section identifies some methods used to protect confidential information. There is no guarantee of nondisclosure protection; therefore, if an organization does not want to automatically disclose its audit findings, it should design its program with nondisclosure as an initial objective.

Trade secrets and confidential business information. The law surrounding the protection of trade secrets and confidential business information is complex; however, a few words of caution are needed for those conducting an environmental audit.

First, if either of these issues is raised as an objection to revealing data needed to complete an audit, refer the issue to the organization's attorney. In this case, the auditor must record the finding "unable to determine." Although the auditor may not be able to see the data, the audited organization can later assure itself of the specific condition related to the protected or confidential data without revealing sensitive data to others.

Second, any information that is sent to EPA, any other agency (such as OSHA), or state agencies can be disclosed under the Freedom of Information Act to anyone who requests the information. Given this, any sensitive or confidential information should be appropriately marked, and specific discussions on how to protect this information from disclosure should be included in planning the audit.

Sensitive information. Sensitive information—other than a trade secret or confidential business information—still needs to be protected in an audit. For example, there may be evidence of violations or conditions that an organization is seeking to correct but does not want to disclose. In this case, two methods may be used, each involving the use of an attorney. The first is the attorney-client privilege and the second is the attorney-work product doctrine. Each is discussed briefly.

Attorney-client privilege. When a client obtains legal advice from an attorney, the substance of their communication, made in confidence, is privileged and immune from disclosure by the attorney or client when the client invokes the privilege. There are limitations on this privilege, however. For example, (1) the attorney must be retained for legal advice, (2) the communication must be made at the direction of the employer, (3) the communication must be within the scope of the employee's duties, and (4) the communication must not be disseminated beyond those involved in the attorney-client relationship. Furthermore, (1) data cannot be protected and (2) the protection is easily, and sometimes inadvertently, waived if the confidential information is disseminated to persons outside the attorney-client relationship.

Attorney-work product doctrine. This privilege may be available through an attorney to provide "work product immunity." This doctrine protects material prepared in anticipation of litigation. Under the general rule, three factors determine the applicability of the doctrine: (1) the material must be prepared for litigation; (2) the material must not be factual or analytical (i.e., data); and (3) the information must be developed by the attorney or under the attorney's supervision. In a recent OSHA case (Steinway, 1993) this doctrine was invoked successfully to protect information in an audit from disclosure.

The preceding discussion of confidentiality issues illustrates the uncertainty of the law and practices in this area. Decisions about protective measures for disclosure should be a top priority. Once the audit is started, it is, in all practical sense, too late to maximize protection. If confidentiality is not a primary concern, some of the other topics in this chapter, such as wording an audit report, will have greater significance in terms of ensuring that inadvertent errors are not made by well-intentioned auditors.

Common Practices

Given all the legal concerns about environmental auditing, practitioners have evolved some practical approaches that are becoming commonly accepted. Some of these are outlined here.

Use internal auditors. Internal auditors may be more desirable to use than external auditors, from a confidentiality standpoint, because they have less opportunity or motivation to "leak" information. External auditors sometimes offer unique skills or knowledge that internal auditors have not had a chance to develop. However, internal auditors should be selected to handle the most sensitive information if that choice is available.

Use attorneys. The specific areas for which attorneys are best employed are planning the audit program, developing the audit protocol, attacking the common law issues, and specifying what parts of consent decrees and orders should be audited. They should also be available for ad-hoc advice on legal interpretations of specific situations that are not clear during the course of an audit. Attorneys could develop and execute the entire environmental audit program, but this option is usually not cost effective.

Use caution in preparing audit reports. Typically, a draft report is prepared before a final report is presented to the audited organization. In the draft report especially, auditors must be careful not to render opinions on compliance status. The report must be factual and present without bias conditions as they were observed. For example, it would be irresponsible for an auditor to report that "Plant X is in violation of Section xxx of RCRA and has been for at least six months." This could be phrased more factually as "Two containers receiving hazardous waste were not labeled. No records available on past practices." Both statements could be attributed to the same situation; however, the latter is more factual and conveys what needs to be done to improve the situation. The first statement is an unsubstantiated legal opinion that audited organizations do not want or need.

Eliminate written reports. A final confidentiality issue relevant to reports is the need for a written report. Occasionally, organizations want to understand compliance status, or just look for areas that need improvement, but do not want the burden of guarding a document from disclosure. These situations are becoming rarer, because the EPA is now seeking documented evidence of an existing audit program to understand corporate attitudes toward environmental issues. In addition, most organizations want the recognition that an audit program provides. However, one way to eliminate the need for protective measures is to avoid producing a written document that must be protected. In these cases, verbal reports are provided on a real-time, daily basis to executives or employees as findings are made. If there is no report, it is difficult to measure progress because a baseline may be missing. However, the value of the audit as a training exercise for affected employees can be enhanced, because they immediately learn of issues and possible solutions.

Because environmental audits focus on compliance with legal requirements, there is a greater need to understand these legal requirements and the implications of noncompliance. Even though there is some judgment applied in developing an environmental audit report, the result of asking a compliance status question is still one of three choices "yes," "no," or "do not know." Admittedly this oversimplifies the outcome, but it is a key characteristic that distinguishes environmental audits from environmental site assessments and is discussed in more detail in the subsequent section.

ENVIRONMENTAL SITE ASSESSMENTS

Environmental site assessments differ from audits in that a greater degree of judgment is typically used by the assessor and the purpose of the activity is different. An assessment is commonly undertaken when there is a real property conveyance. Specifically, when there are stakeholder changes related to ownership, property use, insurance, or financing, an assessment is often conducted to understand whether the value of the property is impaired from environmental issues. Because the stakes are usually high when assessments are conducted for commercial property transactions, it is important to understand the business and legal risks that are associated with assessments.

Business and Legal Risks

Business and legal risks surrounding assessments are large. These are the general categories of risks:

- Failure to detect significant financial liability from environmentally based issues. Thus, a conveyance occurs without an accurate accounting of the property value. This liability may be shared by the assessors as well as the stakeholders.
- The possibility that environmental issues are so large that they will "kill" the deal. Typically, environmental issues are not evaluated until the deal is nearly complete. Consequently, significant environmental issues will not be accounted for until they are detected in the last few moments of "due diligence."
- The possibility that new legal requirements will be identified that require reporting and can possibly delay or "kill" the deal.
- The possibility that the product manufactured at an assessed plant is newly regulated, or even banned, by environmental regulations, thus diminishing the value of the business; however, the value of the property is not affected.

All of these are legitimate business risks that have occurred in a given transaction during the past few years. Based on this situation and the uncertainty associated with assessments, it is likely that there will be real environmental risks, even if an assessment is completed. However, like other business risks associated with a conveyance, there are ways to identify, and therefore minimize, exposure to environmental risks.

Regulatory Drivers

Beyond understanding the business risks associated with impaired property, a few regulatory issues must be considered. Unlike a compliance audit, there is no systematic verification against regulatory standards. Rather, the regulatory issues serve as drivers for motivating stakeholders to undertake an assessment. Some examples are briefly presented.

At the state level, on January 1, 1984, New Jersey passed the Environmental Responsibility and Cleanup Act (ECRA), Public Law 1983, Chapter 330 [effective June 16, 1993, amended as the Industrial Site Recovery Act (ISRA), Senate Bill No. 1070]. ISRA requires that a company notify the New Jersey Department of Environmental Protection within 30 days of an announcement of a pending sale that it has completed an evaluation of its property (i.e., assessment) covering groundwater, surface water, a history of spills, and a detailed description of its testing protocols. The state then reviews the submission, and can hold up any sale until it issues a "Negative Declaration of Toxicity" or an "administrative consent order" for the site. Other states, such as Connecticut, Illinois, and California also have property transfer regulations, but their regulations are less stringent than New Jersey's.

At the federal level, property owners face tremendous liabilities under the Comprehensive Environmental Response, Compensation and Liability Act (CERCLA), otherwise known as Superfund. This legislation essentially requires that anyone who contributed to a hazardous waste at a site, including any financial stakeholders in the site whether old or new, will have liability for cleanup costs. Because the average Superfund site cleanup cost is more than $10 million (U.S.), the liability can be significant. In fact, it is possible to "buy in" to a Superfund site although it is hoped that a properly completed assessment before the purchase would disclose a liability of this magnitude. (See Chapters 3, United States Legal and Legislative Framework; and 8, Hazardous Wastes, for more details.)

Innocent Landowner Provisions of SARA

The Superfund Amendments and Reauthorization Act of 1986 (SARA) provides some protection for property owners. The "innocent landowner" provision of SARA can shield an owner from environmental liabilities if "all appropriate inquiry into the previous ownership and uses of the property" is made prior to the purchase. Unfortunately, the definition of "all appropriate inquiry" has been elusive. Consequently, a large effort has gone into developing standard practices that, if implemented, could help meet this test.

STANDARDS FOR ENVIRONMENTAL AUDITS AND SITE ASSESSMENTS

With today's need to differentiate between audits and assessments in the marketplace, there is clear reason to develop accepted standards for each. From about 1985 to the present, the issue of standards has been one of the most controversial topics in auditing (Priznar, 1993a). At this time the controversy about standards has become international and may result in a mixed landscape of competing standards and auditor certification schemes (Swiss, 1993).

Because of this controversy, it is important that those involved in environmental auditing become aware of some of the fundamental issues and organizations involved in setting standards and certifications.

The Value of Standards

The world of audits and assessments appears to be divided into two positions: those who believe standards are necessary and those who do not. The arguments in favor of standards include these:

- Standards establish a common ground on which expectations can be based. Without standards, audits or assessments conducted for the same purpose, but by different individuals, at different places, or at different times will have nothing in common. If one were to try to evaluate trends or compare one situation with another, a great effort would have to go into understanding the audit or assessment scope of work, qualifications of the individuals involved, and so on. This would add up to wasted effort and a lost opportunity to get more use out of the resulting data.
- Those purchasing the services of an external auditor or assessor cannot get proposals for the same service unless a standard is used as a reference.

Admittedly, the purchaser can develop specifications to meet its own needs to ensure consistency in bids. However, most buyers want to know where they stand with regard to generally accepted standards. That is, they want to know how their programs compare to others. A standard makes this easier because it raises the starting point to a minimum level that industry will accept.

- Insurance companies, lending institutions, and regulatory agencies favor standards so that there is a benchmark. Thus, new insurance products may be developed or regulatory programs implemented, based upon survey results associated with accepted industry standards.

However, legitimate arguments against standards include the following:

- Standards will retard the advancement of the auditing and assessment practice. This assumption is based on the belief that as soon as standards are available, many practitioners will be satisfied to provide the bare minimum effort to achieve that level. There will be no motivation to advance to higher, better, or more innovative practices.
- Standards will prevent organizations from receiving recognition for having invested in programs that achieve results far superior than can be achieved by the standard practice. In fact, superior programs that do not conform to the standard may be perceived to be defective. Thus, the result becomes less important than the process.

Clearly, both positions make legitimate points. The next few years should see a resolution of the concerns in some meaningful direction. In the meantime, one should be aware of these issues and some of the organizations that are setting standards.

Standards-Setting Organizations

Standards-setting organizations involved in audits and assessments are of two kinds: (1) established standards-setting organizations that have recently turned their attention to environmental audit and assessment practices, and (2) newly established organizations formed to fill this specific need. Unfortunately, it can be difficult to differentiate the more credible ones from those that are less valid.

For environmental audits, credible organizations include:

- American Society of Testing and Materials (ASTM), based in Philadelphia, PA. The ASTM is a well recognized and established organization that has started the development of a standard practice for environmental compliance audits. It has also agreed to work with the Institute for Environmental Auditing (IEA) on a number of specialty audit standards. Preliminary indications are that it will have a large base of participation among practitioners both in the United States and overseas.
- Institute for Environmental Auditing (IEA) in Alexandria, Virginia. Formed in 1986 to promote standard setting in environmental auditing, IEA works with any legitimate body that seeks to enhance the fields of environmental auditing, including lawmakers, insurance companies, educational organizations, financial accounting firms, and international concerns. It has developed one specialty standard for environmental management systems audits.

For environmental site assessments, these organizations offer standards:

- ASTM, as just listed, has developed a draft standard for "screening" and Phase 1 assessments with others in progress (ASTM, 1993). This is the most often cited standard of its type.
- Federal National Mortgage Association (Fannie Mae) in Washington, DC, took the lead in establishing a specific policy and protocol for environmental assessments for its multifamily lending program in 1986. Its guidance is still cited by many experts in the field.

The IEA is a good source of the latest information on standards and standards-setting organizations, but several articles offer a more comprehensive list (Bryant, 1993) or advice on how to choose an auditor (Priznar, 1993b).

Several national and international organizations are driven by legal or business requirements to perform audits or assessments. One of the most important international organizations is the International Standards Organization (ISO). ISO standards are most relevant for those conducting audits and assessments outside the United States.

At this time, ISO standards for environmental auditing are being developed by a group of national organizations. United States participation is through ASTM. An internationally accepted standard is expected in 1995. A knowledge of local laws, politics, and standards is critical. However, many countries have adopted U.S. or other standards as their international framework (see Chapter 4, International

Legal and Legislative Framework, for more details on international environmental practices).

ENVIRONMENTAL AUDITING IN PRACTICE

Environmental auditing has evolved and specific practices have been developed. Now practitioners have started to seek improvements to the process. This section describes some of the common practices and tools that are available to improve the efficiency and quality of the audit process.

Automation

Several opportunities exist to automate environmental auditing. For each step of the audit process, automation can provide:

- information on applicable regulations and requirements
- information on the collected findings
- a way to create a standard audit report
- a database of audit findings and site data that can be used for ad-hoc analyses.

Commercially available products are used for all of these processes. In some cases, a product may assist with all of these processes and may be designed specifically for this purpose. Table 7–A lists some of the options and their intended purpose. To keep up with the latest developments in the field, see a member newsletter from one of the auditing associations mentioned previously (e.g., IEA). The newsletter often includes product reviews.

Certification

The qualifications of the auditor are important to anyone planning an audit program. Lists of qualified individuals are published by certifying organizations. Unfortunately, there is no certifying organization that has widespread acceptance or government sanction. Consequently, the value of claimed credentials in the auditing field must be approached with skepticism. Currently, experience and good references are the best indicators of a quality auditor. Once standards are prepared and accepted, there will be some criteria under which individuals can be certified.

Training

There is little formal training available for environmental auditors. Of the programs offered, most are short and intensive, lasting from one to five days. This kind of training does not adequately prepare one to be a good auditor. Good auditors have a combination of knowledge, auditing skills, and personal attributes that make them qualified for the job. Given the need for training, one technique often used today is on-the-job-training (OJT). OJT may come in a variety of ways. One is through a mentoring program for internal and external auditors. This is difficult if the activity is not a full-time or nearly full-time occupation. Many who want to be internal or external auditors have other job duties as well.

A solution may be an audit team approach: An external auditor leads or participates in an audit with members of the facility staff. This allows the internal staff to gain experience, have questions answered, and learn details of the process through mentoring. It also gives management and the external auditor an opportunity to observe the skills of the apprentices while making progress with the audit itself. Variations on this approach seem to be growing in popularity.

Other techniques used in training include cross-training exercises, in which staff from one plant audit the activities of another plant. Some corporations have made this a key part of their auditor development program (McDaniel et al, 1993).

Integrated Audits

Because the process of environmental auditing is so similar to health and safety inspections, the two are sometimes combined into one activity. These are then called *integrated environmental, health, and safety audits*. Many corporations perform this style of integrated audit. For example, the U.S. EPA performs its own internal facilities' audits based on an integrated approach. The downside to integrated audits is that they may take longer and involve a larger team. In addition, the auditors must be proficient with a wider variety of regulations and practices.

On the positive side, the facility becomes disturbed only once by the audit activity, and staff tend to learn more about the relationship between environmental and health and safety issues. Consequently, health, safety, and environmental professionals can widen their professional horizons by participating in integrated audits.

Pollution Prevention Audits

With increasing emphasis on reducing wastes, the concept of pollution prevention audits has emerged as a regular activity (see Chapter 10, Pollution Prevention Approaches and Technologies). Because there are no standards in this area and the process relies substantially on judgment, these are really assessments. Areas such as waste-stream analysis, processes-change analysis, material-substitution analysis, and related investigations are typically part of these audits.

Photo Documentation

Photographs are occasionally included in auditing activities as a way to permanently and clearly document environmental conditions at audited properties. These can be especially useful to help auditors recall specific issues observed during the audit. "A picture is worth a thousand words" certainly is true in audit activities. However, attorneys are reluctant to see photos included in audit reports because they may be subject to differing interpretation and may also unintentionally show evidence of conditions not covered by the audit. Consequently, photographs in audit reports should not be taken or included without approval of either management or legal counsel.

Other photo documentation used includes historical aerial photographs, which can provide a reference of how the property was used. Aerial photographs are generally available through state and county agencies.

ENVIRONMENTAL ASSESSMENTS

The scope for assessments is usually much broader than the scope for audits, because assessments focus on business risks and liabilities, whereas audits focus on regulatory compliance. Consequently, reporting requirements under the Toxic Substances Control Act (TSCA); the Federal Insecticide, Fungicide, and Rodenticide Act (FIFRA); and the Food and Drug Act (FDA) are more important than they would be in audits, because past product stewardship issues are so important (Cahill & Kane, 1989). Independent environmental monitoring and testing are usually conducted if there is any suspicion of contamination. The results of the assessment must also be integrated into financial assessments if there is a conveyance of the property.

As a final issue, assessors require the skills to address questions about potential changes in regulatory requirements, business conditions, and efficiency of existing pollution control equipment for future regulated pollutants and levels. These and some other practical aspects of site assessments are explored in this section. (For a more thorough treatment of this important subject, see Miller, 1993.)

Lenders, equity holders, property managers, developers, and tenants are all stakeholders in a specific property. If their relationship changes, as when a party drops out or a new one is added, a conveyance has occurred. Therefore, conveyance is not just a selling transaction. Whenever a conveyance occurs, typically there is an assessment conducted of financial conditions. Environmental issues are now part of that financial assessment and are separately termed an *environmental site assessment*. In the case of lenders—which form perhaps the most frequent category of stakeholder ordering an assessment—there are some special conditions. These could be escrow accounts, resource coverage requirements, specific contract language, or lender-specified assessments (Weissman & Priznar, 1991).

Environmental Site Assessment Process

Typically, there are three phases to the environmental site assessment process:

- Phase 1: A review of historical records and a visual inspection of the site
- Phase 2: Site characterization by sampling and evaluation of areas suspected of being potential environmental liabilities
- Phase 3: Strategy for site utilization and/or potential remediation or cleanup.

All of these phases are not required for all site assessments. However, at a minimum, a Phase 1 assessment will almost always be completed unless it is clear that there is no need to take even this rudimentary step in screening the site (ASTM, 1993). If further characterization is needed, a Phase 2 is required. Because a Phase 3 assessment is so site specific, only broad generalities can be provided. For example, Phase 3 activities include evaluations of possible groundwater treatment, hazardous waste removal, asbestos removal, or other significant and potentially costly actions designed to eliminate or reduce environmental risk.

Phase 1. A Phase 1 site assessment should quickly and inexpensively determine whether the environmental integrity of a site is impaired by environmental issues. Typically, the environmental issues center on contamination.

The Phase 1 site assessment typically has four parts: records review, interviews, site reconnaissance, and application of judgment resulting in a written report. Each of these parts is further described in Figure 7–3. The information provided in the figure is based on current practices and may not

---

Phase 1 Site Assessment

Part 1. Records Review

Two types of records are generally collected and reviewed:

- *Historical Records:* to understand previous ownership and activities at the site for at least the past 60 years. Examples of records that are typically collected include:
  — Sanborn fire insurance maps
  — Aerial photographs
  — Building/tax department records
  — Cross index/street directories
  — Title and lease records.

An understanding of important historical issues can be derived from interviews as well.

- *Government Environmental Records:* to learn whether there are known environmental risks on the site that may impair the site. Examples of these records fall into two categories: primary sources, which contain information on likely contamination, and secondary sources, which present information on *potential* environmental issues. Primary sources and where to obtain the information include:
  — CERCLIS, EPA's listing of federal Superfund sites
  — Sites of environmental contamination (individual states)
  — Oil and gas contamination sites (individual states)
  — Leaking underground storage tanks (states)
  — County environmental records.

Secondary sources include:

— Hazardous Waste Disposal Management System (HWDMS), EPA's list of licensed facilities to generate, treat, store, or dispose of hazardous waste
— Locations of underground tanks (states)
— Locations of landfills and other waste disposal facilities (states)
— Locations of spills of toxic materials [states, EPA, Department of Transportation (DOT)]
— Areas with high potential for radon contamination (states, EPA)
— Asbestos records (states)
— Location of above-ground storage tanks (fire marshall)
— Discharge permits: air, surface water, groundwater (states, EPA)
— PCBs in transformers and disposal sites (states)
— Community right-to-know disclosures (states, EPA)
— Facilities subject to the TSCA (EPA)
— Compliance inspection reports (states, EPA)
— Interviews
— Security and Exchange Commission (SEC) 10-K disclosure (SEC).

These records are available from the responsible government organization or through environmental information companies that specialize in presenting summaries of these reports.

Part 2. Interviews

The best interviews are structured to collect the same information in all cases. Ideally, then, an interview guide should be used. This guide can be transformed into a questionnaire in case the questions cannot be answered verbally. As a general guideline (Miller, 1993) the following questions should be asked:

1. Do you have any knowledge or indication that hazardous materials have been used on or around the property?
2. Are there any permits affecting environmental conditions or activities issued to the property or its operation?
3. Are any pipes or furnaces wrapped with insulation?
4. When was the building insulated and with what type of material?
5. Are there any underground or above-ground storage tanks on the property?
6. Was there ever an oil-fueled furnace used on the property?
7. Have the ambient radon levels ever been checked?
8. Does the property rely on a private well for water supply? If so, has the water quality been checked?
9. Are any storage drums visible on the property?
10. Name all the parties and the types of businesses occupying the property prior to this transaction.
11. What are the present and intended future uses of the property?
12. Are any areas on the property used to dispose of waste or other materials, or are any areas of uneven settling or unexplainable changes in grade?
13. Does the property now have or did it ever have any oil, gas, water, or injection wells?
14. Are there any areas of unnaturally distressed or dying vegetation?

Figure 7–3. General details of a Phase 1 site assessment.   *(continued)*

15. Are there any areas of soil, asphalt, or concrete that are stained?
16. How close is the property to any industrial areas?
17. Are there any dumps or landfills near the property?
18. Is the property adjacent to a railroad track or underground pipeline?
19. Is the property within one mile of any known environmental problem such as federal or state Superfund sites, leaking underground tanks, and the like?
20. Are there any indications that any type of hazardous materials have been used, stored, or disposed in the neighborhood?

Part 3. Site Reconnaissance

The site reconnaissance consists of a visit to the property to inspect the land and structures built on the land and evaluate environmental issues in the site's vicinity. In general, the following items should be documented:

1. Insulation, flooring, and ceiling materials to identify the presence of asbestos or asbestos-containing materials (ACM) (see also Chapter 13, Indoor Air Quality, asbestos section)
2. Transformers, capacitors, fluorescent light ballasts to identify the presence of PCBs
3. Paint, to determine its current condition and whether it contains lead (see also Chapter 11, Public Health Issues, lead section)
4. Radon level samples
5. Floor/wall staining
6. Presence and condition of underground and above-ground storage tanks (see also Chapter 8, Hazardous Wastes, UST section)
7. Presence and destination of floor drains
8. Presence and condition of storage drums
9. Current and previous waste disposal practices (e.g., waste stream studies)
10. Evidence of soil/pavement staining or unusual cracking
11. Unusual gradient changes
12. Presence of pits, lagoons, ponds, or other surface impoundments
13. Evidence of stressed/unnatural vegetation
14. Evidence of filling/excavation
15. Method of wastewater discharges
16. Method of air emissions
17. Evidence of contamination on neighboring properties
18. Interviews with site operators and neighbors about previous activity on the property or in the area
19. Type and quality of the water supply.

Part 4. Judgment

A report that summarizes and interprets findings of the first three parts must be prepared. In developing the report, the assessor must exercise judgment in determining whether there is significant environmental contamination. The assessor is often presented with an unclear situation, which is why judgment is so critical. In these cases, only an experienced assessor will know how to interpret the entire body of information and understand the subtleties and nuances of clues. In the absence of a clear yes-or-no answer, a Phase 2 site assessment is usually ordered.

Figure 7–3. *(Concluded)*

be consistent with recommendations of all assessors; however, these activities occur in most cases.

Phase 2. The objective of a Phase 2 site assessment is to develop additional information as a result of unclear or ambiguous results in a Phase 1 assessment. The Federal National Mortgage Association (FNMA) has a specific set of assessment procedures that can be used as a guide in conducting a Phase 2 site assessment (FNMA, 1988).

The Phase 2 assessment will involve a more detailed physical site inspection and review of historical records. The purpose of Phase 2 is typically to determine the presence or absence of an uncertain liability (e.g., asbestos, or leaking underground storage tanks) or to quantify the extent of an observed or suspected liability (e.g., soils or groundwater con-

tamination). Because of the specialized nature of the investigations under Phase 2, these assessments must be conducted by a consultant qualified to perform the work.

Examples of the kind of work to be performed in a Phase 2 assessment include:

- bulk asbestos sampling and analysis, and, if required, development of abatement and maintenance and operations programs (see also Chapter 13, Indoor Air Quality)
- underground storage tank leak testing (see also Chapter 8, Hazardous Wastes)
- soil sampling and analysis
- groundwater sampling and analysis
- testing of suspected PCB-contaminated soil and/or facilities (see also Chapter 8, Hazardous Wastes)

• investigation of status of Superfund or RCRA enforcement actions related to neighboring properties.

Lenders should complete and submit the Phase 2 assessments with the consultant's report attached. No specific protocol is mandatory for the Phase 2 assessment consultant's report. However, the report should include a full description of the sampling procedures, the laboratory results, and recommendations. The consultant must certify in the report that the assessment was performed diligently and in accordance with all regulatory and good management standards, and that, to the best of the consultant's knowledge, the results are complete and accurate. The report must be signed by an officer of the consulting firm that performed the work. It is essential that all regulatory standards and good management practices be followed, especially where physical sampling and laboratory analysis is involved.

Although the FNMA guidance is several years old, it is still often cited as a model for conducting site assessments.

Special Issues

Special issues associated with site assessments include:

• *Rapid response.* Most site assessments are completed on a rapid response basis because of an impending sale or transaction. Competent assessors do not let the pressure of large business deals affect their judgment or procedures. Typically, an assessment completed on a rapid response basis will cost up to 100% more than one on a regular work schedule because of overtime expenses.

• *Cost estimating.* Once environmental issues are detected in a site assessment, the assessor is often asked to estimate the cost of cleanup or remediation. This estimate is often beyond the assessor's scope of expertise, and other experts may be needed to assist.

• *Predicting the future.* Assessors are often asked how regulations or other conditions will change in the future and what impact this will have on a site. This is a dangerous practice that could result in assessor liability, so if needed, this should be done in consultation with appropriate experts.

Property Types

It is clear that different property types will generally have different environmental risks and therefore somewhat different site assessment approaches. For the most commonly assessed property types—industrial, commercial, agricultural and undeveloped land, multifamily and single-family residential properties—the following special considerations should be noted.

Industrial. Because legal restrictions on disposal of hazardous waste did not occur until the late 1960s, many industrial sites have become contaminated from storage, spillage, or on-site disposal practices. Also, many industries have old or abandoned underground storage tanks and underground pipelines (Figure 7–4). In general, industrial properties are most likely to have significant contamination issues.

Commercial. A variety of commercial property types exist, but those with potentially greater environmental risk include gas stations, automotive repair shops, laboratories, and dry cleaners. Multiuse office buildings may house tenants with unique environmental issues or, if constructed more than ten years ago, may contain asbestos insulation.

Agricultural. Even proper application of pesticides and/or fertilizers can result in extensive soil and groundwater contamination. Also, improper storage and disposal of these substances can result in highly contaminated areas. Agricultural properties often have above- and below-ground storage tanks for fuel and oil. Finally, agricultural lands may have been used as disposal sites for waste, oil, or septic wastes, or may have been leased for household or industrial waste disposal. A further complication is the possibility of illegal dumping (ABA, 1990).

Agricultural property site assessments should take these unique risks into account. Typically, soil and groundwater testing should be conducted during Phase 1, especially near barns and storage buildings. Interviews should include questions about waste management practices. Record reviews should include a search for oil or gas elements (Olexa, 1991).

Undeveloped. Undeveloped land is categorized either as never developed or as previously developed. Each class presents unique issues. Never developed land may be at risk from illegal dumping or from contamination migrating from neighboring property. Previously developed land presents additional risks from prior users that may have left soil or groundwater contamination (Figure 7–5).

Multifamily residential. Contaminants such as radon or asbestos can be significant because of the potentially large number of exposed individuals. Also, contamination may have resulted because chemicals were improperly stored in the maintenance room or

Figure 7–4. The underground storage tanks at this airpark should be tested for leakage. (Courtesy Frank J. Priznar.)

Figure 7–5. Soil sampling is necessary to determine any risk from illegal dumping or migration of contamination from neighboring property. (Courtesy Dames & Moore.)

because the building was erected on top of old landfills that contained hazardous waste.

Single-family residential. Single-family residences also have radon and asbestos risks. Additionally, they may have storage tanks and improper septic service. Generally, single-family residences have similar but fewer environmental issues than multifamily residences.

These summaries of property types are presented to provide some scope of the magnitude of environmental issues that may occur. The summaries are not meant to be exhaustive.

International Issues

Most multinational corporations that have an audit program maintain similar program goals and processes worldwide. Their specific objectives include (Cahill & Kane, 1989):

- to independently verify compliance with host nation regulations
- to provide assurance that the corporation's policies, procedures, and practices are followed in overseas locations
- to identify and evaluate hazardous conditions, operations, and activities that pose a risk to the plant and to the public
- to ensure that procedures have been established for handling environmental incidents and emergencies.

Multinational environmental audit and assessment programs have unique issues and include:

- Importance of communication in a different language. (This is essential for conducting interviews and reporting progress and findings.)
- Differences in legal requirements that may be more or less strict than those in the United States or require different response strategies.
- Recognition of cultural values such as work hours, dress, and other daily routines.
- Difficulty in estimating costs for remediation or other follow-up actions, if needed.
- Security and safety of any audit team that crosses international borders.

Because the challenges of working overseas are so great, many U.S.-based firms use local auditing firms or develop resident expertise locally for these audits.

SUMMARY

Environmental auditing and assessing are important business techniques that have gained widespread acceptance in corporations, federal agencies, and private institutions. During the past 20 years, environmental auditing and assessing have been likened to

safety inspections. Consequently, safety professionals, who are familiar with safety inspections, should feel comfortable with environmental audits and assessments. Accepted standards have been published, enabling newcomers to learn fundamental expectations and techniques of environmental audits and assessments.

Auditing involves serious legal issues associated with liability, confidentiality, and disclosure. Because legal issues are so complex, and also because financial liability for noncompliance can be large, some attorney involvement is often recommended. Auditors are expected to have recognized credentials at the state or federal level. However, at this time, there is no meaningful registration or certification program available. Automated tools have been helpful for improving auditor productivity and increasing audit quality.

Environmental assessments also have newly set standards. Although there are serious business and legal risks involved in environmental site assessments, most practitioners are technical experts who render judgment and opinion on site conditions. Typically, they are expected to follow standard practices in gathering and interpreting information. Most property types, including undeveloped land, are evaluated for environmental risks through this process. Single-family residential property is just beginning to require environmental assessments.

The need for international environmental audits is growing. The information provided within this chapter will serve as a useful primer for those interested in environmental audits and assessments. It will also be a useful refresher for those with experience.

REFERENCES

American Bankers Association (ABA). *Agricultural Lenders' Guide to Environmental Liability*. Washington DC: ABA, 1990.

Arbuckle JG, Bosco ME, Case OR et al. *The Environmental Law Handbook*. Rockville, MD: Government Institutes, 1991.

American Society of Testing and Materials (ASTM). *Practice for Environmental Site Assessments*. 2 Parts: *Phase 1* and *Transaction Analysis*. Philadelphia: ASTM, 1993.

ASTM. *Draft Standard Practice for Environmental Compliance Auditing*. Philadelphia: ASTM, 1994.

Bryant ME. Site assessment standards emerge, evolve gradually from several sources. *Hazmat World* 29, March 1993.

Cahill L, Kane R. *Environmental Audits,* 6th ed. Rockville, MD: Government Institutes, 1989.

Federal National Mortgage Association (FNMA). Environmental hazards management procedures. In FNMA, *Guide on Conventional Selling.* Washington DC: 61–84, 1988.

Hall RM, Case DR. *All About Environmental Auditing.* Washington DC: Federal Publications Inc., 1987.

McDaniel T, Shih J, Ardiente E. *Environmental Auditing for Continuous Improvement.* Pittsburgh, PA: AWMA Meeting, June 1993.

Miller R. Environmental assessments. In Nanney DC (ed), *Environmental Risks in Real Estate Transactions.* New York: McGraw-Hill, Inc., 1993.

Olexa MT. Contaminated collateral and lender liability: CERCLA and the new age banker. *Am J Agric Econ,* 1388–1393, December 1991.

Priznar FJ. The history of environmental auditing. In Shields J (ed), *The Clean Air Act, Baselines and Environmental Audits.* Von Nostrand Reinhold, 1993a.

Priznar FJ. A guide to environmental auditing. *Scrap Processing and Recycling, J Inst of Scrap Recycling Ind* 50:101, March-April 1993b.

Priznar FJ. Trends in environmental auditing. *Env Law Reporter,* Environmental Law Institute, 20 ELR 10179, Washington DC, 1990.

Steinway D. *Personal Communication.* Washington DC: Kelly Drye Warren, March 1993.

Swiss S. Should environmental auditors be registered individually? *Environmental Risk, Euromoney,* London, U.K., 37, May 1993.

U.S. EPA. *Policy Statement on Environmental Auditing.* Washington DC: 51 Fed. Reg. 25004, 1986.

Weissman R, Priznar FJ. *Lender Directed Property Assessments—Reducing the Loan Risk.* AWMA Meeting, June 1991.

Part 2

Waste Management

Sponges and sorbents are used to contain and absorb an oil-contaminated discharge. Courtesy Breg International.

Hazardous Wastes

Florence Munter
Stephen W. Bell, MBA,
CSP, BSCE
Robert Hollingsworth, MBA,
CSP, SPHR
Joseph W. Gordon
Charles N. Lovinski, MS

Since the mid-1970s, there has been an increasing public demand to clean up environmental contamination from hazardous materials. In the United States, there is a complex framework of legal and regulatory requirements governing the generation, handling, storage, treatment, and disposal of hazardous materials.

In this part, specific chapters are dedicated to a general overview of hazardous wastes; health, safety, and training requirements for hazardous materials workers; and pollution prevention and treatment technology options. Because the hazardous wastes universe is constantly changing, the reader should always be sure the most current regulatory/legal information is consulted.

Hazardous wastes are approximately 1% of the total U.S. solid waste stream. Despite this low percentage, the generation, storage, treatment, disposal, and transportation of hazardous wastes have developed into an entire business sector that materially and financially affects both government and the private sector. This chapter focuses on an analysis of the major U.S. waste laws affecting environmental practice.

The owner or operator of a facility that produces a raw material or manufactures products, by-products, and intermediate goods is responsible for the associated generation of wastes and pollutants from facility activities. The laws and regulations in this country hold owners/operators of wastes and pollutants responsible for ultimate disposition of these materials. Therefore, to owners/operators or employees, the procedures and methods involved in disposition of wastes and pollutants are critical to the survival of their business.

Public perception alone, even if not correct in the assessment of responsibility for wastes and pollutants, can also influence the survival and economics of a business; therefore, it is essential for environmental decision making to occur responsibly. The decisions made on the wastes and pollutants generated and ultimate disposition of materials—disposal, treatment, recycling, or reuse—are as important as the decisions of what materials, goods, or products a business is providing to the customers.

Environmental laws and regulations in recent years are focusing on where wastes and pollutants are generated and methods to reduce them. Regulatory agencies are examining multimedia in attempts to reduce use of the most toxic chemicals. Pollutants exiting stacks or pipes may originate from a different medium (i.e., a wastewater stream in a process may result in air pollutants; see Chapter 10, Pollution Prevention Approaches and Technologies).

Worldwide, there is clear regulatory movement away from treatment, storage, and disposal of wastes and toward prevention and minimization. However, in order to understand the existing U.S. regulatory framework governing hazardous wastes, several concepts and definitions must be clearly understood. Often the practical uses of a term differ from the regulatory definition. It is the regulatory definition that subjects a facility owner/operator or custodian to the requirements of a law or regulatory program. Therefore, understanding the regulatory terms and definitions is a key in the development of environmental management programs and implementation of environmental practices.

As of this writing, *solid waste* means solid, liquid, semisolid, and gaseous material which is discarded or meant for discard [RCRA, Section 1004(27)]. The waste must be a solid waste by definition to qualify as a hazardous waste.

The distinction of a *waste* and a *product* is an important one in the regulations. This can be illustrated by the distinction of a chemical product stored in a tank versus the same material leaking from the tank. For example, if the tank leaks and contaminates soil, a solid waste is produced when the material is disposed. Thus, the "product" chemical in the tank has now become a "waste"—intended for discard unless it could be reused or recycled as an ingredient in another process.

There is also a distinction between a hazardous *waste* and a hazardous *substance*. The RCRA regulations define *hazardous waste* in two distinct categories: (1) listed hazardous waste (i.e., listed in the regulations) and (2) characteristic hazardous waste (i.e., waste that is tested using agency testing methods for certain characteristics or properties). Wastes may be identified in these categories by process knowledge (i.e., the generator knows the chemicals and the process the chemicals are used in to know the resulting waste).

The categories of hazardous waste are more narrowly defined than the categories of hazardous substances. A hazardous substance can be a waste as well as a commercial product. Many of the chemicals considered hazardous substances are ingredients in making a product or intermediate product. The regulatory road is complex and begins in tracking a chemical from the time it is received at a facility through storage, use, and any residue and/or waste produced from use to its ultimate disposition.

To help readers negotiate the convoluted regulatory road, this chapter is organized to systematically review and discuss major U.S. waste laws. The chapter overlaps material presented in several other chapters, particularly Chapters 3, United States Legal and Legislative Framework, and 10, Pollution Prevention Approaches and Technologies; however, the intent of this discussion is to provide a stand-alone analysis of the following topics and laws:

- Solid and hazardous wastes under the Resource Conservation Recovery Act (RCRA) as amended. This discussion includes underground storage tank rules and land disposal requirements under the Hazardous and Solid Waste Amendments of 1984 (HSWA). The HSWA is a revision to the RCRA statues. The section also covers management of hazardous substances in aboveground storage tanks as well as management of hazardous substances during transportation. Transportation of hazardous substances is also regulated by the Hazardous Materials Transportation Act (HMTA), Department of Transportation Rules, and the International Air Transport Association Shipping Requirements.
- Polychlorinated biphenyl (PCB) storage, disposal, and cleanup requirements (Toxic Substances Control Act or TSCA).
- Water quality issues under the Clean Water Act (CWA) and specifically, the National Pollutant Discharge Elimination System (NPDES).
- Air quality issues under the Clean Air Act of 1990 (see Chapter 3, United States Legal and Legislative Framework, for a detailed discussion).
- Comprehensive Environmental Response, Compensation and Liability Act (CERCLA) also known as Superfund.
- Spill assessment and response under CERCLA.
- Emergency Planning and Community Right to Know Act (EPCRA, also referred to as Title III of the Superfund Amendments Reauthorization Act of 1986).
- Regulated medical waste—Medical Waste Tracking Act of 1988.
- Radioactive and mixed wastes.

The U.S. legal framework for many of these regulations was previously discussed in Chapter 3; in addition, the international perspective was presented in Chapter 4. The intent of this chapter is to present a U.S. perspective on hazardous materials that emphasizes practical approaches and detailed descriptions of the regulations. Due to the ever-changing nature of the state and federal regulatory environment, the subsequent discussions must be viewed as general reviews and not as specific compliance manuals. Appropriate legal and/or regulatory compliance guidance should be obtained as needed.

SOLID AND HAZARDOUS WASTES: RCRA

The hazardous waste issue is not new, despite the widespread public perception that hazardous materials problems began with the 1977 Love Canal episode. In fact, industrial activity and its waste by-products are not the only sources of hazardous waste. There are six primary producers of hazardous waste (Dominguez, 1986):

- industrial: manufacturing and formulating processes, oil, gas, and mining
- municipal: power generation facilities and waste treatment facilities
- hospital: biomedical and infectious
- decommissioning: land and buildings
- nuclear
- agricultural.

In the United States, *hazardous waste* has a specific regulatory definition developed by Congress in the passage of RCRA on October 21, 1976:

Hazardous waste is (A) solid waste, or a combination of solid wastes which because of its quantity, concentration, or physical, chemical or infectious characteristics may (B) pose a substantial present or potential hazard ... when improperly treated, stored, transported, or disposed of, or otherwise managed [RCRA § 1004(5), 42 U.S.C. § 6903(5) (1976)].

The RCRA was developed in order to provide a national framework for waste management, and its scope includes solid, liquid, and gas. The U.S. EPA stated, "A fundamental premise of the statute [RCRA] is that human health and the environment will best be protected by careful management of the transportation, treatment, storage, and disposal of hazardous waste, in accordance with standards developed under the Act" [45 Fed. Reg. 12746 (February 26, 1980)]. The clear thrust of the EPA's statement is to create a cradle-to-grave regulatory system.

The RCRA, as an amendment to the Solid Waste Disposal Act, governs solid and hazardous waste management. The RCRA was enacted in 1976 and amended in 1984. The object of this legislation is to regulate both municipal and hazardous waste disposal and to encourage resource recovery and recycling. Most of the regulations developed under the RCRA concern the control of hazardous waste generators, transporters, and treatment, storage, and disposal (TSD) facilities. The RCRA authorizes the EPA to list materials as hazardous wastes and to develop a management system from cradle to grave. This management system includes record keeping, labeling, and handling requirements for these wastes.

The Hazardous Materials Transportation Act (HMTA), referenced by the RCRA, provides for the regulation of hazardous materials that are transported by air, water, rail, or highway. It authorizes the U.S. Department of Transportation (DOT) to issue requirements for the packaging, labeling, and transport of all hazardous materials shipments. The HMTA expands the regulated system to include materials as well as wastes. Thus, products that are toxic or cause a potential safety problem are regulated as well as hazardous wastes. Later sections of this chapter present in-depth discussions of two important RCRA concepts: (1) characteristics of hazardous wastes and (2) generators of hazardous wastes.

RCRA Waste Materials: Characteristics

Regulations require that all wastes be classified as hazardous or nonhazardous before their storage, treatment, or disposal. As a general rule, wastes cannot leave a given facility before the characterization of their hazardous nature. The following are questions that must be addressed in order to determine which type of waste material has been generated:

1. Is the waste material excluded from regulation? (Exclusions are listed in 40 CFR Part 261.4.)
2. Is the waste material a listed waste? (Refers to lists in App. VI or 40 CFR Part 261 Subpart D.)
3. Does the waste material exhibit any of the following hazardous characteristics:
   - Ignitability (I)—liquid with a flash point of less than 140 F (60 C)
   - Corrosivity (C)—pH less than or equal to 2 or pH greater than or equal to 12.5
   - Reactivity (R)—sulfur- or cyanide-bearing wastes or normally unstable or capable of detonating at standard temperature and pressure
   - TCLP toxicity (T)—organic, metal, or pesticide-bearing wastes.

The generator characterizes each waste stream based on the above criteria and generator knowledge of the process from which waste is generated. The generator must conduct sampling and analysis of wastes if changes in a waste stream occur or if the generator is unsure of the characteristics or components of the waste.

Characteristic wastes are no longer considered hazardous if on-site treatment eliminates the hazardous characteristics. Neutralization of acid wastewater in the acid waste storage tank is a good example of a corrosive waste made nonhazardous through treatment. Figures 8–1 and 8–2 provide brief summary flowcharts for defining solid and hazardous waste. Table 8–A is a more detailed presentation of the waste characteristics criteria.

RCRA: EPA Generator Requirements

A *generator* is defined as a person (i.e., owner/operator), by site, whose act or process produces hazardous waste. A generator of hazardous waste must not treat, store, dispose of, transport, or offer for transportation, hazardous waste without having received an EPA identification number. A generator must not offer hazardous waste to transporters or to treatment, storage, or disposal facilities that have not received an EPA identification number. The following delineates the federal requirements; some states have more stringent generator requirements.

Generator categories. A generator's status and specific requirements may change from month to month, depending on chemicals used and subsequent waste generated. The hazardous waste generation rate in a month determines the type of generator and associated regulatory requirements. The three categories of generators are:

Conditionally exempt small-quantity generator
Generates < 220 lb/month (100 kg) (27.5 gal)
Small-quantity generator
Generates ≥ 220 lb/month (100 kg) but < 2,200 lb/month (1,000 kg) (< 275 gal)
Hazardous waste generator (large quantity)
Generates ≥ 2,200 lb/month (1,000 kg) (> 275 gal)

A small-quantity generator may accumulate hazardous waste on-site for 180 days or less (270 days if the disposal facility is more than 200 miles away) without a permit, provided that the following conditions specified by the U.S. EPA are met:

- The quantity of hazardous waste accumulated on-site never exceeds 13,200 lb (6,000 kg) (approximately thirty 55-gallon drums) in a calendar

month and never exceeds 2.2 lb (1 kg) of an acutely hazardous waste. (Note: *Acutely hazardous wastes* are those wastes that, in small doses, are capable of causing death or significantly contributing to irreversible and/or incapacitating illness.) The acute wastes are specifically listed in 40 CFR 261.30.
- The material is temporarily stored in accordance with the regulatory requirements for tanks or containers.
- The date on which each period of accumulation begins is clearly marked and visible for inspection on each container.
- The wastes being accumulated on-site are labeled or marked clearly with the words HAZARDOUS WASTE. This applies to each container and/or tank.
- The facility complies with the requirements to develop a Preparedness and Prevention Plan (40 CFR 265, Subpart C).
- The facility has a Contingency Plan on-site (40 CFR 265, Subpart D): At all times there must be at least one employee either on the premises or on call with the responsibility for coordinating all emergency response measures. This person is the emergency coordinator.
- The generator posts the following information next to the telephone:
  — the name and telephone number of the emergency coordinator
  — location of fire extinguishers and spill control material, and, if present, fire alarm
  — the telephone number of the fire department, unless the facility has a direct alarm
- The generator ensures that the emergency coordinator and all other employees are thoroughly familiar with proper waste handling and emergency procedures. Requirements for training are relevant to their responsibilities during normal facility operations and emergencies.

Generator conditions and procedures. During temporary on-site storage of hazardous wastes, a small-quantity generator must adhere to the following conditions and procedures:

- The floor of the storage area must be impervious to leaks and must be free of drains, cracks, or gaps.
- The waste containers must be in good condition, compatible with the waste material, and inspected at least weekly.
- The containers must be kept closed when not in use.

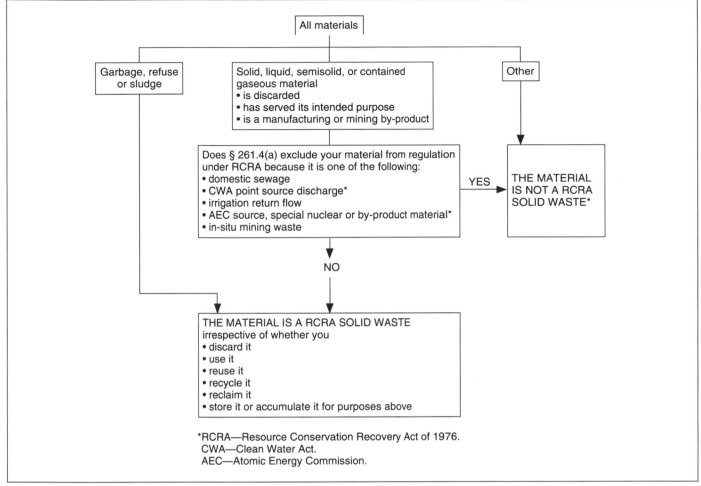

Figure 8–1. Process of defining a solid waste. Source: 40 CFR Part 260 Appendix I.

- The containers holding reactive or ignitable wastes must be stored at least 50 ft (15.3 m) inside the property boundary.
- The incompatible wastes must be separated by a dike, berm, wall, or other device.
- The waste containers must be elevated on pallets to prevent contact with accumulated liquids.

Empty containers. Through standard operating activities, facilities will have empty containers that once held either product or waste. Empty containers consist of steel and plastic drums, drum liners, steel and plastic pails, cardboard drums, and paper bags. Empty containers should be reused and recycled whenever possible. A container or inner liner removed from a container that has previously held any hazardous waste is empty (based on RCRA regulations) if:

- all possible wastes have been removed using the practices commonly employed to remove materials

from that type of container (i.e., pumping and pouring)
- no more than 1 in. (2.5 cm) of residue remains on the bottom of the container or inner liner; no more than 3% by weight of the total capacity of the container remains in the container or inner liner if the container is ≤ 110 gal (416 l) in size; or no more than 0.3% by weight of the total capacity of the container remains in the container or inner liner if the container is greater than 110 gal (416 l) in size.

A container or inner liner removed from a container that has held an acutely hazardous waste is empty if any of the following conditions are met:

- The container or inner liner has been triple rinsed using a solvent capable of removing the commercial chemical product or manufacturing chemical intermediate.
- The container or inner liner has been cleaned by another method that has been proven to achieve

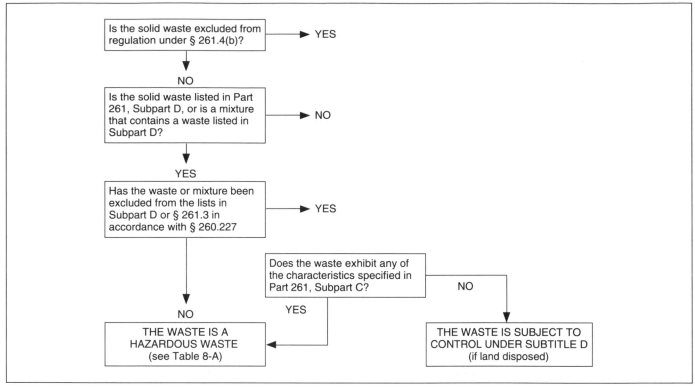

Figure 8–2. Process of defining a hazardous waste.

equivalent removal (the appropriate solvent for rinsing many chemical containers is water combined with nonphosphate detergent).

- The inner liner has been removed (in the case of a container).

A container that has held compressed gas is empty if pressure in the container approaches atmospheric pressure. (See also the section on closing USTs later in this chapter.)

Labeling. Generators must properly label and prepare shipping papers (a hazardous waste label and Uniform Hazardous Waste Manifest) for each hazardous waste container shipped off-site. Samples of each appear in Figures 8–3 and 8–4.

The purpose of a manifest is to keep a record from the time a waste leaves a generator facility to the time it is received at the ultimate disposition facility. Any intermediate storage facility at which waste is stored for more than 10 days or a treatment facility must also be listed on the manifest. The manifest must list an alternate facility if the first choice is unavailable. Each manifest must be retained for three years. Analytical or test results must also be maintained for at least three years. Periods of retention are automatically extended for facilities where enforcement action has occurred. A detailed description of the process and forms used for transporting hazardous

wastes is presented at the end of the RCRA section as part of the discussion of HMTA.

A small-quantity generator is responsible for reporting any manifests of wastes not signed by the disposition facility within 60 days. A legible copy of the manifest with some indication that the generator has not received confirmation of delivery is required to be reported to the regulatory authority. This report must be sent to the EPA regional administrator for the region in which the generator is located or the state regulatory authority. (If the state is authorized to implement the RCRA program, the submission of the exception report and other reports is to be to the state regulatory authority.)

Note: The submission to the EPA need only be a handwritten or typed note on the manifest itself, or on an attached sheet of paper, stating that the return copy was not received.

Large-quantity generators have 35 days in which to receive the returned manifest before contacting the transporter or disposal facility to determine the status of the hazardous waste. They must submit, within 45 days to the regulatory authority a cover letter explaining their attempts to locate the manifest and a copy of the unreturned manifest.

In general, regulatory requirements become stricter and more burdensome as the quantity of hazardous wastes generated increases and the time they

Table 8–A. Characteristics of Waste

Characteristic	Features of the Characteristic
Ignitability	—Liquid other than an aqueous solution containing < 24% alcohol by volume and with a flash point of less than 140 F (60 C) as determined by a Closed Cup Tester, using a specified test method or as determined by an equivalent test approved by an authority. —Nonliquid capable (under standard temperature and pressure) of causing fire through friction, absorption of moisture, or spontaneous chemical changes and, when ignited, it burns so vigorously and persistently that it creates a hazard. —Ignitable compressed gas as determined by test methods described in applicable regulations or equivalent test methods approved by the proper authority. —An oxidizer.
Corrosivity	—Aqueous with a pH $\leq$ 2 or $\geq$ 12.5, as determined by pH meter using approved test method. —Liquid that corrodes steel at a rate greater than 0.250 in (6.35 mm) per year at a test temperature of 130 F (55 C) as determined by an approved test method.
Reactivity	—Solid waste that is normally unstable and readily undergoes violent change without detonating. —Solid waste reacts violently with water. —Solid waste forms potentially explosive mixtures with water. —Mixed with water, solid waste generates toxic gases, vapors, or fumes in a quantity sufficient to present a danger to human health or the environment. —A cyanide- or sulfide-bearing waste that, exposed to pH conditions between 2 and 12.5, can generate toxic gases, vapors, or fumes sufficient to present danger to human health or the environment. —A solid waste capable of detonation or explosive reaction if the waste is subjected to a strong initiating source or if heated under confinement. —A solid waste capable of detonation or explosive decomposition or reaction at standard temperature and pressure.
TCLP Toxicity	The extract from a representative sample of solid waste contains any of these contaminants in a concentration $\geq$ value listed; when the waste contains < 0.5% filterable solids, the waste itself (after filtering) is considered to be the extract:

Contaminant	Maximum Concentration (mg) per Liter
Arsenic	5.0
Barium	100.0
Cadmium	1.0
Chromium	5.0
Lead	5.0
Mercury	0.2
Selenium	1.0
Silver	5.0
Endrin (1,2,3,4,10,10-hexachloro-1,7-epoxy-1,4,4a,5,6,7,8,8a-octahydro-1,4-endo,endo-5,8-dimethoano-naphthalene)	0.02
Lindane (1,2,3,4,5,6-hexachlorocyclohexane, gamma isomer)	0.4
Methoxychlor (1,1,1-Trichloro-2,2-biox(p-methoxy-phenyl)-ethane)	10.0
Toxaphene ($C_{10}H_{10}Cl_8$, technical chlorinated camphene, 67–69% chlorine)	0.5
2,3-D (2,4-Dichlorophenoxyacetic acid)	10.0
2,4,5-TP Silvex (2,4,5-Trichlorophenoxypropionic acid)	1.0

are stored on-site lengthens. Some of the strictest requirements are for facilities that meet the regulatory definition as a treatment, long-term storage, or disposal facility.

Treatment, storage, and disposal facilities—except for totally enclosed treatment facilities and elementary neutralization units—require a Part B permit under RCRA. Permits include detailed information about a facility and define conditions for facility operation. Conditions for facility operation often include monitoring requirements, detailed record-keeping and reporting procedures, and the use of controls to prevent and contain releases from pollution sources. Successful Part B Permit applications require a substantial amount of expertise and time; hence, expert guidance is frequently required.

RCRA characteristics and generators: Summary. The RCRA regulations provide a cradle-to-grave framework for hazardous wastes that covered generators, transporters and treatment, storage, and/or

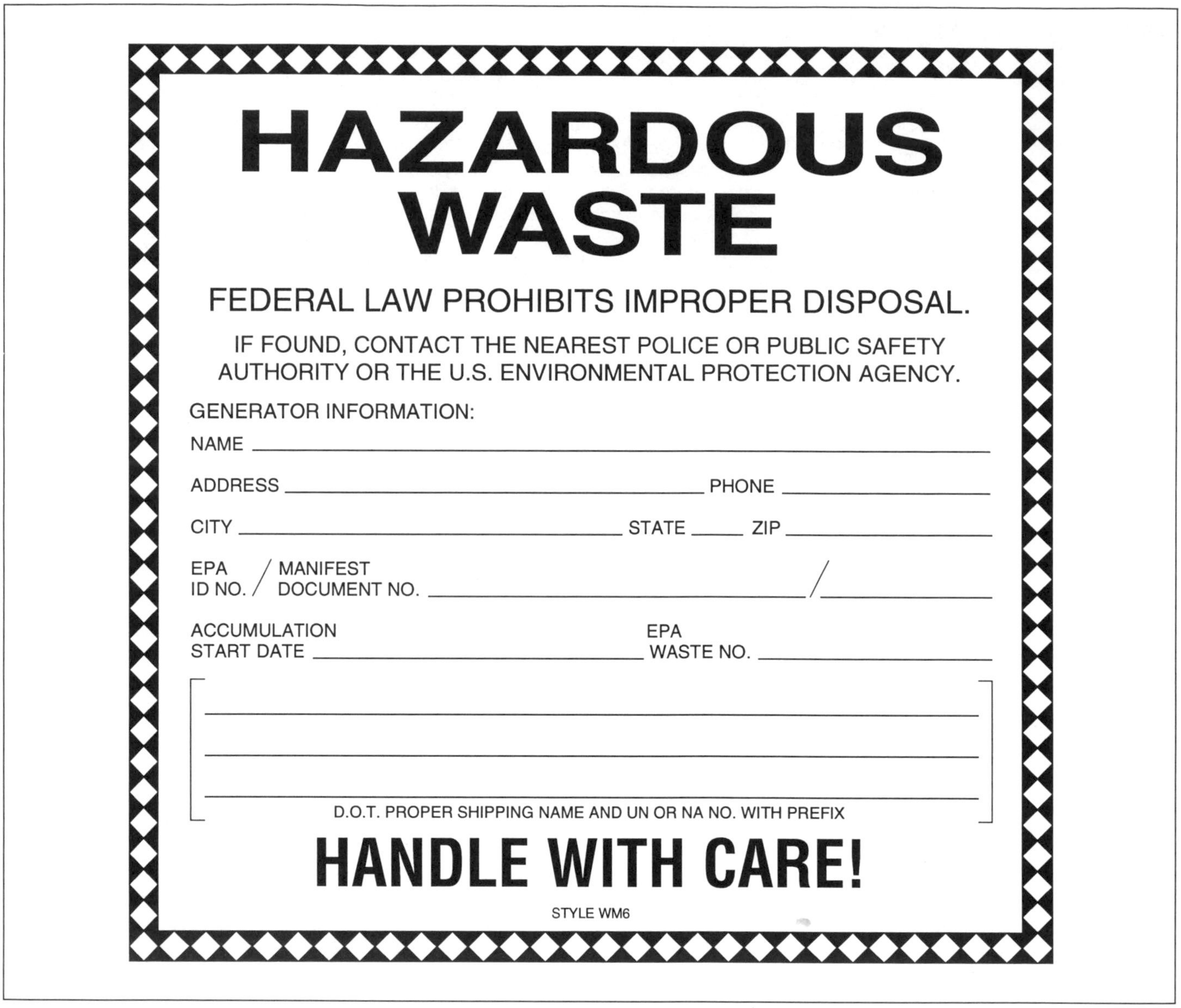
HAZARDOUS WASTE

FEDERAL LAW PROHIBITS IMPROPER DISPOSAL.

IF FOUND, CONTACT THE NEAREST POLICE OR PUBLIC SAFETY AUTHORITY OR THE U.S. ENVIRONMENTAL PROTECTION AGENCY.

GENERATOR INFORMATION:

NAME ______

ADDRESS ______ PHONE ______

CITY ______ STATE ______ ZIP ______

EPA ID NO. / MANIFEST DOCUMENT NO. ______ / ______

ACCUMULATION START DATE ______ EPA WASTE NO. ______

D.O.T. PROPER SHIPPING NAME AND UN OR NA NO. WITH PREFIX

HANDLE WITH CARE!

STYLE WM6

Figure 8–3. Sample of a commercially available hazardous waste label. (Courtesy Labelmaster.)

disposal facilities. The first two sections of this chapter discussed characteristics of an RCRA waste and who is considered a "generator." A brief introduction to the transportation rules was presented; however, this area will be covered in greater detail at the conclusion of the RCRA discussion. The next section describes storage of materials [specifically underground storage tanks (USTs) and disposal]. Treatment options are covered in Chapter 10, Pollution Prevention Approaches and Technologies.

RCRA STORAGE: UNDERGROUND STORAGE TANKS

RCRA was substantially amended in 1984. This amendment, known as the Hazardous and Solid Waste Amendments of 1984 (HSWA), revised the RCRA-related statutes to include notification and technical provisions for USTs. The EPA has devoted considerable attention to the problem of USTs; therefore, this section presents a detailed analysis and discussion of the UST problem and regulations.

An UST is a tank that has 10% or more of its volume (including piping) underground and contains a regulated substance such as:

- petroleum products
- hazardous substances regulated under CERCLA.

History

Since the beginning of this century, it has been the common practice in the United States to store petroleum products and chemicals in buried bare-steel

STATE OF ILLINOIS

ENVIRONMENTAL PROTECTION AGENCY DIVISION OF LAND POLLUTION CONTROL

P.O. BOX 19276   SPRINGFIELD, ILLINOIS 62701-0276 (217) 782-0701

FOR SHIPMENT OF HAZARDOUS, INFECTIOUS AND SPECIAL WASTE.

State Form LPC 62 8/81   IL532-0610

NOTE: FORM DESIGNED TO PRINT 8 LINES PER INCH.

EPA Form 8700-22 (Rev. 6-89)   Form Approved. OMB No. 2050-0039, Expires 9-30-91

UNIFORM HAZARDOUS WASTE MANIFEST	1. Generator's US EPA ID No. GAD123456789	Manifest Document No. 0001	2. Page 1 of	Information in the shaded areas is not required by Federal law, but is required by Illinois law.

3. Generator's Name and Mailing Address   Location If Different:

Waste Maker, Inc.
1372 Big Industry Rd. Stinkeytown, GA 00000

4. Generator's Phone(404) 123-4567

A. Illinois Manifest Document Number
IL 4350464 MANIFEST FEE PAID

5. Transporter 1 Company Name   6. US EPA ID Number
Skip 'n Scam Trucking   GAD234567890

B. Illinois Generator's ID

C. Illinois Transporter's ID
D. ( )   Transporter's Phone

7. Transporter 2 Company Name   8. US EPA ID Number

E. Illinois Transporter's ID
F. ( )   Transporter's Phone

9. Designated Facility Name and Site Address   10. US EPA ID Number

Hump 'n Dump, Ltd.
1 Landfill Rd.
Outasight, GA 00010   GAD345678901

G. Illinois Facility's ID
H. Facility's Phone
( )

11. US DOT Description (Including Proper Shipping Name, Hazard Class, and ID Number)	12. Containers No.	Type	13. Total Quantity	14. Unit Wt/Vol	I. Waste No.
a. Hazardous waste, solid, n.o.s., ORM-E, NA9189	.2 .0	D M	8 8 5 2	P	EPA HW Number X X D 0 0 8 / Authorization Number
b. Waste toluene, Flammable Liquid, UN1294	. .2	D M	9 5	G	EPA HW Number X X U 2 2 0 / Authorization Number
c. Waste trichloroethylene, ORM-A, UN1710	. .1	D M	4 2	G	EPA HW Number X X F 0 0 1 / Authorization Number
d. Waste hydrochloric acid, Corrosive Material, UN1789	. .2	D M	8 9	G	EPA HW Number X X D 0 0 2 / Authorization Number

J. Additional Descriptions for Materials Listed Above

a. Lead-contaminated soil EP-Tox Pb=12.7 mg/l

K. Handling Codes for Wastes Listed Above In Item # 14

1 = Gallons   2 = Cubic Yards

15. Special Handling Instructions and Additional Information

16. **GENERATOR'S CERTIFICATION:** I hereby declare that the contents of this consignment are fully and accurately described above by proper shipping name and are classified, packed, marked, and labeled, and are in all respects in proper condition for transport by highway according to applicable international and national government regulations.

If I am a large quantity generator, I certify that I have a program in place to reduce the volume and toxicity of waste generated to the degree I have determined to be economically practicable and that I have selected the practicable method of treatment, storage, or disposal currently available to me which minimizes the present and future threat to human health and the environment; **OR,** if I am a small quantity generator, I have made a good faith effort to minimize my waste generation and select the best waste management method that is available to me and that I can afford.

Printed/Typed Name   Signature   Date Month Day Year
John Doe   1 0 2 2 8 8

17. Transporter 1 Acknowledgement of Receipt of Materials
Printed/Typed Name   Signature   Date Month Day Year
James Smith   1 0 2 2 8 8

18. Transporter 2 Acknowledgement of Receipt of Materials
Printed/Typed Name   Signature   Date Month Day Year

19. Discrepancy Indication Space

20. Facility Owner or Operator: Certification of receipt of hazardous materials covered by this manifest except as noted in item 19.
Printed/Typed Name   Signature   Date Month Day Year
Steve Jones   1 0 2 4 8 8

This Agency is authorized to require, pursuant to Illinois Revised Statutes, Chapter 111½ Section 21, that this information be submitted to the Agency. Failure to provide the information may result in a civil penalty against the owner or operator of not to exceed $25,000 per day of violation. Falsification of this information may result in a fine up to $50,000 per day of violation and imprisonment up to 5 years. This form has been approved by the Forms Management Center.

COPY 1. TSD MAIL TO GENERATOR

In case of a spill call the Illinois Office of Emergency Response at 217/782-3637 and the National Response Center at 800/424-8802 or 202/426-2675.

Figure 8–4. Sample of a completed Uniform Hazardous Waste Manifest, U.S. EPA Form 8700–22.

tanks. A 1988 review of the type of tanks in use at gasoline service stations (Figure 8–5) shows that 84% were simple bare steel without leak protection. Initially, both federal and state environmental regulations were more concerned with air and surface water contamination than damage to groundwater. However, by the 1980s, concern began to grow around groundwater protection and the potential impact of leaking USTs. Estimates from the EPA indicate that there are 1.6 million regulated USTs in the United States; in addition, somewhere around 25% of these tanks are believed to be leaking. Thus, the potential for significant groundwater impact is quite high.

Gasoline is the commodity stored in approximately two-thirds of the USTs, with diesel accounting for the other one-third. The underground storage mode had been used for fire safety reasons. USTs containing gasoline and diesel fuel may be found on farms; at service stations; convenience stores; public, private, and military motor pools; airports; marinas; and transportation-related companies.

Causes of Leaks

According to EPA studies, the releases from underground storage tanks are typically associated with these problems: (1) piping failure, (2) corrosion, and (3) spills and overfilling. Approximately 80% of the underground releases at gasoline service stations involve improperly installed piping systems. Two types of piping systems are commonly in use. In type 1, a pressure system operates with pressure in the tanks to push the product through the piping and dispensing unit. This type of pressure system is more prone to releases during pumping, because the tanks' contents will continue to be pushed out of any break(s) in the piping. Type 2 piping configuration is a suction system. This system is less likely to experience large volume piping spills because once a leak occurs, the vacuum is lost and the product flows back to the tank.

Corrosion of the tank is another common cause of underground storage tank leakage. Hundreds of thousands of tanks installed in the United States were fabricated of bare steel without adequate leak protection. Many of these tanks have corroded and are leaking.

The third problem, spills and overfills, is the most common cause of releases to the environment from underground storage tanks. Typically, these spills and overfills are small in volume, i.e., less than 20 gal (76 l), and are infrequently reported. Spills often take place during deliveries, when improper hose drainage occurs. Overfills can be attributed in many cases to carelessness during tank-filling operations.

Materials Stored

Both the retail and nonretail sections of business and commerce use USTs for storage of motor fuels, used oil, and hazardous chemicals. Figure 8–6 illustrates the scope of the UST program, i.e., approximately 1.6 million regulated underground tanks, with retail motor fuel tanks representing 39% of the total. Nonretail motor fuel is approximately 38% and includes agriculture, petroleum wholesalers, commercial transportation, and governmental motor pools.

Regulations for USTs

Until 1984, a limited number of federal regulations addressed UST issues. Because underground tanks containing a petroleum product with a capacity in excess of 42,000 gal (159,000 l) could be a direct source of pollution in navigable waters, tank owners/operators have been required, under the Federal Water Pollution Control Act Amendments of 1972 [now the Clean Water Act (CWA)], to prepare and implement when necessary a spill prevention control and countermeasure plan (SPCC). Part of the SPCC plan required corrosion protection and pressure testing of underground storage tanks. However, the CWA impacted few underground tanks, because USTs are generally a threat to groundwater rather than surface water. Furthermore, the RCRA regulations of 1976 had jurisdiction only for tanks that contained hazardous wastes. Thus, there was a regulatory gap: RCRA did not address USTs that contained hazardous products or petroleum.

Further complicating this regulatory omission was a specific CERCLA exemption for petroleum tank leaks. Although CERCLA is specifically directed toward hazardous substance leaks, USTs were not covered. Petroleum tank leaks account for the bulk of reported UST leaks.

To close this obvious regulatory gap, the Hazardous and Solid Waste Amendments (HSWA) to RCRA were signed into law in 1984. The USTs were brought into the mainstream of regulation via Title VI, Subtitle I (Sections 9001–9010) of HSWA. The Subtitle I key provisions include: notification requirements, monitoring and reporting standards, tank standards (including leak-detection equipment), financial responsibility, corrective action, compliance monitoring and enforcement, and state programs approval. HSWA mandated that the EPA establish a

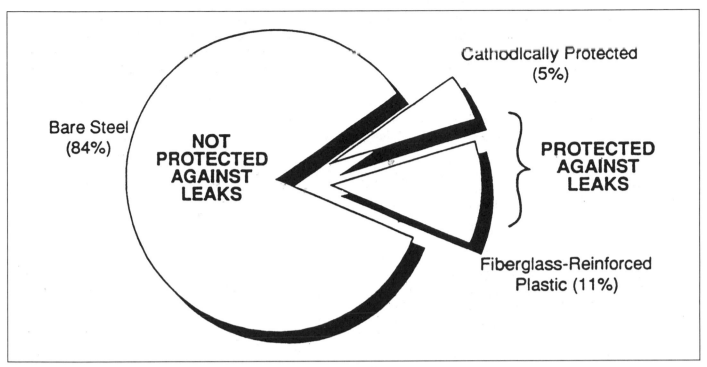

Figure 8–5. Distribution of tank types at gasoline service stations. Source: U.S. EPA, 1989, p. 3.

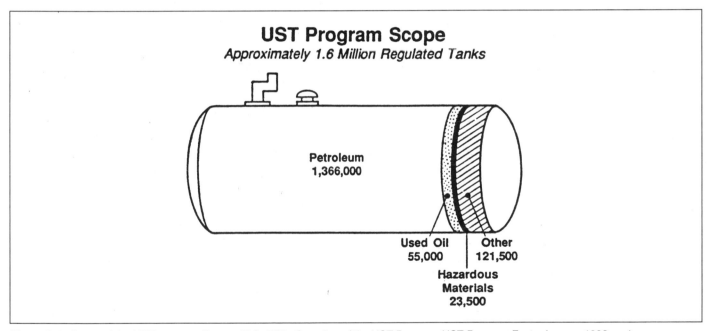

Figure 8–6. Scope of the UST program. Source: U.S. EPA, *Overview of the UST Program: UST Program Facts,* January 1993, p. 1.

full-scale program for UST regulation that would protect human health and the environment. The EPA's Office of Underground Storage Tanks (OUST) developed regulations for technical tank standards, including leak-detection methods and tank construction materials. These standards are set forth in the September 1988 EPA brochure called "Musts for USTs." In October 1988, EPA produced financial responsibility regulations that stipulated the minimum amounts of insurance needed by owners and/or operators to assure their ability to implement corrective action in the event of a leaking UST (LUST).

SARA/UST Trust Fund

The HSWA did not address how to deal with leaking tanks when the owners could not be found or when they were unable or unwilling to take the appropriate corrective action to deal with leakers. Congress again reacted to public concern by generating the Superfund Amendments and Reauthorization Act (SARA) in 1986.

Section 205 of SARA revised RCRA Subtitle I so that federal money became available in order to remediate UST petroleum leaks and spills. A $500 million Leaking Underground Storage Tank Trust Fund was established. The monies were to be generated by a five-year tax on all gasoline sales.

Additionally, the Trust Fund has been made available to the states to increase response time to LUSTs. Cooperative agreements have been signed between the EPA and those states that were interested in accessing the Trust Fund for cleanup purposes. The Trust Fund was not intended for most LUSTs, because the state governments normally direct the responsible party to handle the cleanup. The SARA amendments give the states the authority to direct the responsible party (owner/operator) to:

- test suspected tanks for leaks
- excavate to determine the degree of contamination
- remove the contaminants from water and/or soil
- determine individuals who were exposed to the contamination
- make safe drinking water available to persons who have had their source of water contaminated by the leak
- if conditions dictate, relocate residents who have been adversely affected by the spill or leak.

Congress was aware that cleanups and compensation operations could be costly. Therefore, a minimum insurance coverage of $1 million for each occurrence at locations that refine, produce, and/or market petroleum was established. Thus, the EPA or a state use the Trust Fund only when (1) cleanup cost exceeds the minimum insurance required and additional funds are needed to ensure a successful job, (2) a solvent owner/operator can't be located, or (3) the owner/operator refuses to abide by a cleanup order.

EPA Office of Underground Storage Tanks

In mid-1985, the Office of Underground Storage Tanks (OUST) was established under the auspices of the Office of Solid Waste and Emergency Response (OSWER). The EPA organization was composed of a headquarters unit and 10 regional offices located around the country. With the creation of OUST, a manager and staff were sited at each of the regional offices.

The EPA recognized that the OUST program would require different management techniques than other EPA regulatory programs. The large number of leaking tanks throughout the United States outstripped EPA's ability to successfully deal with the issue. This led EPA to develop a franchise model for the states. This model encourages flexibility and the use of innovative approaches in solving problems at the local level. Once a state or local regulatory agency signs a franchise agreement with EPA, it functions independently. From that point on, the EPA serves as a clearinghouse for ideas and relays information and data to the franchised (authorized) state or local agency. This enables the authorized agency to develop viable UST programs and operate with one set of regulations. Each franchisee must meet, as a minimum, the federal requirements in these areas:

- new UST system design construction, installation, and notification
- upgrading of existing UST systems
- general operating requirements
- release detection
- release reporting, investigation, and confirmation
- corrective action
- out-of-service or closed UST systems
- financial responsibility (U.S. EPA, 1989).

In 1993, 12 states were operating under the franchise provisions, with certain other states functioning with a memorandum of understanding.

New Installation Requirements

A new installation is deemed to be an UST system that has been installed after December 1988. The criteria for new installations are complex and cover five areas: (1) design and installation, (2) corrosion protection, (3) leak detection, (4) corrective measures, and (5) closure procedures and record keeping. New chemical USTs and pre-1980 system requirements and upgrades will also be covered at the end of this discussion.

Design and installation. The design and installation requirements are listed here, along with best management practices to prevent leaks.

- Proper installation—the owner/operator is obliged to follow industry practices and standards. A good reference is the booklet entitled "Installation of Underground Petroleum Storage Systems" by the American Petroleum Institute. A second valuable reference is the EPA's video "Doing It Right."

Improper installation is a significant cause of failure for both fiberglass and steel tanks, particularly for the piping systems. Qualified installers who follow the industry codes are the key to successful installations. During the installation process, attention must be focused on developing good plans, placement of the tank and related equipment, the excavation, depth of burial, assembly of all components, tank anchoring, backfill, and final grading and surfacing.

- Notification—At the time of installation, it is required that the owner/operator inform the state agency on the appropriate notification form that a new UST is being installed. The services of a qualified installer who certifies that the job was done correctly are required.
- Spill and overfill protection—The majority of spills and overfills results from operator error. It is important for the owner/operator to follow correct tank filling practices; in addition, required mechanical devices must be on line. Specifically:
  — Step 1 is to determine whether the capacity of the receiving tank exceeds the volume of material that is being transferred to it.
  — Step 2 is to ensure that the entire transfer operation is constantly watched. For additional literature on this topic the reader is referred to API Publication 1621, 1977, *Recommended Practices for Bulk Liquid Stock Control at Retail Outlets*, 3rd edition; and NFPA 385, 1985, *Standard for Tank Vehicles for Flammable and Combustible Liquids*.
  — Step 3 is to install equipment that prevents both spills and overfills. This is mandatory for new USTs. Overfill devices include automatic flow shutoffs, flow resistors, and full tank alarms that actuate when a tank is nearing capacity during the filling process. Spill prevention devices are items such as catch basins and hose couplings rigged with a dry disconnect. Existing tanks are required to have spill and overfill devices by December 1998.

Corrosion protection. When unprotected steel is buried in the earth, it is exposed to a natural electrochemical environment and/or to artificially generated direct electrical currents in the burial area. When these electrical conditions are combined with moisture, a situation is produced that can corrode steel. Technology to combat this corrosive process has existed for years. Many UST systems have been

well protected from the time of installation. However, in numerous cases, protective measures were not undertaken and bare, unprotected steel USTs were buried. These unprotected tanks are subject to corrosion, pitting, and leakage. Corrosion prevention is required in the EPA regulations, 40 CFR Part 280, for both new and existing tanks. This requirement became effective in December 1988. Another source of potential corrosive leakage is piping.

There are four options for new USTs and piping:

- Use steel tanks that are coated with a protective layer of a corrosion-resistant material and that also have cathodic protection. The cathodic protection can be either in the form of sacrificial anodes attached to the tank and piping or by introducing a direct current into the ground around the UST system via anodes that are not attached to the system.
- Employ the use of fiberglass tanks. When fiberglass is used, it is vital to ensure that the fiberglass is compatible with the stored material. Also, proper bedding of the tank and backfilling of the excavation is crucial in order to avoid damage.
- Use composite tanks constructed with a thick layer of fiberglass bonded to the exterior of a steel tank. Any damage to the fiberglass on a composite tank will negate its corrosion resistance; therefore, careful handling, inspection, and installation are required. Cathodic protection on composite tanks is not required by the federal regulations; however, certain states do require cathodic protection. Therefore, tank installers must check state and/or local regulations prior to installation.
- Use systems that are "no less protective of human health and the environment" than the first three approaches. This option encourages new technology. One technique employed in this category is to install steel tanks and piping with nonmetallic jackets.

Tank leak detection. The criteria for new or upgraded petroleum USTs leak detection are as follows:

- The method(s) chosen must be accurate enough to detect leakage in any part of the tank and piping that contains petroleum on a regular basis.
- The manufacturer's direction must be adhered to relative to the installation, calibration, maintenance, and operation of the equipment.
- Federal regulations Sections 280.43 and 280.44 of Title 40 list the performance criteria for the leak-detection equipment.

Figure 8–7 depicts a number of leak-detection methods. One approach is to drill *monitoring wells* in the vicinity of the UST. These wells are checked regularly to determine whether any product has leaked and is accumulating on or in the water table. The detection within the wells can be either manual or automatic. Another method is to monitor for *vapors* in the soil adjacent to an UST. This method is suitable only where porous backfill material has been used around the tank.

Processes that automatically evaluate the *product volume* and compare it with the inventory figures are an acceptable means of detecting leaks. *Interstitial monitoring* is another technique employed to detect leaks. Instruments monitor the space between the UST and a double wall, partial barrier, or tank liner. With this approach it is possible to detect tank leaks before the contamination has spread to soil and/or groundwater.

Piping leak detection. As previously mentioned, piping systems can either be the pressure or suction type. For a pressure piping system, the owner/operator must use automatic line leak detection. Also, an annual line tightness test or monthly leak-detection monitoring must be employed.

For a suction piping system, leak detection is not required if (1) the buried piping is installed with a downgrade slope to the tank so that the product will automatically run back to the tank if the suction is lost, and (2) the system is designed with only one check valve that is located immediately beneath the suction pump. If a suction piping system does not meet these criteria, the owner/operator must conduct either a line tightness test every three years or monthly leak-detection procedure (groundwater, interstitial or vapor monitoring, or other approved methods).

Corrective measures. If there is reason to believe that an UST system is leaking, prompt action is necessary. If leakage is confirmed, an action plan with appropriate corrective measures is required to be implemented as follows:

1. Take immediate action in order to stop the leak/spill and contain the material.
2. Make certain that the incident will not threaten human life or health by way of explosive vapors or drinking water contamination.
3. Report the incident to the regulatory agency within 24 hours unless it involves less than 25

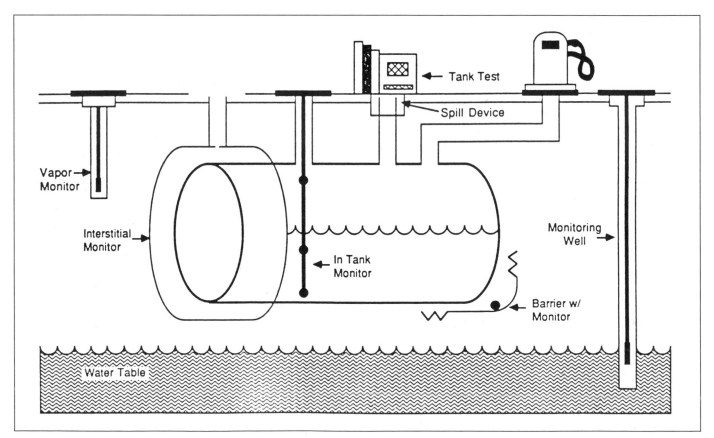

Figure 8–7. Leak detection alternatives for USTs. Source: U.S. EPA November 1990, p. 10.

gallons of petroleum and the spill was promptly contained and cleaned up.

4. Assess the full extent of the leaked material and implement the recovery of spilled material.

5. Report the cleanup progress to the regulatory agency no later than 20 days after the incident is confirmed.

6. Assess within 45 days existing or potential damage to the environment and report to the appropriate regulatory personnel. If groundwater is contaminated, a detailed report of the cleanup action plan must be submitted to the agency.

On a long-term basis, the information transmitted to the regulators could result in an agency-mandated corrective action plan. Repairing of USTs is acceptable if the standard industry codes for repair work are addressed. Within 30 days of a repair, the integrity of the tank must be demonstrated by one of the following procedures:

- Internally inspecting the tank or conducting a tightness test per industry code.

- Applying any of the monthly leak test procedures with the exception of the inventory control/tank tightness test method.

- Employing other techniques approved by the regulatory authority. Repaired USTs that are equipped with cathodic protection must be checked to prove that the cathodic system is correctly functioning. Repair records must be maintained for the duration of time that the UST is kept in service.

- Leaking piping must be replaced, not repaired. Metallic pipe joints that have merely loosened may be tightened. If national codes of practice or the manufacturer's procedures are followed, fiberglass-reinforced plastic piping may be repaired. As in the case of repaired tanks, a test is required within 30 days; however, internal inspections are obviously not appropriate for piping systems.

When considering whether to repair and/or upgrade existing UST systems, the owner/operator may want to employ the services of a corrosion engineer who specializes in this type of work.

Closure procedures and recordkeeping. USTs may be closed on either a temporary or a permanent basis. Tanks that lack corrosion protection and that stay closed in excess of 12 months and tanks that have been selected for closure must adhere to these permanent closure requirements:

- The regulators must be given a 30-day advance closure notice.

- The surrounding area must be checked for leakage. If leakage exists, then the procedures previously discussed are necessary.

- The tank must be emptied, cleaned of any sludge, vapors, and/or liquids. This process can be quite hazardous. A tank that is being decommissioned may present a flammable/explosive condition or may contain toxic residuals.

These jobs should only be performed by personnel who have been properly trained, with the correct equipment and personal protective gear, and who follow all pertinent health and safety practices. Figure 8–8 depicts the American Petroleum Institute's (API's) recommended practice for cleaning a standard UST.

Once the tank is drained and cleaned per the accepted practices, it may be left in the ground or removed. If it remains in the ground, it must be filled with an inert solid material such as sand.

Permanent closure requirements do not apply to an UST if any one of these items is met:

- When the UST in question meets the criteria for a new or upgraded unit, it can be placed in temporarily closed status for as long as it continues to fulfill the temporarily closed requirements.

- USTs that are not protected from corrosion can be granted an extension of the 12-month temporary closure limit by the regulatory authority that has jurisdiction.

- The contents of a tank can be changed to unregulated material. However, the owner/operator must first notify the regulators, empty and clean the tank, survey the surrounding environment for contamination, and perform the appropriate cleanup measures where indicated.

Temporary closure. Temporary closure pertains to tanks that are not used for 3 to 12 months. USTs with leak detection and corrosion prevention in place must be maintained with these systems on-line. In the event of a leak, the response will be in the normal manner. If the tank is empty, there is not a requirement to operate the leak-detection system. While the UST is temporarily closed, all lines, except the vent, must be capped.

Reporting. When an UST is installed, the state in which it is located must be informed by submitting a notification form. All existing USTs should already have been reported by this process.

Procedures for Draining and Excavating an Underground Storage Tank

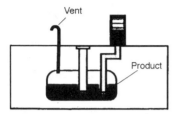

1. Prepare the personnel and area for tank removal.

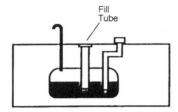

2. Drain the product from the piping into the tank.

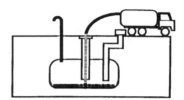

3. Remove liquids and residues from the tank.

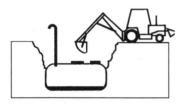

4. Excavate to the top of the tank and remove the piping, pumps, and other fixtures.

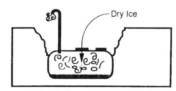

5. Purge the tank of flammable vapors.

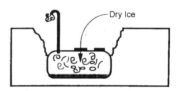

6. Monitor the tank atmosphere for flammable vapors.

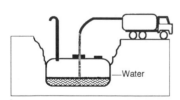

7. Using a minimal volume, triple rinse the tank with water.

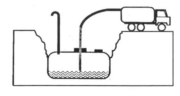

8. Pump out the water.

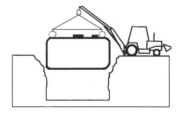

9. Before removal, plug all holes, finish excavating, and attach the proper rigging to keep the tank level.

Figure 8–8. Before cleaning and removing a standard UST, the surrounding area should be barricaded and warning signs placed. Follow applicable trenching and shoring regulations, as well as federal, state, and local safety and health regulations. Source: Adapted with permission from API Recommended Practice 1604.

Releases of both a suspect and confirmed nature must be reported. In the case of confirmed releases that result in damage, the follow-up action plan also must be submitted. Thirty days prior to the permanent closure of an UST, the regulatory personnel must be notified.

Record keeping. The records at a site cover four major sectors: (1) the functioning and maintenance of the leak-detection systems (includes information such as performance specification data provided by manufacturer, the previous year's monitoring results and the most current tightness test, and recent calibration and maintenance records); (2) the availability of the reports from the professionals who performed the last two corrosion system inspections; (3) the documents that show that upgrades and repairs were properly executed; and (4) the records of the site assessments for three years after any permanent closure.

New Chemical USTs/Existing USTs

Hundreds of chemicals are listed as hazardous in the Comprehensive Environmental Response Compensation and Liability Act of 1980 (CERCLA/Superfund). The UST regulations apply to storage of all chemical products and substances. Hazardous wastes regulated under Subtitle C of RCRA are not regulated under the UST program.

New chemical USTs. Systems installed after December 1988 must adhere to the same requirements as apply to new petroleum USTs. These requirements include proper design and installation, corrosion prevention, spill and overflow protection, corrective measures, record-keeping, and closure procedures. In addition, chemical USTs are required to be installed with secondary containment and interstitial monitoring. Secondary containment can be achieved by installing the tank or piping inside a second tank and pipe creating a double-walled system. Another alternative is to place a system inside a concrete vault or third, line the excavation with an impermeable membrane when the tank system is installed. Interstitial monitoring is achieved in the previously described fashion, i.e., monitor the space between the walls.

Existing petroleum UST requirements. Systems that were installed prior to December 1988 are considered to be "existing" USTs. Tank-filling procedures that eliminate spills and overfills were required to be in effect by 1993. By December 1998, spill and overfill prevention devices, such as catch basins and overfill alarms, must be in place. Also by December

1998, existing steel tanks and piping are required to have corrosion protection.

Corrosion protection can be achieved by a corrosion-resistant coating and cathodic protection or by other suitable methods that the regulatory authority approves. The EPA phased in the leak-detection requirement over a five-year period based on the age of the tank. However, the phase-in period ended in December 1993. Compliance with the leak-detection provision is achieved from one of these three options:

- Monthly monitoring of liquids on the groundwater, vapors in the soil, interstitial, or automatic tank gauging. Manual tank gauging inspection is allowed only for small tanks of <1,000 gal (3,785 l).
- Tanks that have a lining, corrosion protection, or spills- and overflow-prevention devices can use monthly inventory control in conjunction with tightness testing on 5-year intervals. This procedure is acceptable for only 10 years after the tank has been lined or rigged with corrosion protection or until December 1998, whichever date is later. Monthly monitoring must be used after 10 years.
- Monthly inventory control in conjunction with annual tank tightness tests are presently acceptable for existing USTs without corrosion control, internal linings, and spills- and overflow-prevention devices. However, this approach is valid only until December 1998. By then, all USTs must have corrosion protection or be lined and outfitted with spills- and overflow-prevention devices. Additionally, the previously discussed leak-detection methods must be used.

Existing pipe leak detection. The choices for pipe leak detection depend on the type of piping system. As of December 1990, existing pressure systems must adhere to the same detection criteria established for new systems. Existing suction systems must follow the same phase-in schedule that has been established for existing tanks. As of December 1993, the existing suction piping systems must meet the leak-detection criteria previously listed for new installations.

RCRA STORAGE: ABOVEGROUND STORAGE TANKS

The only RCRA regulations requiring proper containment of an oily substance are EPA's Title 40, Part

279 regulations on the subject of waste oil. A generator of waste oil has the responsibility under this part to properly store and oversee final disposition of waste oils. No specific storage or tank containment requirements exist for noncommercial waste oil businesses, but improper disposal is prohibited.

Other requirements, such as the Clean Water Act regulations, govern potentially spilled petroleum substances. The EPA's Oil Pollution Prevention Regulations have been in effect since 1973 and are codified in 40 CFR 112. The Oil Pollution Act of 1980 amended these regulations to address nontransportation-related facilities. A spill prevention control and countermeasure plan (SPCC) is required of anyone who stores petroleum substances that could be released into U.S. waters.

Many states require plans for both petroleum substances and hazardous substances when there is a potential of such substances leaking into state waters. Spill response plans related to SPCCs are discussed in the sections of this chapter that address CERCLA. This section is directed toward control strategies that can be applicable to aboveground oil storage facilities:

- **Coverage.** Facilities that drill, produce, gather, store, process, refine, transfer, and/or consume oil products are covered by the SPCC regulations if: (1) the operation is not transportation related; (2) a single aboveground container has a capacity in excess of 660 gal (2,500 l), the aggregate aboveground storage capacity exceeds 1,320 gal (5,000 l), or the total underground storage exceeds 42,000 gal (159,000 l); (3) because of the location of the operation, it is probable for an oil spill to reach U.S. waters or adjoining shorelines. The owners/operators of covered facilities must prepare and maintain on site an SPCC plan that has been certified by a registered professional engineer. The plan must be reviewed every three years.
- **Containment.** Containment systems can be effective in controlling spilled material. Routing of spilled materials to structures such as sumps, dikes, curbed holding areas, and diversion ponds are effective ways to prevent potential pollutant sources from being released to surface waters. Materials collected in these structures can then be treated and disposed of appropriately. Contingency plans are reactive in nature but necessary when positive containment is impractical because of space limitations.

- **Drainage.** Diked storage areas must be equipped with valves or other equipment in order to prevent oil leaks from escaping via drainage systems. General plant drainage systems from areas that are not diked should be designed to retain oil in catch basins or to route it back to the facility for reuse or recycling.
- **Bulk tanks.** Oil storage tanks must be designed and built to handle the anticipated storage material. Compatibility of materials, pressure, and temperature must be considered in such designs. Secondary containment that will hold the contents of the largest tank plus extra capacity for rainwater, etc., is required by regulation.
- **Loading and unloading facilities.** Railway tank car and tank truck loading/unloading facilities must adhere to the U.S. Department of Transportation regulations. Each loading and unloading rack should be built with a containment system that will accommodate the capacity of the single largest compartment in any of the tank cars or trucks that service the facility.
- **Security.** Operations that handle oil and are covered under an SPCC plan should be equipped with fences and lockable gates. All gates should be locked or posted with guards if the plant is not functioning.
- **Barrelled storage.** Areas where materials are stored in barrels or drums should be sited so as to protect the drums from weather and physical damage. Containment is another important criteria. Figure 8–9 shows a storage area for barrels that has a roofed shed for weather protection, a paved yard for the benefit of materials-handling vehicles and collection of spilled material. The entire facility is curbed and sloped to a recovery sump that has a pump. The facility presents an environmentally sound means of dealing with potential spills from barrelled materials.

DISPOSAL: LAND DISPOSAL REQUIREMENTS/SITING STORAGE FACILITIES

As previously discussed, the rules governing the storage of hazardous wastes or potentially hazardous substances in tanks have significantly tightened. Similarly, the ability to use land disposal for hazardous wastes has also been substantially restricted. Typi-

Figure 8–9. This barrelled storage facility handles liquid and solid wastes. The area has sloped concrete paving, curbings, drains, and sumps to prevent any environmental contamination in the event of an inadvertent leak or spill. (Courtesy INTALCO Aluminum Corp. Photo by R. Hollingsworth.)

cally, land disposal is defined to include placement of waste materials in a:

- concrete vault
- injection well
- landfill
- surface impoundment
- waste pile.

The federal government has banned land disposal for certain hazardous wastes that can migrate through soil and pollute groundwater. Hazardous wastes covered by the land ban include liquid metals, free cyanides, dioxin-containing wastes, and discarded chemical products like xylene, formic acid, and methyl alcohol. The ban also prohibits land disposal of diesel fuel, hydrochloric acid, and used solvents without proper treatment or certification of limited recycling options.

The land ban became effective for most spent solvents (i.e., wastes) in November 1986.

Land disposal is restricted unless:

- the wastes meet federal treatment standards
- a quantity of less than 220 lb (100 kg) is generated per month [less than 2.25 lb (1 kg) of acutely hazardous waste]

- the generator obtains a variance from the EPA
- the waste involves soil or debris from a response action under Superfund or corrective action under the RCRA.

Facilities that produce hazardous waste streams clearly have a difficult storage problem. The obvious thrust of current regulations and land ban storage requirements is to emphasize waste minimization and treatment options. Waste minimization and treatment technology options are discussed in Chapter 10, Pollution Prevention Approaches and Technologies. Unfortunately, a variety of processes and daily activities will continue to produce hazardous waste streams requiring adequate treatment, storage, and disposal (TSD) alternatives. The siting of TSD facilities is controversial and highly influenced by the local political climate. A brief synopsis of TSD siting considerations is presented.

Siting of Hazardous Waste Facilities

Several general locational criteria exist when sites for hazardous waste treatment, storage, and disposal (TSD) facilities are selected. In selecting a location, the first concern should be to protect humans and the environment from the adverse effects of toxic or hazardous waste exposure. The area selected for the facility should be a remote area and/or one with a low population density.

The advantages to selecting areas of low population density include reduced land use potential and minimal effect in the event of a release. One disadvantage of selecting areas of low population density is the distance to urban industrial operations where most wastes are generated. The increased distance from generation to TSDs inevitably increases the chances for a transportation accident.

The site considerations should also include areas where the hydrogeology minimizes possible migration and where the slope of the site itself restricts possible migration of leachate or other waste residues. Public health and welfare, water quality standards, and land use plans should all be evaluated. The site should not be close to wetlands, streams, rivers, lakes, ponds, or reservoirs. The site should be at a minimum 0.25 mi (400 m) from any well used to supply water and should not impact any well in a shallow aquifer. A new TSD facility in the United States cannot be within 200 ft (61 m) of a fault in Holocene time, i.e., the most recent geologic period (11,000 B.C. to present). Also, the facility can not be built within 100-year flood plains unless the waste can be removed safely before a flood.

Specifications for container storage of hazardous waste tanks include soil conductivity. The soil must not corrode the tank to any extent that may potentially impact human health and welfare. With surface impoundments, a liner must be placed on a surface that will not uplift, compress, or settle. Double-lined surface impoundments may be exempt from groundwater monitoring if the bottom is above the seasonal high-water table. Landfills should be above the groundwater table, and land treatment units should be 3.28 ft (1 m) above the seasonal high water table.

Complete environmental monitoring procedures are required to be documented in a closure plan to decontaminate and/or decommission facility equipment, buildings, and hazardous wastes. Documentation is required for all soil, groundwater, and surface water sampling results, including effluent quality monitoring conducted at the site during the history of operations. A detailed description of the location, collection, chain of custody, methodology, analysis, laboratory, quality assurance/control procedures, and other applicable records should be retained and available.

Criteria Levels and Rankings

The following levels of criteria and subsequent ranking pertain to selecting hazardous waste sites.

Level-1 criteria. Level-1 criteria are technically defensible criteria that identify and eliminate clearly unacceptable areas and in some instances that identify areas with higher potential for containing suitable sites, including:

- coastal flood hazard areas
- coastal wetland
- diabase or basalt (igneous) bedrock areas
- anthracite mining
- public water supply watersheds
- aquifer/well yield
- critical recharge areas
- aquicludes
- seismic risk zones
- natural areas of designated county, state, regional, or national significance.

Level-2 criteria. Criteria at the second level include land use, zoning, and infrastructure that identify candidate sites suitable for detailed screening, including:

- lands designed for industrial use
- sites of existing facilities

- dedicated lands in public trust
- arterial highways.

Level-3 criteria. Level-3 criteria are technically based and detailed site-specific criteria that exclude outright a site from further consideration and indicate general suitability in terms of "negative," "neutral," and "positive" ratings, including:

- riverine flood hazard areas
- stream proximity and use
- stream flow and quality
- aquifer use
- groundwater flow systems
- groundwater quality
- geologic faults
- unconsolidated deposits
- bedrock
- areas of mineral development
- slope
- prime agricultural land
- soil permeability, soil pH
- solid cation exchanges capacity
- freshwater wetlands
- critical habitat for rare and endangered species
- nearshore commercial shellfish resources
- historic places
- air quality designation
- structures along transportation corridor
- transportation restrictions
- population density in vicinity of facility
- proximity to incompatible facilities/structures
- site ownership.

Ranking system. The following suggested procedure can be used for preliminary site selection:

1. Level-3 criteria ratings are determined for each factor at each site.
2. The number of "negative" and "positive" ratings are totaled for each site.
3. The sites are ranked in two lists, one list for the number of "negative" ratings and the other for the number of "positive" ratings.
4. The quartile (up to a maximum of 15 sites) having the fewest "negative" ratings and the most "positive" ratings is identified.
5. The identified sites from each list are combined, in no particular order, in a new list.

Final site selection. For the final list of suitable sites, field and laboratory investigations are conducted to verify data and to get more detail; public hearings are held to consider political and social concerns. (See also the section on Risk Communication in Chapter 11, Public Health Issues.)

RCRA TRANSPORTATION: MOVING HAZARDOUS MATERIALS

The transportation of hazardous materials impacts virtually all aspects of the business community. When the movement is performed safely and efficiently, activities such as a tank truck distributing motor fuel, a vacuum truck carrying spent solvents, a bulk carrier bringing crude oil from a distant land, or an air parcel service delivering needed chemicals just in time to prevent a shutdown of a production line are routine.

Unfortunately, unsafe acts associated with the transportation of hazardous materials can cause incidents such as oil spills from tanker accidents, chemical releases and fires from train derailments, or aircraft making forced landings because of a spill of improperly packaged chemicals. Fortunately, the transportation industry and regulatory bodies are working together to create standards that, when properly implemented, will prevent such incidents from occurring.

The first legislation regulating the transportation of hazardous materials—the United States Rail Transportation Safety Act of 1906—was enacted to help eliminate accidents caused by hazardous materials transported by rail. After a number of rail incidents involving explosives, the government decided that more extensive legislation was necessary.

As other modes of transport evolved, so did the hazardous material transport regulations, eventually expanding to include vessel (water), highway, and finally air regulations. These regulations were codified under Title 49 of the Code of Federal Regulations (49 CFR), and were the model for most other regulations worldwide. However, the United States regulations have gradually grown apart from those recognized internationally. In an effort to more closely align 49 CFR with the rest of the world's regulations, commonly known as HM-181, recent changes have been made to update 49 CFR. These standards are being phased in gradually and will be completely implemented by the year 2000.

The standards of 49 CFR are administered by the U.S. Department of Transportation (DOT). Within the DOT is the Research and Special Programs Administration (RSPA). RSPA is responsible for developing a national regulatory program to protect against the risks to life and property inherent in the transportation of hazardous material by all modes. Embodied in this regulatory program is the responsibility for the promulgation of Hazardous Materials Regulations (HMR) that are codified at Subchapter C of 49 CFR. Heavy reliance is placed on participation of the four modal administrations (Federal Aviation Administration, Federal Highway Administration, Federal Railroad Administration, and the United States Coast Guard) in the regulatory process.

In a 1985 reorganization of RSPA, the Office of Hazardous Materials Transportation (OHMT) assumed the responsibility for regulating hazardous materials under the Hazardous Materials Transportation Act (HMTA). The regulations address classification, packaging, handling, incident reporting, and hazard communication requirements applicable to the transportation of hazardous materials. These regulations are continually reviewed to address safety concerns and to eliminate obsolete or unnecessary requirements.

Along with regulatory development, enforcement actions, and training, OHMT also publishes the *Emergency Response Guidebook,* which provides guidance for initial actions to be taken by emergency responders to hazardous materials incidents. The *Guidebook* addresses all hazardous materials regulated by the DOT, along with suggested initial response actions in the event of an incident (spill, explosion, fire) involving regulated hazardous materials. The guidebook is updated periodically to accommodate new products and changes in technology. It is available to first responders, police, fire, and other emergency response personnel. OHMT's goal is to have the guidebook in every emergency response vehicle nationwide. To date, more than 2.5 million copies have been distributed, without charge, to the emergency response community.

A related set of transportation regulations is administered by the U.S. EPA for hazardous wastes. These standards, found in 40 CFR 263, address the manifesting, transporting, and tracking of shipments of hazardous waste. The DOT hazardous materials standards for classifying, packaging, and labeling are incorporated into the EPA regulations by reference. The hazardous waste transportation standards are part of the overall EPA cradle-to-grave hazardous waste management philosophy and are in place largely to establish accountability and responsibility for the waste as it is moved from the generator to a treatment, storage, disposal, or final disposition facility.

Significant penalties are associated with breaks in the "chain of custody" as it relates to the proper handling and disposal of hazardous waste. The following discussion presents a more detailed description of the labeling and manifesting requirements for off-site shipments of hazardous wastes.

Procedure for Labeling, Manifesting, and Tracking Hazardous Wastes

Shipment for off-site disposal requires the completion of a hazardous waste label and manifest. Instructions for their completion are described here.

Hazardous waste label. The yellow hazardous waste label shown earlier in Figure 8–3 must be completed for all containers being shipped and must be prominently displayed on the container. The following information is required on the label:

- name of generator
- address
- phone
- EPA ID number
- manifest document number
- accumulation start date
- EPA waste number
- proper DOT shipping name
- UN or NA number.

The accumulation start date should be completed at the time waste is first collected in the drum in the accumulation area. The generator information should be completed on the label. The remaining information—proper DOT shipping name, UN or NA number, and EPA waste number—should also be complete and checked for accuracy by the responsible supervisor/manager.

The manifest document number is from the manifest (shipping paper) under which the container will be shipped. A manifest document number appears on the top line of the manifest. It is a number the owner/operator has assigned to the document for tracking purposes. This connects the container with a particular shipment of waste.

Hazardous waste manifest. Hazardous waste may not be moved outside the immediate vicinity of the source facility without an EPA-approved manifest (Figure 8–4). The manifest is basically a special shipping paper that contains certain important information about the waste:

- where it came from (the "generator" of the waste)
- where it is going (TSD facility)
- who is taking it from the generator to the TSD facility (the "transporter")

- what the waste is
- how much waste is included in the shipment
- what kind of and how many containers are in the shipment
- what kind of hazard the waste can cause.

In addition, personnel representing the generator, the transporter, and the TSD facility must sign special statements (certifications) on the manifest. These statements say that the personnel have examined the drums of waste and found the information to agree with the manifest. This information verifies that the containers are proper for the waste being shipped, the counts or weights match those on the manifest, and the wastes in the containers are the wastes listed on the manifest.

It is important to remember the manifest is the special shipping paper, and that *to manifest* means to transport a shipment with an accompanying special shipping paper, or manifest.

Preparation of a manifest. A sample uniform hazardous waste manifest is shown in Figure 8–4. However, in states that issue their own manifest forms, those forms must be used for wastes being shipped intrastate as well as interstate. Typically, the form has an original (white) with five carbon copies.

Usually, the supervisor/manager or environmental coordinator (EC) prepares the manifest for each off-site shipment. The supervisor/manager is responsible for ensuring that the waste indicated on the manifest (for example, "waste flammable liquid") is the waste being provided to the transporter. The manifest must then be dated and signed. The transporter also signs the manifest and provides one copy of the multicopy form to the shipper's representative. This manifest must be retained by the owner/operator of the generating waste facility.

A manifest must be prepared if a hazardous waste is going to be shipped off-site. If an owner/operator is not shipping the waste off-site, but, for example, moving it from one area of the building to another, a manifest is not required.

Tracking the manifest. The operator of the transportation vehicle will take all copies of the manifest except the shipper's copy, which the owner/operator retains. The shipper's copy must be kept in an appropriate file and maintained by the originator's facility. Once the disposal facility receives the shipment and signs the manifest, the original copy will be returned to the owner/operator. It must be retained along with the shipper's copy in an appropriate file.

The original plus the shipper's copy are considered a *matched pair*. It is an assurance that the waste has been documented from its generation through its transportation to its disposal. If an original is not returned from a disposal facility within 30 days of shipment, the owner/operator must be notified.

Pretransportation procedures. The owner/operator schedules shipments. When the transporter calls to confirm a shipment, the owner/operator must be certain that:

- All containers in the shipment are DOT-specification containers and are in good shape.
- The containers have been securely closed.
- The containers have been properly marked and labeled (the hazardous waste warning sticker has been completely filled out) or confirm who will be responsible for providing completed labels.
- The manifest has been filled out or a designated person will complete the information required by the manifest.

Transfer of shipment to transporter. When the transporter arrives to pick up the shipment of hazardous wastes, a supervisor/manager or EC of the facility should be present. The supervisor/manager or EC is responsible for the proper transfer of the waste to the transporter. A manifest must be completed and signed as described in this section. The operator must be provided with a copy of the signed manifest and the generator's copy must be retained by the owner/operator in the facility files for three years. If the transportation vehicle is in disrepair, not properly placarded, or the "operator" does not appear to be in proper condition, the supervisor/manager or EC should refuse to release the waste to the transporter. Figure 8–10 shows proper U.S. and international placarding. The shipper is responsible for providing the proper placards.

Waste tracking. Generators are required to maintain records of hazardous waste activities for three years. The files to be retained include copies of signed manifests by the designated facility that received the waste; a copy of each biennial report and exception report; inspection logs; and records of any test results, waste analyses, or other hazardous waste determination. All records should be kept for as long as the operator is in business, although the requirement is three years. Figure 8–11 is an example of a RCRA Hazards Assessment Test Report.

The next section moves from the specific U.S. transportation rules to a broader international perspective on hazardous materials transportation.

INTERNATIONAL TRANSPORTATION

By their very nature, air and water transportation of hazardous materials seem to be more conducive to internationally recognized regulations than are rail or highway transport. Currently, no centralized international organization regulates the transportation of hazardous materials by either rail or highway. The increasing cooperation within the European Union may help standardize ground transport within Europe, but it appears that the regulations in other parts of the world will continue to be left to individual governments. However, the international community has taken steps to establish standards for air and water transportation of hazardous materials.

During the 1950s air transport became more common, and international flights bearing hazardous materials became more frequent. The International Air Transport Association (IATA), an industry organization of the world's airlines, concluded it was virtually impossible for the shipping public to comply with each government's different regulations in international shipping. As a result, in 1959 IATA developed the first set of shipping regulations known as the IATA Dangerous Goods Regulations. These regulations are recognized worldwide for the air transport of hazardous materials. Using the input of scientists, chemists, shippers, airline industry experts, and eventually packaging engineers, IATA developed regulations and packaging standards according to a material's properties and its possible hazardous reactions during air transport. IATA's jurisdiction as a regulatory entity was assumed by the International Civil Aviation Organization (ICAO) in 1983. Although no longer considered the governing body for air transport, IATA has grown to include more than 200 of the world's airlines and continues to have a tremendous influence on the air regulations. Published annually, the IATA Dangerous Goods Regulations mirror the ICAO Technical Instructions and continue to be the regulations used by the airline industry and shippers alike, due to their straightforward, easy-to-understand format.

Under the jurisdiction of the United Nations, the ICAO is an organization of government representatives from around the world. In 1983, the ICAO became involved in the rules for the air transport of hazardous materials and began publishing what became the internationally recognized regulations for the air transport of hazardous materials.

The "ICAO Technical Instructions for the Transport of Dangerous Goods by Air" are recognized by most countries and usually may be used in place of

DOMESTIC PLACARDING

Illustration numbers in each square (☐1 through ☐18) refer to TABLES 1 and 2 below.

(WHITE) SQUARE BACKGROUND FOR PLACARD

HIGHWAY
- Used for "HIGHWAY ROUTE CONTROLLED QUANTITY OF RADIOACTIVE MATERIALS". (Sec. 172.507)

RAIL
- Used for RAIL SHIPMENTS-"EXPLOSIVE A." "POISON GAS" and "POISON GAS-RESIDUE" placards. (Sec. 172.510(a))

TABLE 1

HAZARD CLASSES	*NO.
Class A explosives	1
Class B explosives	2
Poison A	4
Flammable solid (DANGEROUS WHEN WET label only)	12
Radioactive material (YELLOW III label)	16
Radioactive material:	
Uranium hexafluoride fissile (containing more than 1.0% U^{235})	16 & 17
Uranium hexafluoride, low-specific activity (containing 1.0% or less U^{235})	16 & 17

NOTE: For details on the use of Tables 1 and 2, see Sec. 172.504 (See footnotes at bottom of tables.)

Guidelines
(CFR, Title 49, Transportation, Parts 100-177)

- Placard *motor vehicles, freight containers,* and *rail cars* containing *any quantity* of hazardous materials listed in TABLE 1.
- Placard *motor vehicles, freight containers* and *rail cars* containing 1,000 pounds or more gross weight of hazardous materials classes listed in TABLE 2.
- Placard *freight containers* 640 cubic feet or more containing *any quantity* of hazardous material classes listed in TABLES 1 and/or 2 when offered for transportation by air or water. Under 640 cubic feet see Sec. 172.512(b).

CAUTION

CHECK EACH SHIPMENT FOR COMPLIANCE WITH THE APPROPRIATE HAZARDOUS MATERIALS REGULATIONS:
Proper Classification   Marking   Placarding
Packaging   Labeling   Documentation
PRIOR TO OFFERING FOR SHIPMENT

TABLE 2

HAZARD CLASSES	*NO.
Class C explosives	18
Blasting agent	3
Nonflammable gas	6
Nonflammable gas (Chlorine)	7
Nonflammable gas (Fluorine)	15
Nonflammable gas (Oxygen, cryogenic liquid)	8
Flammable gas	5
Combustible liquid	10
Flammable liquid	9
Flammable solid	11
Oxidizer	13
Organic peroxide	14
Poison B	15
Corrosive material	17
Irritating material	18

INTERNATIONAL PLACARDING

- Most International placards are similar (color and pictorial symbol(s) to the Domestic placards illustrated above.
- International placards are enlarged ICAO or IMO labels (See International Labeling—Otherside).
- Placard MUST correspond to *hazard class* of material.

- Placard *ANY QUANTITY* of hazardous materials when loaded in FREIGHT CONTAINERS, PORTABLE TANKS, RAIL CARS and HIGHWAY VEHICLES.
- International placards *may be used in addition* to DOT placards for international shipments.

When required, *Subsidiary Risk placards* must be displayed in the same manner as *Primary Risk placards.* Class numbers are *not shown* on Subsidiary Risk placards.

- COMPATIBILITY GROUP DESIGNATORS *must be* displayed on EXPLOSIVES PLACARDS.
- UN CLASS NUMBERS and DIVISION NUMBERS *MUST* be displayed on hazard class placards when required.

UN and NA Identification Numbers

- The four digit UN or NA numbers must be displayed on all hazardous materials packages for which identification numbers are assigned. Example: ACETONE UN 1090.
- UN (United Nations) or NA (North American) numbers are found in the Hazardous Materials Tables, Sec. 172.101 and 172.102 (CFR, Title 49, Parts 100-199)
- Identification numbers may not be displayed on "POISON GAS," "RADIOACTIVE" or "EXPLOSIVE" placards. (Sec. 172.334)
- UN numbers are displayed in the same manner for both Domestic and International shipments.
- NA numbers are used only in the USA and Canada.

When hazardous materials are transported in Tank Cars, Cargo Tanks and Portable Tanks, UN or NA numbers *must* be displayed on:

PLACARDS OR ORANGE PANELS

1090   and

Appropriate Placard must be used.

EUROPEAN NUMBERING SYSTEM—

Top Number—Hazard Index (Identification of Danger, 2 or 3 figures) Example: 33 = highly inflammable liquid.

33
1088

Bottom Number—UN Number of substance Example: 1088 ACETAL

For more complete details on identification Numbers see Sec. 172.300 through 172.338.

Figure 8–10. Hazardous materials warning placards.

RCRA HAZARDS ASSESSMENT TEST REPORT

Activity: _____ Type Sample: _____

Sample ID: _____ Date Received: _____ Sampler (s): _____

Sample No.: _____ Date Reported: _____

261.21 IGNITABILITY

Flash Point: _____ °F (max. allowed 140 F, 75 C)

261.22 CORROSIVITY

pH: _____ (2 < pH < 12.5 allowed)

Halogen: _____

261.23 REACTIVITY

Acid labile cyanide: _____

Acid labile sulfide: _____

261.24 TOXICITY CHARACTERISTIC LEACHING PROCEDURE (TCLP)

SAMPLE TYPE: Solid _____ Semisolid _____ Liquid _____

If liquid or semisolid, nonfiltrable solids = _____

NOTE: If sample contains less than 0.5% nonfiltrable solids, the filtate is the extract.

ANALYTICAL RESULTS

VALUES ARE CONCENTRATIONS OF CONTAMINANT IN EXTRACT

CONTAMINANT	CONCENTRATION	MAXIMUM CONTAMINANT LIMITS, mg/l	EPA HAZARDOUS WASTE NUMBER
INORGANIC COMPOUNDS (METALS)			
Arsenic		5.0	D004
Barium		100.0	D005
Cadmium		1.0	D006
Chromium		5.0	D007
Lead		5.0	D008
Mercury		0.2	D009
Selenium		1.0	D010
Silver		5.0	D011

Figure 8–11. Sample of a RCRA Hazards Assessment Test Report. (Courtesy Dames & Moore.) *(continued)*

CONTAMINANT	CONCENTRATION	MAXIMUM CONTAMINANT LIMITS, mg/l	EPA HAZARDOUS WASTE NUMBER
VOLATILE ORGANIC COMPOUNDS			
Benzene		0.5	D018
Carbon Tetrachloride		0.5	D019
Chlorobenzene		100.0	D021
Chloroform		6.0	D022
1,2-Dichloroethane		0.5	D028
1,1-Dichloroethylene		0.7	D029
Methyl Ethyl Ketone		200.0	D035
Tetrachloroethylene		0.7	D039
Trichloroethylene		0.5	D040
Vinyl Chloride		0.2	D043
EXTRACTABLES—BASE NEUTRALS			
1,4-Dichlorobenzene		7.5	D027
2,4-Dinitrotoluene		0.13	D030
Hexachlorobenzene		0.13	D032
Hexachlorobutadiene		0.5	D033
Hexachloroethane		3.0	D034
Nitrobenzene		2.0	D036
Pyridine		5.0	D038
EXTRACTABLES—ACIDS			
O-Cresol		200.0	D023
M-Cresol		200.0	D024
P-Cresol		200.0	D025
Cresol		200.0	D026
Pentachlorophenol		100.0	D037
2,4,5-Trichlorophenol		400.0	D041
2,4,6-Trichlorophenol		2.0	D042
PESTICIDES			
Chlordane		0.03	D020
Endrin		0.02	D012

Figure 8–11. *(Continued)*

CONTAMINANT	CONCENTRATION	MAXIMUM CONTAMINANT LIMITS, mg/l	EPA HAZARDOUS WASTE NUMBER
Heptachlor (and its hydroxide)		0.008	D031
Lindane		0.4	D013
Methoxychlor		10.0	D014
Toxaphene		0.5	D015
HERBICIDES			
2,4-D		10.0	D016
2,4,5-TP (Silvex)		1.0	D017

Reported By: _____

Figure 8–11. *(Concluded)*

the individual country's regulations. For example, DOT standards permit air carriers to follow either DOT or ICAO criteria. The ICAO representatives meet twice a year, issuing revised ICAO Technical Instructions each time.

The International Maritime Organization (IMO) is the vessel transport equivalent of ICAO. A worldwide organization of government representatives, the IMO began developing the international regulations for hazardous materials transportation by water in 1983. As a "sister" organization to ICAO, the IMO is also under the jurisdiction of the United Nations.

Although not as well-organized or as influential as ICAO, the IMO is the recognized governing body for the vessel transport throughout the world. The IMO publishes the International Maritime Dangerous Goods Regulations (IMDG), which has a classification and numbering system for hazardous materials similar to that found in the ICAO criteria.

HAZARDOUS MATERIALS TRANSPORT REGULATIONS

Most countries still have their own regulations for domestic transport, but the advent of the IATA Dangerous Goods Regulations in 1959 helped to significantly and permanently change the shipping requirements for all types of transport. The IATA (and eventually the ICAO) regulations allowed one common set of rules to be used by shippers worldwide.

After these rules proved to provide a workable solution to a difficult problem, other organizations began to follow suit, including the IMO/IMDG regulations for hazardous materials transported by vessel—which have become closely aligned with ICAO/IATA. Although regulations for highway and rail transport are still generally determined by each country, most countries have—or are beginning to—align their regulations with the ICAO and IMO requirements.

Revisions to DOT Regulations

As noted previously, U.S. DOT regulations are changing and will be in fairly close alignment with other major organizations by the year 2000. This discussion describes the major revisions to these standards.

HM-181. Passed by Congress in 1990, the Hazardous Materials Transportation Uniform Safety Act is an amendment that establishes a phased updating of 49 CFR over a period of time. When these updates are completed in 2000, 49 CFR will be in fairly close alignment with the international regulations. HM-181 will (1) introduce a hazard class number system to 49 CFR, (2) recognize United Nations (UN) packaging (as found in ICAO regulations) as the only acceptable type of packaging to be used, and (3) affect some other requirements as well.

Of particular interest to the environmental remediation industry are the HM-181 requirements for the packaging and shipping of potentially contaminated samples to analytical laboratories. A new Environmentally Hazardous Substance category has been created that encompasses most of these samples, but the more highly hazardous types of samples may fall into more restrictive categories. Restrictions on hazardous wastes require such stringent precautions that packaging costs for samples for air shipment can be prohibitive.

HM-126. Companion legislation that was also passed by Congress in 1991, Hazardous Materials Bill 126 (HM-126) contains two very significant provisions:

- Subpart C requires that an emergency response telephone number be shown on the documentation for any shipment; this number must be monitored 24 hours a day by a person who is knowledgeable of the risks associated with the shipped material, and the proper emergency response procedures. This telephone number may not be for an answering machine or paging device. In addition, Subpart C requires that a copy of emergency response procedures be provided with the shipment.
- Subpart F requires that "Hazmat Employers" provide training to and test their "Hazmat Employees" regarding safe loading, unloading, handling, storing, and transporting of hazardous materials and emergency preparedness for responding to accidents or incidents involving the transportation of hazardous materials. The training categories include General Awareness/Familiarization, which provides instruction in basic HM-181 and hazard communication requirements; Function-specific, in which the Hazmat Employee learns packaging and shipping techniques that are specific to their particular roles in the hazardous materials transportation process; and Safety, which provides information on hazards posed by materials, proper selection and use of personal protective equipment, and use of emergency response information.

In addition, driver training is required for highway transportation and motor vehicle operators. Although transporters will continue to be required to successfully complete training every year (as in the past), all others affected by Subpart F will be required to receive refresher training every two years.

Future of Hazardous Materials Transportation Regulations

The hazardous materials transportation regulations are in a constant state of change, both domestically and internationally. Regulations that are acceptable today may be obsolete tomorrow. Therefore, it is important for all individuals or firms involved with shipping, receiving, handling, and/or transporting hazardous materials to remain constantly alert to potential regulations changes.

In order to stay up-to-date with the latest changes, an owner/operator should stay in close contact with its transportation company, because the company's managers usually know the upcoming rules changes. Owners/operators are advised to get on transportation industry, vendor, and government mailing lists, because they periodically distribute regulatory updates. Owners/operators can also attend a training course from a reputable trainer at least once every two years. Keeping abreast of the current regulations and changes requires some effort, but the alternative is violating the law.

PCB STORAGE, DISPOSAL, AND CLEANUP REQUIREMENTS

Polychlorinated biphenyls (PCBs) present a difficult storage and disposal problem, because the available destruction options are severely restricted—namely, to high-temperature incineration at a TSCA permitted facility. Because of the extremely ubiquitous past usage of these compounds, it is worthwhile to examine the unique set of regulations governing PCB storage, disposal, and cleanup practices.

PCBs are specifically regulated under several programs. They are prohibited from discharge under the toxic pollutant effluent standards of the Clean Water Act and are one of the chemicals regulated under the Safe Drinking Water Act. PCBs are designated hazardous compounds under the Comprehensive Environmental Response, Compensation, and Liability Act and have a reportable quantity (RQ) of 10 lb (4.54 kg). Most importantly, the Toxic Substance Control Act (TSCA) controls the use of PCBs, storage, and disposal requirements, and PCB spill cleanup policy. These regulations are published in 40 CFR 761.

TSCA PCB Regulations

Installation of PCB transformers in or near commercial buildings was banned as of October 1, 1985, except in certain emergency situations or for purposes of reclassification. The U.S. EPA allowed the installation of retrofitted PCB transformers until October 1, 1990. The use of lower secondary voltage network PCB transformers located in the sidewalk vaults near commercial buildings was allowed until October 1, 1993.

PCB storage and disposal. PCB articles or PCB containers (see 40 CFR 761 for definitions) shall not be stored for longer than one year from the date first placed in storage. In order to comply with storage and disposal requirements, an owner/operator needs to know the concentrations of PCB-contaminated articles, items, and liquids. Concentrations of more than 50 parts per million (ppm) PCBs are required to be disposed of in a TSCA-approved facility. Transformers and capacitors are common sources of PCBs.

Except as noted in 40 CFR 761, PCBs and PCB items must be stored in a facility having all of these characteristics:

- adequate roof and walls to prevent rainwater from reaching stored materials
- adequate floor with 6-in (15-cm) curbing having a containment volume of at least twice the largest PCB article or container
- no drain valves, floor valves, expansion joints, sewer lines, or other openings or conduits
- floors and curbings constructed of smooth impervious materials
- not located below the 100-year floodwater elevation.

All PCB stored materials must be marked with the date that the article was placed in storage, and the storage containers must be checked for leaks at least every 30 days. All PCB storage containers, storage areas, transport equipment, and in-use PCB equipment must be labeled with yellow or white labels as described in 40 CFR 761.40.

Disposal requirements. Disposal of PCBs is carefully regulated under TSCA. PCB-contaminated equipment and/or PCBs at concentrations between 50 and 500 ppm can be disposed in a TSCA-approved incinerator or high-efficiency boiler, or alternatively, in a TSCA-approved chemical waste landfill. Concentrations of PCBs of over 500 ppm require destruction in an approved TSCA incinerator. PCBs and PCB contaminated equipment (items, articles, containers) are required to be manifested, using a serial number or other means of identification. Records of the disposal or destruction are required to be maintained as well as an annual document log of all PCB-related activities. These records are required

to be maintained for at least three years after the facility ceases PCB-related activities.

PCB Spill Cleanup Policy

All spills of 10 lb (4.5 kg) or more, or spills directly contaminating surface water, sewers, drinking water, grazing lands, or vegetable gardens must be reported to the U.S. EPA regional office within 24 hours after discovery and cleaned up in accordance with measures outlined in 40 CFR 761.125. Also regulated in this section are spills generating PCB concentrations of 50 ppm or greater. In such a case, soils must be remediated to levels between 1 and 25 ppm as a function of their actual or anticipated land use, i.e., residential (1 ppm), commercial/industrial (10–25 ppm).

Cleanup criteria for nonimpervious surfaces are also provided in this section of the regulations. Residential cleanup criteria are considerably more stringent (1 ppm) and are based on standard EPA risk assessment methodologies (U.S. EPA, 1990). (See also Chapter 12, Risk Assessment.)

Records and Monitoring

The operator of a facility storing at one time at least 99.2 lb (45 kg) of PCBs contained in PCB containers, or one or more PCB transformers, or 50 or more PCB large high- or low-voltage capacitors must develop and maintain records on the storage and disposition of PCBs and PCB items. These records will form the basis of an annual report to be maintained at the storage facility that documents and tracks the disposition of all PCBs and PCB items in storage and disposal.

WATER QUALITY

Surface and groundwater quality is regulated through a number of federal and state programs. The Federal Water Pollution Control Act (FWPCA) was amended in 1977 and renamed the Clean Water Act (CWA). This act is intended to eliminate the discharge of pollutants into navigable waters and to make the nation's waters fishable and swimable. This long-range goal is to be achieved through a combination of industrial discharge regulations, nonpoint source controls, municipal sewage system improvements, and ambient water quality standards. Much of the regulation and implementation of water quality control programs is accomplished through state and/or local governments.

Definitions

The following definitions are relevant to this discussion of water quality:

- **Surface water.** Any body of water with its surface exposed to the atmosphere that flows into waters of the United States. U.S. waters are broadly defined to include lakes, rivers, estuaries, washes, bayous, creeks, conduits, irrigation ditches, and—most recently—wetlands.
- **Point source.** A discharge from a specific point or conveyance such as a pipe, drain, conduit, or ditch that directs or funnels discharge water to surface water. Point sources are regulated primarily through
  - technology-based effluent limitations (how clean the discharge is, using best available control technologies)
  - permits with local authorities, such as pretreatment standards (pretreatment of wastes prior to discharge to the local publicly owned treatment works)
  - permits issued under the National Pollutant Discharge Elimination System (NPDES) for point and nonpoint sources and usually specified limitations on certain constituents.
- **Nonpoint sources.** The 1987 amendments to the CWA give EPA the authority to develop stronger regulations aimed at controlling these secondary or occasional sources of pollution. A nonpoint source is a discharge from an area rather than a specific point. An example of a nonpoint source discharge is contaminated stormwater leaving a facility via overland flow rather than through a specific pipe or point. Soil erosion is another example of a common nonpoint source discharge. Nonpoint sources are regulated primarily through water quality management planning and implementation of best management practices.
- **Stormwater.** Stormwater runoff can originate from a point source or nonpoint source, depending on the activities storm-induced water contacts as well as the way it runs off of the property (through a specific conveyance or widespread overland flow). Stormwater from industrial activity, construction sites, and sources at mining sites is regulated by the state or EPA under the Clean Water Act. The federal regulatory program in 40 CFR 122.26 is implemented through individual, general, or group permits. Focus of the regulations is on pollution prevention through implementation of best management practices. Best management practices are

delineated in stormwater management plans for facilities. If stormwater and/or process water comes into contact with hazardous or regulated materials and runs off-site, then this off-site discharge is regulated under the NPDES program.
- **Groundwater.** Groundwater is water below the land surface in a zone of saturation. There is no umbrella law at the federal level focused on protection of groundwater. Congress has considered various bills aimed specifically at the protection of groundwater supplies; however, as of 1994, none has passed. There are individual state laws and regulations. Most groundwater laws are based on the Safe Drinking Water Act (SDWA) of 1974. This law is intended to provide for the safety of drinking water supplies throughout the nation by establishing and enforcing national drinking water quality standards. SDWA regulates the underground injection of wastewaters and wastes near underground sources of water supplies.

Stormwater Regulations: NPDES Permits

The U.S. EPA has promulgated final regulations for the Water Quality Act of 1987, which requires that industrial facilities apply for a NPDES permit to cover stormwater discharge. The regulations describing the permit application requirements are given in 40 CFR, Part 122.26. The regulations allow for individual industries to submit an application, for group applications for industries that are in the same industrial category, or for industries to be covered under a promulgated general permit. The following describes the classes of facilities that discharge stormwater associated with industrial activity:

- Facilities subject to national effluent limitation guidelines.
- Facilities classified as Standard Industrial Codes (SIC) 24 (except 2434), 26 (except 265 and 267), 28 (except 283), 29, 311, 32 (except 323), 33, 3441, and 373. (These codes include lumber; paper mills; chemicals; petroleum; rubber; leather tanning and finishing; stone, clay, and concrete; metals; enameled iron and metal sanitary ware; and ship/boat manufacturing facilities.)
- Facilities classified as SIC codes 10 through 14, including active and inactive mining and oil and gas operations with contaminated stormwater discharges, except for areas of coal mining operations that have been reclaimed and the performance bond has been released by the requirements after

30 days after publication of the final regulation (see the description of special application provisions for mining operations and oil and gas operations that follow).

- Construction activity (except for disturbances of less than five acres of total land area that are not part of a larger common plan of development or sale).
- Facilities in which materials are exposed to stormwater classified under SIC codes 20, 21, 22, 23, 2434, 25, 265, 27, 283, 285, 30, 31 (except 311), 323, 34 (except 3441), 35, 36, 37 (except 373), 38, 39, and 4221–25. (These codes include food; tobacco; textile; apparel; wood kitchen cabinets; furniture; paperboard containers and boxes; converted paper/paperboard products; printing; drugs' leather; fabricated metal products' industrial and commercial machinery and computer equipment; electronic equipment; transportation equipment; measuring, analyzing, and controlling instruments and photographic, media, and optical goods, and watches and clocks; glass manufacturing; and certain warehousing and storage establishments.)
- Vehicle maintenance, equipment cleaning, or airport deicing areas; railroad, mass transit, school bus, trucking, and courier services; postal service; water transportation; airport facilities; and petroleum bulk stations.
- Treatment works treating domestic sewage or any other sewage sludge or wastewater treatment device or system, used in the storage, treatment, recycling, and reclamation of sewage (including land used for the disposal of sludge located within the confines of the facility) with a design flow of 1.0 million gallons per day (mgd) or more are required to have an approved pretreatment program. This does not include farm lands, domestic gardens, or lands used for beneficial reuse of sludge that are not physically located in the confines of the facility.
- Hazardous waste treatment, storage, or disposal facilities.
- Landfills, land application sites, and open dumps that receive industrial wastes.
- Recycling facilities classified as SIC codes 5015 and 5093. (These codes include metal scrapyards, battery reclaimers, salvage yards, and automobile yards.)
- Steam electric power-generating facilities (including coal-handling sites).

The national policy for the CWA includes providing construction grants for publicly owned treatment works (POTWs), encouraging technological research, implementing area-wide planning and prohibiting discharge of toxic pollutants in toxic amounts. The impact of the significant CWA amendments in 1977 and subsequent court decisions is to incorporate controls to reduce toxic pollutants. Effluent limitations and guidelines are established for specific industries that discharge wastewaters. Water quality programs establish standards and/or controls for toxic chemicals and over 130 priority pollutants.

Sections 301, 304, 306, and 307 of the CWA pertain to technology-based standards for dischargers. These technology-based standards for discharges incorporate industrial direct discharges (source categories) that require existing sources to use best practice technology (BPT) by July 1, 1977, regulated by Section 301(B)(1)(A). For new sources, more stringent requirements are mandated. The technology-based standards for dischargers also require POTWs to use secondary treatment and industrial users of POTWs to use pretreatment according to local standards.

Other provisions to protect the nation's waters include EPA's National Ambient Water Quality Criteria promulgated under Sections 301, 302, and 303 of the CWA. Clean water was to be achieved by July 1, 1977, through treatment standards and application of technology by industries and POTWs under Section 301(B)(1)(C) of the CWA. The Water Quality Management program objectives directed EPA to assist states in establishing further performance-oriented standards and a system to develop stream segment pollutant load allocation for discharges under Section 303(D)(3) and 402 of the CWA. It is important to understand that water quality-based standards can dictate discharge limitations for dischargers more stringently than technology-based standards. Table 8–B lists the mandatory compliance dates for industrial discharges in Sections 301 and 307 of the CWA.

Section 402 of the CWA requires a permit from the National Pollutant Discharge Elimination System (NPDES) for discharges of pollutants. The NPDES is a regulatory mechanism for permitting point source discharges. A permit application and/or notice of intent must be filed 180 days prior to discharge of industrial water or stormwater (Figure 8–12). Point source discharges flow directly into surface waters as opposed to sewer systems. There are different permit applications to file, depending on whether your operation is one of the point source

Table 8–B. Mandatory Compliance Dates for Industrial Discharges (Sections 301 and 307)

Level to Be Achieved	*Date of Compliance*	*Applicable CWA Section(s)*
Best practical technology	July 1, 1977	Section 301(B)(1)(A)
Best available technology	July 1, 1984	Section 301 (B)(2)(C)
Best conventional pollutant control technology	July 1, 1985	Section 301 (B)(2)(E)
Nonconventional nontoxic pollutant limits	Three years after established	Section 301 (B)(2)(F)
Categorical pretreatment standards for POTW users		Sections 307 (B) and 207 (C)

categories with effluent limitations, an existing operation with a permit, or will discharge stormwater only.

To properly manage water quality issues, knowledge of local and EPA environmental compliance guidelines and regulations is a must. The following list is supplied to assist with the initial phases of identifying permit needs. Further investigation of on-site sewer lines and/or identification of liquid waste sources may be needed after the initial information is revised. This information should be gathered for regulated facilities at a minimum and includes:

- For on-site sewer lines:
  — List the known discharge points at the facility (e.g., manholes and drains).
  — Review any construction plans to distinguish between a sanitary sewer and storm sewer lines and discharges.
  — Inspect areas of sewer discharge inlets.
  — Inspect the maintenance garage for floor drains and determine their discharge route.
  — Determine whether there is a local permit for the facility's use of the municipal sewer system. Obtain and review the ordinance or permit regulations.
- For areas of liquid wastes:
  — Conduct an inspection of facility operations (i.e., vehicle wash area, vehicle parking area, maintenance garage cleanup procedures, storage tank's secondary containment structure).
  — Identify any liquids or pollutants that may enter the sewer line or be discharged off-site through a drainage ditch and/or retention pond. These are potential risk areas. Areas of typical pollutant-laden waters are vehicle washwater, liquids from vehicles storing refuse, maintenance garage floor washwater containing oil and grease, antifreeze, hydraulic fluid, motor oil, and transmission fluid leaks.
  — Identify the sewer or drain inlet for these liquids or when liquids may drain off-site.

The following conditions include examples of unacceptable releases. Generally, state and federal law prohibit such discharge. The state and local laws and subsequent regulations should be consulted to understand the facility obligations for (1) notifications of discharge to local government treatment systems, (2) permit applications for discharges into surface waters, and (3) permit applications for installation of wastewater or treatment systems on site.

AIR QUALITY

The Clean Air Act (CAA) of 1970, as amended, is designed to protect the public health and welfare from the harmful effects of air pollution. Subsequent amendments (1990) set definite goals for emission reductions and ambient air quality improvements.

The initial act required the development of primary and secondary National Ambient Air Quality Standards and National Emission Standards for Hazardous Air Pollutants. Under this act and the amendments states are required to develop state implementation plans (SIPs) to implement and enforce these standards. Periodic updates of SIPs are required. In addition, the act's Prevention of Significant Deterioration (PSD) section prohibits degradation of air quality within classes of air quality regions. The most stringent standards are established for certain pristine areas of the country.

For an in-depth discussion of air quality regulations, see Chapter 3, United States Legal and Legislative Framework. It presents a detailed analysis of the 1990 Amendments to the Clean Air Act. In addition, Chapter 14, Global Issues, has a discussion of the CAA amendments applicable to the acid rain problem. Chapter 10, Pollution Prevention Approaches and Technologies, reviews some of the available air pollution control technologies.

CERCLA/SARA

The Comprehensive Environmental Response, Compensation and Liability Act (CERCLA), also known

See Reverse for Instructions

Form Approved. **OMB No. 2040-0086**
Approval expires: 8-31-95

NPDES FORM	

United States Environmental Protection Agency
Washington, DC 20460
Notice of Intent (NOI) for Storm Water Discharges Associated with Industrial Activity Under the NPDES General Permit

Submission of this Notice of Intent constitutes notice that the party identified in Section I of this form intends to be authorized by a NPDES permit issued for storm water discharges associated with industrial activity in the State identified in Section II of this form. Becoming a permittee obligates such discharger to comply with the terms and conditions of the permit.   ALL NECESSARY INFORMATION MUST BE PROVIDED ON THIS FORM.

I. Facility Operator Information

Name: _____   Phone: _____

Address: _____   Status of Owner/Operator: ☐

City: _____   State: ____   ZIP Code: _____

II. Facility/Site Location Information

Name: _____   Is the Facility Located on Indian Lands? (Y or N) ☐

Address: _____

City: _____   State: ____   ZIP Code: _____

Latitude: _____   Longitude: _____   Quarter: ___   Section: ___   Township: _____   Range: _____

III. Site Activity Information

MS4 Operator Name: _____

Receiving Water Body: _____

If You are Filing as a Co-permittee, Enter Storm Water General Permit Number: _____   Are There Existing Quantitative Data? (Y or N) ☐   Is the Facility Required to Submit Monitoring Data? (1, 2, or 3) ☐

SIC or Designated Activity Code: Primary: ____   2nd: ____   3rd: ____   4th: ____

If This Facility is a Member of a Group Application, Enter Group Application Number: ____

If You Have Other Existing NPDES Permits, Enter Permit Numbers: _____   _____   _____

IV. Additional Information Required for Construction Activities Only

Project Start Date: _____   Completion Date: _____   Estimated Area to be Disturbed (in Acres): _____   Is the Storm Water Pollution Prevention Plan in Compliance with State and/or Local Sediment and Erosion Plans? (Y or N) ☐

V. Certification: I certify under penalty of law that this document and all attachments were prepared under my direction or supervision in accordance with a system designed to assure that qualified personnel properly gather and evaluate the information submitted. Based on my inquiry of the person or persons who manage the system, or those persons directly responsible for gathering the information, the information submitted is, to the best of my knowledge and belief, true, accurate, and complete. I am aware that there are significant penalties for submitting false information, including the possibility of fine and imprisonment for knowing violations.

Print Name: _____   Date: _____

Signature: _____

EPA Form 3510-6 (8-92)

♺ *Printed on Recycled Paper*

Figure 8–12. The U.S. EPA's Notice of Intent for Storm Water Discharges Associated with Industrial Activity Form.

(continued)

Instructions - EPA Form 3510-6
Notice Of Intent (NOI) For Storm Water Discharges Associated With Industrial Activity
To Be Covered Under The NPDES General Permit

Who Must File A Notice Of Intent (NOI) Form

Federal law at 40 CFR Part 122 prohibits point source discharges of storm water associated with industrial activity to a water body(ies) of the U.S. without a National Pollutant Discharge Elimination System (NPDES) permit. The operator of an industrial activity that has such a storm water discharge must submit a NOI to obtain coverage under the NPDES Storm Water General Permit. If you have questions about whether you need a permit under the NPDES Storm Water program, or if you need information as to whether a particular program is administered by EPA or a state agency, contact the Storm Water Hotline at (703) 821-4823.

Where To File NOI Form

NOIs must be sent to the following address:

> Storm Water Notice of Intent
> PO Box 1215
> Newington, VA 22122

Completing The Form

You must type or print, using upper-case letters, in the appropriate areas only. Please place each character between the marks. Abbreviate if necessary to stay within the number of characters allowed for each item. Use one space for breaks between words, but not for punctuation marks unless they are needed to clarify your response. If you have any questions on this form, call the Storm Water Hotline at (703) 821-4823.

Section I Facility Operator Information

Give the legal name of the person, firm, public organization, or any other entity that operates the facility or site described in this application. The name of the operator may or may not be the same as the name of the facility. The responsible party is the legal entity that controls the facility's operation, rather than the plant or site manager. Do not use a colloquial name. Enter the complete address and telephone number of the operator.

Enter the appropriate letter to indicate the legal status of the operator of the facility.

> F = Federal   M = Public (other than federal or state)
> S = State   P = Private

Section II Facility/Site Location Information

Enter the facility's or site's official or legal name and complete street address, including city, state, and ZIP code. If the facility or site lacks a street address, indicate the state, the latitude and longitude of the facility to the nearest 15 seconds, or the quarter, section, township, and range (to the nearest quarter section) of the approximate center of the site.

Indicate whether the facility is located on Indian lands.

Section III Site Activity Information

If the storm water discharges to a municipal separate storm sewer system (MS4), enter the name of the operator of the MS4 (e.g., municipality name, county name) and the receiving water of the discharge from the MS4. (A MS4 is defined as a conveyance or system of conveyances (including roads with drainage systems, municipal streets, catch basins, curbs, gutters, ditches, man-made channels, or storm drains) that is owned or operated by a state, city, town, borough, county, parish, district, association, or other public body which is designed or used for collecting or conveying storm water.)

If the facility discharges storm water directly to receiving water(s), enter the name of the receiving water.

If you are filing as a co-permittee and a storm water general permit number has been issued, enter that number in the space provided.

Indicate whether or not the owner or operator of the facility has existing quantitative data that represent the characteristics and concentration of pollutants in storm water discharges.

Indicate whether the facility is required to submit monitoring data by entering one of the following:

> 1 = Not required to submit monitoring data;
> 2 = Required to submit monitoring data;
> 3 = Not required to submit monitoring data; submitting certification for monitoring exclusion

Those facilities that must submit monitoring data (e.g., choice 2) are: Section 313 EPCRA facilities; primary metal industries; land disposal units/incinerators/BIFs; wood treatment facilities; facilities with coal pile runoff; and, battery reclaimers.

List, in descending order of significance, up to four 4-digit standard industrial classification (SIC) codes that best describe the principal products or services provided at the facility or site identified in Section II of this application.

For industrial activities defined in 40 CFR 122.26(b)(14)(i)-(xi) that do not have SIC codes that accurately describe the principal products produced or services provided, the following 2-character codes are to be used:

> HZ = Hazardous waste treatment, storage, or disposal facilities, including those that are operating under interim status or a permit under subtitle C of RCRA [40 CFR 122.26 (b)(14)(iv)];
> LF = Landfills, land application sites, and open dumps that receive or have received any industrial wastes, including those that are subject to regulation under subtitle D of RCRA [40 CFR 122.26 (b)(14)(v)];
> SE = Steam electric power generating facilities, including coal handling sites [40 CFR 122.26 (b)(14)(vii)];
> TW = Treatment works treating domestic sewage or any other sewage sludge or wastewater treatment device or system, used in the storage, treatment, recycling, and reclamation of municipal or domestic sewage [40 CFR 122.26 (b)(14)(ix)]; or,
> CO = Construction activities [40 CFR 122.26 (b)(14)(x)].

If the facility listed in Section II has participated in Part 1 of an approved storm water group application and a group number has been assigned, enter the group application number in the space provided.

If there are other NPDES permits presently issued for the facility or site listed in Section II, list the permit numbers. If an application for the facility has been submitted but no permit number has been assigned, enter the application number.

Section IV Additional Information Required for Construction Activities Only

Construction activities must complete Section IV in addition to Sections I through III. Only construction activities need to complete Section IV.

Enter the project start date and the estimated completion date for the entire development plan.

Provide an estimate of the total number of acres of the site on which soil will be disturbed (round to the nearest acre).

Indicate whether the storm water pollution prevention plan for the site is in compliance with approved state and/or local sediment and erosion plans, permits, or storm water management plans.

Section V Certification

Federal statutes provide for severe penalties for submitting false information on this application form. Federal regulations require this application to be signed as follows:

For a corporation: by a responsible corporate officer, which means: (i) president, secretary, treasurer, or vice-president of the corporation in charge of a principal business function, or any other person who performs similar policy or decision making functions, or (ii) the manager of one or more manufacturing , production, or operating facilities employing more than 250 persons or having gross annual sales or expenditures exceeding $25 million (in second-quarter 1980 dollars), if authority to sign documents has been assigned or delegated to the manager in accordance with corporate procedures;

For a partnership or sole proprietorship: by a general partner or the proprietor; or

For a municipality, state, Federal, or other public facility: by either a principal executive officer or ranking elected official.

Paperwork Reduction Act Notice

Public reporting burden for this application is estimated to average 0.5 hours per application, including time for reviewing instructions, searching existing data sources, gathering and maintaining the data needed, and completing and reviewing the collection of information. Send comments regarding the burden estimate, any other aspect of the collection of information, or suggestions for improving this form, including any suggestions which may increase or reduce this burden to: Chief, Information Policy Branch; PM-223, U.S. Environmental Protection Agency, 401 M Street, SW, Washington, DC 20460, or Director, Office of Information and Regulatory Affairs, Office of Management and Budget, Washington, DC 20503.

Figure 8–12. *(Concluded)*

as "Superfund," in large part traces its lineage back to the 1977 Love Canal episode. The Superfund program was established to identify sites at which releases of hazardous substances into the environment might or have occurred that endanger public health or the environment.

Once sites were identified, the program was further charged with these additional tasks:

- provide a mechanism to clean up the released substances where there is a risk to public health or the environment
- ensure a site is either cleaned up by responsible parties (if they can be identified) or the government, where appropriate
- evaluate damages to natural resources from identified releases
- establish a claims procedure for parties that have either spent money to restore natural resources or have cleaned up sites.

Clearly, a strong emphasis was placed on the identification of potentially responsible parties (PRPs) for cost recovery. Not surprisingly, the first five years of Superfund were hallmarked by substantial litigation about the implementation and constitutionality of the statute.

In 1986, Superfund was reauthorized by Congress, under a five-year extension known as the Superfund Amendments and Reauthorization Act (SARA). The 1993/94 Congress is again faced with reauthorization of Superfund and significant changes have been proposed. Three trends toward streamlining the Superfund process are evident:

- changing the joint and several liability provisions of the law, i.e., everyone is not equally liable financially regardless of the quantity of waste contributed to a site
- modifying the risk assessment process so it is less conservative and produces more technically achievable cleanup levels (see also Chapter 12, Risk Assessment)
- decreasing the amount of time and money directed toward litigation among PRPs and the various federal and state agencies.

Public disenchantment with Superfund is significant, because the perception is that more money has been spent on litigation and study than actual cleanup. In addition, the EPA has become sensitized to the issues of environmental equity and racism. The number of Superfund sites appear to be disproportionately located near or potentially impact poor and/or minority communities. In February 1994, President Clinton signed an executive order mandating greater emphasis on issues of environmental equity. (See also Chapter 2, Economic and Ethical Issues.)

Chapter 3, United States Legal and Legislative Framework, presents a general legal/regulatory framework for Superfund and SARA. This section is directed toward two key parts of environmental practice associated with facility operation under CERCLA/SARA: (1) spill assessment and (2) emergency response and community right-to-know.

Spill Assessment: General Considerations

Spill assessment refers to the steps taken to identify and plan an appropriate response to a spill, and report spills to the regulatory authority. Assessment of spills and releases is initiated as soon as discovery. The scope of an assessment varies with the situation, but the components of the assessment never change. The three steps of a spill assessment are as follows:

1. defining the problem
2. planning appropriate action
3. implementing agency notification and reporting requirements as necessary.

The first two components must be considered prior to controlling the spill to ensure that control activities are conducted safely, quickly, and effectively. The third component is necessary to meet government requirements and must be performed either during spill control for dangerous spills or before spill control for small nonhazardous spills.

Defining the problem or potential problems. Problem definition is completed by identifying the source and size of the spill or threat of a spill and identifying the materials or waste used at the facility and/or spill. An inventory of raw materials, products used, and generated wastes at the facility should be compiled; in addition, material safety data sheets (MSDSs) and information from sampling can be helpful. The toxic, reactive, corrosive, ignitable substances should be identified.

In general, two categories of spills occur at most facilities:

- small nonhazardous or otherwise nonreportable spills that can be cleaned up immediately with no harm to human health or the environment
- spills of hazardous materials that have the potential to harm human health and/or the environment.

Typically, most large facilities have emergency preparedness teams (EPT) and environmental coordinators (EC) who assess these types of spills and determine appropriate response and notification requirements.

The first category of spills or leaks would best fit under good housekeeping procedures. However, it is mentioned here to stress that small leaks and spills can and do result in major environmental problems. Recurring spills or leaks must be reported to the EC so appropriate changes or follow-up action can take place. Examples of this type of spill are:

- fuel spilled at fuel island (i.e., overfilling)
- small spills of dry chemical, cement, and sand that accumulate on the yard
- spills from vendor tankers, fill lines, or discharge hoses
- oil-stained soils around maintenance shop oil tanks.

The second category of spills must be immediately reported to the dispatcher by the person who discovers or causes the spill. Examples of this type of spill include:

- significant leaks of acid due to storage tank valve failure
- field waste drain plugs that cause field wastes to overflow to soil
- acid wastewater or field tank overflows because of rain.

To help an owner/operator define and evaluate the problem, the following list of questions has been developed:

- Is the material spilled hazardous?
- What was spilled?
- Was anyone hurt?
- Is medical assistance needed?
- Is any person, group, or community in danger?
- Is there any danger of fire, explosion, or air quality problems?
- What was the source of the spill?
- Has the source of the spilled material been stopped?
- Is the spilled material flowing off the site?
- Has the spilled material been contained?
- How much material was spilled?

The answers to these questions will be the basis for initial response action. Figures 8–13 and 8–14 present a typical spill response flowchart for a medium or large facility.

Planning appropriate action. The second part of spill assessment involves decisions about how to handle the different types of spills that could occur at the facility. The procedures for internal notification are as important as external notification. Typically, a dispatcher has the first role in this decision-making process. Based on the information provided during the initial spill notification, the dispatcher determines the "first" level of response necessary for each type of incident and immediately contacts the responsible party. The "first" level refers to the "first" call the dispatcher makes after receiving the spill notification. If there is a fire, explosion, injured individuals, or other life-threatening situations, then the "first" call is directed to fire or ambulance services.

In all other cases, the dispatcher initially contacts the EC or the EPT to assist in the spill assessment. Usually, senior supervisory personnel are in the responsible positions. These people are available for guidance in nonemergency situations and are often the most knowledgeable of facility processes, potential problem areas, and substances used at the facility.

For all chemical, fuel, or waste spills except for small nonhazardous spills, the dispatcher should contact appropriate management immediately after the EC or EPT has been contacted.

Assigned personnel are responsible for planning and implementing the appropriate response actions for each reported spill. These actions vary widely based on the size, nature, and complexity of the spill. The response actions may include:

- minor cleanups from small spills with no notification
- calling for additional outside assistance
- notifying federal, state, and local agencies
- isolating the area and setting up work zones
- controlling and cleaning up the spill
- protecting personnel during cleanup
- disposing of waste materials properly
- restoring the spill area and spill equipment.

Notification and reporting requirements. When a spill occurs, the EC or EPT must decide whether agencies are to be notified regarding the type of spill. Reporting requirements are triggered by releasing into the environment, in a 24-hour period:

- a reportable quantity (RQ) or more of a substance listed on the CERCLA hazardous substance list
- one pound (0.45 kg) or more of a substance listed on the SARA Title III Extremely Hazardous

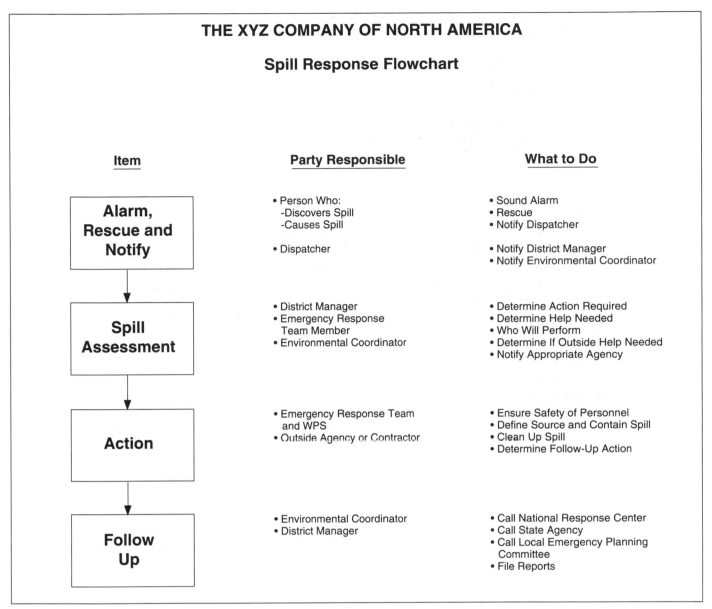

Figure 8–13. Flowchart of typical spill response procedure.

Substance List (EHS) (unless the release resulted in exposure only to persons within the site boundaries where the spill occurred)

- any quantity of a RCRA hazardous waste released from a tank system to the environment (unless it is less than or equal to 1 lb (0.45 kg) and immediately cleaned up)
- any quantity of an RCRA hazardous waste released (regardless of amount) that could threaten human health or the environment
- any release of oil to waters of the United States that causes a film or sheen on or discoloration of the surface of the water or adjoining shorelines or causes a sludge or emulsion to be deposited

beneath the surface of the water or upon adjoining shorelines
- a PCB leak or release into the environment.

The CERCLA and EHS lists should be readily available at the facility. These lists are also printed in 40 CFR, 302.4 (CERCLA list) and 40 CFR, 355, Appendix A & B (EHS list).

The CERCLA list currently includes 721 hazardous substances. This list is a combination of hazardous substances identified in the following environmental laws:

- The Clean Water Act (CWA)
- The Clean Air Act (CAA)

ALARM, RESCUE, AND NOTIFY

Note: Alarm and rescue are usually not required for small recurring spills.

As soon as a spill occurs, you must spread the word or sound the alarm that a problem requiring immediate attention has been identified.

Alarm	Sound the alarm quickly! The person who discovers any spill must notify fellow workers, activate the alarm system, or use the mobile radio to inform the dispatcher that a spill has occurred.
	React promptly to ALL spills no matter how insignificant they may appear.
	Do not be afraid to sound the alarm.
	Give as much information to the dispatcher as possible about the spill location, type of material, approximate quantity, and extent of damage.
Rescue	The person who discovers any spill or the person who causes a spill must make a quick assessment about the need to rescue or assist any person in the area of the spill.
	Wait for assistance before giving medical attention or attempting to rescue any person. Do not take personal risks; heroism can result in serious harm to yourself or others.
Notify	Once the alarm has been sounded, notify the appropriate chain of command.
	The dispatcher will immediately notify the EC and EPT members. The dispatcher will provide information about the spill and indicate whether the discoverer will proceed with cleanup. For small, nonhazardous spills, the EC or an alternate will go to the spill area to review the situation and ensure that cleanup has been performed properly.
	For all chemical, fuel, or waste spills except for small nonhazardous spills, the dispatcher will contact the environmental manager (EM) immediately.
	The EM will in turn notify the company's legal department. Local authorities may also be notified if the EC determines that evacuation is necessary due to imminent threat to human health or environment.
	In all spill cases, the EM will be notified within 24 hours.

Figure 8–14. Response procedure posted for company staff.

- The Hazardous Materials Transportation Act (HMTA)
- The Resource Conservation and Recovery Act (RCRA)
- The Toxic Substances Control Act (TSCA).

The SARA EHS list currently consists of 366 hazardous substances, 134 of which are on the CERCLA list. In order to eliminate the confusion associated with several lists, the EPA has proposed to designate the remaining 232 non-CERCLA EHSs as CERCLA hazardous substances in the near future.

Determining reportable quantities. *Reportable quantity* means, for any CERCLA hazardous substance, the reportable quantity established and identified in the CERCLA list. For extremely hazardous substances (EHSs) the reportable quantity is 1 lb (0.45 kg). RQs are typically listed in pounds.

CERCLA notification and reporting requirements are determined by identifying the chemical specific RQ (from the CERCLA list) and comparing it to the actual quantity of hazardous materials spilled. For each listed chemical contained in a mixture, the RQ must be determined separately by concentration and molecular weight.

The procedure for calculating RQs typically involves the following steps. An Incident Notification and Reporting Data Form like the one in Figure 8–15 must be used to record information collected as a result of these steps:

1. Identify spilled material, including individual components of mixtures. Refer to Material Safety Data Sheets (MSDSs) for percent of hazardous ingredients. List each hazardous ingredient and its respective percentage of the mixture.
2. Match materials and their components to the RQs on the CERCLA list. List the final RQ for each material or hazardous ingredient beside the corresponding material or hazardous ingredient identified in step 1.
3. Estimate total quantity of material that was spilled. Use gallons for liquid spills, cubic feet for gases, and pounds for dry solids. For mixtures, multiply the total quantity spilled by the actual

XYZ COMPANY OF NORTH AMERICA
Incident Notification and Reporting Data Form

Date: _____ Filed by: _____

Incident: _____

PART C – CALCULATING RQs

Instructions: Estimate quantity of materials released, convert to pounds as per flowchart in Figure 8-16, and compare to RQ.

PRODUCTS AND CONSTITUENTS	EST. TOTAL QUANTITY (Qty. Constituents)	UNIT OF MEASURE (gal or ft³)	SG/ GD*	QUANTITY RELEASED (lb)	> RQ? (Yes/No)
1.					
a.					
b.					
c.					
2.					
a.					
b.					
c.					
3.					
a.					
b.					
c.					

PART D – NOTIFICATION AND REPORTING REQUIREMENTS

REGULATIONS	APPLY? (Yes/No)	REQUIREMENT (Verbal/Written)	COMPLETED (Yes/No)
CERCLA			
SARA			
DEQ			
RCRA			
CWA			
OTHER			

* specific gravity or gas density (obtained from Material Safety Data Sheet)

Figure 8–15. Sample Incident Notification and Reporting Data Form.

percentage for each chemical that occurs in the mixture. (Total number of gallons spilled) × (% for each chemical in the mixture) = Number of gallons spilled of each chemical in the mixture. Note: This calculation must be performed separately for each hazardous ingredient contained in a mixture.

4. If the spilled material is a liquid, convert gallons to pounds by using the specific gravity for that particular chemical. (Number of gallons spilled) × (Specific gravity) × (8.3 lb/gal) = Pounds spilled.

5. If the spilled material is a gas, convert cubic feet to pounds by using the gas density, in lb/ft³, for that particular chemical. The specific gravity or gas density is usually specified on the MSDS. If not, refer to a chemical dictionary or call the chemical manufacturer for more information.

6. Compare results with final RQs to determine whether the spill is reportable.

The flowchart in Figure 8–16 illustrates these steps and assists response personnel during the RQ evaluation process.

Spill Response Management Procedures

The information presented in this section can be used as a general guide in preparing facility-specific plans and procedures. Each plan and subsequent procedures will vary depending on location, conditions, and material spilled.

The EC or EPT leader may determine the need for outside assistance at any time following the discovery of a spill or after completing the spill assessment procedure. Outside services include the local fire and police departments as well as vendors that specialize in spill cleanup and disposal. The decision is based on either finding (1) there is an imminent threat to human health or the environment, or (2) the facility does not have sufficient personnel or adequate equipment to control the spill effectively. The dispatcher will make appropriate notification as directed by the EC.

Isolate the area. The spill site should be isolated from operational personnel to reduce the potential for exposure to hazardous substances. Scaffolds or caution ribbon may be placed around the spill site to prevent access. Remove all other hazards, if possible, including:

- electrical hazards
- incompatible chemicals or wastes

- physical hazards
- sources of ignition.

Personnel protection during spill cleanup. The following rules are aimed at reducing the potential for employee exposure to hazards during spill control and cleanup:

- Always wear personal protection to handle spilled materials. A preexisting health and safety plan should specify appropriate equipment to be worn during spill response.
- Provide medical attention as needed for personnel who have come in contact with spilled materials.
- Limit the number of personnel responding to the spill. Allow no unauthorized person to enter the cleanup area.
- Use the buddy system. Two people should conduct the cleanup, using line-of-sight contact with those supervising the activity.

Spill cleanup and control. This discussion provides guidance for the cleanup and control of liquids, solids, and gases.

Liquids. Stop the movement of liquid with a dike of inert absorbent material, such as vermiculite or quick-dry, and cover the remaining liquid with absorbent (Figure 8–17).

Solvent product or waste solvent must not be allowed to flow into shop floor drains. In the event of a spill near the floor drain, cover the drain or place a dike around drain with an inert absorbent material.

Spilled oil should be collected in drums and transferred to the used oil tank. Vermiculite, quick-dry, or a similar product should be used to absorb oil residues remaining on soils or pavement.

Spilled acids and bases should be collected and transferred to the acid wastewater tank. Equipment should be used that is not reactive with the material spilled. For the remaining acid liquids, add sodium bicarbonate and work with a shovel in a circular motion from the perimeter of the spill to the center to form a slurry. Neutralization is achieved when no bubbling is observed after addition of sodium bicarbonate. Use small amounts of dilute hydrochloric acid to neutralize basic residues.

For all liquids, use plastic scoops or shovels to transfer saturated absorbent to the proper waste containers.

Liquids should always be put in compatible containers. A tanker truck may be required for large spills.

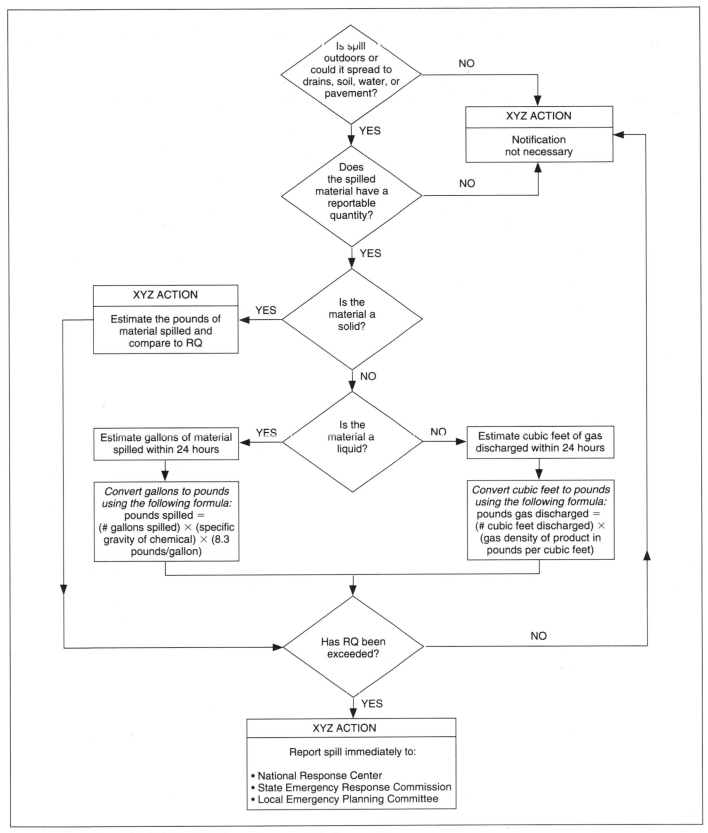

Figure 8–16. Flowchart of typical RQ calculation and agency notification procedure.

Figure 8–17. Sorbent rolls are used for emergency spill containment. (Courtesy Breg International.)

Dry materials. Personnel should cover the spill with a tarp or plastic sheet to prevent the material from blowing away before it is contained.

Small spills of dry material should be swept up and placed in a container for disposal. Large quantities of dry material should be pushed in a pile with a front-end loader and placed into drums or trucks.

Solids and sludges should be placed in drums, plastic bags, or other suitable containers. Trucks may be required for large spills.

Area contamination. Mop the spill area with water, collecting rinse water in properly labeled waste containers. Make sure that the spill site has been completely cleaned (decontaminated) and approved for reentry before unprotected personnel are allowed to enter or resume work in the area.

Gases. Releases involving gases include acid and nitrogen fumes. When a significant (i.e., life-threatening) quantity of gas is released, personnel should be evacuated and local emergency services contacted to control the situation.

Waste disposal. Appropriate procedures must be followed to properly characterize and manage wastes resulting from spill cleanups. Properly label and close all waste containers associated with the spill.

Contaminated soils should be excavated and characterized to determine whether they are hazardous wastes or exhibit hazardous characteristics. Even if the soil is saturated with oil, it is not considered a hazardous waste unless it exhibits at least one of the hazardous waste characteristics. If the soil is determined to be nonhazardous, it may be disposed of in a municipal or industrial solid waste landfill approved to handle such wastes. A check should be made with the disposal facility as to its permit limitations on soils and "special wastes."

Soils contaminated with solvents must be handled and disposed of as hazardous wastes. Soils contaminated with gasoline may be aerated to remove the volatile fraction and may subsequently be disposed of in a municipal or industrial solid waste landfill.

This is usually conducted on a case-by-case basis in conjunction with state agency approval. A land application or treatment permit is usually required of a contaminated solid waste. Air quality permits may be required by state and local agencies to aerate large quantities of contaminated soil.

Follow-up activities. A facility's assessment procedures should include the actions that must be taken after the spill is under control. The procedures include consideration of the following questions during a debriefing:

- Why did the spill occur?
- What can be done to prevent reoccurrence?
- What reports must be submitted internally or to local, state, or federal agencies?
- What facility repairs or equipment replacement is required?

Remedy the cause of spill. Identify what caused the spill. Modify work practices or make necessary repairs. Typical causes of spills include:

- leaking containers
- careless transfer of product
- faulty tank valves or corroded pipes
- eroded berm around surface impoundments or containment areas resulting in overflows to nearby ditch.

Give special attention to the potential spill sources during routine site inspections to ensure that preventative and corrective measures have been implemented. Report discrepancies to the EC.

Decontamination equipment and containers. Clean (decontaminate) all tools and personal protective equipment thoroughly after each use. Collect rinse water in a properly labeled waste container. Inspect tools, measuring instruments, and protective equipment and prepare these items for future use.

Documentation. A Spill Response Form, a sample of which appears in Figure 8–18, should be completed following the control of the spill. This report is useful for tracking environmental problems because it documents the circumstances surrounding the spill and the response action taken.

Emergency Planning and Community Right-to-Know

The amendments to CERCLA, known as the Superfund Amendments and Reauthorization Act (SARA), require owners/operators of facilities to be able to respond in emergency situations. Title III of this act specifically identifies procedures for reporting hazardous substance use and spills to the community. There are three subtitles under Title III:

Emergency Planning and Notification	Subtitle A
—Emergency Response Planning	Section 301–303
—Emergency Release Reporting	Section 304
Reporting Requirements	Subtitle B
—Hazardous Chemical Inventory Reporting	Section 311, 312
—Toxic Chemical Release Reporting	Section 313
General Provisions	Subtitle C

Section 302 of SARA required facilities that handle any of the extremely hazardous substances (EHS) more than the threshold planning quantity (TPQ) to inform state and local officials by May 17, 1987; they must also participate in preparation of community contingency plans for hazardous materials accidents. By September 17, 1987, facilities were required to tell the local committee the name of a designated "facility emergency coordinator" who will work with the local committee in developing emergency plans. These plans were required by October 17, 1988. Figure 8–19 illustrates the emergency response planning requirements.

The emergency release reporting requirements pursuant to Section 304 of SARA (along with 103 of CERCLA) require emergency release notification of leaks, spills, and other releases of specified chemicals (CERCLA hazardous substances and SARA extremely hazardous substances) into the environment. Under CERCLA, those in charge of a facility, including transporters, must report to the National Response Center (NRC) any spill of a specified hazardous substance in an amount equal to or greater than the RQ specified by U.S. EPA. SARA 304 significantly expands these requirements to require reporting of releases of EHS chemicals or CERCLA hazardous chemicals. SARA Title III requires releases to be reported immediately to the state commission and local committee and the National Response Center. Figure 8–20 illustrates the emergency release reporting requirements.

Section 311 of SARA requires facilities that must prepare or have available MSDSs for a "hazardous chemical" under OSHA to submit these MSDSs to the state commission, the local emergency planning

XYZ COMPANY OF NORTH AMERICA

Spill Response Form

TYPE OF INCIDENT _____

_____

DATE OF INCIDENT _____

TIME OF INCIDENT _____

Emergency coordinator's name _____

Emergency coordinator's phone number _____

Location of incident _____

_____

Name of materials involved _____

Quantity of materials involved _____

Extent of injuries _____

Hazards to human health or environment _____

_____

Quantity and disposition of all contaminated or recovered materials, including cleanup debris _____

_____

_____

Cause of emergency _____

_____

Analysis of emergency notification, response, control, and cleanup procedures _____

_____

_____

Full description of incident notification, response, and cleanup, including times, personnel, equipment used _____

_____

_____

_____

_____

_____

_____

Recommended modifications to the Emergency Preparedness Plan _____

_____

_____

_____

_____

_____

Figure 8–18. A typical Spill Response Form.

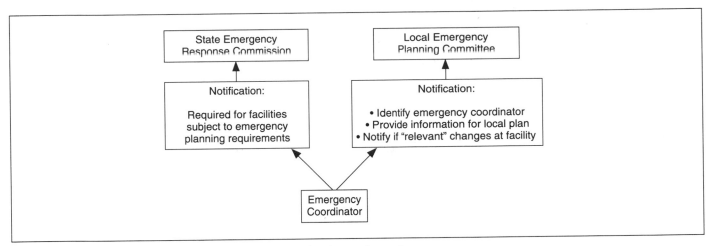

Figure 8–19. SARA Title III overview of emergency response planning requirements.

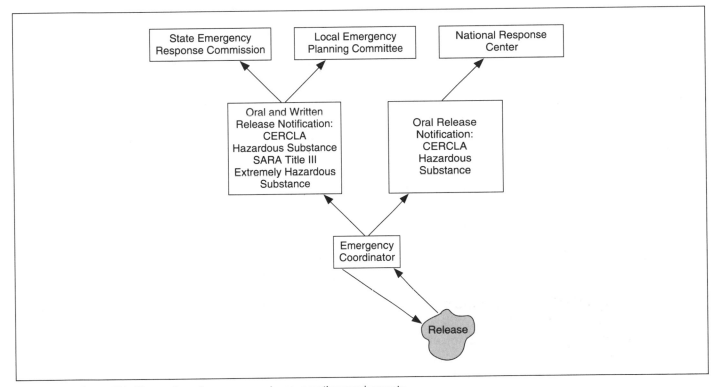

Figure 8–20. SARA Title III overview of emergency release reporting requirements.

committee, and the local fire department. Alternatively, a company may submit a list of hazardous chemicals for which it maintains MSDSs. Reporting is required for a chemical when present in excess of 10,000 lb (4,500 kg) for OSHA hazardous chemicals, and 500 lb (229 kg) on the threshold planning quantity for substances on SARA's EHS list. Known as a Tier 1 Report, this form must be provided annually and contains information on the quantity and location of these hazardous chemicals at the facility aggregated by categories of physical and health hazards. Figure 8–21 illustrates the Hazardous Chemical Inventory Reporting Requirements.

Under SARA Title III, Section 313, certain facilities that manufacture, import, process, or otherwise use a chemical listed on the toxic release inventory must annually report the amount released to the environment. The first toxic chemical inventory report (Form R) was due July 1, 1988, for calendar year 1987. Reports have been due annually thereafter on July 1. Facilities required to submit this report are those that (1) have 10 or more full-time employees; (2) are in Standard Industrial Clarification (SIC) Code 20–39; and (3) manufactured, processed, or used any of the chemicals in excess of the threshold quantity during the preceding calendar year.

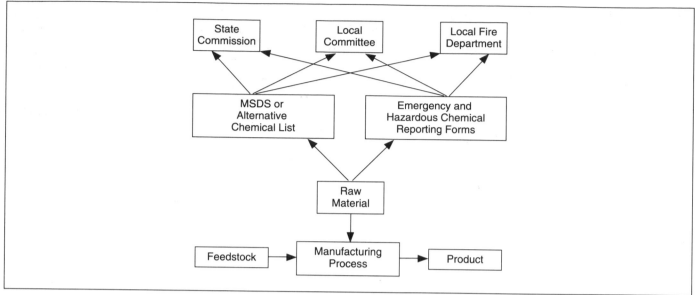

Figure 8–21. SARA Title III overview of hazardous chemical inventory reporting requirements.

The U.S. EPA established phased-in threshold quantities for Section 313 reporting. For manufacturers and processors, the reporting requirement is triggered by the annual amount of a chemical: 75,000 lb (34,040 kg) per year for 1987, 50,000 lb (22,727 kg) per year for 1988, and 25,000 lb (11,363 kg) per year for 1989 and thereafter. For users, reporting is triggered by use of 10,000 lb (4,545 kg) of a chemical per year. A summary of the SARA Title III reporting requirements is provided in Figure 8–22.

Public Access to Facility Information

Title III also requires that facilities provide the public with information about hazardous chemicals in their communities. The state commission and local committee are each required to designate an official to serve as coordinator of information. Facility coordinators must respond to requests for information from state agencies, local officials, the public, and other interested parties.

Section 324 of SARA requires that the following information be made available to the public:

- Toxic Chemical Release Inventory Reporting Forms (Form R)
- MSDSs (or lists) of hazardous chemicals
- Emergency and Hazardous Chemical Inventory Forms (Tier I and Tier II)
- Follow-up Emergency Notification Reports (Section 304)
- local emergency response plan.

Penalties

SARA Title III contains penalties for the violation of its provisions. Civil and administration penalties include fines for companies ranging from $10,000 to as high as $75,000 for repeated violations. Violations subject to fines include the failure of a facility to:

- notify the government that it has extremely hazardous substances on-site above threshold quantities
- provide timely notification of the release of a regulated substance to the environment
- provide hazardous chemical inventory information
- provide the required Toxic Chemical Release Inventory Reporting Forms.

Criminal penalties for the knowing and willful violation of the provisions of SARA Title III can include fines and imprisonment, or both.

MEDICAL WASTES

After repeated episodes of needles, syringes, and other medical paraphernalia appearing on U.S. beaches, Congress passed the Medical Waste Tracking Act (MWTA) of 1988. The MWTA charges the EPA to find an effective means of ensuring that regulated medical waste proceeds from the point of generation to an acceptable point of disposal.

Regulated medical waste (RMW) is defined as waste that is capable of causing disease in humans

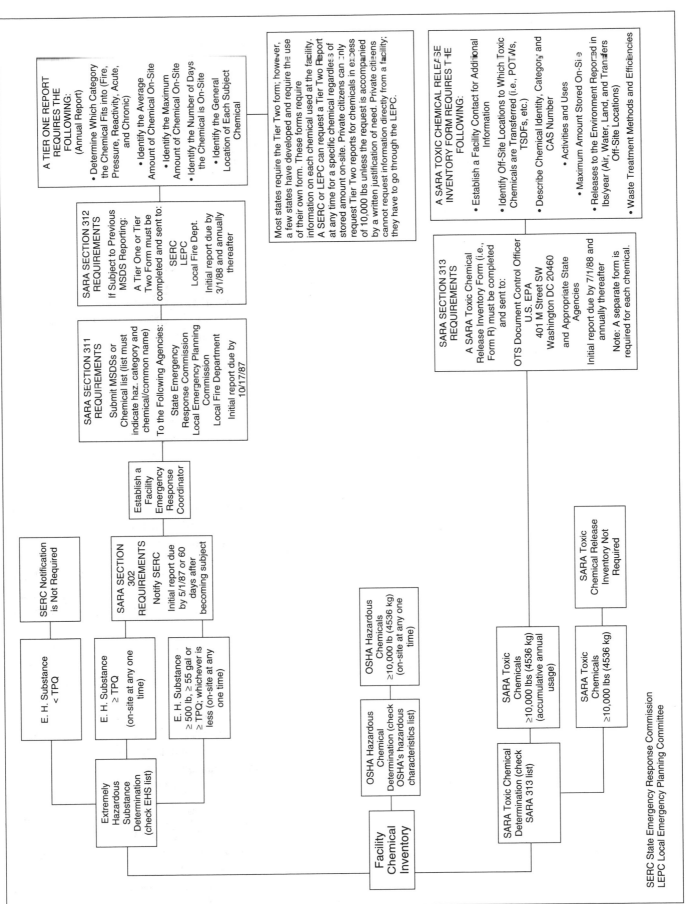

Figure 8–22. SARA Title III chemical reporting requirements.

SERC State Emergency Response Commission
LEPC Local Emergency Planning Committee

A TIER ONE REPORT REQUIRES THE FOLLOWING: (Annual Report)
- Determine Which Category the Chemical Fits into (Fire, Pressure, Reactivity, Acute, and Chronic)
- Identify the Average Amount of Chemical On-Site
- Identify the Maximum Amount of Chemical On-Site
- Identify the Number of Days the Chemical is On-Site
- Identify the General Location of Each Subject Chemical

Most states require the Tier Two form; however, a few states have developed and require the use of their own form. These forms require information on each chemical used at the facility. A SERC or LEPC can request a Tier Two Report at any time for a specific chemical regardless of stored amount on-site. Private citizens can only request Tier Two reports for chemicals in excess of 10,000 lbs unless the request is accompanied by a written justification of need. Private citizens cannot request information directly from a facility; they have to go through the LEPC.

SARA SECTION 312 REQUIREMENTS
If Subject to Previous MSDS Reporting:
A Tier One or Tier Two Form must be completed and sent to:
SERC
LEPC
Local Fire Dept.
Initial report due by 3/1/88 and annually thereafter

A SARA TOXIC CHEMICAL RELEASE INVENTORY FORM REQUIRES THE FOLLOWING:
- Establish a Facility Contact for Additional Information
- Identify Off-Site Locations to Which Toxic Chemicals are Transferred (i.e., POTWs, TSDFs, etc.)
- Describe Chemical Identity, Category and CAS Number
  - Activities and Uses
  - Maximum Amount Stored On-Site
- Releases to the Environment Reported in lbs/year (Air, Water, Land, and Transfers Off-Site Locations)
- Waste Treatment Methods and Efficiencies

SARA SECTION 311 REQUIREMENTS
Submit MSDSs or Chemical list (list must indicate haz. category and chemical/common name)
To the Following Agencies:
State Emergency Response Commission
Local Emergency Planning Commission
Local Fire Department
Initial report due by 10/17/87

SARA SECTION 313 REQUIREMENTS
A SARA Toxic Chemical Release Inventory Form (i.e., Form R) must be completed and sent to:
OTS Document Control Officer
U.S. EPA
401 M Street SW
Washington DC 20460
and Appropriate State Agencies
Initial report due by 7/1/88 and annually thereafter
Note: A separate form is required for each chemical.

Establish a Facility Emergency Response Coordinator

SERC Notification is Not Required

SARA SECTION 302 REQUIREMENTS
Notify SERC
Initial report due by 5/1/87 or 60 days after becoming subject

SARA Toxic Chemical Release Inventory Not Required

E. H. Substance < TPQ

E. H. Substance ≥ TPQ (on-site at any one time)

E. H. Substance ≥ 500 lb, ≥ 55 gal or ≥ TPQ; whichever is less (on-site at any one time)

OSHA Hazardous Chemicals ≥10,000 lb (4536 kg) (on-site at any one time)

SARA Toxic Chemicals ≥10,000 lbs (4536 kg) (accumulative annual usage)

SARA Toxic Chemicals ≥10,000 lbs (4536 kg)

Extremely Hazardous Substance Determination (check EHS list)

OSHA Hazardous Chemical Determination (check OSHA's hazardous characteristics list)

SARA Toxic Chemical Determination (check SARA 313 list)

Facility Chemical Inventory

and that may pose a risk to individual and/or community health if not treated properly. RMW consists of the following classes, as defined in 40 CFR 22 and 259, *Standards for the Tracking and Management of Medical Wastes* (1989):

- Class 1. Cultures, Stocks and Vaccines—Class 1 includes cultures and stocks of infectious agents and associated biologics. This consists of cultures from medical laboratories, discarded live and attenuated vaccines, culture dishes, and devices used to transfer, inoculate, and mix cultures.
- Class 2. Pathological Waste—Class 2 medical wastes consists of human pathological wastes, including tissues, organs, or body fluids that are removed during surgery, autopsy, or other medical procedures. Also included are specimens and body fluids.
- Class 3. Blood and Blood Products—Class 3 wastes consist of free-flowing human blood, plasma, serum, and other blood derivatives that are waste, such as blood in blood bags or bloody drainage in suction containers. Class 3 wastes also include items, such as gauze or bandages, that are saturated or dripping with human blood. Class 3 waste includes items produced during dental procedures.
- Class 4 and Class 7. All Used and Unused Sharps—Sharps that have been used in animal or human patient care, including hypodermic needles, syringes with or without the needle attached, Pasteur pipettes, scalpel blades, blood collection tubes and vials, test tubes, needles attached to tubing, and culture dishes (regardless of the presence or absence of infectious agents) are all RMW of Classes 4 and 7. Other types of broken or unbroken glassware that were in contact with infectious agents, such as used slides and cover slips are also included in this category.
- Class 5 is animal waste and Class 6 is isolation waste.

At the present time, medical waste is regulated in a few states under a demonstration program established by the EPA. These requirements (and EPA's Infectious Waste Guidelines of 1982) are being followed by almost all states at the present time in anticipation of EPA's promulgating medical waste standards for all states. EPA's Report to Congress, "Medical Waste Management in the U.S." (May 1990), states it is EPA's intention to require the standards in 40 CFR 259 for all states, in addition to other requirements. The other requirements mentioned include new source performance standards (NSPSs) for medical waste incinerators under the Clean Air Act, use of Best Available Control Technology (BACT), and development by the facility operator of a site-specific program to handle medical waste.

In order to comply with EPA regulations for "Standards for the Tracking and Management of Medical Waste" (40 CFR 259), several important factors must be analyzed:

- Is the generator employing on-site or off-site disposal of its RMW?
- What is the monthly quantity of waste (pounds per month) generated at the site?
- If on-site disposal is contemplated, will the RMW be commingled with either municipal waste or other regulated hazardous waste?

On-site versus off-site disposal options. Off-site disposal requires that stringent pretransportation and transportation guidelines be followed. On-site treatment and disposal options include autoclaving or incineration. Autoclaving is relatively inexpensive and quick; however, autoclaving does not reduce the total volume of the waste. The advantage of incineration is that the volume of waste is reduced by approximately 90%. However, in order to incinerate RMW in accordance with U.S. EPA regulations, several concerns must be addressed:

- EPA regulations require stringent time and temperature standards for medical waste incineration. In particular, EPA Infectious Waste Guidelines suggest a two-second dwell time at extreme temperatures. To obtain this type of performance generally requires the implementation of costly incinerator-design additions to older existing facilities. Generally, incinerators that are suitable for nonhazardous municipal solid waste do not meet the time and temperature standards set forth by the EPA for biomedical waste incineration.
- Air emissions must be strictly controlled using best available control technology (BACT). Again, incinerators that are appropriate for municipal waste disposal usually are not equipped with the required air emission control devices. Additionally, air emission permits are required for the operations of an incinerator that handles RMW.
- Fly-ash and incinerator bottom ash must be toxicity tested for hazardous waste determination.

The ash is generally considered hazardous waste unless proven otherwise.

As a result of this, it becomes essential for RMW to be segregated from nonhazardous waste at the point of generation and incinerated in a separate hazardous and biomedical waste incinerator.

Hospitals commonly use a combination of both autoclaving and incineration. Autoclaving in itself is relatively inexpensive and renders the waste nonhazardous. Autoclaving prior to incineration alleviates the need for a biomedical classified incinerator. Additionally, an air emission permit for medical waste incineration and toxicity testing, and the potential for hazardous waste disposal of the fly-ash and bed-ash would not be necessary if the RMW was autoclaved prior to incineration. In general, EPA has significantly deemphasized the use of incineration technology (see Chapter 10, Pollution Prevention Approaches and Technologies).

Quantity of waste generated. The EPA has established "small-quantity generator" exemptions from some of the requirements for disposal of RMW for facilities that generate less than 50 lb (22.7 kg) per month.

Mixing regulated medical waste with other waste streams. Segregating RMW from unregulated medical and nonhazardous and hazardous wastes is essential to cost-effective RMW disposal. General nonhazardous waste that is mixed with RMW is regulated as medical waste. If RMW is incinerated with nonhazardous waste, all of the fly-ash and bed-ash must be treated as ash generated from the incineration of RMW; that is, it must all be toxicity tested and possibly treated as hazardous waste. Additionally, if RMW is mixed with other hazardous waste that falls under the small-quantity exemption, the waste mixture must be handled and disposed as RMW under this program.

RADIOACTIVE WASTES

Worldwide, the scrutiny of the proper treatment, storage, and disposal of radioactive wastes ("rad wastes") has dramatically increased since the end of the Cold War. In the United States, the U.S. Department of Energy (DOE) has redirected substantial funding and human resources away from bomb production and toward environmental restoration and waste management. Similar efforts are ongoing in other major nuclear countries as public opinion has focused on the nuclear waste legacy of the Cold War.

Historically, U.S. definitions and requirements for permanent disposal of different classes of radioactive waste were based on the source of the waste and requirements for safe handling and storage rather than requirements for permanent disposal (Kocher, 1990). This section is based on an excellent review by Kocher of the historical, legal, and regulatory requirements associated with the classification and disposal of radioactive wastes. In the United States, the regulation of radioactive wastes involves the overlapping authority of three agencies—Department of Energy, (DOE); Nuclear Regulatory Commission, (NRC); and Environmental Protection Agency, (EPA). Thus, the reader must consult the most current directives and orders from the frequently changed regulatory guidelines and requirements.

This section presents basic information on (1) definitions of principal classes of radioactive waste, (2) framework of the disposal requirements for the principal waste classes, (3) management and disposal of other radioactive (rad) wastes, and (4) decontamination and decommissiong of existing facilities. Finally, an example of the permitting and licensing requirements for a low-level mixed waste facility is included in Appendix 1, Case Studies. This example illustrates the complexity and difficulty of storing and disposing radioactive wastes.

Definitions of Principal Classes of Radioactive Waste

The principal classes and definitions of rad waste are:

- *Spent Fuel:* Nonreprocessed, irradiated nuclear fuel.
- *High-level waste:* Primary waste produced from the chemical reprocessing of spent fuel.
- *Transuranic waste:* Waste that contains more than 100 nCi/g (nanocuries per gram) of a long-lived alpha-emitting transuranium radionuclides. High-level waste is not in this category and is separately defined. One curie (ci) is equal to the activity of 1 g of radium-226. A nanocurie is 10^{-9} ci.
- *Low-level waste:* Waste that is not spent fuel, high-level, or uranium or thorium mill tailings. Transuranic waste is excluded from this category.

As discussed by Kocher, these definitions are generally not quantitative (i.e., expressed as radionuclide concentrations) and do not unambiguously distinguish between the different types of waste. Therefore, under these definitions, it is possible to place rad wastes in different categories despite close similarity in properties. Alternative waste classification schemes have been proposed that attempt to limit or remove ambiguities by increasing the emphasis on quantitative classification based on consideration of risks from waste disposal (Kocher, 1988; Smith & Cohen, 1989).

The responsibility for regulating the previously discussed rad wastes is shared by EPA, NRC, and DOE.

- The EPA establishes the general environmental protection and disposal standards applicable to specified classes of civilian and defense rad waste. The standards for a particular waste class also apply to a given disposal system. EPA standards are enforced by NRC or DOE.
- The NRC establishes the licensing criteria for uses and particular disposal systems regardless of rad waste class. The NRC's licensing authority covers all facilities for disposal of civilian wastes and certain DOE facilities, e.g., geologic repository for civilian or defense spent fuel and high-level waste. The NRC also regulates/permits commercial nuclear power plants.
- The DOE regulates disposal of defense waste that is not subject to NRC licensing, e.g., low-level waste at DOE sites.

Obviously, there are significant overlaps between the three agencies. In addition, the NRC licensing criteria are in exact opposition to the philosophic thrust of EPA standards, i.e., EPA standards are directed toward specific classes of rad waste, whereas NRC licenses particular disposal systems for any rad waste class. This decoupling of waste definition from a specific disposal requirement has one benefit: waste disposal requirements are not affected by the current problems afflicting the classification schemes.

Currently, there are only specific types of disposal systems that are authorized for disposal of some waste classes:

- geologic repository for civilian spent fuel and high-level waste and defense high-level waste
- Waste Isolation Pilot Plant (WIPP) facility, authorized for both defense transuranic and low-level wastes.

There are no authorized intermediate disposal facilities for either dilute high-level waste, transuranic waste, or greater-than-Class C low-level waste. Class C waste is an NRC definition based on radionuclide

half-lives and concentrations. NRC uses this classification for near-surface land disposal. An intermediate disposal system would be a specifically designed facility located a few tens of meters below ground surface.

Management and Disposal of Other Radioactive Wastes

Other significant types of rad waste are:

- uranium or thorium mill tailings
- naturally occurring and accelerator-produced radioactive materials (NORM and NARM)
- mixed waste (radioactive and hazardous chemical wastes).

Mill tailings. Mill tailings were originally regulated under the Atomic Energy Act of 1954 and are not considered a form of low-level waste under current law (Nuclear Waste Policy Act of 1982 and 1987, and the Low-Level Radioactive Waste Policy Act of 1980). The management of the tailings is governed by the Uranium Mill Tailings Radiation Control Act of 1978, also known as UMTRA. UMTRA specifically deals with the control and stabilization of mill tailings in place. If removal of tailings is required because of public health or environmental concerns, permanent disposal and stabilization at non-low-level waste facilities are required. The EPA recently published updated regulations on uranium mill tailings standards, including closure (40 CFR 192; 58 FR 60344).

NORM/NARM waste. This is not a form of low-level waste; however, EPA appears to regulate this material as if it were low-level rad waste. NARM waste would be classified under NRC's waste classification system for near-surface land disposal. For defense waste containing NARM, DOE policies direct management in a fashion consistent with existing mill tailings policies, i.e., stored in place and/or disposed of consistent with EPA requirements for residual radioactive materials.

Mixed radioactive and hazardous chemical wastes. Mixed waste is waste containing regulatory defined hazardous chemical(s) and radioactive materials. The EPA regulates the hazardous portion of mixed waste under RCRA. The RCRA requirements are based on detailed and mandatory technical standards for obtaining permits for treatment, storage, and/or disposal. NRC or DOE regulate the radioactive portion of the mixed waste under the Atomic

Energy Act. Thus, although the definition of hazardous waste in RCRA specifically excludes source, special nuclear, and end by product material (as defined by the Atomic Energy Act), mixed waste must be managed under the authority of EPA's RCRA requirements: (1) EPA and NRC for civilian (commercial) mixed wastes and (2) EPA and DOE for defense waste.

Because of the extreme complexity and possible incompatibility of regulations, RCRA specifically precludes any hazardous waste regulation for mixed waste that is "inconsistent" with the Atomic Energy Act (AEA). Therefore, the AEA takes precedence over an "inconsistent" RCRA requirement. For example, RCRA rules would affect the construction, operation, and closure requirements for a mixed waste facility but not the types of facilities (e.g., geologic repository, near-surface facilities) or the environmental radiation standards applicable to the disposed waste.

Decontamination and Decommissioning of Facilities

After nuclear facilities have reached the end of their useful productive life, a process known as "decontamination and decommissioning" (D&D) must be performed. In this context, *decontamination* is the removal of radioactive contamination that has been deposited on surfaces or spread throughout a given work space. *Decommissioning* is the process of retiring a facility from service in a manner that does not adversely impact human health or the environment. Universities, hospitals, power plants—and now with the end of the Cold War—the U.S. DOE have recognized that D&D will be a major focus of current and future environmental restoration and waste management activities.

Both the techniques and equipment used to decontaminate and dismantle nuclear facilities resemble those used in the nonnuclear industry but have been modified and improved, in many cases, to suit the special nuclear or decommissioning application. Some techniques and equipment have been developed specifically for decommissioning purposes, for example, spalling to remove the surface layer of concrete. Also, because many areas in a nuclear power plant have high radiation fields, the use of remotely operated equipment is required. Although progress has been made in the development of new techniques and equipment in these fields, further work may be desirable in selected cases to improve techniques and equipment, reduce radiation exposures to as low as reasonably achievable (ALARA), and minimize costs.

The objectives of decontamination during decommissioning are to reduce occupational exposure, to permit reuse of the item being decontaminated or to facilitate waste management. In some cases, it may be more beneficial overall not to decontaminate, depending on a comparison of the total radioactive dose and cost involved with reuse of materials.

Radioactive contamination is deposited on concrete and metal surfaces during the operation of all types of nuclear facilities—reactors, examination cells, fuel fabrication plants, etc. The inventory of radioactive contamination can be divided into two categories: (1) radioactivity induced by neutron activation of certain elements in reactor components, such as the pressure vessel and adjacent structures; and (2) the radioactive material deposited on the internal and external surfaces of various out-of-core systems as surface contamination. Contaminants could include activation products, fission products, and fissile metals and oxides.

Neutron-activated material is usually sent for disposal without any decontamination except for superficial surface decontamination to remove loose material in order to allow handling. Core samples of activated concrete including reinforcing bars should be taken to establish the depth of activated material so that it can be separated from nonactivated concrete.

In most nonreactor facilities and for most parts of a reactor complex, contamination occurs either on surfaces or in cracks and crevices. Although this surface contamination contains only a small percentage of the total radioactive inventory in a nuclear power plant, it gives rise to the largest occupational exposures because it is distributed across many readily accessible areas. Before effective decontamination and dismantling work can commence, the location and characteristics of the contamination must be identified. This data is required to determine the decontamination processes and the requirements for shielding, use of remotely operated equipment, waste management, manpower, etc.

Some decontamination techniques recommended for use include high- and low-concentration chemical decontamination, vacuum cleaning, brushing, washing, scrubbing, high-pressure water/steam, abrasive jetting, freon cleaning, ultrasonic cleaning, electrochemical decontamination, scarifying, grinding, drilling, spalling, mechanical breaking, and melting.

The dismantling of nuclear reactors and other facilities contaminated with radioactivity generally involves the segmentation of metal items—reactor vessels, tanks, piping, and other components. Also, in most facilities to be decommissioned, cutting and demolition of concrete components or scarification of the surface to remove contaminated areas is required. A wide variety of processes for the demolition and segmentation of concrete and metal structures have been used, and new processes and techniques are continually being developed. Large decommissioning projects typically require cutting and demolishing of concrete structures such as heavily reinforced, massive concrete; massive, heavy concrete with little or no reinforcement; lightly reinforced or nonreinforced floors and walls; prestressed concrete structures (e.g., reactor buildings). Some of these concrete structures become radioactive during operation. In reactors, concrete adjacent to the core becomes activated as a result of neutron irradiation. This is usually the most difficult removal job because of the relatively high radiation dose rates and poor accessibility for equipment.

The cutting and demolishing techniques recommended for various applications in the nuclear industry include controlled blasting, backhoe mounted ram, flame cutting, wire cutting, wrecking ball, explosive cutting, thermite reaction lance, core drilling, rock splitter, circular carbide or diamond saw, and abrasive water jet.

In addition, a wide range of metal structures and components need to be segmented for easy removal and disposal. These include large items such as reactor vessels, pressure tubes, large and small tanks, and all types of piping and ancillary components. Cutting methods used for highly radioactive components such as pressure vessels or certain reprocessing plant equipment may require remote operation.

The cutting methods used or recommended for use in the nuclear industry include oxygen burning, abrasive water jet, fissuration cutting, plasma arc torch, arc saw cutting, hacksaw and guillotine saw, explosive cutting, circular cutting machines, thermite reaction lance, abrasive cutters, laser cutting, shears, and mechanical nibblers. Selection of the cutting technology needs to account for such factors as need for remote operation, metal type and thickness, accessibility, equipment reliability, reactivity, and waste generation. Many of these techniques have been demonstrated on large-scale decommissioning projects in the United States as well as in Canada, France, Germany, Italy, Japan, and the United Kingdom.

SUMMARY

The rules and regulations governing the treatment, storage, and disposal of hazardous wastes are quite complex and detailed. For example, USTs are a common method for storing either hazardous wastes or potentially hazardous substances. The problem of leaking USTs is significant and facility owner/operators must be prepared to devote the time, energy, and money necessary to meet regulatory requirements. Aboveground storage tanks are regulated under rules governing spills and spill responses.

The labeling, manifesting, and tracking of hazardous wastes is a critical part of RCRA and DOT compliance. The owner/operator must keep meticulous and well-documented records. The penalties for noncompliance are severe and should be taken very seriously.

Co-mingled managing wastewater and stormwater permits can be a complex, time-consuming, and costly undertaking for some facilities. Many times facility process water, sewer water, and stormwater have become co-mingled. Tracking wastewater lines to the point of generation in order to identify pollutants may be necessary in order to manage the stormwaters and wastewaters at a facility in compliance with regulations. Experts in the technical applications and regulatory areas of the water laws and regulations can be useful. A selection of relevant references is included in Appendix 2, Sources of Help.

The regulatory and legal requirements affecting rad waste are voluminous, confusing, and at times contradictory. The current U.S. system for waste classification and disposal is not particularly flexible and is unlikely to accommodate new types of waste produced by evolving technologies. The current state of affairs is unfortunate, because the treatment and disposal of the various types of radioactive wastes and the decontaminating and decommissioning of existing facilities have become an extraordinarily expensive and controversial process for the U.S. DOE and commercial reactor owner/operators.

REFERENCES

Dominguez GS. Hazardous waste management. In Dominguez GS, Barlett, KG (eds), *Hazardous Waste Management: The Law of Toxics and Toxic Substances.* Boca Raton, FL: CRC Press, 1986.

Kocher DC. Classification and disposal of radioactive wastes—Testing and legal and regulatory requirements. *Rad Prot Man* 7(4):58–78, July/August 1990.

Kocher DC. A proposed classification system for high-level and other radioactive waste. *Rad Waste Man Na Cl Fuel Cycle* 11:227, 1988.

Smith CF, Cohen JJ. Development of a comprehensive radioactive waste classification system. In Post RG (ed), *Waste Management '89*. Tucson: University of Arizona, 1989.

U.S. EPA. *Guidance on Remedial Actions for Superfund Sites with PCB Contamination.* EPA/540/G-90/007, August 1990.

U.S. EPA. *Leak Detection Alternatives: Musts for USTs.* EPA/530/UST-88/008. Washington DC: Office of Underground Storage Tanks, 1990.

U.S. EPA. *Tank Tour, Your Guide to the Federal Underground Storage Tank Program.* Office of Underground Storage Tanks, March 1989.

U.S. EPA. *Overview of the UST Program: UST Program Facts,* January 1993.

Health and Safety Training for Hazardous Waste Activities

John G. Danby, CIH

A wide array of field activities associated with environmental practices—environmental site assessments, hazardous waste site cleanups, implementing safe storage practices for hazardous materials, conducting emergency response activities in the event of a spill—have a common goal: preventing potentially harmful materials from reaching the general public. However, the most immediately exposed individuals (receptors) are emergency responders, site investigators, and site cleanup workers. Their protection is critical and begins with health and safety training, i.e., the joining of environmental response with occupational health and safety.

This chapter discusses the various aspects of health and safety training programs for personnel who work with environmental hazards, primarily hazardous waste. The history of hazardous waste training, the training requirements of the Occupational Safety and Health Administration's (OSHA) *Hazardous Waste Operations and Emergency Response* standard, and some methods of providing this training are presented. In addition, three types of operations identified by the standard and the different types of training required for each are also addressed. Finally, the establishment of training goals, the development of practical means of delivering the training, and methods of measuring its effectiveness are discussed.

DEFINITIONS AND EXPLANATIONS

Understanding the training requirements and how they apply to a particular work force requires an understanding of the hazardous waste operations terminology. The terms that follow are used throughout this chapter. Where appropriate, they have been shortened and clarified; complete definitions can be found in the OSHA *Hazardous Waste Operations and Emergency Response* standard (29 CFR 1910.120). For some terms, it is necessary to use the definitions directly from this standard.

▪ **Buddy system:** A system of organizing personnel into workgroups in such a manner that each employee of the workgroup is designated to be observed by at least one other employee in the workgroup. The purpose of the buddy system is to provide rapid assistance to employees in the event of an emergency.

▪ **Decontamination:** The removal of hazardous substances from personnel and their equipment to the extent necessary to preclude the occurrence of foreseeable adverse health effects.

- **Emergency response (ER):** A response effort by employees from outside the immediate release area (such as a facility's designated ER team) or by other designated responders (such as a fire department HAZMAT team) to an occurrence that results, or is likely to result in, an uncontrolled release of hazardous substances (a chemical spill). **Important Note:** Responses to minor spills where the spill can be controlled at the time of release by employees in the immediate area or by maintenance personnel are not considered ER under HAZWOPER.

- **Hazardous materials response (HAZMAT) team:** An organized group of employees, designated by the employer, that is expected to control actual or potential leaks or spills of hazardous substances requiring possible close approach to the substance. The team members perform responses to releases or potential releases of hazardous substances for the purpose of control or stabilization of the incident.

- **Hazardous substance:** Any substance designated or listed as noted in the following, exposure to which results or can result in adverse effects on the health or safety of employees:
  — Any substance defined under Section 101(14) of the Comprehensive Environmental Response, Compensation, and Liability Act (CERCLA), commonly known as Superfund.
  — Any biologic agent and other disease-causing agent that after release into the environment and upon exposure, ingestion, inhalation, or assimilation into any person, either directly from the environment or indirectly by ingestion through food chains, will or can reasonably be anticipated to cause death, disease, behavioral abnormalities, cancer, genetic mutation, physiological malfunctions (including malfunctions in reproduction), or physical deformations in such persons or their offspring.
  — Any substance listed by the U.S. Department of Transportation as hazardous material under 49 CFR 172.101.
  — Hazardous waste as defined in HAZWOPER.

- **Hazardous waste:** A waste or combination of wastes as defined in the Resource Conservation and Recovery Act (RCRA) (40 CFR 261.3), or those substances defined as hazardous wastes in 49 CFR 171.8 (Department of Transportation).

- **HAZWOPER:** An acronym for *Hazardous Waste Operations and Emergency Response,* the OSHA standard.

- **Level A protection:** The highest level of protection from chemical hazards. Known to the media as a "moon suit," it is formally called a "fully encapsulating suit" and includes a supplied-air breathing system. Designed to provide maximum skin and respiratory protection against chemical gases, vapors, and splashes.

- **Level B protection:** The level that uses the supplied-air breathing system for maximum respiratory protection but the protective suit for splash protection; for example, a cloud of chlorine or other gas could pass through seams or openings in the suit and contact the skin.

- **Level C protection:** The Level C protective suit is similar to that for Level B, but the respiratory protection is downgraded to an air-purifying respirator, which has a number of limitations, particularly in unknown environments.

- **Level D protection:** The level consisting of standard work clothing, e.g., a protective helmet, safety glasses, cotton coveralls, and protective boots; only for use where there are no anticipated chemical hazards to skin or respiratory system.

- **Personal protective equipment (PPE):** Specialized clothing, including respirators, that protects the worker from exposure to hazardous chemicals.

HISTORY OF HAZARDOUS WASTE OPERATIONS AND EMERGENCY RESPONSE TRAINING

As it is practiced in the 1990s, most hazardous waste-related health and safety training has its roots in training that was initiated by the Environmental Protection Agency (EPA) in the late 1970s. A significant portion of this training was aimed at the emergency responder. In the early days of environmental awareness, the emergency responder was frequently involved in operations associated with the investigation of hazardous chemicals found at abandoned drum sites, train derailments that included cars of extremely toxic or flammable chemicals, or cleanup of the residue from fires in hazardous waste treatment/storage facilities.

The cornerstone emergency responder course was EPA 165.5, "Hazardous Materials Incident Response Organization" (HMIRO), which focused on abandoned waste site operations. EPA 165.2,

"Personal Protection and Safety," also provided significant health and safety information to the hazardous waste worker. Each of these courses is still taught in a 40-hour program. Many commercial environmental response and cleanup firms modeled their internal training courses after 165.5 and 165.2.

Unfortunately, since these initial EPA courses were primarily for federal, state, and local personnel and only partially available to workers in the private sector, it was relatively difficult for the latter personnel to obtain the training. A variety of OSHA standards required training for specific activities associated with hazardous waste operations and emergency response. However, the emphasis on training present in today's HAZWOPER standard did not exist then, particularly for personnel involved in site investigation and cleanup activities. Most of the personnel who received EPA training tended to work for the emergency response and site cleanup contractors who worked closely with or were under contract to EPA and state environmental agencies. Therefore, less training was available for other personnel working at hazardous waste sites.

The training issue was brought to the attention of Congress, and a response was provided in Section 126 of the Superfund Amendments and Reauthorization Act of 1986 (SARA), which was signed into law on October 17, 1986. OSHA was directed to:

. . . issue an interim final rule within 60 days after the date of enactment, which specified providing no less protection for workers engaged in covered operations than the protections contained in the Environmental Protection Agency's "Health and Safety Requirements for Employees Engaged in Field Activities" manual (EPA Order 1440.2) dated 1981 and the existing OSHA standards under Subpart C of 29 CFR 1926.

In response, OSHA issued *Hazardous Waste Operations and Emergency Response; Interim Final Rule* (29 CFR 1910.120) on December 19, 1986. OSHA borrowed heavily from existing documents, particularly the previously mentioned EPA 1440.2, as well as the joint NIOSH/OSHA/U.S. Coast Guard (USCG)/EPA document entitled *Occupational Safety and Health Guidance Manual for Hazardous Waste Site Activities,* a 1985 manual that continues to be used extensively today. While the final standard was not to be promulgated until March 6, 1989, SARA required the interim final rule to take effect upon issuance; therefore, initial training requirements, as well as several other requirements, were to be fully implemented by March 16, 1987.

When SARA was amended in 1987, Congress directed OSHA to develop specific procedures for the accreditation of hazardous waste operation training programs that are no less comprehensive than those adopted by EPA under the Asbestos Hazard Emergency Response Act (AHERA). (See Chapter 13, Indoor Air Quality, for further details.) Under the proposed rules, which will become 29 CFR 1910.121 when finalized, OSHA would set up a program to review training materials and trainer qualifications, and confer accreditation on those firms that met certain criteria. In addition, the proposed rules specify minimum subject requirements for various types of training courses. As of January 1994, the fate of this standard is in doubt; however, the minimum subject criteria have come into wide usage.

ESTABLISHING THE TRAINING PROGRAM

Personnel cannot take part in any type of hazardous waste-related activity until they have successfully completed the required training. The initial step in establishing a program that will provide the required training is to perform a needs assessment. The employer should first confirm that the personnel will be participating in operations that are covered by the HAZWOPER standard. However, one should resist the temptation to assume that if the particular operation is in a "gray area" not directly covered by the standard, no training is required. Even if the operation does not fall exactly under one of the criteria under 29 CFR 1910.120 (a), "Scope," it is highly likely that it will be impacted by a number of other OSHA standards, such as those listed in Table 9–A.

Employers will frequently use HAZWOPER training as a "catch all" to try to cover all applicable OSHA requirements. Actually, the minimum training under HAZWOPER must often be tailored to include requirements of the other applicable standards, or the initial training must be augmented with additional sessions that include employer-specific information. This is an important concept that must be understood and incorporated into all training, as will be demonstrated in the subsequent sections of this chapter.

Once it has been established that the employer's personnel will be required to participate in activities that are impacted by HAZWOPER, the proper type(s) of training must then be selected. The goals

Table 9–A. Other Potentially Applicable OSHA Standards Under 29 CFR*

Section	OSHA Standard Title
1904	OSHA Injury and Illness Record Keeping
1910.20	Access to Employee Exposure and Medical Records
1910.24	Fixed Industrial Stairs
1910.27	Fixed Ladders
1910.28	Safety Requirements for Scaffolding
1910.38	Employee Emergency Plans and Fire Prevention Plans
1910.94	Ventilation
1910.95	Occupational Noise Exposure
1910.101	Compressed Gases
1910.133	Eye and Face Protection
1910.134	Respiratory Protection
1910.135	Occupational Head Protection
1910.136	Occupational Foot Protection
1910.141	Sanitation
1910.151	Medical Services and First Aid
1910.157	Fire Extinguishers
1910.165	Employee Alarm Systems
1910.181	Derricks
1910.212	General Requirements for All Machines
1910.252	Welding, Cutting, and Brazing
1910.307	Hazardous Locations
1910.1000	Toxic and Hazardous Substances
1910.1200	Hazard Communication
1926.20	General Safety and Health Provisions
1926.21	Safety Training and Education
1926.56	Illumination
1926.58	Asbestos
1926.59	Hazard Communication
1926.151	Fire Prevention
1926.152	Flammable and Combustible Liquids
1926.200	Accident Prevention Signs and Tags
1926.301	Hand Tools
1926.400	Electrical General Requirements
1926.401	Grounding and Bonding
1926.651	Specific Excavation Requirements
1926.652	Trenching Requirements
1926.1000	Roll Over Protective Structures (ROPSs) for Material Handling Equipment
Subpart O	Construction Standards—Motor Vehicles and Mechanized Equipment—Specifically Earth-Moving Equipment.

* Not intended as a complete list.
Source: U.S. EPA, *Standard Operating Safety Guides,* 1992, p. 6.

of training for employees involved in any type of hazardous waste activity are:

- to make workers aware of the potential hazards they may encounter
- to provide the knowledge and skill necessary to perform the work with minimal risk to worker health and safety
- to make workers aware of the purpose and limitations of safety equipment

- to ensure that workers can safely avoid or escape emergencies.

Such goals must be closely aligned with the training course(s) selected for an employer's hazardous waste personnel.

INITIAL TRAINING

There are three distinct types of activities under HAZWOPER that require initial training:

- cleanup operations at hazardous waste sites, including initial investigations
- hazardous waste operations at permitted and interim status RCRA treatment, storage, and disposal facilities (TSDF)
- emergency response operations.

Table 9–B presents specific language from the "Scope" section of HAZWOPER to provide precise definitions of the types of activities under the standard. It is the employer's responsibility to determine which of these definitions addresses its operation.

Cleanup Operations

If the needs assessment indicates that affected personnel will participate in site investigation or cleanup operations, the employer has to consider two more issues: (1) will the employee's job function present potential exposure to health and safety hazards associated with hazardous waste activities? If so, (2) which type of investigation/cleanup activities will each employee be involved in?

OSHA frequently receives inquiries about what activities at a hazardous waste site fall under the HAZWOPER training requirement. A representative interpretive response from OSHA indicated that workers are not covered by HAZWOPER, and, therefore, not subject to the training requirements, if they:

- work exclusively within uncontaminated areas of the hazardous waste site
- do not enter areas where hazardous waste may exist, is stored, or is processed
- are not exposed to health or safety hazards related to hazardous waste operations.

Examples of this situation include clerical staff who work in the site office trailer located outside of the contaminated area, personnel engaged in construction activities in uncontaminated areas of the site, or a truck driver who delivers a load of fill material to a location outside of the contaminated area. Note that OSHA may require the employer to

Table 9–B. Activities Covered Under the HAZWOPER Standard (from 29 CFR 1910.120 (a))

The following operations are covered under HAZWOPER, unless the employer can demonstrate that the operation does not involve employee exposure or the reasonable possibility for employee exposure to safety or health hazards:

1. Cleanup operations required by a governmental body—whether federal, state, local, or other—involving hazardous substances that are conducted at uncontrolled hazardous waste sites (including, but not limited to, the EPA's National Priority Site List (NPL), state priority site lists, sites recommended for the EPA NPL, and initial investigations of government identified sites that are conducted before the presence or absence of hazardous substances has been ascertained)
2. Corrective actions involving cleanup operations at sites covered by the Resource Conservation and Recovery Act of 1976 (RCRA) as amended (42 U.S.C. 6901 et seq)
3. Voluntary cleanup operations at sites recognized by federal, state, local or other governmental bodies as uncontrolled hazardous waste sites
4. Operations involving hazardous waste that are conducted at treatment, storage, disposal (TSD) facilities regulated by 40 CFR 264 and 265 pursuant to RCRA; or by agencies under agreement with U.S. EPA to implement RCRA regulations
5. Emergency response operations for releases of, or substantial threats of releases of, hazardous substances without regard to the location of the hazard.

Hazardous substance cleanup operations within the scope of paragraphs 1–3 must comply with all aspects of the HAZWOPER standard, except paragraphs (p) and (q), which relate to TSDF operations and emergency response.

Operations within the scope of paragraph 4 must comply only with the requirements of paragraph (p) of HAZWOPER, which does include training. However, there are several exceptions to this, many of which impact training requirements; additional information can be found in the "Notes and Exceptions" section at 29 CFR 1910.120 (a)(2)(iii).

Emergency response operations for releases of, or substantial threats of releases of, hazardous substances that are not covered by paragraphs 1–4 must only comply with the requirements of paragraph (q) of HAZWOPER, which includes training requirements for the various classes of emergency responders.

establish that such work areas are in fact free of contamination. Such personnel should be trained in site emergency response procedures in the event that there is a release of hazardous substances requiring an evacuation.

Once it is established that personnel do have to be trained, the employer must decide what type of initial training is required, i.e., 40-hour or 24-hour. In general, 40-hour training is for workers who will be or can be exposed to uncontrolled or uncharacterized hazardous substances and their associated health hazards; 24-hour training is for personnel who work in areas that have been fully characterized as to the type of hazards present, and for which air monitoring has established that it is unlikely that such personnel will be exposed to airborne substances over the occupational exposure limits. Table 9–C provides the HAZWOPER language that addresses this issue.

For all practical purposes, the difference between 40-hour and 24-hour training is that the latter does not provide training in respiratory protection or high-hazard operations. By definition, personnel in the 24-hour training group will not be working in areas of high hazard or chemical exposure potential. If such personnel are later required to work in uncharacterized or high-hazard areas, they must first complete the additional 16 hours of training to bring them up to "40-hour status." Many environmental consulting firms do not use the 24-hour option for their field personnel because it limits flexibility in task assignment. However, 24-hour training may be useful and cost-effective for personnel with task-specific, long-term assignments.

The HAZWOPER standard provides broad, general requirements for both 40- and 24-hour training course contents. HAZWOPER is designed to be a "performance standard;" that is, the employer designs the various aspects of the program within the context of the specific requirements of the standard to address the health and safety hazards particular to the employer's operation. OSHA is emphatic that the level of training provided be consistent with the worker's job function and responsibilities. This is evident when viewing the training elements of HAZWOPER:

- names of personnel and alternates responsible for site safety and health
- safety, health, and other hazards present on the site
- use of personal protective equipment
- work practices by which the employee can minimize risks from hazards
- safe use of engineering controls and equipment on the site
- medical surveillance requirements, including recognition of symptoms and signs that might indicate overexposure to the hazards
- information from the site health and safety plan, including:
  — decontamination procedures
  — emergency response plan

Table 9–C. Initial Training Requirements for 24-Hour Versus 40-Hour Training, from 29 CFR 1910.120(e)

Requirement	Description
1	General site workers (such as equipment operators, general laborers, and supervisory personnel) engaged in hazardous substance removal or other activities which expose or potentially expose workers to hazardous substances and health hazards shall receive a minimum of 40 hours of instruction off the site, and a minimum of three days actual field experience under the direct supervision of a trained, experienced supervisor.
2	Workers on site only occasionally for a specific limited task (such as, but not limited to, groundwater monitoring, land surveying, or geophysical surveying) and who are unlikely to be exposed over permissible exposure limits and published exposure limits shall receive a minimum of 24 hours of instruction off the site, and the minimum of one day of actual field experience under the direct supervision of a trained, experienced supervisor.
3	Workers regularly on site who work in areas that have been monitored and fully characterized indicating that exposures are under permissible exposure limits and published exposure limits where respirators are not necessary, and the characterization indicates that there are no health hazards or the possibility of an emergency developing, shall receive a minimum of 24 hours of instruction off the site, and the minimum of one day actual field experience under the direct supervision of a trained, experienced supervisor.
4	Workers with 24 hours of training who are covered by paragraphs 2 and 3 above, and who become general site workers or who are required to wear respirators, shall have the additional 16 hours and two days of training necessary to total the training specified in paragraph 1 above.

— confined space entry procedures
— spill containment program.

Note that there is an assumption that the training will be designed to be site specific. Such 40- or 24-hour training is the exception rather than the rule, because rarely can an employer anticipate which employees will be assigned to a specific site or in fact anticipate what the employee's first assignment will be. Therefore, most initial training programs are not site specific; rather, most issues are covered in a generic sense, addressing, for example, a model emergency response plan or a typical confined space entry procedure. Under these circumstances, it is the employer's responsibility under this performance standard to see that the workers receive additional site- and task-specific training as part of the three days of on-the-job training prior to commencing site work.

Two examples of the 40-hour training curriculum, one from the EPA's *Standard Operating Safety Guides*, and the other from the Hazardous Waste Action Coalition's Health and Safety Subcommittee, are provided in Tables 9–D and 9–E to illustrate different approaches to identifying initial training requirements for hazardous waste operations.

The proposed training accreditation standard, which was previously discussed, describes minimum criteria and content for training programs that OSHA will require in order to consider them as acceptable for accreditation. These criteria are much more detailed than those provided in HAZWOPER. Various training and hazardous materials consulting firms have adopted these criteria as the working outlines for their training courses. The criteria are provided in Table 9–F.

Treatment, Storage, and Disposal Facilities

Treatment, storage, and disposal facility (TSDF) personnel are already subject to training requirements prescribed by the EPA in 40 CFR 264 and 265; such requirements can be readily woven into the HAZWOPER requirements. As with cleanup operations, the employer must determine whether the particular TSDF is exempted from these requirements; such exemptions are usually associated with "conditionally exempt small quantity generators." Employers should review the "Notes and Exceptions" portion of 29 CFR 1910.120 (a)(2) to see whether this exemption applies. New TSDF employees must complete 24 hours of initial training in order to "perform their assigned duties and functions in a safe and healthful manner so as not [to] endanger themselves or other employees." (See Chapter 8, Hazardous Wastes, for further discussion of TSDFs.)

Current employees of a TSDF for whom the employer can establish previous training and experience equivalent to the 24-hour initial training do not have to take the training. However, "current employee" may be interpreted as being in the position prior to the March 9, 1990, effective date of the HAZWOPER final standard, so one may not wish to read too much into this equivalency provision.

HAZWOPER is vague regarding the specific elements that the 24-hour TSDF course should cover.

Table 9–D. Hazardous Waste Action Coalition Recommended Minimum Contents of the 40-Hour, 24-Hour, and 16-Hour Supplemental Hazardous Waste Operations Health and Safety Courses

Topic	Minimum Time (Hr)*	40-Hour General Site Worker Course	24-Hour Occasional Site Worker Course	16-Hour Supplemental Course
1. Overview of the applicable paragraphs of 29 CFR 1910.120 and its appendixes	0.50	X	X	
2. Overview and explanation of OSHA's hazard communication standard (29 CFR 1910.1200)	0.25	X	X	
3. Rights and responsibilities of employers and employees under OSHA and CERCLA	0.25	X	X	
4. Health hazard recognition • Basic Toxicology • Worker Exposure Limits • Biological Hazards • Radiation Hazards • Corrosive Hazards • Reactive Hazards • Fire and Explosion Hazards	4.00	X	X	
5. Respiratory protection, including limitations, selection, and maintenance. Hands-on training to include fit-testing with air-purifying respirators (APRs), and donning of supplied-air units (SCBA and air-line)	4.50	X		X
6. Use and limitations of personal protective equipment (PPE) including description of levels of protection and a dress-out demonstration	1.50	X	X	
7. Safety hazard recognition • Electrical • Excavation • Powered Equipment • Illumination • Noise • Sanitation • Walking-Working Surfaces	1.50	X	X	
8. Heat stress and cold stress	0.50	X	X	
9. Site control including a description of procedures that will minimize employee exposure to hazards, including work zones, safe work practices, engineering and administrative controls, communications, the buddy system, and security	1.50	X	X	
10. Medical surveillance program elements, including record keeping and employee access requirements of 29 CFR 1910.20	0.25	X	X	
11. Health and safety plan overview	0.50	X	X	
12. Use and limitations of monitoring instruments with a "hands-on" exercise that includes direct-reading instruments capable of measuring toxic gases and vapors, oxygen levels, explosive limits, and ionizing radiation	2.50	X	X	
13. Emergency plans and procedures including spill containment and emergency notification	1.00	X	X	

(continued)

Table 9–D. *(Concluded)*

Topic	Minimum Time (Hr)*	40-Hour General Site Worker Course	24-Hour Occasional Site Worker Course	16-Hour Supplemental Course
14. Use of references and resources including material safety data sheets (MSDSs), hazard coding and labeling systems, and other sources of information that address exposure limits, physical and chemical properties, and health effects	1.25	X	X	
15. Overview of decontamination of equipment and personnel, including a demonstration of personnel decontamination at various levels of protection	1.25	X	X	
16. Overview of container handling, including drums and laboratory containers (not intended to replace site-specific protocols)	0.50	X		X
17. "Hands-on" field simulation, including dressout and skill demonstration in Levels D, C, and B protective equipment by all attendees, and the use of monitoring equipment and a decontamination exercise	4.50	X		X
18. Confined space overview	0.50	X		X
19. Final examination and examination review (must be documented)	1.00	X	X	X
20. Attendees' written evaluation of course		X	X	X

*The amount of time listed represents an HWAC-recommended *minimum* for each topic. Based upon the needs of the specific attendees, the amount of time devoted to a particular topic must be increased, or additional topics included, to satisfy the requirements for 40, 24, or 16 hours of training.
Source: Hazardous Waste Action Coalition. Used with permission from HWAC.

It is implied that it will cover the elements of the required health and safety program: identification, evaluation, and control of safety and health hazards; emergency response criteria; waste-handling procedures; maximum exposure limits; and engineering controls. Also included is hazard communication training, medical surveillance, decontamination, and materials handling. Fortunately, the proposed 1910.121 cuts through all the vagueness and describes what OSHA would really like to see in a TSDF curriculum:

- Overview of the applicable paragraphs of 29 CFR 1910.120 and the elements of an employer's effective occupational safety and health program and those responsible for the program.
- Overview of relevant hazards such as, but not limited to, chemical exposures, biological exposures, fire and explosion exposures, radiological exposures, and heat and cold exposures.
- General safety hazards including those associated with electrical hazards, powered equipment, and walking-working surfaces.

- Confined space hazards and procedures (now addressed by 29 CFR 1910.146).
- Work practices that will minimize employee risk from workplace hazards.
- Emergency response plan and procedures, including first aid meeting the requirements of 1910.120 (p)(8).
- A review of the employer's hazardous waste-handling procedures, including the materials handling program and spill containment program.
- An overview and explanation of the employer's Hazard Communication Program meeting the requirements of 29 CFR 1910.1200 for those chemicals other than hazardous waste in the workplace.
- A review of the employer's medical surveillance programs meeting the requirements of 29 CFR 1910.120(p)(3), including the recognition of signs and symptoms of overexposure to relevant hazardous substances.
- A review of the employer's decontamination program and procedures meeting the requirements of 29 CFR 1910.120(p)(4).

Table 9–E. Recommended Training by Job Category

Training Topic	Emphasis of Training	General Site Worker	On-site Management and Supervisors	Health and Safety Staff
Biology, Chemistry, and Physics of Hazardous Materials	Chemical and physical properties; chemical reactions; chemical compatibilities.	✔	✔	✔
Toxicology	Dosage, exposure routes, toxicity, IDLH values, PELs recommended exposure limits (RELs), TLVs.	✔	✔	✔
Industrial Hygiene	Monitoring workers' need for and selection of PPE.	O	✔	✔
	Calculation of doses and exposure levels; hazard evaluation; and selection of worker health and safety protective measures.	O	✔	✔
Monitoring Equipment	Selection, use, capabilities, limitations, and maintenance.	✔	✔	✔
Hazard Evaluation/ Recognition	Techniques of sampling and assessment.	✔	✔	✔
	Evaluation of field and lab results.	O	✔	✔
	Chemical/physical.	✔	✔	✔
	Risk assessment.		O	✔
Site Safety Plan	Safe practices, safety briefings and meetings, standard operating procedures, site safety map.	✔	✔	✔
Standard Operating Procedures	Hands-on practice.	✔	✔	✔
	Development and compliance.	O	✔	✔
Engineering Controls	The use of barriers, isolation, and distance to minimize hazards.	✔	✔	✔
Personal Protective Clothing and Equipment (PPE)	Assignment, sizing, fit-testing, maintenance, use, limitations, and hands-on training.	✔	✔	✔
	Selection of PPE.	✔	✔	✔
Medical Program	Medical monitoring, first aid, stress recognition.	✔	✔	✔
	CPR and emergency drills.	O	✔	✔
	Design and planning.		O	✔
	Implementation.	✔	✔	✔
Decontamination	Hands-on training using simulated field conditions.	✔	✔	✔
	Design and maintenance.	✔	✔	✔
Legal and Regulatory Aspects	Applicable safety and health regulations (OSHA, EPA).	O	✔	✔
Emergencies/Accidents	Emergency help, self-rescue, drills, alarms, reporting.	✔	✔	✔
	Emergency response, investigation, and documentation.	O	✔	✔
Hazard Communication	Per 29 CFR 1910.1200 and 1926.59 (as applicable).	✔	✔	✔
Employee Rights		✔	✔	✔

✔ = Recommended training   O = Optional
IDLH, Immediately Dangerous to Life or Health; PEL, Permissible Exposure Limits; REL, Recommended Exposure Limits; TLV, Threshold Limit Values. Source: *EPA Standard Operating Safety Guidelines,* 1992.

- A review of the employer's training program and the personnel responsible for that program.
- A review of the employer's personal protective equipment (PPE) program, including the proper selection and use of PPE based upon specific site hazards.
- Safe use of engineering controls and equipment.
- A review of the applicable appendixes to 29 CFR 1910.120.
- Principles of toxicology and biological monitoring.
- Rights and responsibilities of employees and employers under OSHA and RCRA.
- Sources of reference and efficient use of relevant manuals and knowledge of hazard coding systems.
- Hands-on exercises and demonstrations with equipment expected to be used during the performance of work duties.
- Final examination.

Throughout this curriculum description, OSHA refers to the employer's programs or procedures.

Table 9–F. Proposed 29 CFR 1910.121: Required Topics for 40-Hour and 24-Hour Hazardous Waste Operations Training Including the 16-Hour "Upgrade"

40	24	16	Description
x	x		Overview of the applicable paragraphs of 29 CFR 1910.120 and the elements of an employer's effective occupational safety and health program.
x	x	x	Effects of chemical exposures to hazardous substances (such as toxicity, carcinogens, irritants, and sensitizers).
x			Effects of biological and radiological exposures.
x	x	x	Fire and explosion hazards (i.e., flammable and combustible liquids, and reactive materials).
x	x	x	General safety hazards, including electrical hazards, powered equipment hazards, walking-working surface hazards, and those hazards associated with hot and cold temperature extremes.
x		x	Confined space, tank and vault hazards, and entry procedures.
x		x	Names of personnel and alternates, where appropriate, responsible for site safety and health at the site.
x			Specific safety, health, and other hazards that are to be addressed at a site and in the site safety and health plan.
x	x		Use of personal protective equipment and the implementation of the personal protective equipment program.
x	x		Work practices that will minimize employee risk from site hazards.
x	x		Safe use of engineering controls and equipment and any new relevant technology or procedure.
x	x		Content of the medical surveillance program and requirements, including the recognition of signs and symptoms of overexposure to hazardous substances.
x	x		The contents of an effective site safety and health plan.
x	x	x	Use of monitoring equipment with "hands-on" experience and the implementation of the employee and site monitoring program.
x		x	Implementation and use of the informational program.
x		x	Drum and container handling procedures and the elements of a spill containment program.
x		x	Selection and use of materials handling equipment.
x		x	Methods for assessment of risk and handling of radioactive wastes.
x		x	Methods for handling shock-sensitive wastes.
x		x	Laboratory waste pack handling procedures.
x		x	Container sampling procedures and safeguards.
x		x	Safe preparation procedures for shipping and transport of containers.
x	x	x	Decontamination program and procedures.
x	x		Emergency response plan and procedures, including first-aid.
x		x	Safe site illumination levels.
x		x	Site sanitation procedures and equipment for employee needs.
x		x	Review of the applicable appendices to 29 CFR 1910.120.
x		x	Overview and explanation of OSHA's hazard communication standard (29 CFR 1910.1200).
x	x	x	Sources of reference, additional information and efficient use of relevant manuals and hazard coding systems.
x	x		Principles of toxicology and biological monitoring.
x	x		Rights and responsibilities of employees and employers under OSHA and CERCLA.
x	x		"Hands-on" field exercises and demonstrations.
x	x	x	Final examination.

Note: Where "24" and "16" share a topic, the topic is enhanced in the 16-hour course beyond what was presented in the original 24-hour course.
Source: Derived by John Danby from 55 *Federal Register* 2777, Notice of Proposed Rulemaking, 29 CFR 1910.121, January 26, 1990.

More than for the other two types of training, initial training for TSDF personnel can and should be facility-specific. This is because at the completion of the training, employees will be working in one particular facility.

TSDF employees who are to participate in facility emergency response activities are to receive additional training in the elements of the employer's emergency response plan and other issues. However, there are several exceptions to this particular training requirement, so TSDF employers should review 29 CFR 1910.120(p)(8)(iii) to determine the level of emergency response training that may be required for TSDF employees.

Emergency Response Operations

In both hazardous waste cleanup operations and TSDF training, personnel must be oriented to emergency response (ER) procedures in the event that an emergency does occur at their site; however, the emphasis is on health and safety associated with normal site operations. Traditional emergency responders, on the other hand, can face a different situation every time they respond. To address this, OSHA has developed five ER training categories, each accompanied by a different set of training requirements. The employer should consider the following OSHA training categories when developing the emergency response team and assigning personnel to emergency responder tasks.

First responder awareness level. First responders at the awareness level are individuals who are likely to witness or discover a hazardous substance release and who have been trained to initiate an emergency response sequence by notifying the authorities of the release.

First responder operations level. First responders at the operations level are individuals who respond to releases or potential releases of hazardous substances as part of the initial response to the site for the purpose of protecting nearby persons, property, or the environment from the effects of the release. They are trained to respond in a defensive fashion without actually trying to stop the release. Their function is to contain the release from a safe distance, keep it from spreading, and prevent exposures.

Hazardous materials technician. Hazardous materials technicians are individuals who respond to releases or potential release situations for the purpose of preventing or stopping the release. They assume a more aggressive role than a first responder at the operations level in that they will approach the point of release in order to plug, patch, or otherwise stop the release of a hazardous substance.

Hazardous materials specialist. Hazardous materials specialists are individuals who respond with and provide support to hazardous materials technicians. Their duties parallel those of the hazardous materials technician; however, those duties require a more specific knowledge of the various substances they may be called upon to contain. The hazardous materials specialist would also act as the site liaison with federal, state, local, and other government authorities regarding site activities.

On-scene incident commander. Incident commanders assume control of the incident scene beyond the first responder awareness level. OSHA defines the "on-scene incident commander" as the "most senior official on site who has the responsibility for controlling the operations at the site." Such senior officials may be a battalion chief, fire chief, state law enforcement official, or site coordinator. All emergency responders and their communications are to be coordinated and controlled through the on-scene incident commander assisted by the senior official present for each employer.

In contrast to hazardous waste cleanup and TSDF operations, in which the HAZWOPER standard and the proposed training certification standard essentially state what topics must be included in a training program, emergency responders have topic areas in which they must "demonstrate competency." OSHA does not address the training of emergency responders in the proposed 29 CFR 1910.121, stating that it would not be effective to try to certify all the fire agencies throughout the country who provide emergency response services. Clearly, OSHA expects the training curriculum to be tailored to anticipated response activities, and notes that training is required for those who participate or *who are expected to participate* in emergency response.

OSHA's performance goals for the five ER training categories are discussed next. Representative training topics as found in some commercial training programs are included where appropriate.

Performance Goals: Emergency Responders

First responders at the awareness level shall have sufficient training or have had sufficient experience to objectively demonstrate competency in the following areas:

- understanding hazardous substances and the risks associated with them in an incident
- understanding the potential outcomes associated with an emergency created when hazardous substances are present
- recognizing the presence of hazardous substances in an emergency
- identifying the hazardous substances, if possible
- understanding the role of the first responder awareness individual in the employer's emergency response plan, site security and control, and the U.S. Department of Transportation's *Emergency Response Guidebook*
- realizing the need for additional resources and making appropriate notifications to the communication center.

The employer shall certify that first responders at the operational level have received at least eight hours of training or have sufficient experience to objectively demonstrate competency and receive employer certification in the following areas in addition to those listed for the awareness level:

- basic hazard and risk assessment techniques
- selecting and using proper personal protective equipment provided to the first responder operational level

- basic hazardous materials terms
- performing basic control, containment, and/or confinement operations within the capabilities of the resources and personal protective equipment available with their unit
- implementing basic decontamination procedures
- relevant standard operating procedures and termination procedures.

Representative training topics include:

- Regulatory Requirements
- Toxicology
- Industrial Hygiene
- Decontamination
- Spill Control Practicum
- Chemistry of Hazardous Substances
- Hazardous Chemical Workshop
- Personal Protective Equipment
- Spill Control Techniques.

The employer shall certify that hazardous materials technicians have received at least 24 hours of training equal to the first responder operations level and in addition demonstrate competency and receive employer certification in the following areas:

- implementing the employer's emergency response plan
- classification, identification, and verification of known and unknown materials by using field survey instruments and equipment
- function within an assigned role in the incident command system
- selecting and using proper specialized chemical personal protective equipment provided to the hazardous materials technician
- hazard and risk assessment techniques
- performing advance control, containment, and/or confinement operations within the capabilities of the resources and personal protective equipment available with the unit
- understanding and implementing decontamination procedures
- understanding termination procedures
- understanding basic chemical and toxicological terminology and behavior.

Representative training topics include:

- Regulatory Requirements
- Toxicology
- Protective Clothing
- Medical Surveillance
- Physical Hazards

- Site Control Methods
- Contingency Planning
- Incident Command Systems
- Chemical Hazards
- Air Monitoring Instruments
- Respiratory Protection
- Decontamination Procedures
- Safe Work Practices
- Site Safety Plans
- Materials Handling Techniques
- Field Practicum.

The employer shall certify that hazardous materials specialists have competency in the following areas:

- implementing the local emergency response plan
- classification, identification, and verification of known and unknown materials by using advanced survey instruments and equipment
- state emergency response plan
- selecting and using proper specialized chemical personal protective equipment provided to the hazardous materials specialist
- in-depth hazard and risk assessment techniques
- performing specialized control, containment, and/or confinement operations within the capabilities of the resources and personal protective equipment available
- determining and implementing decontamination procedures
- developing a site safety and control plan
- chemical, radiological, and toxicological terminology and behavior.

Training topics for hazardous materials specialists would be similar to those for hazardous materials technicians.

The employer shall certify that incident commanders have received at least 24 hours of training equal to the first responder operations level and, in addition, have competency in the following areas:

- implementing the employer's incident command system
- implementing the employer's emergency response plan
- hazards and risks associated with employees working in chemical protective clothing
- implementing the local emergency response plan
- knowledge of state emergency response plan and of the Federal Regional Response Team
- importance of decontamination procedures.

Representative training topics include:

- Regulatory Requirements
- State ER Plans
- Decontamination
- ICS Operating Requirements
- Organization and Operations
- Personal Protective Equipment
- Emergency Response Planning
- Contingency Planning
- Hazardous Material Recognition
- Factors Affecting Emergency Management
- Media Management
- ICS Workshop.

OSHA provides an interesting training exemption for what it calls "skilled support personnel." These are personnel, such as equipment operators, who have skills that are needed immediately and who will be exposed to site hazards, but who have not had emergency response training. Skilled support personnel do not need to complete the training, but they must have an initial briefing prior to commencing work. It is clear that as soon as properly trained personnel are available to take the place of the skilled support personnel, the latter are to be removed from the hazard area.

REFRESHER TRAINING

Once initial training is completed, additional training requirements must be considered. All three training categories require personnel to complete an annual "refresher" training course. Refresher training provides an opportunity to reemphasize important points and review actual incidents to evaluate lessons learned. Both Hazardous Waste Operations and TSDF personnel must complete an eight-hour refresher course annually. Refresher course requirements and accreditation are specifically excluded from the proposed 29 CFR 1910.121. The HAZWOPER standard specifies that the Hazardous Waste Operations refresher course shall review the topics presented in the initial training course and critique incidents that provide training examples of related work. The standard does not specify curriculum for the TSDF personnel.

Emergency response personnel are to "receive annual refresher training of sufficient content to maintain their competencies, or shall demonstrate competency in those areas at least yearly." The number of hours and type of topics for this refresher course are not specified.

Refresher courses can be accomplished in several ways. Typically, the employer provides an eight-hour classroom session once a year. However, many employers with employees assigned to only one location for the year may schedule periodic "brown bag" training sessions to cover refresher topics. One alternate method is to hold a one-hour session every month, documenting that each employee attends at least eight sessions per year.

The employer must consider what is meant by "annual," for it may vary among federal OSHA and the various state program OSHA agencies. A federal OSHA interpretation of the standard states that "OSHA's intent is that employees should complete their refresher training within twelve months of their initial training," and if a course is missed, the employee should take the next available course. The interpretation notes that some states require the refresher to be completed by the exact anniversary date of the initial training (or previous refresher). At least one state requires initial training to be retaken if the employee does not take a refresher within two years of the anniversary date of the initial training or previous refresher. Note that under similar circumstances federal OSHA requires only that the employee "demonstrate competency" but does recommend retraining if the employee has been away from the field for a significant amount of time.

SUPERVISOR TRAINING

Supervisor training is required of hazardous waste operations on-site managers and supervisors who are directly responsible for, or who supervise employees engaged in, hazardous waste operations. This is generally interpreted to include at least the site manager and the site safety officer (SSO); more people may be included for larger, more complex sites. HAZWOPER requires that supervisors receive the initial training described earlier, as well as eight hours of specialized training on the following topics:

- the employer's safety and health program, and the associated employee training program
- personal protective equipment
- spill containment program
- health hazard monitoring procedures and techniques.

The proposed 29 CFR 1910.121 added:

- management of hazardous wastes and their disposal

- federal, state, and local agencies to be contacted in the event of a release of hazardous substances
- management of emergency procedures in the event of a release of hazardous substances.

Although this is a good beginning for an SSO, it is insufficient for someone who is responsible for health and safety activities at a complex site. Tables 9–G and 9–H describe two approaches to certifying an SSO for various types of hazardous waste site operations.

Supervisor training must also be refreshed annually, which could lead an employer to assume that two separate refresher courses must be conducted each year. A careful evaluation of 29 CFR 1910.120(e)(8), which describes supervisor training requirements, states that employees who have received initial training and supervisors who have received supervisor training have to be refreshed annually on the topics of initial training "and/or" the topics of supervisor training. Therefore, only one annual refresher course need be held, but it must contain components of both initial and supervisor training if supervisory personnel are in attendance.

SUPERVISED FIELD EXPERIENCE

For employees engaged in hazardous waste operations, the initial 24 or 40 hours of off-site (classroom) training must be followed up by supervised field experience. Such field experience must be completed prior to the employee independently engaging in field operations. For personnel with 24 hours of initial training, a minimum of one day of actual field experience is required; for employees with 40 hours of initial training, a minimum of three days of actual field experience must be completed. Such field experience must be conducted under the supervision of a trained, experienced supervisor.

The purpose of supervised field experience is clear. Classroom training, even with a comprehensive field practicum, does not, by itself, adequately prepare personnel for hazardous waste field activities. One must learn to use protective equipment and monitoring equipment under actual field conditions in order to function as an effective team member (Figure 9–1).

The employee's field experience should use the level of protection (A through D) that the employee

---

Table 9–G. Recommended Guidelines for Site Safety Officer (SSO) Training*

Category	Description or Recommended Guideline
Health and safety professional	A Certified Industrial Hygienist, a Certified Safety Professional with demonstrable experience in managing a health and safety program, or a person possessing equivalent credentials and expertise.
Site safety officer	The individual located at a hazardous waste site who is responsible to the employer and has the authority and knowledge necessary to implement the site safety and health plan and verify compliance with the applicable safety and health requirements
	Site safety officers:
	▪ shall at all times be under the supervision of a health and safety professional
	▪ shall be assigned to hazardous waste sites with consideration given to the complexity and level of hazards associated with each site assignment
	▪ shall meet the basic training and medical requirements as defined in paragraphs (e) and (f) of the OSHA *Hazardous Waste Operations and Emergency Response* (HAZWOPER) standard. Training shall be consistent with the worker's job function and responsibilities
	▪ shall meet the additional 8 hours of supervisor-specific training required under paragraph (e)(4) of HAZWOPER
	▪ shall have at least 24 hours of previous experience at the highest level of protection approved in the site safety plan
	▪ shall have at least 30 days of on-site experience in the hazardous waste field
	▪ shall supervise the implementation of the site safety and health plan
	▪ shall make decisions with respect to health and safety based on information and observations made in the field
	▪ shall demonstrate familiarity with the firm's policies, procedures, and health and safety program
	▪ should demonstrate current certification in first aid and cardiopulmonary resuscitation
	▪ should be capable of preparing a site safety and health plan for the review and approval of the health and safety professional.

*These are recommended minimum qualifications to personnel needing to evaluate the qualifications of a person assigned as an SSO during site investigation. This guidance does not, nor is it intended to, reflect the lowest denominator of qualifications required by HWAC member firms.
Source: Developed by the HWAC Health and Safety Subcommittee.

Table 9-H. Hazardous Waste Operations Recommended SSO Qualifications

Activity	40-Hour, Supervisor, and First Aid/CPR	OSHA Construction Safety Course	ASSE Shoring School	EPA Air-Monitoring Course	Training — Heat and Cold Stress Identification and Prevention	4-Hour Air Line/ SCBA Training	3-Day OJT at Level of Protection	Industrial Hygiene Professional	4-Hour Drill Rig Safety	Confined Space Training
Noninvasive Investigation	R									
Drilling/ Noncomplex	R								S	
Drilling/Air Rotary	R			S					S	
Drilling/High Hazard	R			R				S	R	
Trenching with Shoring	R		R							
Heavy Construction	R	R	S							
Level D	R						R			
Level C	R			S	S		R			
Level B	R			R	R	R	R	R		
Level A	R			R	R	R	R	R		
Heat and Cold Stress/ Potential	R				R			S		
Complex Air Monitoring	R			S				R		
Confined Space Entry	R			R	S			R		R

R = Required; S = Suggested; ASSE = American Society of Safety Engineers
Source: Compiled by John Danby from regulatory interpretation and professional experience.

Figure 9–1. Training courses prepare employees to take part in field activities such as this drum-sampling exercise. (Courtesy Gerri Silva.)

will be expected to use during independent project operations. It does no good to complete the supervised field experience in Level D protection, then expect the employee to be able to function effectively in Level B the following week. It is therefore recommended that employees receive three days of supervised field experience in each of the levels of protection in which they will be expected to work.

If an employee with experience in Levels D and C is assigned to a Level B project, then the first three days of work should be supervised field experience. Note that this may not be required if an employee is downgrading the level of protection, such as receiving field experience in Level B, then being assigned to Level D. All supervised field training should be documented in writing and a copy placed in the employee's personnel file.

EQUIVALENT TRAINING

The initial and/or supervisor training requirement is waived if employers of hazardous waste operations

personnel establish that an employee's work experience and/or training is equivalent to the initial training and/or supervisor training required by the standard [29 CFR 1910.120(e)(9)]. The standard states "Equivalent training includes any academic training or the training that existing employees might already have received from actual hazardous waste work experience."

Unfortunately, this paragraph is easily misunderstood and readily misused. Some trainers have given 24 hours of classroom training, then accounted for 16 more hours by certifying various types of field experience as "equivalent" in an attempt to side-step the required additional classroom training. Other similar abuses have occurred. However, in a correction to the HAZWOPER final rule that was published in the *Federal Register* on April 18, 1991, OSHA stated "SARA provides that the required training can result from a course and on the job training or from training and experience *prior to the effective date of the interim standard*" (emphasis added). Because the effective date of the interim standard was April 1987, it would appear that any training or experience obtained after that date may be unacceptable for equivalency.

CONDUCTING THE TRAINING

For most employers, determining whether HAZWOPER training is required for employees is the easy part; developing and implementing the training program is the real work.

The first order of business is to decide whether to do the training "in-house" or to retain an outside consultant to conduct the training. The development of training curriculum, particularly for the longer, more involved 24- or 40-hour course is time-consuming when it is done correctly. For employers with a large number of personnel involved in hazardous waste operations and emergency response, and who have a health and safety professional on staff, it might be reasonable to do some or most of the training in-house. Some firms will use outside consultants for the 24- or 40-hour courses, and do the eight-hour courses, such as the refresher and supervisor courses, in-house. Most small employers, or those with few employees under HAZWOPER, tend to use outside providers exclusively.

IN-HOUSE TRAINING CONSIDERATIONS

If the decision is to do some or all of the applicable training courses in-house, the employer next has to appoint someone to be responsible for developing and conducting the courses. For hazardous waste and TSDF operations, the appropriate choice is an experienced health and safety professional who is familiar with the operations of concern. For example, although fire fighters who are experienced in emergency response can be used for emergency response training, they should be assisted by broadly experienced health and safety professionals. The worst possible approach is to send someone to take the needed course, and after their return have them photocopy the course materials and become the duly appointed trainer.

Logistics is an important issue. If an employer has multiple locations, is it feasible to transport the trainers and/or the trainees to the training site? A significant amount of equipment is used in teaching a 40-hour operations or 24-hour emergency response course. If the employer has a single facility, logistics may not be a deciding factor.

If conducting internal training seems like a good idea, the employer should use the criteria in the following section to fully evaluate the feasibility of this potentially complex undertaking.

COMPONENTS OF A GOOD TRAINING COURSE

There are several issues to consider in the selection of a training course provider. Unfortunately, the criteria that are most readily discernable and that may be of primary interest to the unsophisticated employer may be the least reliable indicators of course quality and applicability. One such criterion is price per student. Although price is an important consideration, it is more important to get a quality training program that fulfills OSHA requirements. Experience has indicated that high prices don't necessarily make a good course, and the low bidder may actually conduct a good course. The higher-priced firm with the glossy brochure may have mediocre training personnel, whereas the low-priced university may be subsidized by a grant. However, to compare several providers, the highest priced and the lowest priced should probably be closely evaluated. Managers should compare the offerings with the criteria listed in this section and, before deciding, ask for and check references regarding course value and quality.

Instructors

The quality of instruction is obviously closely tied to the quality of the course instructors. An employer is well-advised to carefully examine the credentials of the instructors. Not all teachers of HAZWOPER courses are equal in skill level or instructional experience. OSHA requires that, for hazardous waste operations courses,

> Trainers shall be qualified to instruct employees about the subject matter that is being presented in training. Such trainers shall have satisfactorily completed a training program for teaching the subjects they are expected to teach, or they shall have the academic credentials and instructional experience necessary for teaching the subjects. Instructors shall demonstrate competent instructional skills and knowledge of the applicable subject matter.

It may not be easy for the employer to determine whether the instructors meet OSHA's criteria.

For hazardous waste operations and TSDF courses, the course director and/or lead instructor should be a health and safety professional. In the compliance guidelines for HAZWOPER, OSHA recommends that the development and implementation of the site safety and health program be conducted by professional safety and health personnel, such as Certified Safety Professionals (CSPs), Certified Industrial Hygienists (CIHs), and Registered Professional Safety Engineers. This recommendation should be extended to the development and implementation of training courses. Until 29 CFR 1910.121, which certifies training programs, is promulgated, the use of professional safety and health personnel in training is one of the most readily identifiable assurances of trainer competence.

For emergency response trainers, HAZWOPER makes a very specific requirement for trainer qualification:

> Trainers who teach any of the emergency response subjects shall have satisfactorily completed a training course for teaching the subjects they are expected to teach, such as the courses offered by the U.S. National Fire Academy, or they shall have the training and/or academic credentials and instructional experience necessary to demonstrate competent instructional skills and a good command of the subject matter of the courses they are to teach.

Therefore, fire academy credentials would be a good start toward identifying a competent emergency response trainer; beyond that, it becomes a matter of evaluating field and instructional experience in combination with academic credentials in determining trainer competence.

Training experience is not necessarily interchangeable among the three categories of courses. Fire fighters with emergency response experience probably have little experience relative to hazardous waste operations; conversely, the CIH skilled in hazardous waste operations may not have appropriate material for incident command systems or "plug and patch" activities. Unfortunately, some training providers use the same trainers regardless of the course; if the trainers are not thoroughly experienced in all the areas in which they instruct, the credibility and effectiveness of the course suffers greatly.

Although it is not unreasonable to have one instructor teach an 8-hour session, such as supervisor or refresher, it is an unusual individual who can do an adequate job "going solo" for an entire 24- or 40-hour course. It is unrealistic to ask an instructor to teach the entire session. Instructors will fatigue, students will get bored listening to the same person for a week, and the quality of the course inevitably

suffers. A good 40-hour course needs at least two, and preferably three, instructors, whereas a 24-hour offering should have two trainers.

Supporting Materials and Media

Proper use of supporting materials can have a significant impact on the effectiveness of a presentation. Straight lecture for even eight hours will usually result in boredom and poor retention. Eight hours of videos typically creates the same result. Effective instructors will efficiently blend supporting materials into the lecture to deliver a well-balanced course that fosters a high level of retention by the trainees.

Several firms provide high-quality health and safety videos, many of which are specifically designed to address HAZWOPER issues. The typical video runs from 15 to 25 minutes, so it is not unreasonable to show three or four relevant videos during an eight-hour session. For example, in refresher training, a pretest on a subject can be given prior to showing the video, with answers and discussion to follow. The test questions should be geared to the video so the trainees will be "watching" for the correct solutions.

Trainees also respond to well-designed transparencies, such as those that provide topical outlines of the subject under discussion. These can help keep the discussion focused and lead it toward the desired conclusion. Some trainers will include copies of the transparencies in the course manual, with sufficient blank space to permit the trainees to take notes under the topical headings.

Whenever possible, the instructor should have examples of equipment and protective clothing available during the lecture. Few things better help a trainee retain information than being able to visualize the item under discussion. The various types of classes require different types of "show-and-tell" items, such as respirators and cartridges, gloves, chemical-restraint suits, air-monitoring instruments, plug and patch kits, samples containers, labels, and the like. If the items are small and lightweight, they can be passed around the room. Larger items should be placed where trainees can examine them during a break.

Also effective is the "magic show," in which the instructor demonstrates the reactions that can occur when small quantities of incompatible chemicals are mixed (such as may occur in a transportation incident). Popular combinations are combustible liquids with oxidizers and acids with bases. Such exercises must be conducted in a well-ventilated space (or outdoors) using the smallest quantities of chemicals that will create a visible reaction. Instructors and participants must take care to avoid mixtures that will create a toxic reaction by-product and potentially generate a hazardous waste.

The course manual is arguably the most important supporting item, because it will be retained by the trainees for future reference. It is time-consuming to develop a high-quality course manual, and, unfortunately, some commercial courses show their lack of commitment by simply photocopying and binding together various government documents. Although some of these documents are excellent "stand-alone" resources, they tend not to flow well when scrambled together. A good manual augments original narrative text with charts, graphs, and excerpts from the better government documents.

Employers who are designing their own training program should take advantage of the wealth of information available on hazardous waste operations and emergency response. Many of the resources listed in the References section at the end of this chapter are from Appendix D of HAZWOPER.

Exercises

There are three basic types of training exercises that trainers should consider using when developing or evaluating a training course: (1) written classroom exercises, such as the development of a simple contingency or health and safety plan; (2) hands-on classroom exercises, such as monitoring instrument calibration or respirator fit-testing; and (3) field practicums, such as setting up work zones and a decontamination line or suiting up for simulated drum sampling or tank-patching exercises.

A 24-hour or 40-hour course should have each of these types of exercises included in the curriculum. Field practicums are labor-intensive; one instructor can't do it alone. Thus, practicums require a lot of equipment, but there is no better way to equate the classroom lessons with field operations. Some 40-hour courses have one full day dedicated to exercises, starting with the development of a health and safety plan in the classroom, then implementing the plan in field exercises; other courses build through the week to the field practicum as the culminating experience.

Another type of exercise that is effective in shorter courses, particularly in refresher courses in which trainees should already be familiar with the material,

is opening each topic with a short (10- to 15-question) quiz, then holding open discussion to answer the questions. Such discussion encourages participation, and lets trainees bring out their personal experiences for others to learn from.

Measuring Understanding: Exams and Field Practicums

It is essential for employers and trainers to measure the trainees' understanding of the course material. This will help the trainer evaluate the effectiveness of the course, because if a large percentage of the trainees miss a particular test question, there could have been a problem in delivering the material. For the employer, establishing a means of measuring understanding can limit liability exposure in the event of an injury or illness on the job.

All courses should culminate in an examination that reflects the materials presented. Forty-hour courses may have exams of 100 or more questions, whereas the 8-hour courses may feature exams of 40 or 50 questions. Although there is a statistical method for establishing a passing rate for a particular examination, most trainers establish a cutoff at 70% correct answers; a trainee with a score below 70 must either retake the course or undertake remedial studies prior to being certified. Examination cover sheets should have the trainee's name, date, social security number (as a unique identifier), and signature.

Some of the more sophisticated training providers also score the field practicum. Of course, this adds to an already labor-intensive exercise, but objective scoring can be a valuable tool.

Effective training records can be of great assistance in defending against injury and illness claims. For that reason, many employers are creating employee records of training, in which the employee certifies that certain training was received and understood. When used in conjunction with examination results, such records can assist an employer in establishing that an employee received and understood training that was designed to prevent occupational injuries and illnesses. Figure 9–2 shows a sample employee record of training.

Once the classroom sessions, field practicums, and examinations are finished, HAZWOPER requires that successful participants be issued certificates of completion. The certificate must describe the specific type of training provided and state the name of the

Employee Name (print): _____ Date: _____

Job Title: _____ SSN: _____

Course Title: Hazardous Waste Operations 8-Hour Refresher Training _____

Course Version: Version IV _____

Location of Training: _____ Date of Training: _____

Instructor: _____

I have received and understand the training as described below. This training included the following elements:

Subjects Covered	Initials

1. New Information and Control Measures
   a. Occupational Exposure Limits
   b. Interaction with OSHA
   c. Overview of Confined Space Procedures
   d. Health and Safety Plans
   e. Bloodborne Pathogens (Universal Precaution Procedures)
   f. Familiarization with HM 181, 126, and IATA
   g. Interactions with Subcontractors	

2. Review of Actual Incidents	

3. General Principles of Portable Fire Extinguishers (Video)
   a. Fire Classification
   b. Fire Extinguisher Methods
   c. Fire Extinguisher Use
   d. Operating Procedures
   e. Handling Fire Hazards	

4. Review Information
   a. Hazard Communication Standard
   b. Injury, Illness, and Prevention Program (California only)	

5. Course Closure
   a. Questions and Answers
   b. Final Exam	

Instructor's Signature: _____ Employee Signature: _____

Figure 9–2. Sample employee record of training. Source: Used with permission from Dames & Moore.

trainee and course instructor or head instructor. Although not specifically required by the standard, it makes sense to include the date(s) of the training, the name of the firm providing the training, and the location of the course. Many trainers include the trainee's social security number on the certificate as a unique identifier for tracking purposes.

INTERNATIONAL PERSPECTIVE

Hazardous waste activities are not unique to the United States. European nations, for example, are actively identifying and investigating hazardous waste sites; it appears that sites recently revealed in Eastern Europe will be particularly hazardous. Few countries have health and safety regulations similar in scope to HAZWOPER. However, many countries do have standards that reflect those seen in Table 9–A, which provide some impetus to conduct HAZWOPER-type training. Some American-based firms have undertaken training programs for their foreign-based personnel. Although it is possible, and indeed appropriate, for such programs to be conducted, the logistics associated with these activities can be daunting, particularly moving training equipment through customs in a timely fashion. Nevertheless, responsible employers will continue to provide appropriate health and safety training to those employees engaged in international hazardous waste operations.

SUMMARY

Health and safety training for hazardous waste activities has been required by OSHA since 1986, and many contractors have provided such training for their employees since the late 1970s. Properly implementing an appropriate training program remains a challenge for many employers, as turnover and annual refresher requirements create an ongoing need for additional training. By combining needs assessments with appropriate curriculum development and careful selection of instructors, employers can establish a dynamic training program capable of both enhancing regulatory compliance and reducing potential liability exposure.

REFERENCES*

National Institute for Occupational Safety and Health (NIOSH), Occupational Safety and Health Administration (OSHA), U.S. Coast Guard (USCG), and Environmental Protection Agency (EPA). *Occupational Safety and Health Guidance Manual for Hazardous Waste Site Activities.* Cincinnati, OH: NIOSH, 1985.

Occupational Health and Safety Administration (OSHA). *Hazardous Waste Operations and Emergency Response.* 54 FR 9294. Washington, DC: U.S. Government Printing Office, March 6, 1989; corrections in 55 FR 14072, April 13, 1990, and 56 FR 15832, April 18, 1991. Washington DC: U.S. Government Printing Office.

OSHA. *Accreditation of Training Programs for Hazardous Waste Operations; Notice of Proposed Rulemaking.* 55 FR 2776, January 26, 1990. Washington DC: U.S. Government Printing Office.

OSHA. *Hazardous Waste Operations and Emergency Response. Interim Final Rule.* 51 FR 45654, December 19, 1986. Washington DC: U.S. Government Printing Office.

U.S. Department of Transportation. *Emergency Response Guidebook.* Washington DC: U.S. DOT, 1987.

U.S. Environmental Protection Agency, Office of Emergency and Remedial Response, Hazardous Response Support Division, Environmental Response Team. *Standard Operating Safety Guides.* Washington DC: U.S. EPA, 1992.

*See the Training section in Appendix 2, Sources of Help, for recommended readings.

Pollution Prevention Approaches and Technologies

Gary F. Vajda, MS, PE
William L. Hall, PE
Gary R. Krieger, MD, MPH, DABT

Pollution prevention and waste management represent the two ends of the hazardous materials spectrum. Historically, "end-of-pipe" management (waste treatment) has been emphasized over pollution prevention, and substantial energy and expenditures have been devoted to various treatment technologies. Despite this effort, during the late 1980s and early 1990s, there was increasing recognition that pollution prevention emphasizing efficient use of raw materials represents a better long-term economic and environmental strategy. Because environmental spending in the United States is now approaching 2% of the gross national product (GNP), it is critical to develop and implement a strategy that reconciles economic growth and environmental protection. This chapter is dedicated to the two ends of the pollution problem: pollution prevention (waste minimization) and waste treatment technologies.

ROLE OF POLLUTION PREVENTION

As the 1990s progress, the development and implementation of an effective pollution prevention program will likely become the single most essential component of the successful corporate environmental program, with "success" being measured in terms of both compliance and costs. Furthermore, the most successful pollution prevention programs will be those that comprehend the entire manufacturing process, not just the wastes that are generated.

The concept of pollution prevention has been heralded as the environmental focus of the 1990s. But the idea is neither new nor unique. Whether it's called *zero discharge, waste minimization,* or *emission reduction,* the premise of prevention has been part of the environmental world at least since the enactment of the Clean Water Act (CWA) in 1967.

However, the time had come for the ideas of waste minimization and pollution prevention only upon passage of the stiff regulations of the Resource Conservation and Recovery Act (RCRA), its 1984 Amendments, and the liabilities of the Comprehensive Environmental Response, Compensation, and Liability Act (CERCLA). The Pollution Prevention Act (PPA) of 1990, the 1987 stormwater regulations under the CWA, and the Clean Air Act (CAA) Amendments of 1990 completed the statutory basis for a multimedia approach. (See also Chapter 3, United States Legal and Legislative Framework, for additional details.)

Total quality management (TQM) is another concept that can play a significant role in pollution prevention by optimizing the productivity and efficiency of an industry. When waste is recognized as a quality defect that can be reduced or eliminated, the circle is closed. By adapting traditional manufacturing quality control analytical methods to environmental factors, pollution prevention can be accomplished in the same manner as any other operational improvement.

Thus, the scope of pollution prevention goes beyond achieving an environmentally desirable state of affairs; rather it is driven by increasing quality, productivity, and eventually, profits. These objectives can be reached by integrating manufacturing and environmental issues.

DEFINITIONS

In its broadest context, *pollution prevention* is the reduction of a waste's volume and/or toxicity prior to discharge. Although this definition is generally modified to emphasize that pollution control (i.e., an end-of-pipe treatment) is not considered pollution prevention, a thorough and comprehensive evaluation of economic criteria will render this modification a moot point. This is because end-of-pipe treatment is almost always the most expensive alternative when true lifetime costs are considered. When it is not the more expensive option, an end-of-pipe solution should be considered.

Pollution prevention techniques traditionally fall under one of two categories—source reduction or recycling. Source reduction, in turn, consists of product changes or source control. Figure 10–1 illustrates the relationships among these terms.

Product change involves modification of the end product to eliminate a waste. Two examples of past product changes include the replacement of pump sprays for aerosols and a major fast food chain's widely publicized substitution of paper for plastic wrapping. *Source control* comprises modifications to the production process itself. This may involve the technology, the raw materials, the equipment, and/or operating practices. Changing from traditional spray painting to dry-powder painting is an example of technology modification. Closing the covers of unused solvent degreasers is an example of using operating practices to achieve source control. Substituting aqueous cleaners for chlorinated solvents represents a change of raw materials.

Recycling involves the on-site or off-site use, with or without treatment, of a waste. Recycling and re-use can be defined differently as closed-loop recycling versus open-loop recycling. Closed loop implies no further processing of waste material—it is fed directly into the new process. Open loop implies the waste or material must be processed prior to being reused. For example, the smelting of plating sludge in a primary smelter to recover metals is open-loop recycling (i.e., replacing ore).

Reclamation or *recovery* generally involves separating a particular component of a waste stream for reuse. The most common example is solvent recovery through distillation.

Treatment comprises the process(es) by which the characteristics of a waste are changed to reduce the material's volume and/or toxicity. However, treatment takes place after generation, not before, and is generally conducted at the end of pipe. Treatment is not, therefore, generally considered a pollution prevention technique, although the result may be the same.

The ability to recover and recycle wastes has become a major worldwide business. In European countries that are members of the Organization for Economic Cooperation and Development (OECD), approximately 453.5 million tons (500 million tonnes) of industrial and domestic wastes are produced annually. (See Chapter 4, International Legal and Legislative Framework, for further details on the OECD.) Of this total, about one-third is put through some type of recovery operation (Yakowitz, 1993). The remaining two-thirds have typically been disposed of by burial or incineration.

Worldwide, more than 272 million tons (300 million tonnes) of wastes are recovered on an annual basis and represent a multibillion dollar activity affecting 300,000–350,000 jobs (Yakowitz, 1993). Clearly, trade in international wastes represents a major business sector for OECD countries:

- Europe: 50% of the activity
- United States: 33%
- Canada: 10%
- Japan: 2%.

According to Yakowitz (1993), more than 126 million tons (140 million tonnes) of wastes destined for recovery operations (based on internationally agreed upon definitions) crossed European borders in 1990. In March 1994, European Union countries,

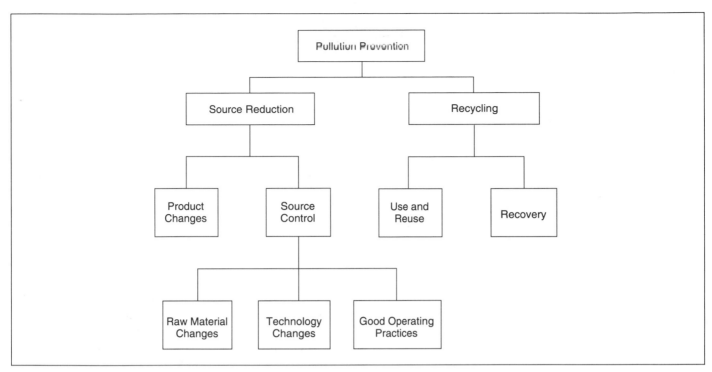

Figure 10–1. Pollution prevention components.

Canada, Japan, and Australia agreed to ban all exports of toxic wastes to developing countries, including exports intended for recycling. Further description of the international rules governing transfrontier movement and definitions of wastes are provided in Chapter 4, International Legal and Legislative Framework. U.S. regulatory definitions of hazardous wastes can be found in Chapter 8, Hazardous Wastes.

HIERARCHY OF WASTE MANAGEMENT

For any given waste stream, there is a general hierarchy of management options (Figure 10–2). These include four major categories:

- source reduction
- recovery/reuse
- waste exchange
- treatment/destruction/disposal.

Source Reduction

As noted previously, source reduction includes those techniques that actually decrease the quantity and/or toxicity of the waste generated by a given process. This should be the preferred management option. The major methods of reduction commonly used include:

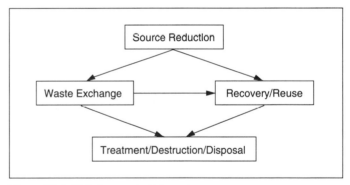

Figure 10–2. Pollution control hierarchy.

- product changes (see examples in Definitions section)
- revising operating practices
- process modifications (changes in raw materials, technology, equipment).

Revising operating practices. A good example of revising operating practices to minimize hazardous waste is *source segregation,* the separation of listed hazardous wastes from nonhazardous wastes. Because of the RCRA "mixture rule," any waste that combines listed hazardous and nonhazardous waste is classified in its entirety as a listed hazardous waste. In practice, this means that if a plant combines a small quantity of electroplating rinsewater treatment sludge (a listed hazardous waste) with a much larger amount of nonhazardous biological sludge, the entire mixture would have to be managed according to RCRA requirements. This is true even if the mixture is itself characteristically nonhazardous. By keeping these two waste streams separate, the facility can significantly reduce the quantity of hazardous waste it produces. As a reduction option, segregation is typically easy to implement and cost-effective. Other practices that should be reviewed include good housekeeping, inventory control, employee training, and spill/leak prevention and control.

Process modifications. Appropriate process modifications might include any of the following:

- technology changes
- equipment changes
- improved control
- procedure changes
- material substitutions.

Examples of process modifications include the replacement of a solvent-degreasing operation with an aqueous degreasing operation, the use of combustion rather than solvents to clean paint racks, and the addition of cooling coils to vapor degreasers to reduce solvent losses. Significant potential reductions can result from fundamental changes in processes, such as using dry-powder paints instead of solvent-based paints. Obviously, a primary consideration in evaluating a process modification is maintaining product quality. This, plus the fact that a substantial investment may be required, generally makes this approach a longer-term consideration.

Recovery/Reuse

Recovery/reuse techniques are most appropriate for two general classes of wastes: (1) relatively concentrated organic waste, such as solvents, coolants, and waste oils; and (2) metal-containing inorganic wastes. There is a growing interest in the use of off-specification materials either as raw materials for similar processes or as feedstocks to the production of the same material. There is little overlapping of the processes applicable for each.

For waste oil, halogenated solvents, and nonhalogenated solvents, four processes have been most commonly identified in conjunction with their potential for recovery. (1) *Distillation* is by far the most prevalent for solvent recovery because of the relative ease of separation by this means. Units of various sizes and temperature ranges are readily available for in-plant use. Distillation is also common for waste oil, although generally at a larger scale than is common to encounter at an individual plant because of the higher temperatures needed. (2) Until recently, *solvent extraction* has been limited to larger-scale uses and to a limited number of organics, particularly phenol. Advances in equipment have expanded this usage somewhat. (3) *Carbon adsorption* is most commonly used for vapor-phase solvent recovery and for low-concentration aqueous streams. (4) *Ultrafiltration*, a low-pressure membrane separation process, is applicable to cleaning coolant oils and recovery of tramp oil from aqueous coolant mixtures.

The most likely candidates among inorganic wastes for recovery/reuse include the caustic and acidic waste streams resulting from metal finishing operations (e.g., caustic cleaners, phosphate solutions, plating solutions, and pickle liquors). They, of course, have some immediate value as neutralizing agents, but other possibilities include recovery of concentrated metal salts from plating baths and the sale of ferric chloride or sulfate pickle liquors to municipal wastewater treatment plants for phosphate removal. When applicable, the advantage of this approach is twofold. Not only is a waste stream recovered for a useful purpose, but the metal content is reduced or eliminated from the wastewater treatment sludge where it would have inevitably ended up.

The processes that can be used for the recovery of metals include: *ion exchange; membrane processes* (reverse osmosis, ultrafiltration, electrolysis); and *evaporation*. All are proven processes, although not necessarily for all applications, and are readily available commercially.

Waste Exchange

A *waste exchange* is an administrative technique based on the premise that "one person's trash is another person's treasure." It is essentially a brokerage that handles wastes that another company could use as raw material. It then becomes identical to recovery/reuse. The feasibility of this operation depends largely on locale, i.e., the distance from a waste generator to a potential waste user. The potential for waste exchange for any specific waste material is a function of several criteria: the value of the raw material, its purity, the quantity available, and the reliability of generation. The most commonly exchanged wastes include solvents, oil, concentrated acids/alkalis, catalysts, and precious metals.

Treatment/Destruction/Disposal

After attempts are made to reduce waste, a residue that cannot be recovered, reused, exchanged, or otherwise managed is likely to remain. At this point, treatment/destruction/disposal options should be examined.

A hierarchy of methods should be considered. For organic materials, incineration or a boiler is an effective, if expensive, means of disposal. Inorganic chemicals and wastes generally require chemical/physical treatment technologies to reduce either the waste's volume or its toxicity. Only for residues that have been rendered essentially inert should landfill options be considered.

IMPLEMENTING A PROGRAM

Companies in every country on the planet earth should be interested in developing a pollution prevention program for several reasons. Economic and political motivations as well as regulatory compliance are major reasons. U.S. regulatory compliance is discussed in the next section and in Chapter 3, United States Legal and Legislative Framework. International regulations are discussed in Chapter 4, International Legal and Legislative Framework.

Economic Motivations

One of the most persuasive reasons to implement a pollution prevention program is economics. In the long term, prevention is less expensive than responding after the fact. These savings might be short term as well. For hazardous waste, there is limited waste management capacity, and as that capacity becomes filled, actual disposal costs will naturally rise. There are also the long-term "hidden" costs associated with the continued liability of generators of waste for any mismanagement (or even proper management in accordance with obsolete rules) of the wastes in years to come. These costs may be orders of magnitude greater than the original disposal costs. The situation is similar for emissions to water and air, primarily because there is an apparent trend to significantly increase permit and other fees on the basis of mass of pollutants being emitted.

There are additional economic pluses, some of which can be substantial. A company can use a properly implemented pollution prevention program to reduce raw material purchases, limit the diversion of productive capacity, and optimize work in progress. In particular, significant cost savings in inventory control can stem from two factors. First, less material has to be purchased and stored; second, disposal of obsolete materials is minimized because the company purchases only what is needed.

Other benefits that can result from an integrated approach to pollution prevention include:

- more effective process control
- increased product quality
- better means of strategic planning to identify high-priority, high-return projects or alternatives.

All of these economic benefits can result from understanding that waste is a quality defect, and then programmatically eliminating that defect in the manufacturing process.

Political Motivations

Pollution prevention is both the politically and economically correct thing to do in the 1990s and should not lose its political properness in years to come. This political correctness is a natural result of two growing trends in public attitude. The first is a desire to avoid dealing with waste in any shape, form, or manner, whether by incineration or landfill. This trend is manifested at one extreme in the series of battles between communities/environmentalists and companies that want to build or operate a waste disposal facility. At the other end of the spectrum, this trend is illustrated by the "greening" of the marketplace, i.e., the use of materials that are supposed to biodegrade or otherwise just disappear when disposed.

The second trend is a general concern for conserving natural resources. The "green market" is also an example of this, as is the current emphasis on recycling, reusing, or recovering household items once simply tossed away. It is not limited to household items: Firms on the cutting edge such as ESPIRIT, a clothing designer and manufacturer, are purposely offering a line of clothes made with recycled goods and manufactured with environmentally friendly substances.

Because pollution prevention directly or indirectly positively deals with both of these trends, a successful program offers substantial public relations value. Although not easily calculable, this value is obvious in better community relations, improved sales figures, and enhanced relations with federal agencies.

REGULATORY COMPLIANCE

During the last two decades, but particularly in the early 1990s, Congress and the U.S. Environmental

Protection Agency (EPA) have developed and begun to implement a number of statutes and regulations that directly or indirectly foster pollution prevention. Many states have followed suit. The means of pollution prevention have included both "large stick" and "carrot" approaches.

Major Regulations

By far the largest "sticks" for hazardous wastes have been the RCRA and CERCLA (Superfund) and their subsequent reauthorizing amendments, the Hazardous and Solid Waste Amendments of 1984 (HSWA), and the Superfund Amendments and Reauthorization Act of 1986 (SARA). The first act decreed what to do with wastes; Superfund indicated what it would cost if wastes were not correctly managed.

During the past several years, HSWA has mandated a number of regulations that significantly impacted how companies manage their hazardous wastes. These regulations resulted in the following: land disposal prohibitions and specified treatment technologies, an expanded listing of wastes categorized as hazardous; strict management and design criteria for treatment and disposal facilities, and increased responsibilities for the waste generator. There is no doubt that the RCRA's regulations will continue and that its emphasis will be on eliminating rather than controlling waste.

The U.S. EPA has completed its mandated schedule to review and determine which hazardous wastes will be permitted to be disposed on land and under what conditions. Experience indicates that problems arise when generators discover that their waste is prohibited from routine land disposal and/or does not meet the specified treatment criteria. Generators rush to find alternative means of disposal and then receive the bill for their waste disposal.

At the same time that the disposal of hazardous wastes is being restricted, the number of wastes identified by the U.S. EPA as hazardous via listings or characteristics is expanding. New information on specific wastes and new characterization parameters and analytic laboratory methods have resulted in the addition of more waste streams to the hazardous category.

Under HSWA, the federal regulatory requirement for waste minimization sounds simple. Generators must certify on their manifests that they have in place a program ". . . to reduce the volume and toxicity of wastes generated to the degree . . . practicable." Generators must also describe in biannual (or annual) reports, the contents and results of their waste minimization programs. The determination of what is achievable is generally made on the basis of short-term economic impacts on the company itself. To date, this requirement has been relatively toothless.

The U.S. EPA is currently reviewing its options for providing increased enforcement and/or encouragement to companies to do more than superficially comply with this requirement. These options have included the inception of an Office of Pollution Prevention, initiation of various research and development programs, and RCRA inspections of waste generators to ensure that they comply with the waste minimization requirement.

CERCLA is responsible for the "hidden" costs of waste management referred to earlier. The Superfund places retroactive responsibility for problems resulting from the past waste management practices on the generator of the waste, regardless of fault in causing the problem. The costs to investigate, negotiate, and remediate these problems have been staggering. (See also Chapter 3, United States Legal and Legislative Framework, for additional information on regulations.)

Emergency Planning and Community Right-to-Know Act

As part of the Superfund Amendments and Reauthorization Act of 1986 under Title III, the Emergency Planning and Community Right-to-Know Act (EPCRA) has specific requirements affecting pollution prevention efforts:

- Congress established the right of communities (state and local) to know what chemicals are used, stored, disposed, and released at facilities. The act and regulations require owner/operators who manufacture, process, use, or store chemicals to report the information on an inventory form to the federal or state government.
- EPCRA and regulations also require annual reporting for certain manufacturing or production facilities of the chemicals used, released to the environment, or recycled. Form R requires facility owners/operators to perform a mass balance in order to report the information.
- The PPA includes reporting requirements linked to the Toxic Release Inventory (TRI) reporting requirements under SARA Section 313 (for certain manufacturing facilities). The reporting practices include the quantity of chemicals entering any waste stream, the amount of chemicals recycled

during the previous year, and the source reduction practices used [PPA (§) 6607(b)].

Under the Section 313 ("Form R") program, companies have been forced to perform at least rudimentary mass balances of their facilities to determine what comes in, what leaves, and where it goes. This regulation includes emissions to all media, not just solid waste. Examples of potential information sources for preparing a material balance inventory include:

- batch make-up records
- design material balances
- emission inventories
- equipment cleaning and validation procedures
- material inventories
- operating logs
- operating procedures and manuals
- production records
- samples, analyses, and flow measurements
- waste manifests.

Compliance with the Form R program can involve substantial effort and may require the use of outside expert consultation.

The U.S. EPA and Congress have also embarked on a program of encouraging and/or mandating the concept of pollution prevention. The PPA of 1990 established a new national policy for environmental protection "that pollution should be prevented or reduced at the source whenever feasible. . . ." The PPA represents a substantial shift in the U.S. EPA's thinking. This paradigm shift is due to a combination of factors:

- Treatment and disposal options are costly.
- The waste problem is perceived as simply shifted from one part of the environment to another.
- Siting and permitting new hazardous waste treatment, storage, and disposal facilities are difficult.

The PPA represents a Congressional desire to establish a multimedia approach to pollution prevention as a national policy. Specifically, the act requires:

- a national policy on pollution prevention
- pollution prevention activities conducted by the U.S. EPA
- matching grants to states for technical assistance programs
- a source reduction clearinghouse

- source reduction and recycling data reported by businesses in their annual toxic chemical release reports
- biennial reports by U.S. EPA to Congress on pollution prevention activities and results.

The PPA established a hierarchy of pollution prevention management options:

1. Pollutants should be prevented or reduced at the source whenever feasible.
2. Pollutants that cannot be prevented should be recycled in an environmentally safe manner.
3. Disposal or other release of pollutants to the environment should be employed only as a last resort and should be conducted in an environmentally safe manner.

Source Reduction

Source reduction is any practice that reduces the amount of hazardous substance or pollutant entering any waste stream or the environment and reduces hazards to the public health and environment that would otherwise be associated with the release [PPA § 6603(5)(A)]. Source reductions can be achieved through a variety of means:

- equipment or technology modifications
- process or procedure modifications
- reformulation or redesign of products
- substitution of raw materials
- improvement in housekeeping and maintenance
- training
- inventory control systems.

In addition to the PPA of 1990, pollution prevention incentives are found in the Clean Air Act Amendments of 1990. It is possible to delay compliance with future standards by agreeing and implementing source reduction now. Pollution prevention plans are also important components of the two most recent regulations from the CWA: stormwater permitting and pretreatment.

Objectives and Goals

The regulatory trends are, therefore, apparent. In the remainder of the 1990s, generators are going to have to concentrate on prevention and not just proper treatment and disposal. The objectives of an effective waste management program should be to:

- minimize waste generation
- minimize off-site disposal

- minimize the long-term liability associated with such disposal.

The goal of a pollution prevention program is to address these objectives in a cost-effective manner. Reducing the generation of wastes at the source is preferred, both from the regulator's and from the generator's perspectives. However, most processes and operations will still produce some waste that must be managed. In many instances, a residue will remain that can only be disposed on land. Off-site disposal, particularly in landfills, should be considered only in the context of a careful review of all other options. The objective should then be to ensure that this residual matter is as environmentally harmless as possible.

POTENTIAL OBSTACLES

Given all of the reasons for implementing a pollution prevention program, surprisingly little has been done. Although almost everyone agrees that pollution prevention is essential, a number of obstacles arise between the plan's intention and successful implementation. In most operations, responsibilities are fragmented. Departments are run with different agendas and objectives, and communications within the facility are often ineffective. Even if the company has an environmental affairs department, responsibilities might be divided by media (i.e., air, water, and waste). This results in potential barriers to a successful pollution prevention program. These barriers can be categorized as follows:

- **Economic**
  - Not enough money is provided to fund the project.
  - Existing stocks of raw materials can delay substitutes that may help the program.
  - Using a new material can mean other production changes.
- **Jurisdictional**
  - Use of a different raw material can have an adverse impact on the process or change how material is handled.
  - Adequate space, utilities, manpower, etc., may not be available.
  - Accepting another plant's feedstock can involve extensive regulatory difficulties.
  - New processes can mean more regulation.
- **Production**
  - Production will be stopped for installation.
  - The new equipment may not work.

- Product quality can deteriorate.
- Customers may not accept the new product characteristics.
- **Attitude**
  - If it isn't broke, don't fix it!
  - A new procedure will be a bottleneck.
  - The changes are not worth the effort by the employees.

These concerns can be valid objections, but they can be overcome. To be effective, a pollution prevention program must put together all of the pieces of this puzzle. One way to do so is by using an interdisciplinary approach to analyze the company's production factors: the integrated manufacturing and environmental (IME) approach to pollution prevention. Its initiation may require training people or obtaining outside assistance.

INTEGRATING MANUFACTURING AND ENVIRONMENTAL ISSUES

For many years, factors attributable to environmental issues were not considered significant components of productivity and operating costs; instead, management focused on labor, materials, and rework. Today, however, environmentally driven costs—ranging from waste disposal to future Superfund liabilities, to the permitting costs of expanding a production operation—must also be considered by plant and corporate management.

An approach to accomplishing this integration is found in several of the techniques and methods used historically to analyze manufacturing operations for productivity and quality improvement. These methodologies include functional systems analysis, statistical sampling programs, design of experiments, traditional alternatives analysis, and structured decision-making techniques. When applied to the objective of pollution prevention, the IME approach enables the company to evaluate the entire manufacturing system, not just the components.

Setting the Stage

It is essential to set the stage for success before the integration team tackles technical aspects. As with any potentially controversial program, good planning and organization can mean the difference between achieving success and confirming the predictions of nay-sayers. Experience has demonstrated that five steps are almost invariably present in a successful pollution program:

1. Obtain the commitment.
2. Establish the goals.
3. Identify the champion(s).
4. Staff the team.
5. Reward accomplishments.

The key to a successful pollution prevention program is a company-wide commitment from the management, implementors, and employees (see Chapter 6, Managing Environmental Resources). It is absolutely essential for all three groups to buy into the program. Management is necessary to provide resources—both dollars and people—and symbolic support. Employees are the ones who actually implement the program and who are most comfortable with the status quo. The most elaborate of plans will not succeed unless the employees on the production floor actively participate. Finally, the implementors are needed to keep the process always moving forward, despite the inevitable glitches that occur.

The second step in gaining commitment is to establish the overall goals of the program. A goals list should be developed first by senior management and then filtered down. The senior management-stated goal(s) may start as relatively broad and generic, then be revised and become more specific. Whether large or small, these goals must be acceptable, achievable, and measurable. People involved must believe that where they are going is reachable, and that there is a verifiable means to document progress. Once set, these program goals can then be incorporated into individual division, department, and/or employee goals.

The third step is to identify one or more individuals to lead the team putting the program in place. The team leader(s) must be familiar with the facility, with waste management and production technologies, and with product requirements. He or she must also have good rapport with both management and employees and be able to bring a diverse group of people together.

Finally, the project team must have input from many departments and specialties, including:

- production
- facilities/maintenance
- process engineering
- quality control
- environmental
- research and development

- safety and health
- marketing
- purchasing
- material control
- legal
- finance
- information systems.

Often, outside assistance can provide expertise and/or a broader and objective perspective. However, it is essential for key input and direction to be provided internally and to be rewarded.

Baseline Characterization

Once the goals and team have been established, the team can perform a baseline characterization of the facility. The purpose of this effort is to define the production process(es), as well as to ascertain the current regulatory status of the facility. This assessment can focus on a particular process or a single medium, but is most effective when it evaluates the facility as a whole in order to highlight potential opportunities. It is essential to realistically characterize the operations and establish an accurate baseline. It is impossible to know how much progress has been made without knowing the starting point.

In the baseline characterization, both environmental and production data is evaluated. The following types of information should be compiled:

- process flowcharts
- piping and instrumentation diagrams
- plant layouts
- equipment design and operating specifications
- labor usage
- labor and utility costs
- raw materials usage and costs
- effluent and emission data
- production rates
- standard operating practices
- reject rates
- regulatory requirements, current and future.

Much of this information should be readily available; some may require sampling or other means of direct measurement.

This information is reassembled in a "model" that integrates and describes the environmental and operational aspects of the facility or process. This model need only be as complicated as is required for a specific circumstance. For example, a complex computer model is neither warranted nor appropriate for a simple plating operation, no matter how large. Conversely, to attempt to assimilate the necessary information for a large, multioperation manufacturing or chemical plant without the aid of at least a simple data management software package is not realistic.

In essence, this model should represent a "resource balance" for the facility or process. It includes not only materials, but also energy, labor, and any design constraints.

For smaller or more structured operations, this should not be too problematic. A good starting point is Section 313, Form R. However, to this must be added the nonmaterial aspects of the balance discussed previously.

For smaller or more focused projects, individual processes and functions can be analyzed using several analytical techniques. Proven methods include:

- statistical sampling programs or charting of historic data

- Taguchi and other design of experiments process diagnostic techniques.

For complex operations, more exotic modeling tools may be appropriate. One useful analytical technique is function analysis. Function analysis permits defining any process by a series of functions or nodes (Figure 10–3). Each node can be described in terms of inputs, outputs, the controls or constraints on the function, and the mechanisms of transformation. An example of a function analysis tool is Integrated Computer-Aided Manufacturing Definition Methodology (IDEF).

In addition to the resource balance, an accurate cost estimate for the product or process is necessary. Any number of techniques can help the team to accomplish this, from simple to complex. Figure 10–4 gives a sample of breaking down components in a product's process. It is important to remember that the accuracy of the result (i.e., the "true" cost) depends on the completeness and accuracy of the input

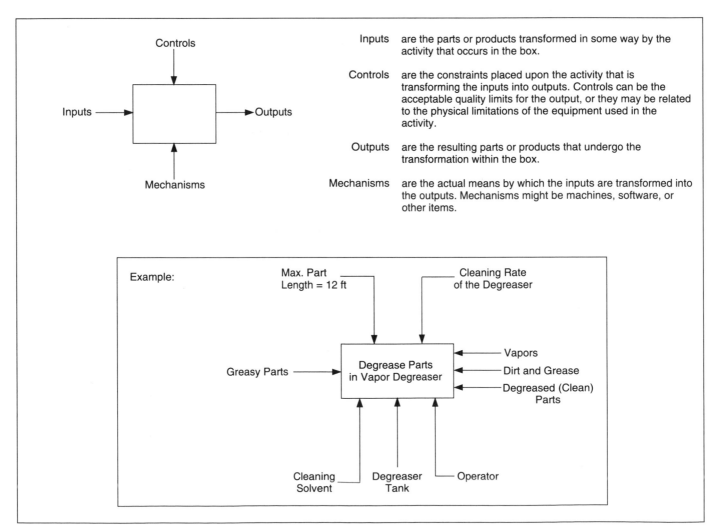

Figure 10–3. Functional analysis diagram.

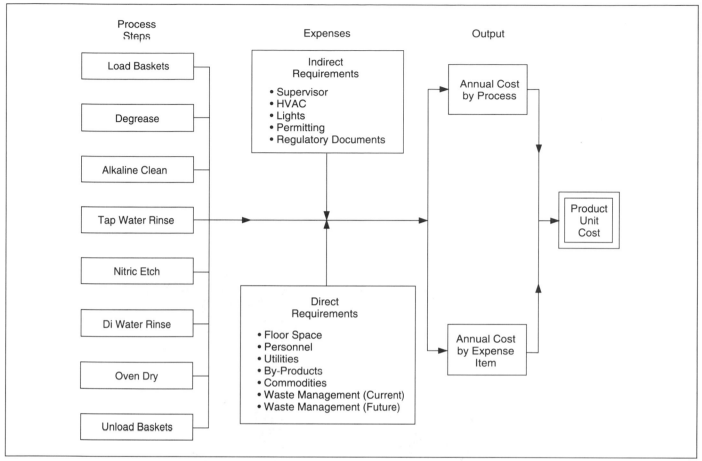

Figure 10–4. Components of a product's cost.

criteria. For example, often left out of traditional product cost analyses are costs for items such as regulatory documentation and compliance (time and materials); waste disposal; disposal liability; and the cost of product rejects (material, labor, lost capacity). Figure 10–5 shows these true costs.

The added value of a baseline cost model is to help identify high-cost operations. Analytical resources can then be focused on those operations or variables that yield the greatest return, environmentally and/or operationally. Among the modeling tools used for complex efforts is a software package called the Standard Assembly Line Manufacturing Industrial Simulation (SAMIS), which calculates a unit production cost for a process or system. The value can then be used to compare costs of the improvement alternatives developed in the next phase.

Alternatives Development

Once the team completes and validates the baseline characterization model(s), team members can develop and evaluate various production improvement and environmental compliance alternatives. Typically, a range of alternatives that partially or completely achieve the previously established pollution prevention goals is identified. Each alternative is then modeled, using the same techniques as in the baseline characterization.

The alternatives can be compared in a number of ways. Ideally, the favored approach to decision making in an uncertain environment is a blend of qualitative and quantitative methods to maximize benefits. Traditional techniques, such as rate of return, minimum capital, unit cost, or life-cycle cost may be appropriate. However, a methodology known as structured decision making can play an important role, particularly when the process or the alternative is complex or the selection cannot be made solely on cost factors.

In general, for structured decision making, an alternative's viability depends on its respective environmental and operational improvement potentials. These can be judged from a number of factors deemed appropriate by the user and may include:

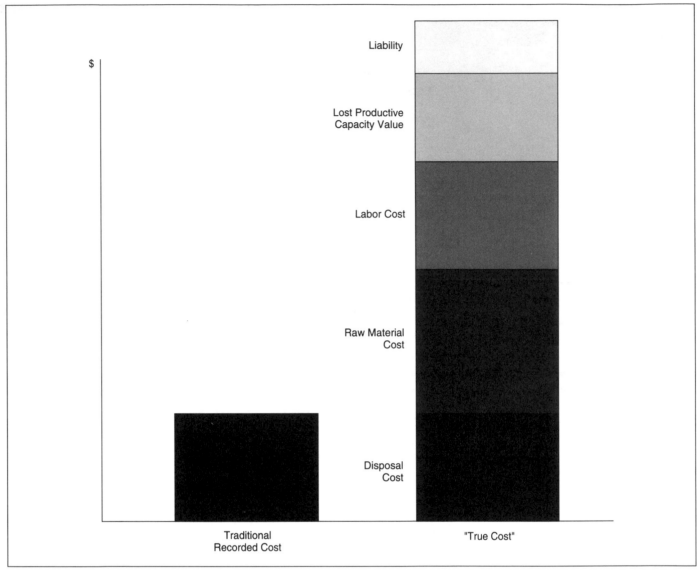

Figure 10–5. Comparing traditional versus true costs.

- Environmental improvement:
  - amount and type of pollutant
  - potential for disruption or legal action
  - level of mitigation technology
  - compliance safety margin provided by control technology
  - permissibility of technology.
- Operational improvement:
  - contribution to increasing inventory turns or reducing lead times (speed)
  - reduction in setup times, product differentiation, and mix (flexibility)
  - reduction of scrap and/or rework (quality)
  - lower operating and capital costs.

Alternatives of these calculations are provided in Figures 10–6 and 10–7.

In Figure 10-8, three alternatives illustrate a structured decision-making process. The initial cost of each alternative is indicated as proportional to the size of the circle. Alternative 1 is the least expensive and provides some environmental improvement but little operational improvement. This alternative represents a typical end-of-pipe pollution control device, such as a solvent fume scrubber.

Alternative 3, on the other hand, involves a significantly higher capital expenditure, which also significantly improves operations as well as the environment. An example might be the replacement of solvent-painting operations at this facility with a dry-powder painting process. Alternative 2, although costing just a little more than Alternative 1, is much better for the environment and improves operations somewhat. This alternative might represent

(1) Criteria	(2) Weight (0–1)	(3) Baseline Performance	(4) Alternatives Degrade — No Charge — Improve −10pts — 0 — +10 pts	(5) Improvement Score [(2) X (4)]
Pollutant	0.2	1,1,1-TCE 2,000 ppm	500 ppm + 7 pts	1.4
Disruption	0.3	Moderate	Low + 5 pts	1.5
Technology	0.1	None	BACT + 9 pts	0.8
Safety Margin	0.1	None	High + 50% + 8 pts	0.9
Permeability	0.3	Yes	High + 8 pts	2.4
Total	1.0	Improvement Index		7.0

Figure 10–6. The data assist in evaluating the potential for environmental improvement if a solvent dip tank is replaced by a vapor degreaser.

(1) Criteria	(2) Weight (0–1)	(3) Baseline Performance	(4) Alternatives Degrade — No Charge — Improve −10pts — 0 — +10 pts	(5) Improvement Score [(2) X (4)]
Cost	0.4	$10/part	$8/part + 20% + 2pts	0.8
Speed	0.2	14 day cue	5 days + 64% + 8pts	1.6
Quality	0.2	10/1,000	10/1,000 + 0% + 0pts	0
Flexibility	0.2	Limited	High + 8 pts	1.6
Total	1.0	Improvement Index		4.0

Figure 10–7. The data assist in evaluating the potential for operational improvement if a solvent dip tank is replaced by a vapor degreaser.

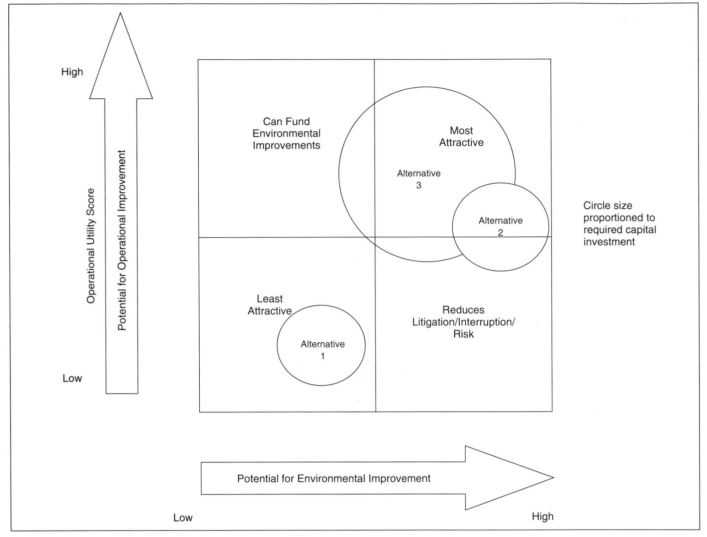

Figure 10–8. Graphical representation of input for structured decision making.

the use of activated carbon adsorbers to recover solvent fumes for reuse in the plant.

If capital cost is the only factor, then the short-sighted selection might be Alternative 1. However, if initial cost is only one major selection criteria, Alternative 2 is the apparent choice. Alternative 3 might be chosen when other factors (such as public relations, life-cycle costs, or unit costs) override initial capital costs.

An end-of-pipe solution to an environmental compliance problem usually achieves no benefit from operational improvement. In addition, negative cash flow is experienced from the purchase and operation of the new control technology equipment. However, a solution combining both operational improvement and environmental improvement can offer increased productivity with a return on invested capital. (See also Appendix 1, Case Studies 6, 7, and 8, for examples of alternatives analysis.)

WASTE MANAGEMENT TECHNOLOGIES

Before the mid-20th century, the natural environment of the United States seemed unbounded in its ability to absorb and cleanse itself of the ever-increasing trash and waste. Today, our ability to produce goods faster and more inexpensively has led to a habit of discarding items that in earlier times would have been repaired, saved for other uses, or returned to the environment in a beneficial manner. By the 1960s reports of fish disappearing from rivers, urban rivers and streams coated with oily sheens, and widespread warning signs prohibiting fishing or swimming began to alert thoughtful Americans to the long-term consequences of carelessness and neglect.

In the late 1970s, several dramatic cases alerted Americans to the potentially disastrous consequences of careless disposal of hazardous waste. New York's Love Canal was a residential area built over buried waste that contaminated the ground and nearby streams. Several hundred unsuspecting homeowners were potentially endangered by the leaking hazardous materials and ultimately were evacuated from their homes. Similar other high-profile cases, such as the Valley of the Drums in Kentucky and the dioxin exposure at Times Beach, Missouri, caught the nation's attention. In all these cases, human health and the environment were potentially threatened and lives were disrupted, in many cases irreparably. A body of federal and local laws and technologies for cleaning and controlling the volumes of waste produced is the legislative result.

The following section discusses wastewater, solid waste, and hazardous waste. For each type of waste, three areas will be covered:

- U.S. federal and state regulations (see Chapter 4, International Legal and Legislative Framework, for discussion of international regulations)
- sources and characteristics
- technologies available for management of wastes produced by both industry and private citizens.

WASTEWATER MANAGEMENT

Wastewater is water that has been changed chemically, biologically, and/or physically by human activity so that it is harmful to humans or natural systems. The United States uses about 400 billion gal (1,600 billion l) of water each day, or about 1,650 gal (6,600 l) for each person each day. The uses are everywhere: industrial processes, personal drinking water, cooking, cleaning, disposal of human and industrial waste, energy production, irrigation of food and nonfood crops, recreation, and livestock. For example, industry annually uses 100 cubic mi (approximately 400 cubic km) of water to manufacture its products. One cubic mile is over a trillion gallons (4 trillion l). According to data published by the U.S. EPA, it takes 1,400 gal (5,600 l) of water just to produce one fast food meal of a hamburger, french fries, and a soft drink (U.S. EPA, 1990a). One ear of corn needs more than 20 gal (80 l) of water before it can find its way to our dinner table.

As water moves through the hydrologic cycle, it is repeatedly used and its quality is changed. As water is used, human waste products, toxic substances from manufacturing, chemicals from agriculture, sediment from urban areas, and waste heat from energy production are potentially added. After each use, the user's challenge is to raise the water quality so it can be repeatedly used for both human purposes and support of the natural environment. (See Chapter 5, Basic Principles of Environmental Science, for more details on ground- and surface-water hydrology.)

Regulatory Framework

The foundation for the U.S. wastewater treatment and control program is the Federal Water Pollution Control Act Amendments of 1972 (USC 92–500), now known as the Clean Water Act (CWA 33 USC/25/a). This act set ambitious goals for the nation's waters. It envisioned that all U.S. waters would be maintained in conditions clean enough for preservation of natural habitat, maintenance of ecosystems, swimming, fishing, and other human recreational usages.

In order to meet the regulatory goals, the act required the U.S. EPA to establish water standards that would be met by all industries and cities. To ensure regulatory compliance, the CWA established a permitting program. This program, known as the National Pollutant Discharge Elimination System (NPDES, 40 CFR 122) prohibits the discharge of any pollutant to the waters of the United States unless the discharge is permitted through the NPDES program (see also Chapter 8, Hazardous Wastes).

Several important terms must be defined in order to understand the NPDES permit system and how it has changed since its beginning in 1972. Perhaps the most important concept is that of a point source. Originally the NPDES program focused on reducing pollutants in wastewater from industrial processes and from municipal sewage treatment plants. The term *point source* was used to define wastewaters that entered a stream or river at a "point," which may be a pipe, ditch, or open canal.

The program emphasis on point sources developed for a variety of reasons. The most important was the belief that the largest volume of contaminants originated from municipal and industrial facilities. Therefore, these facilities had the money and expertise to remove pollutants from their wastewater. As the United States made significant improvement in the quality of discharged industrial and municipal wastewater, it became clear that *nonpoint* sources of water pollution, such as agricultural and urban runoff, significantly contribute to water quality problems.

In response to growing concerns about the environmental impact of stormwater runoff, as part of the Water Quality Act of 1987 (WQA, 33 USC/25/b) the U.S. Congress required the establishment of a two-phased approach for the control of stormwater discharges. Phase 1 was directed at stormwater run-off from industrial facilities and large municipalities, and control was to be achieved by a modification to the NPDES permit system by 1992.

As envisioned by the 1987 WQA, Phase 2 of the stormwater program covers all stormwater discharges not addressed under Phase 1. This would include stormwater discharge from gas/auto service industries, highway systems, large land use developments, agricultural sources, and small municipalities.

In order to obtain a prioritization for dealing with stormwater runoff under Phase 2, the WQA required the states to prepare Nonpoint Source Assessment Reports. These reports identify state waters that cannot be returned to clean conditions without additional control of stormwater runoff. This information guides the U.S. EPA as it establishes an additional body of regulations for the Phase 2 sources by the mid-1990s.

The CWA as well as the WQA of 1987 were federal laws that established national standards for dealing with wastewater management. The federal laws delegated management of the NPDES system to states that wished to maintain control over water quality within their respective borders. By 1993, all but 12 of the states and four territories of the United States had established state programs and were implementing the federal requirements under state law. The various states and territories that still have federal management of their water quality programs are:

- States
  Alaska
  Arizona
  District of Columbia
  Florida
  Idaho
  Louisiana
  Massachusetts
  New Hampshire
  New Mexico
  Oklahoma
  South Dakota
  Texas
- Territories
  American Samoa
  Guam
  Northern Mariana Islands
  Puerto Rico

Sources and Characteristics

Most water pollution comes from three sources: sewage treatment plants; industrial discharges; and stormwater runoff from urban areas, agricultural lands, large developments, and highway systems. Each source's characteristics are unique with regard to the quality of the wastewater and the procedures that are available for management and removal of the water pollutants. An overview of each category is provided next.

Sewage. Sewage consists of wastewater generated from the sanitary wastewater collection systems owned and operated by municipalities. The waste is predominantly domestic wastewater from households and commercial businesses. This domestic wastewater is generated from human waste, cleaning water, and miscellaneous organic and toxic materials disposed of through the sanitary sewer system.

In addition to the domestic waste, sewage usually contains some pollutants generated from industrial facilities. Frequently, industrial facilities discharge into a municipal treatment system rather than directly into a receiving stream or river. To control these discharges, the NPDES program requires industries to "pretreat" the waste before releasing it into a sewage system. This pretreating generally prevents the introduction of toxic pollutants into municipal treatment facilities.

The principal types of pollution found within wastewater are total solids, metals, suspended solids, biological oxygen-demanding material (BOD), chemical oxygen-demanding material (COD), coliform bacteria, acidic or basic pH, oil, grease, and phosphorus. These pollutants, along with the quantity of dissolved oxygen in the wastewater, are used as indicators of the overall quality of the wastewater before and after treatment.

Total solids refers to the residue left in a drying dish after evaporation of a sample of wastewater and subsequent drying in an oven. By contrast, the term *total suspended solids* refers to the residue retained on an ultrafine glass-fiber filter after a sample of wastewater has been passed through the filter.

Biological oxygen demand (BOD) is a measure of the potential impact of wastewater organic matter on a stream or river. High levels of organics deplete the oxygen within a water body. As the oxygen is consumed by biological activity, the ability of a body of water to support healthy populations of fish is diminished. High BOD concentrations also lead to the rapid growth of algae which, in turn, lowers oxygen levels even further.

Chemical oxygen demand (COD) is a measure of the oxygen that would be removed from a water body due to chemical reactions. The COD is a measure of wastewater pollutants that are either not readily biodegradable or that contain compounds that may prevent biological activity from proceeding.

Coliform bacteria counts are an indicator of the disease potential of a wastewater stream. Although coliform bacteria are not necessarily pathogens or they are naturally present in the feces of warm-blooded animals, the presence of large numbers of coliform is an indication that other disease-carrying bacteria can be present.

pH is a measure of the number of disassociated hydrogen ions within a water body. The pH can range from acidic conditions (below 5) to alkaline (above 10). The U.S. EPA criteria for pH is 6.5 to 9. As the pH shifts, compounds such as ammonium may be converted to much more poisonous forms such as un-ionized ammonia.

Oil and grease contaminants encompass a large range of organic compounds. The common sources for oil and grease are petroleum products as well as fats from vegetable oil and meat processing.

Phosphate phosphorus is a nutrient that will stimulate plant growth such as weeds and algae in receiving water bodies. The excessive growth can lead to depletion of oxygen, reduction of sunlight penetration into the water body, and ultimately undesirable changes in the aquatic ecosystem. Spectacular algal blooms have occurred in the Bay of Venice (Italy) because of fertilizer runoff from agricultural sources, and red algal blooms have also episodically appeared in U.S. coastal areas.

Industrial discharges. Industrial discharges include discharges from a wide range of manufacturing facilities. A list of the industry categories as defined by U.S. EPA is as follows:

- adhesives and sealants
- aluminum forming
- auto and other laundries
- battery manufacturing
- coal mining
- coil coating
- copper forming
- electrical and electronic components
- electroplating
- explosives manufacturing
- foundries
- gum and wood chemicals
- inorganic chemicals manufacturing
- iron and steel manufacturing
- leather tanning and finishing
- mechanical products manufacturing
- nonferrous metals manufacturing
- ore mining
- organic chemicals manufacturing
- paint and ink formulation
- pesticides
- petroleum refining
- pharmaceutical preparations
- photographic equipment and supplies
- plastic and synthetic materials manufacturing
- plastics processing
- porcelain enameling
- printing and publishing
- pulp and paper mills
- rubber processing
- soap and detergent manufacturing
- steam electric power plants
- textile mills
- timber products processing.

Some of the categories of waste found in industrial wastewater are discussed below:

- *Spent solvents and solvent still bottoms* result from solvents used in cleaning and degreasing operations. The types of solvents used can include chlorinated solvents (for example, methylene chloride, dichlorobenzene, carbon tetrachloride, trichloroethylene) or nonhalogenated solvents (acetone, xylene, toluene, benzene). Other solvents include kerosene or mineral spirits.

- *Strong acid wastes* are generated wherever any type of metal is formed or processed. Industrial activities that generate acidic waste include metal drawing, rolling, pressing, electroplating, hot-dip galvanizing or hot tinning, anodizing, phosphating, metal coloring, and many more. Acid wastes are also produced from mining operations such as coal and clay. Clay processing can generate large waste streams from the washing of clay to remove impurities.

- *Strong alkaline wastes* are generated from the manufacture of products with aluminum and sometimes zinc. They are also produced from wood-pulping operations that are part of the papermaking process.

- *Plating wastes* generated from electroplating operations may be acidic or alkaline. They generally contain heavy metals. Acid plating solutions generally contain free acids and heavy metals such as copper, nickel, zinc, and possibly tin or cadmium.

Alkaline plating solutions include zinc baths and sometimes tin baths.

- *Heavy metals* are generated from a variety of operations. Contaminants include arsenic, barium, chromium, cadmium, lead, mercury, silver, or selenium. High concentrations of lead are found in battery manufacturing waste streams as well as in ammunition facilities. Other metal waste streams may come from grinding, tank cleanouts, dust collectors, and lead pots.
- *Paint wastes* are generated by paint booths used in the sign and advertising display industry as well as in various wood finishing industries (including furniture) and sports firearm manufacture. Generally, the waste is an organic sludge that contains cadmium, chromium, lead, and/or mercury. Similar wastes are generated in paint and ink formulations.
- *Cyanide wastes* are generated from cyanide plating solutions and simple cyanide solutions. Cyanide plating solutions are used in metal-plating operations. Simple cyanide solutions are used mainly for hardening and metal cleaning. Cyanide baths are common in metal-finishing and heat-treating operations.
- *Ignitable wastes* include any flammable or combustible liquid or combustible solid. Flammable materials have a flash point of less than 100 F (38 C); combustible materials have flash points of 100–200 F (38–93 C).
- *Reactive wastes* are generated primarily by the photographic equipment and supplies industry. Specific wastes may include strong oxidizing agents, such as chromic acid, perchlorates, and permanganates used in metal finishing, and other reactive compounds, such as hypochlorites, peroxides, sulfides, nitrates, cyanides, and sodium hydroxide.
- *Dyes* are discharged in relatively large volumes by the textile industry, especially in the manufacture of carpets.
- *Petroleum wastes* are generated by nearly all industrial operations. Oils may come from cutting, lubricating, and/or quenching as well as from leakage from pumps and motors.

Stormwater runoff. When rain falls on a city, water from streets, buildings, and other commercial areas flows into storm sewers, carrying with it all the oil, dust, and other wastes that have collected on the streets, sidewalks, and rooftops. In the same way, runoff from suburban areas and farms carries previously applied fertilizer, pesticides, and herbicides.

Airborne particulates (e.g., soot and ash) can also enter into water via rainfall or direct precipitation and settling. A prominent example of this phenomenon is acid rain (see Chapter 14, Global Issues, for a discussion of water resources impacted by distant emission sources).

Many of the waste products described for the industrial and municipal waste streams are also present in storm runoff. The effect on streams and rivers in many instances is magnified by the large volumes that enter a stream in short time periods. (See Chapter 5, Basic Principles of Environmental Science, for a discussion of surface water fate and transport.)

Wastewater Management Technologies

Industrial facilities. Wastewater produced by industrial facilities can be treated and disposed of in several ways. The simplest is direct release of the wastewater into the local municipal wastewater treatment facility. This option is becoming less acceptable as U.S. federal and state regulations become more stringent in governing the requirements for water released into a municipal wastewater treatment plant. The second option is pretreatment of the wastewater prior to discharge into a municipal system. Pretreatment usually consists of the removal of contaminants that are particularly difficult for a municipal wastewater treatment plant to manage. This usually includes contaminants such as heavy metals as well as strongly acidic or alkaline waste streams. In most instances oil and grease must also be removed.

The final option is treatment of the waste stream to stringent water-quality standards followed by release directly into a receiving water body. This option is generally the most expensive, but must be used in many cases by industrial facilities that are either not close to a municipal treatment plant or produce wastes that cannot be economically treated by the wastewater treatment plant.

The broad categories of treatment options for industrial wastewater are shown in Figure 10–9. Some or all of the treatment options can be used in a given situation. The treatment necessary for a given industrial wastewater is determined by the characteristics of the contaminants.

Sedimentation is used to treat industrial wastewater when the contaminants are heavier than water and will drop to the bottom if the velocity of the waste stream is sufficiently slowed. Examples include waste streams that contain metal filings or large volumes of grit.

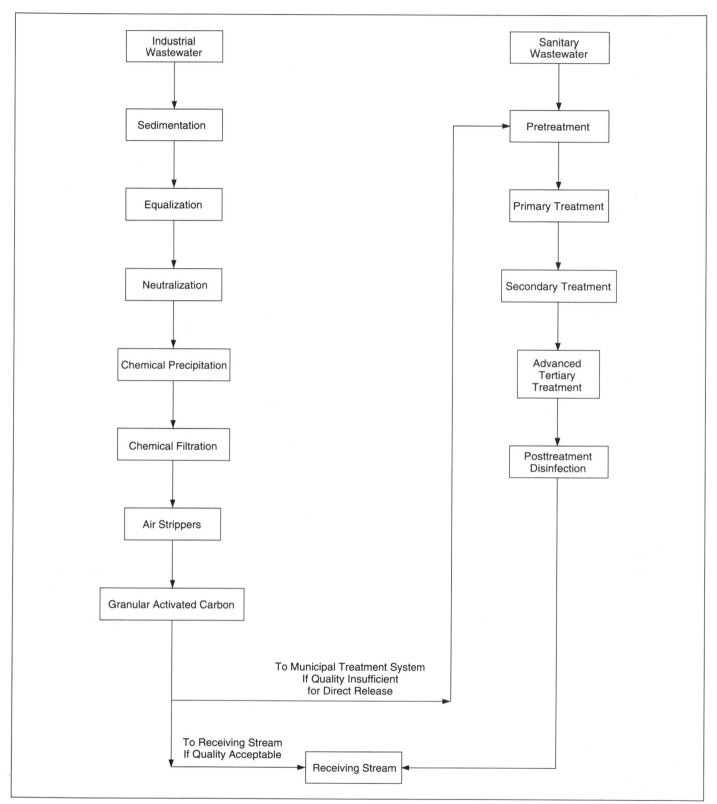

Figure 10–9. Wastewater treatment processes. (Courtesy Dames & Moore. Used with permission.)

Equalization is used by industries that have widely varying waste streams. By combining the waste streams, the toxicity can be reduced through dilution. An example of this technique is textile mill wastewater. This waste stream tends to have high fluctuations in pH and BOD material. Using holding tanks, the waste streams can be combined to provide a more consistent wastewater quality. This higher consistency is essential for subsequent treatment processes.

Neutralization is necessary in waste streams that have a wide range in pH that cannot be controlled through equalization. Waste streams that are either highly alkaline or acidic usually must be neutralized before other contaminants can be removed. This is particularly true of metal contaminants that remain dissolved in water only within limited pH ranges.

Chemical precipitation must be used when contaminants remain suspended or dissolved in water even when the water is completely still. Chemical precipitation removes contaminants through creation of molecules that are heavier than water, elimination of electrical charges that keep particles in suspension, or absorption of contaminants onto the surface of a treatment chemical added to the water. Commonly used chemicals for chemical precipitation are alum, ferric chloride, lime, and powdered activated carbon.

Filtration is used to screen out contaminants from a wastewater stream. Filtration ranges from coarse screens with mesh sizes that can be seen by the naked eye to ultrafine screens that filter molecular constituents.

Air strippers are principally used for light petroleum products such as those associated with gasoline. The process consists of discharging water at the top of a tower while simultaneously blowing air upward through the water stream. The passage of the air removes the volatile contaminants within the wastewater stream.

Granular activated carbon (GAC) is generally used to remove suspended and/or dissolved matter as well as tastes and odors in water. GAC, in many instances, can provide both filtration and absorption of contaminants.

Sanitary waste. Sanitary waste receives up to four stages of treatment, which are generally followed by disinfection prior to release of the wastewater into a receiving stream. Each level cleanses the wastewater to an increased degree. The levels are preliminary, primary, secondary, and advanced tertiary treatment.

Preliminary treatment removes larger objects such as rags, heavy solids, and sometimes excessive volumes of oils and grease. The technique for removal usually consists of trash racks, coarse screens, or grit chambers. The grit chambers are different from screens: large and heavy objects are removed by slowing the water down, allowing material to settle to the bottom where it can be removed.

Primary treatment is generally a mechanical process aimed principally at BOD solids. It consists of sedimentation basins and/or chemical sedimentation. Chemical sedimentation consists of the addition of a coagulant, such as furic chloride, lime, or sulfuric acid, to accelerate the rate at which solids drop out of the wastewater stream.

Secondary treatment uses activity of microorganisms to convert dissolved contaminants and nutrients into a biomass that can then be removed from the wastewater stream. Techniques used in secondary treatment include trickling filters, sand filters, stabilization ponds, and activated sludge. Trickling filters consist of beds of rocks over which wastewater is sprayed. Biological growth attached to the rocks removes wastes from the waste stream as it trickles through the filter. By contrast, sand filters consist of a sandbed up to 3 feet deep. The filter is kept clean by periodically backwashing with clean water and exposing the sand to air to allow the biological growth to decompose. In addition, the upper layer of sand is removed frequently. Stabilization ponds provide treatment by holding wastewater for several months to let aquatic plants, weeds, algae, and microorganisms consume the organic matter. Activated sludge consists of mixing settled sludge material from the wastewater with raw wastewater. Activated sludge contains high concentrations of microorganisms that accelerate the decomposition of the raw wastewater stream.

Advanced tertiary treatment removes fine suspended solids as well as nutrients. Fine suspended solids are removed by ultrafine filtration. The nutrient phosphorus is generally removed by chemical precipitation with aluminum and iron coagulants or lime. The nutrient nitrogen is removed by using bacteria that oxidize ammonium ions to nitrate and nitrite. These constituents can then be converted to nitrogen gas. Finally, tertiary treatment may include removal of inorganic salts by electrodialysis, reverse osmosis, or ion exchange.

SOLID WASTE MANAGEMENT

What is solid waste? A solid waste is any discarded material that is abandoned by being disposed of, burned, or incinerated, or else by being accumulated, stored, or treated before it is abandoned.

Regulatory Framework

The U.S. solid waste management program is built around a body of law and regulations that originate primarily from the Federal Resource Conservation and Recovery Act 1976, amended in 1978 (RCRA, 42 USC 6901). The law delegated the responsibility to the U.S. EPA to develop guidelines for solid waste management. The actual implementation of the standards is largely a function of state law and local municipality implementation.

The criteria for classification of solid waste facilities and practices was first published by the federal government in 1979. These standards provided minimum national performance standards to ensure that "no reasonable probability of adverse effects on the health and the environment will result from solid waste disposal facilities or practices. A facility or practice that meets the criteria is classified as a sanitary landfill" (RCRA, 42 USC 6901).

Early in the 1980s the U.S. Congress recognized the need for modifications to the nation's solid and hazardous waste management practices and passed the 1984 Hazardous and Solid Waste Amendments (HSWA, 42 USC 6901). In response to the requirements of HSWA, the U.S. EPA developed a national strategy for addressing the municipal solid waste management problems, which it published in *The Solid Waste Dilemma: An Agenda for Action* (U.S. EPA, 1989).

The strategy set out three national goals for municipal solid waste management: (1) enhance source reduction and recycling, (2) increase disposal capacity and improve secondary material markets, and (3) improve the safety of solid waste management facilities. To help attain the first goal, the U.S. EPA established a national goal of 25% source reduction and recycling of municipal solid waste by 1992. The second goal is pursued through progressive actions by the U.S. EPA to increase the amount of information available to all parties involved in solid waste management. The third goal is dealt with through federal guidelines for construction and maintenance of solid waste facilities.

Solid waste management facility design and operation is governed by rules and regulations established by the U.S. EPA in 1991 under Subtitle D of RCRA (42 USC 6901). These rules, which must be adopted by the states in a form at least as stringent as the federal standards, establish criteria for landfill liners, contaminated water collection and treatment, daily cover, and landfill closure.

The standards are written specifically to address municipal or sanitary solid waste landfills. Industrial nonhazardous solid waste is not directly addressed.

Currently, in most states, standards for industrial facilities closely resemble the Subtitle D standards. However, there is more flexibility in the design of a nonhazardous industrial waste facility. This greater flexibility results from the more consistent characteristics of an industrial solid waste. This higher degree of consistency enables both the regulatory agencies and the industry to more accurately estimate the potential risk to human health and the environment associated with the waste's contaminants.

Sources and Characteristics

In the 20th century, the United States has seen a steady increase in the volume of material that we find no longer usable or desirable. During the 1950s and 1960s, the U.S. economy was built on products possessing a relatively short life span. This encouraged an ever-increasing volume of production. As a nation, we currently produce about 180 million tons of municipal solid waste a year. The following statistics from the U.S. EPA provide a graphic indication of the nature of solid waste in the United States (U.S. EPA, 1990b):

- Every two weeks we throw away enough bottles and jars to fill the 1,350 ft (412 m) towers of New York's World Trade Center.
- We discard 31.6 million tons of yard waste each year.
- With the aluminum we throw away in three months, the United States could rebuild its entire commercial air fleet. We dispose of 2.5 million plastic bottles every hour (22 billion per year).
- With the office and writing paper we throw away every year, we could build a 12 ft (3.6 m) wall from Los Angeles to New York City.
- We discard more than 200 million tires every year (one for every person in the United States).

Solid Waste Management Technologies

The guiding principle for solid waste management in the United States that has evolved since the 1984 HSWA is integrated waste management. This concept recognizes that communities cannot rely exclusively, as they have in the past, on a single form of waste management. The concept is translated into action through the recognition that there is a hierarchy for waste management. The hierarchy (from most to least desirable) is source reduction, recycling, composting, energy recovery through combustion, and finally, landfilling.

Source reduction. *Source reduction* changes the handling of materials to reduce the volume of discarded materials. On the personal level, this concept is observable in such modern trends as buying food products in bulk rather than in individually wrapped containers. In the manufacturing process, it frequently involves modifying the way raw materials are used and converted into finished products in order to reduce the amount of material that must be disposed. A specific example is modifying the shape of steel sheets so that less material is wasted when patterns are cut from the sheets.

Recycling. Similar to source reduction in its impact on reducing the waste stream, *recycling* involves recovery of materials that have already been discarded or are unusable in their current form. Recyclables include a wide range of household and manufacturing materials that are no longer usable. Examples from the manufacturing industry include paper, metal cuttings, and used oil. Recyclables from households include glass, paper, cans, and plastic.

Composting. *Composting* is the conversion of organic matter to a form that is useful as a soil additive and/or fertilizer. A familiar form of composting is the backyard compost pile. The materials in a backyard compost pile are indicative of the items that can be removed from the solid waste stream. Leaves, grass clippings, and most forms of discarded foodstuffs can be composted successfully.

Combustion. *Combustion* technology can be used to convert portions of the waste stream that have heat value into usable energy. A common approach is the combination of a waste-to-energy facility with an adjacent manufacturing facility that uses steam. Such a combination, called *cogeneration,* takes advantage of the fact that electricity can be produced using steam heated by the combusted wastes. After it is used in production of electricity, the steam can be reused by the manufacturing facility, thereby obtaining double benefit from the solid waste material.

Landfilling. The final element of the integrated waste management approach is *landfilling*. It constitutes the final disposal of material that cannot be otherwise beneficially used, either because of the nature of the material or the local limitations to alternative approaches such as combustion. The technology for construction of landfills is well established. Under the requirements of RCRA Subtitle D, landfills must provide clay and/or synthetic liners to prevent escape of contaminants, leachate systems for collection of contaminated water from within the landfill, gas collection systems for capture of methane gas, and capping requirements following closure of the landfill.

HAZARDOUS WASTE MANAGEMENT

A waste material is considered a hazardous waste if it meets any of the following four conditions (McCoy, 1992):

- The waste exhibits any of the characteristics of ignitability, corrosivity, reactivity, or toxicity.
- The waste is specifically listed as being hazardous in one of the four tables contained in Subpart D of the RCRA.
- The waste is a mixture of a listed waste and a nonhazardous waste.
- The waste has been declared to be hazardous by a generator of the waste.

The materials listed as hazardous waste only become a hazardous waste when they are discarded. Prior to being discarded, the material is considered a raw material, intermediate product, or final product (see also Chapter 8, Hazardous Wastes).

Typical hazardous wastes are discarded pesticide products, solvents used for cleaning, residue containing high concentrations of metals from manufacturing operations that include grinding or cutting of metals products, and sludge from wood preservative operations.

Regulatory Framework

The management of hazardous waste is divided into two broad categories. The first is the management of hazardous waste associated with ongoing active facilities. The second is the management of hazardous waste that has been improperly disposed of or discarded in the past and must be cleaned up to eliminate risk to human health and the environment. This second category is regulated primarily under the Comprehensive Environmental Response, Compensation and Liability Act (CERCLA) of 1980 (42 USC 9601). See Chapter 3, United States Legal and Legislative Framework, for further information on CERCLA.

Active hazardous waste generators are regulated with respect to hazardous waste treatment and disposal under two bodies of regulations: RCRA and HSWA. The RCRA (42 USC 6901) established a program for monitoring and remediating hazardous releases to groundwater from hazardous waste management units. *Hazardous waste management units* include facilities such as surface impoundments, waste piles, land treatment areas, landfill cells, incinerators, tanks, and containment storage areas.

The RCRA regulations were supplemented in 1984 with the passage of the HSWA. The amendment strengthened the requirements for corrective action to clean up and reduce the toxicity of hazardous materials that may be disposed of into or onto the ground.

Hazardous Waste Management Technologies

The last section in this chapter gives detailed descriptions of standard and innovative treatment technologies. Treatment technologies for hazardous waste include both in-situ and ex-situ treatment for both water and soil. Direct treatment processes for hazardous liquid wastes include:

- activated carbon
- activated sludge
- filtration
- precipitation/flocculation
- sedimentation
- ion exchange
- reverse osmosis
- neutralization
- gravity reparation
- air stripping
- chemical oxidation
- chemical reduction.

In-situ refers to treatment technologies that can be applied without large-scale removal of the contaminated material from its existing location. *Ex-situ* refers to technologies that require removal and handling of the contaminated materials. These classes of techniques are described further in the following section based on the U.S. EPA's extensive treatment technologies publications (U.S. EPA, 1991a and b) and illustrated in Figure 10–10.

In-situ treatment. In-situ methods entail the use of chemicals, biological agents, or physical manipulations that detect, separate, or immobilize contaminants (U.S. EPA, 1982). The technologies can be further divided into one of three types: bioreclamation, chemical treatment, and physical methods.

Bioreclamation uses microorganisms that are capable of breaking down many hazardous organic compounds. Microbial activity is classified as either aerobic or anaerobic respiration. *Aerobic respiration*

occurs when oxygen is used to accept electrons. *Anaerobic respiration* refers to conditions in which oxygen is unavailable and sulfate or nitrate serves as an electron receptor.

For the more common hazardous waste compounds such as petroleum hydrocarbons, pesticides, and herbicides, aerobic techniques are the most practical. For lower molecular weight contaminants such as solvents, anaerobic degradation is preferable. The practicality of bioreclamation is affected by the availability of organic and inorganic nutrients to supply the microorganisms, the availability of oxygen, the ability to maintain water saturation, the soil characteristics that affect the water flow, and the types and concentration of contaminants. As experience with bioreclamation has improved, it has been learned that many of the critical factors such as nutrients and oxygen can be manipulated in-situ to achieve environmental conditions that provide rapid degradation of contaminants.

Chemical treatment in-situ consists of immobilizing or detoxifying the contaminants, or flushing them from the soil. Immobilization methods are designed to prevent the contaminants from dissolving into water that flows through the soil. This prevents removal of the contaminants from the soil matrix and thus prevents them from ultimately reaching a human receptor. Detoxification techniques destroy, degrade, or otherwise reduce the toxicity of contaminants in place. Soil flushing, by contrast, consists of washing contaminated soil to remove and collect the contaminants (Figure 10–11).

Immobilization methods currently in use include precipitation, chelation, and polymerization. Precipitation is the most promising method for immobilizing dissolved metals. The technique consists of adding a substance with an opposite electrical charge of the contaminant to be precipitated. The result is creation of a compound that is sufficiently immobile that it cannot move from the site and endanger off-site environments.

Chelating agents can also be effective for immobilizing metals. Stable metal chelates may be strongly bound to the soil to prevent movement.

Polymerization involves the injection of a compound into the groundwater that transforms the groundwater into a gel-like nonmobile mass. The difficulty with this technique is generally the inability to obtain widespread distribution of the compound injected into the groundwater.

Soil flushing consists of a variety of techniques ranging from solvent flushing, ground leaching, and

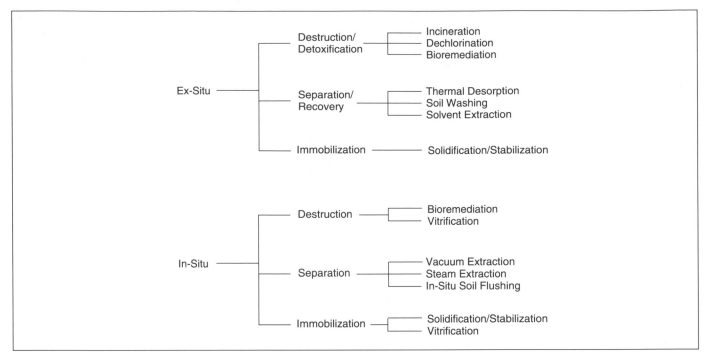

Figure 10–10. Treatment technologies for contaminated soils. Source: U.S. EPA, 1991.

Figure 10–11. Soil-washing unit. (Courtesy Dames & Moore. Used with permission.)

solution mining. The technique consists of injecting water or an aqueous solution into an area of contamination. Contaminants are mobilized into solution either because of their inherent solubility or the increased solubility associated with a chemical additive placed into the injected water. Water can be used to flush water-soluble or water-mobile organics or inorganics. A variation of soil flushing is the use of vapor extraction, in which unsaturated zones in the soil are placed under negative pressure to remove contaminants that are highly volatile.

Detoxification typically consists of neutralization or oxidation/reduction of a contaminant. *Neutralization* involves injecting diluted acids or bases into the groundwater to adjust the pH. The technique is effective in neutralizing acidic or basic groundwater contaminant plumes; often no other treatment is required. *Oxidation* and *reduction* reactions serve to alter the oxidation state of a compound through loss or gain of electrons, respectively. These reactions can serve to change the toxicity characteristics of organics and precipitate or detoxify metals.

Physical in-situ methods involve physical manipulation of the soils in place to immobilize or detoxify waste constituents. These technologies include in situ heating, vitrification, and ground freezing (U.S. EPA, 1985).

In-situ heating has been used to destroy or remove organic contaminants through thermal decomposition, vaporization, and distillation. *Vitrification* consists of converting contaminated soil to a durable glass-like material in which the wastes are crystallized. By contrast, artificial ground freezing has been used through the installation of freezing loops in the ground to immobilize contaminants in an ice matrix.

Ex-situ treatment. Ex-situ methods involve technologies for managing hazardous waste following removal. Generally, treatment technologies can be divided into different types for water and soil. Ex-situ soil treatment generally consists of either solidification/stabilization, thermal destruction, or landfilling in an approved hazardous waste management facility.

The application of these processes can be accomplished through on-site facilities constructed specifically for treatment of the site hazardous waste or by transportation of the removed liquid waste to an existing treatment facility. In addition, a variation that is often used provides partial treatment of the liquid hazardous waste prior to disposal in a municipal treatment facility.

Solidification and stabilization describe treatment systems that accomplish one or more of the following objectives (U.S. EPA, 1982):

- Improve waste handling or other physical characteristics of the waste.
- Decrease the surface area across which transfer or loss of contained pollutants can occur.
- Limit the solidability or toxicity of hazardous waste constituents.

Solidification generally produces a monolithic block of waste with high structural integrity. Typically, the contaminants are bound by being locked within the solidified matrix.

Thermal treatment technologies include incineration, low temperature thermal desorption, and hydrolytic dissipation. Incineration is the most common and most dependable—although usually the most expensive—method. This technology can be applied for organics and consists of burning of the soil to destroy the organic contaminants. *Low-temperature thermal desorption* consists of heating the soil to a temperature that boils the contaminants out of the soil. The contaminants are then captured in the offgases and treated through carbon absorption or incineration. *Hydrolytic dissipation* uses nutrients, organics, and sunlight to accelerate the decomposition and evaporation of organics from the soil. The next section provides detailed information about these different treatment technologies.

INNOVATIVE TREATMENT TECHNOLOGIES

This section describes 10 innovative treatment technologies: incineration, thermal desorption, soil washing, solvent extraction, dechlorination, base-catalyzed decomposition, bioremediation, vacuum extraction, in-situ vitrification, and groundwater treatment technologies. (The material in this section is adapted from U.S. EPA, 1991b.)

Incineration

Incineration uses temperatures ranging from 1,600–2,200 F (871–1,204 C) to volatize and combust (i.e., in the presence of oxygen) organic constituents in hazardous wastes (Figure 10–12). Three common incinerator designs are rotary kilns, infrared furnaces, and circulating fluidized bed incinerators. The destruction and removal efficiency (DRE) for properly operated incinerators often exceeds the 99.99% requirement for hazardous waste. All of these designs can be operated to meet the 99.9999% requirement

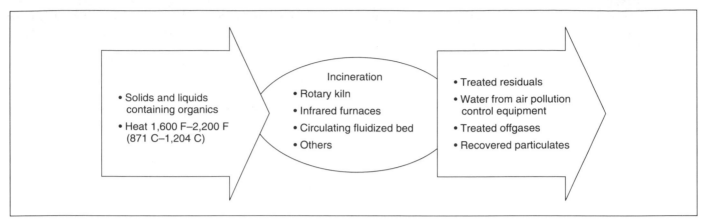

Figure 10–12. The incineration process. Source: U.S. EPA, 1991b.

for PCBs and dioxins. Incinerators usually have primary and secondary combustion units to reach this efficiency.

Incineration is effective in treating soils, sediments, sludges, liquids, and gases. It is used to treat all concentrations of organic constituents. It becomes more competitive when high concentrations are present and other treatments are unfeasible.

The current status of technology limits the capacity for off-site incineration of soils. Rotary kiln, infrared, and circulating fluidized bed incinerators have been successfully demonstrated and used in full-scale remediation (Figure 10–13).

Strengths. Incineration usually destroys organic contaminants in the residual soils with an efficiency greater than 99.9%. The treatment concentrates nonvolatile metal constituents in residues. It may also accomplish some volume reduction of soils, depending on the ash content.

Limitations. Incineration is more expensive than other technologies. Also, public resistance to this method has been high. In 1993, the U.S. EPA began an extensive review of the process of permitting new incinerators and renewing previously licensed units. The long-term implications of the U.S. EPA's combustion strategy are unknown and could significantly affect the long-term economic viability of this technique in the United States. Volatile metals, including lead and arsenic, leave the combustion unit with the flue gases and may have to be removed to avoid excessive emissions. Metals can react with other elements in the feed stream, such as chlorine or sulfur, forming more volatile and toxic compounds than the original. In addition, sodium and potassium in the wastes can attack the brick lining of the thermal unit (a process known as *refractory attack*) and form a sticky particulate that fouls heat transfer surfaces. Finally, acid gases must be removed when wastes containing halogenated, sulfonated, nitrated, or phosphorus compounds are combusted.

Potential materials handling requirements.

- Excavation is required.
- Dewatering may be necessary to achieve acceptable soil moisture content. (The cost of incineration increases as the moisture content increases. In addition, infrared treatment units require more than 22% solids to ensure proper conveyance.)
- The material must be screened to remove oversized particles.
- Size reduction may be needed to achieve the feed size required by the equipment. The rotary kiln has the least stringent requirement (it can accept debris up to 12 in. (30.5 cm) in diameter). Fluidized bed units and infrared units require the waste material to be less than 1–2 in. (2.5–5 cm) in diameter.
- The waste material may require mixing to achieve uniform feed size and moisture content and to dilute troublesome components.
- If the wastes contain a high concentration of clays and fine particles, separation of clays and particles will minimize dust emissions and particulate loadings to the air-pollution control equipment (rotary kiln and fluidized bed). (See Table 10–A.)

Thermal Desorption

Thermal desorption technologies consist of a variety of processes that vaporize volatile and semivolatile organics from soil and sludge (Figures 10–14 and 10–15). The processes are planned and designed to avoid combustion of the contaminants in the primary unit, which offers several advantages. After desorption, the volatilized organics may be treated in an afterburner or condensed for reuse or destruction.

Figure 10–13. A four-module fabric filter system with spray-dry scrubber and electrostatic precipitators controls emissions from a 249 ton/day municipal solid waste incinerator in this waste-to-energy plant. (Courtesy United McGill. Used with permission.)

The types of air pollution control (APC) equipment needed to treat the exhaust gases varies depending on the technology and the nature of the contaminated media. Dust and particulates may be controlled with cyclones, baghouses, or venturi scrubbers. Small amounts of acid vapor may require scrubbing, and residual organics may be condensed and/or captured in activated carbon adsorption units. Air pollution control devices and technologies have been reviewed by the U.S. EPA and others (U.S. EPA, 1991a; Fisher & Deb, 1994).

Although no generally accepted definitions exist for grouping the different types of thermal desorption, the following three terms may be used to describe the different processes:

- directly heated desorption
- indirectly heated desorption
- in-situ steam extraction.

Thermal desorption is appropriate for both high and low concentrations of contaminants. Most wastes require treatability studies to confirm removal levels. Removal efficiencies may vary widely for similar soils. This technology has been selected for many Superfund sites.

Strengths. Lower temperatures eliminate volatilization of some metal compounds (lead, cadmium, copper, and zinc). These processes operate at lower temperatures than incineration and so use less fuel. Concerns with products of incomplete combustion are also eliminated by avoiding combustion in the primary desorbing unit. The technology has the ability to separate and recover concentrated contaminants that may then be taken off-site for treatment. However, the decontaminated soil still retains some organics and soil properties and is, therefore, not ash.

Limitations. The technology is not appropriate for inorganic contaminants. Although thermal desorbers operate at much lower temperatures than incinerators, some metals (e.g., mercury and arsenic) may volatilize during the treatment.

Table 10–A. Waste Characteristics for Soils and Sludges Using High-Temperature Thermal Treatment—General*

Characteristics Impacting Process Feasibility	Reason for Potential Impact	Data Collection Requirements
High moisture content	Moisture content affects handling and feeding and has major impact on process energy requirement.	Analysis for percent moisture
Elevated levels of halogenated organic compounds	Halogens form HCl, HBr, or HF when thermally treated; acid gases may attack refractory material and/or impact air emissions.	Quantitative analysis for organic Cl, Br, and F
Presence of PCBs, dioxins	PCBs and dioxins are required to be incinerated at higher temperatures and long residence times. Thermal systems may require special permits for incineration of these wastes.	Analysis for priority pollutant
Presence of metals	Metals (either pure or as oxides, hydroxides, or salts) that volatilize below 2,000 F (1,093 C) (e.g., As, Hg, Pb, Sn) may vaporize during incineration. These emissions are difficult to remove using conventional air pollution control equipment. Furthermore, elements cannot be broken down to nonhazardous substances by any treatment method. Therefore, thermal treatment is not useful for soils with heavy metals as the primary contaminant. Additionally, an element such as trivalent chromium (Cr^{+3}) can be oxidized to a more toxic valence state, hexavalent chromium (Cr^{+6}), in combustion systems with oxidizing atmospheres.	Analysis for heavy metals
Elevated levels of organic phosphorus compounds	During combustion processes, organic phosphorus compounds may form phosphoric acid anhydride (P_2O_5), which contributes to refractory attack and slagging problems.	Analysis for phosphorus

*Applicable to fluidized bed, infrared, rotary kiln, wet air oxidation, and pyrolytic as well as vitrification processes.
Source: *Technology Screening Guide for Treatment of Soils and Sludges.* EPA/540/2-88/004 (1988).

Potential materials handling requirements.

- Excavation is required for desorber units.
- Dewatering may be necessary to achieve acceptable soil moisture content. (The cost of desorption increases as the moisture content increases.)
- The material must be screened to remove oversized particles.
- Size reduction may be needed to achieve the feed size required by the equipment.
- The pH may need to be adjusted to achieve a pH between 5 and 11.

Several waste characteristics affect performance:

- Temperature and residence time are the primary factors affecting performance.
- Wastes with high moisture content significantly increase fuel usage.
- Fine silt and clay may result in greater dust loading to the downstream air pollution-control equipment, especially for directly heated systems.
- The volatility of the targeted waste constituents will be the primary factor that affects treatment performance. A good indicator of volatility is the pure component boiling point.

Almost all hazardous wastes are mixtures of various organic constituents (both hazardous and nonhazardous), and these other constituents often have significant impact on the actual removal of the specific compound from that matrix. Removal can be achieved at temperatures below the boiling point.

Soil Washing

Soil washing is an aqueous-based technology that uses mechanical processes to separate particles that contain contaminants (Figure 10–16). Because contaminants have a tendency to adhere to the organic carbon and fine-grained soil fraction (i.e., silt and clay) as opposed to the coarse-grained mineral fraction (i.e., sand and gravel), soil washing is a volume-reduction or pretreatment technology. In addition (or in some cases alternatively), contaminants may be removed from the soil as a result of being solubilized in the washwater. Surficial contamination is removed from the coarse fraction by abrasive scouring action.

The washwater may be augmented with a basic leaching agent; surfactant; pH adjustment; or chelating agent, such as ethylene diamine tetra-acetic

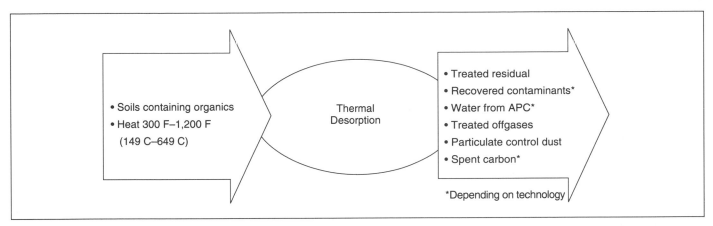

Figure 10–14. Thermal desorption process. Source: U.S. EPA, 1991b.

Figure 10–15. Low-temperature thermal desorption treatment of VOC-contaminated soils. (Courtesy Dames & Moore. Used with permission.)

acid (EDTA) to help remove organics or heavy metals. Treated soil is cleaned of any residual additive compounds. The spent washwater is treated to remove the contaminants prior to recycling back to the treatment unit.

Soil washing may be used for a variety of organic, inorganic, and reactive contaminants. Because extraction additives are selective, the technology is more appropriate for noncomplex wastes contaminated with either metals or organics. Because the technology is primarily a separation and volume-reduction process, it is frequently combined with other technologies. For example, soil washing may be used to separate the highly contaminated "fines" for further treatment. The additional technology is more cost effective when applied to the smaller volume remaining after the soil-washing step. Soil washing is effective on sands and gravel. It is most cost effective when water alone (without additives) is sufficient to achieve cleanup levels (Figure 10–17).

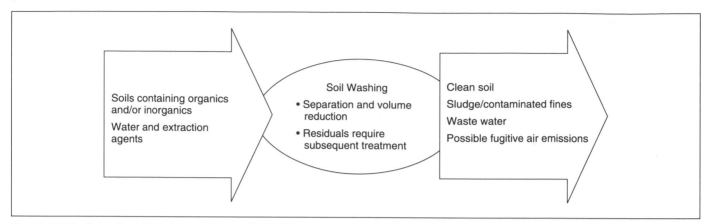

Figure 10–16. The soil-washing process. Source: U.S. EPA, 1991b.

Figure 10–17. Soil washing for a pilot test area. This shows soil staging pad in the foreground, process feed conveyor on the left, and washed soil discard chutes in front of the backhoe. (Courtesy Dames & Moore. Used with permission.)

Soil washing has been selected as a remedial source control technology at many Superfund sites, including eight wood-preserving sites (PAHs, PCP, and metals), one lead-battery recycling site, three pesticide sites, one site with VOCs and metals, and two with metals. This technology is also widely accepted in Europe, especially in Germany, The Netherlands, and Belgium.

Strengths. Soil washing is viewed favorably by the public. It is a relatively low-cost alternative for separating wastes. Furthermore, the technology can use a closed treatment system that permits control of ambient volatile emissions. Testing to date indicates the technology can remove volatile organic contaminants with 90%–99% effectiveness and semivolatile organics with 40%–90% effectiveness.

Limitations. Effectiveness is highly dependent on site conditions. The soil-washing process is relatively ineffective on soils with high silt and clay content. Washing additives (e.g., chelating agents, solvents, surfactants) can be tailored to site, soil, and contaminant conditions; however, these can be hazardous, difficult to recover, and interfere with washwater treatment.

Potential materials handling requirements.

- Excavation is required.
- The wastes can be screened to remove debris and large particles.
- Large particles may need to be reduced in size to achieve a feed size required by the equipment—less than 2 in. (5 cm) in diameter.
- The waste material, washwater, and any additives will require mixing to ensure adequate transfer of the contaminants.

Waste characteristics that limit performance include the complexity of the waste mixture and the content of the soil. Soils with high humic content and fine-grained clay particles are not suitable for this technology.

Solvent Extraction

Solvent extraction uses an organic solvent to separate hazardous organic contaminants from oily wastes, soils, sludges, and sediments, thereby reducing the volume of hazardous waste that must be treated (Figure 10–18). In general, a solvent that preferentially removes hazardous organics is mixed with the contaminated media to transfer contaminants from the media to the solvent phase. The contaminants are then separated from the solvent with a temperature or pressure change and the solvent recycled. Solvent extraction does not destroy wastes but is generally used as one in a series of unit operations. The process can reduce the overall cost of managing a particular site. Solvent extraction separates the waste into its three constituent fractions: concentrated contaminants, solids, and water.

Solvent extraction has been used in Superfund sites. The organic contaminants targeted for solvent extraction include PCBs, VOCs, and pentachlorophenol.

Strengths. Solvent extraction separates organics from the inorganics that remain in the soil. This allows a determination of the most appropriate method for treating, recycling, or disposing of the separate fractions. Solvent extraction produces dry solids as a by-product and can recover oil for reuse. This method is generally not limited by high organics or oil concentrations.

Limitations. Organically bound metals can coextract with organic pollutants, potentially restricting both treatment and recycle options. Also, adding water may be necessary, because wastes must be pumpable. This method is generally least effective on high-molecular-weight organics and highly hydrophilic substances. The presence of water-soluble detergents and emulsifiers can negatively influence extraction efficiency and treatment throughput. Finally, treatment is likely to be necessary for separated water.

Potential materials handling requirements.

- Excavation is required.
- The waste material must be screened to remove debris and large particles.
- Size reduction may be required to achieve a feed size required by the equipment.
- The pH of the waste may require adjustment.
- Wastes may need added water. Solvent extraction processes generally are applied to pumpable wastes. If the contaminated material is not pumpable, water must be added.
- The temperature of the wastes may need adjustment.
- The soil, washwater, and any additives must be mixed to ensure adequate transfer of the contaminants.

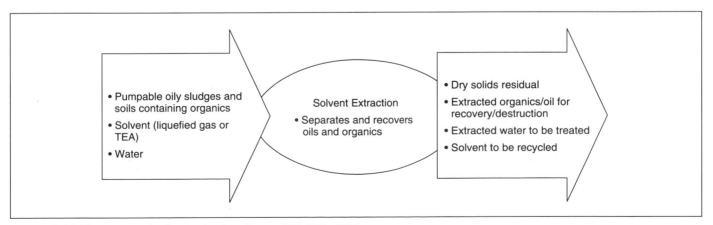

Figure 10–18. The process of solvent extraction. Source: U.S. EPA, 1991b.

Dechlorination

The alkaline metal hydroxide/polyethylene glycol (APEG) dehalogenation technology uses a glycolate reagent generated from an alkaline metal hydroxide and a glycol to remove halogens (e.g., chlorine, bromine, and fluorine) from halogenated aromatic organic compounds in a batch reactor (Figure 10–19). Potassium hydroxide/polyethylene glycol (KPEG) is the most commonly used type of APEG reagent. Potassium hydroxide, or sodium hydroxide/tetraethylene glycol (ATEG), is another variation of the reagent.

APEG processes involve heating and physical mixing contaminated soils, sludges, or liquids with the chemical reagents. During the reaction, water vapor and volatile organics are removed and condensed. Carbon filters are used to trap volatile organic compounds that are not condensed in the vapor. The treated residue is rinsed to remove reactor by-products and reagent and then dewatered before disposal. The process results in treated soil and washwater.

Dechlorination reduces the toxicity of halogenated organic compounds, particularly dioxin and furans, PCBs, and certain chlorinated pesticides. It is appropriate for soils, sludges, sediments, and liquids.

Dechlorination was selected by the U.S. EPA for cleanups at five Superfund remedial sites. The technology has been approved by the U.S. EPA's Office of Toxic Substances to treat PCBs under the Toxic Substances Control Act.

Strengths. Dechlorination has greater public acceptance than incineration. Dehalogenation has been used successfully to treat contaminant concentrations of PCBs ranging from 45,000 ppm to less than 2 ppm. This technology uses standard reactor equipment to mix/heat soil and reagents. Energy requirements are moderate, and operation and maintenance costs are relatively low.

Limitations. This technology is most effective with aromatic halides when APEG and KPEG reagents are used, although ATEG reportedly works with aliphatic halides. The presence of other pollutants, such as metals and other inorganics, can interfere with the process.

Wastewater will be generated from the residual washing process. Treatment may include chemical oxidation, biodegradation, carbon adsorption, or precipitation. Engineering controls, such as a lined and bermed treatment area and carbon filters on gas effluent stacks, may be necessary to prevent releases to the environment.

Potential materials handling requirements.

- Excavation is required.
- The waste material must be screened to remove debris and large particles.
- Size-reduction techniques can be used to achieve a feed size required by the equipment.
- The pH of the waste may require adjustment. The KPEG process operates under highly alkaline conditions. The pH of acidic material must be adjusted to provide an alkaline environment.
- The waste may require dewatering. Very wet (less than 20% water) material requires excessive quantities of reagent.
- The temperature of the wastes may require adjustment. The KPEG process operates at 212–350 F (100–177 C). The waste material must be heated.
- The waste material and reagent may need to be mixed to ensure adequate transfer of the contaminants.
- Waste characteristics can affect the effectiveness of this technology. Treatability studies are necessary to help determine the residence time in the reactor. The treatment process is affected by the type of

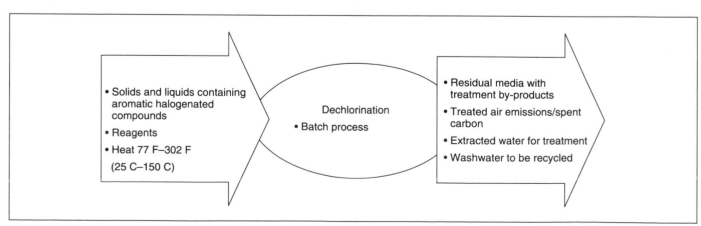

Figure 10–19. The dechlorination process. Source: U.S. EPA, 1991b.

contaminant, initial and desired final concentrations, pH, water content, and the humic and clay content of the soils. The ability to recover and recycle reagents is key to determining the process cost effectiveness. Although individual batch units may have limited capacity, several may be operated in parallel for a large-scale remediation.

Base-Catalyzed Decomposition

Base-catalyzed decomposition (BCD) is another technology for removing chlorine molecules from contaminants such as PCBs, dioxins, and pentachlorophenols. The U.S. EPA and the U.S. Navy are conducting extensive research into this new technology. Like the KPEG process, BCD requires the addition of a reagent to the contaminated soils and heating of the material for the reaction. However, because the reagent is not a glycol reagent, it is significantly less expensive than the KPEG reagent.

Laboratory research indicates that the BCD process is appropriate for PCBs, pentachlorophenol, dioxins, and chlorinated pesticides with a high destruction/removal rate. It also appears to work well on all types of soils. Because this technology has not been widely applied, it is difficult to predict what difficulties will arise at future sites. To be effective, BCD requires the soil to be screened or ground.

Bioremediation

Bioremediation technologies involve enhancing biodegradation of contaminants through the stimulation of indigenous soil and groundwater microbial populations or the addition of proprietary, natural microbial species. Natural biodegradative processes are enhanced by optimizing conditions necessary for microbes to grow and complete metabolic pathways. Bioremediation is applicable only for treating organic contaminants.

Several different types of bioremediation technologies can be used to treat hazardous wastes. These technologies can be classified in two broad categories:

- aboveground (including slurry phase, contained solid phase, land treatment, and composting; Figures 10–20 and 10–21)
- in-situ.

The different approaches vary in terms of the process control, time needed, and cost of remediation. Figure 10–22 shows the general relationship of the different bioremediation technologies with respect to process control, cost, and amount of effort

needed in the engineering/treatability testing phase of the project. Slurry-phased bioremediation, for example, is the most easily controlled process and often the most expensive.

Slurry phase. Slurry-phase bioremediation involves mixing excavated soil with water to create a slurry that is mechanically agitated in an environment (usually a tank, although in-situ lagoon applications are possible) with appropriate ambient conditions of nutrients, oxygen, pH, and temperature. Microorganisms may be seeded initially or added continuously throughout an appropriate residence time. Upon completion of the process, the slurry is dewatered, and the treated soil is disposed.

Land treatment (land farming). This process involves placing contaminated soil in a prepared, lined treatment bed. Wastes may be stockpiled prior to application, which occurs in a series of lifts (several inches thick). Supplements such as manure or nutrients may be added, and the soil is periodically cultivated. Use of standard construction/farm equipment allows management of a large area of treatment.

Contained solid phase. This represents a variety of processes similar to land farming but provides greater process control. Although these systems have been used in Europe, they have not been used in the United States. Excavated soils are mixed; soil amendments (water, nutrients, pH modifiers, bulk modifiers, and microbes) are added; and the conditioned soil placed in an enclosure such as a building, tank, or modified pad. This may improve process control by eliminating water runon/off, moderating temperature, allowing greater moisture control, and controlling VOC emissions. The soil may be several feet deep and require special equipment or processes for reconditioning or aeration. To the extent the building enclosure satisfies the RCRA definition of a tank, the process may be used to satisfy land ban requirements.

Composting is a solid-phase, aboveground biological treatment process in which structurally firm material is added to the contaminated material to enhance the decomposition of organic compounds. Water, oxygen, and nutrients are also added to facilitate microbial growth. This technology has not yet been selected for use at a Superfund site, although the U.S. Army has had some success using composting to degrade munitions wastes in pilot-scale tests.

Composting, particularly if conducted in a closed reactor, is similar to contained solid-phase bioremediation and is included here with the discussion of that technology.

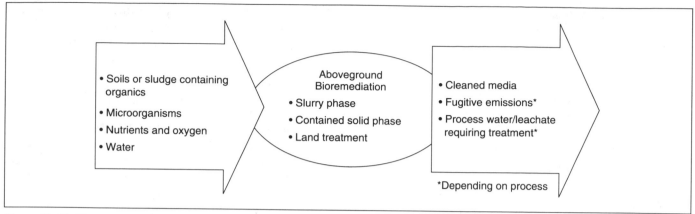

Figure 10–20. Aboveground bioremediation. Source: U.S. EPA, 1991b.

In-situ. This process promotes and accelerates the natural biodegradation processes in the undisturbed soil. Generally, in-situ biodegradation consists of a water recirculation system with aboveground water treatment and conditioning of the infiltration water with nutrients and an oxygen source (Figure 10–23). The system is usually designed to allow uncontaminated groundwater to enter the zone of contamination but not to leave the contaminated zone. Common system design consists of central withdrawal of groundwater and reinfiltration through well injection or infiltration galleries at several locations around the outer border of the treated area.

Biodegradation relies on the contact between contaminants (in the water phase) and microorganisms. In general, subsoil saturation is required. Oxygen is usually the limiting factor. Sources of oxygen include air, pure oxygen, and hydrogen peroxide. Although hydrogen peroxide is commonly used, it can be relatively expensive and its use entails technical problems. Nitrate is being researched with some success as an alternative electron receptor.

The U.S. Army Toxic and Hazardous Materials Agency has pilot tested 40-day composting of soils contaminated with TNT, DNT, and other by-products of munitions manufacture. The original contaminants were successfully degraded and the wastes found to be nontoxic. The U.S. Army is now analyzing the product for the presence of by-products and intermediary products of the degradation of TNT and DNT.

The Superfund sites at which bioremediation has been selected are contaminated with VOCs, PAHs, SVOCs, and creosote. The technology has broad applications for organic wastes.

PAHs with four or more rings are more difficult to degrade than simpler PAHs. The literature reports some success with white rot fungus (Bumpus, 1989) and mycobacterium (Heitkamp et al, 1988).

Recent treatability studies conducted at the American Creosote Works Superfund site (Muller et al, 1990) with creosote-contaminated soils and sediments suggest that:

- *Solid phase* bioremediation strategies may not effectively meet acceptable treatment standards for creosote-contaminated soils and sediments in the time defined by the studies (90 days). Relatively poor removals of pentachlorophenol (PCP) and other multiring, toxic compounds were found (0%–50%).
- *Slurry phase* bioremediation strategies can be effectively employed to remediate creosote-contaminated materials. Within 14 days, two-ring PAHs were totally degraded and the concentrations of PAHs with three, four, and more rings were effectively reduced. The reduction of the heavier PAHs was greatest when there was a significant concentration of phenolics and two-ring PAHs in the contaminated material.
- These studies indicate that PAHs with two rings biodegrade more readily than those with three rings, which, in turn, degrade more readily than those with four rings.

Bioremediation is appropriate for soil, sludges, sediments, groundwater, and surface water. Bioremediation may not be applicable at sites where the contaminated material contains extremely high concentrations of heavy metals, highly chlorinated organics, pesticides, herbicides, or inorganic salts. High concentrations of these contaminants may be toxic to the microorganisms needed for biodegradation.

Figure 10–21. Bioremediation using boom spraying nutrient and bioculture mix. (Courtesy Dames & Moore. Used with permission.)

This technology requires temperature, moisture content, pH, nutrient levels, and oxygen content to be within the limits required by the microorganisms. In-situ bioremediation may not be applicable at sites where these parameters are not within those limits.

Bioremediation has been selected for source control remediation at over 30 Superfund sites. It is currently considered, planned, or operated in full-scale systems in at least 140 CERCLA, RCRA, and UST sites. Five hazardous waste sites have been selected for field evaluation under the Bioremediation Field Initiative program.

Several slurry-phase and solid-phase biological treatment systems are commercially available. Land treatment (farming) has been approved in many administrative records of decision.

A recent international survey found 23 in-situ biodegradation remediation projects in The Netherlands, Germany, and the United States. In-situ treatment has been selected for many Superfund sites. There have been demonstrations that some halogenated compounds may be degraded by bacteria that use methane as their food source.

Strengths. The technologies have fairly broad applicability for organic wastes. Biological technologies are cost effective and likely to be considered as natural processes and supported by the public.

Limitations. Recent treatability studies with creosote-contaminated soils and sediments conducted at the American Creosote Works Superfund site suggest that abiotic losses from bioremediation of creosote-contaminated soils and sediments can be significant. Volatilization is more significant with the low-molecular-weight compounds within creosote. (Volatilization accounted for 12% of the observed losses of naphthalene during solid phase bioremediation of sediments.) Significant quantities of high-molecular-weight compounds were found adsorbed to the bioreactor sludge and residue in slurry-phased bioremediation.

Performance is highly dependent on site conditions. Some wastes are hard or slow to degrade. Complex waste mixtures can inhibit biological activity. A better understanding and optimization of the science is needed. The public can be fearful of microbes suspected of having been "genetically engineered."

Potential materials handling requirements. For slurry-phase, land treatment, and contained solid phase:

- Excavation is required.
- The waste material must be screened to remove debris and large particles.

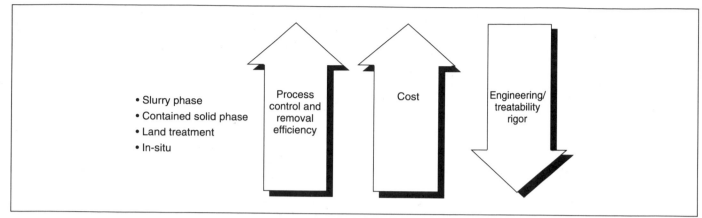

Figure 10–22. Comparison of bioremediation technologies.

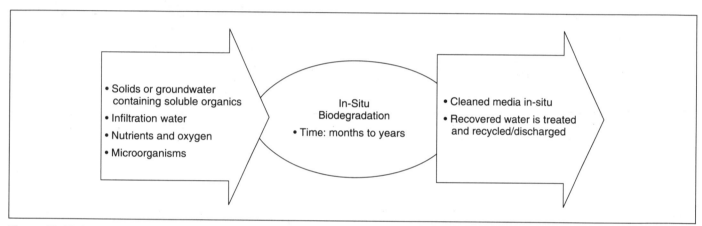

Figure 10–23. In-situ biodegradation. Source: U.S. EPA, 1991b.

- Size-reduction techniques can be used to achieve the feed size required by the equipment. Slurry-phase bioremediation works best if the particles are generally less than ¼ in. (6 mm) in diameter. This small size is needed to ensure adequate contact between the biomass and the contaminants and to keep the particles in suspension in the slurry.
- The pH of the waste may require adjustment. Bioremediation works only for material with a generally neutral pH (> 4.5 and < 8.5). If the contaminated material does not fit within this range, acid or alkali must be added and the material mixed. As the biological processes proceed, the pH of the material may change. The pH may need to be adjusted several times throughout the operation of the technology.
- The waste may require dewatering or wetting. Biological activity is promoted when the moisture content of the contaminated material is between 40% and 80%.

- The temperature of the wastes may be adjusted. Biological activity proceeds more rapidly under warm conditions. Generally, bioremediation projects should be operated between 60 F (15.6 C) and 110 F (43.3 C).
- The waste material, water, added nutrients, and oxygen will need to be mixed to promote microbial growth.

Vacuum Extraction

A vacuum extraction system applies a vacuum to a series of extraction wells to create air flow through the vadose zone. As air moves through the system, volatile contaminants move from the soil and pore water to the air. The contaminated air is withdrawn, often with entrained water, and treated using an emission control system such as activated carbon or catalytic oxidation (Figure 10–24).

In practical terms, the process works well with most soil types. The process demonstrated good performance in removing volatile organics from soil

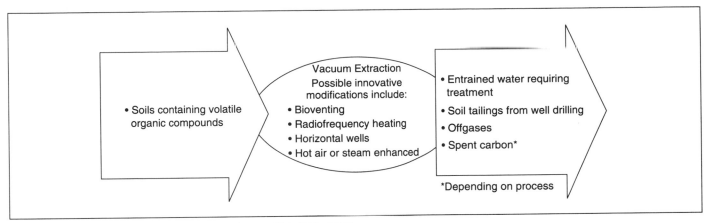

Figure 10–24. The vacuum extraction process. Source: U.S. EPA, 1991b.

with measured permeability ranging from 10^{-4} to 10^{-8} cm/s. Also, the process works well under all weather conditions. It is generally applicable to volatile and moderately volatile organic compounds. The vapor pressure of the compound provides a relative measure of the volatility. Although the success of a vacuum extraction project will depend on site-specific conditions, this technology has been selected for contaminants with vapor pressures as low as 8 torr. The technology also works for removal of volatile, light, nonaqueous-phase liquids floating on the water table or entrained in the capillary fringe. This technology is used at more than 50 sites, including 11 Superfund sites.

Strengths. Vacuum extraction can be used in conjunction with other remedial alternatives. It can be used as pretreatment prior to excavation. It has demonstrated effectiveness in removing VOCs from the vadose zone. It minimally disturbs contaminated soil. In addition, treatment costs are low compared to technologies requiring excavation. Available data indicate that costs are typically near $50 per ton. The economics of this process strongly depend on whether offgas treatment is required and whether wastewater is generated at the site. These factors were considered in the above estimate. The most cost-effective systems have been in homogeneous, sandy soils. The technology is relatively simple and reliable.

Limitations. The process is limited to volatile compounds. Nonhomogeneous soil can result in irregular air flow or "short circuiting," which results in uneven treatment. Entrained soil moisture must be removed from the air stream and treated. High soil moisture can add to the cost of treatment.

Potential materials handling requirements. This technology is used in-situ; no excavation is required.

The contaminated soils may require dewatering. Because this technology works better as the air-filled porosity of the soils increases, performance may be improved and removal of the contaminants maximized early in treatment by dewatering saturated or wet areas before treatment. Because this is an in-situ technology, the material must be dewatered with in-situ methods such as well points and drains.

The major considerations in applying this technology are contaminant volatility, air-filled soil porosity, and site-specific cleanup level.

When soils have low permeability and high moisture content (i.e., low air-filled porosity) a pilot demonstration test should be considered to determine the feasibility of dewatering the soil.

Testing alternatives to activated carbon is ongoing for exhaust air treatment. One such system involves thermal catalytic destruction.

Bioventing. A promising modification of vacuum extraction is its potential to enhance the biodegradation of volatile and semivolatile chemicals in the soil. Vacuum extraction provides air to the vadose zone, thus carrying oxygen that can be used by soil microorganisms to biodegrade contaminants. Field tests conducted by the Air Force in 1990 at Tyndall Air Force Base on soils contaminated with jet fuel suggest that biodegradation becomes increasingly significant as the primary hydrocarbon removal mechanism as the more volatile compounds are stripped from the soil. Under optimal air flow conditions (0.5 air void volumes per day), 82% of the hydrocarbon removal was the result of biodegradation. These field tests also suggest that the effect of temperature on biodegradation rates approximates the effects predicted by the Van't Hoff-Arrhenius equation, i.e., the reaction rate approximately doubles for every 10 C (50 F) rise in temperature.

This modification of the process has two potential benefits. First, it will completely destroy a large portion of contaminants in the ground and minimize the amount of extracted gas requiring treatment, thereby reducing costs. Second, enhanced in-situ biodegradation will also destroy the heavy hydrocarbons that, because of their low volatility, are not removed by conventional vacuum extraction alone.

Radiofrequency heating. Another potential enhancement of the vacuum extraction process is the use of radiofrequency (RF) heating to more rapidly volatilize contaminants and to increase the volatilization of compounds with higher boiling points. The uniform heating that RF energy provides is also expected to improve soil porosity and to improve removal rates in clay and silt soils.

This technology is being demonstrated by the Air Force. Energy is delivered by means of an array of electrodes placed in bore holes. The specific frequency is selected based on dielectric properties of the soil, the depth of treatment desired, and the size of the heated soil volume. Recent tests were conducted on a fire training area at an Air Force base in Wisconsin. A 500 cubic ft (14 cubic m) test volume was heated for 12 days. The first 8 days were required to heat the soil to the 300 F (150 C) target temperature, which was then retained for a period of 4 days. A 97% removal was observed for semivolatile hydrocarbons and a 99% removal for volatile aromatics and aliphatics. Power estimates for full-scale application were quite reasonable at 500 kw-hr/cubic yd.

Steam or hot-air injection. In-situ steam injection facilitates the removal of moderately volatile residual organics, including nonaqueous phase liquids (NAPLs), from the vadose zone. Steam injection technology injects pressured steam to thermally enhance the evaporation rate of the contaminant and its subsequent removal. Injection of steam also can be expected to enhance removal of residual NAPLs in the unsaturated zone by decreasing their viscosities. Steam injection is an emerging technology that appears promising, particularly if used in conjunction with vacuum extraction.

In-Situ Vitrification

In-situ vitrification (ISV) uses electrical power to heat and melt contaminated soils and sludges to form a stable glass and crystalline structure with low leaching characteristics. ISV uses a square array of four electrodes inserted into the ground to establish a current in the soil and heat the soil to a range of 2,900–3,600 F (1,600–2,000 C), well above a typical soil's melting point. As the melt is generated downward from the surface, organic contaminants are destroyed by pyrolysis and the pyrolized products migrate to the surfaces of the vitrified zone, where they combust in the presence of oxygen. Nonvolatile inorganic contaminants are dissolved and incorporated into the melt. The resulting product is devoid of residual organics. A vacuum hood placed over the area collects offgases, which are treated before being released into the atmosphere (Figure 10–25).

In-situ vitrification may be used to destroy, remove, or immobilize all contaminant groups—radioactive, organic, and inorganic. This technology is also appropriate for mixtures of wastes. It can be applied to most soil types. The two limiting factors are: (1) moisture content (excessive moisture must be driven off before vitrification, increasing costs significantly) and (2) presence of glass-forming materials (i.e., silicon and aluminum oxides). If the soil is too wet, it may be dewatered, and if there is insufficient glass-forming material, a fluxing material may be added.

To date, engineering, pilot-scale, and full-scale tests have been conducted on in-situ vitrification of hazardous wastes. A large-scale test has been conducted at Hanford, Washington, on mixed radioactive and chemical wastes containing chromium. Nine full-scale tests have been performed at DOE sites. Only one vendor is licensed by the U.S. DOE to perform ISV. Three U.S. EPA regions have selected ISV to treat contaminated soils at Superfund sites contaminated with PCBs, heavy metals, organics, pesticides, and low-level dioxins.

Strengths. ISV has been successfully tested for the treatment of radioactive and hazardous wastes. It produces a stable crystalline structure with long-term durability, encapsulating the residual inorganic waste.

Limitations. This technology requires potentially hazardous offgas collection and treatment and disposal of spent activated carbon, scrubber water, and other waste materials from the air pollution control equipment. It may require backfilling with clean soil because volume can decrease 20%–40%.

Because contaminants can migrate from the wastes at the periphery of the melt (side migration), a vitrification project may need to include clean material at the edges of the contaminated area to capture migrating contaminants. This will increase the cost of the project by increasing the quantity of material to be vitrified.

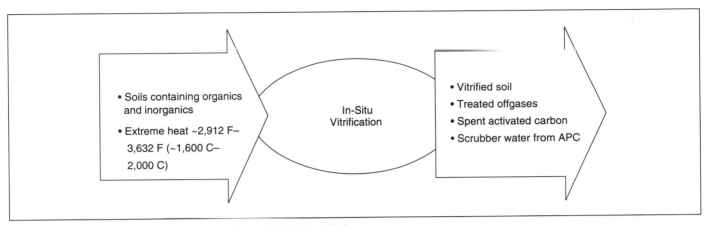

Figure 10–25. The in-situ vitrification process. Source: U.S. EPA, 1991b.

Existing metallic pipes and scrap metals can cause substantial problems in full-scale implementation. To effectively immobilize metals and radionuclides, adequate concentrations of glass-forming elements are required to ensure chemical durability of the product. ISV is not effective below the water table if the hydraulic conductivity is greater than 10^{-4}. The concentration of organic material in the waste must be less than 10% by weight to be accommodated with the existing offgas equipment.

Potential materials handling requirements. This technology is used in-situ; no excavation is required. Some excavation and placement of wastes can be done to stage the waste and lower the cost of the project.

The contaminated material can be dewatered if the groundwater conditions at the site are such that dewatering is cost effective. Because this technology is more efficient with unsaturated soils, costs can be lowered by dewatering saturated or very wet areas before treatment. Because this is an in-situ technology, the material must be dewatered with in-situ methods such as well points and drains.

Several factors affect performance. Soil moisture, which must be driven off before melting occurs, can have a major effect on cost. A second major factor affecting cost-effectiveness is the cost of electricity. The capability of the system depends on two factors: (1) the capabilities of the power supply and (2) the capacity of the offgas system.

The two factors that affect the ability of the power supply system are the presence of saline groundwater and buried metals. Vitrification can occur in saturated soils with low permeabilities if the melting rate is greater than the rate of recharge.

The bulk chemistry of the waste material affects the melt temperature, viscosity of the melt, and durability of the vitrified mass. Miscellaneous buried metal, such as drums, should have little or no effect on the ability to process a site. A conduction path that could lead to shorting between electrodes should be avoided.

The offgas system should maintain a negative pressure to prevent any release of contamination. The concern is with the release of gas from relatively short transient events, such as the release of entrapped air from intrusions into void spaces, penetration of a drum containing combustible material, and intrusion into areas containing solid or liquid combustible materials.

Treatability tests should focus on performance requirements for the offgas treatment system and the type and quantity of secondary waste generated. Almost all soils can be vitrified, and this is generally not a serious consideration during treatability testing.

Groundwater Treatment Technologies

The innovative technologies discussed in the previous sections are primarily source control technologies. These technologies treat hazardous contaminants found in soils, sludges, or debris. Several additional technologies could be considered at sites where groundwater is contaminated. At these sites the nature of the contaminants (concentration, partition, coefficient, solubility, viscosity, and biodegradability) and the hydrogeology (extent of contamination, hydraulic conductivity, storage coefficient, heterogeneity of the aquifer) will determine which treatment technology offers the best possibility of effectively and efficiently treating the groundwater. A thorough investigation of the subsurface conditions is necessary to design a groundwater treatment system, develop appropriate operating and monitoring procedures, and establish realistic termination criteria.

Technologies to clean up groundwater contaminated with hazardous wastes can be considered in two categories: those that rely on pumping followed by treatment above the ground and those that treat the water in-situ. Table 10–B shows the different technologies for groundwater treatment. In many ways, the technologies that are applied aboveground closely resemble the traditional water treatment technologies used to treat industrial and municipal waste waters; they include such approaches as carbon adsorption, air stripping, chemical precipitation, and biological treatment.

The effectiveness of these ex-situ technologies is based on the effectiveness of the pumping system in capturing the hazardous wastes and bringing them to the surface with the groundwater for treatment. If pumping cannot remove the contaminants from the particles within the aquifer, the ex-situ treatment technologies do not have an opportunity to treat them.

Innovative technologies can be used to enhance extraction of the contaminants. These technologies include such techniques as steam extraction and surfactant flushing. Innovative technologies can also be used to treat the groundwater in-situ. As with the technologies to treat soil, these technologies act to separate the contaminants from the groundwater and surrounding aquifer, change the contaminants into a less toxic or less mobile form, or degrade them to eliminate their toxicity.

SUMMARY

The diverse health and environmental problems posed by U.S. wastewater, solid waste, and hazardous waste have resulted in aggressive responses from federal and state governments as well as industry in developing laws and regulations to control the problems. Management of waste in the United States matured during the 1980s as all parties involved grappled with the complex problem through the development of new innovative technologies. Although the United States has only recently begun to make headway in solving the waste problem, the regulatory and technological framework for achieving significant progress before the 21st century is in place.

It is clear that both the general public and the regulatory agencies have begun to switch their focus from "end-of-pipe" waste treatment toward pollution prevention. However, because of the large number of existing historic worldwide waste sites, it is obvious that further development and improvement of treatment technologies will be required. Clearly, a strategy of both prevention and efficient, advanced treatment technologies is required.

REFERENCES

Bumpus JA. Biodegradation of polycyclic aromatic hydrocarbons by phanerochaete chrysosporium. *Appl Environ Microbiol* 55:154–158, 1989.

40 CFR, Subpart 122. "Environmental Protection Agency National Pollutant Discharge Elimination System Permit Regulations." Washington DC: Government Printing Office.

Fisher PW, Deb K. Air pollution control technologies. *Environmental Science and Technology Handbook.* Rockville, MD: Government Institutes, 1994.

Heitkamp MA, et al. Microbial metabolism of polycyclic aromatic hydrocarbons: Isolation and characterization of pyrene-degrading bacterium. *Appl Environ Microbiol* 54:2549–2555, 1988.

Jet Propulsion Laboratory. *Standard Assembly-Line Manufacturing Industry Simulation (SAMIS) PC User's Guide, SAMIS Release 6.0.* Pasadena, CA, December, 1985.

McCoy DE. *RCRA Regulations and Key Word Index.* McCoy & Associates, Inc., 1992.

Muller JG, et al. Bench-scale evaluation of alternative biological treatment processes for remediation of pentachlorophenol- and creosote-contaminated materials: Solid-phase bioremediation. *Environ Sci and Technol* 25(7):1045–1055, 1990.

Table 10–B. Treatment Technologies for Groundwater

Conventional pump and treat	In-Situ	Enhanced Extraction
Standard physical/ chemical technologies (e.g., carbon adsorption, air stripping, chemical precipitation, ion exchange, and reverse osmosis) Standard biological treatment Ultraviolet oxidation	Air purging (possibly with horizontal wells) Bioremediation Dewatering followed by vacuum extraction or bioventing Chemical oxidation	Surfactant flushing Steam extraction

Stouch JC, Schmidt PF. *An Integrated Approach to Pollution Prevention: From Regulatory Requirements to Shop Floor Benefits.* Chicago: HazMat Central, 1991.

U.S. Air Force, Wright-Patterson Air Force Base. *Integrated Computer-Aided Manufacturing (ICAM), Function Modeling Manual (IDEF).* UM110231100, June 1981.

33 USC 1251a. Federal Water Pollution Control Act as Amended by the Water Quality Act, 1977.

33 USC 1251b. Federal Water Pollution Control Act as Amended by the Clean Water Act, 1987.

42 USC 6901. Resource Conservation and Recovery Act of 1976. Amended 1978.

42 USC 9601. Comprehensive Environmental Response, Compensation, and Liability Act of 1980.

U.S. EPA. *1982 Guide to Disposal of Chemically Stabilized and Solidified Waste.* SW-872. Office of Solid Waste and Emergency Response. Washington DC, 1982.

U.S. EPA. *Handbook of Remedial Action at Waste Disposal Sites.* EPA/625/6–85/006. Washington DC, 1985.

U.S. EPA. OSWER. *The Solid Waste Dilemma: An Agenda for Action.* EPA/530-SW-89–019. Washington DC, February 1989.

U.S. EPA. *America's Clean Water Act*, Office of Water Regulations and Standards, CPR, Citation for Clean Water Act. Washington DC, 1990a.

U.S. EPA. *Let's Reduce and Recycle: Curriculum for Solid Waste Awareness.* EPA/330-SW-90–005, Office of Solid Waste and Emergency Response. Washington DC, 1990b.

U.S. EPA. *National Priorities List Sites.* EPA/540/4–90/010, Office of Emergency and Remedial Response, Office of Program Management. Washington DC, 1990c.

U.S. EPA. *Handbook of Control Technologies for Hazardous Air Pollutants.* EPA/625/6–91/014. Washington DC, June 1991a.

U.S. EPA. *Innovative Treatment Technologies.* EPA/540/9–91/002. Washington DC, October 1991b.

Vajda GF. Integrating pollution prevention into the manufacturing process. *Total Quality Environ Man* 1(1):61–70, Autumn 1991.

Vajda GF. Overcoming obstacles to pollution prevention. Paper presented HazTech International Conference, Pittsburgh, PA, 1992.

Vajda GF, Stouch JC. *An Integrated Approach to Waste Minimization.* Paper presented at Air & Waste Management Association Conference, Pittsburgh, PA, 1990.

Yakowitz H. What trade in recoverable wastes? *The OECD Observer* 180:26, February/March 1993.

Part 3

Special Concerns

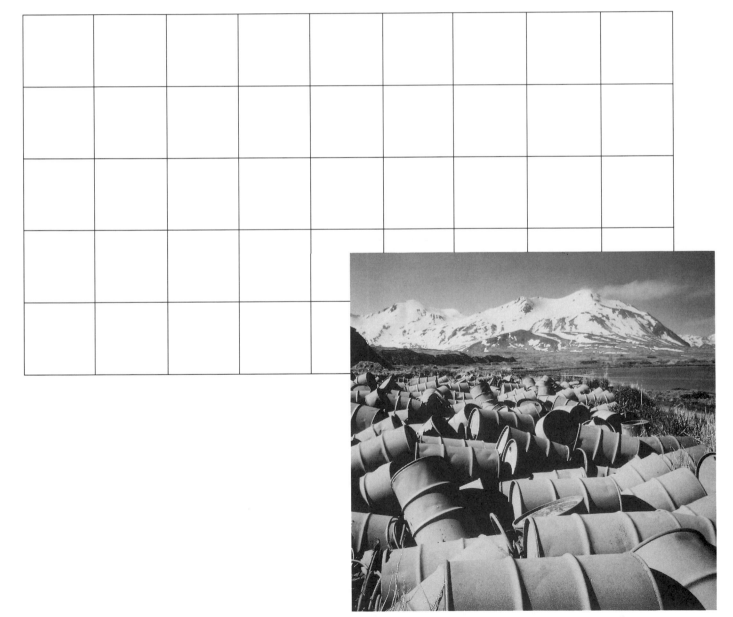

A waste dump in Alaska.

11

Public Health Issues

Gary R. Krieger, MD, MPH, DABT
Christopher P. Weis, PhD, DABT
Martha A. Rozelle, PhD

The previous two sections analyze the ethical, economic, legal, management, and regulatory structure of environmental affairs, particularly those issues that affect potentially hazardous materials. Part 3 is directed toward broader concerns that impact both the external and internal environment. The first two chapters deal with the external sphere of public health, risk communication, and risk assessment. The internal environment—indoor air quality issues—is presented in Chapter 13. Chapter 14, Global Issues, explores concerns that will dominate the environmental debate into the 21st century: population growth, soil degradation and deforestation, biodiversity, water resources, acid rain, ozone depletion, greenhouse gas emissions, and global warming.

Environmental professionals are frequently confronted with the need to evaluate the public health implications of potentially hazardous materials. These evaluations and their subsequent public policy impacts are often based on epidemiological studies and risk assessments. The techniques and vocabulary of public health studies encompass the fields of epidemiology, cancer analysis, and risk communication. This chapter will introduce core concepts in these fields. Chapter 12, Risk Assessment, presents the developing science of human health and ecological risk assessment.

This chapter introduces basic descriptive concepts in epidemiology, followed by a general discussion of the methodological approaches used to analyze possible disease clusters. Specific examples of the epidemiology of electromagnetic fields and lead exposures are presented as brief case studies that illustrate both the techniques of epidemiology and disease cluster analysis. Finally, the science and process of risk communication for public forums are discussed. Risk communication is the crucial link that connects the "science" with the ultimate decision-making audience: the public.

BASIC PRINCIPLES OF EPIDEMIOLOGY

The science of epidemiology has developed from a series of descriptive techniques describing the spread or epidemic nature of disease to a highly quantitative science that formally analyzes and probes the association of disease with a variety of general environmental conditions or exposures. As environmental issues have taken center stage in political and economic debates, the ability to associate various disease outcomes with either old or new potential exposures has become more significant. Multi-million- or billion-dollar economic decisions frequently depend on the attributable risk or calculated strength of association between potential environmental hazards and human and/or ecological receptors.

Epidemiological investigations are keyed to the differences between naturally occurring exposures and experimentally constructed or designed exposures. Environmental epidemiological studies are not in control of the exposure but are in control of the study subjects and the type of analysis. Therefore, these type of studies do not conclusively demonstrate causation but do show strength of association. For example, the association between cigarette smoking and lung cancer has been well demonstrated; however, the specific causative agent(s) is less well defined. It is highly unlikely that anyone or any group would willingly participate in a large-scale designed environment exposure study, so data are usually obtained from three different types of studies: descriptive, analytic, and interventive. Each of these studies has different methodological advantages and problems.

Descriptive studies categorize the various frequencies of disease under a variety of different exposures. An example of this type of study is analysis of different types of cancer found in residents who live near an electrical substation. Descriptive studies generate data that are useful for forming new hypotheses, analyzing trends, and establishing background levels.

Analytic or etiologic studies establish associative relationships between hazard and disease. An example of this type of study design is recent investigations that look for dose-response relationships between increasing electromagnetic field (EMF) strength exposure and childhood leukemia rates.

Intervention studies measure the relative effects of mitigation strategies on previously exposed populations. As an example, an intervention study could analyze the new incidence rate of childhood leukemia in a neighborhood after specific steps were taken to decrease exposure. Clearly, each of the three types of studies can overlap and support the overall goal of describing the relationships between potential hazards and disease or illness endpoints.

Epidemiological Measures of Risk

Two common epidemiological terms must be clearly understood in any discussion of risk: prevalence and incidence. *Prevalence* is the number of existing cases at a specified time divided by the size of the population under consideration. For example, for any given time period, an experimenter could ask, "How many cases of influenza are there in town X?" The prevalence of a disease gives the investigator "background" information on the baseline condition of a population under study. If a new variable or "exposure" is introduced into the population under study, the number of new cases divided by the population at risk is known as the *incidence* rate.

The prevalence rate can be significantly affected by either death or cure. Both of these outcomes can significantly alter the baseline condition of a population under study. Both incidence and prevalence rates are examples of proportions, i.e., the denominator contains the numerator. In general, proportions or rates are more powerful terms than ratios of measured quantities. Thus, it is common for descriptive studies to calculate both incidence and prevalence rates of various disease outcomes.

Furthermore, there is a relationship between the incidence rate and the prevalence rate: prevalence = (incidence × duration). It is clear from this relationship that the knowledge of baseline and new case rates of disease can provide clues in the search for connections between exposure and disease. Later in this chapter, the specific area of disease attributable to EMF exposure will be discussed in detail. Many skeptics of the proposed relationship between EMF exposure and childhood leukemia rates have argued that the overall U.S. prevalence and incidence rates of leukemia have not substantially changed despite massive electrification. Simplistically, the prevalence = (incidence × duration) relationship illustrates that for any condition associated with both relatively constant survival rates and long duration, the prevalence is directly proportional to the incidence rate.

Although significant changes in either the prevalence or incidence rate of a disease can provide insight into the cause(s) of a disease(s), it is also important to estimate the increased risks associated with a given exposure. These measures of risk are known as risk ratios.

The *risk ratio* or *relative risk (RR)* is the ratio between the risk in exposed subjects and the risk in nonexposed subjects. An RR of 1.0 is considered unremarkable. In most studies, RRs are shown within statistical confidence limits. These limits are crucial, because all estimates of risk are associated with variability. The RRs associated with the initial EMF field studies frequently overlapped 1.0 and raised questions about the significance of the findings. In certain circumstances, the true relative risk cannot be calculated, because a sufficient random sample of the study population is not available. In these circumstances, the RR is estimated by obtaining an odds ratio.

The *odds ratio* is the comparison of the number of exposed cases (a) $\times$ the number of nonexposed controls (d) $\div$ by the number of exposed controls (b) $\times$ the number of nonexposed cases (c). This relationship is frequently portrayed in what is known as a 2×2 Table (Table 11–A).

As illustrated by Table 11–A, the mathematical formulation of the relative risk is simply the ratio of incidence rate in the exposed population to incidence rate in the unexposed group. If the disease being studied is relatively rare (e.g., leukemia as opposed to high blood pressure), the total number exposed (a + b) is quite similar to the number of healthy exposed (b). Similarly, the total number not exposed (c + d) is close to the number of healthy nonexposed (d). Therefore, under these conditions, the relative risk can be estimated as the odds ratio ad $\div$ bc. This information is frequently used in case control studies because a, b, c, and d are known but the populations at risk are unknown. Finally, Table 11–A is also useful for illustrating the concept of attributable risk.

The *attributable risk (AR)* is the proportion of events that can be attributed to the exposure under study. The AR is a commonly calculated measure of risk and is used as a benchmark for quantitative risk assessments of carcinogens. This type of risk assessment is discussed in detail in subsequent sections of this chapter. Using Table 11–A, the AR can be defined as $\{[a \div (a + c)] - [b \div (b + d)]\}$ divided by $[1 - b(b + d)]$. Another way to understand the attributable risk concept is to view it as the portion of the total risk of a specified disease in a given population *attributable* to the exposure factor under study. Thus, if the relative risk (r) for a factor (e.g.,

leukemia in a case control study) is known and the frequency of the factor in the population (b) can be reasonably estimated, then the AR $= [b(r - 1)] \div [(b)(r - 1) + 1]$.

A large number of effects can impact the comparability of study populations. For example, a study of lung cancer associated with air pollution exposure must consider the possible compounding effects of cigarette smoking on both the study and the control population. A comparison of the nonadjusted rates would not reveal the true impact of the exposure under study if the rates of cigarette smoking were significantly different in the two populations. Therefore, the rates are *adjusted* to correspond with a standard defined population. Adjustment is not confined to mortality outcomes, although the most commonly encountered ratio is the standard mortality ratio or SMR.

The *SMR* is the observed mortality rate divided by the expected mortality rate. This relationship is frequently presented as SMR = obs $\div$ exp. Similar to the concept of relative risk, an SMR of 100 is expected. SMRs (with associated confidence levels) significantly greater than 100 are generally associated with an excess of disease that may be attributable to a specific exposure. Conversely, SMRs of significantly less than 100 may indicate a protective effect from the exposure under study. Frequently, SMRs from one study are compared to other, "similar" published ratios or rates. These comparisons can be extremely misleading, because the adjustment process associated with each SMR may not be comparable to the process for another SMR. Comparison of SMRs across occupational cancer studies to produce a range of results is a common occurrence; however, because of differences in the age-demographic makeup of the standardizing population, studies may not be directly comparable. A simple example of this problem would be a comparison of workers exposed in plant to chemical "X" versus the general population exposed by stack air emissions.

Workplace SMRs are usually affected by the "healthy worker" phenomenon (e.g., the presence of full-time employment and age range associated with the workplace confers a positive health protective benefit). Conversely, environmentally exposed general populations have different demographics and age/disease distributions that produce SMRs higher than an exposed workplace population despite the differences in exposure concentrations.

Types of Epidemiological Studies

As previously discussed, there are three basic categories of epidemiological studies: (1) descriptive,

---

Table 11–A. Computing Risk

	Cases: Disease Present	Controls: Disease Absent	Total
Exposed	a	b	a + b
Nonexposed	c	d	c + d
Total	a + c	b + d	Total

Odds Ratio = ad $\div$ bc; Relative Risk = $\dfrac{a \div (a + b)}{c \div (c + d)}$

(2) analytic, and (3) interventive. In the environmental epidemiology literature, it is common to review several subtypes of descriptive and analytic studies: (1) prevalence or cross-sectional, (2) case-control, (3) incidence or cohort, and (4) prospective and retrospective (Kahn, 1983; Piantadosi, 1991). Each of these study types is briefly discussed here.

Prevalence or cross-sectional studies examine the relationship between a disease and a variable of interest at one particular time (Friedman, 1974; Last & Tyler, 1992). This relationship can be analyzed as either the prevalence of a disease defined by the presence or absence of the study variable or the presence or absence of the variable in either the diseased or nondiseased group. For example, assume the presence or absence of an EMF is the variable under consideration. A group of childhood leukemia cases can then be analyzed to see whether EMF from power lines is more commonly associated with the diseased group versus a similar nondiseased group. The converse of this approach is also possible: Examine all EMF power lines of a defined strength and then look for the presence or absence of a disease endpoint such as leukemia or brain cancer.

The first EMF example also illustrates a study known as a *case control*. These studies are similar to prevalence studies because they examine the relationship between existing disease (e.g., childhood leukemia) and other potential exposure variables. Case control studies are extremely common in the environmental epidemiology literature and are frequently the first type of study performed in order to attempt to establish an associative link between a disease endpoint and an environmental exposure.

In comparison to prevalence studies, incidence or cohort studies examine the variables related to the development of disease (Friedman, 1974). The *cohort* is a group of individuals under study who are initially disease free. This group (cohort) is then followed through time for the development of the disease under study. The incidence of disease development within the members of the cohort is then calculated and compared within defined subgroups in the original cohort. Cohort studies are very common and can be either *prospectively* or *retrospectively* performed.

Both prospective and retrospective studies are keyed to a given time period: (1) observations have been recorded in the past (retrospective), or (2) observations will begin after the cohort has been defined and will continue into the future (prospective). In general, retrospective studies are cheaper and

quicker to perform, although the data quality can be quite variable. Prospective studies are more expensive, last longer, and usually establish stronger associative links between disease and exposure. Therefore, positive prospective studies usually carry greater "weight" for the environmental regulatory process.

Finally, it is important for a systematic approach to be used when epidemiological studies are evaluated as decision tools for public health policy. Hill (1965) has listed a number of criteria that are useful for judging epidemiological studies:

- **Strength of the association.** The greater the relative risk, the more significant the purported association. Relative risks that are > 1.0 but have statistical confidence intervals that are ≥ 1.0 are less powerful than relative risks clearly elevated above the control. Example: relative risk 1.5, confidence intervals 0.9–2.2, versus relative risk 3.5, confidence intervals 1.9–5.0.
- **Consistency.** Confirmation of the association by other investigators using different types of studies and methods, e.g., case control positive association confirmed by a prospective study.
- **Dose-response relationship.** Quantitative relationship between increasing exposure and disease frequency. The lack of a consistent dose-response relationship raises the question of spurious association, particularly when relative risks are low (i.e., 1.0–1.5 range).
- **Temporal relationship.** Exposure must precede disease. Do disease rates increase or decrease as exposure to the study toxin increases? For example, for EMF exposures, have nationwide leukemia rates increased through the last 50 years as electrification has increased?
- **Specificity.** Are there other factors that may be "operating" in the study population that could also explain the observed disease rates? For example, in studies of electrical workers and brain cancers, exposures to solvents are common and may act as another explanatory or associative variable as significant as the original EMF study variable. In this example, solvent exposure is a *confounder*. Alcohol, smoking, and diet are frequent confounders in many epidemiological studies.
- **Biological plausibility.** Are there reasonable or known physiologic mechanisms that can be evoked to explain the association? In the study of EMF exposure and cancer, the lack of strong biological plausibility relating EMF exposure to irreversible

cellular changes has frequently been seen as a weakness in the EMF epidemiological studies. As new basic mechanistic data are developed, the strength or weakness of this argument could significantly change.

- **Coherence.** The body of positive related facts must be consistent with the new evidence or associations uncovered by a study. New studies should expand and further explain the base of existing confirmed knowledge.
- **Experiment.** Removal of the exposure should decrease the presence of the study disease. This type of data are difficult to obtain unless there is a clear dose-response relationship between disease and a *specific* environmental toxin.
- **Analogy.** If there is a strong association between exposure and disease, then other similar exposures should produce positive associations. For example, if low-level EMF exposures from power lines are a major source of childhood leukemia, researchers would expect to find similar results from other EMF-producing sources (e.g., electric appliances and computers).

CANCER CLUSTERS

Public health officials are often called on to investigate apparent excesses of incidence of cancer among either workers or neighborhoods. These studies are typically known as *cancer cluster investigations*. *Clustering* refers to the discrete space-time aggregation of cancers or other adverse health endpoint (e.g., birth defects) in relatively small areas (Schulte et al, 1987). A small area can mean an office building, an office floor or department, or a neighborhood. Cluster investigations are typically triggered by the belief that an adverse health outcome (e.g., cancer) is occurring with elevated frequency within a small subgroup of the population or work force. Significantly, the distribution of cancer or birth defects can be affected by a wide variety of causal factors, such as: (1) environmental factors, (2) host susceptibilities, and (3) chance.

The analysis and evaluation of cluster events are invariably associated with a climate of fear and skepticism in the study group. These emotional responses are heightened when the group strongly feels that an involuntary, synthetic, environmental (industrial) exposure is the causative agent.

A detailed discussion of the risk communication management of environmental exposures occurs later in the chapter. The emphasis in this section is directed toward the systematic workup of cluster events.

Background

Cluster investigations are extremely common events for most state, local, and federal health agencies. Since the late 1980s, the Centers for Disease Control and Prevention (CDC) have significantly decreased the number of cluster investigations and referred the vast majority of studies to state and local health departments. For example, in Minnesota more than 500 reports led to only five formal studies. Similarly, the Missouri State Health Department, from 1984 to 1989, received 101 disease cluster reports. After initial review and screening, 3 of these 101 reports were formally investigated and only 1 study was able to identify a specific factor associated with the apparent cluster (Raymond, 1989). Equally negative results were found by the CDC in its investigation of 108 cancer clusters from 1961 through 1982.

Schulte et al (1987) reviewed 61 investigations of apparent cancer clusters performed by the National Institute for Occupational Safety and Health (NIOSH) during the years 1978–1984. These investigations were typically triggered by the perception of increased rate of cancer. The questions addressed during the investigations were straightforward:

- Is there an excess of cancer cases?
- Are the cases occurring independently or in some related fashion?
- Were occupational factors causally related?

Interestingly, Schulte et al discovered that relatively few alleged cases could trigger an investigation. Specifically, 45% of the initial cases alleged five or fewer cases. Only 7% of the investigations cited more than 10 cases. A wide variety of anatomical sites were specified (e.g., lung, 21%; gastrointestinal, 13%; breast, 10%; genitourinary, 8%; leukemia, 6%; brain, 5%; melanoma/skin cancer, 5%; Hodgkin's disease, 4%; and all other sites, 28%).

These few alleged case numbers are significant because of the high number of cases above background required to produce statistical elevations in study populations. Neutra (1990), in an analysis of cancer cluster evaluations in the general population, clearly shows that "Most clusters will have to occur at rates of at least eight times their expectation to even qualify as worthy of special investigation." Neutra's quotation refers to a relative risk of 8.0 or greater. Clearly, if a cancer cluster begins with small case numbers and a disease with a high prevalence rate (e.g., cancer in the general population) it will be difficult to demonstrate statistical significance. Neutra (1990) argues that if a cluster investigation had (1) a very high relative risk, (2) at least five cases, and (3) a good estimate of personal exposure, the problem of small case numbers would be less crucial. Unfortunately, as the Schulte et al (1987) investigation demonstrated, case numbers are frequently few and the ability to either quantify exposure or calculate high relative risks is limited.

Approaches

Despite the well-documented negative findings associated with the majority of cluster investigations, public and workplace perceptions will cause researchers to continue to generate studies. Therefore, a clear and systematic approach of these investigations is required. In addition, both the methodological and public relation pitfalls and problems associated with cluster investigations must be appreciated. Armon et al (1991) have produced a clear and well thought out strategy for investigating a cluster in a community. A similar strategy would be effective in a workplace setting.

This approach (Armon et al, 1991) is based on five sequential steps:

1. Case ascertainment
   A. Determine whether the cases are true cases of a specific disease entity. For example, do not mix cancer cases with birth defects. Set a diagnostic criterion.
   B. Define incident cases (i.e., what the period of observation is).
   C. Define the geographical limits and verify that cases belong within these limits. If geography is not a limiting factor, define the subpopulation and verify that cases are part of the subpopulation.
2. Based on results of step 1, calculate the expected frequency and consider two hypotheses:
   A. Null hypothesis—background frequency is being observed.
   B. Causation is attributed to a particular association.
3. If the *expected* number of cases is less than one, ask why since the small number of study cases will produce mathematically elevated rates. Is the reference population appropriate?
4. Compare case ascertainment in the study and reference populations.
   A. Use similar techniques.
   B. Verify that changes in diagnostic criteria have not occurred (i.e., misdiagnosis).
   C. Adjust for age and sex.
5. Calculate the probability that chance alone may explain the findings.

This strategy is useful because it will produce well-documented standardized incidence ratios (SIRs). This type of ratio is similar to the previously discussed standardized mortality rate (SMR) of "observed" versus "expected," where "observed" are based on national data. The SMR uses death (mortality) from a specific disease as an end point, whereas the SIR is based on incidence, or new development of a specific disease. Because it is possible to develop a disease (new incidence) but die from other causes (e.g., accidents or diseases not under study), the incidence ratio attempts to identify and capture all new cases during the study's timeframe. The calculated ratios and their associated confidence intervals must be interpreted based on the criteria discussed for epidemiology in this chapter. Strength of association, biological plausibility, specificity, temporality, etc., are all important factors to consider before a final interpretation is presented. Based on the past track record of cluster investigations, the likelihood of a negative study is quite high.

The investigators must realize that in a general public or workplace forum, emotions are strong, and three general scenarios develop (Health, 1988):

- **Disease in search of pollution.** An individual or group has an unusual health event and seeks an explanatory cause based on a general environmental exposure. An example is cancer cases in a neighborhood located relatively near power lines or substations.
- **Pollution in search of disease.** A major industrial pollution release occurs (airborne) or a long-standing release (contaminated ground water) is discovered. Both preexisting and any future disease will typically be attributed to the pollution source. Temporal association and geographic limits must be evaluated.
- **Pollution and disease in search of each other.** This situation is frequently found in the courtroom. There are numerous examples of large class action toxic tort cases that fit into this category. Examples include asbestos, Agent Orange, and dioxins.

Despite the politics and emotion associated with cluster studies, the investigators must realize that public perception and input are important, relevant, and necessary for the satisfactory outcome of any investigation. Brilliant science will be unsuccessful without adequate workplace or public involvement, because the primary purpose of the investigation is to identify and potentially control causes of the clustering event. If an epidemiological investigation is begun, careful case ascertainment and population definition are required.

The Agency for Toxic Substance Disease Registry has prepared cancer cluster investigation software. This product was released in 1993 (ATSDR, 1993). In addition, many state health departments (e.g., California and Minnesota) have well-conceived written investigation protocols. A poorly designed and executed study is less satisfactory than no investigation because expectations will be raised and not met. Therefore, experienced health professionals should always be consulted and involved.

ELECTROMAGNETIC FIELDS

Previous sections of this chapter illustrated a variety of epidemiological principles with examples taken from the EMF literature. The argument about potential health effects from EMF exposures greatly increased during the 1980s and into the 1990s. There are hundreds of published articles about the effects of EMFs in the scientific and medical literature. In addition, an increasing number of companies have been founded based on the need to measure and analyze the magnetic and electric fields produced by power lines, computers, and the myriad electrical appliances. A number of in-depth literature reviews have been published within the last two to three years, and they should be consulted for further details (Anderson, 1993; Bracken, 1993; Cleary, 1993; Krieger & Irons, 1991; ORAU, 1992; Savitz, 1993).

The debate about the effects of EMF exposures is illustrative of several important epidemiological concepts and principles that were discussed earlier in this chapter. In addition, EMF investigations are frequently triggered by cancer cluster investigations that quickly turn into dramas played out in the media and courtroom. This section briefly provides an update and overview of the EMF literature and a more detailed analysis of the epidemiological evidence that associates EMF exposures with childhood and adult cancers.

Background

Electric power systems produce electric and magnetic fields. Therefore, EMFs are present whenever electricity is generated, transmitted, or used. *Electric fields* represent the forces or strength that electric charges exert on charges at a distance, whereas *magnetic fields* are fields produced by the motion of the charge. Electric fields are measured in volts/meter (V/m) or kilovolts/meter (kV/m, one thousand volts per meter). Magnetic fields (or magnetic flux density) are measured in the System International Units (SIUs) of tesla (T) or microtesla (μT). Many of the published papers, particularly in the U.S. literature, use magnetic field units based on the centimeter-gram-second (CGS) system. These units are the gauss (G) or milligauss (mG). One gauss (1 G) is equal to 10^{-4} T; hence, one milligauss equals 10^{-7} T or 0.1 μT.

As an example, a standard 500 kV transmission line typically has a right-of-way electric field of 1–2 kV/m and a magnetic field or flux density of approximately 10 mG. A wide variety of household appliances also create large magnetic fields that decrease as a function of distance from the source. For example, an electric hair dryer has a magnetic flux density of 60–20,000 mG within 1 cm of the dryer. As distance increases from the dryer, the field strength significantly falls (e.g., it is only 0.1–3 mG at 15 cm). Televisions are also capable of producing magnetic fields of 25–500 mG at 1 cm and 0.1–2 mG at 15 cm.

Electric and magnetic fields can vary according to the frequency or number of times that the field oscillates per second. Power systems produce an oscillating EMF of 50–60 times per second. These EMFs are known as 50–60 hertz (Hz) power. EMFs in the 3–3,000 Hz range are known as *extremely low frequency (ELF)* fields. In the ELF range, the electric and magnetic fields are virtually independent.

The major focus of *in vivo*, *in vitro*, and human studies has been directed toward magnetic field effects. Each of these areas will be briefly reviewed in the following discussions.

In Vitro Studies

Cleary (1993) reviewed and summarized the major physiologic endpoint effects seen in a variety of laboratory test systems:

- effects on DNA synthesis and RNA transcription
- effects on cell proliferation
- modulation of cation fluxes and binding
- effects on immune responses

- membrane signal transduction (i.e., interaction with the response of normal cells to various agents, hormones, enzymes, and neurotransmitters).

These observations are typically produced by short-term exposures to cell test systems at 100 Hz or less and with low field intensities. There has not been a dose-response effect demonstrated; in fact, reported effects depend on field modulation and intensity "windows." A variety of effects are reported *only* within a specific and generally unique frequency window. In addition, there is a significant impact associated with the intermittent application of the field. Hence, the time and fluctuations in exposure are more significant than the magnitude of exposure. A theoretical explanation for these effects has not been clearly articulated. Therefore, the *in vitro* data have not been considered as a powerful biological tool for assessing human health effects.

In Vivo Studies

There is a substantial body of experimental research on the effects of EMF exposure to animals (Anderson, 1993). A large variety of animals, including nonhuman primates, have been exposed to EMFs under disparate conditions. Not surprisingly, results indicated that multiple biological systems can be affected:

- **Neural and neuroendocrine:** specifically, mild and transient behavior alterations, and effects on the timing of circadian rhythms; possibly an inhibition of the nocturnal synthesis of melatonin by the pineal gland.
- **Reproduction and development:** conflicting reports without a clear pattern of reproductive or teratogenic effects.
- **Immunology:** minimal effects on the immune system, positive impacts on immune response to mitogens and antigens.
- **Carcinogenesis and mutagenesis:** no positive studies to date demonstrating spontaneous tumor development. *In vitro* studies have suggested that the intracellular enzyme ornithine decarboxylase (ODC) is induced by specific extremely low frequencies. ODC is induced by a variety of substances, including certain substances that act as cancer-promoting agents. An *in vivo* equivalent of these observations has not been produced. Promoting agents are important substances that further influence the potential malignant transformation of a cell.

Overall, the *in vivo* data demonstrate that a variety of biological effects can be produced with short-term EMF exposure; however, the effects are generally subtle and/or weak and are not consistent with the type of impacts associated with other significant hazards. Further research is required before definite conclusions can be made.

Human Exposure Studies

During the 1980s and early 1990s, more than 40 studies were published that studied the potential relationship between EMFs and cancer. These studies were reviewed and summarized by Savitz (1993). In general, the human epidemiological studies can be divided into four categories (Krieger & Irons, 1991):

- residential studies of children exposed to 50 or 60 Hz magnetic fields
- adult exposure to radio frequency emissions
- residential studies of adults exposed to 50 or 60 Hz fields
- adult occupational exposure studies.

Bracken (1993) clearly articulated many of the exposure assessment problems that confound and confuse these studies:

- EMFs are not detectable by humans at ambient exposure levels; hence, developing self-reporting exposure survey data are extremely difficult.
- There is no accepted definition of dose for an EMF. In addition, there is little or no agreement about which characteristic of the field should be measured: magnitude, frequency, or variation.
- It is unlikely that a true "nonexposed" population exists for control purposes, because almost everyone is exposed to various strength EMFs during activities of daily living.
- Field strength is highly dependent on distance from the primary source. Individuals are not exposed at consistent or uniform levels. Furthermore, variations in shielding and power levels can dramatically affect daily or yearly exposure levels.
- For occupational studies, there is significant source variation and other confounding exposures to a variety of solvents and metals.

Despite these methodological problems, a number of published studies indicate that odds ratios of approximately 1.0–3.0 (95% confidence interval) can be calculated for a variety of childhood cancers (primarily leukemias) and residential EMF exposures. These studies are supported and severely criticized by a variety of well-known and established epidemiologists. A brief historical review of the epidemiological studies is very instructive.

In 1979, Wertheimer and Leeper (1979) published the results of a case-control study that indicated a 2–4 fold increased incidence of childhood leukemia associated with exposure to EMF. The dosimetry model was based on a novel approach. Primary wires were classified as "large-gauge" or "thin-gauge." Large-gauge wires were designed to carry high currents. Thus, study and control homes were classified as either having "high-current configuration (HCC)" or "low-current configuration (LCC)." The wiring code configuration (WCC) was a surrogate for presumed EMF exposures.

Correlations between actual measured EMF exposures and calculated (i.e., calculated dose based on WCC) were not obtained. In addition, there were a number of other significant methodological shortcomings that generated significant scientific controversy. Nevertheless, this paper was clearly a landmark study, because it triggered wide media coverage and resulted in multiple new and sophisticated experimental study designs.

The use of a WCC dosimetry scheme is controversial and requires further amplification, because subsequent studies employed variants of the original Wertheimer and Leeper configuration. Wire code configurations have four general assumptions:

- High-average current flow is associated with high-amp conductors.
- Distance from conductors is linearly related to magnetic field strength.
- A high magnetic field is associated with the first span from a secondary distribution transformer. Spans further downstream have commensurately lower magnetic fields.
- The magnetic field associated with a given distribution line remains constant over time.

These assumptions and subsequent refinements define the exposure unit. Actual measured exposures typically do not correlate with WCCs; however, the key assumption is that the WCC is indicative of the average integrated exposure received by the study subjects. During the 1980s and early 1990s, a variety a studies demonstrated that WCC correlates with childhood leukemia risk better than actual measured exposures. Adult residential exposure studies have produced less potentially significant odds ratios.

Based on the earlier discussion in this chapter surrounding basic principles of epidemiology, it is clear why the entire EMF issue is beset with controversy. Unfortunately, this controversy is not confined to the scientific literature. Several alleged cancer clusters have been blamed on EMF exposures from power lines and electrical substations (Brodeur, 1989). Many people view EMF exposures as a source of environmental pollution that must be minimized and regulated. In this situation, the "pollution" source will clearly attract disease and generate multiple cluster investigations and public demands for regulatory action. The scientific uncertainty fuels the public and litigation passions and sharpens the requirements for accurate and informed risk communication and public participation.

The demand for strong EMF regulatory action was crystallized on September 30, 1992, when the Swedish National Board for Industrial and Technical Development announced that it "will act on the assumption that there is a connection between exposure to power frequency magnetic fields and cancer, in particular, childhood cancer." This dramatic public policy shift was prompted by the release of two major epidemiological studies that assessed the relationship between exposures to weak EMFs and the development of cancer. The Swedish studies covered both residential and workplace exposures. The residential study was performed by Feychting and Ahlbom (1992). This study has been released in detailed abstract form and was presented on November 12, 1992, at the Department of Energy (DOE) annual power line research meeting. Media coverage of the Swedish papers was intense, and extensive print and TV coverage assured wide discussion in both scientific and lay/legal circles.

This 1992 paper was titled "Magnetic Fields and Cancer in People Residing Near Swedish High Voltage Power Lines." The study was designed to test the hypothesis that exposure to magnetic fields produced by high-voltage power lines increases cancer incidence. The study was designed as a case control study that included the population of everyone who lived within 300 meters from any of the 220 and 400 kV power lines in Sweden during the period 1960–1985. For adults, the duration of residence was at least one year. The cases were all instances of cancer diagnosed between 1960–1985. For children, all types of cancer were included; however, the adult study was restricted to leukemia and brain cancers. The selection of adult brain and leukemia cancers is based on the results of multiple published occupational studies on electrical workers.

The controls were matched to the cases on time of diagnosis, age, sex, parish, and power line. Exposure was assessed by several different methods:

1. Spot measurements were performed in the homes of subjects.
2. Power line magnetic fields were calculated by a computer program that analyzed distance, line configuration, and load.
3. Both spot measurements and line loads were simultaneously obtained and magnetic fields were calculated.
4. Historical load records were obtained from station managers.

Overall calculated historical fields were obtained for the various exposure time frames. These calculated fields were the main source of classifying study subjects into different levels of magnetic field exposure. Therefore, the main exposure unit was the annual average of the calculated magnetic field generated by the line. Finally, for a third of the subjects, 24-hour real time measurements were performed.

For childhood leukemia, at the 1 mG and 2 mG (0.1 µT and 0.2 µT) levels the relative risk increased over the two exposure levels and was estimated at 2.7 (95% confidence limit 1.0–6.3) for greater than or equal to 2 mG. When the upper cutoff point was shifted to 3 mG the relative risk increased to 3.8 (1.4–9.3) and the p-value increased from 0.02 to 0.005. Results were not significantly affected by gender, age, time of diagnosis, and area of living. The positive association was only present for one-family homes. There was some relationship with distance but no relationship with spot measurements. There was no significant association for magnetic field exposure and brain tumors or for all childhood cancers.

The authors state that the positive findings are consistent with the assumption that historical calculated field measurements are reasonably good predictors of past fields but that spot measurements are poor predictors of those fields. The positive association for one-family homes may be explained by the limited accuracy of exposure assessments in apartment houses. The authors conclude that the results provide support for the hypothesis that exposure to low strength magnetic fields increases the risk of childhood leukemias.

The Swedish studies are the first published papers that propose a dose-response relationship between EMF exposure and cancer risk (i.e., as exposure increased from 1 mG to 3 mG, the relative risk commensurately increased).

Response to the Swedish studies was immediate. The Oak Ridge Associated Universities (ORAU) panel, a large, multiuniversity scientific panel tasked to report and study EMF issues, produced a detailed and dissenting view of the Feychting and Ahlbom paper (ORAU, 1993). This response pointed out that no association was found for any of the studied malignancies among either children or adults when actual measured magnetic fields were considered. However, there was a positive association with calculated magnetic fields and both childhood lymphatic leukemia and adult myeloid leukemia. Data recalculations by the ORAU panel of the Swedish relative risks produced substantially lower and nonsignificant association for both adult and childhood leukemias when exposure intervals were changed. The ORAU panel concluded that the evidence for EMF and cancer was "empirically weak and biologically implausible." Ahlbom and Feychting (1993) responded vigorously and defended their methods and conclusions.

Hence, despite more than a decade of active and sophisticated research at three levels, *in vitro*, *in vivo*, and human epidemiological studies, the entire EMF issue has not substantially progressed toward a scientific consensus. Earlier in this chapter, criteria for evaluation of epidemiological studies and strengths of association were presented. The strength or weakness of association between EMF and cancer can be analyzed using these criteria. Policymakers, regulators, and the general public will ultimately decide the strength of association. Utility companies and manufacturers must make similar decisions. A central question is: Are exposure avoidance and minimization efforts warranted based on the current scientific evidence? If so, what level of exposure is acceptable and what cost/benefit will be achieved? If not, what further studies are indicated to conclusively demonstrate that there is no significant association between EMF exposures and adverse health impacts? With diligent attention to the issue, these questions will be answered in a reasonable and scientific fashion.

The EMF issue illustrates the difficulty of formulating broad-based public health policy when the scientific database is confusing, contradictory, and controversial. As a counterpoint to the problems of EMF policy issues, the growing public health consensus surrounding the minimization of childhood lead exposures is presented.

LEAD EXPOSURE

Unlike the extreme controversy associated with EMF exposures and public health, the worldwide focus on minimizing lead exposures has continued at an increasingly rapid pace. As an environmental/public health issue, the toxicology of lead exposure and adverse impacts shows the dramatic impact of public health research on environmental policy. A combination of basic toxicology research and long-term epidemiological studies has clearly forced a reconsideration of the potential adverse impacts of low-level lead exposures to sensitive individuals, i.e., children ages 1–3 years.

During the last 20 years, efforts to drastically reduce lead exposures have increased. The net effect has been a significant public health policy aimed at virtually eliminating lead exposures, especially for young children.

From a policy perspective, this is critical because this common, naturally occurring substance is on the verge of severe restriction. With the exception of asbestos, with which lead has many public policy parallels, outright elimination or severe restriction of a substance has usually been restricted to synthetic substances, e.g., DDT and PCBs.

Currently there is a major U.S. public health effort directed toward lead exposure minimization and/or elimination. Increasingly, major lead abatement efforts are directed toward early childhood exposures. These efforts have taken the form of expensive and controversial efforts at lead paint abatement in public and private housing. Just as in the case of asbestos abatement (see Chapter 13, Indoor Air Quality), there are real questions about whether the abatement decreases (via elimination) or increases (via poorly controlled removal) exposure. The toxicology of lead is presented and discussed so that the public health policy implications of lead exposure can be rationally understood.

Health Effects of Lead

The effects of lead on human health have been a subject of medical interest for more than a thousand years. As such, this metal has been studied longer than any other toxicant. Recently, toxicologists and epidemiologists have recognized that lead has central nervous system effects on children at levels below those of previous concern (Figure 11–1). There are indications that these central nervous system effects are persistent, remaining long after exposure to lead has stopped.

The effects of lead on children are distinct from those on adults. Children are at greater risk for a variety of reasons:

- The blood/brain barrier is an anatomical and physiological separation between the fluids in the central nervous system (brain and spinal cord) and the systemic blood. In children this barrier is not fully developed, thus allowing lead in the general circulation into the brain.
- Children are growing far more rapidly than adults. This rapid growth depends on increased uptake of calcium and other minerals and nutrients. It is likely that lead enters the body by mechanisms similar to those used by calcium or other divalent cations, thus increasing the uptake of lead across the intestinal lining.
- Because of their exploratory behavior, children are often exposed to dusts and dirt more frequently and to a greater extent than adults. The presence of lead in house dust, chalking, or peeling lead-based paint, or contaminated residential soil puts this young population at much higher risk of exposure than adults.

Particular concern for increased exposure is focused on children between the ages of one and three years of age. However, children of all ages should be screened for blood lead if environmental lead exposure is a possibility, according to the American Academy of Pediatricians.

The concern for the pediatric population is associated with the central nervous system or brain. Children known to have been exposed to lead through measurements of lead in whole blood exhibit a variety of symptoms. These symptoms range from consistent reductions in learning ability, as measured by the Bailey Mental Development Index (MDI), to coma or death at high levels that may exceed 100 micrograms of lead per deciliter of blood ($\mu g/dL$). In recent years, the Centers for Disease Control (CDC), the U.S. Environmental Protection Agency (EPA), and other federal and state regulatory agencies have focused on a blood lead level of 10 $\mu g/dL$ as a "level of concern." Estimates of the number of children in the United States who currently exceed this level of concern range into the millions.

Although particular attention was devoted to learning deficits, attention span deficits, and hearing

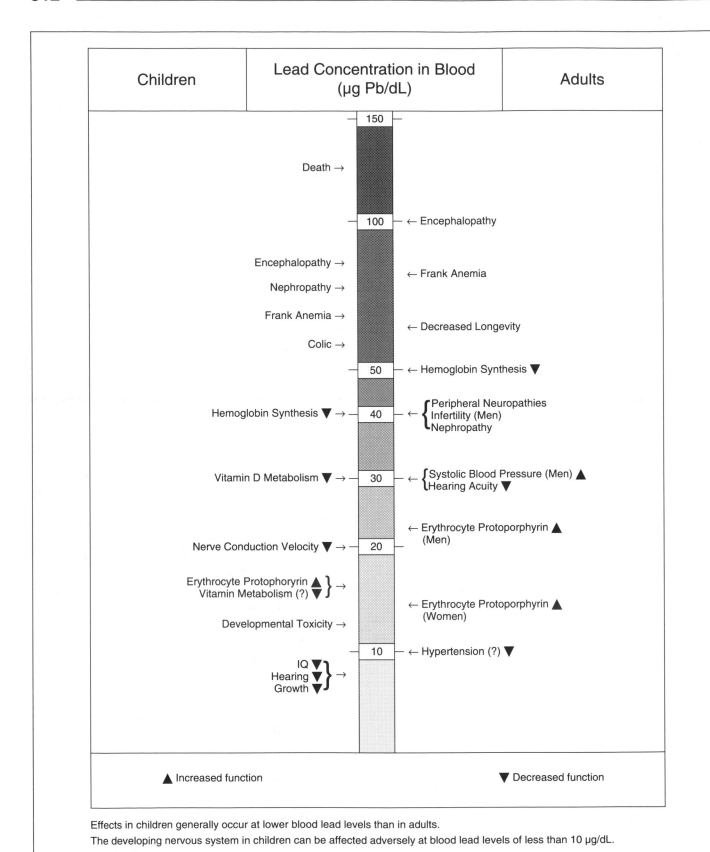

Figure 11–1. The health effects of lead vary widely as a function of dose. In children, the health effects are associated primarily with the central nervous system or brain. In adults, by contrast, observable health effects are expressed in the peripheral nervous system. Source: Adapted from case studies by the Agency for Toxic Substances and Disease Registry (ATSDR), *Environmental Medicine: Lead Toxicity,* 1990.

impairment in citing this level of concern, other hematological, neurotoxic, and blood pressure effects of lead were also considered. A report by Bellinger et al (1989) of an investigation of lead effects on a population of mostly upper-middle-class children demonstrated Bailey MDI deficits at both midrange (6–7 µg/dL) and high (≥ 10 µg/dL) exposures. This work has been substantiated by Wasserman et al (1993). Recent reanalysis of ground-breaking research conducted by Herbert Needleman of the University of Pittsburgh reaffirms public health concerns for children within the target age range for lead exposure (Schwartz, 1993).

Parental attention and adequate nutrition may compensate for small lead-related deficits. However, when such deficits occur in populations of children as occurs in areas of high lead exposure, the effects are clearly significant (Figure 11–2).

The social costs of allowing childhood lead exposure to continue unabated are not trivial. Monetary costs of providing specialized assistance to children below a learning threshold could be prohibitive. Because of the importance of lead exposure for the childhood population, significant efforts have been focused on understanding detailed aspects of lead exposure and uptake. One issue of importance to public health officials is the amount of environmental lead absorbed into the body ("bioavailability").

Bioavailability of Lead

A measure of the amount of toxicant available to cause an effect is referred to as its *bioavailability*. Bioavailability may be defined differently depending on the interest and objectives of the investigator. Generally, the pharmacologist defines it as "the amount of contaminant absorbed into the general circulation and the rate at which the material is absorbed."

The issue of lead bioavailability is particularly important because of the number of residential areas contaminated with soil and house-dust lead and the tendency for children to absorb greater amounts of lead than adults. The wide variety of chemical forms of lead found in paint, water, house dust, and mining/smelting wastes in these areas precludes generalized assumptions about lead absorption. In order

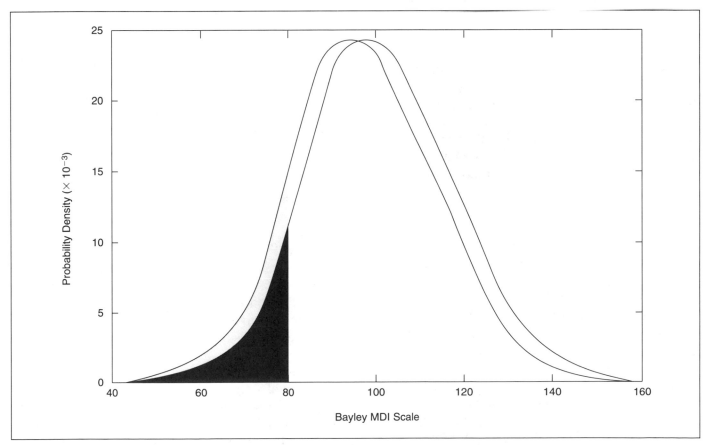

Figure 11–2. The effect of a four-point downward shift in Bayley MDI scale. The curve depicts a normal distribution with a mean of 100 and a standard deviation of 16. The lighter shaded area is approximately one-half the area covered by the darker shading. Source: Davis JM. Risk assessment of the developmental neurotoxicity of lead. *Neurotoxicology* 111:285–292, 1990.

to provide a cost-effective abatement of serious lead exposure problems, an understanding of lead bioavailability in soil, house dust, water, and food is imperative. Assumptions of complete lead absorption would greatly overestimate potential lead exposure hazards, whereas unrealistically low estimates of bioavailability would not be protective of human health.

The absorption of lead is highly dependent on dose. Children tend to absorb a higher proportion of injested lead at low doses than at high doses. These "saturation" effects have recently been confirmed in experimental animals and should be carefully considered when interpreting studies intended to assess lead absorption.

Existing estimates of lead bioavailability span a wide range. It is clear that physiological as well as physicochemical factors may influence the absorption of lead across the gastrointestinal (GI) tract (Weis & LaVelle, 1991). Younger animals and humans are known to absorb greater amounts of lead than do adults (Armbrect, 1979; Mooradian & Song, 1989). Animal studies that do not address the age-dependence of lead bioavailability should be interpreted with caution. Exposure to lead concurrent with a meal greatly reduces the amount of lead taken up by the body.

Additionally, both the chemistry and physical characteristics of the particular lead form may influence absorption. Certain lead forms, such as carbonates and halides of lead found in lead-based paint or automobile emissions, may be absorbed to a far greater extent than lead forms associated with the mineral extraction industries. Sulfides and certain phosphates of lead are relatively insoluble, and their absorption may be measurably lower than other lead forms. However, studies conducted to verify the bioavailability of these materials have netted conflicting results (Freeman et al, 1991; LaVelle et al, 1991; Rabinowitz et al, 1980; Weis et al, 1993).

Particle size and matrix may also be important determinants of lead absorption. Ionic lead found in drinking water or small particulate emissions from automobiles and smelters may be far more bioavailable than lead that is encrusted in a mineral matrix such as iron sulfide. Conscientious environmental analysis of lead for the purpose of risk determination should consider the many characteristics that may influence bioavailability. Studies underway to assess the bioavailability of lead and other metals are likely to improve our understanding of this important issue

in the future (Freeman et al, 1991; LaVelle et al, 1991; Weis et al, 1994).

Health Effects in Adults

With the onset of maturity, the health effects of lead change. Toxic effects of chronic exposure appear in the peripheral nervous system, in the cardiovascular system (hypertension), and, at higher exposures, in the central nervous system (encephalopathy) (Figure 11–2). Gastrointestinal distress (abdominal pain) is a typical early symptom of acute lead exposure in both children and adults. In animal studies, lead has been shown to affect the male reproductive system in a variety of ways, including altered sperm morphology, decreased motility, and reduced number. Such effects may occur in humans following chronic exposure at blood lead levels of 40–50 µg/dL (Wildt et al, 1983). More recently, investigators have suggested that lead exposure in male rats may result in neurological deficits in offspring (Gandley et al, 1992). However, these claims remain unsubstantiated in humans.

Maternal blood contaminated with lead equilibrates readily with blood of the growing fetus. Obvious effects associated with *in utero* exposure to lead include decreased head circumference, decreased chest circumference, and reduced birth weight. Clearly established effects of lead on the developing fetus are derived principally from animal studies. Maternal exposures greater than the level of concern for children should be avoided. Maternal factors such as nutritional state, hypoxia, or concurrent exposure to other contaminants may have a bearing on prenatal lead toxicity.

Assessment of Multimedia Lead Exposure

Among the most useful tools for assessing risk of exposure to lead (or other environmental contaminants) are well-designed environmental monitoring programs. Without well-designed and carefully implemented environmental monitoring, adequate quantification of exposure is not possible. It is important for environmental field personnel and their managers to ensure the collection of environmental data for the purpose of determining multimedia lead exposure. Before conducting any sampling effort, those involved should consider: (1) the specific objectives of the sampling effort, the level of data quality required to meet those objectives, and the usefulness of the measurements from the perspective of childhood lead exposure; (2) the statistical relevance of the sampling design and number of samples to be taken to meet specific statistical criteria; (3) the accuracy and precision of the analytical measurements planned. Following is a brief overview of approaches useful for monitoring lead in residential environments. For more detailed information, the reader is referred to U.S. EPA's *Air Criteria Document for Lead* (1983) and the Society for Environmental Geochemistry and Health Monographs, *Lead in Soil: Issues and Guidelines* (SEGH, 1988) and *The Proceedings of the Symposium on the Bioavailability and Dietary Uptake of Lead* (SEGH, 1991). Much of this information is applicable to the industrial environment as well.

Sampling for lead in the home environment: Lead paint issues. At the outset of any discussion of environmental monitoring one should consider the pertinent exposure pathways. It is generally accepted that exposure risk of children to environmental lead is dominated by the dust pathway. Incidental ingestion of house dust and residential soil by way of hand-to-mouth activity is the predominant lead exposure route for preschool children. It follows that residential monitoring programs that do not address these pathways either directly or indirectly fall short of the needs of those responsible for quantifying potential risk.

A principal source of lead, considering that lead has now been removed from gasoline in the United States, is lead-based paint. Peeling and deteriorating paint can pose an extremely high risk for children dwelling in such environments. The ingestion of a single chip of highly leaded paint may be sufficient to cause serious gastrointestinal symptoms and accompanying neurological symptoms severe enough to warrant hospitalization. A more widespread and less acute problem is the result of chalking of paint suspected of contributing significantly to elevated levels of lead in household dust.

Lead-based paint can be monitored using x-ray fluorescence technology now readily available. The x-ray fluorometer contains an energy source that is of sufficient strength to excite the inner-shell electrons of the target (lead). The relaxation of these electrons results in emission of an energy, the wavelength of which identifies the elements in the target. The intensity of the emission is an indication of the quantity of element present. X-ray fluorescence measurements are highly sensitive to the measurement technique and background matrix of the metal of interest. Great care should be taken to calibrate such instruments against more established analytical methods to ensure accuracy and precision of the

measurements. The presence of lead-based paint, in itself, does not necessarily constitute a threat. Layers of nonlead paint, wallpaper, or other physical barriers may reduce or eliminate exposures. In order for this material to pose a health problem, exposure must occur. Thus, lead-based paint that is in disrepair is likely to pose a much more serious threat than well-maintained paint. Incorrect removal of lead-based paint by untrained individuals can result in extremely serious exposure and toxicity to children in the area and is strongly discouraged.

Sampling for lead in house dust. House dust is likely to be the most significant source of lead exposure for children. Table 11–B demonstrates the potential importance of house dust as an exposure medium by comparing the relationship between blood lead and soil for several areas of known contamination with a similar relationship to house dust. Note that the relationship between house dust and childhood blood lead levels seems to remain similar from site to site. By contrast, the relationship between blood and soil lead can vary widely.

As the most significant source of lead exposure for children, house dust is of particular interest to those responsible for assessing lead risk for the childhood population. Efforts are underway to establish standard methods of measuring lead concentrations in house dust that are meaningful for the purpose of assessing exposure to children. To date, most efforts have involved the measurement of lead concentration (μg lead/gram house dust) and lead loading (μg lead/M^2). Both pieces of information may be collected at the same time if advance preparation is

Table 11–B. The Relationship Between Blood Lead (PbB) and Lead Found in Soil and Dust

Study	Type	Blood-Lead Coefficient (± s.e.) (μg/dL PbB per 1,000 μg/g)	
		Soil Lead	Dust Lead
E. Helena—1983	Active Smelter	2.24 (0.79)	1.54 (0.51)
Kellogg—1983	Unused Smelter	0.34 (0.16)	2.14 (0.85)
Midvale—1990	Mixed	3.05 (0.89)	1.55 (0.58)
Butte—1990	Mining/ Smelter	0.00 (0.12)	1.99 (0.59)

While the slope of the relationship between blood lead and soil lead varies widely from one type of lead-contaminated site to another, the relationship between blood lead and dust lead remains more consistent. Source: Marcus, 1991. © Academic Press.

made to do so. Those investigating exposure should give thought to the specific exposure area(s) of concern within the home. For example, although measurements of lead concentrations in attic dust may provide a useful indication of historical accumulation or potential source, the attic is not likely to be a good indicator for quantification of daily exposure. By contrast, playrooms, living areas, windowsills, and other highly used areas are likely to provide more pertinent information regarding regular exposure. Chalking or peeling paint near the window may be a significant source of lead for measurements taken here.

Sampling for lead in soils. Lead-contaminated soils are present throughout most urban areas of the world. One of the principal sources of this contamination stems from the use of alkyl lead in gasoline. The oxidation of alkyl lead in automobile engines results in the emission of extremely small particulate lead halide, which has found its way into soils adjacent to heavily trafficked streets and highways. Besides lead in auto emissions, lead-based paint on the exterior of homes has resulted in significant residential contamination. Exterior paint contamination is generally localized near the house; however, such contamination may be tracked into the home by residents or pets. In addition to these sources of residential soil-lead contamination, areas in or near mining or nonferrous smelters are often associated with significant soil-lead contamination.

Sampling for lead in soil that is conducted for determining a potential threat should be well planned and executed. Care should be taken to focus sampling on areas likely to be frequented by the population of concern (preschool children) and should be focused such that realistic measures of exposure are obtained. The following objectives should be considered by those responsible for sampling soils for lead contamination:

- Sampling should be focused in areas likely to be frequented by the population of concern. Play areas and bare areas may be particularly important sources of regular exposure.
- The sampling technique should allow for assessment of soil horizons pertinent to contemporary exposure. Samples taken at the surface, such as 0–1 in. (0–2.5 cm) are likely to be more representative of contemporary exposure than samples taken at depth.
- Thought should be given to the particle size of most concern. It has been suggested that a particle

size fraction less than 250 μm (#60 mesh sieve) can adhere to clothes, hands, toys, and pets and thus is both mobile and more likely to be ingested by small children.
- Finally, the diversity of lead types and the wide range of solubilities and potential bioavailabilities suggest that some attention should be paid to the chemical form of lead present at any particular area of concern.

Sampling for lead in water. Although some residential drinking water supplies may have elevated concentrations of lead, most lead found in water derives from lead piping and lead-based solder used within homes and to connect homes to local water distribution systems. As water sits in these pipes, lead leaches into the water as a function of stagnation time. The rate of leaching depends on both the amount of lead in the pipes and the chemistry of the water such as hardness and acidity. Acidity of the water can increase the hydrolysis of lead ions from the piping, thus greatly increasing the leaching rate.

Typically, the lead concentrations are greater following overnight stagnation within the pipe. Samples drawn before first use of a faucet in the morning are likely to produce the highest lead concentrations. By contrast, running the water to completely purge the household plumbing of stagnated water will result in lead measurements indicative of the distribution water only. Actual daily exposures to household water are likely to be a mixture of fully and partially purged water samples.

In the United States, the issue of lead exposure has basically moved from lowering ambient air levels to eliminating or significantly reducing lead house dust, paint, and soil exposures. To address the multimedia nature of residential lead exposure, the U.S. EPA has developed a personal computer-based exposure model that predicts likely distributions of blood lead for children given specific exposure conditions (U.S. EPA, 1994). This integrated exposure, uptake, biokinetic model (IEUBK model) combines modern knowledge of multimedia lead exposure, the biology of lead uptake, and biokinetic information about lead distribution and elimination from the body to estimate likely blood-lead distributions in the childhood population of concern. Major public health programs have been initiated that are directed toward lead paint abatement. Although the scientific rationale is clear, the execution of a general lead paint abatement program is far from simple. The cost and expertise required to perform lead paint abatement are significant, and poorly performed abatements greatly increase exposure and potential adverse health impacts. Finally, the remediation level of lead contamination in soil depends on the bio-availability of the lead material. As discussed, lead bioavailability depends on the form and source of the lead; e.g., mine waste versus lead paint. Bio-availability of lead is also highly dependent upon age and nutritional status.

Legislative efforts to remove lead from gasoline, packaging, paint, and other consumer products are aimed at reducing childhood lead exposure in the United States. The positive results of these important efforts are evident in the overall reduction in child-hood blood-lead concentrations nationwide. How-ever, widespread problems of childhood lead poison-ing remain in underprivileged urban communities where deteriorating housing conditions contribute to exposure through chalking and peeling paint. De-teriorating paint can contaminate house dust, con-tribute to elevated soil-lead concentrations, and pose a serious acute threat to children who may ingest flakes or chips of peeling paint. The Lead-Based Paint Poisoning Prevention Act [42 U.S.C. 4822(d)(1)] has been passed; it addresses exposures to deteriorating lead-based paint.

Homes built using leaded pipes, lead-bearing met-als in plumbing fixtures, or lead-based plumbing sol-der may pose an additional source of lead exposure. The presence of such lead-based construction mate-rials may be of particular concern if the domestic water is slightly acidic or first-draw water is regu-larly consumed. Modern construction materials and practices generally avoid the use of lead in either paint or plumbing. However, the vast number of res-idential dwellings with deteriorating lead-based con-struction materials may pose a significant public health concern for years to come.

Overall, the increasing efforts at severe reduction in allowable lead exposures represent a trend that will persist well into the 21st century. The environ-mental public health policy implications of this elim-ination/restriction will continue to unfold and gen-erate controversy and debate.

RISK COMMUNICATION

Both the lead and EMF issues illustrate the gap that can occur among scientists, policymakers, and the public. During the past 10 years, there has been growing recognition of the need to develop and im-plement a strategy of public involvement for envi-ronmental issues. The practice of risk communica-tion and public involvement continues to develop and evolve. This section presents a brief overview of this new field.

In the past, when government officials spoke in a public meeting, they were usually believed. People left feeling relieved or alarmed depending on the message, but they believed that the official was com-petent and credible. Now the presumption fre-quently is that government bureaucrats are incom-petent, that people have to find the truth themselves, and that the government has a hidden agenda or may even be an obstacle to the truth.

Scientists, by training, tend to focus more on data than on feelings. On the other hand, citizens focus on how they believe a particular situation might af-fect their lives. Unfortunately, neither side listens well to the other. Government and industry repre-sentatives often feel frustrated with communities that do not seem to listen and that appear to be frightened about the "wrong" risks. In response, de-cision makers and their staff can choose to ignore the public, even though it is likely that they may later face increased hostility. Or they can choose to inter-act more effectively with the public.

An important challenge for the government or in-dustry representative or the scientist is to place en-vironmental risks in perspective for the public. Effective risk communication can help decision makers to:

- understand public perception and more easily pre-dict community response to proposed actions
- increase the effectiveness of risk management de-cisions by involving concerned parties
- improve dialogue and reduce tension between communities and government or industry
- explain risks more effectively.

No matter how well it is done, risk communica-tion cannot and should not replace effective risk management or reasonable environmental regula-tion. The field of risk communication, which ex-plores perception and communication of a variety of risks, is growing rapidly. However, the research lit-erature lags significantly behind the wisdom of the best practitioners, who have been experimenting with a variety of approaches to communicating risk. Most importantly, this literature does not immedi-ately translate to practice. Those who are on the fir-ing line and have to answer the mother's "simple" question, "Can you guarantee me that my child will not get cancer by playing near this facility?" will appreciate risk communication techniques that can prove successful.

Studies of risk perception examine the judgments people make when they are asked to respond to potentially hazardous conditions. Perceptions are grounded in people's value systems, not facts and figures. What may sound irrational to some may appear rational to others simply because of differences in individual values and beliefs.

Understanding the way in which the public perceives and accepts risk is key to effective communication about risks. People's concerns are less influenced by what they are told about risk probability than by such factors as whether they consider the risk to be voluntary, controllable, or familiar. The goal of this section is to help the reader better understand public risk perception and become a more effective risk communicator.

How the Experts Perceive Risk

Technical experts view risks as statistical expressions of probability such as "expected annual mortality rates." These same experts are often confused and frustrated by public reactions to risk. People's tempers may flare at a public meeting concerning a risk that might cause less than one in a million chance of an increase in death by cancer. Yet, people smoke during the meeting break and do not wear seat belts when they drive home. These latter risks are far greater than those discussed at the public meeting. However, if scientists point out this contradiction between perception and actual behavior, people become even angrier.

Sometimes, experts dismiss such community reaction as illogical and emotional, and conclude that the public is unable to understand the scientific aspects of risk. Perhaps they believe that communities are in no position to make decisions about how to deal with the risks that confront them and likewise are in no position to make decisions about how their proposed project will affect them. However, when decisions are made that affect communities without involving them first, public outrage can result.

To illustrate how risks are perceived by different groups, Paul Slovic conducted an interesting and highly informative study in 1987 (Slovic, 1987). He used the League of Women Voters and a group of college students to represent the public's opinions and compared those responses with a group of experts in risk-associated professions.

Each group was asked to rate a list of perceived risks, from 1–30, with the most feared or highest perceived risk being number one. The results of this study, presented in Table 11–C, illustrate how perceived risk varies among different groups of people. Why, for example, is there a significant difference of perceived risk between the public and the experts on issues such as nuclear power? Why do the experts perceive this as a much lower risk than do the other groups?

- Is it based on different levels of knowledge between the two groups?
- Or, does the result reflect a recent increase in lack of public trust in the government?
- Is it because of the times? During the 1980s and 1990s people have become more aware of nuclear power, how it is manufactured, how it is handled, and how the wastes it creates are disposed.

Table 11–C. Risk Perceptions by Public and Experts

Activity or Technology	League of Women Voters	College Students	Experts
Nuclear power	1	1	20
Motor vehicles	2	5	1
Handguns	3	2	4
Smoking	4	3	2
Motorcycles	5	6	6
Alcoholic beverages	6	7	3
Private aviation	7	15	12
Police work	8	8	17
Pesticides	9	4	8
Surgery	10	11	5
Fire fighting	11	10	18
Large construction	12	14	13
Hunting	13	18	23
Spray cans	14	13	26
Mountain climbing	15	22	29
Bicycles	16	24	15
Commercial aviation	17	16	16
Electric power	18	19	9
Swimming	19	30	10
Contraceptives	20	9	11
Skiing	21	25	30
X-rays	22	17	7
High school and college football	23	26	27
Railroads	24	23	19
Food preservatives	25	12	14
Food coloring	26	20	21
Power mowers	27	28	28
Prescription antibiotics	28	21	24
Home appliances	29	27	22
Vaccinations	30	29	25

Note: Highest rank 1; Lowest rank 30
Source: P. Slovic, Perception of risk. *Science*, 236 (April 1987). Copyright 1987 by AAAS. Reprinted with permission.

Consider another example. Why do the League of Women Voters view the "risk" of food preservatives so differently from experts and college students? What risk perception factors are at work? A number of different factors might include:

- The age difference between the students and League-members. Does the difference in generations factor into the results? Are preservatives more familiar to League members?
- Does recent information about the chemical content in preservatives influence the groups differently? Does the fact that they are synthetic rather than natural add to the perception of risk?

Many factors play a part in risk perception. The discussion that follows addresses specific factors that influence how people perceive risk.

How the Public Perceives Risk

Although the technical experts define risk by using probability statistics, risk has a different meaning to the public. The statistical probability (what experts mean by risk) can be called a "hazard." The other factors, collectively, can be called "outrage" (Sandman et al, 1987). Risk, then, is the sum of the hazard plus the outrage that people feel. The public pays too little attention to hazard; the experts pay too little attention to outrage.

Conflicting perceptions of risk are demonstrated in the way we choose to live our lives, i.e., voluntary versus involuntary exposure. For example, one individual who spends the week demonstrating against the transport of radioactive waste through his or her neighborhood (an involuntary exposure) may spend the weekend rock-climbing, parasailing, or skydiving (a voluntary exposure). To another, the use of a household product, however toxic, may seem much less risky (a voluntary exposure) than a high-voltage power line in the community (an involuntary exposure).

People perceive risk differently. The factors that influence how an individual will perceive a given risk are discussed in the following paragraphs (Sandman et al, 1987).

- **Voluntary versus involuntary.** EPA research has shown that a voluntary risk is more acceptable to people by three orders of magnitude than a risk to which they are exposed involuntarily. The risk of getting cancer from smoking for 20 years may be statistically greater than the risk from drinking well water contaminated with low levels of trichloroethylene for 20 years. The difference is that a person chooses to smoke in the first instance but may not even know about the contamination in the second.

- **Natural versus synthetic.** Natural risks are less acceptable than voluntary risks but more acceptable than risks created by people. Naturally occurring radon in a person's basement is more acceptable than a radioactive waste disposal facility a mile away from that same person's house.

- **Familiar versus unfamiliar.** If people have experience with a specific risk, their reaction will be stronger than if they don't. Flooding is an example. If people suffer severe damage in a flood, they will be less inclined to stay in their houses when the next flood warning comes along.

- **Not very feared versus very feared.** One of the most feared diseases in the United States today is cancer. Parents may be more concerned about exposing their children to the possibility of cancer from asbestos in the ceiling of their classroom than about exposing them to busy traffic while they walk to school.

- **Diffused in time and space versus focused in time and space.** Many people are more afraid of flying in airplanes than of riding in cars, even though hazard data indicate that cars are more risky. But people in airplanes are clustered in time and space. When an airplane crashes, many people are killed at once. For illustrative purposes, assume that 5,000 people in a given state died from smoking this year. This risk is acceptable because those people died one at a time privately. If they all died on the same day and in the same city, smoking might be outlawed immediately.

- **Controlled by the individual versus controlled by society.** In general, members of the American public consider themselves better-than-average drivers. They feel in control when they are driving. How many people feel comfortable riding while someone else drives? The person with the wheel holds the power. A sense of powerlessness can lead to public outrage.

- **Fair versus unfair.** If the risks and benefits are not distributed evenly, people may perceive the risks as unfair even if the hazard is quite low. A good example that we often see in the western United States involves the siting of a high-voltage transmission lines, often through neighborhoods, parks, and agricultural areas. Although the lines may benefit many people, the neighboring communities may feel that they have to "pay the price" or sacrifice their quality of life for a transmission line from which they receive little or no benefit.

- **Trusted sources versus nontrusted sources.** Do people trust the entity urging them to accept risk? What is their image in the community? A project proponent can also suffer from how much or little trust people have in their industry as a whole. In general, trust is more easily lost than gained. And yet, in the absence of scientific data, being trusted is critical to having the public listen.

Generally, people pay attention when they are told something is a big risk but not when they are told not to worry. People tend to be more trusting when they are told something is dangerous than when they are told something is safe. This is because the public has a greater investment in the dangerous outcome.

The greater the number and seriousness of these "outrage" factors, the greater the likelihood of public concern about risk, regardless of the data. Recalling the results presented in Table 11–C, the risks that provoke the greatest public concern may not be the same risks that scientists have identified as most significant. When scientists dismiss the public's concern as misguided or irrational, the result is anger, distrust, and still greater concern. In other words, outrage increases. This does not suggest that the members of the public disregard the scientific data but that their outrage also matters and should not be ignored.

Scientific experts may say to themselves, "If only members of the public would listen to reason; if only they would understand the facts and benefits." So an outside expert is brought in to offer an explanation. Does this strategy work? Often it does not. The public may bring in an opposing expert—perhaps a neighbor who is a doctor or an agency representative. The debate may focus on personal anecdotes relating to what a neighbor or relative said he or she saw or heard. To the public, personal anecdotes are often more compelling than statistics. The public usually gets the last word in a battle of experts.

Understanding Public Values

When addressing risk-related issues, experts too often try to change people's attitudes. They forget that those attitudes are usually the product of a lifetime of developing one's personal values. Technical experts should understand the importance of values held by potentially affected publics. Experts need to appreciate how these values relate to the ways in which the public perceives environmental risks.

Values are internal standards by which people judge events or behavior to be good or bad, right or wrong, fair or unfair. People derive their values from upbringing and experiences. Parents, religious institutions, and schools can be sources of values. The media are also influential sources.

It is commonly stated that personalities are formed in the first three to four years of life, when life is observed and the foundations of a value structure are laid. What people see and learn is perceived as being "normal." For example, three-year olds in China, Great Britain, and the United States learn to eat their meals holding their utensils differently. Each way is considered normal for them. People also go through intense modeling and socialization phases during their early teens. During this time, a person experiments, modifies, and validates his or her values. Hence, by the time a person reaches the 20s, values become set.

Values are also influenced by the times in which people live:

- People born in the 1920s and 1930s tend to be be intensely patriotic. The Norman Rockwell picture of the family unit was prevalent.
- People who lived through the Great Depression are often very thrifty and security-oriented.
- Those who grew up during World War II tend to be patriotic and have a strong commitment to win.
- In the 1950s, the good life arrived. The baby boomers composed the first generation that was totally indulged.
- Those who grew up in the 1960s tested the basic values of free speech and freedom to choose their way of life.
- The 1970s have been called the "me" generation.
- In the 1980s, people went for the gold: Get as much and as fast as possible.

Once recognizing the role that values play in communication, one may be able to better understand how the public perceives risk.

Communicating to the Public about Environmental Risk

It is difficult to explain risks in ways lay people can understand. But the public's ability to understand the science should not be underestimated. Talking about one in a million increased cancer deaths or comparing one part per billion to one second of time in 32 years or to one sheet in a roll of toilet paper stretching from New York to England leads to two errors: (1) it assumes that low concentration means low risk, and (2) it trivializes the risk. Scientists struggle to convey complex technical information, believing that if they could find a way to explain the data more clearly, citizens would understand the risks the way that scientists do and therefore be less concerned.

One must not mistake the need to explain complex data more clearly as placating the public. The scientist must develop sufficient trust and credibility, place a priority on understanding the public's concerns, and involve members of the public in risk decisions that affect their lives. The most successful communications with the public rely more on mutual respect and understanding and improved interaction with the public than on innovative ways to explain the data.

Nonetheless, it is imperative for agency and industry representatives to try to explain risk information as completely and clearly as possible. Some concrete examples of successful communication may be helpful.

Give the public a mental picture of the parts per billion or tons per day. By using a verbal comparison or graphic representation, communicators can help convey a message that is meaningful and memorable. For example, saying that "Smith County produces 125,000 tons of refuse per day or enough to fill 50 football fields 14 feet deep," will have more effect than simply stating the statistic.

Consider a common question in refuse-to-energy plants, "How much dioxin will come out of the stack?" It may be appropriate to say, "The incinerator will be designed to meet all applicable Resource Conservation and Recovery Act (RCRA) and Toxic Substance Control Act (TSCA) requirements," but that does not explain the possible magnitude of dioxin emissions. Equally unacceptable is the answer, "The design we are currently considering emits 0.11 nanograms of dioxin or less per cubic meter. We expect to achieve even cleaner emissions with further controls." This statement uses the word "nanogram," which is meaningless to the public, yet it still does not explain how much dioxin will fall into community backyards.

Martha Bean suggests that a more satisfactory answer might be:

We believe this plant will not emit more than 0.11 nanograms of dioxin per cubic meter of exhaust gas. We expect that the incinerator will not emit more than X

cubic meters of exhaust gas per day. That works out to about X ounces of dioxin per day falling into the backyards of homes within 500 ft from the stack (Bean, 1988).

The citizen's question was "how much," but an underlying question was "will it hurt me?" By answering the question "how much" with accuracy, the communicator may unwittingly raise the concern of the listener. A possible addendum to the above answer may be:

You may be wondering what X ounces of dioxin means to your health if you live within 500 feet of the stack. If you breathe the air outside your home 24 hours a day for 70 years, you would have slightly less than one chance in a million of contracting cancer from the dioxin that you breathe from the incinerator (Bean, 1988).

The effective communicator must be an insightful listener to identify the real or underlying questions. Effective communication about the magnitude of risk requires giving the listener information about probability. As an example, the scientist is comfortable with the term "excess lifetime cancer risk." But to the layperson, it conveys something unknown and terrible.

Various agencies may require cancer-causing agents to be controlled to the 10^{-6} level at hazardous waste sites. This phrase is unintelligible to the average person even if it is explained by saying, "The excess lifetime risk will be 10^{-6} at this site after cleanup. That means the average incidence of cancer is increased by one per one million people exposed over a lifetime."

Martha Bean gives two other examples of how this information may be communicated more clearly:

At this site we would expect that among one million people drinking two liters (about eight glasses) of water per day for their entire lives, one person may get cancer caused by the contaminant in the water. This person may not die of it.

Let me try to explain what we call a 10^{-6} risk of cancer in personal terms. First, let's assume that you know 200 people in this city. Of these 200 people, 50 will probably contract cancer, regardless of where they live or what water they drink. If all 200 people you know drink water from your well all of their lives, there is a 1 in 5,000 chance that one more of the 200 people you know will contract cancer (Bean, 1988. Copyright © 1988 by National Solid Wastes Management Association. Reprinted with permission.).

Avoid comparisons that seem to minimize or trivialize the risk. If comparisons are used, compare risks that seem similar to the listener. The following guidance from the Chemical Manufacturers Association workshop on risk comparisons may be helpful (Covello, 1988):

- Use comparisons of the same risk at two different times. For example: "The risk of X is 40% less than it was before the latest discharge permit." Or: "We plan to institute new standards next year that will reduce the risk by 10%."
- Compare with a standard. "The emissions are 10% of what is permitted under the old EPA standard and slightly under the level established by the new EPA standard."
- Compare different estimates of the same risk. For example, compare the most cautious "worst case" estimate with the most likely estimate. Or compare the agency's estimate using one method with its estimate using another. Or compare the agency's estimate with an industry's or environmental group's estimate.
- Compare with similar data found elsewhere. For example, compare one community's drinking water data with levels found in other communities in the state.

Thomas Burke of the New Jersey Department of Health says that he tries to put specific risks into context with "the baseline kind of risk . . . the big picture." For example, if you know that the benzene in your air is only slightly higher than the level in the national forests, says Burke, "That's something that's helpful. You're not going to understand whether it's going to cause leukemia in you, but certainly your risks are no higher than anyone else's in the country" (Sandman et al, 1987).

Dealing with the issue of uncertainty is another challenge for the risk communicator. There are at least four types of uncertainty about which scientists typically interact with the public: (1) the uncertainty of science in general, (2) the inexactness of the risk assessment process, (3) the incompleteness of the information the agency has gathered so far, and (4) differences of opinion about the implications of the information and the optimum risk management option.

Acknowledge uncertainty by learning to say "I don't know." This may be difficult to do, but it is a better strategy than claiming to know more than is known. It may even enhance the communicator's credibility because people will recognize his or her honesty and forthrightness. However, citizens need to be given enough background so they do not think something is amiss when the communicator says he or she does not know. People need to understand that uncertainty is part of the process. Be specific about what is being done to find the answers and how research is being conducted. It is important to say "I don't know" so that it does not sound like "I don't care."

One of the predictable things that will happen when one explains a risk number is that someone in the audience will point out that it is not zero, following up with the demand that it should be. The agency or industry must explain that zero risk does not exist. In our daily lives people take quite substantial, yet avoidable risks, so it is not reasonable to demand zero risk from a chemical company. Covello et al (1988) suggest several possible ways to respond to the demand for zero risk:

- The demand for zero risk may be reasonable. The plant may be able to (in effect) eliminate virtually all the risk through some change in procedure. If so, consider the costs and benefits of doing so.
- The demand for zero risk may be an exaggerated way of making the point that the risk is too high. The response to this demand may be complicated by the fact that although further risk reduction is entirely feasible, the remaining risk is insignificant and such action makes no economic sense. However, do not point out the economics. The public will generally reject the argument that the reduced risk to the community does not justify the added costs to the company. If this happens, the discussion has moved from the inherent correctness of a fact to a discussion of how choices based on that fact compare to other choices. In other words, science is left behind and the discussion moves into values.
- The demand for zero risk may be sincere but ill informed. If so, respond gently with some fundamental risk education. Explain that zero risk is nonexistent. Point out that all of life's activities carry some risk, which are usually ignored if the risk is small enough and if the activity is beneficial enough. Be sure to agree that the risk should be made as low as possible and discuss what is being done to achieve the lowest possible level.
- The demand for zero risk may be politically motivated and designed to attract the interest of politicians and the media. The agency or industry is

unlikely to make much headway in public meetings. Try to arrange a private meeting where frank discussion and negotiation might be possible.

- Finally, the demand for zero risk may be a reflection of emotional distress stemming from outrage, anger, and distrust. Risk may not be the central issue. Communicators must address themselves to the underlying antagonism and begin the slow, hard work of building trust.

SUMMARY

The potential public health impacts of lead exposure (particularly childhood exposure) have been well established and form the basis of policy efforts directed toward elimination and/or significant restriction of lead exposures. In the United States, the policy debate has been ongoing for the last 20 years and substantial reductions in general population blood lead levels have occurred. This lowering of total lead body burden has widely been attributed to the removal of lead from gasoline and the subsequent lowering of ambient lead air concentrations.

Unfortunately, a large number of countries worldwide still have leaded gasoline. In particular, the overwhelming majority of developing countries are essentially dependent on leaded gasoline. From a public health perspective, this is quite unfortunate, because developing countries have demographic profiles that are heavily weighted toward children under the age of six. Many environmental/public health advocates feel that an entire generation is potentially irreparably exposed. The attendant public health costs and effects are potentially enormous.

The interactions among epidemiology, toxicology, and risk communication are some of the most significant relationships affecting public health perceptions of environmental hazards, e.g., controversies surrounding EMF and lead impacts. Major public policy decisions on environmental affairs will continue to reflect the interplay of "hard" science and the public's assessment and perception of risk.

REFERENCES

Agency for Toxic Substances and Disease Registry. *Chapter 3.1 Software System for Epidemiologic Analysis*. Atlanta, GA: ATSDR, Division of Health Studies, Feb. 1993.

Ahlbon A, Feychting M. Letter to the Editor. *Science* 260:2, April 1993.

Anderson LE. Biological effects of extremely low-frequency electromagnetic fields: *In vivo* studies. *Am Ind Hyg Assoc J* 54:186–196, April 1993.

Armbrect HJ, Zenser TV, Bruns EH, et al. Effect of age on intestinal calcium transport and adaptation to dietary calcium. *Am J Physiol* 236(6):E769–774, 1979.

Armbrect HJ, Zenser TV, Gross CJ, et al. Adaption to dietary calcium and phosphorous restriction changes with age in the rat. *Am J Physiol* (Endocrinol Metab 2):E322–327, 1980.

Armon C, Daube JR, O'Brien PO, et al. When is an apparent excess of neurologic cases epidemiologically significant? *Neurology* 41:1713–1718, 1991.

Bean M. Nanograms are not the answer to "will I be hurt?" questions. *Waste Age*, March 4, 1988.

Bellinger D, Leviton A, Watermaux C, et al. Low-levels lead exposure, social class, and infant development. *Neurotoxicol Teratol* 10:497–503, 1989.

Bracken TD. Exposure assessment for power frequency electric and magnetic fields. *Am Ind Hyg Assoc J* 54:165–177, April 1993.

Brodeur P. *Currents of Death: Powerlines, Computer Terminals and the Attempt to Cover Up Their Threat to Your Health*. New York: Simon & Schuster, 1989.

Cleary SF. A review of in vitro studies: Low-frequency electromagnetic fields. *Am Ind Hyg Assoc J* 54(4):178, April 1993.

Covello VT, Sandman PM, Slovic P. *Risk Communication, Risk Statistics, and Risk Comparisons: A Manual for Plant Managers*. Washington DC. Chemical Manufacturers Associations, 1988.

Davis JM. Risk assessment of the developmental neurotoxicity of lead. *Neurotoxicology* 11:285–292, 1990.

Feychting M, Ahlbom A. Magnetic fields and cancer in people residing near Swedish high voltage power lines. IMM rapport 6/92, Karolinska Institute, Stockholm, Sweden, 1992.

Freeman GB, Johnson SC, Liao P, et al. Effect of soil dose on bioavailability of lead from mining waste soil in rats. *Chem Speciation and Bioavail* 3:121–128, 1991.

Friedman GD. *Primer of Epidemiology*. New York: McGraw-Hill, 1974.

Gandley RE, Couture M, Silbergeld EK, et al. Paternal exposure to lead (Pb) alters initial genomic expression in offspring. *The Toxicologist* 12:213, 1992.

Health CW. Uses of epidemiologic information in pollution episode management. *Arch of Environ Health* 43(2):75–80, March/April 1988.

Hill AB. The environment and disease: Association or causation? *Proc Royal Soc Med* 58:295–300, 1965.

Kahn HA. An introduction to epidemiologic methods. In Lilienfeld AM (ed), *Monographs in Epidemiology and Biostatistics*. New York: Oxford University Press, 1983.

Krieger GR, Irons R. Non-ionizing electromagnetic radiation. In Sullivan, JB, Krieger, G (eds), *Hazardous Materials Toxicology*. Baltimore: R. Williams and Wilkins, 1991.

Last JM, Tyler CW. Epidemiology. In Maxcy-Rosenau-Last. *Public Health and Preventative Medicine,* 13th ed. pp. 11–41. East Norwalk, CN: Appleton and Lange, 1992.

LaVelle JM, Poppenga RH, Thacker BJ, et al. Bioavailability of lead in mining wastes: an oral intubation study in young swine. *Chem Speciation and Bioavail* 3:105–111, 1991.

Marcus AH. Comparative approaches to Superfund site assessments for young children exposed to lead. Presented before the Society of Toxicology, 1991.

Mooridian AD, Song MK. Age related alterations in duodenal calcium transport in rats. *Mech of Aging and Dev* 47:221–227, 1989.

Neutra RR. Counterpoint from a cluster buster. *Am J of Epidemiol* 132(1):1–8, July 1990.

ORAU, 1992. *Health Effects of Low-frequency Electric and Magnetic Fields*. ORAU Oak Ridge Associated Universities, Oak Ridge, TN 92/F8. June 1992.

ORAU, 1993. Letter to the Editor. *Science* 20:13, April 2, 1993.

Piantadosi S. Epidemiology and principles of surveillance regarding toxic hazards in the environment. In Sullivan JB, Krieger GR, *Hazardous Materials Toxicology*. Baltimore: Williams and Wilkins, 1991.

Rabinowitz MB, Kopple JD, Wetherill GW. Effect of food intake and fasting on gastrointestinal lead absorption in humans. *Am J Clin Nutr* 33:1784–1788, 1980.

Raymond C. Nagging doubt, public opinion offer obstacles to ending 'cluster' studies. *JAMA* 261(16):2297–2298, April 28, 1989.

Sandman PM, Caron C, Hance BJ. Improving dialogue with communities: A risk communication manual for government, submitted to the New Jersey Department of Environmental Protection. New Brunswick, NJ: Rutgers University Press, 1987, p. 67.

Savitz DA. Overview of epidemiologic research on electric and magnetic fields and cancer. *Am Ind Hyg Assoc J* 54:197–204, April 1993.

Schulte PA, Ehrenberg RL, Singal M. Investigation of occupational cancer clusters: Theory and practice. *Am J Pa* 77(1):52–56, January 1987.

Schwartz J. Beyond LOEL's, p valves, and vote counting: Methods for looking at the shapes and strengths of associations. *Neurotoxicology* 14(2–3):237–246, 1993.

Society for Environmental Geochemistry and Health. Davies, BE, Wixon, BG. Lead in soil: Issues and guidelines. Northwood, U.K.: Science Reviews Limited, 1988.

Society for Environmental Geochemistry and Health (1991), Proceedings of the Symposium on the Bioavailability and Dietary Exposure of Lead. *Chem Speciation and Bioavail* 3, Science and Technology Letters.

Slovic, Paul. Perception of Risk. *Science* 236:280–285, April 17, 1987.

U.S. Department of Health, Education and Welfare: Smoking and Health: A Report of the Surgeon General. Washington DC: U.S. Government Printing Office, 1964.

U.S. Environmental Protection Agency (EPA). *Guidance Manual for the Integrated Exposure Uptake Biokinetic Model for Lead in Children*. EPA/540/R-93/081, 1994.

U.S. EPA. *Air Criteria for Lead*. EPA 600/8–83/028[a-d]F, 1983.

Wasserman GA, Factor-Litvak P, Shrout P, et al. Prospective study of the effects of lead exposure on intelligence at 4 years. *The Toxicologist* 13:350, 1993.

Weis CP, Henningsen GH, Poppenga RH, et al. Pharmacokinetics of lead in blood of immature swine following acute oral and IV exposures. *The Toxicologist* 13:175, 1993.

Weis CP, LaVelle JM. Characteristics to consider when choosing an animal model for the study of lead bioavailability. *Chem Species and Bioavail,* 3:113–119, 1991.

Weis CP, Poppenga RH, Thacker BJ, et al. Design of pharmacokinetic and bioavailability studies of lead in an immature swine model. In Beard M E, Iske SA (eds), *Lead in Paint, Soil, and Dust: Health Risks, Exposure Studies, Control Measures, Measurement Methods, and Quality Assurance, ASTM STP 1226.* Philadelphia: American Society for Testing and Materials, 19103–1187, forthcoming 1995.

Wertheimer N, Leeper E. Electrical wiring configurations and childhood cancer. *Am J of Epidemiol* 109:273–284, 1979.

Wildt K, Eliasson R, Berlin M. Effects of occupational exposure to lead on sperm and semen. In Clarkson TW, Nordberg GF, Sager PR (eds). *Reproductive and Developmental Toxicity of Metals.* [Proceedings of a joint meeting, May 1982; Rochester, NY.] New York: Plenum Press, 1983, pp. 279–300.

12

Risk Assessment

Janet E. Kester, PhD
Holly A. Hattemer-Frey, MS
Joseph W. Gordon
Gary R. Krieger, MD, MPH, DABT

The development of formal, quantitative human health and ecological risk assessment represents a major advance in environmental management techniques. Quantitative risk assessment allows for a numerical analysis of current and future risks associated with a variety of industrial/commercial processes and activities. The use of formal risk assessments has steadily increased since the 1980s. A wide variety of both U.S. and international regulatory agencies base their management and administrative decisions on the numerical output derived from the risk assessment process.

Not surprisingly, risk assessment is not without its flaws and problems. This chapter focuses on the three major types of risk assessment commonly practiced in environmental management today: (1) human health, (2) probabilistic, and (3) ecological. This chapter is long and detailed; however, the risk assessment is frequently the major document that summarizes and presents the formal findings of many environmental investigators. Increasingly, risk assessment is the decision "driver"; therefore, managers and policy makers must be conversant with the essential terminology and basic methods of quantitative risk assessment.

HUMAN HEALTH RISK ASSESSMENT

The risk assessment process synthesizes available data on exposure of specified receptors and the toxicity of potential chemicals of concern (COCs or PCOCs) to estimate the associated risk to human health. The characterization of potential adverse human health effects resulting from exposure to chemicals involves five basic steps:

1. **Data collection:** All available information pertaining to a site must be carefully reviewed to determine basic site characteristics, identify potential exposure pathways and points, and determine data needs for risk assessment. Necessary information includes:
   - physical characteristics of the site
   - chemicals and processes historically used at the site
   - levels of contaminants present in various environmental media
   - potentially exposed populations
   - reliable site-specific data for various human activities and behavior patterns.
2. **Data evaluation:** Available site data are reviewed to characterize the site, define the nature and

magnitude of chemical releases to environmental media (soil, air, and water), and identify potential site-related COCs, exposure media, and receptor populations (see also Chapter 5, Basic Principles of Environmental Science). Key questions are identified and plans for addressing them are made at this stage.

3. **Exposure assessment:** Assessment of the degree of exposure involves estimation of the amount, frequency, duration, and routes of human exposure to site-related chemicals. The exposure assessment should consider both current and likely future site uses. Identifying exposure routes (such as inhalation and ingestion) and receptors (i.e., the person(s) who could come in contact with a chemical) are crucial to determining the validity of an exposure pathway. After potentially complete exposure pathways are identified, exposure point concentrations and chemical intakes by exposed humans are calculated.

4. **Toxicity assessment:** Assessing the degree of toxicity involves review of available information to identify the nature and degree of toxicity of the COCs and to characterize the dose-response relationship (the relationship between magnitude of exposure and magnitude of adverse health effects).

5. **Risk characterization:** Characterizing risk to the identified receptors involves combining exposure and toxicity information to determine the degree of potential risks at a site, and to estimate what residual levels of chemicals may *not* pose unacceptable risks to receptors. This step also includes an analysis of the uncertainties associated with the risk estimates to put them and potential target criteria into appropriate perspective for decision making.

This sequence of events applies to human health risk assessments for both carcinogenic and noncarcinogenic chemicals. Therefore, the risk assessment is frequently used as a basis for regulatory policy decisions because a variety of chemicals can be evaluated.

The information generated in the risk assessment can be used to:

- determine whether remedial action (cleanup) is necessary from a human health standpoint (i.e., if risks exceed some preestablished criterion)
- identify the contaminated media and specific areas that may require remediation

- provide a basis for establishing cleanup levels that must be achieved in order to adequately protect human health, if remediation is required.

Typically, U.S. EPA-style risk assessments are done in accordance with guidelines outlined in the *Risk Assessment Guidance for Superfund, Volume I, Human Health Evaluation Manual (Part A)* (U.S. EPA, 1989b), *Part B* (1991b), *Part C* (1991c); the Office of Solid Waste and Emergency Response (OSWER) Directive 9285.6–03, *Supplemental Guidance: Standard Default Exposure Factors* (U.S. EPA, 1991d); *Exposure Factors Handbook* (U.S. EPA, 1990a); and *Guidance for Data Usability in Risk Assessment* (U.S. EPA, 1990b).

A characterization of the physical setting should summarize information on climate, hydrogeology, geologic setting, and soil type, as well as the location and description of surface water and groundwater bodies on or near the site. An understanding of historical operations at the site can help guide the site characterization process by suggesting what chemicals might be found and in what form they might exist.

Chemicals and Processes Used On-Site

Identification of chemicals present on-site that may also be present in the background (i.e., unrelated to operations on-site) is an important part of the site characterization process. Because the risk assessment typically quantifies health effects from potential exposures to on-site chemicals exceeding background concentrations, careful evaluation of background concentrations can justify the exclusion of certain chemicals from the quantitative component of the risk assessment process. This is especially true for metals, which can be toxic under certain conditions but are natural components of the earth's crust (e.g., arsenic, cadmium, lead, and mercury).

In addition, the widespread use and disposal of organochlorine pesticides (e.g., DDT and dieldrin) and polychlorinated biphenyls (PCBs) have resulted in their detection in virtually all media (including human tissue worldwide). Other ubiquitous anthropogenic compounds, including polycyclic aromatic hydrocarbons (PAHs; e.g., benzo[a]pyrene) and chlorinated dibenzodioxins and -furans, derive from combustion processes.

Levels of Contaminants in Various Media

Based on the review of available information, a risk assessor can formulate a *conceptual model* representing his or her understanding of the (1) sources of site-related COCs, (2) potentially affected media, (3) the means by which they are released and transported within and among media, and (4) the exposure pathways and routes by which COCs may contact human receptors.

Potentially Exposed Populations

A description of potentially exposed populations should include information on sensitive subgroups (e.g., children, pregnant women, and the elderly) as well as activity patterns that could affect exposures. Such information can be obtained from local agencies and site-specific survey results. This section of the assessment should also include information on individuals who could be exposed to site-related contaminants even if they do not live on or near the site (e.g., a nearby resident who consumes fish from a river or creek that traverses the site).

The risk assessment typically identifies potentially exposed populations associated with the following land-use categories:

- residential
- commercial/industrial
- recreational
- trespasser—a term used in a nonlegal sense, referring to episodic exposures at a site.

Activity Patterns and Behavior

To accurately define current land uses, assessors often must gather information on activity patterns (e.g., determining the amount of time residents, workers, or trespassers spend in potentially contaminated portions of the site, and assessing how activities and exposures can change with the seasons). Information on the proximity of individuals to potentially contaminated areas can be obtained from zoning maps, census data, and population surveys conducted on or near the site. Although future land use of the site is typically characterized by reviewing zoning plans, preparing census projections, and making reasonable assumptions about the probable future use of the area, information on how long residents have lived on or near the site can aid in making realistic judgments about future uses of the site. For example, if the site is in the middle of an industrial area, assessors can judge how likely it is to be developed into a single-family residential area.

The collection of site-specific data enables assessors to more accurately estimate risk in its geographical context and precludes the need to use default assumptions from agency guidelines. For instance, if

warranted, assessors could conduct a survey of the people living on a site or who use a site to reduce the uncertainties of assuming default values for many exposure factors. The goal of such an exposure assessment survey would be to obtain site-specific information on behavioral patterns and frequencies so that accurate, realistic exposure parameters could be incorporated in the risk assessment. In addition, this survey would permit the risk assessment to delineate the factors that contribute most to exposure (U.S. EPA, 1992b). Conducting a site-specific exposure survey can be a relatively inexpensive means of achieving more realistic risk estimates.

Data Evaluation

A principal objective of the data evaluation task is to organize the chemical data into a form appropriate for the risk assessment. This includes the following steps:

1. Evaluate the data sampling plan for adequacy in characterizing the amounts and distributions of site-related chemicals.
2. Qualify the analytical methods with regard to their appropriateness for use in the risk assessment.
3. Evaluate the quality of data with respect to sample quantization and detection limits.
4. Examine laboratory qualifiers assigned to monitoring data and evaluate potential quality assurance/quality control problems.
5. Evaluate data with respect to blanks and tentatively identified compounds (TICs).
6. Summarize information on background concentrations of chemicals and compare background levels with site-related levels.
7. Test the statistical distribution of the data. Most environmental data are lognormally rather than normally distributed. The effect of distribution type is important because the calculation of the mean and upper confidence limit (UCL) is distributionally sensitive. The assumption of normal distribution may overestimate the "true" mean of the data set.
8. Identify COCs.

The analytical methodology employed should be appropriate for the medium and chemicals being tested and, in most instances, be approved by a regulatory agency.

Evaluate sample quantization and detection limits. Detection limits vary depending on the chemical analyzed, the analytical instrument used, and the characteristics of the medium being tested. Before any chemicals can be eliminated from consideration in the risk assessment, it is necessary to consider how detection limits can vary between sampling events. Some chemical concentrations may not be detected in samples from some sampling events because of varying detection limits. Because detection limits can vary, assessors must establish that the reported detection limit does not exceed levels practical for use in a quantitative risk assessment. If a detection limit concentration is greater than twice the most commonly observed detection limit, the elevated detection limit should be eliminated (Gansecki, 1992). Chemicals not detected in any of the samples that have appropriate detection limits can also be eliminated from further consideration.

Results classified as nondetects are typically treated as equal to one-half the detection limit and included in the calculation of mean and upper-bound (95% UCL) estimates (U.S. EPA, 1992b, 1992d). This method tends to bias estimates of the mean and standard deviation (Porter et al, 1988; U.S. EPA, 1992d). Nondetect data from a site should therefore be handled, if possible, using standard, EPA-approved statistical techniques for left-censored data sets (U.S. EPA, 1992d). A left-censored data set has a high proportion (10%–50%) of samples testing below detection limits. The calculated exposure point concentration has important ramifications on cleanup levels, so a substantial amount of effort has been devoted to this problem.

Log-probit analysis has been shown to be a robust method for estimating the geometric mean and standard deviation of samples with values in the nondetectable range (Gilliom & Helsel, 1984). In a log-probit analysis, all measurements (both detects and nondetects) are assumed to be samples taken from the same lognormal probability distribution. When samples from a lognormal distribution are plotted on a probit scale, they tend to lie on a straight line (Grossjean, 1983). In a probit analysis involving both detects and nondetects, the nondetects are treated as unknowns, but their percentile values are taken into account. Thus, if there are 100 samples, 30 of which are nondetects, the first detectable data point is plotted at the 31st percentile. If sufficient detectable values exist, they can be used to establish the line, using linear regression, that characterizes the entire data set. The geometric mean for the data set (both detects and nondetects) is the 50th percentile value. Thus, a probit analysis allows the geometric mean to be extrapolated from a data set even if values are below detection limits (if there are sufficient detected values to define the probability distribution). Available probit tables (Finney, 1952) and computerized programs (Abou-Setta et al, 1986; Lieberman, 1983; Sette et al, 1986; Woodfork & Burrell, 1985) facilitate the use of log-probit or other statistical analyses in the risk assessment process.

Test statistical distribution. For evaluation of the likelihood of increased human health risk from a site, it is important to define study areas that represent potential exposure units (EUs), i.e., geographic areas or sets of chemical concentrations that could be contacted by receptors during the exposure period. The appropriate definition of exposure units depends on the spatial distribution of COCs as well as on site-specific land use and exposure conditions. Mapping methods such as kriging and triangulation are established methods for accounting for the spatial distribution of chemicals and are likely to be recommended in future EPA guidelines for exposure characterization (U.S. EPA, 1993).

Assessors can also use these methods for defining potential exposure units. In general, it is recommended that:

- Surface soil across the site should be evaluated as a single exposure unit where contact with soil is spatially random, i.e., in a trespasser scenario.
- Subareas of the site where a specific activity is expected to occur should be evaluated as *hot spots* when contamination is unevenly distributed across the site.
- Groundwater influenced by site-related COCs should be evaluated as a possible future source of drinking water, if groundwater is shown to be potable.

The U.S. EPA has encouraged the use of geostatistical techniques by releasing a public-domain software package (GEO-EAS, Ada, OK) that facilitates the actual calculations. Spatial statistical analysis of a site can potentially provide greater flexibility and "realism" in site assessment. These techniques avoid portraying the data set as a unique point estimate. Hence, with spatial analysis techniques, specific exposure patterns or contamination sources (e.g., hot spots or plumes) can be more realistically mapped and analyzed. In addition, appropriate statistical techniques can be employed to optimize final sample location and sizes so that redundant and expensive data collection is avoided. Geostatistical techniques are not perfect and require a high degree of technical expertise and experience. Any mathematical technique must be consistent with the reality on the ground.

Identify COCs. The COC selection process accounts for (1) the concentration of contaminants present in various environmental media; (2) their predicted mobility, persistence, and potential transformation/degradation in the environment; and (3) their observed toxicity.

Screening of chemicals. Selection of COCs involves several steps. The first step includes the screening of chemicals after detection limits, detection frequencies, and blank samples have been examined.

Important considerations in the screening process include:

- Detection limits vary depending on the compound, the analytical instrument and method used, and the characteristics of the matrix. If the detection limits are acceptable for risk assessment purposes (i.e., if methods are sufficiently sensitive to detect potentially significant concentrations), target analytes not detected in any samples may be eliminated from further consideration.

- Infrequently detected chemicals may be artifacts of sampling, handling, or analytical errors. EPA guidance allows for the elimination of a chemical if (1) it is detected in low concentrations in few samples of only one or two media, and (2) there is no reason to believe that the chemical is site-related (U.S. EPA, 1989b). If a chemical is detected in less than 5% of samples and is not a carcinogen, it can be eliminated (U.S. EPA, 1989b). Chemicals not detected in any samples are eliminated from further consideration, whereas any carcinogen detected at an order of magnitude or more above the detection limit should be retained regardless of its frequency of occurrence.

- Blank samples provide a measure of accidental contamination of the sample set. To prevent inclusion of on-site-related chemicals in the risk assessment, concentrations of all chemicals detected in blanks are compared with concentrations detected in site samples. A chemical can be eliminated from further consideration if its concentration in site samples is less than 10 times the blank concentration for common laboratory contaminants or 5 times the blank concentration for other chemicals (U.S. EPA, 1989b).

Chemicals considered essential human nutrients (zinc, iron, sodium, calcium, etc.) can also usually be eliminated (U.S. EPA, 1989b).

Background comparisons. Chemicals not eliminated during these steps of the screening process are compared to background levels. The purpose of background comparisons in conducting baseline risk assessments is to determine whether chemical concentrations detected at the site are distinguishable from concentrations originating from causes other than activities that have occurred at the site. Chemical concentrations that are indistinguishable from the background levels may not have resulted from site activities.

Also, in characterizing health risks, the risks associated with concentrations indistinguishable from background would have less significance than concentrations that are clearly associated with site activities. Statistically, this is a test of the null hypothesis that the mean concentration of a chemical at the study area is not significantly different from the mean value at the background (or remote, uncontaminated) location (U.S. EPA, 1990b).

Site concentrations should be compared to data collected from background areas that are otherwise comparable to study areas defined for the site. Factors that determine comparability include natural features, such as soil types and mineralized zones in the subsurface. Anthropogenic (synthetic or induced) influences derive from current and past land use, e.g., agriculture. Where background concentrations are unevenly distributed, background data should be grouped into geographic areas of similar size for comparison to study areas on the site.

If local background data for a given medium or chemical are unavailable, U.S. or regional background data, such as that reported by Adriano (1986) or Shacklette & Boerngen (1984) may be used. The choice of statistical methods for comparison of mean concentrations in the defined site and background study areas depends on the distribution of the data and the frequency of nondetected levels in the site and background sample sets. Nonparametric tests are often useful, because these tests do not require data to follow the normal distribution or any other specific distribution.

In addition, these tests can accommodate a moderate number of missing data or concentrations reported as trace or nondetect (Gilbert, 1987). If the mean concentration of a chemical in a given medium is statistically higher (at the 95% level of confidence) than its mean level in background samples as determined by a one-tailed t-test, that compound should be considered a potential COC. As a result of this process, only chemicals likely to be site-related and detected at statistically significant concentrations are ultimately selected as COCs.

Multiple comparisons of several independent data sets are required where site-related contamination could occur in only some subset of exposure units or source areas, and/or where background concentrations are unevenly distributed. This situation commonly occurs where groundwater is monitored at multiple upgradient (background) and downgradient (potentially impacted) locations.

Another example is the case of focused soil surveys surrounding each of several separate potential source areas. Pairwise statistical tests for each potentially impacted area and comparable background data set would result in an unacceptably high false-positive rate for the site. For example, if 20 potential source areas are evaluated separately using a 5% significance level for each test, the chance of incorrectly concluding that site-related contamination is present in at least one of the areas is $1 - (0.95)^{20} = 64\%$.

Multiple comparisons procedures. Statistical procedures that control the overall false-positive rate when several independent data sets are compared are called *multiple comparisons procedures*. Examples of multiple comparisons procedures include the parametric analysis of variance (ANOVA) and the nonparametric Kruskal-Wallis test.

Toxicity concentration screen. If a further reduction of COCs is required (e.g., to reduce the number of chemicals that must be quantified to a more manageable number), a toxicity-concentration screen can be used to focus the quantitative risk assessment on those contaminants posing the greatest risk for the exposure scenarios being considered. Two of the most important factors determining the potential toxicological significance of a chemical are (1) its measured concentrations in site media and (2) its toxicity. Therefore, in the toxicity-concentration screening procedure, each chemical (whether non-carcinogen or carcinogen) identified in each medium is combined with appropriate toxicity constants such as *reference doses* (RfDs) or *cancer slope factors* (CSFs):

$$R_{ij} = C_{ij} \times T_{ij} \qquad \text{(Equation 12–1)}$$

where:

R_{ij} = risk factor for chemical i in medium j
C_{ij} = concentration of chemical i in medium j
T_{ij} = toxicity value for chemical i in medium j (i.e., either the CSF or 1/RfD).

Chemical-specific risk factors are summed to obtain the total noncarcinogenic and carcinogenic risk factors for all chemicals of potential concern in each medium. Risk factor ratios are calculated as the ratio of chemical-specific risk factors to the total risk factor. Chemicals with relatively low ratios (e.g., less than 0.01) can be eliminated from further consideration.

Exposure Assessment

The objective of the exposure assessment is to estimate the amount, frequency, duration, and routes of human exposure to site-related chemicals. The exposure assessment should consider both current and future risks to human health under land-use scenarios consistent with known or likely activity.

This task involves the following steps:

1. Characterizing the exposure setting
2. Characterizing pathways of exposure
3. Identifying potentially exposed populations
4. Estimating exposure point concentrations
5. Estimating total contaminant intake by potentially exposed individuals for all relevant pathways of exposure.

Characterizing the exposure setting. In response to concerns that the traditional human health risk assessment process can be needlessly conservative for specific sites, recent guidance prescribes use of both individual and populational descriptors of risk. For example, risk estimates calculated using both average or median and upper-bound parameter values may be presented (U.S. EPA, 1992b). Future guidance is likely to formally endorse the use of distributions for exposure parameters, and definition of default distributions for exposure parameters unlikely to vary among sites is under consideration (U.S. EPA, 1993).

Although a well-designed survey is the best way to characterize behavior patterns of potential receptor groups living, working, or otherwise contacting chemicals related to a site, the information content of more readily available data should be thoroughly evaluated first. National survey information can be altered to account for known significant differences between a site and the nation in general (e.g., climatic conditions). Local data on population distribution, economic level, types of work, other activities in the area, possible future development, etc., may be readily obtainable from chambers of commerce, planning commissions, departments of motor vehicles, utility company records, or other local bodies. An evaluation of the adequacy of such information can determine whether a local survey is necessary.

If collection of site-specific data is considered necessary to accurately define current land uses and potential receptor groups, appropriate survey instruments must be designed. Information to be gathered might include

- characterization of typical activity patterns (e.g., the amount of time residents, workers, or trespassers spend in contaminated portions of the site, and seasonal differences in activities and exposures)
- information on the proximity of individuals to potentially contaminated areas (obtainable from zoning maps, census data)
- population surveys conducted on or near the site.

The results of appropriately designed surveys or reviews of national survey information can be used to construct site-specific probability distributions describing behavioral patterns and frequencies for incorporation in the risk assessment. In addition, such surveys enable the risk assessor to delineate the factors that contribute most to exposure (U.S. EPA, 1992b).

Characterizing pathways of exposure. Environmental fate and transport modeling and/or monitoring data can be used to estimate chemical concentrations in the transport medium (i.e., air or groundwater) at the point of contact with the receptor. Such contact constitutes human exposure. Because the movement of chemicals within and among various environmental media can substantially influence how far the chemical can migrate and in what form it might exist at the exposure point, it is necessary to employ measurement or predictive techniques to evaluate the fate and transport of COCs.

Many chemical, physical, and environmental parameters, such as water solubility, vapor pressure, octanol-water partitioning, and bioaccumulation, influence the behavior and fate of organics released into the environment. The importance of these factors and how they influence each other is often inadequately understood. However, the examination of a few basic physicochemical properties can provide insight into the behavior and fate of chemicals released into the environment. (See also Chapter 5, Basic Principles of Environmental Science.)

Water solubility is the maximum amount of a chemical that will dissolve in pure water at a specific temperature and pH. The solubility of a chemical in water affects its fate and transport in soils and could significantly influence human exposure to organics through aquatic pathways. Highly soluble compounds tend to leach rapidly from soil into groundwater and surface-water supplies. In addition, they tend to be less volatile (Menzer & Nelson, 1980), more biodegradable (Lyman et al, 1982), and more mobile (Briggs, 1981) than less soluble chemicals.

Vapor pressure measures the relative volatility of a chemical in its pure state and is useful for determining the extent to which a chemical will be transported into air from soil and water surfaces. Volatilization is a major route for the distribution of many chemicals in the environment (Dobbs & Cull, 1982). The volatility of a chemical is affected by its solubility, vapor pressure, and molecular weight, as well as the nature of the air-to-water or soil-to-water interface through which the chemical must pass (Lyman et al, 1982). Chemicals with a low vapor pressure and a high affinity for soil or water are less likely to vaporize than are chemicals that have a high vapor pressure and a weak affinity for soil or water.

As mentioned earlier, the migration of chemicals to groundwater and surface water is largely controlled by aqueous solubility. However, soil particles bearing sorbed chemicals may also be transported by flowing water and eventually be deposited as sediments. The release of chemicals from soil to groundwater may occur through diffusion or through leaching and percolation. Gases and vapors forming in the vadose zone (the unsaturated soil generally above the water table) are transported by diffusion to the surface, where they may escape into the atmosphere.

Both diffusion and percolation through the soil column may be slowed by sorption of a contaminant to the surface of soil particles. The interactions of chemicals with soils are complex, varied, and poorly understood. It is generally acknowledged that adsorption of organic chemicals to soils is primarily related to organic matter content (represented by the organic carbon:water partition coefficient, Koc). High Koc values predict decreased mobility in the soil because of sorption. Under equilibrium conditions, the distribution of a chemical between the soil and infiltrating groundwater can be estimated using the partition coefficient, Kd, which may be estimated as

$$Kd = foc \times Koc \qquad \text{(Equation 12–2)}$$

where foc is the fraction of organic carbon in the site soil.

Chemicals present in various environmental media can undergo a number of physical and chemical processes that may alter their chemical structure or properties. For example, chemicals exposed to direct sunlight may be degraded by photolysis. Chemicals sorbed to soil particles may be effectively immobilized. However, changes in soil moisture content or chemistry can serve to release adsorbed chemicals. With the exception of sorption, chemicals present in soil, surface water, and sediments are exposed to similar fate mechanisms. These mechanisms include chemical reactions, biotransformation, and bioaccumulation.

Fate and transport predictions can be physically tested or monitored in the field for model calibration and/or to guide additional sampling efforts. Because of the complexity of chemical fate and transport properties, computer models are often used to estimate migration patterns. At the present time, however, no models are appropriate to demonstrate all aspects of chemical movement and interaction in the subsurface. Models may be either conceptual (providing a qualitative description of the characteristics and dynamics of the geologic/hydrogeologic/chemical system) or mathematical (translating the conceptual model into mathematical equations representing the system properties and processes). Types of mathematical models include:

- analytical models—set(s) of mathematical equations that enable "exact" solutions
- numerical models—one-dimensional/single phase/equilibrium, one-dimensional/multiphase/nonequilibrium, multidimensional, or any combination that offers "approximate solutions"
- combinations of analytical and numerical models.

A complete site assessment must be performed and option selection discussed with relevant regulatory agencies prior to the initiation of fate and transport modeling. Well-documented, field-validated, and scientifically peer-reviewed models are preferred. The model selection process should include:

- Statement of the objectives of the fate and transport analysis, and discussion of whether the proposed model(s) is likely to accomplish these objectives under site conditions.
- Description of the concepts and calculations used in the model(s). Can they accurately simulate significant mass transport, partitioning, and transformation processes at the site?

- Identification of the processes (e.g., biodegradation, gaseous diffusion/volatilization, advection, sorption, etc.) that the model(s) is expected to simulate.
- Summary of the strengths, weaknesses, assumptions, and uncertainties in each component of the model(s). A model sensitivity analysis should be performed and documented to identify the most sensitive input parameters. A discussion of how variations in sensitive parameters are likely to affect model results, and the status of knowledge regarding these parameters at the site, should be included.
- Description of how the model will be calibrated and applied.
- Description of the methods (e.g., analytical, experimental, physical) that will be used to validate modeling results.

If the model has been validated elsewhere, discuss the applicability of the validation techniques used to the site of interest.

Selection of complete exposure pathways. For risk assessment purposes, an important objective in evaluating the environmental behavior and fate of various chemicals is predicting the major pathways and extent of human exposure. An exposure pathway provides the mechanism by which a COC comes in contact with a receptor. A complete exposure pathway has the following elements:

- a source and mechanism(s) of chemical release to the environment
- an environmental transport medium for the released chemical (groundwater, soil gas, soil, or ambient air)
- an exposure point (defined as a spatial area where a receptor could contact the affected medium)
- an exposure route (ingestion, inhalation, or dermal contact) at the exposure point.

If any of these components is missing, the pathway is incomplete and does not contribute to receptor exposure.

The relative importance of an exposure pathway depends on the concentration of a chemical in the relevant medium and the rate of intake by exposed individuals. Each land-use category (residential, industrial, or recreational) is defined by a distinct set of exposure pathways. Exposure pathways that might be evaluated at a hypothetical site are shown in Table 12–A.

Table 12–A. Potential Exposure Pathways that Could Be Evaluated During a Human Health Risk Assessment[1]

Pathway[a]	Current and Future Residents	Current and Future Workers	Current and Future Trespassers	Rationale
Soil				
Dermal Absorption	Maybe	Maybe	Maybe	Only if COCs will be absorbed through the skin.
Ingestion	Yes	Yes	Yes	Incidental ingestion of outdoor soil and indoor dust is possible.
Air				
Inhalation (Vapors)	Maybe	Maybe	Maybe	Only if the COCs volatilize at ambient temperatures.
Inhalation (Particulates)	Yes	Yes	Yes	Resuspension of particulates via wind or other mechanical processes is possible.
Food				
Ingestion of Produce Grown On-Site	Yes	No	No	Trespasser and workers do not live on-site or consume food items originating from the contaminated area. Residents could consume produce grown in a backyard garden.
Ingestion of Meat and Milk/Dairy Products	No	No	No	Trespasser and workers do not live on-site or consume food items produced on-site. It is unlikely that residential receptors raise their own cows.
Ingestion of Fish	Yes	No	Yes	The resident and trespasser could catch and consume fish from local water bodies.
Ingestion of Game	No	No	No	Intake of COCs from ingesting game from the contaminated area is assumed to be low relative to other exposure pathways.
Surface Water				
Dermal Contact While Swimming	Maybe	No	Maybe	Residents and trespassers could use potentially affected surface water bodies for recreational purposes. This pathway would apply only if the COCs will readily absorb through the skin.
Ingestion While Swimming	Yes	No	Yes	Residents and trespassers could use potentially affected surface water bodies for recreational purposes and ingest water while swimming.
Groundwater				
Ingestion	Maybe	No	No	The trespasser does not live on-site or consume local groundwater. Worker is expected to consume municipal water while at work. This pathway is applicable only if residents could use the groundwater now or in the future.
Inhalation During Showering	Maybe	No	No	The trespasser and worker do not live on-site or use the local groundwater for domestic purposes. This pathway is applicable for residents only if COCs are likely to volatilize during showering.
Dermal Absorption During Showering	Maybe	No	No	The trespasser and worker do not live on-site or use the local groundwater for domestic purposes. This pathway is applicable for residents only if COCs are readily absorbed through the skin.
Irrigation of Crops with Ground or Surface Water	Maybe	No	No	The trespasser and worker do not live on-site or use the local groundwater for agricultural purposes. This pathway is applicable for residents only if sufficient rain does not fall annually to support agricultural production.

[1]Some or all of these pathways may exist at a given site. This table illustrates possible exposure routes.
[a]All pathways apply to current and future child and adult receptors except for the worker, who is assumed to be an adult.

Identifying potentially exposed populations. Generally, one or more of four distinct receptor groups could be exposed to COCs at a given site: residents, workers, recreational visitors, and trespassers. Residential exposures should be evaluated whenever individuals currently live on or near a site, or future use of the site includes residential development. Residential receptors are assumed to come in frequent and repeated contact with site-related contaminants and generally experience the highest exposures.

Under the commercial/industrial land-use scenario, employees who work on or near the site could be exposed to site-related contaminants. Exposures are typically lower than those estimated for residential receptors, because worker exposure is generally limited to a maximum of 8 hours a day, 250 days a year. The trespasser or recreational land-use scenario addresses individuals who do not live on-site but who spend a limited amount of time at or near the site playing, fishing, hunting, or engaging in other recreational activities. Recreational exposures are expected to be higher at inactive versus active sites, because there is generally less surveillance at inactive sites.

Bioaccumulation. Assessing the potential toxicological significance of environmental chemicals also depends on being able to predict the extent to which they tend to *bioaccumulate* in organisms that are part of the human food chain, including edible plants, fish, cattle, dairy products, etc. Aquatic organisms can concentrate certain chemicals in their tissues at levels substantially higher than the concentration in water. Bioaccumulation is generally less marked in terrestrial organisms, which obtain chemicals principally via the food chain.

Nonionic organic chemicals such as PAHs, PCBs, dioxins, and organochlorine pesticides tend to accumulate in the lipid (fatty) portions of biotic tissues. Thus, one way to predict the bioaccumulation potential of a chemical is to measure how lipophilic it is. It is difficult to directly measure the lipophilicity of a chemical in a living organism, so researchers typically use the octanol-water partition coefficient (Kow) as a surrogate to predict its tendency to partition between an octanol component (a surrogate for fat) and water. Kow is defined as the ratio of the concentration of a chemical in the octanol phase to its concentration in the aqueous phase (Lyman et al, 1982). It has been shown to correlate well with the tendency of a chemical to concentrate in biota (Chiou et al, 1982; Geyer et al, 1982, 1986, 1987; Kenaga, 1980; Kenaga & Goring, 1980). Nonionic organic chemicals with large Kow values thus tend to accumulate in soil, sediment, and biota, and can enter human tissues through the agricultural food chain. Conversely, chemicals with low Kow values tend to partition mostly into air or water. For example, volatile organics such as trichloroethylene and tetrachloroethylene tend to be widely distributed in air with inhalation being the primary pathway of human exposure.

Estimating exposure point concentrations. Accurate estimates of chemical concentrations at points of potential human exposure are a prerequisite for evaluating chemical intake in potentially exposed individuals. Actual concentrations as determined from sampling data should be used whenever possible. In some cases, direct measurement of chemical concentrations and chemical contribution to another medium may not be feasible, accurate (e.g., health effects may occur below limits of detection), or cost-effective.

EPA defines the exposure point concentration used to estimate reasonable maximum exposure (RME) as the 95% upper confidence limit (UCL) on the arithmetic mean. An important complication in calculating the RME concentration is that a statistical distribution (usually either normal or lognormal) must be assumed. Supplemental guidance for risk assessment (U.S. EPA, 1992d) cites EPA's experience that most large environmental data sets are lognormally rather than normally distributed. In this case, the UCL can be calculated using a method presented in Gilbert (1987). However, UCL values based on the lognormal model are often an order of magnitude (10 times) or more above the range of observed concentrations, for data sets that exhibit high variability. This is because the UCLs are extrapolations representing the average concentrations associated with extreme long-tailed (highly skewed) distributions that could have generated the observed sample data with 5% probability. Therefore, where the UCL is higher than any measured value, a more reasonable approach may be to consider the maximum concentration detected for any composite area (U.S. EPA, 1989b, 1992d).

To provide some estimate of what more common exposures and risks might be, an average or most likely exposure (MLE) scenario using mean exposure assumptions should also be considered. Mean chemical concentrations are used to estimate exposure point concentrations. The advantage of evaluating both exposure scenarios is that they provide decision makers with a broader perspective on the true range of risks likely to be experienced by individuals exposed to site-related chemicals. The use of a point estimate of risk, especially an upper-bound estimate, does not provide adequate information to individuals who must decide if the health risks are excessive and therefore if the site warrants remediation.

Estimating chemical intakes. Chronic daily intakes are estimated using standard U.S. EPA (1989b) exposure equations. Calculated intakes are expressed as the amount of chemical that is taken into the body (not the amount that is absorbed through the lung or gut once the chemical has been ingested or inhaled). This method of calculating exposures is appropriate if the toxicity data used in the assessment are based on administered (versus absorbed) dose, such as the majority of numerical toxicity values recommended by the EPA. In the case of dermal contact, an absorbed dose must be estimated. Therefore, the toxicity values used should be based on absorbed versus administered dose. Because such values are largely unavailable, the comparison of absorbed doses with RfDs and CSFs based on administered dose introduces additional uncertainty.

Estimation of a 30-year residential exposure may be approached by (1) calculating two distinct exposures: a 6-year exposure for children and a 24-year exposure for adults, or (2) calculating a 30-year, time-weighted average exposure including exposures during both childhood and adulthood. In the first method, intakes for children and adults are calculated separately (using different ingestion rates, body weights, and exposure durations). Exposures are calculated assuming a 6-year exposure duration (ED) for children, a 24-year ED for adult residents (U.S. EPA, 1989b). Default body weights (BW) of 33 lb (15 kg) and 154 lb (70 kg) are typically assumed for children and adults, respectively (U.S. EPA, 1991d). Carcinogens and noncarcinogens have different mechanisms of toxicity, so an averaging time of 25,550 days [365 days a year × 70 years (lifetime)] is used to model exposures to carcinogens, while an averaging time equivalent to the ED (days) × 365 days per year is used for noncarcinogens.

In the second approach, intake parameters reflecting different soil ingestion rates, exposure duration, and body weights are integrated, resulting in age-adjusted average exposures (U.S. EPA, 1991b). The same default values for exposure duration and body weights are used. For example, the age-adjusted soil intake parameter is calculated as

Age-adjusted soil intake parameter

$$= \frac{IR_{soil/age\ 1-6} \times ED_{age\ 1-6}}{BW_{age\ 1-6}}$$

$$+ \frac{IR_{soil/age\ 7-31} \times ED_{age\ 7-31}}{BW_{age\ 7-31}}$$

(Equation 12–3)

where

$IR_{soil/age\ 1-6}$ = ingestion rate of soil for children (200 mg/day)

$IR_{soil/age\ 7-31}$ = ingestion rate of soil for adults (100 mg/day)

$ED_{age\ 1-6}$ = exposure duration for children (6 years)

$ED_{age\ 7-31}$ = exposure duration for adults (24 years)

$BW_{age\ 1-6}$ = mean body weight of children, 33 lb (15 kg)

$BW_{age\ 7-31}$ = mean body weight of adults, 154 lb (70 kg).

Note that the resultant ingestion factor (114) is in units of mg-year/kg-day and is not directly comparable to simple soil ingestion rates in units of mg soil/day or mg soil/kg body weight/day.

Modeling exposures from soil-related pathways. Individuals who come in contact with contaminated soils may be exposed via three exposure routes: ingestion, dermal absorption, and inhalation. The relative importance of these exposure routes depends on the concentration of chemical in soil, and its physicochemical characteristics.

Incidental ingestion of soil. Individuals can be exposed to chemicals by intentionally eating soil (e.g., pica behavior in children) or by inadvertent hand-to-mouth transfer of soil during gardening, cleaning, and recreational activities. Chronic daily intake from ingesting contaminated soil is calculated using the following equation (U.S. EPA, 1989b):

(Equation 12–4)

$$I = \frac{C_S \times IR \times CF \times FI \times EF \times ED}{AT \times BW}$$

where

I = intake from ingestion contaminated soil/dust (mg/kg/day)

C_S = concentration of chemical in soil (mg/kg)

IR = ingestion rate (mg soil/day)

CF = conversion factor (10^{-6} kg/mg)

FI = fraction ingested from contaminated area (unit less)

EF = exposure frequency (days/year)

ED = exposure duration (years)

AT = averaging time, or the period over which exposure is averaged (days)

BW = body weight (kg).

As explained earlier, estimates of soil intake for residential scenarios may be age-adjusted. The resulting estimate of COC intake is less conservative (lower) than if only childhood exposure was considered but is considered to be adequately protective of the entire population (U.S. EPA, 1991b).

Dermal contact with soil. Because dermal contact with chemicals in soil will not result in an adverse effect unless chemicals are absorbed through the skin, the following equation is used to estimate the amount of chemical absorbed through the skin (U.S. EPA, 1989b):

(Equation 12–5)

$$AD = \frac{C_s \times CF \times SA \times AF \times ABS \times EF \times ED}{BW \times AT \times 365 \text{ days/year}}$$

where

AD = absorbed dose (mg/kg-day)
C_s = concentration of chemical in surface soils (mg/kg)
CF = conversion factor (10^{-6} kg/mg)
SA = skin surface area available for contact (cm²)
AF = soil-to-skin adherence factor (mg/cm²-event)
ABS = chemical-specific absorption factor (unitless)
EF = exposure frequency (events/year)
ED = exposure duration (years)
BW = body weight (kg)
AT = averaging time (years).

Modeling exposures via inhalation. Exposure from inhalation of chemicals sorbed to resuspended dust is calculated as a function of the concentration of contaminants in soil, the concentration of respirable soil particles in air, respiration rate, and body weight as follows:

(Equation 12–6)

$$I = \frac{C_s \times EF \times ED \times IR_{air} \times 1/PEF}{BW \times AT \times 365 \text{ days/year}}$$

where

I = intake from inhaling contaminated particles (mg/kg-day)
C_s = concentration of COC in soil (mg/kg)
IR_{air} = respiration rate (m³/hour)
EF = exposure frequency (days/year)
ED = exposure duration (years)
PEF = particulate emission factor (m³/kg)
BW = body weight (kg)
AT = averaging time (years).

The particulate emission factor (PEF) relates the COC concentration in soil with the concentration of respirable particles (PM_{10}) in the air due to fugitive dust emissions from affected soils. It is recommended by U.S. EPA (1991b) because it represents a conservative estimate for inhalation of respirable particles.

Using default values (cf. U.S. EPA, 1991b), the PEF is equal to 4.63×10^9 m³/kg. If site-specific information is available, it may be used to calculate a modified PEF for use at that site.

Inhalation of chemical vapors. Intake of chemical vapors is quantified as

(Equation 12–7)

$$I = \frac{C_s \times EF \times ED \times IR_{air} \times 1/VF}{BW \times AT \times 365 \text{ days/year}}$$

where

I = intake from inhaling chemical vapor (mg/kg-day)
C_s = concentration of COC in soil (mg/kg)
IR_{air} = respiration rate (m³/hour)
EF = exposure frequency (days/year)
ED = exposure duration (years)
VF = chemical-specific volatilization factor (m³/kg)
BW = body weight (kg)
AT = averaging time (years).

The chemical-specific volatilization factor VF defines the relationship between the concentration of a chemical in soil and in air.

Ingestion of groundwater. Even if residents do not currently use groundwater for domestic or agricultural purposes, there are often no constraints on the use of such resources as a potable supply in the future. For groundwater to be considered as a potential water source in the future, the water must be both potable and have sufficient production capacity to support domestic or agricultural uses. Similarly, it should also be assumed that local ground or surface water will not be used by future residents for irrigation if sufficient rain falls yearly to support agricultural production without irrigation. Again, this pathway is typically limited to residential receptors, because workers are assumed to consume water from a municipal source.

Daily chemical intake from ingestion of contaminated water can be calculated using the following equation (U.S. EPA, 1989b):

(Equation 12–8)

$$I = \frac{C_w \times IR_{water} \times EF \times ED}{AT \times BW \times 365 \text{ days/year}}$$

where

I = intake from ingestion of groundwater (mg/kg-day)

C_w = concentration of COC in drinking water (mg/L)

IR_{water} = water ingestion rate (L/d)

EF = exposure frequency (days/year)

ED = exposure duration (years)

AT = averaging time (years)

BW = body weight (kg).

Modeling food chain intakes. Assessing the magnitude of human exposure to contaminants through the food chain depends largely on being able to predict the bioaccumulation of contaminants in the terrestrial and aquatic food chains. For most contaminants, estimating human exposure to chemicals present in the food chain is accomplished through the use of various predictive equations that estimate exposure from ingestion of contaminated produce (fruits and vegetables), meat, milk/dairy products, and fish (Shor et al, 1982; Fries, 1987; Hattemer-Frey et al, in press). These equations are conservative and tend to overestimate exposure. Because estimating the concentration of chemicals in various food items is highly uncertain and the EPA does not provide guidance for quantitatively assessing plant or animal uptake of inorganic compounds, the U.S. EPA (1991d) recommends that on-site fruits, vegetables, and fish be sampled to determine the concentration of COCs in food items of the human diet. If site-specific data cannot be obtained because of scheduling constraints, intakes from food chain-related pathways should be discussed qualitatively in the risk assessment.

Toxicity Assessment

Toxicity assessment evaluates the nature and severity of health effects associated with exposure to each site-related COC. It consists of identifying the *intrinsic* toxicity of a chemical (i.e., a hazard evaluation), and determining the frequency of an adverse health effect (e.g., cancer) associated with a unit daily dose (dose-response assessment).

The hazard evaluation involves a comprehensive review of available toxicity data to identify the severity of toxic properties associated with the COCs. Once the potential toxicity of a chemical has been established, the next step is to determine the amount of chemical exposure that may result in adverse human health effects.

COCs are classified into two broad groups based on their toxic endpoint: carcinogens and noncarcinogens. These classifications are selected because certain chemicals can have both properties, and health risks are calculated differently for carcinogenic and noncarcinogenic effects.

Carcinogenic health effects. Carcinogenic effects are expressed as the probability that an individual will develop cancer from a lifetime of COC-specific exposure. This probability is based on projected intakes and chemical-specific dose-response data. These factors are developed by the EPA and are available on the EPA's Integrated Risk Information System (IRIS).

Evidence of chemical carcinogenicity originates primarily from two sources: (1) lifetime studies with laboratory animals, and (2) human epidemiological studies. For most chemical carcinogens, animal data from laboratory experiments represent the primary basis for the extrapolation. Federal regulatory agencies have traditionally estimated human cancer risks associated with exposure to chemical carcinogens on the basis of administered dose according to the following approach:

- The relationship between administered dose and cancer incidence in animals is based on experimental animal bioassays.
- The relationship between administered dose and cancer incidence in the low-dose range is based on mathematical models.
- The dose-response relationship is assumed to be the same for both humans and animals, if the administered dose is measured in the proper units [typically, in milligram of chemical per kilogram body weight per day of exposure (mg/kg/day)].

Major assumptions arise from the necessity of extrapolating experimental results (1) across species (from laboratory animals to humans), (2) from high doses (used in animal studies) to low doses (levels to which humans are likely to be exposed in the environment), and (3) across routes of administration (e.g., projecting inhalation risks from data derived from an ingestion study).

The uncertainty involved in extrapolating from experimental animals receiving relatively high doses to humans receiving much lower exposures is qualitatively addressed by classifying chemicals into groups based on the weight of evidence from epidemiological and animal studies that they are actually carcinogenic in humans. The EPA's weight-of-evidence classification system is shown in Table 12–B. A quantitative risk assessment is generally performed for all Group A and B carcinogens, whereas Group C carcinogens are considered for inclusion on a case-by-case basis.

Table 12–B. U.S. EPA's Weight-of-Evidence Carcinogenicity Classification Scheme

Group	Description
A	Human carcinogen
B1	Probable human carcinogen—limited human data are available
B2	Probable human carcinogen—sufficient evidence in animals and inadequate or no evidence in humans
C	Possible human carcinogen
D	Not classifiable as to human carcinogenicity
E	Evidence of noncarcinogenicity for humans

Source: U.S. EPA, 1989b.

The U.S. EPA assumes that a small number of molecular events can evoke changes in a single cell that can lead to uncontrolled cellular proliferation and tumor induction. It is assumed that there is no level of exposure to a given chemical that does not pose a small, but finite, probability of generating a carcinogenic response. Although this is the prevailing paradigm, it should be put into perspective by noting that the background incidence of cancer in the United States is approximately 33%. This can be discussed in the uncertainty analysis section of the risk assessment. Also, other uncertainties that pertain to how the dose-response relationships are calculated should be discussed.

Because risk at low exposure levels cannot be measured directly either in laboratory animals or in human epidemiology studies, mathematical models have been developed to extrapolate from high to low exposures. Currently, CSFs are estimated with the *linearized multistage model*, which is based on the theory that multiple events may be needed to yield tumor induction (Crump et al, 1977). The linearized multistage model incorporates procedures for estimating the 95% upper confidence level on the slope of the dose-response curve extrapolated to doses lower than those used in experiments. The probability that the true risk is higher than that estimated is thus only 5%, and it may be zero. The animal data used for extrapolation are taken from the most sensitive species studied, based on the assumption that humans are at least as sensitive as the most sensitive animal species.

The risk estimates made with this model should be regarded as conservative, representing the upper limit of risk. Actual risk is likely to be lower, and it could even be zero. Cancer dose-response relationships can also be derived from human epidemiological studies, but such data are seldom adequate for regulatory purposes.

Other available low-dose extrapolation models produce quantitatively similar results in the range of observable data but yield estimates that can vary by three or four orders of magnitude at lower doses. Animal bioassay data are not adequate to determine whether any of the competing models is better than the others. Moreover, there is no evidence to indicate that the precision of low-dose risk estimates increases through the use of more sophisticated models. Thus, if a carcinogenic response occurs at the exposure level studied, it is assumed that a similar response will occur at all lower doses. Although the output of such models is a single point estimate of cancer potency, the magnitude of uncertainty surrounding these estimates is unquantifiable but probably quite large. Thus, to enable informed risk management, this source of uncertainty must be clearly presented in the risk assessment document.

Noncarcinogenic health effects. Potential effects from chronic exposure to noncarcinogenic compounds are assessed by comparing exposure levels to chronic reference doses (RfDs). Reference doses were previously termed acceptable daily intakes. However, the U.S. EPA changed terminology in order to avoid sanctioning some level of chemical exposure as "acceptable." Unlike carcinogenic compounds, chemicals that cause toxic effects other than cancer appear to do so through mechanisms that exhibit a physiological threshold. Thus, a certain dose of the chemical must be present in the body before toxic effects are observed. The approach used to estimate the likelihood that exposed individuals will experience noncarcinogenic effects assumes that there is some level of exposure (i.e., the RfD value) that individuals can tolerate without experiencing adverse health effects. If exposure exceeds this threshold level, there may be some concern that exposed individuals could experience noncarcinogenic health effects. Thus, an RfD is defined as a level of intake that is not expected to produce adverse effects, even in sensitive subpopulations, over a lifetime of exposure. The uncertainty in RfDs is said to span about an order of magnitude.

RfDs are calculated by dividing an effect level—typically NOAEL or LOAEL doses (in units of mg/kg/day)—by an *uncertainty factor* (UF) that typically ranges from 10 to 10,000. NOEL, NOAEL, and LOAEL are defined as follows (U.S. EPA, 1989b):

- **NOEL: No Observed Effect Level**—The dose at which there are no statistically or biologically significant increases in the frequency or severity of the effects observed in the study between exposed and control populations (i.e., effects of concern in the study are not produced at this dose).

- **NOAEL: No Observed Adverse Effect Level**—The dose at which there are no statistically or biologically significant increases in the frequency or severity of adverse effects observed in the study between the exposed population control populations. Effects are produced at this dose, but they are not considered adverse.
- **LOAEL: Lowest Observed Adverse Effect Level**— The lowest dose of a chemical in a study that produces statistically or biologically significant increases in the frequency or severity of adverse effects observed in the study between the exposed population and its appropriate control.

RfDs are commonly derived from NOAELs or LOAELs for the critical toxic effect by the application of UFs and a modifying factor (MF), if necessary. The NOAEL is considered the most appropriate basis for RfD development, because it represents a level at which effects are observed, but they are not deemed to be toxicologically significant. Where only LOAELs are available, an additional uncertainty factor is used.

UFs generally consist of multiples of 10 (although values less than 10 are sometimes used), with each factor representing a specific area of uncertainty inherent in the extrapolation from the available data to the potentially exposed population. Because the data are typically lognormally distributed, a UF of 3 still represents a substantial mathematical adjustment.

(Equation 12–9)

$$RfD = \frac{NOAEL \text{ or } LOAEL}{(UF_1 \times UF_2 \times UF_3 \times U_4)}$$

- UF_1 accounts for the variation in sensitivity among the general population, and is intended to protect subpopulations such as children, pregnant women, and the elderly that may be more sensitive than the average person.
- UF_2 accounts for the interspecies variability between the test species and humans.
- UF_3 is used to extrapolate data from subchronic studies to estimate risks from lifelong (chronic) exposures.
- UF_4 is used to extrapolate LOAELs to NOAELs.

The maximum multiplicative UF from this scheme is 10,000 ($10 \times 10 \times 10 \times 10$). It is conservative

because it assumes that humans are more sensitive than laboratory animals. If the laboratory animal is, in fact, more sensitive to a given chemical than humans are, the adverse health impact estimated using that RfD could be overestimated. For example, based on a study evaluating 490 LD_{50} studies, Weil (1972) showed that a factor of 3 to 6 was adequate to adjust for intraspecies differences. Extrapolation from a subchronic to a chronic NOAEL showed that a factor of 5 was adequate about 90% of the time (Weil & McCollister, 1963), whereas a factor of 10 was adequate 97% of the time.

Lewis et al (1990) have considered other uncertainty or safety factors that could also be applied and examined available data to estimate reasonable values for quantifying the uncertainty involved in animal to human extrapolations. Their results indicated that uncertainty factors need not be as large as 10 to provide values protective of human health. Lewis et al. (1990) suggested that a composite adjustment of 250 is sufficient to be protective of human health versus the maximum value of 10,000 sometimes invoked by the EPA.

Inhalation RfDs are derived from available effects levels by applying UFs similar to those used for oral RfDs. Inhalation toxicity data can be listed as a *reference concentration* (RfC; expressed in units of mg/m³) or as an inhaled intake (in mg/kg-day). RfCs are the concentrations of chemical in air (in mg/m³) that are not expected to cause adverse health effects in exposed populations assuming continuous chronic exposure. The default adult body weight and inhalation rate of 70 kg and 20 m³/day, respectively, can be used to convert an RfC to an inhaled intake.

Oral and inhalation RfDs are typically expressed as one significant figure in units of mg/kg-day. Most RfDs developed by EPA are based on the administered versus absorbed doses. The administered dose approach for dermal and inhalation routes is also conservative; it does not account for the fact that some fraction of the dose the animal receives may not be absorbed across the lung or skin, making the dose that actually affects a target organ or system lower. It is generally agreed, however, that the administered dose approach is reasonable and provides results that are protective of human health (OSTP, 1985).

Chemicals without toxicity values. If a COC lacks EPA-derived toxicity values, then available toxicity information and EPA methodology may be used to develop appropriate values. The agency recommends that this effort be undertaken in conjunction with regional and national EPA personnel.

Risk Characterization

Risk characterization involves estimating the magnitude of the potential adverse health effects of the chemicals under study. It combines the results of the toxicity and exposure assessments to provide numerical estimates of health risk. Risk characterization also considers the nature and weight of evidence supporting these risk estimates as well as the magnitude of uncertainty surrounding such estimates. Risks are typically calculated for both current and future land-use scenarios and both the RME and MLE exposure scenarios. This allows for a comparison of the range of risks between average (likely) and upper bound (highly unlikely), allowing for better risk management decisions.

Carcinogenic health effects. CSFs and the estimated daily intake of a chemical, averaged over a lifetime of exposure, are used to estimate the incremental risk that an individual exposed to that compound may develop cancer using the equation:

$$\text{Risk} = \text{Intake} \times \text{CSF} \qquad \text{(Equation 12–10)}$$

Cancer risks from exposure to multiple carcinogens and multiple pathways are assumed to be additive (U.S. EPA, 1989b). To obtain an estimate of total risk from all carcinogens at the site, cancer risks are summed across all exposure pathways for potential carcinogens of concern. Risks are deemed acceptable if they are within the 10^{-6} to 10^{-4} *target risk range,* or a possible excess case of cancer in 1 million exposed individuals to 1 excess cancer in 10,000 exposed individuals (U.S. EPA, 1989b). In practice, EPA frequently uses the 1 per 1 million (1 $\times 10^{-6}$) level as the *point of departure* for regulatory consideration.

Because EPA has not derived CSFs for the dermal route of exposure, oral slope factors are often used to quantify risks associated with exposure to carcinogens via this route. It is inappropriate, however, to use oral slope factors to quantify risks for chemicals that might cause skin cancer through direct action at the point of application or contact (U.S. EPA, 1989b). This class of skin carcinogens should be evaluated differently than those that cause cancer through a systemic rather than local mechanism (e.g., arsenic).

As mentioned previously, most RfDs and CSFs are based on administered versus absorbed doses. Thus, to quantify risks associated with dermal exposures,

the assessor must adjust oral toxicity values from an administered to an absorbed dose using the following equations for noncarcinogens and carcinogens, respectively (U.S. EPA, 1989b):

(Equation 12–11a)

$$\text{Absorbed RfD} = \text{Oral Administered RfD} \\ \times \text{Oral Absorption Efficiency}$$

(Equation 12–11b)

$$\text{Absorbed CSF} = \frac{\text{Oral Administered CSF}}{\text{Oral Absorption Efficiency}}$$

Noncarcinogenic health effects. Potential health effects of chronic exposure to noncarcinogenic compounds are assessed by calculating a hazard quotient (HQ) for each chemical of concern. An HQ is derived by dividing the estimated daily intake by a chemical-specific RfD, as shown in this equation:

$$\text{Hazard Quotient} = \frac{\text{Intake}}{\text{RfD}} \qquad \text{(Equation 12–12)}$$

An HQ greater than one indicates that exposure to that chemical may cause adverse health effects in exposed populations. It is important to note, however, that the level of concern associated with exposure to noncarcinogenic compounds does not increase linearly as HQs exceed one. In other words, HQ values do not represent a fixed probability of an adverse effect occurring. It should also be noted that RfDs, because they are based on a single, often equivocal, point on the dose-response curve, do not directly indicate a chemical's potency. Thus, HQs for different chemicals cannot be assumed to be equally meaningful from a risk perspective. All one can conclude is that HQ values greater than one indicate that noncarcinogenic health impacts are possible and that the more the HQ exceeds unity, the greater the potential severity of the adverse health effect.

The characterization of potential risks from exposure to chemicals at a hazardous waste site is complicated by the fact that humans are rarely exposed to only one chemical at a time. Ideally, in assessing the potential effects of a chemical mixture, the RfD for that mixture would be determined. However, as "custom blend" RfDs are not available, the typical approach is to sum chemical-specific HQs for each pathway to calculate a hazard index (HI). This approach can result in a situation where HI values exceed 1.0 even when no chemical-specific HQs exceed unity (i.e., adverse systemic health effects would be expected to occur only if the receptor were exposed to several COCs simultaneously). In this situation, chemicals are segregated by similar effect on a target organ, and a separate HI value for each effect/target organ is calculated (U.S. EPA, 1989b). If the HI value for any effect exceeds 1.0, then adverse, noncarcinogenic health effects are considered possible.

Because the toxicological significance of HQs is uncertain, the significance of HIs is also uncertain. A major source of uncertainty is the fact that the toxicological effects a chemical may have in isolation can be affected by the concurrent responses to other chemicals. Chemical interactions may be *additive, synergistic, antagonistic,* or *potentiating.* An *additive* effect occurs when the combined effect of two or more chemicals is equal to the sum of the effect of each chemical administered alone (2 + 4 + 8 = 14). A *synergistic* effect occurs when the combined effect of two or more chemicals is much greater than their individual effects when given alone (2 + 4 + 8 = 140). *Potentiation* occurs when one chemical is nontoxic alone, but increases the toxicity of another chemical (0 + 2 = 10). Chemicals are *antagonistic* when they interfere with one another's toxic effects (2 + 8 − 4), or one reduces the toxicity of another (4 + −4 = 0).

In the absence of direct knowledge about the possible interactive effects of simultaneous exposure to multiple compounds, simple additivity is typically assumed if similar target organs or similar mechanisms of toxicity exist. If mechanisms of toxicity and/or target organ effects are different, it is not appropriate to calculate HIs. Risk estimates would be lower if the effects of COCs are antagonistic but higher if they are synergistic or potentiating.

Uncertainty analysis. Like other forms of mathematical modeling, risk assessment relies on certain assumptions that have varying degrees of accuracy and validity. Variations in the data and models selected are influenced by several factors including data availability, differences in judgment concerning the balance of conservatism and realism, and agency requirements. Risk assessments typically use *deterministic* models that produce single-value estimates of risk. These single-value estimates based on EPA's recommended default parameter values combine a series of average, conservative, upperbound, and worst-case assumptions. As such, they do not consider the inherent variability in exposure assessment input factors and major data gaps. As a result, they provide limited information to decision makers who

must determine whether estimated health risks warrant remediation. An uncertainty analysis accounts for the variability in measured and estimated parameters, which allows decision makers to better evaluate risk estimates in the context of the assumptions and data used in the assessment.

Qualitative uncertainty analyses. Qualitative uncertainty analysis is useful in determining the extent, consistency, completeness, and quality of the parameter values and assumptions used in the risk assessment. A simple approach to qualitative uncertainty analysis is to develop a qualitative or quantitative description of the uncertainty for each parameter and indicate the possible influence of these uncertainties on the final risk estimates. General causes of uncertainty can be categorized as follows:

- **Measurement error:** Uncertainty arises from random and/or systematic error in the measurement technique.
- **Sampling error:** Uncertainty arises from the degree of representativeness of the data of the actual population being sampled.
- **Variability:** The natural variability in environmental and exposure-related parameters are a major source of uncertainty.
- **Dose-response extrapolations:** Uncertainty arises from the extrapolation of high-dose results in experimental animals to low-dose characterizations and from extrapolations from particular experimental animal species to humans.
- **Professional judgment:** Data gaps must often be filled based on engineering or scientific assumptions, which have inherent uncertainty.

Some major sources of uncertainty and the effect they may have on estimated risks are outlined in Table 12–C.

Quantitative uncertainty analyses. Although qualitative uncertainty analysis is helpful for evaluating the general quality of available information and ranking sources of uncertainty, it is difficult for risk assessors, risk managers, and the public to put exposure and risk estimates and resultant target criteria into appropriate perspective for decision making in the absence of quantitative estimates of the magnitude of uncertainty. As a result, the usual risk assessment process provides a sense of security if the estimated risks fall beneath the point-of-departure risk level but offers no indication of where the calculated risk falls in the distribution of actual risks, and no context for interpretation of results that exceed this arbitrary benchmark.

Table 12–C. General Uncertainty Factors

Uncertainty Factor	Effect of Uncertainty	Comment
Use of cancer slope factors.	May overestimate risks.	CSFs are upper 95% confidence limits on dose-response slopes derived from the linearized model. Considered unlikely to underestimate true risk (which may be zero).
Risks/doses within an exposure route assumed to be additive.	May over- or underestimate risks.	Does not account for synergism or antagonism.
Toxicity values derived primarily from animal studies.	May over- or underestimate risks.	Extrapolation from animals to humans may induce error due to differences in pharmacokinetics, target organs, and population variability.
Toxicity values derived primarily from high doses; most exposures are at low doses.	May over- or underestimate risks.	Assumes linearity at low doses. Tends to have conservative exposure assumptions.
Toxicity values.	May over- or underestimate risks.	Not all values represent the same degree of certainty. All are subject to change as new evidence becomes available.
Effect of absorption.	May over- or underestimate risks.	The assumption that absorption is equivalent across species is implicit in the derivation of the critical toxicity values. Absorption may actually vary with species and age.
Effect of excluding chemicals without toxicity data.	May underestimate risks.	These chemicals are not addressed quantitatively.
Home-grown produce pathway.	Lack of site-specific data prohibits accurate plant uptake; models tend to over-predict concentrations in vegetables.	Uptake factors are soil-, chemical-, and plant-specific; percent amount in regular diet is unknown.
Effect of applying toxicity values to soil exposures.	May overestimate risks.	Assumes bioavailability of chemicals sorbed onto soils is the same as in matrices administered in lab studies. Chemicals administered in lab studies may be more bioavailable.
Exposures assumed constant over time.	May over- or underestimate risks.	Does not account for environmental fate, transport, or transfer that may alter concentration.

Source: U.S. EPA, 1992c.

Quantitative uncertainty analysis consists of a *sensitivity analysis* and a *probabilistic simulation* (e.g., Monte Carlo or Latin Hypercube Simulation). A sensitivity analysis identifies the parameters that contribute most to overall model output. Specifically, a sensitivity analysis defines the magnitude of the effect a given parameter has on model output and ranks each parameter according to its influence on model output (e.g., cancer risks). In a sensitivity analysis, each parameter is varied across its range of plausible values while all other model variables are held constant to observe the effect each parameter has on model outcome (Hoffman & Gardner, 1983).

The most influential parameters on model output are defined as the most sensitive. Additional research may then be justified to either find more realistic or accurate values for these parameters or to develop *probability distributions* or *probability density functions* that describe the range of reasonable values for the parameter. For parameters that have a strong influence on model output, it might be worthwhile to obtain site-specific data. Performing a sensitivity analysis thus allows cost-effective use of limited resources so that the uncertainty associated with these parameters (as well as overall uncertainty) can be more accurately described or, if possible, reduced.

Because of these advantages, recent EPA draft guidelines for risk assessment recommend development of probability distributions for exposure parameters (U.S. EPA, 1993).

Monte Carlo simulation is a powerful analytical tool for evaluating the joint effects of all the input factors working together. The advantages of using a Monte Carlo analysis include ease of use, applicability to a broad range of model equations, a general lack of restraints on the form of the input distributions used or the functional form of the model equations, and the ability to characterize the output distribution. A model exists as a series of input parameters, each of which can be described as a random variable with an associated probability distribution. Thus, the first step is to obtain data on the range across which each parameter used in the risk assessment (or each sensitive parameter) is reasonably likely to vary. In this way the proper distribution of values for each parameter can be determined. Most input parameters for estimating chronic daily intake or absorbed dose are available in the technical literature or in EPA guidance documents, and values for these inputs can be expressed as ranges.

For some parameters, however, values are not available in the technical literature. For these parameters, estimates are made from existing data or are based on professional judgment and immediate knowledge of the site. Random samples are taken from these distributions and entered into the exposure model to yield a single value estimate of exposure or health risk. This process is repeated many times to produce a distribution of model predictions. Hence, as opposed to a single-value estimate that has no characterization of model variability, a Monte Carlo analysis defines a range of possible values for model output from which a best estimate (e.g., mean or median) can be selected. For example, 50th percentile values can be used as a best estimate of health risks, whereas 95th percentile values can be used recognizing that they represent an upperbound estimate.

PROBABILISTIC RISK ASSESSMENT

Probabilistic risk assessment (PRA) techniques can be used to address a wide variety of process safety and reliability issues. The specific complement of techniques used depends largely on the aims of the study. PRA techniques have evolved in response to specific needs and specific applications. Informal PRAs have long been performed as people have decided how they can best meet their needs without subjecting themselves to injury or loss.

Although the concept of PRA is not new, the development of formal, disciplined techniques with which to evaluate risk and reliability are recent phenomena. The fault tree methodology, for example, was developed by Bell Laboratories and applied by the Boeing Company in the early 1960s to the Minuteman missile program. This application of methodology represented the first step toward disciplining, codifying, and documenting an analysis technique that addressed itself to systems analysis. For the nuclear power industry, which has been the leader in the development and application of PRA techniques in recent years, formal risk assessment activities began in the mid-1950s. In 1956, WASH-740, a PRA for the operation of nuclear power plants, was published. This risk assessment was used to help determine the level of indemnification needed for the operation of nuclear power plants in the utility environment. No systematic methods for the systems analysis or for risk assessment had been developed at the time of WASH-740. Therefore, the analyses were deterministic ones of the sort that one often encountered in a safety analysis review.

From the beginning of activities within the nuclear power industry until the emergence of the WASH-1400 report, systems analysis was applied in isolated cases, including the evaluation of probabilities of catastrophic events. WASH-1400 was a landmark in the nuclear power industry and was the first major, relatively complete probabilistic assessment ever made for the nuclear industry. Its publication in 1975 brought confidence and support from the nuclear industry because the study presented data that the development of nuclear power plants in our society did not represent a major risk. However, intervenors and opponents used the same document to show that the risk associated with nuclear power plants was unacceptable. Opponents pointed out that this was a new technology, that many mistakes could have been made, that assumptions were likely to have been incorrect, and that uncertainties in the data were potentially significant.

A number of insights were developed as a result of the WASH-1400 study and have generally been incorporated into subsequent PRAs. These insights are somewhat different from those presented in safety analysis reviews (SARs). SARs historically have examined large events but not small ones. Insights from the WASH-1400 study include:

- In PRAs, human error and common cause failure mechanisms have been found to be major contributors to risk.
- Dominant accident sequences have been determined to vary from one facility to another, primarily because of differences in design and operational practices. Thus, the calculated risks of the same large event can be substantially different across superficially similar facilities.

Since the WASH-1400 study, and particularly subsequent to the Three Mile Island accident in 1979, the U.S. Nuclear Regulatory Commission (NRC) and the nuclear industry have initiated (and completed in many cases) several programs and studies that further broaden the base of PRA applications in the nuclear industry. Table 12–D provides a partial chronological listing of PRA developmental milestones.

Figure 12–1 shows the organization of the analysis and the flow of information in NUREG-1150, *Severe Accident Risks: An Assessment for Five U.S. Nuclear Power Plants.* NUREG-1150 is a relatively recently completed study that employed a traditional PRA approach to the analysis of risk of severe accidents. Figure 12–2 provides examples of a fault tree and an event tree, common tools used in PRA.

A PRA consists of an analysis of the design, operation, and maintenance of a plant/facility that focuses on the accident sequences that could lead to the occurrence of some specified undesired event. This analysis includes an investigation of the basic causes and frequencies of the accident sequences. A PRA often serves the purpose of investigating four fundamental categories of questions:

- What can happen? What combinations of equipment failures or other events must occur for a specified undesired event to occur?
- What is the probability that these particular combinations of events will actually occur?
- How can the undesired event be prevented from occurring, or, if it does occur, how can it be mitigated?
- What are the potential health, economic, and environmental related effects if specified quantities of certain radiological and hazardous materials escape?

Table 12–D. Partial Listing of PRA Milestones

Year	Milestone Event
1962	Development of Fault Tree Analysis
1969	Development of KITT Codes
1974	Conference on Reliability and Fault Tree Analysis
1975	Reactor Safety Study (WASH-1400) Development of CRAC Code
1978	ANS Topical Meeting on Probabilistic Analysis
1979	Initiation of the NRC's Interim Reliability Evaluation Program Results from Oyster Creek PRA, the first major utility-funded PRA Fault Tree Handbook (NUREG-0492)
1980	Data Summaries of Licensee Event Reports (Several NUREGs) Development of MARCH Code Handbook on Human Reliability (NUREG/CR-1278)
1981	Zion Probabilistic Safety Study
1983	PRA Procedures Guide (NUREG/CR-2300)
1984	Probabilistic Safety Analysis Procedure Guide (NUREG/CR-2815) Industry Degraded Core Rulemaking Program (IDCOR) Report on Nuclear Power Plant Response to Severe Accidents
1986	Individual Plant Evaluation Methodology developed by the IDCOR Program
1988	Individual Plant Examination for Severe Accident Vulnerabilities (NRC's Generic Letter 88-20)
1989	Severe Accident Risks; An Assessment for Five U.S. Nuclear Power Plants (Formerly the Reactor Risk Reference Document) (NUREG-1150)

The answers to these questions are used by management personnel to support the resolution of certain safety issues and facilitate planning and operational decisions.

A potentially important use of risk assessment methodology is in the regulatory process, i.e., the ranking of safety issues. Many safety-related issues exist, but not all are equally important. Regulatory requirements are sometimes imposed without comprehensively evaluating their contribution to safety or cost. PRA can be used as a cost/benefit tool to evaluate proposed regulations. The assessment of design or operational adequacy is an obvious application of risk assessment technology. There are myriad other applications that can be made, including the optimization of spare parts inventories and the prioritization of those systems or components that should be inspected, tested, and maintained on a routine basis.

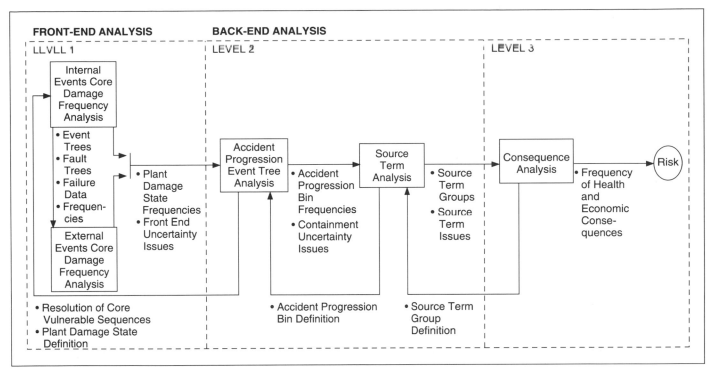

Figure 12–1. Organization of the analysis and the flow of information in NUREG-1150.

PRA techniques can play a beneficial part in the evaluation, understanding, and safety of waste management facilities. The performance of a PRA provides a comprehensive decision-making tool by which safety can be measured and evaluated. The PRA methodology provides an excellent method for ranking safety issues in order of their significance. Design weaknesses may be located by using PRA techniques even though the design meets traditional licensing requirements. A PRA plays a unique function in linking the design, operation, maintenance, testing, and safety of a facility. The PRA activity not only enhances safety but also provides the basis for increasing plant productivity.

Risk Assessment Techniques

There are several accepted PRA methodologies:

- **Fault Tree Analysis (FTA):** FTA is currently the most commonly used method for assessing system availability. Numerous in-depth studies of nuclear power plants have used and are using FTA techniques. Using FTA, a system is analyzed in the context of its environment and actual operation to find all credible ways in which the system can fail. A graphical model is generated of the various parallel and sequential combinations of faults that could occur. FTA is a deductive process, whereby the top event is postulated and the possible means for that

event to occur are systematically deduced. Boolean logic is used to interconnect the various events in a fault tree model.

Several modeling considerations exist when assessors perform FTA. The boundary conditions, system success criteria, operational state of the system, level of detail in the models, logic loops, sharing of a system, and preferential alignments are some of the topics that must be considered in fault tree construction. A human reliability analysis and common cause analysis are usually involved also. Documents exist that provide detailed guidance for constructing fault trees, or, if necessary, a guideline can be created for evaluating a specific facility. Numerous computer codes (for example, GRAFTER, CAFTA, IRRAS, and SETS) exist to construct and quantify fault trees.

- **Event Tree Analysis (ETA):** Scenario development is often achieved by using event trees. Event trees are simple inductive logic models for identifying the various possible outcomes of a given initiating event. For a nonreactor facility, an initiating event is often defined as one that could initiate an accident situation involving a hazardous material. For a reactor, an initiating event is typically a system or component failure that induces a reactor trip. Events subsequent to the initiating event on an event tree are determined by the specific characteristics of the facility being evaluated. The safety and

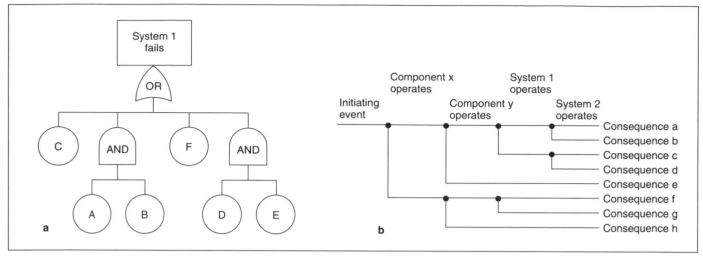

Figure 12–2. Examples of PRA tools: *(a)* fault tree and *(b)* event tree. Source: Developed by Joseph W. Gordon.

nonsafety systems that are capable of mitigating the accident must be identified. These systems are then structured in the form of headings for the event tree model. The order in which the systems appear is based on the sequence of events following the initiating event and the interdependencies among the systems. Once the required systems have been identified and ordered for a given event, the set of all possible failure and success states for each of the systems must be defined and enumerated. Binary logic is often employed, i.e., one failed and one success state for each system. Once the system failure and success states have been defined, the states are then combined to obtain the various accident sequences that are associated with the given initiating event. Conditional effects must be considered during event tree construction. For example, System 2 may depend on the success of System 1.

Fault tree analysis is generally used to calculate the conditional (failure/success) probabilities needed for each system. The conditional probabilities associated with each sequence are then multiplied together, thereby producing the probability of any given sequence occurring. Documents commonly used in event tree construction include design literature, system descriptions, and normal and emergency operating procedures. Several computer codes (for example, SUPER and NUPRA) exist to construct and quantify event trees.

- **Failure Modes and Effects Analysis (FMEA):** The emphasis of FMEA is on identifying failure modes for the components of concern. MIL-STD-1629, *Failure Modes and Effects Analysis,* specifies the

process. FMEA provides for an orderly examination of possible hardware failures and is relatively simple to apply because it considers only one failure at a time rather than multiple, preexisting, or common-cause failures. FMEA is a qualitative, inductive identification of system hazards and does not usually contain quantitative analyses. It is most effectively used as input information to fault tree models.

- **Reliability Block Diagrams (RBDs):** RBDs divide a system into blocks that represent distinct elements according to system success pathways. RBDs are generally used to represent active elements in a system in a manner that allows an exhaustive search for, and the identification of, all pathways for success. Reliability block components can be successively decomposed until the desired level of detail is obtained, with subsequent numerical calculations of system reliability made at this level.

- **GO Method:** The GO method is a success-oriented, systems analysis technique. It consists of an arrangement of GO symbols that represent the engineering function of each of the components, subsystem, or system. GO models are generally constructed from engineering drawings by replacing the engineering symbols with one or more GO symbols. The symbols are combined to represent system functions and logic. The GO computer code uses the GO model to quantify system performance. GO charts tend to become inscrutable for modeling certain facilities (e.g., power plants) and are generally best suited for modeling complex electrical-type systems.

- **Markov Analysis:** The Markov analysis is an analytical technique applied to both nonrepairable and repairable systems. The repair process is not assumed to be instantaneous or negligible compared with the operating time. The Markov approach models the time dependence associated with component failures.

Consequence Assessment

At the endpoint of each accident sequence identified during the scenario development process, assessors must evaluate the possible harmful effects on property, personnel, and the environment. Many sequences on an event tree will have "success" for endpoints, i.e., the accident was successfully mitigated. Endpoints that are assessed to be of similar consequence are often aggregated to determine a total probability of occurrence for a specific consequence. For example, all sequences that lead to an off-site impact might be binned together to determine the (total) probability that operation of a given facility will have an off-site impact.

Analysis of exposure through multiple exposure pathways is necessary to accurately model impacts from waste management activities. For example, releases from waste management activities would likely include particulate, volatile chemical, and radionuclide releases. These releases would follow airborne and possibly waterborne dispersion pathways and expose human receptors through direct and resuspended inhalation, water and food ingestion, dermal exposure, ground shine, plume shine, and submersion. All pathway analyses require site-specific parameters or the appropriate selection of generic model parameters. Software algorithms are used to describe and calculate the potential impacts of accidental releases on local populations. Most programs of this type have been developed for description of nuclear facilities (e.g., weapons or power facilities). Therefore, dosimetry and health effects are calculated secondary to radiological releases. The algorithms vary in their features and in their flexibility and accuracy of application. Some air release packages offer plume modeling capabilities, consideration of complex meteorological and demographical scenarios, and even evacuation. Groundwater pathway programs generally consider the movement of released radionuclides through groundwater transport mechanisms to various off-site locations. For hazardous chemicals, the algorithms from some air release modeling software can be adapted for estimating changes in concentrations from point of release to any other downwind location. Health effects from hazardous chemical releases are often evaluated by comparing concentrations determined in the accident analysis to important threshold values determined by regulatory agencies.

For radiological releases, previous studies have demonstrated that inhalation is the mechanism for giving the largest dose to local and on-site workers. To determine the grams or curies deposited in an individual via inhalation, the following equation is used:

[Release rate (curies or grams per unit time)] × [χ/Q (time per unit volume)] × [Breathing rate (volume per unit time)] × [Time of exposure (time)] = Curies or grams deposited in an individual's body via inhalation.

The expression χ/Q (meaning the Greek letter chi over Q) accounts for the dilution of a release at some specified distance downstream of the release under given meteorological conditions. Numerous textbooks discuss the derivation of an expression to calculate χ/Q. The Radiological Safety Analysis Computer (RSAC) code is often used to calculate χ/Q values.

Release fraction/radioactive release phenomena modeling computer codes represent a class of tools developed for evaluating complex and energetic release events that could result in a release of radioactive material and a compromise in the extent to which it can be contained. Many of these programs (e.g., CRAC and CORRAL) were developed for reactor accidents, and their capability for evaluating the release potential at nonreactors is limited. The types of phenomena considered, however, could result in any situation where thermal effects may become a strong driving force, and knowledge of such programs is helpful for understanding and adding credibility to release fraction calculations.

ECOLOGICAL RISK ASSESSMENT

Whereas human health and probabilistic risk assessments have developed during the last 20 years, ecological risk assessment is in its relative infancy. This is ironic because concern about the possible ecological effects of chemicals and anthropogenic activities became widespread in the 1950s and 1960s, when some agricultural pesticides were found to harm wildlife (Hoffman et al, 1990). When Rachel Carson wrote in *Silent Spring* (1962) ". . . We have put poisonous and biologically potent chemicals indiscriminately in the hands of persons largely or wholly ignorant of their potentials for harm, . . ." the subject that had been quietly discussed among scientists erupted into a noisy public debate. This debate resulted in the formation of the U.S. EPA, the ban of DDT (discussed in a monograph by Dunlap, 1981) and other pesticides, and the commencement of an era of environmental regulation and risk assessment in the United States.

Despite the dramatic impact of the 20th century environmental movement on U.S. societal values and regulatory practice, the EPA—mandated to protect human health *and* the environment with limited funds and personnel, and equipped only with crude, unproven scientific and regulatory tools—opted to concentrate on human health first. As discussed in the first section of this chapter, much progress has been made in developing standard methods for characterizing human exposure and risk. It has become clear, however, that protection of human health does not ensure protection of environmental health, and that additional emphasis must be placed on environmental issues (U.S. EPA, 1987, 1990c). However stringent criteria for protection of human health may be, they are based on *human* exposure patterns and responses. In some cases, nonhuman organisms may be more exposed or more susceptible to chemicals (or other anthropogenic disturbances) than humans are (Suter, 1993). For example,

- Nonhuman species are more sensitive than humans to many common environmental chemicals, both naturally occurring (e.g., chlorine, ammonia, and sulfur dioxide) and anthropogenic (e.g., pesticides). An example is the organochlorine pesticide DDT, which is relatively nontoxic to humans but had devastating effects on predatory bird populations.
- Nonhuman exposure can occur by routes irrelevant for humans, e.g., respiration of water (fish) or consumption of soil (worms).
- Nonhuman organisms may experience greater exposure than do humans to a contaminated source. For example, whereas humans usually eat a variety of foods from diverse sources, a fish-eating bird may dine daily on fish from a contaminated water body.
- Humans usually have access to alternative sources of food and shelter in case of emergency, but nonhuman organisms exist as components of interdependent communities and ecosystems. Even if an animal is itself unaffected by a toxic chemical, it will suffer secondary impacts if organisms on which it feeds are affected.

Although a wide range of validated toxicological methodologies exist to examine the effects of chemicals on individual organisms, tissues, cells, and (with the advent of molecular biology) molecules, the impacts of chemicals on the environment remain far more difficult to evaluate. This difficulty arises not only from inadequate methods for measuring environmental impacts but also from inadequate scientific basis and conceptual framework for developing the questions to be investigated. It is compounded by legal, financial, and regulatory constraints.

The current status of interdisciplinary efforts to develop a sound process for assessment of risk to ecological systems is the subject of this section.

Definition and Scope of Ecological Risk Assessment

> We have to accept that prediction of ecological effects by pollutants is unlikely, for the foreseeable future, ever to be very precise (Moriarty, 1988).
>
> We need to return to common sense and basics (i.e., a recognition of complexity rather than further efforts to oversimplify) in order to avoid the increasing use of bad science as a basis for regulation and management (Chapman, 1991).

Ecological risk assessment (ERA) is a process that evaluates the likelihood that adverse ecological effects may occur or are occurring as a result of exposure to one or more *stressors* (U.S. EPA, 1992a). The term *stressor* describes any chemical, physical, or biological entity or condition that has the potential to induce adverse effects on any *ecological component*, which might be individuals, populations, communities, or ecosystems (Norton et al, 1992). Definitions for these and other terms are presented in Table 12–E. Figure 12–3 represents the major variables involved in ERA, and how they relate to one another.

Before this discussion of the principles and practices of ERA begins, it is important to emphasize that risk assessment is not itself a science but rather a decision-making process informed, guided, and limited by current scientific knowledge. In addition, political considerations also play a major role in ERA.

Ecology. *Ecology* can be described as scientific natural history. It is an essentially holistic science

Table 12–E. Definition of Ecological Terms

Component	Definition
Environment	The whole world: both inanimate (air, soil, water) and animate (plants, animals, fungi).
Habitat	The inanimate world: air, water, soil.
Individual	Single member of a species.
Species	Groups of genetically similar organisms that are capable of interbreeding.
Niche	The position within an ecosystem occupied by a given species.
Population	Members of a single species that occur within a defined area.
Community	Populations of different species that coexist and interact with one another in the same area.
Ecosystem	A community and its habitat.

concerned with the structure and dynamics of *ecosystems,* functional systems of complementary relationships between communities of organisms, and their physical environment. Ecosystems can be divided into three general types:

- *Terrestrial* ecosystems are classified according to the dominant vegetation types (primary producers). Examples are various kinds of forests, grasslands, and deserts.
- *Aquatic* ecosystems may be primarily freshwater (lakes, ponds, streams, rivers) or marine (ocean, sea coast, estuaries).
- *Wetland* ecosystems are areas of transition between terrestrial and aquatic environments, e.g., bogs and salt marshes.

Nutrients, matter, and energy flow through ecosystems via food chains, which consist of a number of trophic levels (a functional classification of types of organisms based on feeding relationships). An example food chain is shown in Table 12–F. In all food chains, the first trophic level consists of primary producers (green plants that transform radiant solar energy into chemical energy via photosynthesis). Higher trophic levels of all food chains obtain their energy from organic food (i.e., members of lower trophic levels). In practice, feeding relationships are more complex than the simple food chain suggests. For example, some species are omnivores (consuming both plants and animals) and thus feed on species in more than one trophic level. These interconnecting food chains form a food web, depicted in Figure 12–4.

Plant-eating organisms at the base of a food web tend to be relatively numerous, whereas carnivores in the upper trophic levels are relatively few, forming a pyramid of numbers from bottom to top of the food web. Besides fewer species and smaller populations, lower reproductive rates, increased body size, and increased foraging range are generally found as trophic level increases (Moriarty, 1988). For example, nesting bald eagles may cover large areas in search of diverse prey such as fish, rodents, and other birds to feed their clutch of two or three young, whereas the fish that teem in one of the water bodies they hunt may lay millions of eggs in a breeding season.

The ultimate goal of ERA is to protect *ecosystem health,* a relative concept subject to various interpretations. Some view a healthy ecosystem as one "having a high level of biodiversity, productivity, and habitability which lead naturally to the endpoint of diversity, productivity, and habitat preservation" (Chapman, 1991). Because an ecosystem is a dynamic and variable assemblage, effects on its health status can be perceived as changes in these parameters. The standards by which assessors attempt to measure these changes and judge the effects of chemical stressors on ecosystems are provided by the new and rapidly developing field of *ecotoxicology,* the study of the fate and effects of toxic agents in ecosystems.

Ecotoxicology. As its name implies, ecotoxicology combines concepts and methods of both ecology and toxicology. In contrast with ecology, toxicology is a reductionist science that is principally concerned with effects on single organisms and their individual systems, organs, cells, biochemical processes, and molecules. As such, the study of toxic perturbations of ecological systems is beyond the scope of this field. Unlike individual organisms, ecosystems are not highly homeostatic; they are only loosely interconnected, with many different compensatory and feedback mechanisms affecting response. Thus, the response of an organism (or species) to a chemical cannot be simply predicted from traditional laboratory toxicity testing methods, even with "environmental" organisms like fathead minnows or water fleas. (Some of the reasons for this and strategies for toxicity testing are discussed later in this chapter.)

The state of the science at present combines field and laboratory approaches to determine what adverse effects may be caused by exposure to chemical stressors:

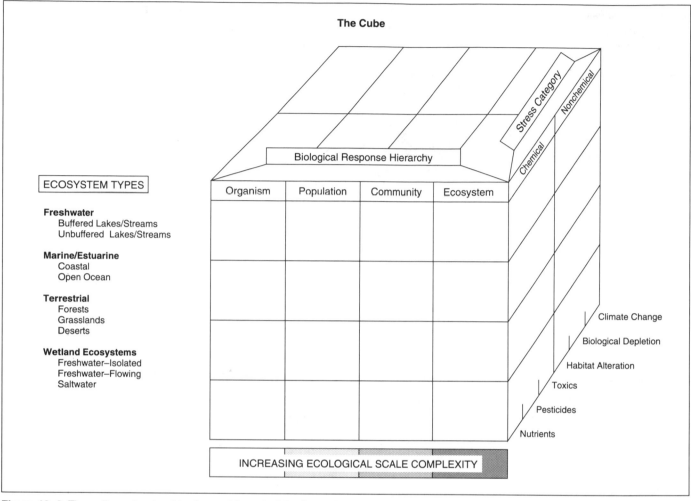

Figure 12–3. Three-dimensional matrix of organizing principles for ecological risk assessment.

Table 12–F. Example of a Food Chain

Group Name	Trophic Level	Example
Primary producer	1	Pine trees
Primary consumer	2	Aphids (herbivores)
Secondary consumer	3	Spider (carnivores)
Tertiary consumer	4	Song birds (carnivores)
Quaternary consumer	5	Hawks (carnivores)
Decomposer		Bacteria, fungi (recyclers)

Source: Adapted from Moriarty, 1988.

- **Field surveys.** Visual inspection of a site may reveal impacts on ecosystem parameters such as the abundance, diversity, or density of certain species.
- **The toxicity quotient (TQ).** Analogous to the hazard quotient (HQ) (ratio of daily intake to reference dose or risk-specific dose) used in human health risk assessments, the TQ is calculated as the ratio of measured or modeled media concentrations with concentrations of a chemical (in the same media) that are thought to have no adverse effects. Calculation of TQs is usually used to screen out relatively "clean" areas from further consideration in an ERA.
- **Toxicity testing.** Laboratory or *in situ* bioassays are conducted with site media to measure the effects of site-related chemicals on laboratory or resident test organisms.

All of these approaches are complementary, and each may be useful in different contexts. To adequately and reliably characterize potential chronic adverse effects on exposed populations, guidance suggests that at least two (and preferably all three) should be performed.

Comparison with human health risk assessment. Although the paradigm being developed for ERA

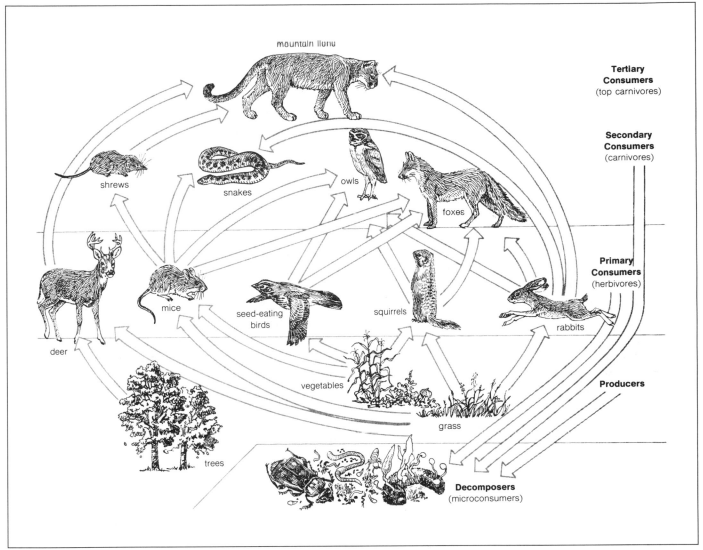

Figure 12–4. A simplified food web. Source: *Living in the Environment,* 3rd ed. by G. Tyler Miller, Jr. Copyright © 1982 by Wadsworth, Inc. Reprinted by permission of the publisher.

(described subsequently) borrows much from the relatively well-validated human health risk assessment (HRA) process, the two also differ in significant (and necessary) ways:

- The subject of HRA is the human individual, but ERAs may focus on any one or any combination of ecological components.
- They focus on different *endpoints* (defined as characteristics of an ecological component that may be affected by exposure to a stressor). Although the endpoints of HRAs are relatively limited and well-defined (e.g., cancer, systemic toxicity, developmental or reproductive effects), there are no universally appropriate indexes of ecosystem "health" that can be applied in all ERAs. As a result, assessment endpoints must be selected for each site

on the basis of both scientific and political considerations (which may have little in common).

- Ecological risk assessors must be aware of the potential effects of not only chemicals but also of physical and biological agents on ecological components. Physical stressors include global phenomena such as ozone depletion as well as local and regional phenomena such as habitat destruction or alteration by natural events (drought, fire) or human activities (construction, farming), and extremes of natural conditions (e.g., temperature, moisture, water level, and flow rate). Potential biological stressors include disease and predation. Although HRA focuses primarily on chemical stressors, recent studies have shown that the scale of environmental problems is increasing from local to regional to global, and that nonchemical stressors may be more significant than chemical (U.S. EPA, 1992a).

Ecological Risk Assessment Process

As previously mentioned, the mandate of the EPA to regulate potentially harmful chemicals in the environment relates not only to humans but also to animals and plants; not only to individuals but also to populations; not only to single species but also to ecosystems. However, although the EPA has provided agency-wide guidance for performing human health risk assessments since 1986, attempts to develop rationale and consensus for a set of definitive guidelines for the protection of the environment have begun only recently.

Guidance for ecological assessment of hazardous waste sites was presented in 1989 (U.S. EPA, 1989a and 1989c), and a proposed framework for general ecological risk assessment was published in 1992 (U.S. EPA, 1992a). In its present state of development, this framework does not define regulatory requirements or even provide procedural guidance. Rather, it presents a suggested paradigm intended to foster a consistent approach to ecological risk assessment, identify key issues, and define terminology. Additional guidance documents will be published in the next few years.

The proposed framework is shown in Figure 12–5. It consists of three major phases:

- problem formulation
- analysis
- risk characterization.

These main components are conceptually similar to those developed for human health risk assessment, as shown in Table 12–G. Although this step is not part of the ERA framework, Figure 12–5 also illustrates the importance of discussion between risk assessors (those who perform the risk assessment) and risk managers (those who apply the results of the risk assessment to decide any remedial actions that should be taken) at both scoping (problem formulation) and risk characterization stages. The risk manager can advise the risk assessor on what information is necessary for informed decision making, whereas the risk assessor must provide the risk manager with a complete understanding of the assumptions, conclusions, uncertainties, and limitations of the risk assessment.

In addition, a role for verification and monitoring in the risk assessment process is indicated in Figure 12–5. *Verification* refers to validation of the process,

as well as confirmation of specific predictions made on the basis of the risk assessment. *Monitoring* of site conditions can aid in the verification process and may identify other areas or ecological receptors in need of risk assessment. Verification and monitoring actions thus permit evaluation of the accuracy of the risk assessment conclusions and the effectiveness and practicality of the selected remedy.

Problem formulation. The ecological risk assessment process begins when an adverse ecological effect (e.g., fish kill, population decline) is observed or when a stressor of potential concern is identified (e.g., dredging of a harbor or use of a new pesticide). The first phase of ecological risk assessment is *problem formulation* (Figure 12–6), a systematic planning and scoping process to establish the goals, breadth, and focus of the risk assessment. This process incorporates regulatory and policy issues as well as scientific considerations. The final product of the problem formulation process is a *conceptual model* that identifies the assessment endpoints, the data needed, and the analyses to be used.

Careful planning is required in the process because of the complexity of ecological systems and their differential and inconstant "value" in terms of public opinion and policy. The risks of multiple stressors to many species as well as risks to multiple ecological components may need to be considered. The objectives of the ecological assessment must be clearly defined and should reflect both primary ecological concerns and the information required for effective decision making.

The initial steps in problem formulation are the identification and preliminary characterization of stressors, the ecosystem potentially at risk, and the ecological effects observed or expected.

Stressors. As mentioned previously, stressors may be chemical, physical, or biological in nature. Sample stressor characteristics are shown in Table 12–H. For chemical stressors, a phased selection process can be used to identify COCs, those chemicals most likely to contribute significantly to risk. The process of identifying COCs for ecological receptors accounts for (1) chemical concentrations in relevant environmental media relative to background or baseline concentrations; (2) their predicted mobility, persistence, transformation, and bioaccumulative potential (tendency to accumulate to high concentrations in animal tissues) in the environment; and (3) their observed toxicological hazards to biota.

A three-phase screening process can be used to identify COCs that may cause adverse impacts on

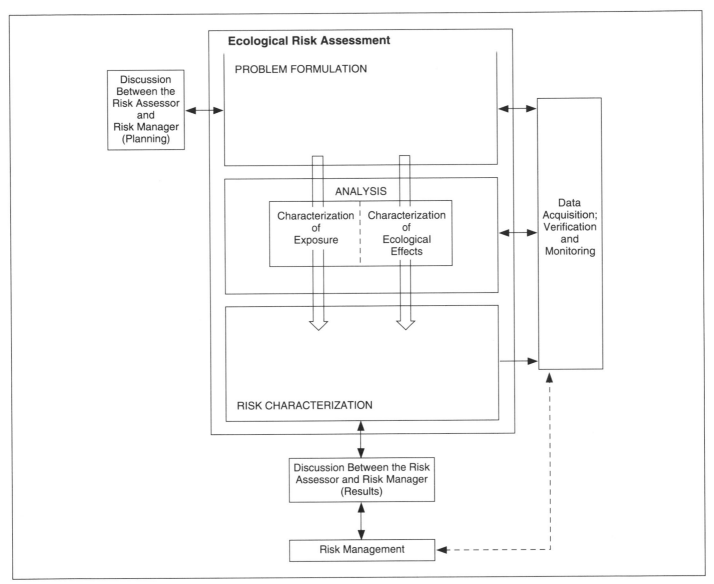

Figure 12–5. Framework for an ERA. Source: U.S. EPA, 1992a.

Table 12–G. Relationship of Human and Ecological Risk Assessment Paradigms

Human	*Ecological*
	Problem formulation
Hazard identification Dose-response assessment }	Characterization of ecological effects
Exposure assessment	Characterization of exposure
Risk characterization	Risk characterization

environmental receptors (and humans ingesting them, if relevant). In Phase I, detected compounds are screened with respect to detection limits, detection frequency, and concentrations in blank samples.

- Detection limits vary depending on the compound, the analytical instrument and method used, and the characteristics of the matrix. If the detection limits are acceptable for risk assessment purposes (i.e., if methods are sufficiently sensitive to detect potentially significant concentrations), target analytes not detected in any samples may be eliminated from further consideration.

- Infrequently detected chemicals may be artifacts of sampling, handling, or analytical errors. EPA guidance allows for the elimination of a chemical if (1) it is detected in low concentrations in few samples of only one or two media, and (2) there is no reason to believe that the chemical is site-related (U.S. EPA, 1989b).

- Blank samples provide a measure of accidental contamination of the sample. To prevent inclusion of nonsite-related chemicals in the risk assessment, concentrations of all chemicals detected in blanks

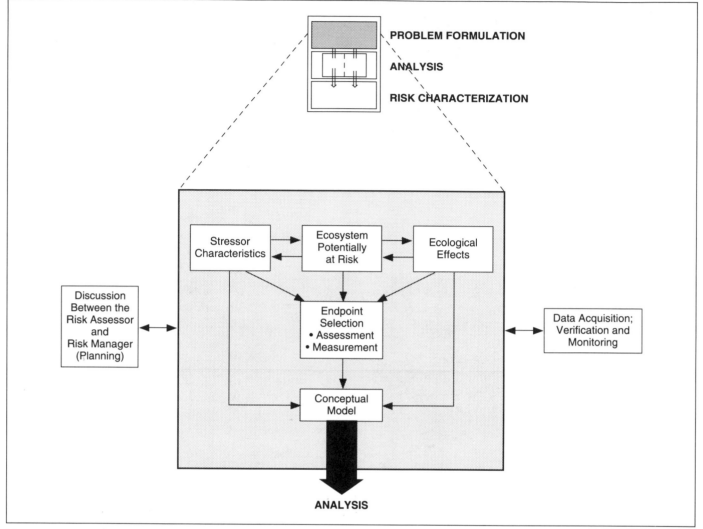

Figure 12–6. Problem formulation is the first phase of an ERA. Source: U.S. EPA, 1992a.

Table 12–H. Examples of Stressor Characteristics

Characteristic	Descriptor
Type	Chemical, physical, or biological
Intensity	Concentration or magnitude
Duration	Short- or long-term
Frequency	Single event, episodic, or continuous
Timing	Occurrence relative to biological events or cycles
Scale	Spatial extent and heterogeneity

Source: Adapted from U.S. EPA, 1992a.

are compared with concentrations detected in site samples. A chemical can be eliminated from further consideration if its concentration in site samples is less than 10 times the blank concentration for common laboratory contaminants, or 5 times the blank concentration for other chemicals (U.S. EPA, 1989b).

The second phase involves comparing estimated concentrations in samples to media-specific background data. If the mean concentration of a compound in a given medium is statistically higher than mean background levels, it will be considered a COC. As a result of this process, compounds likely to be site-related and detected at statistically significant concentrations will be selected as COCs.

In the third phase of COC selection, concentrations of those contaminants identified in the first two steps are compared to reference toxicity data (i.e., ambient water quality criteria) found in the literature to calculate TQs. Those contaminants whose concentrations exceed their respective reference criteria (i.e., have TQs > 1) are considered COCs.

Ecosystem potentially at risk. Knowledge of the nature of the ecosystem potentially at risk provides the context for the assessment and permits assessors to identify appropriate subjects for the ERA. The process of identifying the ecosystem at risk is influenced by how the risk assessment was initiated. If adverse effects were observed first, those effects can indicate areas and potential *receptors of concern* (ROC), species that play key roles in ecosystem structure and/or function. If a potential stressor was identified first, information on its temporal and spatial distribution patterns and expected effects (if known) may be useful in selecting appropriate ecological ROCs.

Potential ROCs include:

- *Species that are vital to the structure and function of the food web* (i.e., principal prey species or species that are major food items for principal prey species). In general, loss of a few individuals of a species is unlikely to significantly diminish the viability of the population or disrupt the community or ecosystem of which it is a part. As a result, the fundamental unit for an ERA is generally the population rather than the individual, with the exception of threatened and endangered (T&E) species (U.S. EPA, 1992b; this concept is discussed further later in this chapter). However, significant impacts on species that occupy critical positions in the food web structure may ramify throughout the ecosystem, potentially disrupting higher trophic level populations that depend on the affected population for survival and/or stability.
- *Species that exhibit a marked toxicological sensitivity to the COCs.* Ecosystem function can be impaired if certain component species are particularly vulnerable to chemical exposure. Selection of ROCs is thus designed to ensure that benchmark criteria are protective of the most sensitive organisms actually present at the site.
- *Species that have unique life histories and/or feeding habits.* Significant impacts on such species might eliminate unique ecological niches, with unpredictable results on the ecosystem as a whole.
- *Species for which toxicological data are readily available in the scientific literature.* Although such species may not be "key" in the sense of occupying a critical ecological niche, the availability of data on their responses to COCs reduces uncertainty in the evaluation as a whole.

Ecological effects. Data on ecological effects can be obtained from a number of sources. In order to optimize the acquisition process, data are often collected in phases or tiers (usually of increasing complexity and cost). The decision to terminate the ERA or to proceed to the next tier is based on objective decision points set at certain levels of exposure or effects.

A sample framework is shown in Figure 12-7. Tier 1 may use available data and relatively conservative assumptions, e.g., calculation of a TQ. If a TQ > 1 is noted, the assessment would advance to more sophisticated and site-specific analyses. Tiers may also be designed to guide the investigation through a series of questions regarding ecological effects and potential causes. For example, Tier 1 might be a field survey to determine the reproductive success of predatory birds nesting at a Superfund site. If any adverse effects are evident, subsequent tiers would focus on identifying causes for the observed field effect. For example, eggs might be collected and analyzed for pesticides, heavy metals, or other chemicals of potential concern. The optimal approach to the assessment must be decided on a site-by-site basis, depending on the nature of the site, the information required to support the decision-making process, and the time and funds available.

Selection of endpoints. Once the ecosystem at risk, ecological effects, and characteristics of potential stressors are known, assessors can select ecologically based endpoints that are relevant to the decision making. Two types of endpoints are defined: assessment and measurement. An *assessment endpoint* is "an explicit expression of the actual environmental value that is to be protected" (U.S. EPA, 1992a). As the name implies, *measurement endpoints* are quantitative expressions of a measurable effect or characteristic designed to provide insight into potential impacts on the corresponding assessment endpoints.

Both assessment and measurement endpoints may relate to ecological components at any level of ecological organization, from individual organisms to the entire ecosystem. Because *individual-level endpoints* refer to stressor effects on single organisms, they are not usually considered as assessment endpoints (however, they compose the largest and best understood class of measurement endpoints, as will be evident). *Population-level endpoints* can be used to determine risks to an entire population of a species. *Community-level endpoints* can be used to evaluate whether the composition (diversity and/or richness) and abundance of species have been altered as a result of exposure to the stressors. *Ecosystem-level endpoints* are used to evaluate the impacts of stressors associated with various environmental media on ecosystem function.

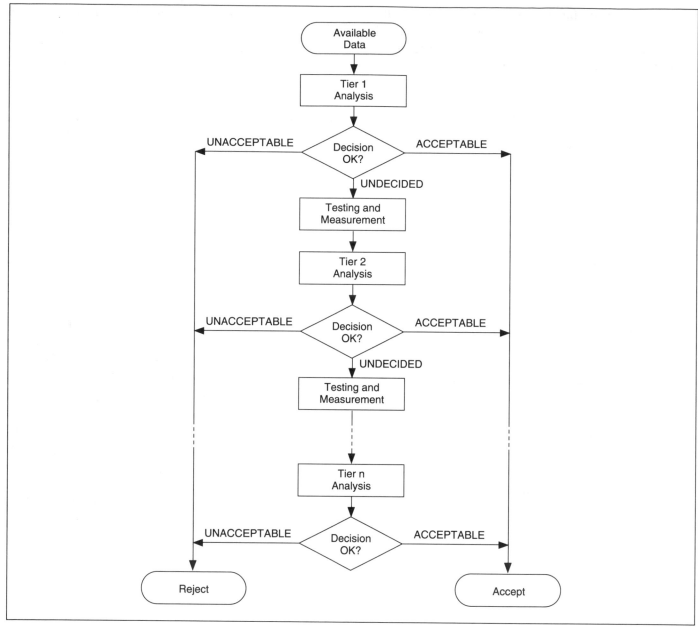

Figure 12–7. A diagrammatic paradigm of hazard assessment. Source: Cairns et al, 1978. Copyright ASTM. Reprinted with permission.

At sites where contamination is relatively widespread, found in several habitat types, and/or may affect organisms in multiple trophic levels, consideration of endpoints at several levels is recommended to ensure that no significant component is overlooked. Important considerations in selecting endpoints are summarized in Table 12–I, and examples of endpoints at the various organizational levels are shown in Table 12–J. Endpoint evaluation is discussed further later in the chapter, and more information on endpoint types and selection is provided in EPA guidance (U.S. EPA 1989c, 1992a, 1992c).

Conceptual model development. Available information on the ecosystem, stressor characteristics, and ecological effects are combined to define possible *exposure scenarios* (descriptive hypotheses of how exposure to ROCs may occur). Each scenario includes the ecosystem and its principal components, the stressor, its source, and how it may contact or affect the ecosystem, and the spatial and temporal scales of the interaction.

Those scenarios considered most likely to contribute to risk are incorporated into a *site conceptual model* that

- describes the ecosystem potentially at risk
- summarizes reasonable hypotheses of how the stressor(s) might affect ecological components

Table 12–I. Considerations in Selecting Ecological Endpoints

Assessment Endpoints	*Measurement Endpoints*
Ecological relevance. Relevant endpoints are those that reflect important characteristics of the system and are functionally related to other endpoints.	*Relationship to assessment endpoint.* A measurement endpoint should correspond to or be predictive of an assessment endpoint.
Societal and political relevance. Important societal concerns may not coincide with ecological concerns. For example, bald eagles may not be a critical component of the food web in certain locations, but their great symbolic importance (and protected status) requires that they be a focus of ecological risk assessment wherever they occur.	*Readily measurable.* Ideal measurement endpoints are cost-effective and easily measured.
Unambiguous definition. Clearly defined assessment endpoints have specified subjects (e.g., spotted owls) and characteristics of interest (e.g., extinction).	*Standardized.* The use of standardized methods and endpoints ensures replicability of results and facilitates comparison with other sites and development of predictive models.
Predictability. Assessment endpoints should be directly predictable from measurement endpoints.	*Broadly applicable.* Measurements on widely occurring species can be compared among sites and regions, providing insight into stressor responses and site-specific characteristics that may influence responses.
Susceptibility to the stressor. Assessment endpoints should be sensitive and significant signals of exposure to the stressor. For example, if a stressor decreases the prey base, an appropriate assessment endpoint might be predator population viability.	*Appropriate to the scale of the site.* Measurement endpoints should be appropriate to the scale of the stressed site. For example, the impacts of deforestation of a one-acre plot would not be reflected by assessment of the productivity of local deer herds.
	Consistency with assessment endpoint scenarios. The measurement endpoint must correspond to the assessment endpoint in terms of exposure pathway, magnitude, and temporal dynamics.
	Low natural variability. Stressor effects on a highly variable measurement endpoint may be difficult to distinguish from "noise."
	Diagnostic ability. Unique responses to a stressor can be valuable in defining causal relationships. For example, induction of certain enzyme activities is diagnostic of exposure to dioxin-like chemicals.

- illustrates the relationship between assessment and measurement endpoints.

Hypotheses considered most likely to contribute to risk are evaluated further in the second phase of the ERA, analysis.

Analysis. The *analysis* phase, based on the conceptual model developed in problem formulation, is a site characterization process. In this phase, assessors identify ecosystems in which impacts caused by stressors may occur, the biota species that may be exposed, and the magnitude and spatial and temporal patterns of exposure (Figure 12–8). Data on the effects of the stressors are summarized and related to the assessment endpoints selected in the problem formulation phase. The summary consists of two main interactive activities:

- characterization of exposure
- characterization of ecological effects.

Exposure characterization. The purposes of exposure characterization are to (1) measure or predict the distribution of a stressor in space and time (*stressor characterization*), and (2) determine whether the stressor is likely to contact ecological ROCs (*ecosystem characterization and exposure analysis*).

For chemical stressors, a combination of modeling and monitoring data commonly determines the magnitude and extent of contamination. For nonchemical stressors, the pattern of change may depend on season or land-use practices. These stressors can be characterized by ground reconnaissance, aerial photography, or satellite imagery. The spatial and temporal distributions of the ecological components are

Table 12–J. Potential Measurement Endpoints at Standard Levels of Ecological Hierarchy

Level	Assessment Endpoint	Measurement Endpoint
Individual	Organism health	Death Overt toxicity Growth Reproduction Tissue concentrations Behavior
Population	Viability	Occurrence Abundance Age-size-class structure Reproductive performance Birth rate/death rate Frequency of morbidity or mortality
Community	Difference in structure and function from a reference community	Number of organisms present Number of species present Species diversity
Ecosystem	Difference in structure and function from a reference ecosystem	Biomass Productivity Nutrient dynamics Material and energy flow

Source: Adapted from U.S. EPA, 1992b.

discussed in more detail and combined with the spatial and temporal distributions of the stressor to evaluate potential exposure.

The final product of the exposure characterization is the *exposure profile,* which quantifies the magnitude of exposure through space and time. For chemical stressors, it is usually expressed in terms of intake per unit time (e.g., mg chemical/kg body weight/day). For nonchemical stressors, it may be expressed as the percentage of habitat removed per year or event.

Characterization of ecological effects. The goals of the second major component of the analysis phase, ecological effects characterization, are to (1) identify and quantify the adverse effects potentially caused by the stressor(s), and (2) evaluate cause-and-effect relationships to the extent possible. Measurement endpoints are evaluated in terms of their relationships to corresponding assessment endpoints selected during problem formulation. A variety of statistical and modeling methods may be used to quantify the *stressor-response* relationship (the relationship between the magnitude of exposure and the magnitude of response). If the measured response is not the assessment endpoint, a series of extrapolations is necessary to relate the exposure to the assessment endpoint. Some of the general types of extrapolations are listed in Table 12–K.

The product of the ecological effects characterization is a *stressor-response profile* that describes the stressor-response relationship and reviews available evidence of causality. When the stressor-response relationship is based on field surveys as opposed to controlled toxicity tests, the causal factors responsible for observed effects may not be evident. Satisfaction of some or all of the following criteria supports a presumption of causality (see also the epidemiology discussion in Chapter 11, Public Health Issues):

- **Strength:** Exposure to the stressor causes an observable reaction.
- **Consistency:** The association is repeatedly observed.
- **Specificity:** The effect is frequently associated with a particular stressor (or combination of stressors).
- **Temporality:** The effect only occurs after exposure to the stressor.
- **Presence of a biological gradient:** The degree of response depends on the degree of exposure.
- **Plausible mechanism of action:** The proposed interaction between stressor and ecological component is feasible.
- **Coherence:** The hypothesis is consistent with knowledge of ecology and ecotoxicology.
- **Experimental evidence:** The stressor-response relationship can be replicated experimentally.
- **Analogy:** Similar stressors cause similar responses.

Risk characterization. In *risk characterization,* the final phase of both HRA and ERA, the stressor-response and exposure profiles developed in the analysis phase are integrated to evaluate the likelihood of adverse ecological effects associated with exposure to a stressor(s) (Figure 12–9). The purpose of risk characterization is to provide as complete and accurate a picture as possible of the potential risks associated with contact or co-occurrence with stressors. Risks are described in terms of the assessment endpoint(s), and can be expressed qualitatively or quantitatively, depending on available data. The ecological significance of potential effects is discussed in terms of their types and severity, spatial and temporal patterns, and the likelihood of recovery.

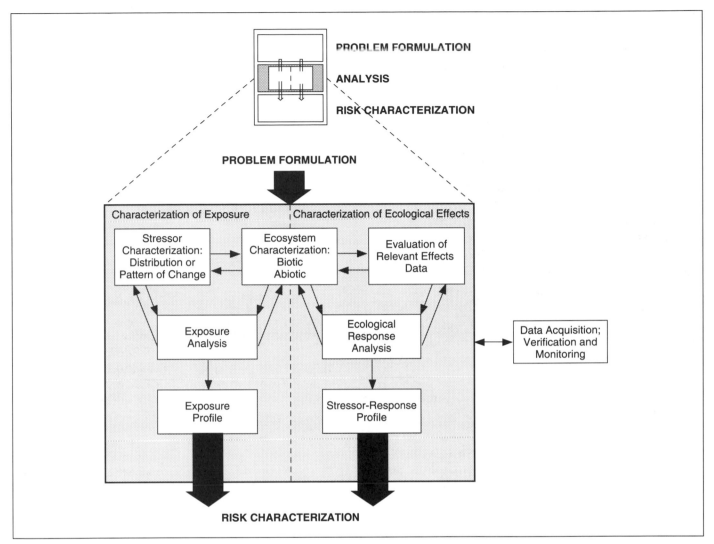

Figure 12–8. The analysis phase. Source: U.S. EPA, 1992a.

Uncertainty. The conclusions of any ERA, and the decisions and actions based on them, are only as sound as the assessment and measurement endpoints selected, the data collected and interpreted, the observations made and recorded, and the quality of the professional judgment applied (i.e., garbage in, garbage out). It is important to recognize that each step of the ERA process has some associated degree of uncertainty. The risk assessor therefore provides the risk manager with a sense of the overall level of confidence that can be placed in the assessment by specifying and justifying all the assumptions made and parameter values used, analyzing the strengths and weaknesses of the analyses used, and estimating (qualitatively or quantitatively) the uncertainties associated with their application to the subject site. To compensate for the substantial inherent uncertainty of ERA processes, risk assessors and managers

generally choose conservative assumptions and parameters, and accept that results will err on the side of overprotectiveness.

Unique Ecological Assessment Issues

Although there are many similarities to the HRA process, several key issues are unique to ERA. These differences have substantial cost and time implications for all sites undergoing the ecologic risk assessment process. In general, ERAs are more costly and time consuming than human assessments due to the complexity of the laboratory and technical analyses.

Population versus individual risk. The subjects of human health risk assessment are individuals of one species, *Homo sapiens*. In contrast, the fundamental unit of ecological risk assessment is generally considered to be the population rather than the individual (U.S. EPA, 1992a). Although the immediate effects of stressors in the environment are on individuals, the ecological significance of these effects (or lack thereof) is determined by the impact on the population of the affected species, and the ramifications of such impacts on the ecosystem as a whole. For example, the death of an organism is obviously catastrophic for the deceased individual but unlikely to impact the viability of the population of which it was a member. On the other hand, a chemical that kills few organisms but reduces the reproductive success of a key species may have considerable ecological consequences.

Ecotoxicity testing. Although the primary interest in ERA is the population or some other ecological component greater than the individual, we are limited in evaluating effects on higher levels of ecological organization by a lack of both assessment and measurement endpoints that can provide information relevant for decision making. As mentioned previously, the study of toxic perturbations of ecological systems is beyond the scope of classical toxicology for a number of reasons:

- Whereas classical human toxicology relies on extrapolation of toxicity data from a handful of mammalian species, ecotoxicology must rely on extrapolation from a few test species to an enormous number of species with a tremendous taxonomic range (e.g., from algae to eagles) and spanning vast differences in size, physiology, and native habitat.

- The spatial and temporal heterogeneity of exposure and conditions in natural systems can cause large variations in the doses and responses observed.

- Organisms in the environment are rarely (if ever) exposed to pure compounds alone but rather to complex mixtures of chemicals whose effects in combination are unknown.

- Chemicals may be volatilized, transformed to more or less toxic products, and/or sequestered in the environment.

This lack of knowledge of environmental variables and limited ability to replicate them in the laboratory or control them in the field results in a high level of uncertainty in assessors' predictions of the effects of stressors on any given ecosystem component from laboratory toxicity tests. Despite this uncertainty, however, toxicity testing can provide valuable information as part of a tiered approach to ERA.

The risk assessor must use professional judgment to decide (1) which types of toxicity tests most effec-

Table 12–K. Some Extrapolations Required to Relate Measurement and Assessment Endpoints

Extrapolation	*Example*
Between taxonomic groups	From laboratory mouse to field mouse
Between responses to stressor	From mortality in laboratory rats to a no-observed-effect-level in prairie dogs
Between laboratory and field conditions	From fish tank to lake
Between one field site and another	Benthic community structure in a swamp versus a bayou
Between individual animals to population	From decreased growth rate in captive individuals to effects on a wild population
Between short- and long-term exposure conditions	From acute or subchronic toxicity tests to lifetime exposure
Between laboratory and natural exposure media	Percent uptake of chemical mixed with laboratory diet versus adsorbed to soil
Between spatial scales	Evaluation of the impact of exposure to a contaminated field on predators whose foraging range is 50 times as large

Source: Adapted from U.S. EPA, 1992a.

tively reduce the inherent extrapolation uncertainty (and are most cost-effective), and (2) at what point the uncertainty is reduced enough to allow a decision to be made regarding the potential or actual impact of a chemical stressor(s) on an ecosystem. Because the physical, chemical, and biological factors involved in the exposure and response of a receptor to a chemical are so complex and variable, uncertainty is best addressed by using standardized, site-appropriate, state-of-the-science procedures whose results can be readily verified and compared with results reported elsewhere.

Types of toxicity tests. Standard test methods are published by the EPA, the American Society for Testing and Materials, and other governmental agencies and standard-setting organizations. Toxicity tests can provide data on the *acute* (short-term) and *chronic* (long-term) effects of chemicals on organisms. The choice of test species is a critical part of this process. Because different types of organisms vary widely in their responses to chemicals (and

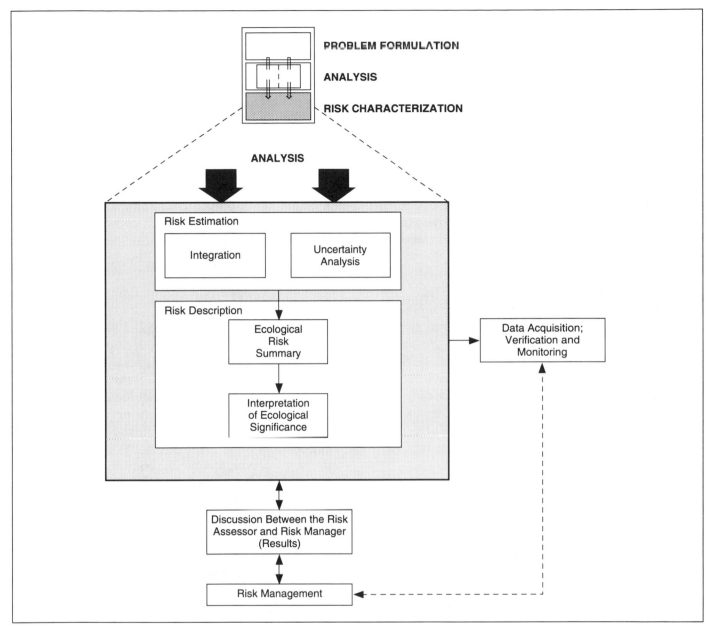

Figure 12–9. The risk characterization phase. Source: U.S. EPA, 1992a.

other stressors), toxicity tests should use species (or combinations of species) representative of the ecosystem under study. An appropriate test species is one that is:

- easily obtainable and easily maintained in the laboratory, with well-defined growth patterns and nutritional requirements
- of economic, recreational, and/or ecological importance, or similar to such species
- widely distributed in the environment, allowing comparison of data obtained from different areas
- highly sensitive to the COCs.

In some cases, tests of species that live at a site may be appropriate. However, most toxicity tests are conducted under controlled conditions in laboratories with samples of contaminated soil, water, or sediment collected from the site being assessed. The use of site media accounts for the site-specific characteristics (e.g., water hardness, pH, temperature, particle size, organic carbon and acid volatile sulfide content) that constitute the habitat and control chemical bioavailability. *In situ* toxicity tests, conducted with either laboratory or resident species at the site itself, can provide information or organism responses to actual field conditions.

Common endpoints examined in acute and chronic toxicity tests are (1) percent response in a group of test organisms exposed to site media, and (2) concentration-response relationships for tests run at several different proportions of site media. The responses observed may be lethal or sublethal (e.g., effects on growth, reproduction, behavior). The statistic usually calculated from acute lethality studies is the LC_{50} or LD_{50} (the concentration or administered dose of chemical that caused death in 50% of exposed organisms; the median lethal concentration or dose). Where sublethal endpoints are observed, the EC_{50} or ED_{50} (median effective concentration or dose, at which 50% of exposed organisms displayed the effect) is calculated.

Chronic test results are usually expressed as estimates of concentrations below which no adverse effects are expected. The *maximum acceptable toxicant concentration (MATC)* is presented as two test concentrations: (1) the *no-observed-effects-concentration (NOEC)*, defined as the highest test concentration that caused no significant toxic effects, and (2) the *lowest-observed-effect-concentration (LOEC)*, defined as the lowest test concentration at which any adverse effect was observed. The geometric mean of these two values is an estimate of the threshold for chronic toxicity.

Future directions in ecotoxicity testing. It is widely agreed that ecosystem-level analyses are most relevant for ERAs. An active area of current research is therefore identifying and improving methods of extrapolating results observed in single organisms to the population level. Many researchers use population models for this purpose. However, extrapolation of the endpoints examined in single-species studies to multiple species at the community or ecosystem level is much more difficult (reviewed by Suter, 1993).

One strategy being developed to examine endpoints at the level of the ecosystem is toxicity testing with multiple species. Multispecies test systems include model ecosystems such as *microcosms* (laboratory systems intended to stimulate all or part of an ecosystem), *mesocosms* (partially confined outdoor experimental systems), and *field tests* (tests of the effects of a treatment in an open ecosystem). Reviews of these and other methods can be found in EPA (1989a) and Suter (1993). Although such tests are promising, the lack of well-defined ecosystem-level assessment endpoints presently limits the applicability of their results to ERAs.

Sediments. As part of PL92–500, the EPA was mandated by Congress to develop federal water quality criteria (WQC). The federal criteria serve as a basis for each state's own water quality standards. Based on a broad range of toxicological studies using a wide range of water column organisms and life stages, these criteria were intended to ensure protection of 95% of all aquatic organisms from adverse effects. However, significant environmental damage has occurred even in areas where WQC are met. In such cases, the primary source of exposure is *sediment* (Table 12–L).

Aquatic sediment can be defined as a collection of debris—coarse to fine particles of mineral and organic matter deposited at the bottom of water bod-

Table 12–L. Some Approaches to Development of Chemical-Specific Sediment Quality Criteria

Approach	Description	Major Assumptions
Bulk sediment chemistry approach	Chemical concentrations in contaminated sediments are compared to concentrations found in reference sediments, which form the basis of the regulatory criteria.	▪ There is no threshold sediment contaminant concentration below which no adverse effects are expected. ▪ Sediment characteristics (e.g., grain size, organic carbon content) do not significantly influence partitioning of chemicals from sediment to water (i.e., differences in chemical bioavailability are negligible).
Equilibrium partitioning approach (presently applicable only to nonionic organic chemicals for which WQC exist)	Sediment quality criteria (SQC) are calculated from water quality criteria (WQC) using Koc (chemical-specific) and foc (site-specific): $SQCoc = Koc \times WQC$ $SQC \text{ (site-specific)} = SQCoc \times foc.$	▪ Partitioning of chemicals largely depends on organic carbon content. ▪ Benthic and water column organisms are equally sensitive to contaminant effects. ▪ Soluble and adsorbed chemical are in equilibrium.

ies. Fifty to 90% of the volume is "interstitial" or "pore" water, depending on depth and degree of compaction. Organic matter content typically ranges from 0 to 10%. Sediments are an important component of aquatic ecosystems because of the habitat they provide for *benthic* (sediment-dwelling) organisms, including commercially important organisms such as shrimp, crayfish, lobster, crab, mussels, and clams, as well as species important in the environmental food web, such as many species of worms, amphipods, bivalves, and insects.

Because of the tendency of both organic and inorganic molecules to adsorb to particles, sediments are also repositories for a wide variety of potentially toxic chemicals, both naturally occurring and anthropogenic. Chemicals undetected or found in only trace amounts in the water column can accumulate to high levels in sediments. These chemicals can enter the water from point sources (e.g., municipal and industrial effluents), nonpoint sources (e.g., agricultural and urban runoff), and from spills, leaks, or dumping of wastes. Common sediment contaminants include:

- organic chemicals such as halogenated hydrocarbons (e.g., PCBs, dioxins and furans, organochlorine pesticides)
- PAHs such as benzo[a]pyrene

- solvents (e.g., benzene, trichloroethylene), as well as metals (e.g., arsenic, lead, cadmium, mercury).

Although this "sink" effect serves to isolate chemicals from aquatic organisms, the sediment can also serve as a reservoir, slowly releasing chemicals from sediment to pore water to organisms. Thus, for the large number of commercial and food-chain organisms that spend a portion of their life cycles close to the sediment, sediment provides chemical exposure as well as nurture (Table 12–M).

In fact, direct transfer of chemicals from sediments to organisms is now considered to be a major route of exposure for many benthic species. Further, consumption of benthic organisms by higher trophic level animals, such as fish-eating birds and humans, can be a major pathway of exposure to chemicals otherwise isolated in the sediment sink. Water quality criteria may not be protective of benthic species (and ultimately of predators that feed on them) because they are principally based on studies with water column organisms that live and feed in the overlying waters as opposed to benthic organisms that live and feed in sediments. As a result, attention has shifted from point to nonpoint sources of pollution, and protection of sediment quality has emerged as a necessary extension of water quality protection.

To evaluate the threat posed by contaminated sediments, several questions must be answered:

Table 12–M. Some Approaches to Development of Chemical Mixture Sediment Quality Criteria

Approach	Description	Major Assumptions
Sediment bioassay approach	The responses of test organisms to field-collected sediments (and/or interstitial water) from contaminated and reference areas are compared to determined potential problem areas. Sediments are spiked in the laboratory to determine the effects of different concentrations of chemicals (or mixtures of chemicals) on test organisms.	Sediment assays performed in the laboratory are representative of effects in the field. Spiked sediments have the same properties (e.g., contaminant bioavailability) as field sediments.
Benthic community structure and function approach	Communities of benthic organisms from contaminated and reference areas are compared in terms of species composition, relative abundance, and community function.	The selected measures of benthic community structure are adequate indicators of adverse effects.
Sediment quality triad approach	Independent measures of sediment chemistry, sediment toxicity, and benthic community structure in contaminated versus reference areas are combined to form an integrated picture of chemical concentration and effects at a site.	Because the sediment quality triad is a combination of sediment chemistry, bioassay, and benthic community analysis approaches, the same assumptions are made.

- What concentrations of contaminants cause adverse effects?
- What proportions of adsorbed contaminants are bioavailable (capable of dissociating from sediment into pore water and thence into an organism)?
- What are the factors that mediate the toxicity and bioavailability of contaminants in sediment?

Development of sediment quality criteria. Several environmental statutes—the Clean Water Act; the Toxic Substances Control Act; the Federal Insecticide, Fungicide, and Rodenticide Act; as well as RCRA and CERCLA—give EPA the authority to address these questions and develop methods for management of contaminated sediments. The agency is authorized under Section 304(a) of the Clean Water Act to develop and implement sediment quality criteria (SQC). Because (1) the extent and significance of the problem are not well defined, and (2) scientific approaches to the assessment and remediation of sediments are relatively new and unproven, this is a difficult charge.

The EPA has conducted an analysis of 11 of the currently available methods for sediment quality assessment (U.S. EPA, 1989e). These approaches may be divided into two general types:

- approaches that provide criteria for individual chemicals
- approaches that consider the effects of the mixtures of chemicals that are usually encountered in the environment.

Examples of each are briefly described in Tables 12–L and 12–M. Additional information on these and other approaches can be found in excellent reviews by Adams et al (1992) and Chapman (1989).

The chemical-specific approaches yield numerical criteria representing "safe" levels of individual chemicals, whereas chemical mixture approaches are used to identify sediments that cause adverse biological effects without specifying or quantifying the contributions of individual chemicals to these effects. Both approaches have strengths and weaknesses, but neither is adequate for development of generally applicable sediment criteria. Although the chemical-specific approaches are generic and easy to apply, they do not account for the effects of chemical interactions within the complex chemical mixtures found in most sediments. The EPA has proposed the equilibrium partitioning approach for derivation of

chemical-specific SQC for nonionic organic chemicals (U.S. EPA, 1991a).

As indicated in Table 12–L, this method calculates an organic carbon-normalized SQC (SQCoc) from WQC (the assumed pore water concentration) using the organic carbon:water partition coefficient Koc. It allows for differences in chemical bioavailability at different sites by multiplying the generic SQCoc by the site-specific fraction of organic carbon (foc). Because organic carbon does not control the adsorption of metals or ionic organic chemicals to sediments, this approach is only applicable to nonionic organics for which WQC have been developed. Efforts to develop other normalization methods for these other chemical groups are currently in progress.

The interpretive difficulty posed by the presence of multiple contaminants is avoided in the chemical mixture approaches, because they are based on observed effects on organisms rather than chemical concentrations. However, they are quite expensive to perform and difficult to interpret, and results may not be applicable to any other sites. One of the most promising of these methods is the sediment quality triad approach, which combines independent measures of sediment contamination levels, sediment toxicity (e.g., media toxicity tests), and *in situ* measurements of benthic community health (e.g., species abundance, diversity) (Chapman, 1990, 1991; Chapman et al, 1992). Examples of possible results and interpretations of sediment quality triad data are shown in Table 12–N.

Additional efforts needed. Greater understanding of the role of sediments as a reservoir of potentially toxic chemicals is necessary to protect public health and the environment. Unfortunately, development of sediment quality criteria is considerably more difficult than development of water quality criteria because of the many site-specific factors that influence the toxicity of chemicals in sediment but not in water. These factors derive from the nature of sediment. Unlike the overlying water, sediment is relatively structured and immobile, and consists of a complex array of particles of different sizes and chemical characteristics. These characteristics influence the degree and affinity of chemical adsorption, and hence bioavailability. Adsorbed contaminants do not mix freely in sediment as they would in water, but depend on sediment deposition and erosion patterns for movement. Thus, the concentration and distribution of sediment-associated chemicals can be

Table 12–N. The Sediment Quality Triad Model

Sediment Contamination	Sediment Toxicity	Benthic Community Alteration	Possible Conclusions
+	+	+	Strong evidence for pollution-induced alteration of the community.
−	−	−	Lack of evidence of pollution-induced community alteration.
+	−	−	Contaminants are not bioavailable.
−	+	−	Unmeasured chemicals/conditions exist with the potential to cause alteration of the community.
−	−	+	Alteration in the community is not caused by toxic chemicals.
+	+	−	Toxic chemicals are stressing the system.
−	+	+	Unmeasured toxic chemicals are altering the community.
+	−	+	Alteration of the community is not caused by toxic chemicals, *or* chemicals are not bioavailable.

Positive (+) and negative (−) signs indicate whether statistically significant differences from the reference conditions exist.
Source: Adapted from Chapman PM, Power EA, Burton GA. Integrative assessments in aquatic ecosystems. In *Sediment Toxicity Assessment*. Burton GA (ed). Chelsea, MI: Lewis Publishers, 1992, a subsidiary of CRC Press, Boca Raton, Florida. Used with permission.

highly heterogeneous in both space and time. Because no existing assessment methodologies have undergone extensive review or field validation, the few criteria presently available cannot be regarded as more than preliminary. Immediate efforts are needed to:

- identify the extent of sediment contamination and prioritize areas of greatest potential concern
- incorporate existing approaches into a flexible tiered approach
- expand, improve, and validate existing assessment methodologies
- develop new approaches as conceptual and technological advances allow.

Future Directions in Ecological Risk Assessment

The purpose of ERA and the legislation that mandates it is to protect the "environment" from physical, chemical, and biological "stressors" of all descriptions. This is a necessary goal, but it is also substantially beyond researchers' (or regulators') current capabilities. This is partially because funds are lacking to support and skilled personnel to perform research and testing on the tens of thousands of extant chemicals and the physical changes occurring on a global scale. The other reason is the absence as yet of an intellectual framework to guide such efforts. This framework must be grounded in a new, multidisciplinary science melding concepts and approaches of ecology, toxicology, chemistry, physics, engineering, and other sciences.

Development of an ecosystem-level perspective in ecotoxicology and ecological risk assessment is considered to be the first priority for advancing researchers' understanding of the impacts of anthropogenic (and natural) stressors on ecosystems. The major challenges facing risk assessors are (1) to define what should or must be preserved in the environment; (2) to identify explicit, rational assessment endpoints reflective of these values; (3) to develop corresponding measurement endpoints for quantitative evaluation of their responses to (and recovery from) contact with stressors; and (4) to develop accurate and cost-effective methods of stressor evaluation that will reduce the uncertainty involved in extrapolation of results from laboratory to natural conditions or from one site to another.

Because of the enormous complexity of the environment, its potential stressors, and their interactions, it is unlikely that numerical criteria can be developed to ensure environmental health. Even if they could, chemicals are not the only or even the most serious anthropogenic threat to the environment. Rather than resort to generalizations that can be misleading when applied to particular sites and situations, some experts (e.g., Chapman, 1991) recommend that environmental decision-making be based on flexible environmental quality guidelines developed from the best available science and a reasoned compromise between cost and benefit.

SUMMARY

A review of site-specific information is required to identify the physical characteristics of the site, potential COCs, and potentially exposed populations.

Current and future land uses should be defined as clearly as possible. The collection of site-specific survey data on behavior patterns and the likelihood of exposure can yield more accurate risk estimates.

Selection of COCs involves several phases. The first phase includes the screening of chemicals after detection limits, detection frequencies, and blank samples are examined, whereas the second phase involves comparing chemical concentrations in various environmental media (e.g., air, soil, and water) to medium-specific background data. A concentration-toxicity screen can also be used to further reduce the number of COCs.

The detailed formal and rigorous statistical analysis of data sets represents a substantial methodologic enhancement of the risk assessment process. Because obtaining site-specific chemical data sets is so expensive and time consuming, a proper and defensible method for collecting and analyzing data is mandatory. Tremendous cost and effort are associated with the site investigation and cleanup process; thus, environmental program managers need to understand both the risk assessment process and the problems and pitfalls associated with site data collection and analysis.

Because organic chemicals tend to accumulate in the media in which they are most soluble, a few basic physicochemical properties can be used to predict the behavior and fate of chemicals released into the environment. Highly lipophilic compounds tend to sequester in soil, sediment, and biota, and the food chain is the primary pathway of human exposure. Highly soluble compounds are found in water, and ingestion of contaminated water is usually the primary pathway of human exposure. Volatile compounds tend to partition into air, and inhalation is the major source of human exposure.

Generally, four distinct receptor groups could be exposed to COCs at a given site: residential, commercial/industrial, and trespassers/recreational receptors, each of which is defined by a distinct set of exposure pathways. The exposure point concentration used to estimate RME exposures is the upper-bound confidence limit on the arithmetic or geometric mean or the maximum value detected, whereas mean data are used to quantify average exposures.

Toxicity assessment evaluates the nature and extent of health effects from exposure to site-related chemicals. EPA-derived toxicity criteria are available for many chemicals. Cancer slope factors represent the 95% upper confidence limit on the probability of a carcinogenic response; hence, there is a 95%

chance that the true risk is lower than the projected risk. Because carcinogenic risks at low exposure levels cannot be measured directly either in laboratory animal or human epidemiologic studies, mathematical models are used to extrapolate from high to low exposure levels. Most models produce quantitatively similar results in the range of observable data but yield estimates that can vary by three or four orders of magnitude at lower doses.

RfDs represent the level of exposure not expected to cause adverse health effects even in sensitive subpopulations, with variability typically spanning one order of magnitude. Given the EPA's conservative method of deriving RfDs, they are likely to overestimate the possibility that exposed individuals could experience adverse health effects. Lewis et al (1990) evaluated numerous studies regarding toxicological extrapolations from animals to humans in an effort to determine more representative uncertainty factors and concluded that uncertainty factors need not be as large as EPA's recommended defaults to provide values protective of human health.

The use of a conservative dose-response cancer model has historically provoked a substantial outcry from the regulated community, because highly conservative CSFs result in conservative risk estimates. There is a substantial financial increment associated with each additional magnitude (factor of 10) of calculated risk. Thus, based on typical EPA practice, if risks are considered for regulation beginning at one per million (1×10^{-6}), then an overestimation of 3 or 4 orders of magnitude can result in a site moving from no cleanup (no or minimal cost) to multiyear, multimillion dollar efforts.

The process of risk characterization combines information from the exposure and toxicity assessments to estimate the magnitude of potential adverse health effects due to site-related COCs. Cancer risks from exposure to multiple carcinogens and multiple pathways are assumed to be additive. Risks are frequently acceptable if they are within the 10^{-6} to 10^{-4} target risk range; however, EPA wields considerable latitude in their interpretation of "acceptable."

Potential health effects of chronic exposure to noncarcinogenic compounds are assessed by calculating a hazard quotient (HQ) for each chemical of concern. Although an HQ greater than one does indicate that exposure to that contaminant may cause adverse health effects in exposed populations, the level of concern associated with exposure to noncarcinogenic compounds does not increase linearly as HQs exceed one. If mechanisms of toxicity and/or end-organ effects are similar, individual HQs for chemicals that affect the same target organ or system are summed to form an HI.

Use of single-value estimates for input parameters—especially upperbound estimates—provides limited information to decision makers who must determine whether estimated health risks warrant remediation. A quantitative uncertainty analysis provides decision makers with an estimate of the range of possible health impacts so that the optimal, risk-based decision can be made.

Overall, the human health risk assessment process has undergone significant refinement and revision in recent years. The human health risk assessment process is still controversial and either too conservative (a view held by the regulated community) or too inexact and liberal (a view frequently held by nongovernmental organizations). Despite these divergent viewpoints, human health risk assessment techniques do provide a consistent, uniform process for risk managers who must evaluate a wide variety of simple and complex sites and pathways. Continued evolution of the risk assessment process is to be expected; however, wholesale revolution in the process is unlikely in the near future. Risk management judgment will remain the final arbiter. The next section of this chapter describes a family of risk assessment techniques that have gained increasing importance for industrial systems analysis.

Probabilistic risk assessments are a family of specialized risk analysis techniques that developed in conjunction with the nuclear power industry. However, these techniques have broad application for many types of industrial process safety and reliability studies. As public pressure has focused on the performance and safety of large-scale industrial facilities, the requirement for formal PRAs has increased. The Occupational Safety and Health Administration (OSHA) has recently reinforced this trend by promulgating its *Process Safety Rules and Regulations* (CFR 1910.119). Although OSHA's rules do not require formal quantitative PRAs, it is not difficult to envision the time when probabilistic safety analyses are required. California's Risk Management Prevention Plan (RMPP) currently requires a formal risk analysis prior to industrial process facility modification. The use of PRAs will clearly increase during the 1990s and into the 21st century.

REFERENCES

Abou-Setta MM, Sorrell RW, Childers CC. A computer program in BASIC for determining probit and log-probit or logit correlation for toxicology and biology. *Bull Environ Contam Toxicol* 36:242–249, 1986.

Adams WJ, Kimerly RA, Barnett, Jr, JW. Sediment quality and aquatic life assessment. *Environ Science and Tech* 26:1864–1875, 1992.

Adriano DC. *Trace Elements in the Terrestrial Environment*. New York: Springer-Verlag, 1986.

Briggs GC. Theoretical and experimental relationships between soil adsorption, octanol-water partition coefficients, water solubilities, bioconcentration factors, and the parachor. *J Agric Food Chem* 29:1050–1059, 1981.

Cairns, Jr, J, Dickson DL, Maki AW (eds). *Estimating the Hazard of Chemical Substances to Aquatic Life*, STP 657. Philadelphia: American Society for Testing and Materials, 1978.

Carson R. *Silent Spring*. Boston: Houghton Mifflin, 1962.

Chapman PM. The sediment quality triad approach to determining pollution-induced degradation. *Science of the Total Environ* 97–98:815–825, 1990.

Chapman PM. Environmental quality criteria: What type should we be developing? *Environ Science and Tech* 25:1353–1359, 1991.

Chapman PM. Current approaches to developing sediment quality criteria. *Environ Toxicol and Chem* 8:589–599, 1989.

Chapman PM, Power EA, Burton GA. Integrative assessments in aquatic ecosystems. In *Sediment Toxicity Assessment*. Burton, GA (ed). Chelsea, MI: Lewis Publishers, 1992.

Chiou CT, Schmedding DW, Manes M. Partitioning of organic compounds in octanol-water systems. *Environ Sci Technol* 16(1):4–10, 1982.

Crump KS, Guess HA, Deal KL. Confidence intervals and tests of hypotheses concerning dose response relations inferred from animal carcinogenicity data. *Biometrics* 33:437–451, 1977.

Dobbs AJ, Cull MR. Volatilization of chemicals—Relative loss rates and the estimation of vapor pressures. *Environ Poll* (Series B) 3:289–298, 1982.

Dunlap TR. *DDT: Scientists, Citizens, and Public Policy*. Princeton, NJ: Princeton University Press, 1981.

Finney DJ. *Probit Analysis: A Statistical Treatment of the Sigmoid Response Curve*, 2nd ed. London: Cambridge University Press, 1952.

Fries GF. Assessment of potential residues in foods derived from animals exposed to TCDD-contaminated soil. *Chemosphere* 16(8/9):2123–2128, 1987.

Gansecki M. Comments on the "Background Geochemical Characterization Report for 1989—Rocky Flats Plant." Mike Gansecki, August 1, 1992.

Geyer HJ, Scheunert I, Korte F. Correlation between the bioconcentration potential of organic environmental chemicals in humans and their n-octanol/water partition coefficients. *Chemosphere* 16(1):239–252, 1987.

Geyer HJ, Scheunert I, Korte F. Bioconcentration potential of organic environmental chemicals in humans. *Regul Toxicol Pharmacol* 6:313–347, 1986.

Geyer HJ, Sheehan P, Kotzias D, et al. Prediction of ecotoxicological behavior of chemicals: Relationship between physicochemical properties and bioaccumulation of organic chemicals in the mussel. *Mytilus edulis. Chemosphere* 11(11):1121–1134, 1982.

Gilbert RO. *Statistical Methods for Environmental Pollution Monitoring*. New York: Van Nostrand Reinhold, 1987.

Gilliom RJ, Helsel DR. *Estimation of Distributional Parameters for Censored Trace-Level Water Quality Data*. U.S. Geological Survey Open File Report 84–729, 1984.

Grossjean D. Distribution of atmospheric nitrogenous pollutants at a Los Angeles area smog receptor site. *Environ Science Technol* 17(1):13–19, 1983.

Hattemer-Frey HA, Lau V, Krieger GR. An evaluation of the soil, plant, and chemical parameters that influence root uptake of metals. *Environ Geochem Health*, in press.

Hoffman DJ, Rattner BA, Hall RJ. Wildlife toxicology. *Environ Science and Tech* 24:276–283, 1990.

Hoffman HO, Gardner RH. Evaluation of uncertainties in radiological risk assessment models. In *Radiological Assessment*, NUREG/CR-3332. Till JE, Meyer HR (eds). Washington DC: U.S. Nuclear Regulatory Commission, 1983.

Kenaga EE, Goring CAI. Relationship between water solubility, sorption, octanol-water partitioning, and concentration of chemicals in biota. In *Aquatic Toxicology*. Eaton JE, Parrish PR, Hendricks AC (eds). Philadelphia: American Society for Testing Materials STP 707, 1980.

Kenaga EE. Correlations of bioconcentration factors in aquatic and terrestrial organisms with their physical and chemical properties. *Environ Sci Technol* 14:553–556, 1980.

Lewis SC, Lynch JR, Nikiforov AI. A new approach to deriving community exposure guidelines from "No-Observed-Adverse-Effect Levels." *Reg Toxicol Pharmacol* 11:314–330, 1990.

Lieberman HR. Estimating LD_{50} using the probit technique: A basic computer program. *Drug Chem Toxicol* 6(1):111–116, 1983.

Lyman WJ, Reehl WF, Rosenblatt DH. *Handbook of Chemical Property Estimation*. New York: McGraw-Hill, 1982.

Menzer RE, Nelson JO. Water and soil pollutants. In Doull J, Klaassen CD, Amdur MD (eds). *Toxicology*. New York: MacMillan Press, 1980.

Moriarty F. *Ecotoxicology: The Study of Pollutants in Ecosystems*, 2nd ed. New York: Academic Press, 1988.

Norton SB, Rodier DJ, Gentile JR, et al. A framework for ecological risk assessment at the EPA. *Environ Toxicol and Chem* 11:1663–1672, 1992.

Office of Science and Technology Policy (OSTP). Chemical carcinogens: A review of the science and its associated principles. *Fed Reg* 50:10372–10422, 1985.

Porter PS, Ward RC, Bell HF. The detection limit. *Environ Sci Technol* 22(8):856–861, 1988.

Sette A, Adorini L, Marubini E, et al. A microcomputer program for probit analysis of interleukin-2 (IL-2) titration data. *J Immunolog Methods* 86:265–277, 1986.

Shacklette HT, Boerngen JG. *Elemental Concentrations in Soils and Other Surficial Materials of the Conterminous United States*, U.S. Geological Survey Professional Paper 1270. Washington DC: U.S. Government Printing Office, 1984.

Shor RW, Baes III CF, Sharp RD. *Agriculture Production in the United States by County: A Compilation of Information from the 1974 Census of Agriculture for Use in Terrestrial Foodchain Transport and Assessment Models*, ORNL-5763. Oak Ridge, TN: Oak Ridge National Laboratory, 1982.

Suter II GW. *Ecological Risk Assessment*. Chelsea, MI: Lewis Publishers, 1993.

U.S. Environmental Protection Agency (EPA). *Unfinished Business: A Comparative Assessment of Environmental Problems*. Washington DC: EPA, 1987.

U.S. EPA. *Ecological Assessment of Hazardous Waste Sites: A Field and Laboratory Reference*. EPA/600/3–89–913. Washington DC: EPA, 1989a.

U.S. EPA. *Risk Assessment Guidance for Superfund. Volume I. Human Health Evaluation Manual, Part A*. Washington DC: Office of Emergency and Remedial Response, 1989b.

U.S. EPA. *Risk Assessment Guidance for Superfund, Volume II. Environmental Evaluation Manual*. EPA/540/1–89/001. Interim Final. Washington DC: EPA, 1989c.

U.S. EPA. *RCRA Facility Investigator's (RFI) Guidance*. Vols. I-IV. EPA 530/SW-89–031. Washington DC: EPA, 1989d.

U.S. EPA. *Sediment Classification Methods Compendium*. Washington DC: Draft Final Report, Watershed Protection Division, 1989e.

U.S. EPA. *Exposure Factors Handbook*, EPA/600/8–82/043. Washington DC: Office of Health and Environmental Assessment, 1990a.

U.S. EPA. *Guidance for Data Usability in Risk Assessment*. EPA/540/G-90/008. Washington DC: EPA, 1990b.

U.S. EPA. *Reducing Risk: Setting Priorities and Strategies for Environmental Protection*. SAB-EC-90–021. Washington DC: Science Advisory Board 1990c.

U.S. EPA. *Handbook for Remediation of Contaminated Sediments*. EPA/625/6–91/028. Washington DC: EPA, 1991a.

U.S. EPA. *Risk Assessment Guidance for Superfund. Volume I—Human Health Evaluation Manual, Part B: Development of Risk-Based Preliminary Remediation Goals*, OSWER Directive 9285.7–01B. Washington DC: Office of Solid Waste and Emergency Response, 1991b.

U.S. EPA. *Risk Assessment Guidance for Superfund. Volume I—Human Health Evaluation Manual, Part C: Risk Evaluation of Remedial Alternatives*. OSWER Directive 9285.7–01C. Washington DC: Office of Solid Waste and Emergency Response, 1991c.

U.S. EPA. *Risk Assessment Guidance for Superfund, Volume 1, Human Health Evaluation Manual, Supplemental Guidance "Standard Default Exposure Factors*, Draft Final, March 25, 1991, OSWER Directive 9285.6–03. Washington DC: Office of Solid Waste and Emergency Response, 1991d.

U.S. EPA. *Framework for Ecological Risk Assessment*. EPA/630/R-92–001. Washington DC: EPA, 1992a.

U.S. EPA. Guidelines for exposure assessment. *Fed Reg* 57(104):22888–22938, Washington DC: 1992b.

U.S. EPA. *Report the Ecologic Risk Assessment Guidelines Strategic Planning Workshop*. EPA/630/R-92/002. Washington DC: EPA, 1992c.

U.S. EPA. *Supplemental Guidance to RAGS: Calculating the Concentration Term*. Publication 9285.7–081. Washington DC: Office of Emergency and Remedial Response, 1992d.

U.S. EPA. *Superfund Site Health Risk Assessment Guidelines* (Review Draft). Washington DC: Environmental Health Committee, Science Advisory Board. January 1993.

Weil CS, McCollister DD. Relationship between short and long-term feeding studies in designing an effective toxicity test. *Agric Food Chem* 11:486–491, 1963.

Weil CS. Statistics versus safety factors and scientific judgment in the evaluation of safety for man. *Toxicol Appl Pharmacol* 21:454–463, 1972.

Woodfork K, Burrell R. A basic computer program for calculation of CH_{50} values by probit analysis. *Computers Biol Med* 15(3):133–136, 1985.

13

Indoor Air Quality

D. Jeff Burton, MS, PE, CSP, CIH
Roger Greenway, MS, MBA, CCM
Gary R. Krieger, MD, MPH, DABT
Matthew D. Ziff, MS
William J. Keane, PhD, CIH

The previous chapters have been directed toward the "external" world of human interaction with the environment. Depending upon the age of those surveyed, up to 90% of human activity is spent indoors (U.S. EPA, 1989). Therefore, the "internal" or indoor environment can represent the largest single potential exposure source.

Throughout the last 20 years, the attention directed toward issues of indoor air quality (IAQ) has significantly increased. This new scrutiny is largely because of both real and perceived changes in commercial and home building design and construction. As buildings have increased their energy efficiencies, various IAQ problems have been described and attributed to poor ventilation, lack of fresh air exchange (inadequate dilution of existing indoor contaminants), and poor design and layout of interior spaces (e.g., cubicles and partitions).

This chapter explores the analysis of IAQ problems based on four integrated approaches:

- architectural and design characteristics—location, form, functions, and human factors for interior environments
- medical and epidemiological—the types of medical problems and complaints associated with IAQ problems, and the strategies that are useful for their investigation and resolution
- engineering—evaluation and testing of heating, ventilation, and air conditioning systems (HVAC)
- monitoring and measuring potential indoor air contaminants.

In addition, a special section on asbestos, a controversial indoor air contaminant, is presented at the end of this chapter. The asbestos section focuses on the evaluation and control of asbestos-contaminated materials in the indoor environment.

ARCHITECTURAL AND DESIGN CHARACTERISTICS

Issues in IAQ relating to the architectural design of spaces involve relationships between the people occupying a space and the physical characteristics of that space. A physical space (a room, a house, a building floor area, or an entire building) acts as a modulator between an interior environment and surrounding environments, including adjacent and nearby interior environments and the exterior or natural environment. The modulation is typically designed by architects, engineers, interior designers, and other design consultants to provide an appropriate interior climate for the activities that will occur within the space. This environmental modulation is achieved through architectural and engineering proposals that are typically reviewed and approved by the future users of the spaces.

The interior environment of a space or building is often quite different from the simultaneously existing environment outside the space. The environment in one space can even be substantially different from an environment in a nearby interior space. An interior environment is created and maintained by architectural and design-related components and materials. The building envelope, fenestration and glazing, interior materials and finishes, and environmental support systems such as the HVAC equipment and lighting elements all affect the characteristics of an interior environment.

The architectural design of spaces can influence both the actual and perceived IAQ, as defined by the presence of smoke, pesticides, fungi, volatile organic compounds (VOCs), etc., and an accompanying range of health problems. Perceptual characteristics that are not necessarily health threatening include inappropriate air temperature, stale air (often a result of insufficient air exchange rates within a space), airborne odors, inappropriate humidity levels, proximity to fenestration and glazing (a location for convective air currents and extreme shifts in indoor air temperatures), and proximity to HVAC supply and return registers.

Architectural factors can influence both the actual and the perceived IAQ of a space. These factors include the size and shape of the space, its orientation to sunlight, the materials used to construct and finish it, its type or function, its relationships to adjoining or nearby spaces, and its location within its geographic and climatic context. In well-planned architecture and interiors, the relationships between all of these factors are understandable and important. The size and shape of a space should be a result of or at least contribute to the desired function of the space. Similarly, the materials used in a space should have an appropriate relationship to the function as well as to the larger geographic and climatic environment of the space. The relationships of these architectural characteristics and the quality of indoor air are discussed next.

Size and Shape of Spaces

The size and shape of a space is a result of architectural design considerations that attempt to integrate the function and the physical, spatial layout of the space. These design considerations include the number of people who will use the space, the kinds of behaviors that will be performed within it, its physical orientation relative to sunlight, and the physical relationships required between the people using it.

The size and shape of spaces, whether small rooms of a house or large auditoriums, affect both the actual physical characteristics and the perceived characteristics. For example, placement of HVAC components, such as supply and return registers and the resulting patterns of air movement, can affect IAQ within the space. Stagnant air or disturbing air motion, excessive heat or cold, inappropriate humidity levels, and excessive deposits of dust and dirt all can result from the size and shape of a space. Often HVAC systems are designed based on standard distribution patterns related to the square footage or meters of a space, cubic feet or meters per minute of air distributed, and standard humidity and temperature levels for various types of spaces. Spaces that do not conform to the standards can be fitted with a system that does not adequately provide the environmental characteristics required by the inhabitants.

Orientation of a Space

The amount of sunlight that enters a space influences the temperature levels, the perception of airborne particulate matter such as dust, the humidity level, and the amount of surface mold and fungus growth. An office building floor layout usually tries to maximize efficiency in terms of square footage. Materials, functions, and sizes of spaces that occur on the north side of an office floor are often exactly the same as those occurring on the south side of the same office floor. Architects and designers have rediscovered the virtues of design that is responsive to solar orientation, thanks largely to the oil embargo of 1973–1974. The presence of direct sunlight is well known as a desired characteristic of many types of spaces, from kitchens and breakfast rooms to executive offices.

Construction and Finishing Materials for Interior Spaces

Materials used to construct building envelopes and create the interior furniture, finishes, and partitions also can decrease IAQ. *Offgasing* is a well-known phenomenon of many materials used in building construction, interior spaces, furniture, and finishes. Chemicals used in bonding, covering, stabilizing, and insulating elements of the spaces we live and work in give off formaldehyde and VOCs. Studies of offgasing materials have suggested that the most significant amounts of released chemicals occur within the first three to six weeks of installation of the materials. After that time, the release of chemicals occurs at a steady but low level for quite a long time. Indoor air that is continuously recirculated by adding only a small percentage of fresh outside air will be affected by these chemical releases.

Function of a Space and Air Quality

The function of a space influences the design of the architecture, the interior layout, the engineering of environmental control systems, the contents of the space, and the behavior of the people using the space. The type and intensity of chemical contaminants in the air are directly related to the kind of activities that occur within a space. In newly constructed or renovated spaces, after the offgasing from construction materials, furniture, and finishes has abated, the air quality within the space is primarily the result of mechanical and chemical processes taking place in the space. Also, environmental control systems such as HVAC components can promote the growth of microorganisms and the accumulation of dust, dirt, and debris.

Impact of Nearby Spaces on Air Quality

Where does the air circulating within an interior space come from? Depending on the design of the HVAC system, fresh air intake ducts with louvered or screen-covered grills may be located at or just outside the building's exterior. The location of these fresh air intake grills is often dictated by the design of the interior components of the HVAC system and the mechanical rooms that are dedicated to that equipment, by the immediate surroundings of the building, or by cost-driven choices that may not be relevant to system efficiency or air quality.

A fresh air intake is supposed to supply a calculated amount of fresh air to the building's HVAC system during the course of daily operation and use of the building. The amount of fresh air required is calculated by establishing air exchange rates based on the number of people using the space and their activities in the space. The fresh air intakes tend to be located close to the mechanical equipment rooms and other areas remote from public spaces.

One adverse situation found in large, multifunctional buildings is the location of main fresh air intake grills in an adjacent parking garage, which may

be above or below ground level. The air drawn into the HVAC system from such a location can contain many unpleasant and potentially toxic contaminants from vehicular exhaust. Filtering systems can eliminate many of these contaminants, but that defeats part of the purpose of bringing in fresh air and is another costly component in operating a building's ventilation system.

Geographic and Climatic Factors in Architectural Design

One of the reasons architects study indigenous or "native" buildings is to understand fundamental principles in environmentally sensitive design responses. Indigenous buildings tend to be extremely responsive to climatic conditions within their geographic region. Historically, it was possible to look at the materials, methods of construction, and overall design of a building and be able to tell quite accurately where it was located. This was possible because the materials, methods of construction, and design were responses to particular climatic, geographic, and cultural conditions. Buildings of this type did not create interior environments that established contradictory or illogical situations.

Today, we regularly design buildings in hot, humid climates with large wall surfaces of south-facing fixed glass. With developments in technologies that enabled building systems to override or ignore environmental limitations, indigenous buildings, with their careful response to climate, became anachronistic or "inflexible" in the eyes of designers and their clients. They saw advantages in disregarding climatic characteristics which could be expressed in short-term or local profitability. This design directly contradicts a logical environmental response and represents a decision to ignore straightforward, responsive, and user-controlled design in favor of a design that relies on the intervention of high-technology and high-energy use for environmental control. It is likely that all subsequent design decisions, including interior systems decisions, will be required to use similar technological and energy-demanding responses to keep the interior environment in control. In a struggle for control, systems created by human beings are invariably eroded and, given enough time, destroyed by the natural world. Not surprisingly, high-technology, energy-demanding buildings soon develop "problems" that require even greater levels of intervention and control. Mold, fungi, bacterial growth, humidity, and moisture trapped in walls all can lead to rot, rust, decay, and general deterioration of a building's exterior and interior components. Problems with the interior air that is blown through the ducts, plenum spaces, rooms, and hallways will accompany this general erosion of the environment.

The role of the natural environment on materials, human health, and human social structure has been studied for thousands of years. The effects of living within the artificial environment of contemporary buildings on the physical and social characteristics of humans are not fully understood. Unfortunately, some evidence suggests a negative health effect: Some individuals now state that they suffer from a new medical malady called *multiple chemical sensitivity* (Sullivan & Brooks, 1991). It is not uncommon to discover this real or perceived problem as part of IAQ investigations. The next section discusses the medical evaluation of IAQ problems. More investigation is needed to develop better environmental control and response systems.

Investigation of the air quality inside our homes, offices, institutions, and shopping centers requires an understanding of the fundamental decisions that create these physical environments. The problem of IAQ exists because humans live in indoor environments. These indoor environments are initially the result of architectural, engineering, and construction design decisions. The effects of these decisions upon IAQ must be evaluated at the earliest design stage.

MEDICAL APPROACHES TO IAQ PROBLEMS

The health impacts of IAQ problems were graphically illustrated by the mass outbreak of building-related pneumonia at the 1976 American Legion convention in Philadelphia. This episode was associated with 182 cases of illness and resulted in 29 fatalities. The combination of the 1976 *Legionella pneumonia* epidemic and the 1970s energy crisis focused new attention on the significant health impacts associated with the indoor environment.

Various medical complaints were quickly attributed to IAQ problems. Further complicating the diagnostic situation, several descriptive terms were used to describe and diagnose IAQ problems: (1) building-related illness (BRI), (2) tight-building syndrome (TBS), and (3) sick-building syndrome (SBS). All of these terms imply that a defined pathophysiology or true clinical syndrome can be attributed to a commercial or residential building environment.

The nomenclature problem of describing IAQ effects is not insignificant: There appear to be many potential etiologic sources of "problem buildings," such as ventilation, building materials, and infectious agents. J.E. Woods has proposed two fundamental types of building problems: (1) building-related and (2) sick- (tight-) building syndrome (Woods, 1991). Under this classification scheme, BRI is typically associated with objective clinical and/or laboratory findings, including fever, infection, blood serology conversions, reproducibly identifiable indoor air pollutants, and prolonged recovery times after exposure terminates (Woods, 1991). SBS or TBS is considered as a working hypothesis when a variety of self-reported, nonobjective symptoms are reported for more than two weeks at frequencies significantly greater than 20% of occupants in an area of the building or in the entire building (Woods, 1991; Quinlan, 1989). Nonspecific symptoms include headaches, fatigue, nausea, and skin irritation without persistent rash. The discussion of SBS lists common health complaints associated with SBS. Although there can be overlap between BRI and SBS, the SBS symptoms tend to be self-reported and not easily confirmed by independent evaluation. Discussions later in the chapter cover each of these building "diagnosis" categories.

Building-Related Illness

Five diseases involve clearly recognized pathophysiologic mechanisms that can be attributed to IAQ problems and BRI (Hodgson, 1993):

- *Asthma*—reversible narrowing of airways typically triggered by a single source, such as chemicals like formaldehyde or isocyanates.
- *Hypersensitivity pneumonitis*—fever, cough, chest tightness, fatigue, and pulmonary infiltrates on a chest X ray. These findings are also associated with changes in lung function related to a specific workplace exposure. Exposure sources include organic dusts; endotoxins; and aerosols from cooling towers that are contaminated with bacteria, molds, and fungi, such as *Alternaria aspergillus, Cladosporium, Fusarium,* and *Penicillium.*
- *Allergic rhinitis*—nasal discharge, congestion, and interior swelling associated with a positive challenge test to a specific agent found in the building.
- *Infections*—myriad viruses, bacteria, fungi, and rickettsial organisms that cause human disease and can be transmitted within the indoor environment. These diseases include *Legionella,* tuberculosis, Pontiac fever, and *Coxiella burnetti* (Q fever).

- *Dermatitis*—skin rashes attributable to irritants, light, and allergies. Various types of fibers (e.g., fiberglass) may be released from ventilation dust linings and produce strong irritant reactions. Rashes have also been attributed to relative humidities below 30%; however, these rashes may not be associated with visible medical findings.

These disease categories are distinguished by their objective and generally reproducible clinical findings. Unfortunately, most IAQ investigations are not triggered by disease findings that lead to an obvious source. According to Woods, most IAQ investigations lead to findings consistent with SBS rather than BRI (1991). Therefore, the next section is an analysis of the problems and pitfalls associated with the clinical evaluation of SBS.

Sick-Building Syndrome

Because the reporting of medical symptoms typically triggers an IAQ investigation, it is worthwhile to carefully examine these symptoms and the etiologic clues that they may indicate.

The common health complaints associated with SBS are:

- Irritation
  - mucous membrane—eye, ears, nose, throat
  - skin rashes—allergic, eczema, dyshydrotic
  - contact lens problems
  - sinus congestion
- Respiratory
  - cough, wheezing, hoarseness
  - difficulty breathing, chest tightness, shortness of breath
- Miscellaneous
  - fatigue, headache, dizziness, muscle and joint aches
  - headache
  - dizziness
  - lethargy, poor concentration, forgetfulness, nausea
  - heightened sensitivity to odors (NIOSH, 1989).

These typical symptoms reported in SBS investigations have several common characteristics: (1) self-reported, i.e., not obtained by an independent examiner, (2) nonspecific and not clearly attributed to a known pathophysiologic mechanism, and (3) difficult to obtain objective, independent confirmation. For example, it is not possible to independently confirm or deny the presence or absence of a headache.

These common denominators present a significant diagnostic/evaluation problem because baseline comparative rates for nonspecific, self-reported symptoms are not readily available. Thus, if a building owner/manager is confronted with symptom reports (e.g., headache and fatigue), it is extremely difficult to know whether these findings are symptomatic of a clear building problem or people's discovery that others are reporting of health impacts.

Differentiation is critical because IAQ problems can quickly degenerate into major confrontations between "staff" and "management." It is not uncommon for nonspecific medical complaints to follow an epidemic curve and produce new "cases" at an astonishing speed. In many of these situations, a set of symptoms is rapidly spread by conversation between affected and nonaffected employees. It is important, however, to temper initial skepticism by realizing that many occupational and environmental studies have used symptom prevalence rates as a possible indicator of low-level environmental chemical exposure.

In a 1992 study of three neighborhood communities 2–10 mi (3.2–16 km) from a hazardous waste site, Lipscomb et al (1992) found the symptom prevalence rates for skin irritation, eye irritation, fatigue, and sleep disturbance were extremely variable (10%–49%) and highly dependent on several factors: (1) self-administered questionnaire versus trained interviewer surveys, i.e., a two- to fivefold higher symptom level in the self-reporting group; (2) sex differences—female rates were higher than male rates; (3) ethnicity—Caucasians had greater rates than other ethnic groups; (4) degree of preexisting environmental worry, i.e., irritant symptoms such as eye complaints were strongly correlated with preexisting levels of environmental concern.

The importance of symptom prevalence rates and their correlates should not be minimized because most investigations of SBS do not find a clear explanatory source such as a single chemical exposure. Invariably, symptom surveys are performed in a self-reporting fashion and under conditions in which the building occupants generally perceive that significant potential health threat exists. Therefore, it is highly likely that high (greater than 20%) symptom prevalence rates will be found. The presence of "high" rates typically confirms the presumptive diagnosis of SBS and leads to further expensive and time-consuming engineering and/or industrial hygiene evaluations. Figures 13–1 and 13–2 present sample symptom evaluation forms. These and any other health impact forms should be used carefully.

Sample Office Staff Health Questionnaire

Some people are concerned about the office environment. In order to investigate these complaints, we would like your cooperation in filling out this questionnaire.

1. Problems. (Please select those which apply to you. Not all complaints have been registered—it is a complete list of all possible complaints.)

- ☐ Aching joints
- ☐ Hearing problems
- ☐ Discolored skin
- ☐ Heartburn
- ☐ Sinus trouble
- ☐ Congestion
- ☐ Problems wearing contact lenses
- ☐ Sore throat
- ☐ Sleepiness
- ☐ Too humid

- ☐ Musty smell
- ☐ Muscle pain
- ☐ Dizziness
- ☐ Skin itching
- ☐ Nausea
- ☐ Sneezing
- ☐ Chest tightness
- ☐ Fatigue
- ☐ Air too dry
- ☐ Frequent colds or flu
- ☐ Dirty air

- ☐ Back pain
- ☐ Dry, flaking skin
- ☐ Skin irritation
- ☐ Noticeable odors
- ☐ Watering eyes
- ☐ Eye irritation
- ☐ Headache
- ☐ Drowsiness
- ☐ Temperature too cold
- ☐ Temperature too hot
- ☐ Excessive noise

2. When do these complaints occur?

- ☐ Morning
- ☐ Daily
- ☐ No noticeable trend or time

- ☐ Afternoon
- ☐ Specific days of week (Specify:)

- ☐ All day

3. When do you experience relief of these complaints? _____

_____

4. Do you have any of the following?

- ☐ Hay fever
- ☐ Other allergies

- ☐ Skin allergies, skin irritation
- ☐ Colds/flu

- ☐ Sinus problems

5. Do you smoke tobacco?

☐ No   ☐ Yes   ☐ Number of cigarettes/cigars/pipes per day: _____

6. Please note your location. _____

7. Comments or suggestions: _____

_____

8. Name (optional): _____

Figure 13–1. Sample office staff health questionnaire for gathering general health data. Source: Adapted by Burton DJ from OSHA, TI Course No. 233 Material, 1991. Used with permission.

These observations have generally been confirmed in various U.S. and European surveys of SBS symptoms (Kreiss, 1989). A nationwide, random stratified sample of office workers reported by Woods et al (1987) found that of 600 office workers sampled, 24% perceived air quality problems in their office environment. Similarly, a 1985 Danish study reported a 22% prevalence of work-attributed mucous membrane irritation within a statistical sample of the Danish population (Andersen, 1985). Burge et al (1987) reported even higher rates in a study of 4,373 English office workers in 42 buildings: (1) lethargy—57%, (2) blocked nasal passages—47%, (3) dry throat—46%, (4) headache—46%.

Guidotti et al have noted these types of findings and subsequent diagnostic dilemma. They have proposed several epidemiological features that may distinguish SBS because of chemical exposure from outbreaks of mass psychogenic illness (Guidotti et al, 1987):

- Shifting incidence of symptom complaints not consistent with ventilation patterns. This presumes

---

Sample Office Staff Health Questionnaire

This questionnaire is to determine the cause of complaints expressed by employees. Your cooperation is essential. Your responses will be confidential.

Age: (   )   Sex: (   )   Location (   )

Which of the following symptoms, if any, do/did you have?

Yes	No	Not Sure	
☐	☐	☐	Irritated or watering eyes
☐	☐	☐	Sneezing
☐	☐	☐	Stuffy nose
☐	☐	☐	Postnasal drip
☐	☐	☐	Sore throat
☐	☐	☐	Cough
☐	☐	☐	Difficulty breathing
☐	☐	☐	Headache
☐	☐	☐	Wheezing
☐	☐	☐	Rash
☐	☐	☐	Fever
☐	☐	☐	Vomiting
☐	☐	☐	Diarrhea
☐	☐	☐	Cold symptoms
☐	☐	☐	Sinus trouble

General questions

Yes	No	Not Sure	
☐	☐	☐	Do you know the cause of symptoms?
☐	☐	☐	Do you wear contact lenses?
☐	☐	☐	Do you smoke tobacco?
☐	☐	☐	If yes, how many cigarettes/pipes/cigars per day? ( )
☐	☐	☐	Does tobacco smoke bother you at work?
☐	☐	☐	Do you sit near people who smoke?
☐	☐	☐	Do you know the source of the complaints you were experiencing?

For those symptoms you marked above,

Yes	No	Not Sure	
☐	☐	☐	Do you have a medical condition that would explain them?
☐	☐	☐	Do your symptoms improve when you leave the office?
☐	☐	☐	Can you wear your contact lenses at work?
☐	☐	☐	Have all your symptoms disappeared?
☐	☐	☐	Is there a time of day your symptoms are worse? If so, what time: ( )

How many days were you bothered by these symptoms? (   )
How soon after arriving at work do you notice the symptoms? (hours) (minutes)

Figure 13–2. Sample office staff health questionnaire for gathering data about office IAQ complaints. Source: Adapted by Burton DJ from OSHA, TI Course No. 233 Material, 1991.

that the ventilation system is serving as the transmission vector.

- Time sequence not consistent with the ventilation flow rates.
- Absence of consistent (or objective) medical findings compatible with a single chemical exposure source.
- Epidemic curve (the case production rate over time) consistent with "person-to-person" (conversation) rather than "common-source" transmission.

Kreiss has well summarized the epidemiological and diagnostic problems that plague IAQ investigations (Kreiss, 1989):

Our understanding of indoor air quality complaints and their prevalence will increase only when we systematically study buildings and their occupants without regard to complaints, in order to see what parameters of buildings, their furnishings, and their ventilation systems are associated with healthy, happy occupants.

Thus, although the medical/epidemiological approach to IAQ problems has many strengths, it does not provide stand-alone answers. The next section examines the role of engineering evaluations in the workup of IAQ.

ENGINEERING APPROACHES TO IAQ PROBLEMS

During the period 1971–1988, the National Institute for Occupational Safety and Health (NIOSH) conducted more than 1,100 investigations of IAQ problems (NIOSH, 1989). The vast majority of these investigations were triggered by worker symptoms or illness reports that were felt to be building related. The composite results of these investigations is highly illuminating, because the relative percentages of each diagnostic category have not appreciably changed over time:

- inadequate ventilation—52%
- outside contamination—17%
- microbiological contamination—5%
- unknown—12%.

Figure 13–3 illustrates these findings. Based on the results of the NIOSH studies and his own investigations of 30 buildings, Woods concluded that "air quality problems in residential and commercial buildings are nearly always associated with inadequacies in design and methods of operation" (Woods, 1991). Furthermore, Woods argued that 20%–30% of existing buildings have sufficient problems to indicate that either SBS or BRI findings are present and that another 10%–20% of buildings are

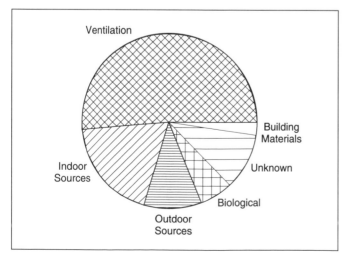

Figure 13–3. NIOSH sick-building syndrome investigations: Factors contributing to IAQ problems. Source: NIOSH, September, 1989.

undiagnosed and would be "abnormal" if investigated (1991). Based on these figures, only 50%–70% of buildings would qualify as "healthy."

The NIOSH investigations and Woods's work clearly imply that ventilation problems are a major contributor to IAQ complaints. In the United States, the American Society of Heating, Refrigeration, and Air Conditioning Engineers (ASHRAE) is the organization that has historically produced HVAC standards. Building ventilation adequacy has typically been measured and compared to ASHRAE standards. Several ASHRAE standards are important for purposes of understanding IAQ problems:

- ASHRAE 55–1992: *Thermal Environmental Conditions for Human Occupancy*
- ASHRAE 62–1981: *Ventilation for Acceptable Indoor Air Quality*
- ASHRAE 62–1989: *Ventilation for Acceptable Indoor Air Quality.*

Documents 62–1981 and 62–1989 make specific recommendations on the acceptable amount of airflow per occupant in an indoor setting. The 62–1981 standard recommended a minimum of 5 cubic feet per minute (cfm; 0.14 m/min) of outdoor air per person in an office area to prevent excessive buildup of carbon dioxide (CO_2). This minimum was increased to 20 cfm (0.56 m/min) if smoking was permitted in the office. The 62–1989 standard, which is still in effect, recommends a 15–25 cfm range regardless of smoking status. The recommendation for office space is 20 cfm, whereas the 15 cfm figure applies to classrooms and the 25 cfm to patient rooms (hospitals and nursing homes). The thermal environment was addressed in the 55–1992 standard, which recommends a 68–74.5 F (20–23 C) temperature range in winter and 73–79 F (22–26 C) in summer.

Both standards now recommend 60% relative humidity as an upper limit. In order to understand the development, basis, and implications of these figures, a working knowledge of HVAC engineering is necessary.

BASIC PRINCIPLES OF HVAC ENGINEERING

HVAC Systems

HVAC is the generic term for ventilating, heating, cooling, humidifying, dehumidifying, and cleaning air for comfort, safety, and health. HVAC also concerns itself with odor control and maintaining CO_2 levels at acceptable concentrations. Carbon dioxide is a normal constituent of exhaled breath and is frequently used as an indirect measure of whether adequate quantities of fresh outdoor air are being introduced into a building or work area (NIOSH, 1989). Normal outdoor concentrations of CO_2 range from 330–360 ppm, depending on location, time of day, and other factors. ASHRAE 62–1989 suggests that indoor CO_2 concentrations exceeding 1,000 ppm are suggestive of inadequate ventilation. The use of CO_2 measurements will be discussed in subsequent sections.

Mechanical air-handling systems (AHSs) range from a simple fan to complex central air-handling units. All of these systems distribute air to meet odor, temperature, humidity, and air quality requirements established by the user and by appropriate regulatory agencies. Individual units may be installed in the space they serve, or central units can be installed to serve multiple areas or zones.

Zoning, for HVAC engineers, defines which areas are to be served by an AHS. The smaller the zone, the better chance there is to provide satisfactory conditions. However, supplying air to multiple small zones can increase operating costs. Although most zones have one thermostat, some systems are designed to provide individual control of room air temperature in a multiple-zone system.

In designing an AHS, the user must balance combinations of volume flow rate, temperature, humidity, and air quality that will satisfy the needs of the space. Figure 13–4 is a schematic of a simple air volume commercial HVAC system.

This system could exist as either a large central unit for an entire building or as a small unit serving a specific zone within a building. This type of system heats or cools the air at a central location and then distributes it among various zones. There are several standard variations of the air volume HVAC system (Hughes & O'Brien, 1986):

- **Constant or variable air volume control.** The constant system operates at a fixed flow rate and only varies the temperature. The variable system can alter the air volume as well as temperature. These systems recirculate the return air as an energy conservation measure; the percentage of air recirculated can be varied from 10%–100%.
- **Air and water systems.** This is similar to all air systems, but prior to discharge into the zone, the air passes through a fan coil unit that is supplied with heating/cooling water.

- **All-water systems.** These systems empty self-contained fan coil units that distribute the heated or cooled air to the work space. All of the heating or cooling airflow is recirculated from the work space as opposed to the previously described air and water systems.

Humidification and dehumidification operations can be performed within all three systems. In addition, dust-particle filtration and charcoal beds for odor control are fairly easy to incorporate into these units.

Every HVAC system should have an outdoor air (OA) intake, which is usually a louvered opening on the top or side of the building. As air enters the intake, pushed by atmospheric pressure, a damper regulates the amount of OA taken into the system. The damper should never close completely (except in rare weather conditions, e.g., subzero temperatures.)

The incoming air moves to a point where return air (RA) from the occupied space mixes with the OA (forming mixed air, MA). The MA usually passes through a prefilter that provides good control of bees, flies, bird feathers, leaves, and larger dust particles. Small particles usually pass through to a more efficient filter. After exiting the more efficient filter, air enters a centrifugal fan.

Exiting the fan outlet, the air is now under positive pressure, and is pushed toward the coils. These coils heat and cool the air, depending on the temperature of the air and season. Winter air is typically heated to 68–80 F (20–27 C). Summer air may be cooled to 55–70 F (13–21 C). Below the coils is a drain pan. If water condenses on the cooling coils, it will drip and collect in the pan, which can be a potential source of bioaerosols. Leaving the coils, the air can pass through a humidifier or dehumidifier to adjust relative humidity to 40%–60%.

After the air is suitably conditioned, it moves along in metal ductwork (sometimes insulated with a fiberglass lining) at about 10–20 mph (1,000–2,000 ft/min) or 16–32 km/hr and quickly reaches a distribution box. From there it travels through smaller, flexible ducts to the supply terminals, or diffusers.

Once in the room, the air typically hugs the ceiling and walls, and its velocity slows to a "terminal velocity" of about 40–50 ft/min (12–15 m/min). The air should not blow on people until after it reaches terminal velocity.

Slowly (in about 5–10 min) the air migrates through the space to the return air register, which is

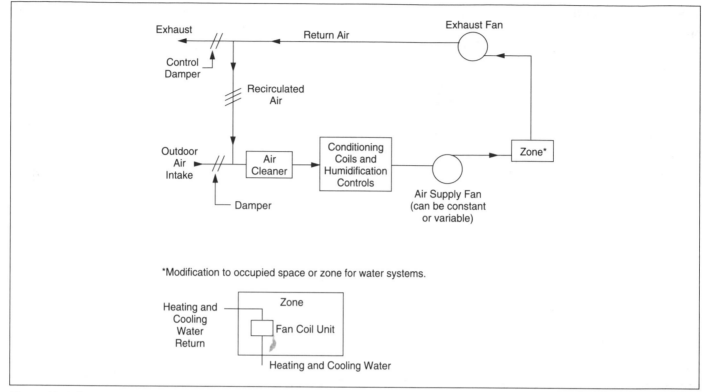

Figure 13–4. Schematic of a simple air volume HVAC system. Source: Adapted by Krieger GR from NIOSH, 1989.

usually a louvered opening into the crawl space above the ceiling tiles. Finally, the air moves to the return air duct, where it can either proceed around again (recirculation) or be exhausted to the outdoor exhaust air (EA). The building is usually maintained under positive pressure and air can exfiltrate anywhere along the building periphery. Sophisticated controls manage the system and determine: (1) the amount of OA that will be used, (2) the percent of RA that will be exhausted, and (3) temperature and humidity that will be maintained.

Although this system appears very uncomplicated, other mechanical issues can significantly affect HVAC performance:

- **Air mixing in the occupied zone.** The delivery of air to a space does not guarantee that proper mixing will occur. Offices with partitions and constantly changing floor plans are particularly susceptible to this problem and create microenvironments, i.e., locations of variable airflow within a zone. The placement or movement of bookshelves, furniture, windows, and walls will change the pattern of air movement, often for the worse. This problem is often quantified as the *mixing efficiency* or *ventilation effectiveness*. The *mixing factor* (K_m) is the ratio of the amount of air required to dilute

a contaminant to the ideal amount of air that should reduce it.

$$K_m = \frac{Q_{actual}}{Q_{ideal}} \qquad \text{(Equation 13–1)}$$

In well-planned buildings served by good HVAC systems, K_m is often close to 1.0 (perfect mixing), although it can never actually reach 1.0 because of the randomness of air behavior. There are several approaches for maximizing the mixing factor, i.e., keeping K_m near 1.0:

— Provide and properly position an adequate number of supply and return registers so that the registers are not mismatched with the locations of office occupants.

— Provide freestanding fans for persons located in areas of possible poor mixing. This approach is often cost-effective and satisfying to building occupants, although it does not address the underlying ventilation performance issues.

- **Controlling airflow with dampers in the HVAC duct system.** Damper positions can be set automatically or manually, depending on the type and sophistication of the HVAC control system. Fire and smoke dampers may be used to restrict the spread of heat and smoke during a fire. Dampers, actuators, and control systems should always be checked and maintained regularly to ensure proper flow of air through the system. Improperly set dampers are a common finding in many IAQ investigations.

- **Controlling distribution of the air supply with terminal devices.** Terminal devices consist of supply diffusers, return and exhaust grills, and the associated dampers and controls. The number, location, and type of terminal devices determine the air distribution in the occupied space. Improper devices that lower the ventilation effectiveness can create stagnant areas, drafts, odor buildup, uneven temperatures, short-circuiting of the air, and air stratification.

  Occupants who complain of draftiness, odors, stagnation, and uneven temperatures may adjust or block supply diffusers. This can disrupt the proper distribution of air to other areas of the building.

- **Providing return air systems.** These are typically located at the crawl space above the ceiling (return plenum). Air enters the return plenum and is ducted to the air-handling unit. In addition, some systems use return fans to ensure proper pressure relationships between the supply and return ducts. The space should be kept clean and dry to reduce bioaerosol formation (molds, fungi, bacteria, viruses). Morey & Shattuck (1989) provide additional detailed information on RA system requirements for maintenance personnel and contractors.

The following paragraphs explore commissioning, testing, and evaluating HVAC system performance.

Commissioning, Testing, and Evaluating Ventilation System Performance

Commissioning is a process in which a new HVAC system's performance is identified, verified, and documented to ensure proper operation and compliance with codes, standards, and design intentions. Commissioning often requires tests and demonstrations to verify that the system operates properly. Troubleshooting and maintenance activities also require testing. Objectives of this process include:

- establishment of proper baseline or startup conditions
- determination of the continued effectiveness of the HVAC system, i.e., comfort maintained, compliance with standards, specifications, and design criteria

- maintenance of performance throughout the life of the system.

Another important function that must be performed on HVAC systems is testing and balancing (TAB or T&B). This involves testing and adjusting system components (e.g., dampers) to establish adequate air distribution to the occupied spaces. Commissioning, testing and balancing, troubleshooting, and maintenance are required on every HVAC system.

Most building staff cannot conduct in-depth testing of HVAC systems because specialized knowledge of testing and balancing engineering is required for modern HVAC systems. However, the following guidelines for simple testing and troubleshooting should be considered by facility managers:

- Develop sufficient familiarity with the HVAC system characteristics.
- Determine the intended or desired operating parameters.
- Perform cursory ventilation checks with smoke tubes, balometers, velometers, and pressure-measuring equipment.

IAQ investigations should involve the building engineer or a person who knows the HVAC system intimately. Frequently, building engineers make modifications that are not specified on system plans but that significantly impact HVAC performance.

A non-HVAC engineer can easily ask these critical questions:

- Does the room have a supply diffuser? A return?
- Is air moving through diffusers and return grills?
- Are air diffusers and grills open? Blocked? Attached to ductwork?
- Is supply air distributed throughout the occupied space?
- Do people actually feel air moving?
- Are there "dead air spaces" in the office or room?
- Do printers, copiers, and other equipment have adequate ventilation?
- Are mixing fans or portable heaters used by occupants?
- Does the HVAC system always operate when people are in the building?
- Is the air too hot? Too cold? Too humid?
- Do people actually detect odors?

Table 13–A illustrates the basic measurement data to obtain during the evaluation of HVAC performance.

Table 13–A. HVAC Evaluations

What to Measure	Typical Equipment
In occupied space: • Dry bulb temperature • Wet bulb temperature • Relative humidity	Thermometer, psychrometer
System temperatures: • Supply air (SA) • Outdoor air (OA) • Mixed air (MA) • Return air (RA)	Thermometer
CO_2 measurements: • Supply air (SA) • Outdoor air (OA) • Return air (RA) • Occupied space air	Detector tubes, CO_2 monitors
Air movement: • Any location	Smoke tubes, velometers

A simple approach to evaluating the HVAC system is to follow it from start to finish. The investigator should first go to the air intakes. The airflow should be followed through the dampers, filters, fans, coils, ductwork, terminal boxes, and supply registers. The return grills should be identified and the air followed back to the air handler. The investigator should note the following:

- equipment that is and is not operating
- closed or jammed dampers
- clogged or misplaced filters
- presence of moisture pools
- thermostat locations and temperature levels
- supply air quality (clean, odorless, humidity)
- drafts and/or stuffiness in occupied space
- supply and return register locations and damper settings
- potential sources of contaminants (microbial, chemicals)
- provision for OA
- positioning of OA and room air dampers
- controls operation
- variable air volume (VAV) system air delivery schedule, minimums.

In conjunction with the building engineer, the IAQ investigator should review the "as built" and "as modified" drawings to become familiar with the HVAC system and the building. Original (or modified) specifications should be read to learn the intended operating parameters so that comparison to current performance can be made.

Troubleshooting HVAC systems. Invariably, something goes wrong with ventilation systems. Simple troubleshooting usually involves three phases of study:

Phase 1. Data collection and building inspection/ walkthrough.

This phase involves the following activities:

- characterizing complaints and gathering background data (see the Medical Approaches to IAQ Problems section)
- checking performance of ventilation systems and their controls, including the following:
  — insufficient OA introduced to the system
  — poor distribution/stratification of supply air in occupied space
  — draftiness—too much supply air or improper terminal settings
  — stuffiness—not enough air delivered or air delivered improperly
  — pressure differences—doors hard to open
  — temperature extremes—too hot or too cold
  — humidity extremes—too dry or too humid
  — poor filtration—dirt, insects, or pollen in air delivery system
  — poor maintenance
  — increased recirculation and decreased fresh air intakes because of energy conservation efforts
  — stagnant water in system
  — visual evidence of slime or mold
  — imbalance of distribution system
  — dampers at incorrect positions
  — terminal diffusers not at correct positions
  — VAV systems in nondelivery or low-delivery mode
  — CO_2 measurement.

(See Figures 13–5 through 13–7 for sample checklists.)

Phase 2. Analysis.

The following are examples of typical occupant complaints and possible explanations for these complaints:

- The temperature is too warm (or too cold).
  Potential problems: Thermostats are misadjusted; supply air temperature settings are too high or low; too much or too little supply air; supply diffuser blows air directly on occupants; temperature sensor malfunctioning; cold air not mixing with occupied space air; HVAC system defective or undersized; building under negative pressure which causes infiltration of air at the building perimeter.

Occupants' Room Ventilation Checklist

☐ Does each room have a source of air?
☐ Is air moving through diffusers and return grills?
☐ Are diffusers and grills open? Blocked?
☐ Is the air distributed throughout the space where people are?
☐ Do people actually feel air moving?
☐ Are there "dead spaces" in the office or room?
☐ Do printers, copiers, and other equipment have adequate ventilation?
☐ Does the ventilation system operate when people are in the building?
☐ Is the air too hot? Too cold? Too humid?
☐ Does the air smell bad?
☐ Does the air make anyone uncomfortable? Sick?
☐ What contaminates the air?

Figure 13–5. Checklist for use by an environmental or building manager to gather data from a room's occupants. Source: Burton DJ, *IAQ and HVAC Workbook.* Salt Lake City, UT: IVE, 1993. Used with permission.

Moisture Control Checklist for Building Occupants and Operators

☐ Avoid spilling water at sinks and coffee areas.
☐ Clean up spills until dry.
☐ Always use exhaust fans in bathrooms, showers, and tubs.
☐ Avoid the use of room humidifiers.
☐ Repair roof leaks.
☐ Remove standing water on roofs, in basements, in crawl spaces, and on internal surfaces.
☐ Provide adequate drainage of flat roofs.
☐ Do not allow water to condense in air-handling systems.
☐ Do not allow water to condense on perimeter windows and basement walls.

Figure 13–6. Periodic monitoring for moisture control to help reduce bioaerosol levels. Source: Burton DJ, *IAQ and HVAC Workbook.* Salt Lake City, UT: IVE, 1993. Used with permission.

- The air is too dry (or too humid).
  Potential problems: Humidity controls not operating correctly or undersized. Normally, relative humidity should range between 30%–60%.
- The air is stuffy or stagnant, or there is no air movement.
  Potential problems: Nondelivery or low delivery rate of air to space; filters overloaded; variable air volume dampers malfunctioning; restriction in ductwork; ductwork disconnected from supply diffusers; duct leaking; inadequate delivery of OA.
- There is no air movement when it gets cold.
  Potential problems: VAV system set to deliver no air when system is not calling for cooling. This is a common problem in older VAV systems.

Checklist for Reducing Microbial Problems

☐ Prevent buildup of moisture in occupied spaces.
☐ Prevent moisture collection in HVAC components.
☐ Remove stagnant water and slime from mechanical equipment.
☐ Use steam for humidifying.
☐ Avoid use of water sprays in HVAC systems.
☐ Maintain relative humidity less than 60%.
☐ Use filters with a 50%–70% collection efficiency rating. (ASHRAE Dust-spot rating.)
☐ Find and discard microbial-damaged furnishings and equipment.
☐ Remove room humidifiers.
☐ Provide preventive maintenance.
☐ Provide pigeon screens on intakes and exhausts (this will prohibit the contamination of the system by bird droppings, feathers, nesting materials, food, etc.

Figure 13–7. A monitoring and maintenance program for reducing microbial problems should be in place. This checklist helps the environmental manager check important environmental factors on a regular basis. Source: Burton DJ, *IAQ and HVAC Workbook.* Salt Lake City, UT: IVE, 1993. Used with permission.

- There are too many drafts.
  Potential problems: Supply diffuser set to blow air directly on occupant; occupant near open door or window; freestanding fan blowing on occupant.
- Air smells like diesel exhaust.
  Potential problems: Air intake near loading dock or the building has a negative pressure relative to the fume source and pulls in outdoor odors.
- Air smells and has a musty, dirty sock smell.
  Potential problem: Microbiologic contamination. This problem is complex and controversial, and requires an experienced investigator's judgment. The investigation of bioaerosol problems is discussed in the next two sections.

Phase 3. Correction of problem and maintenance improvement.

Duct cleaning is discussed in detail in the next section because this has been proposed as a major source of IAQ problems and bioaerosol formation.

Duct Cleaning

In recent years the number of commercial duct-cleaning companies has multiplied. Many are actively marketing duct cleaning as a solution to IAQ problems. When should duct cleaning be performed? What are the advantages and disadvantages?

Ducts can become both the source and the pathway for dirt, dust, and biological contaminants that spread through the building. ASHRAE Standard

62–1989 suggests that efforts be made to keep dirt, moisture, and high humidity from ductwork. Filters must be used and kept in good working order to keep contaminants from collecting in the HVAC system. To some IAQ investigators, this implies that periodic duct cleaning may be beneficial; however, duct cleaning is controversial. Duct contaminants can include an enormous variety of chemicals and materials: hair and dander, skin particles, insects and insect parts, organic and inorganic dust, carbon and oil particles, glass fibers, asbestos, pollen, mold, mildew, bacteria, leaves, dirt, and paper. All of these could create IAQ problems. However, the mere presence of these contaminants in the ductwork may have no effect on people if they are not carried to the occupied zone or do not emit offensive odors. Indeed, there have been cases where inert and inactive dusts were aerosolized during duct cleaning, resulting in occupant complaints. Most ducts have small amounts of dust collected on their surfaces—a common occurrence which almost never requires duct cleaning.

Duct cleaning or replacement is warranted when there is:

- permanent water damage
- slime growth
- debris that restricts airflow
- dust actually seen emitting from supply registers
- an offensive odor originating from the ductwork.

Not all dirt originates in the duct. Some of the gray or black streaks seen around supply registers are created by small particles in the room air that are deposited by thermal and inertial forces at the register, and are not dust particles originating in the duct. Dirt, debris, and microbiological growth in ductwork can be minimized by the following:

- well-maintained filter systems
- regular HVAC maintenance
- good housekeeping in the occupied space
- location of air intakes in noncontaminated locations
- keeping all HVAC system components dry or drained.

If ductwork becomes damaged or water-soaked, cleanup and/or replacement is required. When sound or thermal liners become water soaked, they almost always require replacement. Failure to adequately manage system moisture is a significant contributor to bioaerosol formation and propagation.

In addition, if glass-fiber duct liners are used, special attention must be paid to air filtration and keeping the liner dry. The large porous surface can trap organic dusts and water, which then provide an excellent location for the growth of microorganisms.

Duct cleaning must be performed by experienced personnel in a manner that protects building occupants (e.g., cleaning during nonoccupied hours). Additionally, the problem that originally caused the duct contamination must be corrected.

As of early 1994, the efficacy and need for regular duct cleaning has not been established. However, there are a few circumstances that warrant actions. Positive responses to the following questions provide guidelines for action:

- Are there contaminants in the ductwork?
- Has testing confirmed their type and quantity?
- Do they (or their odors or by-products) leave the duct and enter the occupied space?
- Is the source of these contaminants known? Can the source be controlled? (If not, cleaning is only a temporary control measure.)
- Do the contaminants actually cause IAQ problems?
- Will duct cleaning effectively remove, neutralize, or inactivate the contaminant?
- Is duct cleaning the only or most cost-effective solution?
- Has a qualified and reputable duct-cleaning firm been identified?
- Have the firm's references been checked?
- Does the duct-cleaning firm have a sensible, sound approach? Does the company have the right kind of equipment? Will the cleaning process protect the HVAC equipment and the occupants of the space during cleaning?
- Will the duct-cleaning company give a guarantee?

Any "no" answers should delay duct cleaning and prompt reconsideration of the rationale for cleaning.

The duct-cleaning process can consist of contaminant removal (e.g., through brushing/vibration plus vacuum cleaning), encapsulation (e.g., spraying a sealer into the duct), disinfection (e.g., using a fungicide to inhibit the growth of mold), and duct replacement. Sometimes one or more methods may be employed. Encapsulation is a particularly controversial technique and may not be effective.

If duct cleaning is performed, these suggestions should be considered:

- Keep ducts under negative pressure during the cleaning operation. This will minimize the discharge of dirt and dust into the occupied space.
- Protect the duct system. Avoid unnecessarily cutting holes in the duct or duct liner.
- Schedule the cleaning when the building is not occupied.
- Provide at least 10 air changes in the building after duct cleaning and *before* occupants are readmitted to the building.
- Vacuum cleaning and collection equipment for the ductwork should be located outside the building. When vacuum collection equipment must be inside, high-efficiency particulate air (HEPA) filtration should be provided for vacuum discharge.
- Vacuum cleaning should be used in conjunction with gentle brushing of settled materials. Vacuum cleaning alone is not likely to be enough to clean the ductwork.
- If biocides are used, select only products approved by the U.S. Environmental Protection Agency (EPA) and follow the manufacturer's instructions carefully.
- If sanitizers, deodorizers, or chemicals are used, be sure they are rinsed and removed from system equipment before occupants return to the building.
- Sealants should not be used to cover interior-contaminated ductwork. Sealants have not been shown to be effective as a barrier to microbiological growth. In addition, their long-term health effects are unknown, and they can void fire safety ratings.
- Where water-damaged or biocontaminated porous materials are encountered, they should be removed, not cleaned.
- Institute a contamination-prevention program so ducts do not have to be cleaned again.

MONITORING AND MEASURING

Bioaerosols

Bioaerosols present a particularly difficult problem in IAQ investigations. Several methodological problems are associated with sampling for bioaerosols, i.e., viable versus nonviable material, bacterial, fungal, molds, or products of microbial growth (Burge et al, 1987). In addition, bioaerosols are present to some degree in most indoor environments, so it is difficult to differentiate between normal "background" and abnormal levels of biological material.

Routine cultures in air filters and ductwork invariably produce an uninterpretable list of fungi, bacteria, and molds. Colony-forming units or plaque counts are similarly difficult to interpret, because there are no standards currently available for comparison or reference. The lack of reference standards is unfortunate, because bioaerosols have become more popular as an explanation for nonspecific symptoms of fatigue, headache, dry skin, and mucous membrane irritation (Kreiss, 1989).

Most buildings contain many potential multiple sources of bioaerosol production and odors that require investigation:

- contaminated cooling coils, humidifiers, and air washers
- high-velocity air passing through wet cooling coils
- inadequate preventive maintenance
- improperly designed or inaccessible air-handling equipment
- porous synthetic fiber insulation inside air-handling and fan coil ventilation equipment that has become moist
- contaminated acoustical lining in ductwork
- stagnant water in drain pans
- excessive humidity (greater than 70%)
- the recirculation and buildup of human-shed bacteria and viruses
- transfer of viruses and bacteria by human-origin aerosols (e.g., coughing and sneezing)
- water-contaminated furnishings, particularly organic fibers (e.g., wool and cotton)
- use of cool-water room humidifiers
- water runoff from windows into unit ventilators
- flooding of any type
- locating OA intakes near external bioaerosol sources (e.g., cooling towers).

Singly or in combination, these sources can provide the necessary elements for the nourishment, growth, and dissemination of significant quantities of potentially harmful bioaerosols. A thorough understanding of the mechanisms by which bioaerosols evolve is critical to effectively evaluating and eliminating this source of IAQ problems. Sampling for bioaerosols will be discussed in the section on Measurement of Indoor Air Contaminants.

Carbon Dioxide

Carbon dioxide is a normal component of the atmosphere, with typical outside levels ranging from 250–350 ppm. In the indoor environment, the exhaled breath of the building occupants is an important continuous source of CO_2. The CO_2 in an occupied space will gradually build up over time if inadequate dilution air is provided. Consequently, in some buildings, measurement of indoor CO_2 levels may provide a useful indication of the adequacy of ventilation to an occupied space. For example, comparison of indoor and outdoor levels, peak concentration variations at different locations in the building, and measurement of changes in levels during a 24-hour period can all provide useful information. Sampling methods for CO_2 are listed in Table 13–A.

New Approaches for Assessing Ventilation Performance

The data obtained from the previously discussed assessment technique should provide potential solutions to most IAQ problems attributable to HVAC performance. However, in some difficult-to-resolve problems, more sophisticated approaches must be considered. Recent work at the National Renewable Energy Laboratory (NREL, previously known as the Solar Energy Research Institute) has been directed toward developing techniques to provide more accurate measurements of ventilation and pollutant removal rates (Anderson, 1988). Specific performance problems that can be diagnosed include:

- flow recirculation—microenvironmental problems of poor air mixing due to partitions, furniture, or other objects
- flow short-circuiting—part of the air supply bypassing a room due to poor system design
- duct leakage.

The NREL system uses approaches that were originally developed for remote-sensing applications and provides a digitized image of more than 360,000 points at which optical density can be measured to determine pollutant concentration. Figures 13–8 through 13–10 illustrate this system.

The careful evaluation of the HVAC system is likely to reveal the source of most IAQ complaints. It must be emphasized that analysis of the entire HVAC system is necessary to ensure a thorough understanding of the problem. Simply increasing the supply of outdoor air may not solve the problem if, for example, the issue is poor distribution, contaminated ductwork, polluted OA, or a source unrelated to the HVAC system (Menzies, 1993; Farnham, 1993).

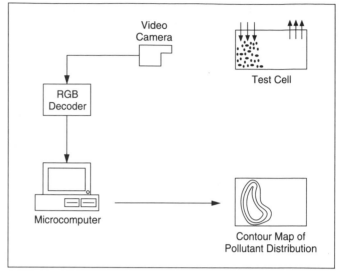

Figure 13–8. Image processing for pollutant measurements. Source: Anderson, 1988.

Measurement of Indoor Air Contaminants

Most investigations of IAQ complaints involve gathering information about the following sources:

- building occupants
- HVAC system
- pollutant pathways
- pollutant sources.

The majority of complaints can be resolved by correcting problems revealed by information gathered from these sources. However, some situations require sampling the air to determine the concentration of specific contaminants. For example, if the information gathered indicates that a specific contaminant is the cause of the complaint, it may be appropriate to collect air samples to confirm the suspected source before any attempt is made to resolve the problem. Once a specific source is confirmed, sampling before and after remediation can provide an indication of whether the attempted solution has been effective.

Although sampling for airborne contaminants may sound like an easy way to assess IAQ, if no specific contaminant source is suspected, air sampling may not provide any further useful information. Without a suspect contaminant, the usefulness of collecting air samples is limited, because the range of potential sources is enormous. In addition, the concentrations at which the contaminants are present are often so low as to make detection difficult. To further complicate the interpretation process,

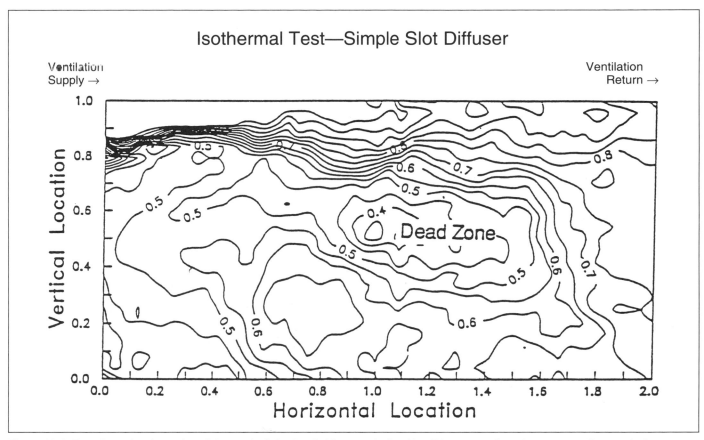

Figure 13–9. Two-dimensional mapping of the result of simulated airflow monitoring, identifying areas of no air movement. Source: Anderson, 1988.

there is no exposure standard for many contaminants to help determine whether the concentration measured is hazardous.

If air sampling becomes necessary, it should be performed by an experienced professional familiar with the necessary techniques, including their applicability and limitations.

Common sources of indoor chemical contaminants include (Sullivan et al, 1991):

- Environmental tobacco smoke (ETS)—composed of a vapor phase (300–500 compounds) and particulate phase (more than 3,500 compounds). ETS was probably the most common source of indoor pollutants and is considered a human carcinogen by the U.S. EPA. A substantial number of VOCs are present in ETS, such as benzene, toluene, formaldehyde, styrene, and acetic acid.
- Building materials—sealants, caulking compounds, paints, woods, plastics, vinyl products, foams, ceiling tiles, and insulation material. The chemicals associated with these materials have been extensively reviewed by Levin (1989).

- Cleaning agents—these materials can be both odor-producing and irritants because they contain a variety of amines, alcohols, acids, phenols, and caustics. Typically, cleaning agents are used on a relatively predictable schedule so that a large temporal component should be present between application and symptom complaint if these materials are the source of the problem.
- Carpets, fabrics, and adhesives—a variety of VOCs can offgas from these sources; however, the compound of greatest interest is 4-phenylcyclohexane (4-PC). 4-PC is the presumed source of "new carpet odor." The source of this compound appears to be the styrene-butadiene rubber latex that is used to bind the backing of new carpet (Van Ert et al, 1988). The toxicology of 4-PC is not well known, although some anecdotal information indicates symptom production when concentrations were in the low parts per billion (Sullivan et al, 1991).
- Outside sources brought inside—this can include vehicle exhaust, pollen, dust, industrial contaminants, roofing fumes, oxides of nitrogen, and other sources or products of incomplete combustion.

Concentration/Distribution from Image Analysis System

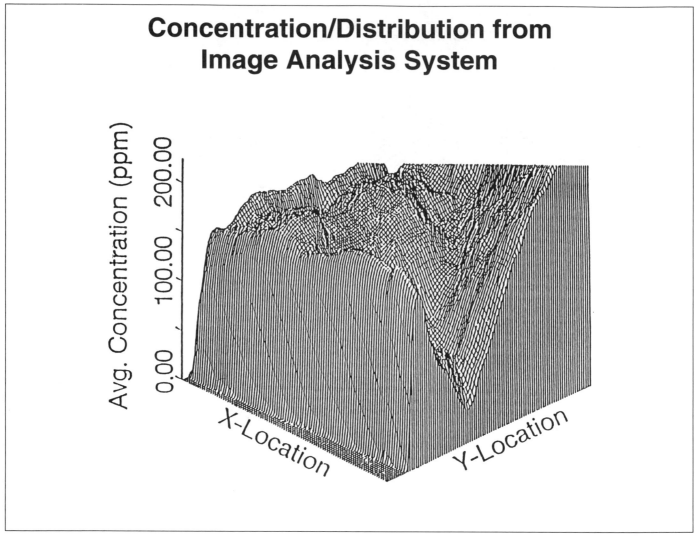

Figure 13–10. Three-dimensional model showing "dead" space area representing little or no effective air movement (areas of depression). Source: Anderson, 1988.

- Bioaerosols—these include microorganisms (culturable, nonculturable, and dead microorganisms) and fragments, toxins, and particulate waste products from all varieties of living things. Examples include bacteria, fungi, and infectious agents.
- Radon (radon-222)—this naturally occurring radioactive gas is produced as trace amounts of uranium decay in soil. As radon decays, it produces alpha particles before it reaches a stable end product, lead-206. The amount of uranium in soil is highly variable; however, the indoor concentration of radon can be higher than outdoor levels. Testing for radon is inexpensive and simple. In addition, control methods are straightforward and usually include some form of local exhaust ventilation.
- VOCs—these are extremely common in the indoor environment and are produced by numerous sources, e.g., plastics, adhesives, copiers, and wallpaper. Usually VOCs are found in low concentrations (low ppm or ppb) by using gas chromatography-mass spectrometry (GC/MS). VOC levels are typically well below Occupational Safety and Health Administration (OSHA) standards. It is unclear whether these low-concentration findings can account for occupant symptoms. One recently developed method for investigating VOC levels is GC/MS "fingerprinting" that clearly documents postremediation improvement. Although quite impressive, this technique is expensive and requires an experienced and skilled hygienist to perform it. Another technique is the "bake-out." This approach calls for heating the building to 90 F (32.5 C) for one or more days (Girman, 1989). The results of this strategy have been mixed, i.e., some VOC levels decreased and others (formaldehyde) remained unchanged. This technique also

cracks paint and causes wood warping and expansion of joints.

- Asbestos—asbestos-containing materials (ACM) are frequently found in many buildings and homes. ACM have been the subject of extensive controversy and significant regulatory activity. The rules and regulations covering ACM and its abatement are discussed in a subsequent section. Tables 13–B and 13–C provide a brief summary of important facts and regulations. See also Chapter 7, Environmental Audits and Site Assessments, for additional information.

- Physical sources and thermal comfort factors— these include lighting, noise, humidity, and temperature. The latter two were discussed previously. Lighting and noise are treated as human factor issues and should be considered along with other psychological aspects of IAQ.

- Psychological factors—the role of factors such as supervisor-employee relations, boredom because of highly repetitive work, deadline pressure, and lack of communication in the workplace should not be minimized as contributing factors for IAQ. Nevertheless, IAQ problems require a systematic and thorough approach; therefore, a diagnosis of psychogenic illness should not be assumed before the investigation is complete. Figures 13–11 through 13–14 provide environmental testing checklists.

ASBESTOS IN BUILDINGS

Asbestos has been used in building products for many years. Asbestos-containing building materials (ACBMs) include roofing felt, roofing tar, boiler insulation, pipe and pipe-fitting insulation, vinyl asbestos floor tile, fireproofing, acoustic plaster, ceiling panels, hard transite panels, decorative architectural panels, and other structural and architectural products.

ACBMs in and of themselves do not pose a significant hazard to building occupants or to maintenance personnel. When the surface of ACBMs is broken, however, there is the potential, depending on the type and extent of disruption, for ACBMs to become damaged. Breaking apart friable asbestos can release asbestos fibers into the air. Again, releasing

---

Table 13–B. Asbestos Facts

Classification
Fibrous mineral silicates

Other Names
Actinolite, amosite, anthophyllite, chrysotile, crocidolite, and tremolite

Flammability
Nonflammable; supresses flame and smoke generation

Primary Applications
Fireproofing; heat, cold, and sound insulation; acoustic plaster

Forms of Asbestos-Containing Materials (ACM)
Sprayed- or troweled-on surfacing materials; pipe, boiler, duct, and flue insulations; ceiling tiles; floor tiles; wallboards; water and sewer pipes; automobile and elevator brake pads

Health Hazards
Human carcinogen; known cause of asbestosis and mesothelioma

Routes of Entry into Body
Inhalation or ingestion

Extent of Usage
The U.S. EPA has estimated that approximately 31,000 schools and 733,000 public and commercial buildings nationwide contain asbestos.

Current Usage
Asbestos is no longer being installed in buildings for insulation or decorative purposes. However, many asbestos-containing products are still being manufactured, including asbestos-reinforced water and sewer pipes, and some states require the use of asbestos-composite brake pads on elevators.

Table 13–C. Enforcement of Asbestos Regulations

- 1973—the U.S. EPA partially banned the application of sprayed-on asbestos-containing materials (ACM) in new buildings and implemented methods of managing ACM during demolitions.
- 1975 and 1978—the U.S. EPA regulations were revised to encompass building renovations, all applications in new buildings, and asbestos emissions monitoring during waste disposal procedures.
- 1986—Congress passed the Asbestos Hazard Emergency Response Act (AHERA), which ordered the U.S. EPA to develop asbestos regulations for schools. The act also ordered the EPA to study the extent of asbestos contamination in public buildings.
- 1989—the U.S. EPA prohibited most remaining uses of ACM, with this ban to be implemented over the next seven years.
- The Occupational Safety and Health Administration (OSHA) regulates occupational exposure to asbestos in all industries covered under the Occupational Safety and Health Act, including general industry, the maritime industry, and the construction industry. OSHA standards cover permissible exposure limits, ventilation requirements, respiratory protection, personal protective equipment, sanitation, medical services and first aid, and eye and face protection. OSHA PEL is 0.2 fibers/cc and the action level is 0.1 fibers/cc.
- State asbestos regulations vary and, in some cases, may be more stringent than federal regulations.

Building Information Checklist

Building: _____   Contact person: _____
Address: _____   Telephone: _____
Date: _____   Investigator: _____

Building Description

Year built _____	HVAC type _____	Interior layout _____
Occupants _____	Owner _____	Date occupied _____
Construction _____	No. of floors _____	Neighborhood type _____
Emission sources _____	Traffic pattern _____	Location of garages _____
Interior construction _____	Tightness of doors, windows ____	Insulation type _____

Occupant Space Description

Number of people _____	Sq ft/person _____	Type of activity _____
Smoking policy _____	No. of smokers _____	Chemicals present _____
Cleaning materials _____	Furnishings _____	Construction material _____
Recent construction _____	Recent changes _____	Freestanding fans _____
Temperature, humidity _____	Mold/dirt _____	Wet surfaces _____
Adjacent space _____	Room pressure _____	Carbon dioxide _____
Drafts _____	Stuffiness _____	Interviews _____

HVAC Systems

Type of system _____	Condition _____	Windows _____
Type of fuel _____	Type of diffuser _____	Location of intakes _____
Location of exhaust _____	OA provisions _____	Distribution _____
Terminal velocities _____	Noise _____	Dust/dirt _____
Economizer cycle _____	Controls _____	Zones _____
Total cfm _____	Total OA _____	Heat exchanger _____
Local exhaust _____	Makeup air _____	Duct type _____
Water in system _____	Type of humidifier _____	Restroom exhaust _____
Air cleaner type _____	Air cleaner efficiency _____	Person in charge _____

Figure 13–11. This building information checklist should be periodically updated and accessible to environmental and building managers. Source: Burton DJ, *IAQ and HVAC Workbook*, Salt Lake City, UT: IVE, 1993. Used with permission.

some fibers from ACBMs does not necessarily pose a hazard to building tenants or maintenance workers. One means of measuring whether fiber levels from the removal or disruption of ACBMs is to monitor airborne levels and to compare monitored levels to regulatory standards. OSHA has established a permissible exposure level (PEL) of 0.2 f/cc, an action level of 0.1 f/cc, with a short-term excursion limit (STEL) of (1 f/cc). These are levels that OSHA deems acceptable for people who handle ACBMs. Appropriate respiratory protection is recommended to limit the exposure to no more than the PEL or STEL levels.

Building Assessments

Building owners and potential buyers sometimes want to know what ACBMs, if any, are present in a building. A common way of determining this is to conduct a building assessment. The building assessment involves reviewing plans and specifications for the building (if available) and conducting an inspection of the building.

Plans and specifications are reviewed to determine whether ACMs were specified or called for in constructing the building. For example, structural steel drawings for the building may indicate that asbestos fireproofing of a certain rating, e.g., two-hour or four-hour, is required on structural steel in the building. Likewise, specifications for the building may call for vinyl asbestos tile as a floor covering. It is also common that alternate materials are permitted, and therefore, the presence or absence of a note indicating ACMs cannot be taken as proof of the presence or absence of asbestos.

Checklist for In-Depth On-Site Investigation

At the site of the problem or complaint.
☐ Conduct interviews, or distribute questionnaires.
☐ Look for obvious sources of the problem.
☐ Check temperature, humidity, odors, and carbon dioxide levels.
☐ Become familiar with air-handling systems.

Parking garages, boiler rooms, incinerators, emergency generator rooms, motor pool areas.
☐ Check for sulfur dioxide, carbon monoxide, carbon dioxide, and nitrogen dioxide with direct reading and inexpensive indicator tubes.

Locations of building air intakes and exhausts
☐ Check for blockages, debris, nests, etc.

General office spaces.
☐ Check carbon dioxide, temperature, humidity, and formaldehyde with direct reading instruments.
☐ Look for visible mold, wet or discolored ceiling tiles, and wet or damp carpet and furnishings.
☐ Look for obvious chemical sources (copiers, blueprint machines, industrial processes, cleaning solutions).

Mechanical rooms and duct chases.
☐ Look for visible mold, standing wet water, and chemical sources.

Computer rooms.
☐ Check for the presence of ozone.

Laboratories, photo-processing rooms, and chemical storage areas.
☐ Check for VOCs, spills, open containers, inoperative lab hoods, improper storage practices, air supplies, and returns.

Cafeterias and food service areas.
☐ Look for decaying organic materials, grease, and other potential sources of odors.
☐ Check for carbon monoxide, carbon dioxide, and others as necessary.

Restrooms, locker rooms, shower rooms.
☐ Look for microbial growth, settled water, air handling.
☐ Check for carbon dioxide.

Attics, penthouses, cooling towers, air intake locations.
☐ Look for slime, mold, dead birds, settled water, maintenance, and distance from contaminant sources.

Designated smoking areas.
☐ Look at air-handling systems.
☐ Check carbon monoxide and carbon dioxide.

Figure 13–12. This checklist can help environmental and building managers to identify problems. Source: Adapted from Burton DJ, *IAQ and HVAC Workbook*, Salt Lake City, UT: IVE, 1993. Used with permission.

An inspection of the building involves a walk-through by a trained inspector. The inspector normally is accompanied by a representative of the building who is familiar with various areas of the building and who can allow the inspector into areas that are normally locked.

The result of the assessment is identification of materials that are suspected of containing asbestos. These materials are candidates for sampling to determine the presence or absence of asbestos.

Sampling

Having identified potential ACBMs in the building, sampling can be undertaken. Sampling involves taking bulk samples of suspected material for laboratory analysis to identify the presence or absence of asbestos. Additionally, the condition of the material should be noted and data on which to base recommendations for possible response actions can be gathered. Along with this information, homogeneous areas are delineated. Homogeneous areas are areas in which the potential ACM appears to be the same. For example, a floor of a building may be considered a homogeneous area with respect to vinyl asbestos tile (VAT). A particular run of pipe may be considered a homogeneous area. A sufficient number of bulk samples is taken of each homogeneous area to characterize the given material.

Chemical Contaminants and Potential Sources

Formaldehyde (tobacco smoke, insulation, paneling, furnishings, carpet, stay-pressed cloth, deodorizers, paper products)

Ozone (photocopiers, electrical equipment, ozone generators, electrostatic air cleaners)

NO_2 (vehicles, gas heating and cooking, industrial processes)

CO_2 (people density, flame operations)

Organic chemicals (photocopiers, industrial processes, labs, new furniture, furnishings, building materials, cleaning materials, paint)

Allergens (pollen, fungi, bacteria, mold, mites, dust, ETS)

Fibers (insulation, fireproofing, equipment)

SO_2 (industrial processes)

CO_2 (people, open flames, industrial processes, OA, malfunctioning heating equipment, vehicles)

ETS (tobacco smoke)

Figure 13–13(a). This checklist identifies some chemical contaminants and their potential sources. Source: Burton DJ, *IAQ and HVAC Workbook*, Salt Lake City, UT: IVE, 1993. Used with permission.

A licensed laboratory can determine the presence or absence of asbestos using optical microscopy. The procedure is to pick apart the material under a stereo microscope to determine the percentage of the material that is fibrous. Next, the fibers are placed under a polarizing microscope and stained. The fibers will turn color, depending on the mineral composition. A trained microscopist can determine asbestos from nonasbestos fibers. Generally speaking, 1% asbestos is considered a threshold for defining a material as asbestos-containing. This is the cutoff for defining an ACM under federal regulations, and most state regulations follow this convention.

Operation and Maintenance Plans

Most buildings have operation and maintenance (O&M) plans that detail procedures for maintaining building systems such as HVAC, electrical, lighting, and other systems. The U.S. EPA recommends that O&M plans include appropriate procedures with respect to ACBMs. Components of an asbestos O&M plan may include notifying occupants and maintenance workers of the presence of ACBMs in the building, training maintenance personnel in proper procedures for conducting activities that require handling or removing ACMs, providing equipment and supplies for the asbestos maintenance personnel, providing respirators for the O&M staff, providing fitness-for-duty physicals for the employees who will participate in the program, and conducting inspections to track the condition of the material.

Although some form of O&M is desirable, recent studies have shown that building occupants, custodians, and even maintenance workers are highly unlikely to be exposed to airborne asbestos fibers exceeding the OSHA action level, PEL, or STEL (Price et al, 1992). This study showed that even during maintenance activities involving disturbing ACBM, airborne levels do not (at the 90th percentile) exceed the PEL.

What O&M components may be appropriate? Notifying occupants, custodial staff, and maintenance workers of the type and location of ACBM is a good idea. It is also recommended to train maintenance workers to know what ACBMs are present in their buildings and how to most appropriately deal with any planned handling or removal.

The U.S. EPA provides general guidance for conducting operations and maintenance plans in "the green book" (U.S. EPA, 1990). The U.S. EPA had previously provided guidance on managing ACM in the "orange" (U.S. EPA, 1984) and "purple" (U.S. EPA, 1985) books. In addition, the National Institute of Building Sciences has provided guidance on O&M plans in its document, *Guidance Manual, Operations and Maintenance Plans* (NIBS, 1992). These resources provide a general overview of the principles of operations and maintenance. They contain lists of possible components of asbestos management programs; however, not all of the components in these references may be necessary or appropriate for any particular building. What makes sense in terms of operations and maintenance is a function of the

IAQ Contaminant Sources, by Groups

Major Category	Major Source Type	Contributing Sources (Examples)
Outside sources	Contaminated OA	Pollen, dust, mold spores industrial pollutants general vehicle exhaust, smoke
	Nearby emission sources	Parking garages, lots freeway traffic loading docks, truck parks dumpster odor, debris and garbage building exhaust reentrainment
	Soil gases	Radon leakage from HC storage tanks soil contamination (e.g., landfill) pesticides, herbicides
	Standing water	Rooftop crawl spaces
Building equipment	HVAC system	Leaking refrigerant dirt, dust, debris in ductwork microbiologic growth biocides, sealants, water treatment cooling towers
	Other Equipment	Elevator motors office equipment (e.g., copiers) office cleaning supplies ships, labs, closets, kitchens
Human activities	Personal	Smoking perfume, cosmetics body odor cooking
	Housekeeping	Cleaning chemicals deodorizers clean activities (e.g., dusty sweeping)
	Maintenance	Paint, caulk, adhesive, oil, solvent pesticides cleaning materials
Building furnishings and materials	VOC, ROC emitters	Textured surface, carpet, textiles, fleecy particle board, shelving
	Inert aerosol emitters	Old asbestos old furnishings
	Microbiologic origins	Water-damaged furnishings, mtls surface condensation standing water (e.g., clogged drains)
Other	Accidents	Spills, floods, leaks
	Special use areas	Smoking area, exercise room, kitchen
	Repair and remodeling	New furnishings demolition (dust, bioaerosols) paint, caulk, sealants, adhesives

Figure 13–13(b). This checklist identifies IAQ contaminant sources by category and source type. Source: Burton DJ, *IAQ and HVAC Workbook*, Salt Lake City, UT: IVE, 1993. Used with permission.

material in the building, its condition, its potential for disturbance, and activities that may require handling or removal of the material.

Training options for custodians and maintenance workers can involve two-hour awareness training to make them aware of what types of ACBMs are in the building and what to do if they note damage (generally, notify a trained maintenance person). Training for maintenance workers who may clean up damage or for those who will deliberately handle or remove material—for example, by conducting small spot scrapes or by drilling through material to attach hangers or to run wires—would include 14- to 16-hour training.

When operations and maintenance activities require removing or drilling through ACBMs, workers should have respiratory protection available. Whether respiratory protection is required is a

Contaminant Concentrations Checklist		
Chemical Concern	Typical Concentrations*	Concentration of Concern**
CO_2	350–1,000 ppm	>1,000 ppm
TVOC	1–2 ppm	Depends on compound
Formaldehyde	0.04–0.1 ppm	>0.1 ppm
CO	1–5 ppm	>9 ppm
NO_2	<0.03	>0.05 ppm
Ozone	0.01–0.02 ppm	>0.05 ppm
Particles	<0.075 mg/m³ (total) <0.050 mg/m³ (PM-10)	>0.075 mg/m³ >0.050 mg/m³
Bioaerosols	Varies with site	(Other information more important)
Asbestos	<0.01 f/cc <2 µg/m³	>0.1 f/cc
Radon	<0.5 pCi/l	Canisters
Odors	"none"	Any detectable for long periods of time

*Nonindustrial environments; concentrations in excess of these values may warrant investigation.
**ASHRAE 62-1989; WHO *Indoor Air Quality Research,* Report 103, 1984; U.S. EPA; "Concern" refers to limited health effects and/or comfort.

Figure 13–14. This checklist identifies typical concentrations of some common contaminants. Concentrations in excess of these values warrant investigation. Source: Burton DJ, *IAQ and HVAC Workbook*, Salt Lake City, UT: IVE, 1993. Used with permission.

function of the ambient airborne asbestos fiber levels. OSHA sets levels above which respiratory protection is required. As noted previously, O&M personnel are unlikely to be exposed to airborne fiber levels requiring respiratory protection. The recommendation to have a respirator available is a precaution, and respirator use should be left up to the discretion of the trained maintenance worker.

Routine inspection of the building is also recommended. Typically, personnel in the O&M program will conduct an inspection of the building on a twice-a-year basis. The purpose of this inspection is to look for evidence of damage to ACBMs that might require repair or removal. Additionally, every three years, an outside inspection may be undertaken. Costs for O&M programs can range from negligible (for notification) to several thousands of dollars per year.

Table 13–D shows the components of a typical O&M plan and the general level of costs associated with initial training and equipment and costs typical for individual O&M activities.

Abatement Planning

If a building owner decides that a response action to the presence of ACBMs (other than operations and maintenance) is necessary, several options are available.

First, ACBMs can be encapsulated. One form, *bridging encapsulation,* involves coating an ACBM with a film similar to latex paint that effectively seals the material. A second form of encapsulation involves using a penetrating encapsulant. Encapsulation is often the response of choice for surfacing material such as acoustical plaster. This material is thin, hard, and often painted anyway in its use as a ceiling material. As long as the material is in good shape and is not delaminating, encapsulation may be an effective response action. Other potential response actions include building airtight enclosures and removal.

Removal Planning

Once an owner decides to undertake abatement and chooses removal of the ACM as the method, a plan for the removal must be developed. Many states and some cities require abatement designers and contractors to be licensed. In areas where abatement planners must be licensed, owners sometimes obtain the license themselves; however, most commonly, owners will retain an outside firm to design the removal and to conduct removal oversight and air monitoring. Typically, the owner and removal firm create a direct contract to remove the ACBMs.

Table 13–D. Typical Operations & Maintenance Costs

Initial Year Costs

Item	Average Cost $	Number	Total $
O&M plan	1,500	1	1,500
O&M training	500	3	1,500
Awareness training	75	1	75
Medical monitoring	200	3	600
Equipment	1,500	1	1,500
Expendable supplies	750	1	750
Disposal costs	50	3	150
Personal air sampling	350	3	1,050
		Total	$7,125

Subsequent Year Costs

Item	Unit Cost $	Number	Total $
Eqpt. repair & replacement	350	1	350
Training (new employee)	500	0.333	166
Awareness training	75	1	75
Medical monitoring	200	3	600
Reinspection	4,000	0.333	1,332
Expendable supplies	75	3	225
Disposal costs	50	3	150
		Total	$2,898

Table 13–E. Outline of Asbestos Removal Specification

Notice to Abatement Contractors
Quotation Forms
Work Items—Scope of Work
Certificate of Attendance
Language of Detail Specifications
Notifications and Permits
Documentation
Temporary Facilities
Scope of Work
Air Monitoring and Quality Control
Contractor OSHA Monitoring
Protective Equipment and Respiratory Protection
Supplies and Equipment
Decontamination
Barrier Wall Construction for Bag-Out
Work Area Preparation
Interior Demolition
Negative Air Pressure Differential System
Asbestos Removal
Final Cleaning of Work Area
Application of Sealant
Disposal of Asbestos Contaminated Waste
Site Responsibility
Acceptance of Work
Refireproofing

APPENDIX A—DRAWINGS
APPENDIX B—HVAC DRAWINGS
APPENDIX C—U.L. INFORMATION FOR
    FIREPROOFING

One of the first decisions to make is whether all ACBMs in a given area (say a floor) should be removed at the same time. Where a planned abatement involves removing asbestos-containing fireproofing, acoustical plaster, or vinyl asbestos floor tile covering in an entire area, it is normally cost-effective to remove other ACBMs at the same time.

The abatement planner designs the removal project. The planner generally prepares drawings indicating the containment (work area) on each floor and prepares specifications describing the specific work areas, approximate quantities of material to be removed, and what work practices are to be followed. The specifications will also describe laws to be complied with and monitoring to be undertaken.

Planning for removal involves coordinating the abatement work with other planned construction activities. Removals are sometimes undertaken at the same time as a major renovation. Therefore, the removal must be coordinated with building tenants and with other construction and renovation activities. In a typical office building, and asbestos removal project may add four to six weeks to the renovation of a floor. An outline for an asbestos removal specification is shown in Table 13–E.

Removal Design

Designing the removal of ACM on a floor of a building will involve preparing a work area or "containment." The containment will be constructed by hanging one or two layers of polyethylene sheeting on the walls and two layers on the floor. All windows, HVAC outlets, and other points where air may enter or leave the area will be further "critical barriered" with an additional layer of poly, and, in some cases, a fixed barrier of plywood or dry wall will be installed before covering with the polyethylene. Negative airflow is established by installing HEPA-filtered air-handling machines in the work area. The number of machines required is a function of the volume of the work area. Sufficient airflow must be maintained to achieve at least four air changes per hour. Additionally, the HVAC system must be blocked off to the work area to ensure that negative pressure in the work area or positive pressure in the HVAC system does not cause airflow from the work area.

Selecting an Abatement Contractor

Asbestos abatement contractors should be prequalified before being permitted to submit bids for an

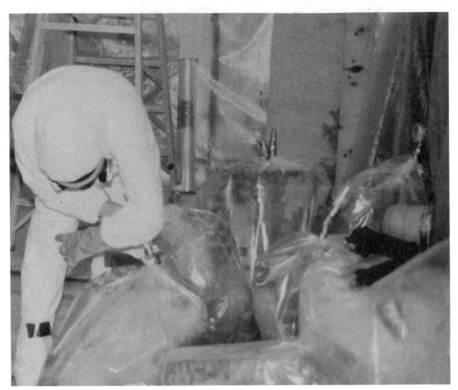

Figure 13–15. In an asbestos removal enclosure, an employee double-bags ACM, taping each bag securely before disposal.

asbestos-removal project. Prequalification involves checking the contractors to ensure that they are: (1) licensed to perform work in the area, (2) experienced in the type of project contemplated, (3) of sufficient size to carry out the project in the required time schedule, and (4) staffed with the appropriate number of full-time employees to carry out the project. References are normally required of abatement contractors prior to their being prequalified. References should be contacted.

The process for retaining an abatement contractor to do a removal project involves the following:

1. Prepare removal specifications that detail work practices, approximate material quantities, and disposal requirements.
2. Perform prequalification of various contracting companies to develop a short list of four to six contractors from which to obtain bids.
3. Do a walkthrough, at which time the prospective contractors are shown the removal area and given an opportunity to ask questions and to see for themselves the specific features of the building and the material involved that might affect their bids.
4. Receive bids from contractors, review, and recommend to the owner which bidder should be awarded the contract. Normally, the low bidder is awarded the contract. However, under certain circumstances a low bidder is not awarded the contract, for example, if the low bid is based on an unusual approach or technology that is unproven and that may not be acceptable to regulatory agencies.

Awarding the bid to a contractor offering a very low price based on an innovative technology exposes the owner to the risk that the regulatory agencies may not approve the approach or that the approach may not work as anticipated. In either case, the contractor would then be forced to use a traditional abatement approach, presumably at higher cost, leaving the owner vulnerable to a high-priced change order for any extra work that may be identified.

One way to safeguard against expensive change orders is to build into the bid process unit prices for any additional material which may be added to the work.

Project Monitoring

During any removal project, an outside firm is usually retained to serve as the owner's on-site project monitor. The monitor serves a number of functions. First, the monitor is present throughout the work to observe the contractor's work practices. It is the monitor's function to check the contractor's conformance to the removal specification and to local laws and regulations. This involves a number of activities:

1. The monitor performs a visual inspection several times a day to ensure that the containment is holding up and that there are no gaps in the seals.
2. The monitor checks the credentials of each worker. If new workers are added to a work shift, the monitor requires proof of the workers' training and medical surveillance status.
3. The monitor performs visual inspections at critical points in the work. Before bulk removal can begin, the monitor performs a visual inspection of the containment to ensure that the containment is properly installed, that a sufficient number of negative airflow machines are present and operating, and that negative air has been attained.
4. Next, the monitor performs a visual inspection before allowing encapsulation of the substrate. Having passed this visual inspection, the contractor is allowed to spray the lock-down encapsulant on the substrate used to contain the ACBM.
5. After the encapsulant has dried, the monitor does a postencapsulation inspection. This inspection is aimed at identifying any flaws or missed spots in the encapsulation.
6. Once the contractor passes the encapsulant inspection, final clearance air samples are taken. These air samples are taken aggressively: The air inside the containment is stirred up using electric leaf blowers, and fans are used to continually keep air circulating during the time that the final clearance air samples are taken. Final clearance air samples are laboratory analyzed to compare with the agreed-upon clearance criteria, usually 0.01 f/cc.
7. When the air quality has been shown to meet the final clearance level, permission is given for the contractor to break down the containment and demobilize.
8. After the containment is broken down, the monitor conducts one final visual inspection to identify any debris that may need to be vacuumed.

A note on clearance air samples: All samples can be taken for analysis by phase contrast microscopy (PCM) or by transmission electron microscopy (TEM). The TEM samples cost significantly more than PCM samples for two reasons. First, TEM samples are more expensive to analyze. As of early 1994, they cost approximately $80–$150 each, compared to $10–$20 each for PCM. Additionally, more TEM samples must be taken. This is because TEM assesses the amount of asbestos fibers above a background reference concentration (usually an outdoor reading). To do this, five samples inside the containment and five samples outside the containment are needed, as well as field and trip blanks. Typically, 10 to 13 samples are analyzed by TEM in order to obtain clearance. PCM, by contrast, requires only three to five samples inside the containment. A limitation of PCM is that it is not specific for asbestos. This, however, makes PCM a conservative measure, because it counts all fibers, whether they are asbestos or not.

The Asbestos Hazard Emergency Response Act (AHERA) of 1986 requires TEM for clearance in school buildings. This requirement does not extend to private buildings, however, and in practice, private building owners do not use TEM unless required to. One instance in which TEM is used is when final clearance cannot be obtained using PCM. Because PCM counts all fibers in buildings, buildings with cloth or paper fiber problems or with heavy dust levels may not obtain clearance with PCM even though the level of asbestos is acceptable. If the monitor or contractor suspects that this is the case, TEM samples are sometimes used to resolve the issue.

After successful clearance samples have been obtained and the contractor demobilizes, the monitor performs one final task. At this point, a final report is prepared documenting the work that was done, the results of all inspections and air sampling, and the summary of the project activities. A certificate or letter of completion is provided to the owner for submission to the regulatory agency, documenting that work has been completed in accordance with the specifications and local laws and regulations.

SUMMARY

IAQ is a rapidly emerging field of study for environmental practice. Because most people spend a high percentage of time indoors (up to 90%), the internal environment represents a tremendous potential exposure. The design and layout of interior spaces and buildings is a critical factor for IAQ investigations. As designers, architects, and engineers have become more aware of the hazards of the indoor environment, significant improvement has occurred. Nevertheless, the environmental manager should not be surprised if he/she is confronted by the need to evaluate buildings associated with medical symptoms and perform a thorough HVAC and industrial hygiene investigation. Many of the approaches, problems, and pitfalls of IAQ investigations were presented and discussed. It is anticipated that IAQ problems will continue to be a major area of investigation, litigation, and research into the 21st century.

Asbestos may present a problem in IAQ investigations because it is a known human carcinogen at high levels, widely present in many buildings, and subject to significant regulatory action (for example, in school buildings). The mere presence of ACBMs does not constitute a health threat and should not be responsible for occupant symptoms. However, high levels of airborne asbestos fibers in the indoor environment may present a potential long-term health problem and must be carefully handled within the confines of existing regulations. These regulations are controversial and have been the subject of an ongoing scientific debate (Mossman et al, 1990).

REFERENCES

Andersen I. Scandinavian experiences with indoor air pollution. *Tokai J Exp Clin Med* 10:309–314, 1985.

Anderson R. *Determination of Ventilation Efficiency Based Upon Short-Term Tests.* Washington DC: U.S. Department of Energy, Solar Energy Research Institute (SERI), 1988.

ASHRAE. *Guideline for the Commissioning of HVAC Systems,* Guideline 1–1989. Atlanta: ASHRAE, 1989.

Burge HA, Catigny M, Feeley J, et al. Guidelines for assessment and sampling of saprophytic bioaerosols in the indoor environment. *Appl Ind Hyg* 2:R 10–16, 1987.

Burge S, Hedge A, Wilson S, et al. Sick building syndrome: A study of 4373 office workers. *Ann Occ Hyg* 31:493–504, 1987.

Burton DJ. *IAQ and HVAC Workbook.* Salt Lake City, UT: IVE, Inc., 1993.

Burton DJ. *Industrial Ventilation Workbook,* 2nd ed. Salt Lake City, UT: IVE, Inc., 1992.

Farnham JE. Levels of October air supply and symptoms of sick building syndrome. *OEM Report* 7(11):93–94, November 1993.

Girman JR. Volatile organic compounds and building bake out. *Occ Med: State of the Art Rev* 4(4):695–712, October–December 1989.

Guidotti TL, Alexander RW, Fedorak MJ. Epidemiologies: Features that may distinguish between building-associated illness outbreak due to chemical exposure or psychogenic origin. *JOM* 29(2):148–150, February 1987.

Hodgson MJ. Buildings and health. *Health & Environ Dig* 7(2):1–3, May 1993.

Hughes RT, O'Brien DM. Evaluation of building ventilation systems. *Am Ind Hyg Assoc J* 47(4):207–213, 1986. Also located in NIOSH 1989 reference.

Kreiss K. The epidemiology of building-related complaints and illness. *Occ Med: State of the Art Rev* 4(4):575–592, October–December 1989.

Levin H. Building materials and indoor air quality. *Occ Med: State of the Art Rev* 4(4), October–December 1989.

Lipscomb IA, Satin KP, Neutra RR. Reported symptom prevalence rates from corporation populations in community-based environmental studies. *Arch of Environ Health* 47(4):263–267, July/August 1992.

Menzies R, Tamblyn R, Farant JP, et al. The effect of varying levels of outdoor air supply on the symptoms of sick building syndrome. *NEJM* 328:8231–827, March 25, 1993.

Morey PR, Shattuck DE. Role of ventilation in the causation of building-associated illness. *Occ Med: State of the Art Rev* 4(4):625–642, October–December 1989.

Mossman BT, et al. Asbestos: Scientific developments and implications for public policy. *Science* 247: 294–301, 1990.

National Institute of Building Sciences. *Guidance Manual: Asbestos Operations and Maintenance Work Practices.* Washington DC: NIBS, 1992.

National Institute of Occupational Safety & Health (NIOSH). *Indoor Air Quality: Selected References.* Cincinnati: NIOSH, September 1989.

Price B, Crump K, Baird E. Airborne asbestos levels in buildings: Maintenance worker and occupant exposures. *J of Exposure Anal and Environ Epidemiol* 2(3):215–220, November 3, 1992.

Quilan P, Macher JM, Alevantis LE, Gene JE. Protocol for the comprehensive evaluation of building-associated illness. *Occ Med: State of the Art Rev* 4(4):771–797, October-December 1989.

Sullivan JB, Brooks B. Multiple chemical sensitivities. In Sullivan JB, Krieger GR (eds). *Hazardous Materials Toxicology*. Baltimore: Williams & Wilkins. pp. 215–220, 1991.

Sullivan JB, Van Ert M, Krieger GR. Indoor air quality and human health. In Sullivan JB and Krieger GR (eds). *Hazardous Materials Toxicology*. Baltimore: Williams & Wilkins, pp. 667–689, 1991.

U.S. EPA. *Asbestos-Containing Materials in School Buildings: A Guidance Document*. Part 2. EPA-451/2-78-014. Washington DC: U.S. GPO, June 1984.

U.S. EPA. *The Exposure Factor Handbook*. EPA/600/8-89-1 043. Washington DC: U.S. GPO, 1989.

U.S. EPA. *Guidance for Controlling Asbestos-Containing Materials in Buildings*. EPA 560 15-85-024. Washington DC: U.S. GPO, June 1985.

U.S. EPA. *Managing Asbestos in Place*. 20T-2003. Washington DC: U.S. GPO, July 1990.

Van Ert MD, Clayton JW, Crabb CL, et al. Identification and characterization of R-phency—An emission product of new carpeting. Presented at the American Industrial Hygiene Conference, San Francisco, May 1988.

Woods JE. An engineering approach to controlling air quality. *Environ Health Persp* 95:15–21, 1991.

Woods JE, Orewry GM, Morey PR. Office worker perceptions of indoor air quality effects on discomfort and performance. In Seifert B, Esdern H, Fischer M, et al (eds). *Indoor Air '87, Proceedings of the 4th International Conference on Indoor Air Quality and Climate*. Berlin: Institute for Water, Soil, and Air Hygiene, 1987.

14

Global Issues

James M. Ohi, PhD
Gary R. Krieger, MD, MPH, DABT

During the last 25 years, there has been increasing recognition that certain types of environmental problems transcend an individual country's political and scientific ability to formulate an adequate response. Beginning with the 1972 UN Conference on the Human Environment in Stockholm, the prominence and importance of global environmental initiatives have steadily increased.

This chapter will examine environmental problems that are global in scope and have been core issues for two UN programs [United Nations Environment Programme (UNEP) and United Nations Development Programme (UNDP)] and the Organization for Economic Co-operation and Development (OECD):

- population growth
- soil degradation and deforestation
- biodiversity
- water resources
- energy
- atmosphere and climate.

Much of the material covering these topics is based on the World Resources Institute's *World Resources 1992–93, A Guide to the Global Environment, Towards Sustainable Development.* Each of these issues is briefly reviewed and potential solutions are presented.

Because the debate surrounding these global issues is frequently couched in terms of "sustainable development" language, the concepts and theory of this paradigm will also be briefly reviewed. (See also Chapter 2, Economic and Ethical Issues.) Although many of the global issues have generated debate, an in-depth analysis will be presented on three significant issues: (1) acid rain, (2) ozone depletion, and (3) greenhouse gas emissions. Each of these areas has passionate advocates and critics; hence, special examination is warranted.

Finally, it is critical to realize that a central driver of all of the major global issues is the rapid expansion in worldwide population. Therefore, as a prelude to the main discussion areas, it is worthwhile to review the trend in global population during the last half of the 20th century.

POPULATION GROWTH

The population of the world in 1950 was approximately 2.5 billion. By 1987, world population had increased to more than 5 billion. Current projections predict a global population of 11 billion by the year

2100. Can the planet sustain this level of population growth? The rapid emergence of environmental issues in the developing nations is partially attributable to the rapid population growths that have occurred since 1950 and are projected to occur by the year 2025. Between 1990 and 2025, 95% (3 billion) of the worldwide population expansion will be found in developing countries (Table 14–A).

The rapid deforestation of the developing nations is driven in large measure by expanding populations. In addition, high birth rates in the countryside have forced subsistence farmers into tropical rainforests and onto increasingly marginal farmlands. The resulting deforestation and soil degradation (secondary to intensive pesticide and fertilizer applications) have produced flooding and detrimental impacts on local flora and fauna.

Unfortunately, the effects of population growth are not confined to rural areas. The population in urban areas of the developing nations has also dramatically expanded. In 1950 approximately 17% of the population in developing countries was urban; however, by the year 2020, this figure will have increased to more than 50% (Keyfitz, 1989).

The dramatic increase in air pollution in urban areas of developing countries has been well documented (WRI, 1992). For example, whereas the concentrations in airborne lead have dramatically declined over the last 15 years in the developed (industrialized) world, the reverse has occurred in developing country urban areas (WRI, 1992). In Mexico City, 7 out of 10 newborns had blood lead levels greater than World Health Organization norms (WRI, 1992).

Obviously, each country must make choices between sufficient levels of economic development and

environmental protection. The rapid expansion of the human population is a major political, social, and economic issue that defies simple solutions; however, the environmental impacts are becoming increasingly obvious and require multidimensional solutions. Unfortunately, it is easier to document adverse effects than to develop sustainable solutions. Nevertheless, potential solutions can be developed based on a more complete understanding of the depths of the problems.

SOIL DEGRADATION AND DEFORESTATION

Since 1970 global food production has generally met the demands of world population. However, it is uncertain whether this trend can indefinitely continue and match future global population projections. According to the World Resources Institute (WRI), most agricultural production in the world uses farming practices that are environmentally unsustainable (WRI, 1992). Specifically, soil erosion and deforestation secondary to agricultural activities have occurred at alarming rates. According to the International Soil Reference and Information Center, approximately 2 billion hectares (4.9 billion acres) of soils or more than 11% of the earth's vegetated surface have suffered some degree of degradation, and 300 million hectares (750 million acres) of this total have suffered strong to extreme degradation (see Table 14–B; WRI, 1992). Agricultural activities were implicated in 28% of the degradation, overgrazing in 34%, and deforestation in 29%. Most of the damage has been done by wind and water erosion, although salinization, loss of nutrients, compaction, overusage of pesticides and fertilizers, and waterlogging are also contributors.

Table 14–A. Population 1950–2025

Region	Population or Percent Share of World Population				
	1950	*1970*	*1990*	*2000*	*2025*
World Total (billions)	2.5	3.7	5.3	6.3	8.5
Industrialized countries	33.0%	28.4%	22.8%	20.2%	15.9%
Europe	15.6%	12.4%	9.4%	8.1%	6.1%
North America	6.6%	6.1%	5.2%	4.7%	3.9%
Developing countries	66.9%	71.6%	77.2%	79.8%	84.1%
Central and South America/Mexico	6.6%	7.7%	8.5%	8.6%	8.9%
Asia	55.0%	57.0%	58.8%	59.3%	57.8%
Africa	8.8%	9.8%	12.1%	13.8%	18.8%

Source: Compiled from WRI, 1992.

Table 14–B. Moderate, Severe, and Extreme Land Degradation

Region	Aprox. Percent of Vegetated Land
World	10%
Oceania	< 5%
North America	4–5%
South America	8–10%
Asia	10–15%
Africa	12–15%
Europe	15–18%
Central America/Mexico	22–25%

Source: Compiled from WRI, 1992.

The majority of agricultural and overgrazing damage has been found in Asia and Africa. In the United States, approximately 800 million metric tons of eroded agricultural soils annually run into reservoirs (WRI, 1992). Damage attributable to waterborne sediments is estimated to cost more than $10 billion per year. The WRI estimates that about 36% of the eroded soils are from croplands.

Parallel with the increase in soil degradation, estimates of deforestation rates have also accelerated. A 1991 UN global assessment of deforestation found tropical deforestation to be almost 17 million hectares per year (42.5 million acres) (FAO, 1991). The impacts of deforestation are far-reaching:

- decreased biodiversity
- potential impacts on climate change by releasing stored carbon into the atmosphere
- soil degradation
- impacts on human habitats for indigenous peoples (local forest dwellers)
- diminished repositories for natural resources.

The potential solutions to the problems of soil degradation and deforestation are controversial and complex. The current paradigm in environmental management is based on concepts of *sustainable development* (see Chapter 2, Economic and Ethical Issues). Sustainable development has four interlocking dimensions: economic, human, environmental, and technological. The World Commission on Environment and Development defined sustainable development as development that "meets the needs of the present without compromising the ability of future generations to meet their own needs" (WCED, 1987).

Unfortunately, this definition can mean anything to anyone and provides little concrete guidance.

Therefore, more precise definitions of sustainable development have been attempted but even these provide minimal direction:

- *Physical definition:* Use of renewable natural resources in a manner that does not eliminate or degrade them, i.e., diminish renewability for future generations while maintaining adequate resources such as soil, groundwater, and biomass (Goodland & Ledee, 1987).
- *Economic definitions:* Maximizing the net benefits of economic development, subject to maintaining the services and quality of natural resources (Barbier, 1989); "Our economic systems should be managed so that we live off the dividend of our resources, maintaining and improving the asset base." (Repetto, 1986).
- *Social definition:* Development that improves health care, education, and social well-being and involves local peoples in decisions that impact their lives (UNDP, 1991).

Not surprisingly, sustainable development "solutions" for soil degradation and deforestation are based strongly on economic analysis. Many environmental economists believe that the industrialized economies have agricultural policies that produce serious environmental consequences:

- direct subsidies that encourage maximum use of pesticides and fertilizers
- crop selection driven by distorted price structures, i.e., crops or products (such as milk) are grown or produced on the basis of high support prices rather than "rational" land usage
- subsidization of irrigation water, e.g., the western United States, with increased production of relatively low-value alfalfa and subsequent secondary decline in groundwater aquifer levels and increased accumulation of salts and minerals in land and surface waters (WRI, 1992).

Economic "solutions" are based on reversing these policies and providing financial incentives that promote sustainability and renewability. Similarly, policies directed toward tropical forest management have heavy economic and political emphasis (WRI, 1992):

- *Change natural resource accounting so that natural forests are not undervalued.* Only the value of timber is currently considered versus the nontimber goods: fruits, nuts, resins, oils, etc.

- *Reform land ownership, tenure, and distribution.* Provide forest residents with secured right of use or access versus the government control that emphasizes "improvement" via cleavage for pasture or plantation agriculture.
- *Agrarian reform.* The inequitable distribution of agricultural land is a major issue in developing countries, because landless subsistence farmers are frequently attracted to government controlled tropical forests.
- *Adjust trade structure.* The price of commercially traded tropical woods does not include the true environmental externalities (cost). See also Chapter 2, Economic and Ethical Issues, for a detailed discussion of trade issues and environmental externalities.
- *Decrease demand for tropical timber.* A variety of countries (the United States and European Union) have taken action to significantly reduce the importation of tropical woods.
- *Use debt to finance conservation.* A variety of "debt-for-nature" swaps have been proposed between the industrialized nations and developing countries.
- *Rights of indigenous forest dwellers.* Many indigenous forest dwellers are increasingly demanding a dominant role in decisions affecting development of their traditional domains. An example of this actively is the *Coordina dora de las Organizaciones Indígenas de la Cueca Amazonica* representing 1.2 million local peoples in the Amazon basin.

Currently, a variety of these strategies are being selectively implemented. The long-term results and best approaches are still unknown.

BIODIVERSITY

Biodiversity is the "the variety among living organisms and the ecological communities they inhabit" (WRI, 1992). Although the total number of living species is unknown (estimates range from 5 to 30 million), the WRI estimates that an extinction rate of 15% during the next 25 years is possible. The major driver of species extinction is destruction or degradation of habitat: (1) tropical deforestation, (2) wetlands elimination, (3) coral reef impacts, and (4) temperate forest alterations.

Species extinction rates are at best an uncertain calculation. Calculations are based on the relationship between the known geographical size of a habitat and the number of known species *within the defined area.* This "species area curve" allows approximate calculations of percentages of species lost if a given habitat area of known size is eliminated. Because tropical forest habitats are estimated to contain 50%–90% of the world's species, deforestation is the dominant biodiversity threat (WRI, 1992).

Although absolute loss of tropical forest acreage is critical, various other direct and indirect forces have accelerated the adverse impacts on biodiversity:

- *Population growth with subsequent increased resource consumption.* There is increased pressure to convert species-intense habitat areas into agricultural and/or urban usage.
- *Knowledge deficiency about species and ecosystems.* Most species have not been studied in detail, and there is limited information on their potential beneficial uses (e.g., pharmaceuticals). In the United States, however, there is increasing emphasis on ecologic impacts and analysis (see Chapter 12, Risk Assessment, specifically the Ecological Risk Assessment section).
- *Ineffective and/or uncoordinated governmental efforts.* The emphasis on increased agricultural output is frequently at variance with the desire to maintain unaltered habitats. This tension is frequently heightened by the global trading system, which positively rewards national economies based on comparative advantage and specialization. This observation is particularly true for countries that emphasize large-scale monoculture farming (e.g., coffee and cocoa).
- *National accounts valuation.* It is difficult to "value" biodiversity versus the price quantification that can be placed on direct farm output.
- *Issues of social equity, resource distribution, and the rights of indigenous peoples.* Chapter 2, Economic and Ethical Issues, discusses these issues in detail. Resource control, land rights, and ownership issues are critical but highly sensitive in terms of individual political and social policies in a given country.

Obviously, these political, social, and economic forces can interact in a synergistic fashion. Therefore, the potential solutions to the "biodiversity problem" are unlikely to be simple or uniform. A variety of international forums and programs have been specifically directed toward developing biodiversity preservation (see Chapter 4, International Legal and Legislative Framework). In particular, the June 1992 UNEP Rio meeting specifically developed a Biological Diversity Convention. This convention was signed by the United States in June 1993 and stipulates that:

- Industrial countries must help developing nations technically and financially *over* and *above* current assistance levels.
- Rural communities and indigenous people must be first beneficiaries of profit based on products from wild plants (e.g, pharmaceuticals) and animals.
- The Global Biodiversity Strategy (developed under the auspices of the WRI, the World Conservation Union, and the UNEP) recommends four major areas of policy reform:
  — Reform existing policies that invite the loss of biodiversity.
  — Adopt new policies and accounting methods that promote conservation.
  — Reduce demand for biological resources.
  — Integrate biodiversity conservation into national planning.

Although these recommendations appear laudable, the "devil is in the details." Thus, it is far from certain what specific concrete actions will be produced by either the Biodiversity Convention or Strategy.

WATER RESOURCES

Evaluation of global water issues is a two-part resource problem: (1) freshwater and (2) oceans and coasts. These two areas will be analyzed separately; however, there is significant overlap in both their general problems and proposed solutions.

Freshwater

The cycle of freshwater "production" is inextricably linked with many other aspects of the environment. Chapter 5, Basic Principles of Environmental Science, presents the essential concepts governing the hydrologic cycle as well as the differences and interconnections between surface water and groundwater. The primary *supplies* of freshwater for most of humanity are rivers, lakes, and reservoirs. However, the primary *source* of freshwater is precipitation. Unfortunately, neither the source nor the supplies are evenly distributed across the planet.

This maldistribution is the source of significant political and economic tension. For example, most freshwater in the world is in river basins, and approximately one-half of these basins are shared by two or more countries (WRI, 1992). Obviously, unilateral action by one country has the potential for severe political, social, and economic impacts. The Middle East illustrates this problem. Egypt receives 86% of its domestic water resources from the Nile; however, the water of the Nile originates and is controlled by eight upstream countries (Starr, 1991). Similarly, the major aquifer that provides 25%–40% of Israel's water lies beneath both pre-1967 Israel and the West Bank (Starr, 1991).

Across the world, 3,240 cubic kilometers of freshwater are withdrawn from various sources and used on an annual basis (WRI, 1992). Approximately 69% is for agriculture, 23% for industry, and 8% for domestic uses (Table 14–C). Just as the water sources and supplies are unevenly distributed, it is not surprising to find that annual average per capita water use also shows tremendous geographic variation: North and Central America, 1,692 cubic meters (m^3); Europe, 526 m^3; Asia, 526 m^3; South America, 476 m^3; and Africa, 244 m^3. Within these geographic regions, agricultural and industrial usage account for the substantial observed annual usage differences, e.g., Asia accounts for 86% for agriculture; Europe, 54% industrial; North America, 42% industrial (WRI, 1992).

Based on these figures, it is easy to predict that the major sources of freshwater pollution would fall into three broad categories:

- *Agricultural.* Soil erosion, nitrate and pesticide pollution, and poor irrigation practices that increase salinity
- *Industrial.* Wastewater discharges of heavy metals and various synthetic organic compounds
- *Human domestic activities.* Pathogenic contamination via untreated or inadequately treated human/animal sewage.

These pollution sources are unlikely to diminish on a worldwide basis because water withdrawals have been increasing approximately 4%–8% per year, particularly in the developing countries (WRI, 1992). The rapid expansion of human population in the developing nations will inevitably lead to severe pressure in the food production sector and subsequent increase in agriculture associated water usage and contamination.

The developing nations do not have an exclusive franchise on nonsustainable agricultural practices. For example, the western United States has traditionally depended on subsidized water for agricultural activities that would probably be uneconomical

Table 14–C. Three Major Uses of Freshwater

Region	Percent of Total Withdrawals		
	Domestic	Industry	Agriculture
World	5–10	20–25	60–70
North and Central America	5–10	40	45
Europe	8–12	50–55	30–35
Africa	3–7	1–3	80–90
Asia	3–7	5–10	80–85
South America	18–20	18–22	58–60

if market prices were applied. However, the most spectacular example of water resource mismanagement and subsequent ecologic calamity has occurred in the former Soviet Union. Diversion of water from the Amu Daiya and Syr Darya rivers was begun in order to provide additional irrigation for cotton production despite the fact that agricultural practice in this region was already heavily directed toward cotton production. The river diversion had a catastrophic impact on the size of the Aral Sea (it shrank to less than one-sixth of its 1960 area) and produced tremendous increases in salinity such that all native fish species disappeared. Before the diversion, the Aral Sea was the fourth largest freshwater lake in the world.

Not surprisingly, economic and technical solutions to freshwater issues are focused in these areas:

- Price water resources correctly so that environmental externalities are considered.
- Improve water resource management for agriculture.
- Control point (industrial) and nonpoint (agriculture) sources of pollution.
- Integrate industrial and agricultural water resources management plans with sanitation efforts.
- Revitalize the efforts to improve and provide secure domestic sources of water. The 1981–1990 International Drinking Water Supply and Sanitation Decade was not an unqualified success since most regions (except West Asia) barely kept pace with or fell behind water demands attributable to population growth.

Oceans and Coasts

Traditionally, the oceans have been viewed as an unlimited resource that could not be significantly environmentally degraded. However, during the last 20 years there has been a real increase in coastal pollution and destruction of coastal marine habitats. In

addition, in 1990, for the first time in 13 years, the total global fish catch decreased and has continued to decline (*Economist*, 1994).

In response to these observations, substantial international efforts have been directed toward controlling both dumping of hazardous and radioactive materials at sea and from ships. Under the UNEP, more than 10 regions have formed Regional Seas Programmes to protect coastal areas.

Pollution control efforts are directed toward two areas: (1) direct dumping and (2) land-based point sources, e.g., industrial and municipal discharge pipes. *Direct dumping* includes sludge, radioactive and toxic wastes, and dredge spoils (i.e., silt from harbors and rivers). The UNEP estimates that 80%–90% of the material dumped at sea consists of dredge spoils, of which 10% is contaminated with oil, heavy metals, nutrients, and chlorinated hydrocarbons (UNEP, 1990).

Land-based point sources can also affect coastal areas. Major pollutant sources include (1) logging, (2) agriculture, (3) dam construction and irrigation, (4) urbanization adjacent to coastal areas, i.e., partially treated or untreated sewage discharges, and (5) deposition of airborne pollutants such as nitrogen from power plants, industrial facilities, and agriculture.

The ability of an individual country to cope with internal, land-based pollution point sources is highly variable. The cost of point source "end-of-pipe" treatment is high; few countries have aggressive collection and treatment of sewage. For example, as of the late 1980s, 72% of European sewage was treated (at any level), and less than 5% was treated in developing regions. Although dumping from ships can be banned, the political and economic will to control onshore and atmospheric sources of marine pollution is currently lacking. Unfortunately, for many waterways such as the Mediterranean, Black Sea, or

Maryland's Chesapeake Bay, the problems are significant and the long-term outlook uncertain.

ENERGY

Energy production and use are critical components of any country's economy and environment. The selection and mix of energy sources—coal, wood, oil, natural gas—have a profound effect on both overall environmental quality and the ability of a country to sustain adequate economic development. Worldwide, energy production has increased by approximately 50% during the last 20 years. This increase has been overwhelmingly concentrated on fossil fuels (coal, oil, gas). Significantly, although industrialized countries use 10 times more energy on a per capita basis, traditional fuels (firewood and animal wastes) are not considered in these statistics (WRI, 1992). Traditional fuels are significant sources of energy for developing countries, particularly for oil-importing African nations. The environmental impacts of these fuel sources are important because of problems with deforestation and degraded indoor/outdoor air quality.

The development of new energy resources is currently split between two goals that frequently clash:

- economics, i.e., price viability of different energy sources, particularly nonfossil fuels such as solar or wind
- use of renewable and sustainable sources.

The fossil fuels (oil and coal) have clearly been the most economical energy resource based on typical economic analyses. These analyses do not include environmental externalities such as air pollution, impacts of spills, and land-usage impacts. Throughout the early 1990s, fossil fuel prices either remained stable or declined. In addition, the political viability of carbon content taxes has been highly variable (e.g., not accepted in the United States).

Nevertheless, environmentally "friendly" energy resource allocation can only occur if prices actually reflect true costs. Subsidized energy sources invariably result in severe and unintended environmental impacts, e.g., deforestation of CO_2-absorbing rainforests in order to produce hydroelectric power for potentially pollution-generating heavy industry, or use of high-sulfur coals with subsequent acid rain production and high atmospheric particulate counts. The current industrial dislocation in Central Europe is partially attributable to massive energy subsidies for fossil fuels that favored gigantic, inefficient, and highly polluting state-owned heavy industries.

Energy projects that do not consider environmental impacts simply reflect the errors of the past and do not allow developing countries the ability to "leap frog" the mistakes of the industrialized world by using environmentally sound and energy efficient technology. The long-term cost of not incorporating realistic pricing and new technology is far greater than the immediate short-term, onetime gain in industrial production. Reversible technologies, e.g., biogas and biomethization, are potentially viable alternatives for developing countries, whereas new technologies (e.g., fluidized techniques and cogeneration) are available for countries that wish to efficiently use indigenous coal sources. The ability of developing and industrializing countries to maximize fossil fuel efficiency while minimizing environmental impacts is critical because these energy sources are major producers of air and water pollution emissions. Future energy development projects must combine local needs with larger regional global imperatives.

ATMOSPHERE AND CLIMATE

The potential that human activities may seriously alter the earth's atmosphere and climate has become a major environmental issue in the 1990s. Because the atmosphere and climate represent a true "commons," any anthropogenic alteration in atmosphere/climate stability inevitably leads to concern, controversy, and a call for some type of action. For many in both the political and scientific communities, this atmospheric environment clarion call is premature and unwarranted. Within the general topic of atmosphere and climate issues, three major topics continue to receive intense scientific and media scrutiny:

- acid rain
- ozone depletion secondary to chlorofluorocarbons
- greenhouse gas emissions and global warming.

These topics will be presented and discussed next in detail.

Acid Rain

The process of acid deposition begins with the interaction of sulfur dioxide (SO_2) and nitrogen oxides (NO_x) with sunlight and water vapor in the upper atmosphere. SO_2 emissions are primarily derived from coal-burning power plants. The net results of the SO_2 and NO_x interaction and chemical conversion is the formation of sulfuric and nitric acids.

Significantly, acid rain is not solely attributable to 19th or 20th century emission sources. Sulfuric and nitrogen compounds are also released by natural processes such as volcanism and the activity of soil bacteria.

The SO_2–NO_x reaction cycle occurs in the lowest 6–7.5 mi (10–12 km) (troposphere) of the atmosphere. Figure 14–1 illustrates this cycle. A key step in the process is the formation of the hydroxyl radical (HO). Each HO radical can oxidize thousands of sulfur-containing molecules; therefore, the overall reaction is only limited by the amount of pollution initially present in the air.

As Figure 14-1 shows, the sulfuric and nitric acids can easily become incorporated into existing cloud formations. The subsequent acid deposition then occurs in two forms: (1) wet (e.g., rain or snow) and (2) dry. *Dry deposition* refers to both direct settling of sulfate particles and to capture of sulfur dioxide gas by vegetation.

The presence of sulfuric and nitric acids in cloud droplets can produce very low pH water. Figure 14–2 illustrates the pH scale and describes how "acid" is acid rain. Typically, acid rain refers to pH below 5.6. Significantly more acidic water has been associated with the base of clouds (Mohnen, 1988). Although it is unusual to find low-pH droplets at the upper levels of clouds, high-altitude forests can be directly exposed to acidic cloud bases. Thus, mountain soils and vegetation can be exposed to high acidic precipitation. Figure 14–2 illustrates areas where precipitation in the eastern United States has a pH below 5. The eastern United States has typically been more vulnerable to the potential effects of acid rain because of the large number of coal-burning plants that use high-sulfur coals.

Once the acid rain has fallen on land, it is subject to a variety of chemical and physical transformations

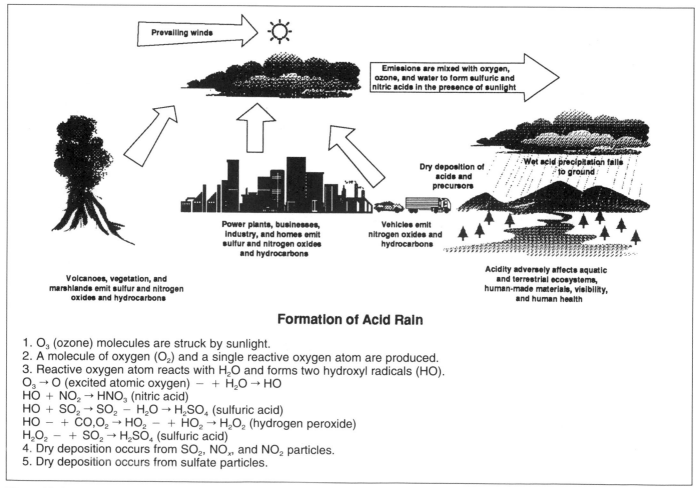

Formation of Acid Rain

1. O_3 (ozone) molecules are struck by sunlight.
2. A molecule of oxygen (O_2) and a single reactive oxygen atom are produced.
3. Reactive oxygen atom reacts with H_2O and forms two hydroxyl radicals (HO).

$O_3 \rightarrow O$ (excited atomic oxygen) $- + H_2O \rightarrow HO$
$HO + NO_2 \rightarrow HNO_3$ (nitric acid)
$HO + SO_2 \rightarrow SO_2 - H_2O \rightarrow H_2SO_4$ (sulfuric acid)
$HO - + CO, O_2 \rightarrow HO_2 - + HO_2 \rightarrow H_2O_2$ (hydrogen peroxide)
$H_2O_2 - + SO_2 \rightarrow H_2SO_4$ (sulfuric acid)

4. Dry deposition occurs from SO_2, NO_x, and NO_2 particles.
5. Dry deposition occurs from sulfate particles.

Figure 14–1. Acid rain reaction cycle. Source: U.S. EPA, 1990.

How "Acid" is Acid Rain?

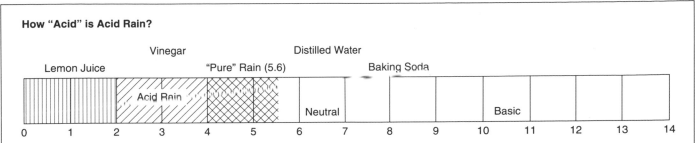

The pH scale ranges from 0 to 14. A value of 7.0 is neutral. Readings below 7.0 are acidic; readings above 7.0 are alkaline. The more pH decreases below 7.0, the more acidity increases.

Because the pH scale is logarithmic, there is a tenfold difference between one number and the one next to it. Therefore, a drop in pH from 6.0 to 5.0 represents a tenfold increase in acidity, while a drop from 6.0 to 4.0 represents a hundredfold increase.

All rain is slightly acidic. Only rain with a pH below 5.6 is considered "acid rain."

Source: U.S. EPA 1990.

Areas Where Precipitation in the East Is Below pH 5:

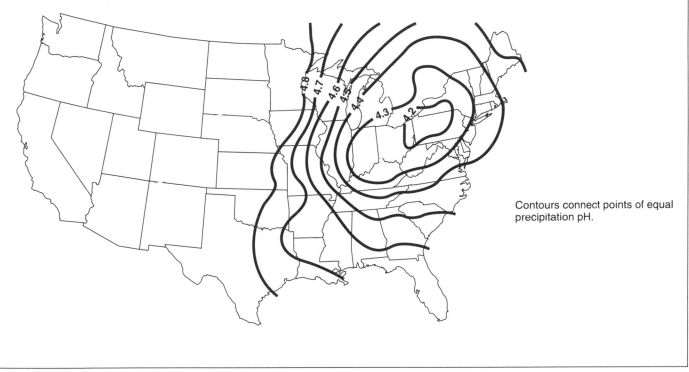

Contours connect points of equal precipitation pH.

Figure 14–2. Acid rain concepts and distribution of effects. Source: *National Acid Precipitation Assessment Program Interim Report,* 1987, U.S. EPA.

that alter the acidity and chemical characteristics of the water. Generally, three processes affect the final composition of the water that reaches lakes and streams:

- Direct neutralization via alkaline soils (i.e., soils containing significant amounts of limestone).
- Immobilization of the sulfate and nitrate ions by soil and/or vegetation.
- Cation exchange, in which soil ions such as calcium and magnesium replace the acid's hydrogen

ions. Metal ions are generated by rock weathering; however, rocks such as quartz (i.e., quartz sands) are resistant to weathering and do not provide a significant buffering effect.

Once the acid water reaches lakes or streams, it can be further buffered by bicarbonate ions. Thus, there are a variety of "self-protective" mechanisms that tend to mitigate the initial pH of acid rain. The content of neutralizing ions in a body of water is known as the *acid-neutralizing capacity (ANC)*. The higher the ANC, the less likely a lake will be susceptible to the effects of acid deposition. Of course, the converse proposition is also true. Therefore, a lake with an ANC of zero is highly susceptible to the potential impacts of acid rain.

Typically, an acidified lake has a pH below 5–6 and contains high levels of sulfate and ions like aluminum. Aluminum ions are mobilized when acidic water percolates through soil. The formation of acid lakes or streams is a dynamic rather than a static process (i.e., the ANC capacity of a water body can be replenished by weathering of minerals associated with materials surrounding the lake or stream).

Therefore, the speed of acidification cannot be easily predicted; however, water bodies subject to heavy acid rain and associated with soils poor in weatherable materials will tend to acidify. National surface water surveys and acid precipitation assessment programs (U.S. EPA, 1987) have mapped significant percentages of acid lakes in the Adirondacks (New York), Pocono Mountains (eastern Pennsylvania), and Michigan's Upper Peninsula. In addition, portions of southern and central New England may also be affected.

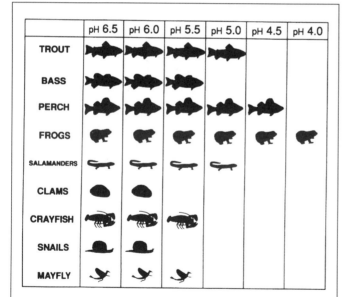

	pH 6.5	pH 6.0	pH 5.5	pH 5.0	pH 4.5	pH 4.0
TROUT	🐟	🐟	🐟	🐟		
BASS	🐟	🐟	🐟			
PERCH	🐟	🐟	🐟	🐟	🐟	
FROGS	🐸	🐸	🐸	🐸	🐸	🐸
SALAMANDERS						
CLAMS						
CRAYFISH						
SNAILS						
MAYFLY						

This chart shows that not all fish, shellfish, or their food insects can tolerate the same amount of acid. Fish like trout, bass, and perch are affected at different pH levels. Which type of fish are the most sensitive to acid? Generally, the young of most species are more sensitive than adults. Frogs may tolerate relatively high levels of acidity, but if they eat insects like the mayfly, they may be affected because part of their food supply may disappear.

Figure 14–3. Effects of acid rain on aquatic species. Source: U.S. EPA, 1990.

Similar effects have also been noted in Europe, particularly in Germany. The Germans have coined a specific term, *Waldsterben* or "forest death," for the adverse impacts of acid rain on forest ecosystems.

Ecosystem effects. Generally, trees are affected by direct damage to leaves and the indirect effects of uptake of toxic substances (e.g., aluminum released from soil by the acidic water). Affected forests are also susceptible to damage by diseases and insects. In addition, weakened trees are more likely to be injured by cold weather.

Table 14–D. Significant Acid Rain Accords, Agreements, and Cooperative Studies

- 1972 Stockholm U.N. Conference on Human Environment
  - Swedish study asserted that acid rain was responsible for both adverse ecologic and human health impacts.
- 1972–1980 Initiation of Norwegian Interdisciplinary Research Program (SNSF)
  - Effects of acid precipitation on forests and fish.
  - 1976 Telemar, Norway Conference.
  - 1980 Sovde Fjord, Norway Conference.
- 1972–1977 OECD Collaborative Study
  - Eleven European nation collaborative effort.
  - Measured the contribution of local and transboundary sources of sulfur deposition.
  - Reinforced the conclusion that significant portions of acid deposition was due to transboundary pollution.
- 1978 Great Lakes (United States, Canada) Water Quality Agreement
- ECE Convention on Long Range Transboundary Pollution
- 1980 United States and Canadian Transboundary Memorandum of Intent
  - Established joint coordinating committee.
  - Regulation development, increased enforcement.
- 1980 Acid Precipitation Act (United States)
  - Established National Acid Precipitation Assessment Program (NAPAP) and Interagency Task Force on Acid Precipitation.
- 1984 Canada–Europe Ministerial Conference on Acid Rain
- U.S. EPA Aeration of State Acid Rain Program (STAR)
  - States have utilized program money to develop emission inventories or evaluate economic implications of control strategies.
- 1990 U.S. Clean Air Act Amendments: Title IV: Acid Deposition Control
  - SO_2 allowance program.
  - Excess emissions penalties.
  - NO_x reduction program.
  - Permit and compliance program.
  - Monitoring and reporting requirements.

Source: Compiled by Gary R. Krieger.

Aquatic environments are sensitive to pHs below 6.0. Typically, the total numbers of fish and aquatic plants and animals decrease as water pH decreases below 6.0. At pH 5.0, most fish eggs cannot hatch. Increasing levels of aluminum are also responsible for fish toxicity. Figure 14-3 illustrates the relative pH water sensitivity of a variety of aquatic species.

Domestic and international agreements. Since 1972, several national (United States) and international agreements concerning acid rain have been negotiated. The 1972 Stockholm United Nations Conference on Human Environment was a significant initial clarion call and triggered subsequent accords and cooperative studies. Table 14–D lists the major international agreements.

Acid deposition control. Strategies for acid deposition control have developed along two tracks: (1) formal regulatory pressure directed at reducing SO_2 and NO_x emissions and (2) technologic improvements that lower SO_2 and NO_x emissions from power plant sources. Regulatory action is typically directed toward emission limitations for specific fossil fuel-burning electric utility sources. These limitations may be met by alternative methods of compliance provided by an emission allocation and

transfer system. The Title IV of the 1990 United States Clean Air Act Amendments were specifically directed toward acid deposition control compliance requirements. These requirements are listed in Table 14–E.

In the United States, regulatory compliance requirements have also produced a system of trading "offset" of emission quantities for economic gain. Thus, within a given geographic area, an efficient plant could sell its permitted SO_2 and NO_x emissions capacity to a less efficient plant for a direct monetary gain. The emphasis of the trading system is to provide an economic incentive for investment in efficient control technologies. Table 14–F lists a variety of existing and experimental technologies for controlling SO_2 and NO_x emissions.

To sum up the recent history, in the United States, the 1990 Clean Air Act Amendments formally specified acid rain mitigation actions. Similar steps have occurred in Canada and Europe. Although there have been a variety of opposing interpretations of existing scientific knowledge, the net thrust of worldwide political initiatives has clearly been directed toward reducing SO_2 and NO_x emissions. Both direct

Table 14–E. Acid Deposition Control Compliance Requirements

	Affected Sources*	Scheduled Compliance Date
SO_2 Emissions Reduction		
Phase I	110 power plants with emissions > 2.5 lb mmBtu**	January 1, 1995 January 1, 1997†
Phase II	Most fossil-fueled power plants > 25 MW with emission > 1.2 lb SO_2/mmBtu	January 1, 2000
NO_x Emissions Reduction	Tangentially fired boilers, with emissions of 0.45 lb NO_x/mmBtu	January 1, 1995
	Dry bottom, wall-fired boilers with emissions of 0.5 lb NO_x/mmBtu	January 1, 1995
	Wet bottom, wall-fired boilers; cyclone boilers; boilers with cell burner technology; all other types of utility boilers	January 1, 1997
Continuous Emissions Monitoring System		
Phase I	110 power plants, all units	November 15, 1993
Phase II	All power plants, all units > 25 MW with emissions > 1.2 lb SO_2/mmBtu	January 1, 1995
	All new generation units	Upon commencement of operations

Source: Public Law 101-549, Title IV.
Note: Because of the complexity and level of detail included in the amendments for Phase II requirements, this table is only intended to address the majority of those power plants that will be affected.
*Affected source or affected unit is a source or unit that is subject to emission reduction requirements under this program.
**mmBtu = million Btu heat input (using the nomenclature of the amendments).
†Affected units that install qualifying Phase I control technology (that is, a 90% reduction in emissions from the emissions that would have resulted from the use of fuels that were not subject to treatment prior to combustion) will be granted a two-year extension until 1997.

Table 14–F. Technologies for Controlling Acid Deposition

- Physical Coal Cleaning
  - Separate coal from impurities (sulfur).
- Fuel Oil Desulfurization
- Wet Flue Gas Desulfurization
  - Wet scrubbers, use either limestone or lime.
- Dry Flue Gas Desulfurization
  - Cheaper, simpler, and easier to recycle waste stream.
- Regenerable Flue Gas Desulfurization
- Staged Combustion
  - Nitrogen oxides control technology.
- Flue Gas Recirculation
  - NO_x control technology.
- Dual-Register Burner
  - NO_x control technology.
- Low Excess Air
  - Reduces formation of thermal and fuel NO_x.
- Low Nitrogen Oxide Burners
  - Useful for retrofit purposes.
- Flue Gas Treatment
  - Post combustion alternative, selective catalytic reduction removes 90% of NO_x.
- Chemical Coal Cleaning
  - Chemical reagents used to remove both inorganic and organic impurities.
- Limestone Injection Multistage Burner
  - Reduces both SO_2 and NO_x.
  - Useful for retrofits.
- Atmospheric Fluidized Bed Combustion (AFBC)
  - High (90%) SO_2 removal rates.
  - Easily disposed dry waste is produced.
  - Can burn high or low sulfur coal.
- Pressurized Fluidized Bed Combustion
  - Similar to AFBC but higher NO_x removal but lower SO removal.
  - High efficiency units.
- Integrated Gasification Combined—Cycle Technology
  - SO_2 removal has high as 95%.
  - Significant NO_x emission reduction.

Source: Reprinted from *The Acid Rain Controversy*, by James L. Regens and Robert W. Rycroft, by permission of the University of Pittsburgh Press. © 1988 by University of Pittsburgh Press.

regulatory and economic incentives have been developed and implemented. In 1980, Howard and Perley wrote, "Acid rain is an environmental crisis with sweeping financial implications, but its solution is political" (Howard & Perley, 1980). The policy initiatives of the 1990s indicate that while a political consensus has been reached, the economic and scientific costs and benefits will require long-term study and monitoring.

Ozone Depletion

Chlorofluorocarbons and the ozone layer. The vastness of the sky disguises a delicately balanced and vulnerable life-sustaining envelope. The atmosphere filters biologically harmful ultraviolet radiation, maintains the heat balance of the earth, and transports energy and moisture to create climatic patterns that help to determine the character and distribution of life on earth. The global energy balance is maintained at a nearly constant level through a complex feedback system of "ocean and air currents, evaporation and precipitation, [and] surface and cloud reflection and absorption" (SCOPE, 1970). The delicacy of this balance is demonstrated by oxygen isotope analyses of deep cores in the tropics that suggest that differences of only a few degrees Celsius marked changes from glacial to postglacial climates (Bryson, 1977).

Variables regulating global heat balance are the intensity of sunlight, the transmittance of the atmosphere, the albedo (reflectivity) of the earth-atmosphere system, and the "greenhouse" control of infrared fluxes from the earth (Bryson, 1977). Only the first of these variables, called the solar constant, is unaffected by human activities.

Gases in trace amounts also play an important role in regulating the global heat balance. One of these trace gases is *ozone*, a toxic and highly reactive gas that is used in industry as a disinfectant. The ambient concentration of ozone is one component of the "pollution standards index" used by health agencies to indicate the level of air pollution. In the stratosphere, however, a minute concentration of ozone, which does not exceed 10 parts per million by volume (ppmv) (NRC, 1979a), helps maintain the temperature structure of the atmosphere while screening biologically harmful ultraviolet radiation. Ozone may be as essential as oxygen in maintaining life on earth. The presence of the "ozone layer" is generally considered a critical factor in the interplay of energy and matter that allowed life to evolve on earth. [A notable dissenter is James Lovelock, who states that "the theory of the ozone shield as an essential prerequisite of surface life is . . . almost certainly plain wrong" (Lovelock, 1980). Lovelock developed an instrument sensitive enough to measure CFCs in the atmosphere (Lovelock, 1971).]

Chlorofluorocarbons (CFCs) are synthetic hydrocarbon compounds containing chlorine and fluorine. They are inert stable compounds with other chemical properties that make them well-suited for propellants, refrigerants, solvents, and other industrial uses. For example, they are commonly used as a refrigerant in refrigerators and air conditioners. Because they are chemically inert, CFCs accumulate in the troposphere and slowly migrate to the stratosphere, where they decompose and release chlorine atoms. Chlorine atoms play an important role in complex chemical reactions that regulate the production and removal of ozone in the stratosphere. Synthetic sources of chlorine can upset the chemical equilibrium of the stratosphere and reduce the amount of ozone in the stratosphere.

Historical background. In 1974 two chemists at the University of California at Irvine, Mario Molina and F.S. Rowland, raised the possibility that CFCs could deplete ozone in the stratosphere (Molina & Rowland, 1974). Molina and Rowland argued that the consequence of releasing CFCs (CF_2Cl_2 and $CFCl_3$) into the atmosphere was destruction of ozone in the stratosphere. The argument presented by Molina and Rowland is based on the following reasoning:

1. The atmospheric concentration of CFCs corresponds roughly to their total worldwide industrial production to date.
2. CFCs are insoluble, chemically inert, and there are no obvious sinks for their removal in the troposphere.
3. Photolysis of CFCs does not occur in the troposphere, because CFCs are transparent to wavelengths of light longer than 290 nanometers (nm), and ultraviolet radiation below the tropopause is longer than 290 nm.
4. Photolysis in the stratosphere is the most likely sink for CFC releases to the atmosphere.

Molina and Rowland concluded that chlorine atoms are dissociated from CFCs and trigger a catalytic chain reaction that results in a net destruction of ozone in the stratosphere.

The method used by Molina and Rowland was to search first for tropospheric sinks that could remove CFCs before they percolated to the stratosphere (Dotto & Schiff, 1978). Through a process of elimination they concluded that there were no likely removal processes in the troposphere and that CFCs would slowly migrate to the stratosphere. Once in the stratosphere where they are no longer shielded by ozone, CFCs will be split apart by the sun's ultraviolet radiation and will release chlorine atoms.

Molina and Rowland argued that "odd-chlorine" atoms form catalytic chain reaction analogues to that of the odd-atom species of hydrogen and nitrogen. The odd-nitrogen chain reaction was known to be important in the regulation of ozone in the stratosphere (Johnston, 1976), and the chlorine-chlorine

oxide (Cl-ClO$_x$) chain was thought to be considerably more efficient in catalytically converting ozone to oxygen (Molina & Rowland, 1974). In their paper Molina and Rowland set out the major issues that would characterize the ensuing controversy:

- The lifetime of CFCs in the troposphere is in the range of 40 to 150 years, and decades will be required to approach a steady-state concentration even if CFC releases are immediately halted.
- Because of this long lifetime, CFCs can be expected to accumulate in the troposphere, and any detrimental effects will not be obvious immediately but only later.
- The atmosphere has a finite capacity for absorbing chlorine atoms produced in the stratosphere, and although this capacity is not known, it will probably be exceeded when steady-state conditions based on 1974 release rates of CFCs are reached.
- Estimates of the formation rates of chlorine atoms at different altitudes and the lifetime of CFCs in the troposphere depend on the photodissociation rates and the "eddy diffusion coefficient" used in calculations and mathematical models.

Almost at the same time of the Molina-Rowland paper, R.S. Stolarski and R.J. Cicerone (1974) also proposed that the oxides of chlorine may constitute an important sink for stratospheric ozone. Stolarski and Cicerone examined the oxides of chlorine (ClO$_x$) and proposed that they formed a catalytic ozone destruction cycle analogous to that of the oxides of nitrogen (NO$_x$). The approach of Stolarski and Cicerone was to estimate the ratio of each chlorine species to the total chlorine species from their local steady-state chemistry, because their reaction rates, as known at the time of their paper, were rapid and atmospheric transport processes could be ignored.

By comparing the relative speed of the reactions, Stolarski and Cicerone were able to estimate the relative importance of each reaction and derive a model of the odd-oxygen cycle in the stratosphere. By examining local steady-state chemistry, they concluded that the ClO$_x$ cycle is not the dominant mechanism for odd-oxygen destruction but was significant at total chlorine concentrations near the 1 part per billion (ppb) range. The ClO$_x$ cycle is dominant but only at concentrations 10 ppb that are typical at altitudes greater than 20 km which are most important for ozone destruction. Thus, the ClO$_x$ cycle is less significant but more efficient for ozone destruction.

Stolarski and Cicerone mentioned industry, sea salt spray, fumaroles (holes in a volcanic area from which volcanic smoke and ashes arise), volcanic eruptions, and solid rocket fuels as likely sources of chlorine but not CFCs. They also pointed out that their model was incomplete and that the ClO_x cycle should be coupled with NO and HO cycles for a better understanding of ozone destruction processes in the stratosphere.

The papers by Molina and Rowland and by Stolarski and Cicerone are only two of many on the chemistry of the stratosphere that appeared in the scientific literature about the same time. For example, the Stolarski-Cicerone article appeared in a special issue of the *Canadian Journal of Chemistry* devoted to the proceedings of the International Association of Geomagnetism and Aeronomy (IAGA) meeting in Kyoto in September 1973, the first time that chlorine in the earth's atmosphere was seriously discussed by experts in stratospheric chemistry and physics (Dotto & Schiff, 1978). The Stolarski-Cicerone paper is important because it introduced the issue of chlorine chemistry in the stratosphere to the IAGA gathering of leading atmospheric chemists and physicists. The Molina-Rowland paper is important because it was the first to link CFCs released at the earth's surface to the presence of chlorine in the stratosphere.

The Molina-Rowland paper was perceived as a bombshell by the press and news media, but the capacity of chlorine to destroy ozone was common knowledge among researchers specializing in the stratosphere (Dotto & Schiff, 1978). Both the Molina-Rowland and the Stolarski-Cicerone teams built their findings in part on the work done by other scientists under the Climate Impact Assessment Program (CIAP) sponsored by the U.S. Department of Transportation (DOT) to assess the atmospheric effects of a fleet of supersonic transport aircraft (SST). The CIAP was a three-year study that involved nearly 1,000 scientists and cost $20 million (Glantz et al, 1982). A significant part of the research sponsored under CIAP concerned the impacts of SST exhaust gases on the stratosphere, and the focus of the research concerned with the ozone layer was the NO_x cycle, due primarily to Harold Johnston's contention that NO_x rather than water vapor was the critical constituent of SST exhaust (Broderick & Oliver, 1982).

Most of the research on NO_x conducted under CIAP was applicable to the ClO_x problem identified by Molina-Rowland, Stolarski-Cicerone, and others. The CIAP study and a parallel study by the National Academy of Sciences (NRC, 1975) commissioned by the CIAP-DOT management to maintain objectivity of the CIAP study meant that the scientific community was well-prepared to respond when the CFC/ozone-depletion controversy immediately followed. The enormous effort to assess the environmental effects of stratospheric flight created an international network of scientists, which was quickly mobilized when the CFC/ozone-depletion controversy arose, to compare theoretical calculations, atmospheric measurements, and laboratory experiments. Public awareness about the ozone layer and teamwork among U.S. governmental agencies involved in the issues was extraordinarily high.

The SST legacy, however, influenced the CFC/ozone issue in a less positive way as well. When the photochemistry of NO_x was adapted to illuminate the photochemistry of ClO_x, a problem-solving approach as well as analogous chemical reaction paths were borrowed. The dominant environmental issue in the SST controversy was originally noise. Ozone depletion by water vapor, and later NO_x in SST exhaust gases, was a concern only in the closing hours of the public debate. The SST controversy was initially a political issue (government funding for a technological advancement with dubious social value) to which scientific issues (noise and ozone depletion) were secondary. The CFC controversy, on the other hand, started out as a scientific issue and only later became a political issue. The SST/ozone-depletion connection, however, left a lasting mark on the CFC/ozone-depletion issue.

In 1970 James MacDonald, testifying before a subcommittee of the U.S. House of Representatives, argued that one consequence of ozone reduction could be a greater incidence of skin cancer because more ultraviolet radiation was reaching the earth's surface (Broderick & Oliver, 1982). MacDonald's testimony changed the stratospheric ozone-depletion issue from a scientific debate among technical experts to a public and media event (Dotto & Schiff, 1978). Unlike climate change, the threat of skin cancer was perceived as immediate, direct, and universal, even though the causal predecessor of both effects—stratospheric ozone depletion—was problematic and uncertain. In other words, public reaction to a possible consequence affected the scientific approach to understanding the cause. Once skin cancer and ozone depletion were linked, it became almost impossible to separate scientific research on stratospheric ozone depletion from the demands of public health and welfare (Dotto & Schiff, 1978).

(The following three paragraphs are based on H. H. Lamb, *Climate: Present, Past and Future*, vol. 1: *Fundamentals and Climate Now*. London: Metheun, 1972.)

Scientific background. This discussion of the scientific background of the ozone issue encompasses ozone photochemistry and the fluctuations in stratospheric concentration and distribution of ozone that are not caused by human uses of ozone-interfering chemical substances.

Overview of ozone photochemistry. The production and removal of ozone in the atmosphere is shown in Figure 14–4. The temperature of the atmosphere decreases as altitude increases until a minimum, the *tropopause*, occurs. The height of the tropopause (and the temperature minimum) varies with latitude and occurs at about 5 mi (8 km) near the poles and about 10 mi (16 km) at the tropics. The region of the atmosphere below the tropopause is the *troposphere*; that above it is the *stratosphere*.

In the stratosphere, temperature rises gradually with altitude until a maximum, the stratopause, is reached at about 31 mi (50 km). The positive temperature-altitude profile means that cooler, denser air lies beneath warmer, less dense air and that vertical mixing is minimal. Unlike the troposphere,

where descending cold air and rising warm air create vigorous circulation patterns, the stratosphere is stable and almost cloudless.

Atmospheric pressure decreases with altitude from about 1,000 millibars (mb) at the earth's surface to about 100 mb at the top of the tropopause. At the top of the stratosphere, atmospheric pressure is only 1 mb. Circulation in the stratosphere is caused by large-scale, nearly horizontal movement of air masses. With little gravity to cause heavier elements to settle out, the rate of transport does not depend on the nature of the molecule being transported.

The *mesosphere* lies above the stratopause. The temperature-altitude profile again becomes negative and reaches a minimum, the mesopause, at 52.8 mi (85 km). Above the mesopause, temperature rises strongly with altitude as gas molecules are ionized by solar radiation, and these charged ions and free electrons characterize the ionosphere. The outermost regions of the atmosphere from 52.8 mi (85 km) to 435 mi (700 km) are known as the *thermosphere* and *exosphere*.

In the stratosphere ozone is a trace gas present in concentrations of a few parts per million. [Unless otherwise noted, the following discussion on stratospheric chemistry is based on National Research

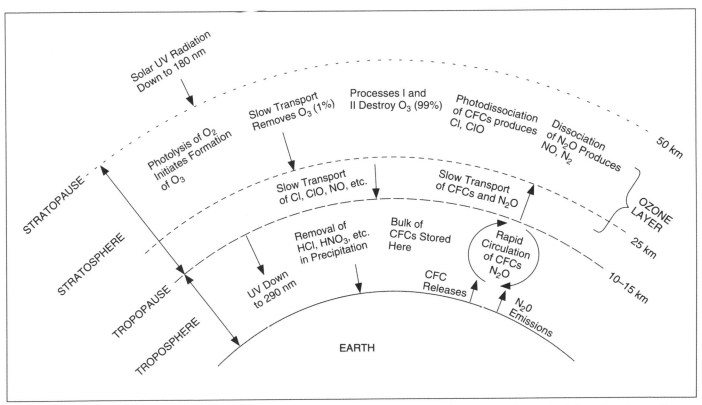

Figure 14–4. Production and removal of ozone in the atmosphere. Source: Reprinted with permission from *Ozone Depletion, Greenhouse Gases, and Climate Change*. Copyright © 1989 by the National Academy of Sciences. Courtesy of the National Academy Press, Washington DC.

Council, *Halocarbons: Effects on Stratospheric Ozone* (NRC, 1976), and on NRC's *Cause and Effects of Stratospheric Ozone Reduction: An Update* (NRC, 1982b).]

Ozone is formed by the photodissociation of oxygen molecules by solar radiation (hv) in the ultraviolet portion of the spectrum with wavelengths less than 242 nanometers (nm):

$$O_2 + hv = O + O \qquad \text{(Equation 14–1)}$$

$$O + O_2 + M = O_3 + M \text{ (twice)} \qquad \text{(Equation 14–2)}$$

Net: $3O_2 = 2O_3$

where M is another molecule that removes the excess energy of ozone to increase its stability.

Through this photodissociation or photolysis, about 50,000 million metric tons per year of ozone are produced, with 90% of this amount produced at altitudes above 15.5 mi (25 km).

Ozone is formed by photolysis and is "destroyed" by the same process:

$$O_3 + hv = O_2 + O \qquad \text{(Equation 14–3)}$$

It is this absorption of solar energy (hv) that shields the earth from harmful ultraviolet radiation. The photolysis of ozone shown in equation 14–3 is not a true destruction mechanism, because almost all of the oxygen atoms produced recombine rapidly with molecular oxygen to reform ozone (equation 14–2). The two reactions (equations 14–1 and 14–3) are also important; they convert solar energy into heat and produce the temperature inversion that characterizes the stratosphere.

The formation of ozone in the stratosphere is balanced by several processes that destroy ozone. About 1% of the ozone is removed by transport to the troposphere. The remaining 99% of the destruction is caused by chemical reactions that reform oxygen. As a result of these chemical reactions, ozone is destroyed by one of two reactions that produce molecular oxygen as follows:

$$O_3 + O = O_2 + O_2 \qquad \text{(Equation 14–4)}$$

$$O_3 + O_3 = 3O_2 \qquad \text{(Equation 14–5)}$$

Most of the ozone is removed by reactions represented by equation 14–4. Reactions in equations 14–1 and 14–4 are much slower than reactions in equations 14–2 and 14–3 and control the amount of "odd oxygen" (O and O_3) in the stratosphere. The more rapid processes, reactions (equations 14–2 and

14–3), control the distribution of odd oxygen between O and O_3. Ozone is the dominant form of odd oxygen, but the slowness of reactions (equations 14–1 and 14–4) make ozone vulnerable to other removal processes. The cycle involving species derived from oxygen was proposed by Chapman in 1930 and accounts for about 20% of the natural destruction of ozone.

Destruction processes involving other chemical species have been discovered during the past 25 years. The most important of these processes is the catalytic cycle involving nitric oxide (NO) and nitrogen dioxide (NO_2), which collectively (as NO_x) account for 45% of the natural destruction rate of ozone at altitudes between 15.5 mi (25 km) and 28 mi (45 km). The dominant reactions in this cycle are as follows:

$$O + NO_2 = NO + O_2 \qquad \text{(Equation 14–6)}$$

$$NO + O_3 = NO_2 + O_2 \qquad \text{(Equation 14–7)}$$

Net: $O + O_3 = O_2 + O_2 \qquad \text{(Equation 14–8)}$

This catalytic cycle produces the same effect as equation 14–4 without NO_x being consumed.

The distribution of NO_x between NO_2 and NO is determined by the following cycle:

$$NO_2 + hv = NO + O \qquad \text{(Equation 14–9)}$$

$$NO + O_3 = NO_2 + O_2 \qquad \text{(Equation 14–7)}$$

$$O + O_2 + M = O_3 + M \qquad \text{(Equation 14–2)}$$

This cycle competes with the catalytic cycle but does not remove either ozone or NO_x. It is important at lower altitudes during daylight hours.

About 10% of the removal of ozone by natural processes involves catalytic cycles containing hydrogen species: hydrogen atoms (H), hydroxyl (HO), and hydroperoxyl (HO_2). The "odd hydrogen" family of reactions is similar to those of oxygen and nitrogen as follows:

$$O + HO_2 = HO + O_2 \qquad \text{(Equation 14–9)}$$

$$HO + O_3 = HO_2 + O_2 \qquad \text{(Equation 14–10)}$$

Net: $O + O_3 = O_2 + O_2$

The "ozone war" was sparked by the contention that chlorine (Cl_2) forms an ozone-destruction cycle involving "odd chlorine" species analogous to those of oxygen, nitrogen, and hydrogen as follows:

$$Cl_2 + hv = Cl + Cl \qquad \text{(Equation 14–11)}$$

$$Cl + O_3 = ClO + O_2 \qquad \text{(Equation 14–12)}$$

$$O + ClO = Cl + O_2 \qquad \text{(Equation 14–13)}$$

Net: $O + O_3 = O_2 + O_2$

According to the chemical model just described, ozone in the stratosphere is formed and destroyed as oxygen atoms (O) are alternately attached to oxygen molecules (O_2) to form ozone (O_3) and detached to reform O and O_2. The chain reaction is terminated when two oxygen atoms combine to form an oxygen molecule. Because the oxygen atom spends far more time as O_3 than as O, the O_3 concentration depends directly on the length of the chain reaction. The problem of ozone depletion in the stratosphere arises when other chemical processes, described earlier, shorten the length of this chain (NASA, 1984).

The reactions involving NO_x, HO_x, and ClO_x are catalytic because they are not consumed in the reactions and remain available to continue converting O and O_3 to O_2. For example, the active chlorine species, ClO_x, is generated many times over before chlorine, as hydrochloric acid, is removed from the stratosphere to the upper troposhere, where it is washed out by rainfall. The average Cl atom and NO molecule have residence times of about two years in the stratosphere and can each remove between 10,000 and 100,000 molecules of ozone during this time (NRC, 1976). A process analogous to that described for chlorine removes nitrogen from the NO_x cycle as nitric acid. The significance of the catalytic cycles is that minute concentrations (in the part per billion range) of these species in the stratosphere can affect the concentration of ozone present at a concentration of no more than 10 ppm.

Natural fluctuations in stratospheric ozone concentration and distribution. To understand the scientific complexity of the CFC/ozone-depletion issue, it is necessary to review what is known about the natural fluctuations in stratospheric ozone concentration and distribution. These natural fluctuations are themselves complex and only incompletely known. The rate at which ozone is created and destroyed varies with the wavelength of ultraviolet radiation. Ozone is created by the absorption of energy of wavelengths less than 200 nm and is destroyed by energy of wavelengths 250 nm and 320 nm as well as by longer wavelengths around 600 nm.

Equilibrium concentrations at any time depend on the ratio of the intensities of solar radiation at these shorter and longer wavelengths. Because solar energy at wavelengths less than 300 nm varies and increases considerably in the shortest wavelengths during solar disturbances, the amount of ozone produced should vary over a period of years (Lamb, 1972). This variation has been linked with the 11-year cycle in sunspot activity (NRC, 1982a).

Although 90% of the ozone in the stratosphere is produced at altitudes above 15.5 mi (25 km), only 30% of global ozone is stored above this altitude. Below 15.5 mi (25 km) the abundance of ozone is determined by a balance between ozone transported down from higher altitudes and ozone lost to the troposphere.

There is less variation in the rate at which ozone is destroyed, but this, too, affects the natural distribution of ozone in the stratosphere. At the top of the stratosphere, the creation and destruction of ozone attains an equilibrium. Ozone descending below this equilibrium layer is shielded and is less likely to dissociate (equation 14–3). Short-wave radiation is also depleted by scattering at lower altitudes. Below about 18.6 mi (30 km) ozone is formed and destroyed more slowly; below 12.4 mi (20 km) ozone is almost stable (Lamb, 1972).

The major catalytic cycles that control the chemistry of ozone production and removal in the stratosphere provide a microscale understanding of the CFC/ozone-depletion problem. Certain large-scale meteorological processes make the traditional scientific procedure of empirical validation difficult. These difficulties of relating the microscopic perspective of photochemistry to macroscopic meteorological phenomena become evident when scientists try to differentiate the depletion caused by human activities from the natural fluctuations of ozone concentration and distribution that are themselves complex and difficult to quantify.

Total ozone in the stratosphere (that above a unit area of the earth's surface and also called the *ozone column*) varies by latitude and longitude and by season. The global distribution of ozone is tied to the general circulation patterns of the atmosphere, and the concentration of ozone reaches a minimum over equatorial regions even though most of the ozone production, because of photochemical reasons, takes place in these regions. Ozone is transported toward the poles, and its concentration increases with latitude in each hemisphere, reaching maximum 60° to 70° in the southern hemisphere (London & Angell, 1981). The seasonal variation of total ozone shows a maximum in the spring in each hemisphere with the intensity of the maximum increasing toward the poles. Little seasonal variation occurs between 25° north and 25° south.

The distribution of ozone with altitude also varies with season, latitude, and height. Largest variations occur during late winter and early fall, and the amplitude of these variations increases with latitude (London & Angell, 1981). Similar to total ozone variation, maxima occur in the spring and minima in the fall. Below 15.5 mi (25 km) seasonal variations are determined primarily by circulation processes in the lower stratosphere. The altitude of maximum ozone concentration also varies by latitude and is highest over the equatorial regions. This altitude generally follows the variation of the height of the troposphere with latitude, but it is usually about 6.2 mi (10 km) higher (London & Angell, 1981).

Empirical confirmation of ozone depletion by CFCs is also complicated by the long-term variation in total ozone. The broad pattern computed from data weighted by area and averaged across the observation stations shows ozone in the northern hemisphere declining in the late 1950s, increasing in the 1960s, and declining slightly or remaining steady in the first half of the 1970s (London & Angell, 1981). Interannual variation in total ozone in the late 1970s appeared to be increasing in north temperate latitudes, similar to interannual variation in temperature during the same period. London and Angell (1981) suggest long-term variation in ozone concentration is associated with that of temperature in temperate latitudes and also suggest this relationship will complicate the CFC/ozone-depletion issue.

The long-term variation of ozone distribution by height also complicates detection of depletion caused by CFCs. Ozone concentrations are determined for nine pressure layers (from 500 to 0.98 mb) with each layer corresponding to a vertical depth of about 4 km (London & Angell, 1981). The upper layers, from 19.8 mi (32 km) to 28.5 mi (46 km), are especially important because of the sensitivity of ozone concentration to variations in ultraviolet radiation at these heights and to anthrogenic perturbations. Data suggest a significant increase in ozone at these heights between 1964 and the early 1970s, but measurements showing this increase may be unreliable due to the interference from aerosols introduced into the stratosphere by volcanic eruptions. The same volcanic eruptions have also interfered with the study of the relationship between ozone concentration and sunspot activity.

The natural variations in ozone concentration indicate the difficulty of obtaining data sufficient in quantity and quality to establish a base against which ozone depletion by CFCs can be measured.

The quality of the data available does not permit a trend analysis of ozone concentration sensitive enough to detect perturbations caused by human activity (NRC, 1984). Measurements of chemical species that control the formation and removal of ozone are also subject to the same problem. It is not always clear whether differences in measurements of the same species at different times are caused by measurement inaccuracies or to atmospheric variability (NASA, 1984).

Ozone in the lower atmosphere has been measured for about 140 years, or since it was first clearly identified. The ozone column has been measured for about 60 years. Data from a worldwide network of measuring stations are available only for the past 20 years. These data are subject to errors ± 3% to 5% that are not all random and therefore do not cancel when averaged across time or space. Errors from station to station differ according to lack of proper instrument calibration, meteorological conditions under which measurements were made, and the care with which measurements were made (London & Angell, 1981).

The available data are also nonuniform across space. More data are available for total ozone than for its vertical distribution. As with other topics of scientific inquiry (for example, soils), more data are available for the northern hemisphere than for the southern. About 80% of the observations for total ozone are in the northern hemisphere, and most of the observation stations are located in the midlatitudes of each hemisphere. Almost no observations are available in the equatorial regions even though the equatorial midstratosphere is a major ozone-producing region (London & Angell, 1981). The uneven distribution of observation stations may insert a strong geographic bias into the calculations of total ozone concentrations based on measurement made at these stations.

The difficulties inherent in detecting change against a shifting background are obvious. In the case of stratospheric ozone depletion, there has been a heroic effort to improve researchers' understanding of the chemistry, physics, and meteorology of the stratosphere and thereby to resolve the CFC/ozone-depletion controversy. The effort has deepened scientific understanding without resolving the controversy. Indeed, the more that is understood about the complexity involved, the more difficulty of resolving the issue is appreciated. (A similar situation exists for the CO_2-global warming issue, for which a signal of temperature increase is difficult to detect above the "noise" of the system (Madden & Ramanathan, 1980).

The Montreal Protocol. An international convention banning nonessential uses of CFCs was first proposed by Sweden and Finland in January 1982. It was proposed to a working group of technical and legal experts formed by the United Nations Environment Programme (UNEP) to develop a global framework convention to protect the ozone layer. This proposed convention, which called upon signatories to "limit, reduce, and prevent activities under their jurisdiction or control which have or are likely to have adverse effect upon the stratospheric ozone-layer" (UNEP, 1982), underwent numerous revisions, and it took the working group three years to devise a consensus framework.

In March 1985 UNEP convened an International Conference on the Protection of the Ozone Layer. The conference adopted the Vienna Convention for the Protection of the Ozone Layer that, among other things, called upon the executive director of UNEP to convene a diplomatic conference in 1987 for the purpose of adopting a protocol to enforce control of global production, emissions, and use of CFCs. Elements of many proposals to limit emissions of ozone depleting chemicals were eventually incorporated in the Montreal Protocol, which was signed in September 1987 (OTA, 1988). On January 1, 1989, the Montreal Protocol was ratified by enough countries to enter into force (Brown, 1992).

The protocol is a complex document that seeks to limit production, consumption, and trade of certain CFCs (Group I compounds) and specific bromine-containing chemicals (Group II compounds) commonly known as halons. By 1999, developed countries must reduce production and consumption of Group I compounds to 50% of 1986 levels while freezing Group II compounds at 1986 levels (NRC, 1989). Developing countries, which produce and consume relatively low amounts of both categories of compounds, have a 10-year delay, beginning in 1989, during which they can increase production and consumption with certain limits. Beginning in 1999, these countries must also cut production and consumption of Group I compounds over a 10-year period and freeze production and consumption of Group II compounds. These restrictions are, however, pegged to consumption levels during 1995–1997 rather than 1986 (OTA, 1988).

With evidence (as discussed later in this chapter) of more rapid ozone depletion than was evident during the negotiations over the protocol, the international community moved quickly to amend the protocol. In June 1990, after a series of meetings, 93 nations agreed to phase out use of CFCs altogether by the year 2000 and extend the provisions of the protocol to other previously nonregulated, ozone-depleting chemicals (Brown, 1992). As part of these later negotiations, the United States agreed to establish a fund of up to $240 million to help developing countries purchase substitutes for CFCs (Brown, 1992).

This worldwide concern about ozone depletion has also prompted industry to accelerate the phasing out of CFCs. For example, many American, European, and Japanese automakers are already using non-CFC compounds in air conditioning units in certain high-end vehicle models.

Recent developments. The "ozone hole" over Antarctica was discovered during deliberations regarding the Montreal Protocol (NRC, 1989). By October 1987, the amount of ozone over Antarctica was reduced by more than 50% of its amount in 1979, and local depletion of between 9.3 mi (15 km) and 12.4 mi (20 km) altitude was as high as 95%. Not only was the ozone level the lowest on record, but the seasonal period of depletion also lasted the longest (NRC, 1989). The amount and distribution of atmospheric total ozone column has been observed by the Nimbus-7 Total Ozone Mapping Spectrometer (TOMS) since November 1978. From 1979 to 1991, the amount of total column ozone has decreased over most of the globe with small (3%–5%) losses at middle latitudes, larger (6%–8%) losses at high latitudes, and no losses at near the equator (Gleason et al, 1993). The average area-weighted loss rate for ozone for all seasons from 1979 to 1991, after corrections for the solar cycle and other effects, is estimated at 2.7% + 1.4% per decade, with most of the loss occurring at middle and high latitudes (Gleason et al, 1993). The global average total ozone in 1992 measured by the TOMS was 2%–3% lower than in any year measured by the TOMS, with largest decreases in the regions from 10° S to 20° S and 10° N to 60° N (Gleason et al, 1993). It is also significant that the 1992 measurements showed for the first time a simultaneous and sustained decrease of ozone across a wide range of latitudes in both hemispheres (Gleason et al, 1993). The reason for the low ozone values in 1992 are uncertain, but Gleason et al (1993) make a "first guess" that the decrease is related to the presence of aerosol from the Mount Pinatubo eruption.

It has been almost 20 years since Molina and Rowland first put forward their theory that CFCs attack stratospheric ozone. During this timeframe there have been detailed and intensive scientific debates relative to both the theory and supporting experimental data. In addition, substantial international accords have been signed and implemented that phase out the use of CFCs. Thus it is ironic that at a scientific forum preceding the environmental 1992 summit in Rio de Janeiro, a Brazilian meteorologist would declare the ozone depletion a "sham" (Tauber, June 11, 1993). Additional critics include (1) talk-show host Rush Limbaugh, who is quoted as declaring the ozone depletion theory "balderdash" and "poppycock," (2) former governor of the state of Washington, Dixie Lee Ray (in her book *Trashing the Planet*), and (3) commentaries by S. Fred Singer, former chief scientist for the Department of Transportation, in *The Wall Street Journal* and *National Review*.

Despite this frontal assault on the ozone-depletion theory, most mainstream atmospheric scientists appear to support the basic tenets and findings of the theory. In the Department of Energy's (DOE's) 1989 Ozone Project Review, the essence of the argument was well stated:

> On the one hand, recent evaluations of stratospheric and global tropospheric ozone trends indicate substantial on this progressive impacts that, if allowed to continue, could result in widespread and unacceptable damage. On the other hand, current and proposed remediation effects have resulted and will result in severe and potentially unacceptable, socioeconomic impacts.

Finally, perhaps the best answer and direction was provided by Ari Patrinos, director of the DOE Ozone Project: "There's only one atmosphere and sometimes we have to be very conservative in the actions we take."

GREENHOUSE GAS EMISSIONS/ GLOBAL WARMING

The controversy associated with the ozone-depletion theory pales in comparison to the ongoing debate about global warming. From 1984 through 1994 the global warming issue rapidly evolved from an interesting scientific debate to a major issue on the international political agenda. In 1988, the Intergovernmental Panel on Climate Change (IPCC) was impaneled by the UN to "study the scientific basis of climate change, to access its potential impacts, and to explore appropriate policy responses" (IPCC, 1992; Jones, 1993). In 1992, 154 countries signed a Framework Convention on Climate Change at the UN Conference on Environment and Development in Rio de Janeiro. What prompted this breathtaking political response, and what are the long-term economic and political implications?

In 1992, the IPCC concluded, if current emission trends continue, mean global temperatures will rise at the rate of 0.3°C per decade, based on *current models* (WRI, 1992, emphasis added). Mean sea levels were also predicted to rise by 0.23 in. (0.6 cm) per decade or approximately 2.53 in. (6.4 cm) by the year 2100 (IPCC, 1992).

Because of both the long-term scientific importance and potential economic impacts, the OECD "has focused its efforts on 'translating' knowledge on the science and potential impacts of global warming into the economic policy terms that decision-makers require" (Jones, 1993). Specifically, the OECD program tries to answer questions:

- Are abatement policies necessary?
- What are the potential costs, and can predictive financial and economic models be developed?
- What are the consequences of using market incentives, i.e., emission taxes or a system of tradeable emission permits, to achieve abatement objectives?
- What are the economics of negotiation, i.e., how much do nonsignatories benefit from international agreements?

In order to begin to answer these questions and evaluate the ongoing debate, a detailed understanding of the scientific issues surrounding greenhouse gas emissions is necessary. Therefore, this subject is presented and discussed in detail.

Historical Background

The "greenhouse" function of certain trace gases in the atmosphere was well established by the end of the 19th century. In 1827 Fourier compared the atmosphere to the glass of a hothouse that let through the "light rays of the sun" but retained the "dark rays from the ground" (Arrhenius, 1896). The selective absorption of the atmosphere in the long-wave portion of the spectrum, primarily by water vapor and "carbonic acid" (CO_2), was established by the research of eminent scientists such as Tyndall, Roentgen, Langley, and Angstrom. For example, Tyndall (1863) showed through laboratory experiments that molecules of "aqueous vapor" in air behaved as "barriers to waves of heat" and that absorption of heat was directly proportional to the quantity of water vapor present. It was obvious to Tyndall that this property of water vapor exerted an immense influence on meteorology and served as "a blanket more necessary to the vegetable life of England than clothing is to man."

By the end of the 19th century, Chamberlin (1899), among others, had drawn the broad outlines of the global carbon cycle and its relation to the regulation and change of climate. Chamberlin proposed a "working hypothesis" in which depletion of atmospheric CO_2 alternates with its reenrichment and causes periods of glaciation and warming. Callendar (1930) inquired whether human activity, particularly the demand for heat and power, had reached a level that could disturb the "carbon-balance" of nature. Callendar made rough estimates of the amount of "fossil carbon" released to the atmosphere by the burning of coal and oil and calculated what should be the corresponding increase in atmospheric CO_2 content from the preindustrial level. His calculated values of atmospheric CO_2 concentrations were lower than those observed, and Callendar concluded that other human activities such as clearing forests and cultivating land were adding significant amounts of CO_2 to the atmosphere. Callendar also pointed out that the ocean had a large capacity to absorb excess CO_2 and would delay any noticeable warming of the globe.

According to Ausubel (1983), public attention and concern about global warming caused by CO_2 emissions probably can be traced to a scientific conference convened by the Conservation Foundation in 1963. The Conservation Foundation (1963) published a "statement of trends and implications of carbon dioxide research" that described in nontechnical terms the potential problems of accumulating CO_2 in the atmosphere and the need for better understanding of biogeochemical cycles in general and the carbon cycle in particular. The global warming-CO_2 emissions issue became a part of the national political agenda in 1965, when it was included in *Restoring the Quality of Our Environment*, the Report of the Environmental Pollution Panel of the President's Science Advisory Committee (Ausubel, 1983).

Scientific Background

Earth's radiation balance. The greenhouse effect is created by the radiative energy balance of the earth-atmosphere system and is best illustrated by considering the global average radiative energy budget of this system (Ramanathan et al, 1987). Radiation budget measurements (incoming solar radiation, reflected solar radiation, and outgoing long-wave radiation) made by satellites at the top of the atmosphere show that the amount of solar radiation absorbed, after reflective losses are subtracted, is about 236 W/m² and is balanced by the same amount of energy emitted to space, primarily in wavelengths longer than 4 μm. The mean annual radiation and heat balance of the atmosphere-earth system is shown in Figure 14–5.

The global mean altitude at which long-wave emission to space is calculated is about 3.7 mi (6 km) above the earth's surface (Hansen et al, 1981). The surface temperature of the earth is about 288°K (15°C) and, at this temperature, the earth emits about 390 W/m² in long-wave radiation to the atmosphere. The difference between the energy emitted at the surface of the earth and at the top of the atmosphere, 154 W/m², is the greenhouse effect, the long-wave energy that is intercepted by the atmosphere and that would otherwise be returned to space (Ramanathan et al, 1987).

Radiatively active gases both absorb and emit long-wave radiation, but because the air temperature in the troposphere decreases with altitude, these gases absorb radiation emitted by a warmer surface but emit this radiation to space at a colder atmospheric temperature. In other words, radiatively active gases in the troposphere absorb more upward radiative flux than they emit (Dickinson & Cicerone, 1986). The net effect of the process is to reduce radiation to space at the top of the atmosphere (Ramanathan et al, 1987). If the increase in radiatively active gases does not alter the amount of solar radiation absorbed, there will be more radiative energy available to the surface-atmosphere system. To maintain radiative energy balance with the net incoming solar radiation, the surface-troposphere system would warm and radiate more long-wave energy to space until the radiative energy balance at the top of the atmosphere is restored (Ramanathan et al, 1987).

Greenhouse gases. The term *greenhouse gas* generally describes gases that are highly transparent to incoming solar radiation but are relatively opaque to long-wave radiation, similar (for descriptive purpose) to glass in a greenhouse (Luther & Ellingson, 1985). Atmospheric gases differ greatly in their absorptive properties. The most important atmospheric constituent of the greenhouse effect is water vapor (H_2O). Water vapor, together with CO_2 and clouds, create about 90% of the total effect (Ramanathan et al, 1987). The remaining 10% of the greenhouse effect is due to the presence of trace amounts

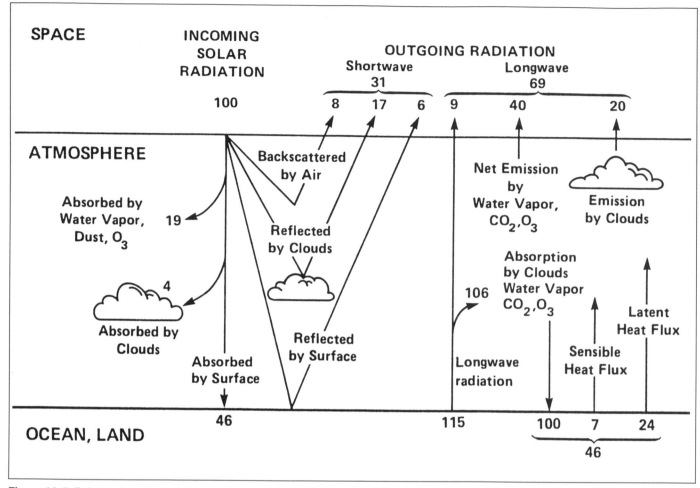

Figure 14–5. Relative sizes of the global average components of the earth's energy balance expressed as percent of incoming solar radiation at the top of the atmosphere. Source: Reprinted with permission from MacCracken et al, *Energy and Climate Change: Report of the DOE Multi-laboratory Climate Change Committee.* Chelsea, MI: Lewis Publishers, 1990. Copyright Lewis Publishers, a subsidiary of CRC Press, Boca Raton, Florida.

of ozone (O_3), methane (CH_4), and nitrous oxide (N_2O).

Unlike H_2O, which varies in concentration according to temperature and weather, CO_2 is mixed almost uniformly throughout the atmosphere to an altitude of 62.1 mi (100 km) (Dickinson, 1982) and thus plays an important role in controlling the vertical temperature distribution of the stratosphere (the portion of the atmosphere above the troposphere) and not just the temperature of the lower troposphere and at the surface of the earth.

Although the role of CO_2 in controlling the temperature of the troposphere is well known, CO_2 plays an equally critical role in cooling the middle and upper stratosphere and in balancing heating due to absorption of solar radiation by O_3. Net radiative heating of the troposphere caused by an increase of atmospheric CO_2 can be accompanied by significant enhancement of the cooling rate in the middle and upper stratosphere. Because the photochemical rates of O_3 formation and destruction in the stratosphere are dependent on temperature, cooling of the stratosphere can lead to an increase in stratospheric O_3 concentration that, in turn, can act as a negative feedback on cooling due to CO_2 (Ramanathan et al, 1987).

The other important greenhouse gases are CH_4, N_2O, and chlorofluorocarbons (CFCs). These gases behave similarly to CO_2 by reducing the outgoing long-wave radiation to space and thereby causing the surface-troposphere system to warm until radiation balance is restored (Luther & Ellingson, 1985). The combined greenhouse effect of present-day atmospheric concentrations of CH_4 and N_2O is comparable to the projected surface temperature warming caused by a doubling of CO_2 (Ramanathan et al, 1987). Both CH_4 and N_2O are chemically as well radiatively active and influence the greenhouse effect indirectly through chemical reactions with other gaseous species, especially by influencing the concentration and vertical distribution of O_3 that play an important role in the radiative energy balance of the stratosphere and troposphere.

CFCs (primarily F-11 [$CFCl_3$] and F-12 [CF_2Cl_2]) are important greenhouse gases because their absorption band strengths are several times stronger than those of CO_2. Furthermore, the greenhouse effect of CFCs scales linearly with concentration rather than logarithmically as with CO_2 (Ramanathan et al, 1987). These differences make the greenhouse effect of one molecule of CFC equivalent to 10^4 molecules of CO_2 (Ramanathan et al, 1987). CFCs are chemically inert and do not decompose in the troposphere. These compounds eventually percolate up to the stratosphere, where they are dissociated by ultraviolet (UV) radiation and release chlorine atoms which, in turn, catalyze the destruction of O_3.

Of the greenhouse gases described so far, O_3 presents a more complicated case, because it absorbs strongly in the solar portion of the spectrum, is chemically reactive, and plays an important role in the radiative balance of both the stratosphere and troposphere. O_3 is formed in the middle and upper stratosphere by the photolysis of molecular oxygen (O_2) by UV radiation. The oxygen atoms formed by this photolysis recombine with O_2 to form O_3. Little UV radiation reaches the surface of the earth, because it is absorbed by O_3 in the process of its formation and destruction. Almost all of the solar absorption in the stratosphere is caused by absorption by O_3 (Luther & Ellingson, 1985). The absorption of UV radiation by O_3 maintains the stable thermal structure of the stratosphere and may also be responsible for the existence of the tropopause that separates the stratosphere and troposphere (Wang et al, 1985).

Aerosols inhibit the transmission of solar radiation to the surface and can scatter a fraction of the incident radiation back to space, depending on the local concentration and optical properties of the aerosols and on the surface albedo (reflectivity) and solar zenith angle (Luther & Ellingson, 1985). Aerosols can increase or decrease planetary albedo, depending on whether the albedo of the surface over which they are suspended is lower, such as the oceans, or higher, such as deserts or snow (Luther & Ellingson, 1985). Aerosols can be soot from urban and industrial sources, dust from wind-blown soil and sand, water-soluble sulfates and sea salt particles from maritime sources, and ash from volcanic sources.

Trace gases that are not radiatively active but that react chemically with those that are also influence the radiative energy balance of the atmosphere-surface system. These gases include carbon monoxide (CO) and oxides of nitrogen ($NO_x = NO_3 + NO_2$). Although CO has no direct effect on the radiative balance of the atmosphere-surface system, CO reacts with hydroxyl ions (OH) to form CO_2 and, by removing tropospheric, OH^-, indirectly increases concentrations of CH_4 and O_3 (Ramanathan et al, 1987). In the troposphere, NO_x is an important source of O_3 and plays an indirect but significant role in the radiative balance of the atmosphere-surface system. Ramanathan et al (1987) estimate that from 1970 to 1980 anthropogenic emissions of NO_x have caused a 5% increase in tropospheric O_3 in the northern hemisphere.

In addition to emissions of greenhouse gases, the effects on the atmosphere-surface system of increased aerosol loading of the atmosphere from natural and artificial sources are extremely complicated and add significantly to the uncertainty surrounding climate change. According to Ramanathan et al (1987), even after decades of research it is still not known whether the general effect of aerosols is to cool or to warm the climate. Aerosols affect climate directly by altering the radiation budget in both the solar and long-wave portions and indirectly by altering the optical properties of clouds (Luther & Ellingson, 1985).

Aerosols can produce considerable "noise" in the meteorological data and make it difficult to establish a direct relationship between increases in atmospheric trace gas concentrations and climate change (Ramanathan et al, 1987). The effects of increased aerosol loading on the radiation budget of the atmosphere-surface system can be comparable in magnitude to that of a doubling of CO_2, and it is not known whether or how aerosol loading will change in response to climate change induced by increased CO_2.

Clouds also complicate our understanding of the effect of greenhouse gases on global climate. On the average, clouds cover about half the earth and play a major role in regulating the radiative heating of the atmosphere-surface system (Dickinson, 1982). Clouds create a cooling effect by reflecting a large part of incoming solar radiation and by increasing the albedo of the earth to twice what it would otherwise be in the absence of clouds. Clouds create a warming effect by absorbing long-wave radiation emitted by the surface and by reemitting this radiative energy to space at colder cloudtop temperatures. In other words, clouds warm the atmosphere-surface system much in the same way as greenhouse gases. The greenhouse effect of clouds, however, is larger than what could result from a hundredfold increase of atmospheric CO_2 (Ramanathan et al, 1989). These competing effects of cooling and warming vary with time, geography, cloud type, and cloud structure in a complex manner that is not well understood.

Sources and sinks of greenhouse gases. Anthropogenic emissions of CO_2 disrupt the natural circulation of carbon among the atmosphere, ocean, and biosphere by returning to the atmosphere, in a few decades, carbon stored in sediments for over 500 million years. Figure 14–6 shows recent estimates of the major reservoirs of carbon and fluxes of carbon between these reservoirs. Determining the influence of CO_2 emissions on climate change is difficult, because of the natural variability of both climate and atmospheric concentration. Although precise measurements of the atmospheric concentration of CO_2 were not available before 1957—when reliable instrumental measurements were begun at Mauna Loa, Hawaii, by Charles Keeling of the Scripps Institution of Oceanography (SIO)—it is indisputable that atmospheric CO_2 has steadily increased since then from about 316 ppmv to 351 ppmv in 1988, an 11% increase (MacCracken et al, 1990). The annual rate of increase CO_2 is about 0.4% (Wuebbles & Edmonds, 1988). The absolute annual increase is about 1.5 ppmv, an imbalance of about 3 GT (10^9 metric tons) C per year in the global carbon budget (Woodwell, 1988).

Figure 14–7 shows the annual increase in the monthly mean concentrations of CO_2 compiled by the SIO at Mauna Loa. The year-to-year increase in CO_2 concentration is roughly in step with CO_2 emissions from combustion of fossil fuels (CDAC, 1983). These measurements of atmospheric CO_2 levels by the SIO are corroborated by measurements made since 1970 by the National Oceanic and Atmospheric Administration (NOAA) at four remote measuring stations located at Pt. Barrow, Alaska; Mauna Loa, Hawaii; American Samoa; and the South Pole (Gammon et al, 1985).

Present annual emission to the atmosphere from fossil fuel combustion is about 5.0×10^{15} g carbon (Mooney et al, 1987). The partitioning of anthropogenic carbon emissions among reservoirs in the atmosphere, ocean, and biosphere is the subject of a major scientific controversy between terrestrial ecologists and geochemists (Detwiler & Hall, 1988). Although researchers agree that industrial emissions of CO_2, primarily fossil fuel use and kilning of limestone in cement production, resulted in a total release in 1980 of about 5.3×10^{15} g of C to the atmosphere, measurements of atmospheric concentrations of CO_2 at Mauna Loa since 1958 indicate that only 55% of the CO_2 released from these activities (the "airborne fraction") remains in the atmosphere (Detwiler & Hall, 1988). The bulk of the remaining 45% of CO_2 emissions was most likely absorbed by the ocean.

Geochemists use models of carbon cycling to estimate uptake of CO_2 by the ocean because it is not possible to measure the increase in inorganic carbon in seawater directly. Present versions of these models provide estimates that only 35% of CO_2 released by industrial activities is stored in the ocean (Detwiler & Hall, 1988).

The total airborne fraction will determine when a doubling of CO_2 will occur and, hence, the timing of climate change under equilibrium modeling assumptions. The transient response of the climate system is intimately connected to the cycling of carbon, particularly between the atmosphere and ocean. Uptake of CO_2 by the ocean is complicated by thermal stratification of the ocean, which varies with latitude and by a complex chemical buffering process involving carbonate and bicarbonate ions. The CDAC (1983) estimates the annual ocean uptake of CO_2 is about 2 GT (10^9 tons) of carbon, or approximately 40% of fossil fuel emissions.

The possible biotic sources of CO_2 emissions are not well understood. Although there is agreement that terrestrial storage of carbon has been greatly reduced during the past century, researchers substantially disagree on the timing and amount of this depletion (and emission to the atmosphere) (CDAC, 1983). Estimates of the net release of carbon from terrestrial ecosystems to the atmosphere during the past 100 years or so based on $^{13}C/^{12}C$ records in tree rings range from about 70×10^{15} to more than 200×10^{15} g C (Houghton et al, 1985). Woodwell (1987) reports a current release rate in range of 0.5 to 4.7 GT (1 GT = 10^{15} g) of carbon per year from deforestation alone.

Uncertainty surrounding biotic emission sources may be further complicated by the potential effects of temperature increase on the metabolism of terrestrial ecosystems. According to Woodwell (1987), temperature increase may have little direct effect on

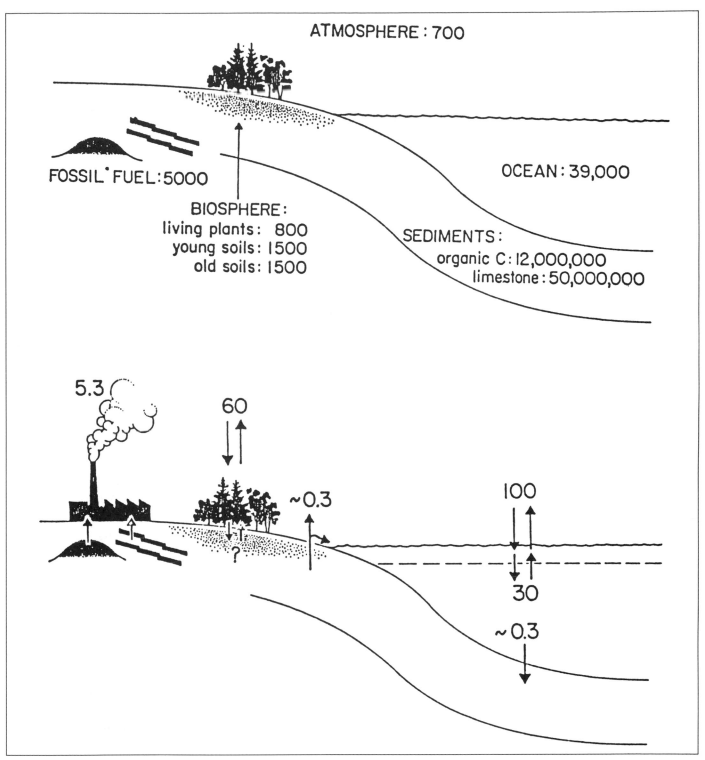

Figure 14–6. Major reservoirs of the global carbon cycle in gigatons (10^9 metric tons) of carbon [GT (C)] (top). Fluxes of carbon in Gt (C)/yr between the reservoirs (bottom). Source: Keeling, 1983.

rates of photosynthesis but commonly induces an increase in rates of respiration. Woodwell (1988) further estimates that the approximate global warming of 0.5° C during the past century may have increased respiration rates of soils in high latitudes by 5%–15%. Woodwell (1987) questions the "naivete" of

assuming that the growth rate of natural populations of terrestrial biota will be enhanced by increased CO_2 in the atmosphere. This assumption is based on agricultural studies in which other essential growth factors such as water and nutrients are maintained. Even under these ideal conditions, the stimulation

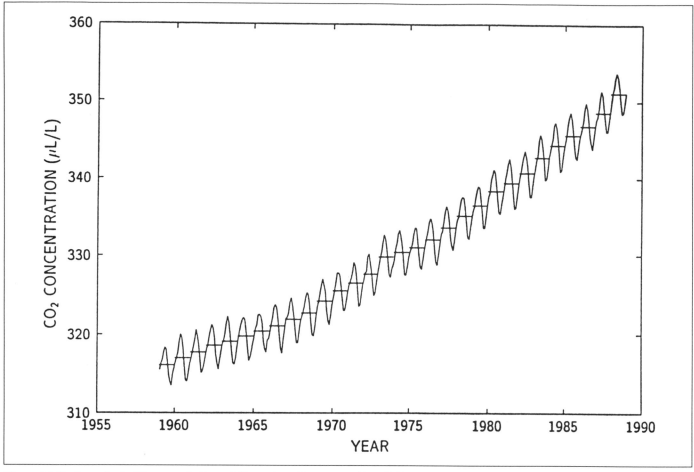

Figure 14–7. Atmospheric CO_2 concentrations at Mauna Loa Observatory, Hawaii. Lines connect monthly average values to form the oscillating curve. Horizontal bars indicate annual average concentrations. Source: Reprinted with permission from MacCracken et al, *Energy and Climate Change: Report of the DOE Multi-laboratory Climate Change Committee.* Chelsea, MI: Lewis Publishers, 1990. Copyright Lewis Publishers, a subsidiary of CRC Press, Boca Raton, Florida.

growth from 25%–30% increase in CO_2 is not more than a few percent (Woodwell, 1987). Carbon uptake and biotic storage through CO_2 fertilization is also questioned by ecologists who point out that growth of terrestrial ecosystems is probably limited by factors other than the CO_2 content of air (Detwiler & Hall, 1988).

The natural cycling of carbon is complex and shows the same kind of interacting feedback mechanisms that characterize the climate system. The difficulties geochemists and terrestrial ecologists face in trying to balance the global carbon cycle and to separate the effects of anthropogenic carbon sources on this cycle mirror the difficulties faced by climate scientists attempting to determine the effects of human activities on the ever-changing context of the climate system.

The presence of methane (CH_4) in the atmosphere was detected in the 1940s, and *in situ* measurements of atmospheric CH_4 were first made in the 1960s

(Bolle et al, 1986). The present-day worldwide concentration of tropospheric CH_4 is about 1.68 ppmv; it increased 11% between 1978 and 1987 (Blake & Rowland, 1988). Analyses of air extracted from polar ice cores show that the tropospheric concentration of CH_4 before 1700 was about 0.7 ppmv and has increased exponentially since then in remarkable correlation with the growth in world human population. This correlation indicates that the increase of tropospheric CH_4 is most likely caused by human activities, primarily agriculture (Bolle et al, 1986).

Methane is produced by microbial action during the mineralization of organic carbon under anaerobic conditions found, for example, in waterlogged soils and in the intestines of herbivorous animals (Bolle et al, 1986). The major sources of tropospheric CH_4 are enteric fermentation in livestock and insects, rice fields and natural wetlands, biomass burning, landfills, and gas and coal fields (Pearman & Fraser, 1988). The CH_4 budget is highly uncertain. Pearman and Fraser (1988) give an annual emissions range of 275–790 terragrams (Tg) (10^{12} g) for all sources, whereas Bolle et al (1986) quote sources in the range of 225–1,210 Tg/yr. This order-of-magnitude uncertainty is primarily caused by uncertainties in production rates, by uncertainties in the size of individual ecosystems that produce CH_4, and by the complexity of methanogenic food chains (Bolle et al, 1986).

The tropospheric concentration of nitrous oxide (N_2O) increased slowly from about 299 ppbv (parts per billion by volume) to 302 ppbv in the northern hemisphere between 1976 and 1980, a rate of increase of about 0.2% per year (Bolle et al, 1986). The major source of N_2O emissions is microbial processes in soil and water, and *both* denitrification (reduction of NO_3 to N_2) and nitrification (oxidation of NH_4 to NO_3) are sources of N_2O (Bolle et al, 1986). The total emission of N_2O from all source is estimated to be about 12–15 Tg N per year (Bolle et al, 1986).

The principal atmospheric sink for N_2O is photolysis and reaction with oxygen atoms in the stratosphere to form NO, which in turn plays a major role in controlling the concentration and distribution of O_3 in the stratosphere (Bolle et al, 1986). Nitrous oxide is also destroyed in soil and water, and reliable flux rates to the atmosphere from ocean and soil can be determined only by *in situ* measurements under natural conditions (Bolle et al, 1986).

Tropospheric sources of O_3 are transport from the stratosphere and *in situ* production involving nitrogen oxides (Wang et al, 1985). The lifetime of O_3 in the troposphere is on the order of a few weeks. Consequently, there is considerable spatial variation in its concentration, and it is difficult to measure global trends (Bolle et al, 1986). There is clear evidence of increase in the middle and high latitudes of the northern hemisphere during the last two to three decades, and the present rate of increase is about 1%–2% per year (Bolle et al, 1986). Areas in which extensive biomass burning occurs also show an upward trend in tropospheric O_3.

General circulation climate models. The radiative, chemical, and dynamic processes that control climate are complex and operate over a large range of spatial and temporal scales. Although statistical analysis of geological records provides evidence of a series of cyclic glacial (cooling) and interglacial (warming) periods during the past million years, the details of the physical processes and specific causes responsible for these periodic changes in climate cannot be deduced from diagnostic studies alone (Gates, 1985). Analyses of the climate of the past 100 years reveal both short-term variations and long-term trends but do not provide a cause-effect relationship between climate change and forcing mechanisms such as volcanic emission of aerosols, variations in atmospheric CO_2 concentration, or variations in solar radiation.

Because the specific causes of past climate changes remain uncertain, it is not possible to predict future climate by simply extrapolating recent or past trends (Luther & Ellingson, 1985). Moreover, empirical studies of the CO_2 climate change issue would require analogues of large climate change with a comparable space and time domain, and one of the major concerns raised by climatologists about present CO_2 emission rates is that projected temperature increases will be larger than any experienced during the era of human civilization (Schneider, 1989). For these reasons and others, the only known way to study the details of the many physical processes controlling climate is to construct mathematical models based on fundamental physical principles that govern these processes (Gates, 1985).

Climate models are mathematical representations of physical laws governing the behavior of the climate system and provide a rational physical basis for studying climate and climate changes. Mathematical climate models can simulate in a physically consistent manner not only the present climate but also how this climate will change in response to a change in a given forcing mechanism, such as a change in atmospheric composition. Such models, however, are inherently imperfect and cannot simulate the actual climate system (Schlesinger, 1983). This should not be surprising because, as Gates (1985) points out, there is as yet no unifying general theory of climate. The physical mechanisms of climate processes are imperfectly known, so mathematical models and numerical simulations of these processes are also imperfect and projections of climate change based on these models uncertain. Because models provide the only way to study and understand the intricacies of the climate system in a comprehensive way, an examination of climate modeling provides valuable insight into scientific knowledge underlying and the uncertainty surrounding projections of climate change.

Dickinson (1982) identified the following major elements of the climate system:

- radiative fluxes and budgets
- the atmospheric general circulation, including vertical convective heat transfer
- the atmospheric and surface hydrological cycle
- the ocean as a reservoir and conveyor of heat
- air-sea and air-land energy and momentum transfer processes.

Each of these elements, in turn, consists of many subcomponents that are in themselves complex processes. The major shortcoming of climate modeling so far has been the inability to link all of the elements into a comprehensive model of the climate system. Some of the elements have been modeled with varying degrees of simplification, and some initial attempts have been made to couple models of two or more elements.

The most progress has been in modeling radiative fluxes and the general circulation of the atmosphere, and this modeling has provided the basic understanding that eventual global warming at the surface by increasing concentrations of CO_2 is controlled by the interaction of radiative fluxes and processes regulating the vertical distribution of water vapor, clouds, and temperature in the atmosphere (Dickinson, 1982). Progress in modeling this aspect of the climate system may have placed undue emphasis on average global warming that according to Dickinson (1982) is "unlikely to be the most important consequence of future increases of CO_2." Average global warming has been emphasized "primarily because it is the only aspect of CO_2-induced climate change that can be modeled with some degree of confidence and also because other consequences are believed to depend on how much warming would occur" (Dickinson, 1982).

According to Dickinson (1986), general circulation models (GCMs) simulate the atmosphere in three dimensions and time by jointly solving hydrodynamic equations of motion for atmospheric winds and conservation of thermal energy and water vapor, including transport by the calculated atmospheric motions. The equation for thermal energy computes vertical radiative transfer of solar and long-wave radiation and relies on parameters for the redistribution of thermal energy and moisture by convection and turbulence. The equation for water vapor includes a source of evapotranspiration at the earth's surface and a sink through the formation of rain and snow.

Although the basic physical laws governing the time-dependent behavior of the atmosphere are well known and can be expressed in terms of equations for the conservation of momentum, heat, and moisture in a GCM (Gates, 1985), the atmosphere is only one component of the climate system, and a GCM depends on stipulated boundary conditions at the earth's surface (or coupling to other models that simulate these conditions) to simulate climate (Dickinson, 1986). Models providing the surface temperature of the ocean as well as models for sea ice and land-surface processes such as snow cover, soil moisture, and evapotranspiration are needed to augment atmospheric GCMs (Dickinson, 1986). The uncertainty associated with forecasts of climate change based on GCM simulations can be assessed by separating sources of practical and theoretical limitations that are internal and external to GCMs.

The most immediate practical limitation of GCMs is their computational complexity. The governing equations of GCMs are nonlinear, partial differential equations that can be solved only by numerical methods using the largest and fastest available computers (Schlesinger, 1983). In order to solve these equations numerically, the atmosphere is subdivided vertically into discrete layers and horizontally into discrete grid points. Conditions are specified at each point, and the governing equations are solved at each point using a time step approach and an interpolation scheme between grid points (Henderson-Sellers & McGuffie, 1987).

As a consequence of the computational complexity, the spatial resolution of GCMs is constrained by the speed and memory of available computers. To compute each of the basic atmospheric variables at each grid point requires storing, retrieving, recalculating, and restoring roughly 10^5 numbers (Henderson-Sellers & McGuffie, 1987). Increasing spatial resolution requires an increase in memory (linearly for vertical resolution and quadratically for horizontal resolution) and usually a reduction in the integration time step (Schlesinger, 1983). In current GCMs, horizontal resolution is limited to grid points spaced between 3° and 5° of latitude and longitude apart, vertical resolution to between 6 and 15 layers, and time steps to about 30 minutes (Henderson-Sellers & McGuffie, 1987).

The limitation of horizontal resolution to a few hundred kilometers has important theoretical consequences because it means that GCMs cannot resolve important physical climatic processes that occur on smaller spatial scales. These processes include turbulent transfer of heat, moisture, and momentum within the "boundary layer" between the earth's surface and the atmosphere and within the atmosphere by dry and moist convection; condensation of water vapor; formation of clouds and their radiative interaction; formation and dissipation of snow; physics of soil heat and moisture; and transfer of solar and long-wave radiation (Schlesinger, 1983).

These subscale processes are incorporated in the GCMs by the technique of *parameterization,* whereby they are resolved in terms of larger-scale, resolved behavior. For example, the processes that determine the net atmospheric heating rate (radiation and latent heating and the effects of conduction, convection, and diffusion) occur on scales too small to be resolved by GCMs and are represented in terms of large-scale distributions of temperature, pressure, and water vapor (Gates, 1985). Radiation is the most complex of these parameterizations because of the presence of radiative, chemical, and dynamic interactions among solar and long-wave radiation, greenhouse gases, clouds, and aerosols as described earlier. According to Gates (1985), the treatment of sub-scale processes is perhaps the most important feature of a climate model.

In addition to parameterizing subscale processes, boundary conditions and physical constants for the atmosphere-surface system must be stipulated. These boundary conditions and constants include the radius, surface gravity, and rotation speed of earth; the solar constant and orbital parameters for earth; total atmospheric mass and composition; thermodynamic and radiation constants of atmospheric gases and clouds; surface albedo; and surface elevation (Schlesinger, 1983). Although a significant share of the practical limitations of GCMs to stimulate climate change can be attributed to computational burden, a realistic portrayal of climate processes is also hindered by a lack of understanding about fundamental physical processes of the climate system and lack of observations and data.

Aside from the difficulty inherent in modeling the nonlinear fluid dynamics of the atmosphere along with its chemical and radiative interactions, a realistic climate model must also incorporate linkages and interactions among the atmosphere, oceans, sea ice, and the earth's surface albedo and hydrology (Dickinson, 1986), because these components are closely coupled in the climate system. These linkages and interactions are known as "feedbacks," and ability to realistically treat feedbacks determines the sensitivity of a model (how it responds) to external forcing, such as a doubling of the atmospheric CO_2 content.

Climate modeling results. The reliability of the model is tested by validation against present and past climates and against high-resolution submodels and observational data. Methods used to improve confidence in the results of model simulations of climate change include intercomparison of modeling results, sensitivity analysis, and uncertainty analysis.

The capability of GCMs to simulate the seasonal cycle is a critical test, because seasonal temperature changes on a hemispheric average are several times larger than the projected greenhouse warming and the temperature change from an ice age to an interglacial period (Schneider, 1989). GCMs are able to reproduce the large seasonal variations in surface air temperatures of the northern hemisphere continents as well as global average surface temperatures and pole-to-equator temperature differences (Dickinson, 1986). Verifications of GCMs against seasonal variations in precipitation, relative humidity, and other variables remain to be checked (Schneider, 1989). Validation of GCMs against seasonal changes does not address their adequacy in modeling essential but slower climate processes, such as deep ocean circulation or ice cover, which may have important effects on time-scales of decades to centuries during which the atmospheric concentration of CO_2 is projected to double (Schneider, 1989).

Validation of GCMs by simulating earlier climates is limited by the lack of sufficient data about these climates to allow unambiguous testing. Although GCMs have been used to simulate the last ice age (which occurred some 18,000 years ago), these simulations used mostly available data for boundary conditions and do not provide model validation (Dickinson, 1986). Hansen et al (1981) used a GCM simulation of the last ice age to measure, among other things, the feedback factor for such components of the climate system as water vapor and clouds, sea and land ice, and vegetation. Simulations of interglacial climates are even less developed, and the methods and criteria to verify the results are still being developed. The causal factors that influence past climates and estimates of the differences between past and present climates are too uncertain to permit judgment of model reliability by simulations of past climates (Webb & Wigley, 1985).

Intercomparisons of modeling results have been used as an implicit measure of consensus among atmospheric scientists on the probable effect of increased concentrations (usually a doubling from pre-industrial estimates) of atmospheric CO_2. The often-quoted conclusion by the Climate Research Board (NRC, 1979b) of the National Academy of Sciences (NAS) that the "equilibrium surface global warming due to doubled CO_2 will be in the range of 1.5 C to 4.5 C, with the most probable value near 3 C," is based on a comparison of results of GCM simulations that are "basically consistent and mutually supporting." The board's conclusion was reaffirmed in another influential NAS study by the CO_2/Climate Review Panel (NRC, 1982b) that did not find "any new results that necessitate substantial revision of the conclusions of the [Climate Research Board] study."

Comparisons of model results also allow comparisons of how physical climate processes are incorporated in the models and provide a way to evaluate sources of uncertainty in the modeling. As Hall (1985) points out, analyzing and reconciling the disagreement among modeling results are essential to establishing the reliability of climate models. Because GCMs are the most complete analytical tools available to study climate change, many comparisons of the results of GCM simulations of CO_2-induced climate changes have been made. Dickinson (1986) compared the results of three of the "more realistic" GCMs, those that have annual cycles, "realistic continents and orography," and "adequate horizontal and vertical resolution." GCMs fulfilling Dickinson's criteria are those developed by the National Oceanographic and Atmospheric Administration's Geophysical Fluid Dynamics Laboratory (GFDL), the National Center for Atmospheric Research (NCAR), and the National Aeronautics and Space Administration's Goddard Institute for Space Sciences (GISS). All three models show a global average surface temperature increase of about 4° C, the NCAR (Washington & Meehl, 1984) and GISS (Hansen et al, 1981) models for a steady-state doubling of CO_2 but the GFDL (Manabe & Stouffer, 1980) model for a steady-state quadrupling of CO_2.

The GFDL and NCAR models, and—to a lesser extent—the GISS model, show the largest warming in high latitudes of the northern hemisphere in winter. The models, however, show significant differences in the geographic distribution of summer and winter maxima in surface warming. The geographic and seasonal extent of sea ice is an important determinant of model response, because sea ice persists into spring and summer when solar irradiance is largest and largely determines the ice-albedo feedback in a GCM. All three models show significant spatial and temporal differences between observed and simulated extent of sea ice as well as differences in the models' simulations of sea-ice extent. Another important climate feature that is also difficult to model is cloudiness. The GFDL model does not calculate clouds but prescribes them as independent of longitude and variable with latitude based on observation. The GISS model calculates larger changes in cloudiness than does the NCAR model, particularly low-level clouds in high latitudes and high-level clouds in middle latitudes; clouds are responsible for a large amplification of warming in the GISS model.

Changes in soil moisture represent the difference between precipitation and losses caused by runoff and evaporation. The changes are an important feature of climate change calculated by GCMs. All three models show a significant increase in precipitation in high latitudes but differences in the geographic and season gains and losses of soil moisture. The GFDL model shows summer drying of soils in the middle and high latitudes of the northern hemisphere. The NCAR model generally shows an increase in soil moisture for northern hemisphere areas outside of the tropics with the largest increase occurring in spring and winter. The GISS model does not show substantial drying in the middle-to-upper latitudes in spring and summer.

Schlesinger (1983) compared the results of eight GCMs, including the GFDL model assessed by Dickinson and discussed earlier, and GCMs developed at Oregon State University (OSU), the fourth of the four U.S. institutions that have developed the principal GCMs (and various refinements of these GCMs) used for climate change forecasting (Gates et al, 1981). An important conclusion reached by Schlesinger is that the "control" climate (based on the reference CO_2 concentration that is doubled for the simulated climate change) is the critical determinant of model-dependent results. His general conclusions from comparing the results of these models are that the warmer the simulated control climate: (1) the larger the CO_2-induced warming and increase in precipitation rate, and (2) the smaller the control precipitation rate and, therefore, the larger the percentage increase in CO_2-induced precipitation rate. Schlesinger also shows that it is essential that the control and CO_2-doubling simulations reach equilibrium before averages are taken.

According to Schlesinger and Mitchell (1985), the state-of-the-art in GCMs still means imperfect simulation of the present climate, because these models incorporate dubious treatments of climate processes and simulations of climate change by different models show many quantitative and even qualitative differences. Not all of the simulations could be correct, and, indeed, they could all be incorrect. For Schlesinger and Mitchell, attaining confidence in GCM projections of CO_2-induced equilibrium climate change requires that these models "correctly simulate at least one known equilibrium climate, with the present climate being the best choice because of the quantity, quality, and global distribution of contemporary instrumental observations." However, evaluating the accuracy of simulations of the present climate is difficult because, among other things, the observed climate as well as the simulated climate may not represent their corresponding equilibrium climates, and many of the measurements of the observed climate—such as precipitation over oceans and soil moisture—are of poor quality. Even if GCMs could simulate the present climate perfectly, doubts would remain whether GCMs could simulate a climate different from the present one because the few reconstructions of past climates available may not be of sufficient quality to provide a meaningful test of the capabilities of GCMs.

Schlesinger and Mitchell (1985) conclude that there is an "inherent limitation" on the capability of scientists to validate the accuracy of GCM simulations of CO_2-induced climate change, which in turn means there will always be doubts about the accuracy of simulations of future climate change. Confidence in GCM simulations can be increased by better understanding of the dissimilarities, particularly feedbacks and parameterizations, among the state-of-the-art GCMs; by systematically validating the parameterizations of subgrid physical processes; by developing physically based parameterizations of clouds; and by developing coupled atmosphere-ocean GCMs. GCMs will remain critically dependent on parameterizations of physical climate processes, because they span 14 orders of magnitude from the planetary (10^7 m) scale to the cloud microphysical (10^{-6} m) scale. Schlesinger and Mitchell point out that even a thousandfold increase in computer speed would increase the spatial resolution of GCMs by only one order of magnitude.

Detecting climate change. Isolating and identifying climate change induced by emissions of CO_2 present both theoretical and practical difficulties. Climate can be defined as "the statistical description of the mean state (including the variability) of the atmosphere, ocean, ice, and land surface in a specific region over a specified time period" with the conventional time period for averaging being at least 30 years (Gates & MacCracken, 1985). Climate *variability* is generally the variations of climatic parameters *within* a typical averaging period and can be measured by the standard deviation of continuous elements, such as temperature and pressure or the frequency spectrum of observed variations (Hare, 1985). Climate *change* is defined by Gates and MacCracken (1985) as the "difference between climate statistics evaluated over two similar time intervals." These definitions reveal some of the difficulties of detecting CO_2-induced climate change, of which the following four are the most salient:

- A "change" in climate is as much a problem of definition as detection, because it depends on the two time intervals selected for comparison.
- If climate is a statistical description of a mean state, researchers' understanding of climate change depends on the availability, quality, temporal extent, area covered, and interpretation of data.
- Because climate itself is variable, detecting climate change requires the difficult task of isolating and identifying change as distinct from climate variability.
- Attributing a particular change in a changing system to a particular forcing mechanism among many other interacting mechanisms requires a detailed understanding of climate processes.

The observation record of past climate provides a baseline against which a future, CO_2-induced change can be measured. This baseline provides information about the natural variability of climate, the "noise" level above which anthropogenic effects must be detected and measured. The record of past climates also provides insights into the way complex feedback mechanisms respond to forcing as well as the way these mechanisms may have responded to past change in atmospheric CO_2 (Wigley et al, 1986).

On the time scale of 10^3–10^5 years, a fairly regular series of glacial periods becomes apparent in the climate record about 1.7 million years before the present (Wigley et al, 1986). Cycles of glacial/interglacial periods occur approximately every 100,000 years, caused primarily by changes in earth's radiation budget (amount and/or spatial distribution) due to changes in earth's orbital parameters. Evidence from deep sea sediments links changes in earth's surface temperature to periodicities in earth's orbital parameters.

Large-scale changes in CO_2 concentration at the glacial/interglacial time scale have been measured using fossil CO_2 in bubbles from ice cores and inferred from carbon isotope data from deep-sea sediments (Wigley et al, 1986). At the peak of the last glaciation (20,000–15,000 years ago), atmospheric CO_2 concentration was about 200 ppm, which indicates that CO_2 concentration—meditated by ocean processes—may be an important positive feedback that amplifies climate change caused by orbital effects (Wigley et al, 1986).

During the last 2000 years, the largest changes in the climate record occurred during the Medieval Warm Epoch (about 800–1200 AD), which may have been restricted to the North Atlantic Basin region, and during the Little Ice Age (about 1400 = n1800 AD), which seems to have been more global in effect. The cause of the Little Ice Age is not known, although changes in solar activity and in the frequency of explosive volcanic activity have been proposed, and some evidence exists for depressed atmospheric concentrations of CO_2 and CH_4 during this period (Wigley et al, 1986).

On the 100-year time scale, many warming and cooling episodes have occurred, but these events are observable only through indirect or proxy evidence that is local and often poorly dated. Evidence provided by vegetation changes and glacial fluctuations and isotopic data from ice cores and other proxy climate indicators show that climate change on a regional scale has been substantial. This evidence and data, however, do not show whether the global mean climate changed. According to Wigley et al (1986), climatologists do not even know whether mean global temperature has changed since about 6000 years ago, the end of the early Holocene (most recent geologic time) warm period.

Only after 1850 did instrumental data become widely available. Data on precipitation, temperature, and mean sea level pressure are available, but confidence in such data depends on the particular variable, size of the area covered, and on location. Even today, spatial coverage for these variables is poor over the oceans, especially in the southeastern Pacific and in the high southern latitudes. Upper air data, which are essential because climate is a three-dimensional process, exist only for the period since the late 1940s (Wigley et al, 1986). Because the uncertainty in the expected signal from CO_2 and other trace gases decreases with increasing spatial scale, large area averages are most useful for both temperature and precipitation data.

Precipitation is an important climate parameter and is a determinant of soil moisture that is much more important than temperature in determining where plant life can grow (Kellogg & Zhao, 1988). Changes in precipitation amounts and patterns that may accompany a change in mean global surface temperature is a critical concern for agriculture and other human activities. There is, however, little agreement on the normal level of mean global precipitation, let alone on any trend revealed in the past climate record (Ellsaesser et al, 1986).

Another indicator of climate change is the response of the cryosphere, whose components include snow cover, sea-ice, land ice, and permafrost (Barry, 1989). Records for average snow cover in the northern hemisphere are quite limited, and year-to-year changes of snow cover in the northern hemisphere can vary by as much as 15%, which makes it difficult to determine any recent trend or to use changes in snow cover as an indicator of future climate regimes (Ellsaesser et al, 1986).

Sea level change should accompany global warming because of thermal expansion of seawater and melting of land ice. A uniform warming of a 2.5 mi (4 km) deep ocean by 1° C should increase sea level by about 2 ft (60 cm) (Ellsaesser et al, 1986). Records from tidal gauges available from 1880 onward indicate that the global mean sea level has risen between 1.0 and 1.5 mm/yr, which is consistent with estimates of alpine glacier melting and thermal expansion of the upper layers of the oceans (Gornitz & Solow, 1989). As with other climate indicators, however, sea level measurements show great variability in the pattern of change across relatively small distances, and there is strong evidence of nonlinear behavior in the data from stations with the longest continuous records (Gornitz & Solow, 1989).

Surface air temperature is the best-documented climate variable, and the signal for temperature is most likely to be better established than for other variables. The first reliable estimates of large-scale area-average temperatures were made during the 1960s. More recent estimates were published during the late 1970s and early 1980s. Most of these estimates are based on data from individual land stations and have little coverage across the ocean. Individual station records may be influenced by various sources of nonclimatic inhomogeneities resulting from changes in measurement location, instruments and/or exposure of these instruments during measurements, observation time, and computational methods to produce daily and monthly averages.

Land-based records for the northern hemisphere from 1880 to 1980 show the general trends of warming to about 1940, cooling to the mid-1960s, and warming since around 1970. Temperature data taken over the oceans in both hemispheres show general warming from the late 19th century to around 1940, cooling to the mid-1960s, and warming since the mid-1960s with this warming delayed in the northern hemisphere compared to the southern hemisphere. The reason for this hemispherical difference in warming is not known (Wigley et al, 1986). However, Wigley (1989) explores the possibility that increasing emissions of sulfur dioxide in the northern hemisphere from industrial and urban sources could lead to increases in nonsea-salt sulfates, which in turn may act as cloud condensation nuclei. These nuclei affect the albedo of clouds and could account for delayed warming of the northern hemisphere relative to the southern hemisphere.

Although there have been significant global cooling and warming trends during the past century, regional trends tend to differ markedly from each other and from global or hemispheric mean trends (Wigley & Schlesinger, 1985). For example, between 1917 and 1939, most land areas in the northern hemisphere showed strong warming, especially in high northern latitudes, but central northern Canada and part of central Asia cooled. During the subsequent hemispheric cooling period between 1940 and 1964, cooling was strongest in high latitudes, but again there were large areas in which temperatures increased. In addition, these areas were not the same areas that showed cooling during the preceding hemispherical warming trend. During the most recent warming, the pattern of regional differences shows little similarity to the earlier warming, and a significant part of the Arctic has shown a cooling trend (Wigley et al, 1986).

Even if the problems of spatial inhomogeneity and gaps in the spatial coverage of the data are put aside, noise in the record of past temperature on the time scale of decades or more makes it difficult to detect an unequivocal signal of mean surface temperature increase caused by CO_2 emissions. The observed data show marked hemispheric-scale fluctuations in both directions on all time scales. The most striking fluctuation is the cooling episode between about 1940 and 1970, during which the northern hemisphere cooled by about 0.2° C (Wigley & Schlesinger, 1985). An explanation of this cooling anomaly is important, because if estimates CO_2-induced warming by climate models are correct, this cooling must be caused either by an external factor other than greenhouse gases or to a major internal climate oscillation. Changes in stratospheric aerosol loading and/or changes in solar irradiance have been suggested as causal factors, but not enough is known about these two forcing mechanisms to prove these suggestions. According to Wigley et al (1986), it is unlikely that this cooling episode will be satisfactorily explained, because the appropriate data are not available.

Climate variability. *Climate variability* is the variations of climatic parameters within a typical averaging period (Gates & MacCracken, 1985). Variability is measured as deviations from the mean within a given averaging period, whereas change is measured in terms of a difference of mean values between averaging periods. Historical observations of climate parameters, such as mean annual surface temperature and annual total precipitation, for a given area within a given period on the time scale of a decade to a century, typically show large variations in values.

Changes in regional climate variability rather than changes in the global mean are more readily perceived and felt. The impacts of changes in climate variability may be as great if not greater than the impacts of changes in climate means (Katz, 1988). In practical terms, the impacts of climate change that will be more immediately experienced will be changes in climate variability—extreme events such as droughts, heat waves, cold spells, and flooding occurring more frequently, lasting longer, and affecting larger areas.

In addition to changes forced by external factors, climate may vary on monthly, seasonal, interannual, decadal, or longer time scales without apparent changes in external forcing factors (Gates & MacCracken, 1985). This "natural" variability is caused by interactions among many physical processes in the atmosphere, ocean, and land surface that are "internal" to the climate system. These interactions represent feedback loops, such as those among clouds, radiation, and temperature, that maintain climate variables within certain ranges. The variability of each climate variable has a characteristic amplitude that depends on how the variable is involved in the major feedback processes and how sensitive those processes are to other internal factors (Gates & MacCracken, 1985). On time scales of up to hundreds of years, variability of climate caused by internal factors can be comparable in magnitude to changes caused by external forcing factors (Gates & MacCracken, 1985).

Scientific consensus. Scientific consensus on detection of climate change induced by an increasing atmospheric concentration of CO_2 hinges on two issues: (1) the capability of current state-of-the-art climate models to define the CO_2-induced signal of climate change, and (2) the adequacy of climate data to enable a statistically significant identification of change in a climate parameter (Gates & MacCracken, 1985). Consensus among climate scientists depends on achieving a satisfactory degree of correspondence between what is understood of the physical processes of climate as embodied in climate models and what the observed data reveals about climate processes.

The observed temperature record shows that the global mean temperature has increased about 0.5° C since the late 19th century, with large fluctuations on the decadal and shorter time scale superimposed on this overall warming trend (Wigley et al, 1986). If the results from climate modeling are correct, there should have already been a substantial increase in global mean surface temperature caused by increasing emissions of CO_2 and other greenhouse gases. Wigley et al (1986) estimate that between 1900 and 1985, CO_2 and other trace gases should have warmed the globe by 0.3–1.1° C. Although the observed temperature trend is in agreement with the lower end of temperature increases projected by climate models (1.5–4.5° C), Wigley et al (1986) state that a short answer to the question of whether a statistically significant anthropogenic greenhouse effect has been detected is "no."

Given the statistical uncertainties in estimating temperature trends, uncertainties in data quality, and gaps in data coverage, the post-1900 warming could be anywhere within the range of 0.3–0.7° C, which may include a substantial component due to natural variations (Wigley et al, 1986). Given the limitations of historical observations and uncertainties in modeling climate change, it is not surprising that most climate scientists choose to err on the side of caution. The prevailing opinion among climate scientists is that although the observed global mean temperature record during the past 100 years or so is not inconsistent with climate change attributable to increasing atmospheric concentrations of CO_2 and other greenhouse gases, it is not yet possible to confirm a causal relationship between increases of these gases and changes in the climate record from the pre-industrial period to the present.

Impacts, Mitigation, and Adaptation

Impacts of climate change. In assessing the potential impacts of climate change, as in detecting the effects of greenhouse gas emissions on average global surface temperatures, one must first consider the background "noise" within which additional incremental changes may take place. In the case of impacts of climate change, this noise stems from the global-scale alteration of terrestrial, biological, hydrological, and atmospheric systems by humans during the last 10,000 years. For example, the human population, which is increasing by 1 million every 5 days, uses more than 35% of the total land surface of the earth to harvest—directly or indirectly—plant-derived materials. Agriculture now accounts for about 10% of primary terrestrial production (IPCC/IGBP, 1992). Furthermore, given the uncertainties implicit in climate models and the lack of spatial resolution in these models as discussed earlier, assessing impacts caused by climate change (in turn due to anthropogenic emissions of greenhouse gases) quickly becomes a speculative enterprise.

With explicit understanding of the high degree of uncertainty involved, MacCracken et al (1990) made some qualitative estimates of the potential effects of climate change (caused by a doubling of CO_2) for agriculture, water resources, forestry, energy, air quality, human health, and other sectors. Even qualitative assessments of the potential impacts of climate change on all of these sectors are uncertain because of the lack of understanding and information about the regional consequences of global climate change. For example, the impacts on agriculture, water resources, and forestry sectors all are uncertain due to the uncertainties in the potential regional effects of climate change on temperature, precipitation, and other weather variables. Impacts on all three sectors will depend greatly on the rapidity of climate change, the effect of climate change on the seasonality of rainfall, and the human and societal response to climate change.

The potential impacts on energy, air quality, and human health are equally uncertain. In the energy sector, climate change will directly affect the amount and way energy is used. Renewable energy (e.g., hydroelectricity, wind, biomass, photovoltaic, etc.) will be more fundamentally affected, compared to fossil and nuclear energy because these energy sources depend on solar energy in both its direct (as solar radiation) and indirect (as stored solar energy in biomass and as hydroelectric energy from the hydrological cycle that is driven by solar energy) forms. Air quality, especially in nonattainment areas where meteorological conditions already play a strong role in determining the severity of air pollution, will be affected in yet unknown ways by climate change.

Scientists have not established direct links between potential climate change and human health although potential impacts can be inferred from established correlations between health conditions and weather variables or seasonality. For example, the severity and frequency of summer and winter temperature extremes due to climate change could increase the risk to those already prone to heat stress and cardiovascular problems. The incidence and severity of diseases linked to plants, fungi, insects, and climate and seasons in general (influenza, asthma, encephalitis, etc.) could be affected by climate change. In all of these areas, the uncertainties are better known than the potential impacts.

Mitigation and adaptation strategies. Mitigation strategies to address the potential effects of climate change are often distinguished from "adaptation" strategies. For example, the National Academy of Sciences (NAS) (1991) defines *mitigation options* as those that "can take effect prior to the onset of climate change and slow its pace" and *adaptation options* as those whose "effects come into play after climate has changed." The NAS report (1991) lists two categories of mitigation options: (1) those that eliminate or reduce greenhouse gas emissions, and (2) those that compensate for increased greenhouse gas emissions by altering the energy flows of the earth-atmosphere system. Some of the options listed by the NAS in category 2, such as placing mirrors in orbit to reflect incoming sunlight or placing soot in the lower stratosphere, seem impractical and environmentally risky. Of the category 2 options, only reforestation and, perhaps, stimulation of marine phytoplankton to absorb CO_2 by iron fertilization, seem practical. On the other hand, many category 1 options are feasible, particularly attaining greater energy efficiency throughout our national economy and in making greater use of nonfossil forms of energy, especially renewable energy. The NAS (1991) report discusses these options in detail.

Adaptation strategies as defined by the NAS require behavioral, institutional, and infrastructure changes after the onset of climate change. Such changes in just the agricultural sector would include modification of farming practices to accommodate changes in temperature average, seasonality, and extremes; modification of agricultural markets and crop insurance policies; and building of reservoirs and changes in water release practices. Adaptation strategies also require foresight and planning for climate change. For example, again in just the agricultural sector, more research may be needed to understand tolerance of crops to climate change and to develop varietals that are more resistant to changes in temperature averages, extremes, and seasonality as well as to changes in soil moisture. These considerations are discussed in detail in the NAS (1991) report.

Economic Strategies and Responses

The beginning of this section posed four economic questions relative to global warming/greenhouse gas emission. It is worthwhile to reconsider these questions and provide preliminary answers.

Are abatement policies necessary? As the previous discussion of the science of global warming illustrates, the need for abatement policies does not inexorably follow from the known scientific facts. Furthermore, the potential mitigation and adaptation strategies involve substantial political will to force the required behavioral, institutional, and infrastructure changes. It can be (and has been) argued that many of the proposed changes, e.g., greater energy efficiency and increased use of nonfossil (renewable) energy source, are meritorious and cost-effective *unrelated* to the greenhouse gas emission controversy (OECD/IEA, 1992).

What are the potential costs of abatement measures? The OECD has developed a multisector, multiregion, dynamic model for the evaluation of emission abatement strategies. This model is known as GREEN (General Equilibrium Environmental Model). A summary of the predictive output from GREEN (Burniaux et al, 1992) is:

- Global emissions will grow by 2% per year during 1990–2050 if no action is taken.
- Global emissions are unlikely to fall by the year 2000 (assuming stabilization of OECD emissions at the 1990 level) unless non-OECD countries also stabilize emissions at the 1990 level.
- OECD countries reduce emissions to 80% of 1990 volumes to the year 2000 and then stabilize emissions; non-OECD countries increase emissions to 150% of 1990 volumes by 2010 and then stabilize; the required carbon taxes necessary to achieve this theoretical scenario are $250 per ton in 2005 based on 1985 prices (exchange rates) and decreasing to $116 per ton by 2050. Non-OECD carbon tax

rates would vary from $450 to $50 per ton in the years 2005 and 2050, respectively. Significantly, the average annual real income loss to the world economy between 1995–2050 would be 2.1% (Burniaux et al, 1992).

- Elimination of all existing energy subsidies would probably lower global CO_2 emissions by 20% of 2050 baseline projections. This reduction would be attributable to non-OECD regions.

What are the consequences of using market incentives? The two major market incentives for controlling CO_2 emissions are (1) emission taxes and (2) a system of tradable permits that allow the purchase of rights to emit various amounts of greenhouse gases. The U.S. EPA currently encourages the tradable permits system within the framework of the 1990 Clean Air Act Amendments (see Chapter 3, United States Legal and Legislative Framework).

Preliminary GREEN results indicate that the tradable permits system would achieve significant improvement in emissions at a substantially lower economic cost than specified mandatory regional reductions.

The use of carbon taxes is complex and politically sensitive. In the U.S. 1993 budget debate, carbon taxes were politically unacceptable and were withdrawn by the Clinton administration. Worldwide, carbon taxes have had variable impacts that typically provide (1) exemptions for key industrial sectors and (2) attempt to offer tax relief (i.e., revenue neutrality) to offset the negative impacts of the carbon taxes. Countries with carbon taxes includes Denmark, Finland, Netherlands, Norway, and Sweden.

What are the economics of negotiation? As indicated by the predictive output from GREEN, the "free rider" effect is substantial. The free rider effect for nonsignatory or nonparticipatory countries manifests itself in two ways: (1) direct effect—free-riders incur no costs but receive the benefits produced by others and (2) price effect—abatement/energy conservation measures by signatories causes a fall in fossil fuel prices (secondary to increased supply) and produces an incentive for free-riders to increase fossil fuel consumption (and thereby increase emissions). Unfortunately, the solution to the free-rider problem is not self-evident and requires further study (OECD, 1992b).

The potential economic impacts of abatement/mitigation strategies are substantial. Nevertheless, a large number of countries have made commitments

to reduce or stabilize emissions. Within the OECD members, these approaches have been undertaken:

- Unilateral, unconditional commitments to targets—Austria, Canada, Iceland, Luxemburg, and Switzerland—elaborated by funded plans of action and supported by carbon or CO_2 taxes—Netherlands, Denmark
- Unilateral nonbinding commitments to targets—Austria, Germany, Italy, New Zealand, United Kingdom—and two supported by carbon tax—Finland, Norway
- Target adjusted for need for economic growth—Spain
- Conditional targets based on per capita emissions—France and Japan
- Commitment to a set of policies which will stabilize emissions—United States
- Regional targets balanced out by allowing economic-growth-adjusted targets with offsets from other member countries—European Union
- Targets not specified but implicit in European Union—Belgium, Greece, Iceland, Ireland, Portugal—and supported by carbon tax—Sweden (OECD/IEA, 1992).

The long-term economic impacts and emission reductions of these commitments are unknown. However, it is clear that some combination of economic incentives (carbon taxes, tradable permit systems) and reduction in energy subsidies is required if the 1992 Framework Convention in climate change is to have any long-lasting substance.

The global warming/greenhouse gas debate is complex and emotional. Because the science does not provide an unambiguous direction, the debate has quickly become framed in political-economic terms that frequently pit the developed against the developing nations. This is unfortunate, because everyone lives and relies on the atmosphere "commons." The resolution of the global warming controversy is unlikely; however, the debate has moved in the direction of action and intervention. The type, structure, and results of these actions/interventions are clearly in evolution.

SUMMARY

The scope and potential impact of the global issues discussed in this chapter are truly revolutionary. The advocates of the "sustainable development" paradigm are putting forth an agenda for ". . . fundamental change in existing policies and practices"

(WRI, 1992). The sustainable development movement will require simultaneous changes in economic, human, environmental, and technological activities. The breadth of the proposed change is of such great magnitude that it is difficult to foresee complete adoption of the "sustainable development" paradigm in the immediate future. However, the extent of the global environmental problems and the lack of easy short-term fixes illustrate the need for examining and rethinking previous practices. To paraphrase DOE's Patrino, there's only one earth and sometimes humans have to be conservative in the actions they consider and take.

REFERENCES

Arrhenius S. On the influence of carbonic acid in the air on the temperature of the ground. *Phil Mag* 41 (April 1896):237–276.

Ausubel JH. Historical note. *NRC* (1983): 488–491.

Barbier EB. *Economics, National Resources, Scarcity and Development: Conventional and Alternative Views*. London: Earthscan Publications, Ltd., 1989.

Barry RG. Observational evidence of changes in global snow and ice cover. In U.S. Department of Energy, *DOE Workshop on Greenhouse-Gas-Induced Climate Change: A Critical Appraisal of Simulations and Observations*, Amherst: Univ. of Massachusetts, May 8–12, 1989.

Blake DR, Rowland FS. Continuing worldwide increase in tropospheric methane, 1978 to 1987. *Science* 239 (March 4, 1988):1129–1131.

Bolin B, Doos BR, Jager J, Warrick RA (eds). *The Greenhouse Effect, Climatic Change, and Ecosystems*. SCOPE 29. Chichester, U.K.: John Wiley & Son, 1986.

Bolle H-J, Seiler W, Bolin B. Other greenhouse gases and aerosols. In Bolin B, Doos BR, Jager J, Warrick RA (eds), *The Greenhouse Effect, Climatic Change, and Ecosystems*. SCOPE 29. Chichester, U.K.: John Wiley & Son, 1986, 157–203.

Broderick AJ, Oliver RC. The supersonic transport. In Bower FA, Ward RB (eds). *Stratospheric Ozone and Man*, 2 vols. Boca Raton, FL: CRC Press, 1981, 2:14.

Brown LR. *State of the World 1992*. New York: W.W. Norton, 1992.

Bryson RA. A perspective on climate change. *Science* 184 (17 May 1974):753.

Burniaux JM, Martin JP, Nicoletti G, et al. *The Costs of Reducing CO₂ Emissions: Evidence from GREEN*. Working Papers Nos. 115 and 116. Paris: OECD Economics Department, 1992.

Callendar GS. Can carbon dioxide influence climate? *Weather* 4 (1930):310–314.

Carbon Dioxide Assessment Committee (CDAC). Synthesis. *NRC* (1983):5–86.

Chamberlin TC. An attempt to frame a working hypothesis of the cause of glacial periods on an atmospheric basis. *Journal of Geology* 7(1899):545–584.

Conservation Foundation. *Implications of Growing Carbon Dioxide Content of the Atmosphere*. New York: The Conservation Foundation, 1963.

Detwiler RP, Hall CAS. Tropical forests and the global carbon cycle. *Science* 239 (January 1, 1988): 42–47.

Dickinson RE. Modeling climate changes due to carbon dioxide increases. In Clark WC (ed). *Carbon Dioxide Review: 1982*. New York: Oxford Univ. Press, 1982, 101–142.

Dickinson RE, Cicerone RJ. Future global warming from atmospheric trace gases. *Nature* 319 (January 9, 1986):109–115.

Dickinson RE. How will climate change? In Bolin B, Doos BR, Jager J, Warrick RA (eds), *The Greenhouse Effect, Climatic Change, and Ecosystems*. SCOPE 29. Chichester: John Wiley & Son, 1986, 207–270.

Dotto L, Schiff H. *The Ozone War*. Garden City, NJ: Doubleday, 1978.

Ellsaesser HW, MacCracken MC, Walton JJ, et al. Global climate trends as revealed by the recorded data. *Rev of Geophys* 24 (November 1986):745–792.

Food & Agriculture Organization (FAO) of the United Nations, "Second Interim Report on the State of Tropical Forests." 10th World Forestry Congress. 1991.

Gammon RH, Sundquist ET, Fraser PJ. History of carbon dioxide in the atmosphere. In Trabalka JR (ed), *Atmospheric Carbon Dioxide and the Global Carbon Cycle*. Washington DC: U.S. Dept. of Energy (DOE/ER-0239), 1985, 25–62.

Gates WL. Modeling as a means of studying the climate system. In MacCracken MC, and Luther FM (eds), *Projecting the Climate Effects of Increasing Carbon Dioxide*. Washington DC: U.S. Dept. of Energy (DOE/ER-0237) (1985), 57–79.

Gates WL, Cook KH, Schlesinger ME. Preliminary analysis of experiments on the climatic effects of increased CO_2 with an atmospheric general circulation model and a climatological ocean. *J of Geophys Res* 86 (1981):6385–6393.

Gates WL, MacCracken MC. The challenge of detecting climate change induced by increasing carbon dioxide. In MacCracken MC, Luther FM (eds), *Detecting the Climatic Effects of Carbon Dioxide*. Washington DC: U.S. Dept. of Energy, DOE/ER-0235, (1985), 1–12.

Glantz MH, Robinson J, Krenz ME. Climate-related impact studies: a review of past experiences. In Clark W (ed), *Carbon Dioxide Rev*. New York: Oxford Univ. Press, 1982.

Gleason JF, Bhartia PK, Herman JR. Record low global ozone in 1992. *Science* 260 (23 April, 1993): 523–526.

Goodland R, Ledee G. Neoclassical economics and principles of sustainable development. *Ecologize Modeling* 38 (1987):36.

Gornitz V, Solow A. Observations of long-term tide-gauge records for indications of accelerated sea-level rise. In U.S. Department of Energy, *DOE Workshop on Greenhouse-Gas-Induced Climate Change: A Critical Appraisal of Simulations and Observations*. Amherst: Univ. of Massachusetts, May 8–12, 1989.

Hall MCG. Estimating the reliability of climate model projections—steps toward a solution. In MacCracken MC, Luther FM (eds), *Projecting the Climate Effects of Increasing Carbon Dioxide*. Washington DC: U.S. Department of Energy (DOE/ER-0237), 1985, 337–366.

Hansen J, Johnson D, Lacis A, et al. Climate impact of increasing atmospheric carbon dioxide. *Science* 213 (28 August 1981):957–966.

Hare FK. Climate variability and change. In Kates RW, et al (eds), *Climate Impact Assessment: Studies of the Interaction of Climate and Society*. SCOPE 27. New York: John Wiley & Sons, 1985.

Henderson-Sellers A, McGuffie K. *A Climate Modelling Primer*. Chichester, U.K.: John Wiley & Sons, 1987.

Houghton RA, Schlesinger WH, Brown S, et al. Carbon dioxide exchange between the atmosphere and terrestrial ecosystems. In Trabolka JR (ed), *Atmospheric Carbon Dioxide and the Global Carbon Cycle*. Washington DC: U.S. DOE (DOE/ER-0239), 1985:113–140.

Howard R, Perley M. *Acid Rain: The Devastating Impact on North America*. New York: McGraw-Hill, 1980.

Intergovernmental Panel on Climate Change (IPCC)/International Geosphere-Biosphere Program (IGBP), *Global Change: Reducing Uncertainty*. Stockholm: Royal Swedish Academy of Sciences, 1992.

IGBP. *Climate Change: The IPCC Scientific Assessment*. Cambridge, UK: Cambridge University Press, 1990.

IPCC. *Scientific Assessment of Climate Change 1992 Supplement*. UNEP/World Meteorological Organization, Geneva, 1992.

Johnston H. Reduction of stratospheric ozone by nitrogen oxide catalysts from supersonic transport exhaust. *Science* 17 (6 August 1976):517–522.

Jones T. The economics of climate change. *OECD Observer* 179 (December 1992/January 1993):22.

Katz RW. Statistics of climate change: implications for scenario development. In Glantz MH (ed), *Societal Responses to Regional Climate Change: Forecasting by Analogy*. Boulder, CO: Westview Press, 1988, 95–112.

Keeling CD. The global carbon cycle: what we know and could know from atmospheric, biospheric and oceanic observations. In *Proceedings: Carbon Dioxide Research Conference: Carbon Dioxide, Science and Consensus*, September 19–23, 1982, Berkeley Springs, WV: DOE, CONF-820970, 1993.

Kellogg WW, Zhao Z. Sensitivity of soil moisture to doubling of carbon dioxide in climate model experiments. Part I: North America. *J of Climate Change* 1 (April 1988):348–366.

Keyfitz N. The growing human population. *Sci Am* 261 (3) (1989):118–128.

Lamb HH. *Climate: Present, Past and Future*, vol. 1: *Fundamentals and Climate Now*. London: Metheun, 1972.

London J, Angell JK. The observed distribution of ozone and its variations. In Bower FA and Ward RB (eds), *Stratospheric Ozone and Man*, 2 vols. Boca Raton, FL: CRC Press, 1981, 1:19.

Lovelock J. A policy for fluorocarbons. *Sci and Pub Pol* (June 1980):203.

Lovelock J. Atmospheric fluorine compounds as indicators of air movement. *Nature* 230 (9 April 1971):379.

Luther FM, Ellingson RG. Carbon dioxide and the radiation budget. In MacCracken MC, Luther FM (eds), *Projecting the Climate Effects of Increasing Carbon Dioxide*. Washington DC: U.S. Dept. of Energy (DOE/ER-0237), 25–56 (1985).

MacCracken MC, Aronson E, Barns D. *Energy and Climate Change: Report of the DOE Multi-laboratory Climate Change Committee*. Chelsea, MI: Lewis Publishers, 1990.

Madden RA, Ramanathan V. Detecting climate change due to increasing carbon dioxide. *Science* 209 (15 August 1980):766.

Manabe S, Stouffer RJ. Sensitivity of a global climate model to an increase of CO_2 concentration in the atmosphere. *J of Geophys Res* 85 (1980):5529–5554.

Mohnen VA. The challenge of acid rain. *Sci Amer* 259 (2), (1 August 1988):30–39.

Molina MJ, Rowland FS. Stratospheric sink for chlorofluoromethanes: chlorine atom-catalysed destruction of ozone. *Nature* 249 (29 June 1974):918–812.

Mooney HA, Vitovsek PM, Mason PA. Exchange of materials between terrestrial ecosystems and the atmosphere. *Science* 238 (13 November 1987):926–932.

National Academy of Sciences (NAS). *Policy Options of Greenhouse Warming*. Washington DC: National Academy Press, 1991.

National Aeronautics and Space Administration (NASA). *Present State of Knowledge of the Upper Atmosphere: An Assessment Report*, Washington DC: Government Printing Office, January 1984.

NRC. *Carbon Dioxide and Climate: A Scientific Assessment*. Washington DC: National Academy Press, 1979b.

NRC. *Carbon Dioxide and Climate: A Second Assessment*. Report of the CO_2/Climate Review Panel, National Research Council. Washington DC: National Academy Press, 1982b.

NRC. *Causes and Effects of Changes in Stratospheric Ozone: Update 1982*. Washington DC: National Academy Press, 1984.

NRC. *Causes and Effects of Stratospheric Ozone Reduction: An Update*. Washington DC: National Academy Press, 1982a.

NRC. *Changing Climate: Report of the Carbon Dioxide Assessment Committee*. Washington DC: National Academy Press, 1983.

NRC. *Environmental Impact of Stratospheric Flight: Biological and Climatic Effects of Aircraft Emission in the Stratosphere*. Washington DC: National Academy of Sciences, 1975.

NRC. *Halocarbons: Effects on Stratospheric Ozone*. Washington DC: National Academy of Sciences, 1976.

NRC. *Ozone Depletion, Greenhouse Gases, and Climate Change*. Washington DC: National Academy Press, 1989.

NRC. *Protection Against Depletion of Stratospheric Ozone by Chlorofluorocarbons*, Washington DC: National Academy of Sciences, 1979a.

OECD. *A Convention on Climate Change: Economic Aspects of Negotiations*. Paris: OECD Publications, 1992b.

OECD. *Global Warming: Estimating the Economic Benefits of Abatement*. Paris: OECD Publications, 1992a.

OECD/IEA. Climate Change Policy Initiatives. *Energy and the Environment Series*. Paris: International Energy Agency, OECD/IEA, 1992.

Office of Technology Assessment (OTA). *An Analysis of the Montreal Protocol on Substances That Deplete the Ozone Layer*, Staff Paper, rev. February 1988.

Pearman GI, Fraser PJ. Sources of increased methane. *Nature* 332 (7 April 1988):489–490.

Ramanathan V, Callis L, Cess R, et al. Climate-chemical interactions and effects of changing atmospheric trace gases. *Rev of Geophys* 25 (August 1987):1441–1482.

Ramanathan V, Cess RD, Harrison EF, et al. Cloud-radiative forcing and climate: Results from the earth radiation budget experiment. *Science* 243 (January 6, 1989):57–63.

Regens JL, Rycroft RW. *The Acid Rain Controversy*. Pittsburgh: Univ. of Pittsburgh Press, 1989.

Repetto R. *World Enough and Time*. New Haven, CN: Yale Univ. Press, 1986.

Schlesinger ME. *A Review of Climate Model Simulations of CO$_2$-induced Climatic Change*. Climate Research Institute Report No. 41. Corvallis, OR: Oregon State Univ., 1983.

Schlesinger ME, Mitchell JFB. Model projections of the equilibrium climatic response to increased carbon dioxide. In MacCracken MC, Luther FM (eds), *Projecting the Climate Effects of Increasing Carbon Dioxide*. Washington DC: U.S. Dept. of Energy (DOE/ER-0237) (1985), 81–148.

Schneider SH. The greenhouse effect: Science and policy. *Science* 243 (1989):771–781.

SCOPE. *Man's Impact on the Global Environment*. Report of the Study of Critical Environmental Problems. Cambridge, MA: MIT Press, 1970.

Starr JR. Water wars. *Foreign Policy* No. 82 (Spring) 1991.

Stolarski RS, Cicerone RJ. Stratospheric chlorine: A possible sink for ozone. *Can J of Chem* 52(1974): 1610–1615.

Tauber G. The ozone backlash. *News and Com Science* 260 (11 June 1993):1580–1583.

The tragedy of the oceans. *The Economist* 330 (7855) (March 19, 1994):21–24.

Tyndall J. On radiation through the earth's atmosphere. *Phil Mag* 4 (1863):202ff.

UNDP. *Human Development Report, 1991*. New York: OUP, 1991.

United Nations Environment Programme (UNEP). *The State of the Movie Environment*. Group of Experts on the Scientific Aspects of Movie Pollution. Nairobi, 1990.

UNEP. *Draft International Convention for the Protection of the Stratospheric Ozone Layer*. UNEP/WG.69/3, January 1, 1982.

U.S. Department of Energy. *DOE Workshop on Greenhouse-Gas-Induced Climate Change: A Critical Appraisal of Simulations and Observations*. Amherst: University of Massachusetts, May 8–12, 1989.

U.S. EPA. *Acid Rain: A Student's First Sourcebook*. EPA/600/9–90/027, July 1990. Washington DC: U.S. EPA, 1990.

U.S. EPA. *National Acid Precipitation Assessment Program: Interim Report*. Washington DC: U.S. EPA, 1987.

Wang WC, Wuebbles DJ, Washington WM. Potential climate effects of perturbations other than carbon dioxide. In MacCracken MC, Luther FM (eds), *Projecting the Climate Effects of Increasing Carbon Dioxide*. Washington DC: U.S. Dept. of Energy (DOE/ER-0237) (1985), 191–236.

Washington WM, Meehl GA. Seasonal cycle experiment on the climate sensitivity due to a doubling of CO$_2$ with an atmospheric general circulation model coupled to a simple mixed layer ocean model. *J Geophys Res* 89 (1984):9475–9503.

Webb III T, Wigley TML. What past climates can indicate about a warmer world. In MacCracken MC, Luther FM (eds), *Projecting the Climate Effects of Increasing Carbon Dioxide*. Washington DC: U.S. Dept. of Energy (DOE/ER-0237) 1985.

Wigley TML. Possible climate change due to SO$_2$-derived cloud condensation nuclei. *Nature* 339 (June 1, 1989):365–367.

Wigley TML, Jones PD, Kelly PM. Empirical climate studies. In Bolin B, Doos BR, Jager J, Warrick RA (eds), *The Greenhouse Effect, Climatic Change, and Ecosystems*. SCOPE 29. Chichester, U.K.: John Wiley & Son, 1986, 271–322.

Wigley TML, Schlesinger, ME. Analytical solution for the effect of increasing CO$_2$ on global mean temperature. *Nature* 315 (20 June, 1985):649–652.

Woodwell GM. Forests and climate: Surprises in store? *Oceanus* 29 (Winter 1986/1987):71–75.

Woodwell GM. Letters: The global carbon cycle. *Science* 241 (30 September 1988):1736–1737.

World Commission on Environment and Development. *Our Common Future*. New York: OUP, 1987.

World Resources Institute. *World Resources 1992–93, A Guide to the Global Environment Toward Sustainable Development*. New York: Oxford Univ. Press, 1992.

Wuebbles DJ, Edmonds J. *A Primer on Greenhouse Gases*. Washington DC: U.S. DOE (DOE/NBB-0083), 1988.

Appendixes

Air quality monitoring at an oil and gas production facility. Courtesy Dames & Moore.

Appendix 1

Case Studies

CASE 1
PHASE I ENVIRONMENTAL ASSESSMENT RESULTS

John Г. Montgomery, PhD, ASP, CHMM

As part of its corporate philosophy, the XYZ Remanufacturing Company conducts in-house Phase I Environmental Assessments of its properties. Currently XYZ has 10 remanufacturing facilities throughout the United States and has two subsidiaries that manufacture boxes and packing for shipping of remanufactured parts.

As the new environmental manager at XYZ, you have been requested to conduct a Phase I Environmental Assessment of the Eminence facility, located in the Missouri Ozarks. As part of your Phase I walkthrough, you have identified the following suspicious situations:

1. The Eminence facility was built 10 years ago on property that was formerly used for industrial purposes.
2. Adjacent upstream property has recently had several underground storage tanks removed by a local rental and service company.
3. A landfill located approximately 1 mi (1.6 km) to the north of your facility sits beside a small stream that flows through your property.
4. Oils and solvent have been removed from your site by the ABC Company, a local company from Rolla.
5. Appropriate copies of manifests for hazardous waste disposal are incomplete and are uncategorized.
6. HAZWOPER and Hazard Communication Training for employees working with solvents, acids, and other hazardous materials are not recorded. Discussions with employees indicate that training has not been completed.
7. Outside areas of the building show signs of staining, with several areas of barren soil with grass growing on the perimeter.

Discussion Questions

As the corporate environmental manager, you have determined that for some of your discoveries a Phase II Environmental Assessment must be completed by an environmental consulting firm. You are in the process of determining which discoveries will be included in the scope of work for a request for proposal (RFP).

1. Your company can institute various remediation activities to meet OSHA and EPA regulations. Identify and discuss your strategy for each of these in-house "fixes."
2. What should you do concerning discoveries 1 and 7?
3. What should you do concerning discoveries 2 and 4?
4. Which of these discoveries constitutes the most immediate EPA/OSHA regulatory threat to your facility?

Recommendations

Question 1. Your company can institute various remediation activities to meet OSHA and EPA regulations. Identify and discuss your strategy for each of these in-house "fixes."

a. #4. Oils and solvent have been removed from your site by the ABC Company, a local company from Rolla.

You and your facility manager should visit the ABC Company and conduct an audit of the station to determine if they have appropriate state and EPA numbers, permits, licenses, and certificates. Additionally, the cleanliness and orderliness of the facility will provide an overview of the manner in which the managers run their business. A copy of audits of the facility conducted by insurance companies, state EPA, or by other companies should also be requested. If these are unavailable, request the name of the company that performed the audit. Local, state, and federal governmental agencies should be visited to determine whether the facility is in compliance with the EPA, OSHA, or other regulatory agencies. Their liability insurance policy should be substantial enough to meet potential Superfund liabilities. Request a copy of the annual report.

Any irregularities or inconsistencies may indicate that the company may be in noncompliance and may lead to future troubles for your company.

b. #5. Appropriate copies of manifests for hazardous waste disposal are incomplete and are uncategorized.

If the signed final copy of the manifest has not been received from the TSD or Transporter, a request should be made immediately for the signed copy to be returned to your company as the generator (60 days for small quantity generators and 45 days for large quantity generators). If these are not returned to you within the required time, you must file an Exemption Report with the EPA or authorized state agency. The Exemption Report will consist of a cover letter from your facility explaining the steps to ensure that the failure to receive the completed manifest does not recur and a legible copy of the initial manifest (40 CFR 262.42).

c. #6. HAZWOPER and Hazard Communication Training for employees working with solvents, acids, and other hazardous materials are not recorded. Discussions with employees indicate that training has not been completed.

Training classes for these two OSHA requirements, as well as others that may be subsequently discovered, should be conducted at the earliest opportunity. Outside consultants may be required to conduct these training courses; however, a large corporation such as XYZ should have corporate programs that should be used.

Question 2. What should you do concerning discoveries 1 and 7?

a. #1. The Eminence facility was built 10 years ago on property that was formerly used for industrial purposes.

Following your investigation of the former use of the property and the confirmation that possible soil contamination may have occurred under the previous owner, a scope of work should be developed to solicit bids for a study of the drainage system (leaks, egress, stormwater drainage, industrial waste) to typify the soil found around the exterior of the building and at the parameters of the plant fence line. The depth and number will depend on the geological formation of the property.

b. #7. Outside areas of the building show signs of staining, with several areas of barren soil with grass growing on the perimeter.

An investigation should be conducted to determine the plant's practices for disposal of solvent, oils, or waste into the property's earth. Following the investigation, sampling—both vapor and soil—should be conducted with appropriate instrumentation and provided to a licensed AIHA laboratory (industrial hygiene) that also meets EPA protocols. Following receipt of the results, appropriate strategies should be developed to complete further sampling, if appropriate, or to begin remediation.

Question 3. What should you do concerning discoveries 2 and 4?

a. #2. Adjacent upstream property recently had several underground storage tanks removed by a local rental and service company.

An investigation concerning the local rental and service company's reputation, licensing, and past conduct may be implemented. If warranted, continued investigation with local and/or state EPA regulators concerning the removal and closure of the particular site may be requested under the Freedom of Information Act.

b. #3. A landfill located approximately 1 mi (1.6 km) to the north of your facility sits beside a small stream that flows through your property.

Several samples of water flowing from the landfill, through XYZ property, and continuing into the nearby lake should be taken to ensure that water is not contaminated by activities at your facility. Water samples should be taken at the point of entry onto XYZ property as well as the point of egress from the property. Several samples should be taken, including those following unusual landfill activities as well as following rain episodes. Contrast between entry and egress results will indicate potential contamination by your facility.

Question 4. Prioritize in order of immediate, intermediate, and long-term threats to your company as a result of your Phase I Assessment.

a. Immediate—#4, #5, and #6
b. Intermediate—#1 and #7
c. Long-term—#2 and #3.

CASE 2
SELECTING A CONSULTANT TO PERFORM DUE DILIGENCE ASSESSMENTS

John F. Montgomery, PhD, ASP, CHMM

Your company, XYZ Remanufacturing, has just notified you that it is in the process of negotiating for a parcel of property in a nearby industrial park. As the corporate manager—environmental compliance, you have been asked whether you believe that the property is suitable for purchase, and if not, what steps the company should initiate to identify potential noncompliance and/or contamination.

The property is located in an old section of the Loma Vista Industrial Park and had formerly been used as an automobile parts manufacturing facility for the DEF Company. You have visited the site and have identified a foul smell and staining in and about several of the grates as well as around the rear of the building. The building was constructed in 1959, and various parts of the building may contain sprayed-on asbestos or asbestos mastic for floor tiles or other parts.

You have decided that a due diligence environmental assessment should be conducted by a third-party environmental consultant. You have solicited bids from several environmental consultants and are in the process of reviewing their proposals as well as their credentials.

Discussion Questions
As your company's manager responsible for the selection of the environmental consultant/company, what criteria will you use to address each of the following topics?

1. professional expertise and education
2. professional certification
3. past experience, references, and sample product
4. liability insurance.

Recommendations
Question 1. Professional expertise and education

The prospective environmental consultant or environmental consultant company should have professionals on staff with degrees in such fields as engineering, environmental science, hydrogeology, industrial hygiene, chemistry, safety engineering or management, or similar vocations. Employees with graduate degrees—master's (various) and doctorates in these fields—add to the company's credentials and competencies. Years in service as a professional environmental consultant with a good track record may be substituted for university degrees. Professional resumes should include solid technical expertise, education, and professional experience.

As the environmental manager, your visit with the professional responsible for conducting the due diligence assessment will provide a quick snapshot of the company and its ability to perform the requested work. The project manager is the responsible individual for the conduct of the assessment and will control its management, progress, and budget, and is your key point of contact.

Question 2. Professional certification

Professional certification has become a "hot topic" as increased legislation from both the EPA and OSHA and litigation require proof of competence.

Liability and civil penalties for noncompetence have also highlighted the need for proof of competence from accepted safety, industrial hygiene, and environmental certification boards. Prospective consultants should have certifications from boards that require a combination of degrees, years in service, and the successful completion of a competency test. Certification such as professional engineer (PE), professional geologist (PG), certified safety professional (CSP), certified industrial hygienist (CIH), certified hazard control manager (CHCM-Safety), certified hazardous materials manager (CHMM-Environment), certified environmental professional (CEP), and others offer tested proof of competence.

Question 3. Past experience, references, and sample product

The reputation of the prospective company drawn from former clients is the best method for judging the ability of the consultant to accomplish your desired scope of work. A consultant company that has been in business for a number of years, that provides a list of clients, and provides a work sample is a good candidate.

A history of successful, on-time completion of work, accomplished within budget and with a complete, clear description and understandable discussion of discoveries, is the most impressive resume.

Needless to say, the prospective environmental consultant or consultant company should not appear on any "lists" for noncompliance or failure to complete assignments and should not be listed by business bureaus for unethical behavior.

Question 4. Liability insurance

Insurance carried by the company should match the requirements of the job and include workers' compensation, automobile insurance, and liability insurance. Liability, the most important for the due diligence assessment, may also be the most controversial. Environmental consultants will quite naturally elect to limit the amount required, while the client will want the maximum amount available. The right insurance for the job is the best rule of thumb. If the consultant's contract is for $30,000 and the property will eventually earn your company $20 million, just how much insurance will you require? All insurance requirements must be clearly stated in the request for proposal (RFP).

CASE 3
DETERMINING THE REPORTABILITY OF A CHEMICAL SPILL

John F. Montgomery, PhD, ASP, CHMM

During painting activities at your facility, a fork lift driven by a new employee inadvertently comes into contact with two tanks used in the painting process. One of the tanks is filled with acrylic enamel thinner and the second with blue enamel. Both of the tanks have been punctured, and before your emergency response team arrives, have leaked their entire 100 lb (37.3 kg) contents onto the floor of your main painting facility.

As the environmental coordinator at your facility, you are concerned about your responsibility to report the two spilled materials to local, state, or federal EPA. You pull a copy of the material safety data sheets (MSDSs) and summaries of their characteristics (Tables A1–A to A1–D) for the two products to determine how best to respond.

Discussion Questions

From the information provided on the MSDSs determine:

1. Does the acrylic enamel thinner spill reach the reportable quantity level?
2. How did you determine if the spill should or should not be reported?
3. Does the blue enamel spill reach the reportable quantity level?
4. How did you determine if the spill should or should not be reported?
5. If the spills are to be reported, to whom should they be reported? Why?
6. Under what criteria did you make your reporting decision?

Recommendations

Question 1. Yes

Question 2.
a. Determine the amount spilled in gallons.
b. Convert to pounds: (gal × specific gravity (SG) from MSDS × weight) = P(lbs product)
c. Find the pounds of each constituent chemical: % of constituent chemical from MSDS × P = C (amount per constituent chemical).
d. Compare the pounds of each constituent chemical to the reportable quantity for each.
e. If C per constituent chemical is greater than or equal to the reportable quantity, *the spill must be reported.*

Table A1–A. Material Safety Data Sheet—Acrylic Enamel Thinner

Product	CAS PEL*	OSHA Product	%	RQ*-Pounds
Hydrocarbon Solvent	64742-89-8	100 ppm	0–35	10,000
Toluene	108-88-03	100 ppm	25–45	1,000
Ethyl benzene	100-41-4	100 ppm	0–5	1,000
Xylene	1330-20-7	100 ppm	5–45	1,000
Light aromatic naphtha	64742-95-6	100 ppm	0–5	10,000
Methanol	67-56-1	200 ppm	<5	5,000
2-Butoxyethanol	111-76-2	25 ppm	0–5	1
2-Butoxyethyl	112-07-2	50 ppm	0–15	10,000
Acetone	67-64-1	750 ppm	5–15	5,000

*CAS PEL = Chemical Abstract Service permissible exposure level
+RQ = reportable quantity

Table A1–B. Material Safety Data Sheet—Blue Enamel Paint

Product	CAS	OSHA PEL	% Product	RQ-Pounds
Acetone	67-64-1	750 ppm	30–40	5,000
Xylene	1330-20-7	100 ppm	5–10	1,000
Ethyl alcohol	64-17-5	1,000 ppm	5–10	10,000
Diacetone alcohol	123-42-2	50 ppm	5–10	1,000
Methyl ethyl ketone	78-93-3	200 ppm	<5	10,000
Propellant—propane	N/A	N/A	23	

Table A1–C. Physical/Chemical Characteristics of Acrylic Enamel Thinner

Characteristic	Value
Boiling point	132 to 384 F (55.6 to 196 C)
Vapor pressure (mm/Hg)	From 0.6 for 2-Butoxyethanol to 180.0 for acetone
Vapor density: Air = 1	Not given
Specific gravity: (H_2O = 1)	Not given (assume 1)
Melting point	Not given
Volatile organic compound	Not given
Evaporation rate	Slower than ether
Weight/gallon	6.5–7.2
pH	None
Solubility in water	Not given
Appearance and odor	Not given

Example: # gal × SG × weight in lbs × % spilled = quantity spilled
Acetone (RQ = 5,000 lbs) = 100 gal × 1 × 7.2 × 15% = 108 lbs *Not Reportable*
2-Butoxyethanol (RQ = 1 lb) = 100 gal × 1 × 7.2 × 5% = 36 lbs *Reportable*

Question 3. No

Table A1–D. Physical/Chemical Characteristics of Blue Enamel Paint

Characteristic	Value
Boiling point	133 F (56.1 C)
Vapor pressure (mm/Hg)	188 mm Hg @ 20 C vapor
Density: air = 1	N/A
Specific gravity: (H_2O = 1)	0.90
Melting point	Not given
Volatile organic compound	Not given
Theoretical	5.1 lbs/gal
Percent volatile	85%–90%
Solubility in water	Not given
Appearance and odor	Paint/solvent

Question 4.
a. Determine the amount spilled in gallons.
b. Convert to pounds: (gal × SG × weight) = P
c. Find the pounds of each constituent chemical: (each constituent % from MSDS) × P = C (amount per constituent chemical)
d. Compare the pounds of each constituent chemical to the reportable quantity for each.
e. If C per constituent chemical is greater than or equal to the reportable quantity, *the spill must be reported.*

Example: # gal × SG × weight in lbs × % spilled = quantity spilled

Acetone (RQ = 5,000 lbs) = 100 gal × .90 × 5.1 × 40% = 183 lbs *Not Reportable*

Xylene (RQ = 1,000 lbs) = 100 gal × .90 × 5.1 × 5% = 10 lbs *Not Reportable*

Question 5. The spills are not required to be reported because neither left the facility and both were cleaned up and waste materials disposed as required by RCRA. If the spills meet reporting criteria, they should be reported to the Local Emergency Planning Committee (LEPC), the State Emergency Response Commission (SERC), and the National Response Center (NRC).

Question 6. The spills did not enter water (streams, stormwater, lakes), air, or dirt.

CASE 4
SPILL INCIDENT

Susan Gesoff, CHMM

You are conducting a Phase I Assessment of one of your company's field stations and are reviewing the facility environmental documents. In the files you discover an account of a spill incident that occurred two weeks ago. After conversations with facility personnel, you piece the story together as it appears here. Please read carefully and consider what correct and incorrect decisions were made, and what actions should now be taken.

The Discovery

John Smith's shift was finished for the day, and he headed for the parking lot gate to begin his weekend. As he walked between the automotive maintenance shop and the fueling area, something caught his attention. A stream of liquid was running along the curb. It was a bright sunny day, so the trickle of liquid running down into the storm sewer seemed strange. He followed the stream of fluid back up to a huge dumpster beside the auto maintenance building. Upon further investigation, it was obvious that the leak was not from the dumpster, but from the green storage shed behind it.

John put down his coat and bag and walked over to the wooden shed. He opened the door and was instantly aware of a pungent odor. There was only one small window in the shed, but the sunlight made it possible for John to see that the entire floor of the shed was a large pool of liquid. In order to determine what the liquid was, John stepped up into the shed. There was a long work bench against one wall, three storage cabinets, and two full 55-gallon drums in the corner.

John was able to determine the source of the liquid. A spot on the side of one of the drums had rusted through and liquid was leaking. Reacting quickly to stop the leak, he grabbed the drum and tipped it enough to stop the flow of liquid. He then wedged it, in this tipped position, between the other drum and the one of the cabinets. He tried to find a label on the drum to determine the nature of the liquid, but it was badly deteriorated. He was unable to make out the product name, but a manufacturer's name, JKL, was legible, and the side of the drum was stenciled with the word WASTE.

Noticing the time, he quickly wiped his hands on his jeans and decided to head home. It was already after 5:00 p.m., so he decided to leave the drum for the weekend and mention it to his supervisor on Monday.

Subsequent Events

Leaving the scene. As he was leaving the shed, John noticed water dripping down the inside wall of the shed behind the 55-gallon drums. He went outside the shed and found that the source of the water was condensation dripping onto the shed's roof from the air-conditioning unit on top of the auto shop. As he closed the shed door, he thought to himself that this dripping condensation was probably the cause of both the rust and the water damage to the label.

Spreading the word. When John arrived at work on Monday morning, he found his supervisor and told him about the leak and the actions he had taken. The supervisor asked John to go to the general manager's office and look through the MSDSs to see whether he could find some specific information about the product in the shed. John didn't know what an MSDS was, but he walked to manager's office to look for them. He asked the secretary where they were, and she pulled out a binder labeled Material Safety Data Sheets. Unfortunately, they were not in any particular order so he was forced to page through about a 100 MSDSs. He failed to locate an MSDS for any product made by JKL.

The cleanup. In the interim, the supervisor called Facilities Maintenance to inform that supervisor of the leak. One of the supervisors on duty, Steve, went to examine the situation. Everything in the shed remained in the condition discovered the previous Friday. Steve had brought several rags and absorbent pads with him and quickly spread them on the floor of the shed. He used clay litter to absorb the spilled liquid in the corners and near the door.

When most of the spilled liquid was absorbed, Steve put on his work gloves and began to remove the soaked pads and rags. He stacked them in a pile on the grass outside the shed. He used a shovel to scoop the clay litter into the trash can. After he finished this cleanup, he returned to the office to ask what the product might be.

One of the supervisors remembered that the shed was used as a paint room for auto maintenance a few years back. He explained that the general manager had asked him to clear out the shed months ago, but he had not yet found the time. He guessed that the liquid might be some type of waste solvent.

Reporting. The facilities manager overheard the conversation as he walked into the office and suggested reporting the spill to the EPA. Because the environmental coordinator was out of town and no one could find instructions on spill reporting in the files, Steve decided to document the incident and left a report for the general manager, who filed it in his environmental files.

Discussion Questions

1. List the environmental/safety violations that occurred during this event.
2. What recommendations will you list in your assessment report?
3. Who may be liable for any or all of these environmental violations?
4. What actions should now be taken?

Recommendations

Question 1. List the environmental/safety violations that occurred during this event.

a. John was not properly trained in spill response and therefore should not have responded to the spill, but he should have notified his superior.
b. John took no action to stop the spilled liquid from entering the storm sewer.
c. John entered a dangerous confined space with unknown hazards and had no training or knowledge of how to monitor the atmosphere before entering.
d. The fact that a dripping compressor had resulted in corrosion highlights the lack of weekly facility inspections and proper equipment maintenance.
e. The drums were labeled as both new product and waste.

f. Assuming the drums contained spent solvents, they were stored in an improper facility, had no secondary containment, listed no start accumulation date, and (because they were full) should have been removed to a permanent hazardous waste storage area and disposed within the 90 days permitted to generators of hazardous waste.
g. No MSDSs were available for the original product specified on the drum. In addition, John had not been properly trained to understand the need for MSDSs and the hazards associated with chemicals in the workplace. Furthermore, the MSDSs were not "readily available" to employees.
h. Cleanup of the hazardous materials was performed by untrained personnel.
i. Hazardous waste—spent absorbents—were disposed of illegally in nonapproved containers as regular waste.
j. Soil was contaminated by soaked rags being stacked on the ground, and this soil was not removed and disposed of as hazardous waste.
k. The spill entered the storm drain and no assessment was made regarding the reportability of the spill to local, state, and federal agencies.

Question 2. What recommendations will you list in your assessment report?

a. Employee training, including OSHA Hazard Communication, OSHA HAZWOPER, and RCRA waste disposal.
b. Contaminated soil should be tested, remediation activities should be conducted, and/or soil should be removed.
c. An analysis should be performed of the waste to determine its exact composition. The drums of waste solvent should be properly manifested and shipped off-site as hazardous waste immediately.
d. Weekly inspections of the facility should be conducted to assess environmental and maintenance issues.
e. A proper waste storage area should be established, and proper labeling and handling requirements per RCRA should be strictly followed.
f. An analysis should be performed to determine whether any spilled chemical constituent exceeded the reportable quantity. If so, the spill should be reported immediately to state and federal agencies.
g. MSDSs for every product used at the facility should be made available in all workplaces using

chemical products, and these sheets should be up-to-date and well organized.

Question 3. Who may be liable for any or all of these environmental violations?

- the facility's supervisor
- the general manager
- the corporate environmental department
- the senior officers of the corporation.

CASE 5
PLASTIC CONTAINER MANUFACTURING FACILITY

Gary R. Vajda, MS, PE

Alternative analysis has been successfully used in both large and small firms, and in industries as diverse as metal finishing to consumer product packaging. This and the following two case studies are examples of alternative analysis projects that can be used to test pollution prevention and waste minimization techniques.

The objective of this project was to reduce the scrap rate associated with the production of plastic containers. The study had two major components: (1) to identify which of the five primary systems along each of four production lines was the source of the majority of the scrap losses and (2) to identify which variable(s) were causing the losses. Statistical sampling was performed in an attempt to assess any positional (along the line), temporal (time dependent), or cyclical (beginning, middle, or end of a production run) tendencies. This was performed during the course of two weeks, on two shifts, on four production lines for five products, and at five stations on each line.

A valuable aspect of the sampling phase was time spent observing the operations, procedures followed, and reactions to problems. The data were plotted and several conclusions drawn about where further effort was to be focused and the range of variables contributing to scrap loss.

Approximately 30 variables were identified as contributors to waste generation. It was impractical to test a range of settings for each variable, so a survey/feedback approach was used to narrow the list of 30 to 8 primary variables that would be the focus of the design of experiments. Another constraint on

experimentation was to minimize the amount of downtime experienced by the line on which experiments were conducted. Therefore, production scheduling, plant engineering, maintenance, and operations were asked for close cooperation on factors such as test sequences and test set points for each variable.

The designed experiments were run with a minimum of downtime on the line, less then one-third of the time that was budgeted. Data were analyzed for both primary and interactive effects of the variables. This analysis of variance showed three variables as primary contributors to waste generation.

A written report was prepared to document what was done, how it was done, and the conclusions reached. Most importantly, however, oral presentations and discussions were held for management, engineering personnel, and key operators on all four lines on all shifts.

The methodologies used included statistical sampling, multivariant analysis, survey feedback, design of experiments, analysis of variance, and direct operations and maintenance involvement.

The key results included minimal downtime spent to allow sampling and testing to be carried out and a recommendation resulting in an immediate 70% reduction in waste material generated. The simple payback period for this was less than 60 days, including the cost of the study and the value of lost production during the downtime. Although this project did not initially have an environmental goal, the result certainly was environmentally beneficial, and the company did take some bows for it.

CASE 6
AIRCRAFT COMPONENT MANUFACTURING COMPLEX

Gary R. Vajda, MS, PE

This project involved a rather old manufacturing facility that had reached the breaking point with the local air regulatory agency. The company needed to decide whether to close, move, or modernize.

The approach was to first quantify the environmental compliance concerns, then develop a plan to identify and recommend measures (equipment, facilities, procedures, systems, substitute materials, etc.) that addressed both compliance and factory modernization. The scope of work involved completing an integrated manufacturing and environmental evaluation of this 13,000-employee aircraft component fabrication and assembly complex.

More than 400 production functions were evaluated for environmental concerns and operational improvement potential by adapting a structured system analysis methodology for use by a combined team of environmental, health and safety, manufacturing, and facilities engineers. The types of operations assessed included painting, curing, bonding, composite materials fabrication, foundries, degreasing, metal cleaning (acid and alkaline etch), chemical milling, metal removal, metal cutting, metal joining, and electroplating.

Efficient data analysis, decision making, and prioritization processes were keys to managing this effort. Data were gathered and entered or referenced into a PC-based relational data base. The key identifier for all data gathered was a numbering system that assigned a unique number to each of the more than 400 operating functions. And, because of the relational capabilities of the data base, queries were made and reports produced on the basis of regulation, department number, category of emission or discharge, mitigative action strategy, etc. Categories of data gathered included permits, Material Safety Data Sheet (MSDS) information, applicable regulations (federal, state, and local), production throughput and operating cost data, facility data such as process tank surface area and paint booth size, and cost estimates for compliance measures.

The methodologies used included: IDEF function analysis methodology; SAMIS cost modeling (environmental and production operating costs); a relational data base for environmental, operational, and alternative action strategy; cost data; and structured decision making.

The key results included a total estimated installed cost of implementing needed control technology that was one-third that of previous internal estimates, and development of a decision model to assist in ranking prospective projects on the basis of potential improvements to both environmental and manufacturing operations. This model would allow screening of low or negative project alternatives that might otherwise be equal in their ability to mitigate an environmental, health, or safety problem. In addition, a group of 24 "fast-track" projects was identified and defined to the point that further engineering definition was not needed prior to the submission of a capital expenditure request. An operating cost model was also developed to serve as a baseline against which future improvements would be measured.

CASE 7
METAL PRODUCTS MANUFACTURING FACILITY

Gary R. Vajda, MS, PE

The objectives of this effort were to establish for a multifacility metal products manufacturing company the potential air emission limits under the maximum achievable control technology (MACT) provisions of the Clean Air Act Amendments and then to estimate the cost of compliance. A secondary objective of the project was to identify activities that could be conducted now that would result in significant cost reductions.

After completing a general survey of the facilities with respect to operations, emissions, and materials, it was determined to concentrate on one unit process, vapor degreasing with 1,1,1-trichloroethane (TCA), because it was common to most of the facilities and resulted in a large proportion of the VOC emissions.

It was also decided to focus on one typical plant. This facility operated 35 degreasers with individual capacities of 5 to 300 gallons. These degreasers were used across three shifts, seven days a week. This resulted in over 300 tons (270 metric tons) of TCA being emitted per year.

Because of the focused nature of this effort, the baseline characterization was performed using a questionnaire, operator interviews, vendor information, and observations. Information was compiled on the design and operating characteristics of each unit and its operators, emission rates and solvent usage, anomalies between recommended and actual operations, and operator and vendor recommendations for improvement.

The list of recommendations ranged from the elimination of certain units, to operator training to operational modifications (regulate crane speed, improve instrumentation, regular maintenance), to unit modifications (install covers, increase cooling capacity, disconnect fans, increase freeboard). To implement all of these changes was estimated to require a capital cost on the order of $1 million and would result in an approximately 50% reduction in TCA emissions. This was too costly. An alternative analysis focused on limiting equipment changes to the most used units. It indicated that approximately the same TCA reduction could be achieved for one-quarter less initial cost (i.e., about $750,000). Moreover, when the life-cycle true cost of the nonemitted TCA was calculated, the payback period was slightly longer than one year.

CASE 8
CONDUCTING A RECYCLING AUDIT

Beth Hayes

Airport XYZ has a 32-gate terminal and currently has a white paper recycling program in two administration areas with different employee groups collecting aluminum cans for personal programs.

Recently, the environmental coordinator at XYZ airport received two new pieces of legislation regarding the reduction of the solid waste stream. One piece of legislation was from the city, the other from the state.

The City's Ordinance

The City's Municipal Waste Reduction and Recycling Ordinance states that beginning January 1, 1994, as a condition of receiving, renewing, and maintaining any city license, each licensee shall be required to certify in the license application that an effective recycling program will be conducted on the premises during the license period. By January 1, 1994, at least four items on the following recyclable materials list must be recycled:

- corrugated cardboard
- computer paper
- white ledger paper
- mixed paper
- magazines
- metal containers, aluminum, steel
- newspaper
- clear glass containers
- colored glass containers
- plastic containers.

The goal of this ordinance was to implement effective recycling programs, to enable each establishment to recycle at least 15% of its waste stream by January 1, 1995, and 25% of its waste stream by January 1, 1997.

The State's Ordinance

The state's ordinance says that by March 1, 1991, each county, with a population of greater that 100,000 must submit a waste management plan that includes a recycling program. Such plans shall be designed to recycle, by the end of March 1994, 15%, and by March 1996, 25% of the waste generated. The plan should identify potential markets for at least three recyclable materials and the program shall include education and notification programs to encourage compliance.

Waste Audit

XYZ Airport performed a waste audit and found that they currently recycle about 5% of their solid waste stream. During the solid waste stream audit the following was determined:

- Terminal. A large amount of paper is generated from the office areas; however, the majority of the paper is 100% recycled and is not considered a white paper. Most of the cardboard is generated at the Cabin Service operation, smaller amounts are also generated at Flight Services, and the Inbound Bagroom. Magazines are generated at Cabin Service from overstocked magazines. Newspapers and magazines are generated by employees and passengers in the concourse, plastics are generated in the employee cafeteria, and glass is generated in small amounts by the Admiral's Club.
- Off the aircraft. Aluminum and plastics are the two main source generators onboard the aircraft. Currently, aluminum is being recycled by the caterer; however, recycling of plastics (formerly recycled by the caterer) has been discontinued because of the poor regional plastics market. Twice a month, a magazine exchange occurs on all aircraft during the overnight cleaning. Newspaper wastes are generated by passengers, mostly on the morning flights.

Two 6-yard compactors are used for the current white paper recycling program, located at each end of the terminal. Space available around the terminal for extra rolloffs and compactors is extremely limited. There is space for one 8-yard rolloff near Cabin Service next to the current trash compactor.

Discussion Questions

1. What four recyclables would you collect at the airport?
2. How would you collect each of the four recyclables?
3. How would you implement the program?
4. How would you promote the program and provide user information?

Recommendations

Question 1. What four recyclables would you collect at the airport?

- mixed paper
- cardboard
- magazines
- newspapers.

The white paper program should be changed to a mixed paper program, because most of the paper coming from the facility is not considered white quality. Cardboard, magazines, and newspapers are chosen because of the amount generated and availability of recycling markets in the region.

Glass was not included because of the limited amounts generated by the airport. Plastic was not chosen because of the poor plastic recycling markets in the area. Aluminum was not chosen because the caterers have a program that recycles aluminum off the aircraft and employee groups are currently involved in their own programs throughout the terminal.

Question 2. How would you collect each of the four recyclables?

- mixed paper: The present program should be expanded and put into place in all administration areas and an employee education program should describe methods to accommodate a mixed paper recycling program. Employees should continue to empty their recycling containers into the large recycling containers at the copier areas, and building cleaners should continue to empty large recycling containers in the 6-yard compactors. These compactors should now be designated as mixed paper compactors.
- cardboard: An eight yard rolloff should be placed outside Cabin Service for cardboard materials. Cabin Service employees should continue to break down the cardboard, and dispose of it in the recycling container instead of in the trash dumpster. Utility people will pick up cardboard at Flight Service and the Inbound bagroom during regular nightly trash pickup. The cardboard from these areas will be combined in the Cabin Service cardboard recycling container.
- magazines and newspapers: Small 2-yard wheeled containers may be strategically placed throughout the ramp for magazines and newspapers. Cabin Service cleaners will collect magazines in bags on magazine exchange nights. Newspapers will be collected by Cabin Service cleaners during cleaning of early morning flights. Building cleaners will separate magazines and newspapers from other trash as they police the concourse.

Question 3. How would you implement the program?

a. The first step is to secure upper management support, form a recycling committee, and designate a program coordinator.
b. Conduct a waste audit to determine products that may be recycled and the availability of markets for these recyclables.
c. Determine the benchmark for all solid waste streams before the program is started.
d. Conduct a study to determine the quantities of the generated recyclables.
e. When quantities are determined, the size and type of internal and external collection containers must be identified.
f. Obtain janitorial approval, set up contracts with the recyling haulers and vendors, conduct a financial analysis, and schedule accounting procedures and reporting procedures.
g. Develop a training and promotional program to introduce the initiative to the facility.
h. Document all recyclables and solid waste stream reductions.

The recycling program should begin with all four recyclables at the same time and plan to achieve a 25% reduction regulation within the first month.

Question 4. How would you promote the program and provide user information?

The program can be promoted and information provided to employees in the following ways:

- posters
- flyers
- letters
- internal newspaper articles
- electronic mail
- videotapes
- paycheck stuffers
- recycling fairs
- invitations
- drawings
- contests
- giveaways.

CASE 9
ALPHA LOW-LEVEL WASTE (ALLW) AND ALPHA LOW-LEVEL MIXED WASTE (ALLMW) TREATMENT FACILITY

Florence Munter

This case study presents a prototypical private sector low-level mixed waste treatment facility. This case study highlights the jungle of regulatory licensing and permitting requirements that must be analyzed and acknowledged in order to develop a mixed waste treatment facility. This example is only for illustrative purposes and is directed toward RCRA and DOE compliance. Many other regulations that affect siting, clean air, NPDES, and risk assessment are not presented.

This review of the applicability of RCRA to ALLW and ALLMW treatment assumes that storage of hazardous wastes will be in containers and approval for storage greater than 90 days will be necessary. Management units requiring groundwater monitoring are not anticipated at this facility. Units requiring groundwater monitoring include certain waste piles, landfills, and surface impoundments.

- Source material: Stored materials from DOE facilities
- Regulatory bodies: EPA, NRC, DOE, state and local authorities
- Assumption: An incinerator is built as part of the facility and equipped with appropriate best available technology under the Clean Air Act.

Significant elements of the RCRA permit are shown in Table A1–E. Part A of the RCRA Permit Application can be submitted as soon as sufficient information to identify the waste has been acquired to characterize the waste in accordance with 40 CFR 261.

The RCRA facility location requirement relating to seismic considerations is met provided that the political jurisdiction is not listed in 40 CFR 264, Appendix VI. If the facility location is in a political jurisdiction listed in Appendix VI, the seismic standard can still be met if no faults or lineaments of Holocene or more recent ages are identified within 200 ft of the facility. However, if PCBs are present in concentrations of wet 50 ppm, then under TSCA, the facility could not be sited if it is listed under 40 CFR 264, Appendix VI, regardless of actual fault locations. Similarly with the floodplain standard, engineering design modifications may allow siting in a 100-year floodplain under RCRA but not under TSCA.

The RCRA regulations for incinerators provide for preliminary approval of the Part B Permit Application contingent upon a successful trial burn. Principal organic hazardous constituent (POHC) limitations are initially established by the regulatory authority after review of the trial burn and waste analysis plans. Final approval to operate the incinerator is provided after the results of the trial burn are evaluated.

Certain elements of the RCRA program overlap with elements of TSCA, AEA, and the CAA. These elements include the following:

- Container storage: RCRA, TSCA, and AEA all have their own requirements.
- Contingency plan: RCRA requires a contingency plan to address sudden accidental releases. NRC requires an accident risk evaluation based on potential exposure.
- Financial assurance and liability requirements: RCRA, TSCA, and NRC all have their own requirements. Some states may also have additional requirements.
- Closure documentation: RCRA and TSCA regulations require written closure plans and have different requirements for documentation of clean closure. AEA regulations require a decommissioning plan. Standards have been proposed for incorporation into a generic environmental impact statement to document termination of the license.
- Trial burn plan and trial burn: RCRA, TSCA, and the CAA require successful trial burns under potentially different incinerator operation conditions for final permit approvals.
- Final approval: the regulatory authority staff indicate that the RCRA final incinerator permit and the operating permit required by the CAA are reviewed together so that a joint RCRA/CAA permit is issued by the EPA or state regulatory authority. If the state has authorization under the Clean Air Act and RCRA, it is assumed that the EPA TSCA permit approval process is separate.

Nuclear Regulatory Commission Requirements

If the state is not an agreement state in terms of NRC licensing, the NRC would be the regulatory authority. A potential commercial operator of treatment services would require a radioactive materials license for a facility operating in the state. A radioactive materials license would have to be granted before any radioactive materials could be received/handled. This license would be the only major NRC requirement for "permitting" the proposed facility. The assumption behind this statement is that the treatment facility would not be considered a production or utilization facility as defined in 10 CFR 50, *Domestic Licensing of Production and Utilization Facilities*. This assumption, which is supported by the NRC's

Table A1–E. State RCRA Hazardous Waste Facility Permit Application Requirements

Requirement	Regulatory Citation	Requirement Trigger	Scope	Comments
Part A of Application	40 CFR 270.10(f) (requirement) 40 CFR 270.13 (contents)	New hazardous waste management facility	Summarizes proposed activities.	Complete federal form supplied by EPA region. Include general facility information for owner/ operator; SIC code(s); topographic map; list of wastes designated under §261; other permits; hazardous debris (y/n). Make any confidentiality claims with Part A.
Part B of Application				
Waste analysis plan	40 CFR 270.14(b)(3) 40 CFR 264.13	Treatment, storage, of hazardous waste	Characterizes waste/ develop waste acceptance criteria (WAC). Characterizes residues and methods per 268.	Chemical and physical analyses of representative samples of hazardous waste and hazardous debris; representative sampling and testing methods; procedures for compliance with WAC. Procedures for compliance with land disposal restrictions.
Inspection schedule	40 CFR 270.14(b)(5), 264.15, and 264.174 (containers) 40 CFR 264.347 (incinerators)	Facility owner/ operator	Prevents releases to environment, threats to human health.	Include monitoring, inspection necessary to comply with permit, i.e., with container and incinerator requirements, and equipment leaks.
Manifest system	40 CFR 264.1050 et seq. (equipment leaks)	Organic concentration of wastes >10 percent	Maintains designed operation.	
Contingency plan	40 CFR 270.14(b)(7) 40 CFR 264.50–56	Facility owner/ operator	Addresses hazards from fire, explosion or release of hazardous waste, constituents.	Obtain agreements from local agencies/police, fire, hospitals, contractors, and state and local emergency response teams pursuant to 264.37. Include evacuation plan for facility.
Traffic control plan	40 CFR 270.14(b)(10)	Facility owner/ operator	Prevents accident.	Include vehicle type(s); volume; access road(s).
Facility location information	40 CFR 270.14(b)(11) 40 CFR 264.18(a) 40 CFR 264 App. VI	Facility owner/ operator	Identify physical hazards from seismicity and floods.	1. Determine applicability of seismic standard by reference to 264 App. VI. 2. Determine whether facility is within 100-year flood plain. If either criterion is applicable, the facility may be suitable under RCRA but not under TSCA.

(continued)

Table A1–E. *(Continued)*

Requirement	Regulatory Citation	Requirement Trigger	Scope	Comments
Personnel training program	40 CFR 270.14(b)(12) 40 CFR 264.16	Hazardous waste facility	Promote safe operation of facility, reduce hazard error, improve response to accidents.	Detail training of personnel in: • emergency response systems and shutdown procedures • maintenance and inspection procedures under normal operation • sampling and analysis • record keeping and reporting
Closure plan	40 CFR 270.14(b)(13) 40 CFR 264.10–115 40 CFR 264.178 (containers) 40 CFR 264.351 (incinerators)	Hazardous waste management unit(s)	Documents closure, obtain release from financial assurance requirements.	Identify off-site hazardous waste management units and decontamination procedures; develop waste management plan for waste generated by closure.
Closure cost estimate and financial assurance	40 CFR 270.14(b)(15) 40 CFR 264.142, 264.143	Hazardous waste management unit(s)	Document that owner/operator will have means to close facility.	Provide cost estimate to close facility according to approved plan (above) in current dollars. Select mechanism for documenting financial assurance and establish mechanism. Note that documentation for 270.142 and 143 may be submitted separately but not later than 60 days before waste is received.
Liability requirements	40 CFR 270.14(b)(17) 40 CFR 264.147(a)	Hazardous waste management unit(s)	Insurance coverage required for sudden accidental occurrences.	Liability >$1 m/occurrence, >$2 m/year aggregate, exclusive of legal costs. Select mechanism for assurance. If a state mechanism is required under 270.14(b)(18), it may be supplemented.
Other federal laws	40 CFR 270.3 40 CFR 270.14(b)(20)	Activities required by any laws listed under 270.3	Document compliance with other federal laws.	Evaluate applicability of other federal laws listed under 270.3, for example, Wild and Scenic Rivers Act, Endangered Species Act and others.
Specific information for containers	40 CFR 270.15 40 CFR 264.170–178	Hazardous waste storage >90 days in containers	Meet design specifications for container storage and management.	Container and secondary containment design, as applicable. Address free liquids management. (Note also NRC Regulations for mixed waste storage, and TSCA Regulations for storage of PCB items [40 CFR 761.65].)

(continued)

Table A1–E. *(Concluded)*

Requirement	Regulatory Citation	Requirement Trigger	Scope	Comments
Specific information for incinerators: trial burn plan	40 CFR 270.10 40 CFR 264.340–351 and 270.62 40 CFR 264.343–345	Owner/operator of hazardous waste incinerator	Determine permit conditions necessary to satisfy performance standard(s) (264.343), select POHCs.	Trial Burn Plan to be submitted with Part B application. • Analysis of each waste mixture • Engineering design of proposed incinerator • Test schedule for each waste included in trial burn • Statement of conditions necessary to operate within performance standards before final approval is received. Regional Administrator will designate POHCs.

definitions, eliminates the need for either a preliminary or final safety analysis report, and for NRC-specific construction or operation permits.

The radioactive materials license application for this facility would be primarily concerned with special nuclear material (plutonium), although another assumption in this analysis is that the proposed treatment facility may need to be licensed for handling some by-product and source materials. Other requirements that are triggered by granting an NRC license for handling radioactive materials are the radiation safety criteria of 10 CFR 20, *Standards for Protection Against Radiation,* and the packaging and transportation requirements found in 10 CFR 71. Details concerning applicable requirements are presented in Table A1–F.

DOE and NRC Regulatory Differences

This discussion documents the review conducted to determine the regulatory differences between DOE and NRC requirements that would be applicable to a ALLMW treatment facility. The identification of potential regulatory gaps between NRC licensing for operation and DOE-owned operations focuses on the DOE requirements that are more conservative or are in addition to the NRC requirements. However, determination of the legal applicability of these requirements to a private mixed waste treatment facility contracted for DOE waste treatment would be necessary.

By identifying the differences between DOE and NRC, a requirements baseline has been established. The review is applicable only to general operational requirements. If DOE requirements are applicable to the private sector contractor, a thorough review of additional DOE requirements such as design and detailed operational requirements should be performed. Such a detailed review would ensure all applicable DOE requirements are implemented.

To initiate the process of defining differences between the DOE and NRC requirements, a number of DOE Orders were reviewed for consistency with either requirements. Table A1–G identifies the DOE Orders that were initially reviewed.

Comparison of DOE, NRC, and EPA Requirements

Each DOE order was reviewed for additional, or more conservative DOE requirements as related to published NRC or EPA requirements. When such a requirement was identified it was documented in Table A1–H. The table provides the general requirements or applicable subject matter; the specific DOE requirement and citation; a paraphrase of the requirement; and a comparison to NRC or EPA requirements.

Three general requirement differences were discovered among the DOE, NRC, and EPA:

• determination of maximum effective dose equivalent (EDE)
• treatment of liquid radioactive waste streams
• mission standard.

The first difference is the method by which the maximum EDE is determined. Both DOE and NRC use 100 millirem (mrem) as the maximum EDE for

Table A1–F. NRC Licensing Requirements

Requirement	Requirement Trigger	Citation	Scope	Comments
Radiation Protection Programs	Activities licensed under 10 CFR Parts 30–35, 39, 40, 50, 60, 61, 70, and/or 72.	10 CFR 20, Subpart B	Requires that licensee must develop, document, and implement a program sufficient to ensure compliance with 10 CFR 20.	Discussed as part of license application.
Waste Disposal Plan	Waste disposal via incineration requires specific authorization in accordance with 10 CFR 20.2002, Method for obtaining approval of proposed disposal procedures.	10 CFR 20, Subpart K	Includes proposed procedures to dispose of licensed material, including the following: analysis of environment, description of waste and disposal process, nature and location of other facilities affected, ALARA analysis and procedures.	Submitted as part of license application.
License Application (By-Product Material)	Manufacture, produce, transfer, receive, acquire, own, possess, or use by-product material.	10 CFR 30, 33	Same as for Special Nuclear Material.	For the purposes of licensing, requirements are similar to those for special nuclear material.
License Application (Source Material)	Receive title to, own, receive, possess, use, transfer, provide for long-term care, deliver or dispose of by-product material or residual radioactive material as defined . . . or source material.	10 CFR 40	Same as for Special Nuclear Material.	For the purposes of licensing, requirements are similar to those for special nuclear material.
License Application (Special Nuclear Material)	Receive title to, own, acquire, deliver, receive, possess use, or transfer special nuclear material.	10 CFR 70	Includes background information, data on SNM, technical qualifications, description of health and safety equipment/ facilities, proposed procedures for health and safety, SNM control and accounting program, decommissioning funding plan.	License application must be submitted at least nine months prior to construction of facility if filed with an environmental report.

members of the public. However, the DOE considers "multiple sources" of radiation in the EDE determination, whereas the EDE for the NRC is specific to the licensee.

The second difference is the treatment of liquid radioactive waste streams. The DOE requires incorporation of the best available technology (BAT) policy as the appropriate level of treatment. A requirement to identify BAT for any waste stream was not found in the NRC documentation. However, the NRC does require the use of the "as low as reasonably achievable" (ALARA) policy that could effectively implement the BAT policy.

The third difference is specific requirements associated with the CAA emission standards. Title 40 CFR, Part 61, Subparts H and I address air emission standards for DOE facilities and those licensed by the NRC, respectively. Specific Subpart H

Table A1–G. DOE Order Review

DOE Order	Title	General Subject
5400.3	Hazardous and Radioactive Mixed Waste Program	Radiation Protection Standards
5400.5	Radiation Protection of the Public and the Environment	Radiation Protection Standards
5440.1	National Environmental Policy Act	Radiation Protection Standards
5480.11	Radiation Protection for Occupational Workers	Radiation Protection Standards
5820.2	Radioactive Waste Management	Radiation Protection Standards
5480.1	Environment, Safety, and Health Program for Department of Energy Operations	Radiation Protection Standards
5480.5	Safety of Nuclear Facilities	Radiation Protection Standards
5500.3	Reactor and Nonreactor Nuclear Facility Emergency Planning, Preparedness, and Response Program for DOE Operations	Radiation Protection Standards
5632.10	Safeguards and Security (S&S) Equipment Standardization	Radiation Protection Standards
5634.1	Facility Approval, Security Surveys, and Nuclear Materials Surveys	Radiation Protection Standards
5660.1	Management of Nuclear Materials	Radiation Protection Standards
DOE/EH-0256T	Radiological Control Manual	(Chapter 2) Radiation Protection Standards (Chapter 4) Packaging and Transportation
5820.2	Radioactive Waste Management	Radiation Protection Standards
5633.2	Control and Accountability of Nuclear Materials, Responsibilities, and Authorities	Special Nuclear Material
5633.3	Control and Accountability of Nuclear Materials	Special Nuclear Material
5633.4	Nuclear Materials Transactions: Documentation and Reporting	Special Nuclear Material

Table A1–H. DOE Requirements That Are More Restrictive Than NRC and EPA Requirements

General Requirement/ Subject	DOE Citation	DOE Requirements	NRC/EPA Comparison
100 mrem effective dose equivalent	DOE 5400.5	Specifies that reasonable efforts shall be made to limit dose to members of the public from multiple sources of radiation to the maximum effective dose equivalent level (100 mrem).	10 CFR 20 implements the 100 mrem effective dose equivalent requirement but is specific to the licensee, (i.e., does not consider multiple sources of radiation).
Treatment of liquid radioactive waste streams	DOE 5400.5	The order adopts the BAT as the appropriate level of treatment for liquid radioactive waste streams.	No NRC citation was found that implements the BAT policy.
Clean Air Act emission standards	40 CFR 61, Subpart H	For purposes of compliance, the maximally exposed individual living within 1.86 mi (3 km) of all sources of emissions in the facility may be modeled using EPA's COMPLY. Submittal of an annual report documenting the highest calculated EDE to any member of the public is required.	40 CFR 61 Subpart I addresses National Emission Standards for Radionuclide Emissions from Facilities Licensed by the NRC. Subpart I does not use a 1.86 mi (3 km) restriction. There are exemptions for an annual report as identified in 61.104.

requirements (DOE facilities) that differ from Subpart I (NRC) are as follows: the DOE identifies monitoring and test procedures for the maximally exposed individual that lives less than 1.86 mi (3 km) from the facility; and an annual report must be completed that identifies the highest calculated EDE to any member of the public. NRC regulations do not impose these requirements.

Mixed Waste Treatment Facility Summary

As illustrated by this case, the regulatory analysis of applicable EPA, DOE, and NRC orders, rules, and regulations are exceedingly complex and detailed. This case study did not deal with the siting issues required by the National Environmental Policy Act (NEPA). NEPA requirements are crucial and would impact any permit or license decision. (See Chapter 3, United States Legal and Legislative Framework, for a brief discussion of NEPA.) The case study demonstrates why the problem of mixed waste treatment and disposal facilities is likely to persist well into the 21st century.

CASE 10
DETERMINING VENTILATION REQUIREMENTS

John F. Montgomery, PhD, ASP, CHMM

During daily painting activities at the RST facility, a general shop area painting operation results in the emission of three pints of methyl ethyl ketone (MEK) (specific gravity = 0.81, molecular weight = 72.12) released each hour of the operation. The TLV for this substance is 200 ppm, the STEL is 300 ppm, and IDLH is 3,000 ppm.

As your company's environmental manager, you are concerned that the concentration level of MEK may be above the TLV. You conduct a series of industrial hygiene samplings and determine that the concentration in the spray room is 300 ppm.

Discussion Questions
From the information just provided, determine:

1. What should be the proper ventilation, in cfm, for your paint booth operation?
2. What is the ventilation, in cfm, for the paint booth being used in your current painting operation?
3. What is the increased amount of ventilation required to lower the TLV to OSHA regulations?

Recommendations
Use the following formula (K Factor is 4):

$$\frac{403 \times SG \times 10^6 \times \text{pints/hour} \times K\ Factor}{MW \times TLV \times 60\ \text{min.}}$$

1. $\dfrac{403 \times .81 \times 10^6 \times 3 \times 6}{72.12 \times 200 \times 60} = \dfrac{5,875,740,000}{865,440}$

   $= 4,526$ cfm

2. $\dfrac{403 \times .81 \times 10^6 \times 3 \times 6}{72.12 \times 300 \times 60} = \dfrac{5,875,740,000}{1,298,160}$

   $= 3,017$ cfm

3. An increase of 1,509 cfm

Appendix 2

Sources
of Help

Robert J. Marecek, MALS

The environmental professional frequently needs highly specialized or up-to-the-minute, unpublished information. The sources for obtaining this information are numerous. Professional societies and trade associations are excellent sources of help; however, their charters of responsibility are varied. To aid the environmental professional's search for information, this appendix presents selected sources and defines their functions.

On a particularly difficult problem, it may be necessary to contact a number of sources before an effective solution can be obtained. Governmental agencies can help with information about regulations or applicable standards. Insurance companies or their associations may offer assistance through their knowledge of a similar problem. The trade association in the industry may have developed materials and aids in solving the problems. The National Safety Council, through its resources and membership, can provide added input to the development of an effective countermeasure.

SERVICE ORGANIZATIONS

United States Environmental Protection Agency
401 M Street, SW
Washington DC 20460
202/260-2090

The U.S. Environmental Protection Agency (EPA) was established in 1970 as an independent agency within the executive branch of the federal government. The agency was created to permit coordinated and effective government action on behalf of the environment. Its mission is to control and abate pollution in the areas of air, pesticides, radiation, solid wastes, toxic substances, and water. It does this through an integrated, coordinated attack on environmental pollution in cooperation with state and local governments.

The EPA performs a variety of research, monitoring, standards setting, and enforcement roles. It coordinates and supports research and antipollution activities by state and local governments, private and public groups, individuals, and educational institutions.

U.S. EPA Regional Offices

The 10 regional offices are the EPA's principal representatives for contacts and relationships with federal, state, interstate, and local agencies; industry; academic institutions; and other public and private groups. They are responsible for accomplishing

465

within their regions the national program objectives established by the agency.

This list gives the telephone numbers and addresses for each regional EPA office, as well as the group of states each office serves.

Region 1 (CT, ME, MA, NH, RI, VT)
John F. Kennedy Federal Building
One Congress Street
Boston, MA 02203
617/565-3420
617/565-3232 (indoor air)
617/565-4502 (radon)
617/565-3744 (asbestos)
617/565-3265 (NESHAP)

Region 2 (NJ, NY, PR, VI)
26 Federal Plaza
New York, NY 10278
212/264-2657
212/264-4410 (indoor air)
212/264-4410 (radon)
212/264-6671 (asbestos)
212/264-6770 (NESHAP)

Region 3 (DE, DC, MD, PA, VA, WV)
841 Chestnut Building
Philadelphia, PA 19107
215/597-9800
215/597-8322 (indoor air)
215/597-4084 (radon)
215/597-3160 (asbestos)
215/597-1970 (NESHAP)

Region 4 (AL, FL, GA, KY, MS, NC, SC, TN)
345 Courtland Street, NE
Atlanta, GA 30365
404/347-4727
404/347-2864 (indoor air)
404/347-3907 (radon)
404/347-5014 (asbestos)
404/347-5014 (NESHAP)

Region 5 (IL, IN, MI, MN, OH, WI)
230 South Dearborn Street
Chicago, IL 60604
312/353-2000

Region 5 Environmental Hotline:
1-800/572-2515 (IL)
1-800/621-8431 (IN, MI, MN, OH, WI)
1-312/886-7930 (outside Region 5)

Region 6 (AR, LA, NM, OK, TX)
1445 Ross Avenue
Dallas, TX 75202-2733

214/655-6444
214/655-7223 (indoor air)
214/655-7223 (radon)
214/655-7223 (asbestos)
214/655-7223 (NESHAP)

Region 7 (IA, KS, NE, MO)
726 Minnesota Avenue
Kansas City, KS 66101
913/551-7000
913/551-7020 (indoor air)
913/551-7020 (radon)
913/551-7020 (asbestos)
913/551-7020 (NESHAP)

Region 8 (CO, MT, ND, SD, UT, WY)
999 18th Street, Suite 500
Denver, CO 80202-2405
303/293-1603
303/293-1440 (indoor air)
303/293-0988 (radon)
303/293-1442 (asbestos)
303/294-7611 (NESHAP)

Region 9 (AZ, CA, HI, NV, AS, GU)
75 Hawthorne Street, A-1-1
San Francisco, CA 94105
415/744-1305
415/744-1133 (indoor air)
415/744-1045 (radon)
415/744-1136 (asbestos)
415/744-1135 (NESHAP)

Region 10 (AK, ID, OR, WA)
1200 Sixth Avenue
Seattle, WA 98101
206/442-1200
206/553-2589 (indoor air)
206/553-7299 (radon)
206/553-4762 (asbestos)
206/553-1757 (NESHAP)

State Environmental Protection Agencies

The responsibilities of state governments are similar to those of the federal government. The agencies listed coordinate and manage the state's pollution control programs. The state environmental protection agencies improve the overall quality of the environment by planning, granting permits, and regulating standards.

Alabama
Department of Environmental Management
1751 Congressman Dickinson Drive
Montgomery, AL 36130
205/271-7761

Alaska
Department of Environmental Conservation
P.O. Box 0
Juneau, AK 99811
907/465-2600

Arizona
Department of Environmental Quality
2006 North Central Avenue
Phoenix, AZ 85004
602/257-2300

Arkansas
Pollution Control and Ecology
8001 National Drive
Little Rock, AR 72209
501/562-7400

California
Environmental Protection Agency
555 Capitol Mall
Sacramento, CA 95814
916/445-3846

Colorado
Health and Environmental Protection Office
Department of Health
4210 East 11th Avenue
Denver, CO 80220
303/331-4510

Connecticut
Division of Environmental Quality
Department of Environmental Protection
165 Capitol Avenue, Room 161
Hartford, CT 06106
203/566-2110

Delaware
Natural Resources and Environment Department
89 Kings Highway
P.O. Box 1401
Dover, DE 19903
302/739-4403

Florida
Department of Environmental Regulations
Twin Towers
2600 Blairstone Road
Tallahassee, FL 32399
904/488-4805

Georgia
Environmental Protection Division
Department of Natural Resources
205 Butler Street, SW
Atlanta, GA 30334
404/656-4713

Hawaii
Environmental Health Administration
Department of Health
1250 Punchbowl Street
Honolulu, HI 96813
808/548-4139

Idaho
Division of Environment
Department of Health and Welfare
1410 North Hilton
Boise, ID 83706
208/334-5840

Illinois
Environmental Protection Agency
2200 Churchill Road
Springfield, IL 62708
217/782-3397

Indiana
Department of Environmental Management
105 South Meridian Street
P.O. Box 6015
Indianapolis, IN 46225
317/232-8162

Iowa
Environmental Protection Division
Department of Natural Resources
Wallace State Office Building
Des Moines, IA 50319
515/281-6284

Kansas
Division of Environment
Department of Health and Environment
Forbes Field, Building 740
Topeka, KS 66620
913/296-1535

Louisiana
Department of Environmental Quality
P.O. Box 44066
Baton Rouge, LA 70804
504/342-1266

Maine
Environmental Protection Department
State House Station #17
Augusta, ME 04333
207/289-2812

Maryland
Department of Environment
2500 Broening Highway
Baltimore, MD 21224
301/631-3084

Massachusetts
Department of Environmental Management
1 Winter Street
Boston, MA 02108
617/292-5856

Michigan
Environmental Protection Bureau
Department of Natural Resources
P.O. Box 30028
Lansing, MI 48909
517/373-7917

Minnesota
Environmental Quality Board
State Planning Agency
658 Cedar Street, Suite 300
St. Paul, MN 55155
612/296-0212

Mississippi
Division of Environmental Quality
Department of Natural Resources
P.O. Box 20305
Jackson, MS 39289
601/961-5000

Missouri
Division of Environmental Quality
Department of Natural Resources
P.O. Box 176
Jefferson City, MO 65102
314/751-4810

Montana
Environmental Sciences Division
Cogswell Building
Helena, MT 59620
406/444-2544

Nebraska
Department of Environmental Control
301 Centennial Mall South
P.O. Box 94877
Lincoln, NE 68509
402/471-2186

Nevada
Division of Environmental Protection
123 West Nye Lane
Carson City, NV 89710
702/687-4670

New Hampshire
Department of Environmental Services
6 Hazen Drive
Concord, NH 03301
603/271-3503

New Jersey
Department of Environmental Protection
401 East State Street, CN 402
Trenton, NJ 08625
609/292-2885

New Mexico
Department of Environment
P.O. Box 26110
Santa Fe, NM 87502
505/827-2850

New York
Department of Environmental Conservation
50 Wolf Road
Albany, NY 12233
518/457-3446

North Carolina
Division of Environmental Management
Environment, Health and Natural Resources
512 North Salisbury Street
P.O. Box 27687
Raleigh, NC 27604
919/733-7015

North Dakota
Environmental Health Section
Department of Health
1200 Missouri Avenue
P.O. Box 5520
Bismarck, ND 58502
701/224-2374

Ohio
Environmental Protection Agency
1800 Watermark Drive
P.O. Box 1049
Columbus, OH 43266
614/644-2782

Oklahoma
Environmental Health Services
Department of Health
1000 NE 10th Street
Oklahoma City, OK 73117
405/271-8056

Oregon
Department of Environmental Quality
811 SW Sixth Avenue
Portland, OR 97204
503/229-5300

Pennsylvania
Department of Environmental Resources
Fulton Building, 9th Floor
P.O. Box 2063
Harrisburg, PA 17105
717/787-5028

Rhode Island
Department of Environmental Management
9 Hayes Street
Providence, RI 02908
401/277-2771

South Carolina
Environmental Quality Control Office
Department of Health and Environmental Control
2600 Bull Street
Columbia, SC 29201
803/734-5360

South Dakota
Department of Water and Natural Resources
523 East Capitol Avenue, Joe Foss Building
Pierre, SC 57501
605/773-3151

Tennessee
Bureau of Environment
Department of Environment and Conservation
150 Ninth Avenue, North, 1st Floor
Nashville, TN 37247
615/741-3657

Texas
Environment and Consumer Health Protection
Department of Health
1100 West 49th Street
Austin, TX 78756
512/458-7541

Utah
Division of Environmental Health
Department of Health
288 North 1460 West
P.O. Box 16690
Salt Lake City, UT 84114
801/538-6121

Vermont
Department of Environmental Conservation
Agency of Natural Resources
103 South Main Street, 1 South Building
Waterbury, VT 05676
802/244-8755

Virginia
Council on the Environment
903 Ninth Street Office Building
Richmond, VA 23219
804/786-4500

Washington
Department of Ecology
M/S: PV-11
Olympia, WA 98503
206/459-6168

West Virginia
Division of Natural Resources
Commerce, Labor and Environmental Resource Department
Capitol Complex, Building 3
Charleston, WV 25305
304/348-2754

Wisconsin
Environmental Equality Division
Department of Natural Resources
P.O. Box 7921
Madison, WI 53707
608/266-1099

Wyoming
Department of Environmental Quality
Herschler Building
122 West 25th Street
Cheyenne, WY 82002
307/777-7938

District of Columbia
Housing and Environmental Regulation Administration
Consumer and Regulatory Affairs
614 H Street, NW, Room 505
Washington DC 20001
202/727-7395

American Samoa
Environmental Quality Commission
Office of the Governor
Pago Pago, AS 96799
684/633-2304

Guam
Environmental Protection Agency
130 Rojas Street
Harmon, GU 96911
671/646-8863

Northern Mariana Islands
Environmental Quality Division
Public Health and Environmental Services
P.O. Box 409
Saipan, MP 96950
670/234-6114

Puerto Rico
Environmental Quality Board
P.O. Box 11488
Santurce, PR 00910
809/767-8056

U.S. Virgin Islands
Department of Planning and Natural Resources
Nisky Center, Suite 231
St. Thomas, VI 00802
809/774-3320

PROFESSIONAL AND STANDARDS SETTING ORGANIZATIONS

Organizations

Air and Waste Management Association
Box 2861
Pittsburgh, PA 15230
412/232-3444
A not-for-profit technical and educational organization that collects and disseminates information about air pollution control and hazardous waste management.

Air Conditioning and Refrigeration Institute
4301 North Fairfax Drive, No. 425
Arlington, VA 22203
703/524-8800
An organization consisting of manufacturers of air conditioning, refrigeration, and heating products and components. The organization develops and establishes equipment and application standards and certifies the performance of certain industry products.

American Academy of Environmental Engineers
130 Holiday Court, #100
Annapolis, MD 21401
301/266-3311
A professional organization consisting of about 2,600 sanitary and environmental engineers. The academy publishes *Who's Who in Environmental Engineering* and the *Environmental Engineering Selection Guide*.

American Association for the Advancement of Science
1333 H Street, NW
Washington DC 20005
202/326-6400
The largest general scientific organization (135,000 members) representing all disciplines of science. Its objectives are to further the work of scientists and to improve the effectiveness of science in promoting human welfare.

American Board of Toxicology
4208 6th Fork Road, Suite 220G
Raleigh, NC 27609
919/782-0036
Conducts a certification program in toxicology; also holds annual certification and recertification exams.

American Chemical Society
1155 16th Street, NW
Washington DC 20036
202/872-4600
A national association of chemists that supports all branches of chemistry. The society publishes *Environmental Sciences & Technology*.

American Conference of Governmental Industrial Hygienists
6500 Glenway Avenue, Building D7
Cincinnati, OH 45211
513/661-7881
Society of persons employed by governmental units responsible for programs and research dealing with industrial hygiene. Chief concern is with worker health protection.

American College of Toxicology
9650 Rockville Pike
Bethesda, MD 20814
301/571-1840
A multidisciplinary society whose members have a common interest in toxicology. It publishes the *Journal of the American College of Toxicology.*

American Ground Water Trust
6375 Riverside Drive
Dublin, OH 43017
614/761-2215
This organization is dedicated to the optimal utilization and protection of groundwater.

American Industrial Health Council
1330 Connecticut Avenue, NW, Suite 300
Washington DC 20036
202/659-0060
The Council consists of chemical and related industries working to promote scientific methods to identify potential industrial carcinogens, mutagens, and teratogens.

American Industrial Hygiene Association
2700 Prosperity Avenue, Suite 250
Fairfax, VA 22031
703/849-8888
This association is made up of occupational and environmental health professionals. It promotes the study and control of environmental stresses in the workplace and the effects of products on the health of workers and the public.

American Institute of Chemical Engineers
345 East 47th Street
New York, NY 10017
212/705-7338
The national society for chemical engineers, with over 51,000 members. Publications include *AICHE Journal, Chemical Engineering Progress,* and *Plant/Operations Progress.*

American Mining Congress
1920 N Street, NW, Suite 300
Washington DC 20036
202-861-2800
An industry association of producers of metals, coal, industrial and agricultural minerals, manufacturers of mining processing and supplies, and mining engineering firms.

American Petroleum Institute
1220 L Street, NW
Washington DC 20005
202/682-8000
The petroleum industry trade association consisting of major oil companies, independent oil producers and fuel distributors, and owners of service stations. It produces publications in the area of health and the environment.

American Society of Agricultural Engineers
2950 Niles Road
St. Joseph, MI 49085
616/429-0300
A professional and technical organization interested in the science and application of engineering in the numerous aspects of farming.

American Society of Civil Engineers
345 East 14th Street
New York, NY 10017
212/705-7496
A national association of engineers with more than 110,000 members. Publications include the *Journal of Environmental Engineering.*

American Society of Sanitary Engineers
Box 40362
Bay Village, OH 44140
216/835-3040
A national association of engineers who work on sanitary projects dealing with liquid waste collection, processing, and disposal.

American Society for Testing and Materials
1916 Race Street
Philadelphia, PA 19103
215/299-5000
A standards-setting organization for materials, products, systems, and services. Has developed more than 9,000 standard test methods, specifications, classifications, definitions, and recommended practices. Publishes an annual book of ASTM standards.

American Society of Heating, Refrigeration and Air Conditioning Engineers
1791 Tullie Circle, NE
Atlanta, GA 30329
404/636-8400
Society composed of heating, ventilating, refrigeration, and air-conditioning engineers. Sponsors

research projects dealing with the effects of air-conditioning, quality of indoor air, heat transfer and flow, and cooling processes. Publishes the *ASHRAE Journal.*

Association of American Pesticide Control Officials
P.O. Box 1249
Hardwick, VT 05843
802/472-6956
This organization is composed of state, municipal, and federal officials interested in a uniform approach to the enforcement of laws controlling the proper handling and use of pesticides.

Association of Ground Water Scientists and Engineers
6375 Riverside Drive
Dublin, OH 43017
614/761-1711
This organization of about 15,000 members is devoted to communicating information about the occurrence, development, and protection of groundwater.

Association of Metropolitan Sewerage Agencies
1000 Connecticut Avenue, NW, Suite 1001
Washington DC 20006
202/682-5886
Composed of municipal sewerage agencies, this association serves to exchange technical data and deal with the federal government on environmental and regulatory matters.

Association of State and Territorial Solid Waste Management Officials
444 North Capitol Street, NW, Suite 388
Washington DC 20001
202/624-5828
A nonprofit trade association that represents waste management program officials in states and territorial agencies who are involved in the management of solid and hazardous waste, and remediation programs.

Cement Kiln Recycling Coalition
1101 30th Street, NW, 5th Floor
Washington DC 20007
202/625-3440
This coalition was created to promote the use of wastes as fuel in cement kilns while promoting the environmental health and safety of the process. Its membership includes major cement producers and companies that handle wastes to use as fuel in cement kilns.

Center for Environmental Management
Tufts University
Curtis Hall
474 Boston Avenue
Medford, MA 02155
As a research, education, and training institute, the center focuses on emerging environmental issues such as solid wastes, prevention of pollution, risk communication, environmental management in corporations, and worldwide environmental problems.

Chemical Manufacturers Association
2501 M Street, NW
Washington DC 20037
202/887-1100
This organization consists of companies that manufacture chemicals in the United States and Canada. The association publishes *ChemEcology*, a monthly newsletter.

Chemical Producers and Distributors Association
1430 Duke Street
Alexandria, VA 20314
703/548-7700
A voluntary, not-for-profit association consisting of companies that either manufacture, formulate, distribute, or sell products used in the production of food, feed, and fiber crops, and for home lawn and garden care.

Chemical Waste Transportation Council
1730 Rhode Island Avenue, NW, Suite 1000
Washington DC 20036
202/659-4613
This Council represents the interests of companies that haul chemical wastes. It is affiliated with the National Solid Wastes Management Association.

Chem Trec Center-Nonemergency Services
2501 M Street, NW
Washington DC 20037
202/887-1100
800/CMA-8200
Created by the Chemical Manufacturer's Association, the Center responds to requests for information on chemicals in nonemergency situations. It is supported by more than 130 companies, many of which can provide lists of their products.

Environmental Business Association
1150 Connecticut Avenue, NW
Washington DC 20036
202/862-4363
This national organization consists of members who represent all segments of the environmental industry. Its services include an information exchange, sponsoring educational services, and tracking industry trends, technology changes, and legislation.

Environmental Industry Council
1825 K Street, NW, Suite 210
Washington DC 20006
202/331-7706
The council consists of companies that manufacture environmental protection equipment and materials.

Hazardous Materials Advisory Council
1110 Vermont Avenue, NW, Suite 250
Washington DC 20005
202/728-1460
A national organization representing the entire hazardous materials industry. It deals with safety in the domestic and international transportation of hazardous materials and wastes.

Hazardous Materials Control Research Institute
9300 Columbia Boulevard
Silver Springs, MD 20910-1702
301/587-9390
The institute is a national membership organization consisting of engineers, scientists, government administrators, and corporations interested in the problems of managing and controlling hazardous materials. It publishes the bimonthly magazine *Hazardous Materials Control*.

Hazardous Waste Treatment Council
1440 New York Avenue, Suite 310
Washington DC 20005
217/783-0870
The council is a trade association of waste disposal firms that make use of high-technology processing rather than land disposal of wastes.

Industrial Gas Cleaning Institute
1707 L Street, NW #570
Washington DC 20036
202/457-0911
The institute is composed of manufacturers of industrial air pollution control equipment for stationary sources.

Institute for Environmental Auditing
P.O. Box 23686
L'Enfant Plaza Station
Washington DC 20026 3686
This nonprofit organization promotes the development of guidelines for environmental audits and programs, and encourages its use as a tool for business planning management.

Institute of Hazardous Materials Managers
11900 Parklawn Drive, #220
Rockville, MD 20852
301/984-8969
The institute certifies environmental engineers and scientists leading to the credential of Certified Hazardous Materials Manager.

Institute of Noise Control Engineering
Box 3206, Arlington Branch
Poughkeepsie, NY 12603
914/462-4006
This professional organization is concerned with noise control technology to solve environmental noise problems. The institute publishes *Noise Control Engineering Journal*.

Institute of Scrap Recycling Industries, Inc.
1325 G Street, NW, Suite 1000
Washington DC 20005
202/466-4050
This trade association represents processors, brokers and consumers of scrap commodities, as well as suppliers of equipment and services to the industry. It publishes the bimonthly magazine, *Scrap Processing and Recycling*.

International Labour Organization (ILO)
(Organisation Internationale du Travail—OIT)
4, rue des Morillons
CH-1211 Geneva 22, Switzerland
22 7996111
A specialized agency associated with the United Nations. Exists to promote the voluntary cooperation of nations to improve labor conditions and raise living standards, thereby improving prospects of peace by fostering economic and social stability throughout the world. Sets standards covering all aspects of work life, including human rights, through the adoption of conventions and recommendations. Aids developing nations through a technical cooperation program.

Publishes *Bulletin of Labour Statistics* (in English, French, and Spanish) quarterly, *ILO Information* (in English, French, and Spanish), 5/year, *International Labour Documentation* (in English, French, and Spanish), monthly, *International Labour Review* (in English, French, and Spanish), bimonthly.

International Organization for Standardization (ISO)

1, rue de Verembe
Case Postale 56
CH-1121 Geneva 20, Switzerland
22 749 0111
An organization composed of national standards bodies united to promote standardization worldwide. Facilitates exchange of goods and services and fosters mutual cooperation in intellectual, scientific, technological, and economic spheres of endeavor. Develops and publishes standards, including the ISO 9000 series.

National Agricultural Chemicals Association

1155 15th Street, NW
Madison Building, Suite 900
Washington DC 20005
202/296-1585
A nonprofit organization of manufacturers, formulators, and distributors of pest control products used in agriculture.

National Association of Environmental Professionals

P.O. Box 15210
Alexandria, VA 22309-0210
703/660-2364
This interdisciplinary society consists of people dealing with all aspects of the environmental process—administration, planning, engineering, law, research, and education. It awards the CEP (Certified Environmental Professional) designation.

National Association of Environmental Risk Auditors

4211 East Third Street
Bloomington, IN 47401
812/333-0077
This organization trains and certifies environmental professionals on three levels: Certified Environmental Risk Screener (CERS), Certified Environmental Audit Reviewer (CEAR), and Certified Environmental Risk Auditor (CERA).

National Association of Chemical Recyclers

1875 Connecticut Avenue, NW
Washington DC 20009
202/986-8150
This association represents companies interested in the environmentally sound recycling of solvents, waste water, used oil, certain metals, and organic compounds.

National Conference of States on Building Codes and Standards

505 Huntmar Park Drive, Suite 210
Herndon, VA 22090
703/437-0100
State building code officials, building-related manufacturers, associations, educators, and consumer groups seeking solutions to the many problems in the building regulatory system.

National Council of the Paper Industry for Air and Stream Improvements

260 Madison Avenue
New York, NY 10016
212/532-9000
This technical organization deals with environmental problems in the manufacture of pulp, paper, and wood products.

National Environmental Health Association

720 South Colorado #970, South Tower
Denver, CO 80222
303/756-9096
Members of this national professional and educational organization deal with all areas of the environmental health field. The NEHA tests and certifies environmental health technicians and hazardous waste specialists, and provides continuing education. It publishes the *Journal of Environmental Health*.

National Safety Council

1121 Spring Lake Drive
Itasca, IL 60143-3201
708/285-1121
The National Safety Council is the largest organization in the world devoting its entire efforts to the prevention of accidents and illnesses. It is nongovernmental, not-for-profit, and nonpolitical. Its staff members work as a team with more than 2,000 volunteer officers, directors, and members of various divisions and committees to develop accident prevention material and programs in specific areas of safety, health, and environment. These areas include industrial, traffic, home, recreational, and public

safety. The Council headquarters includes one of the largest safety libraries in the world.

At the Itasca, Illinois, headquarters, a staff of more than 250—about half of whom are engineers, editors, statisticians, writers, educators, data processors, librarians, and other specialists—carry out the major activities. In addition to its main office and Distribution Center, the council has regional offices in California, New York, Virginia, Georgia, and the United Kingdom as well as a Public Policy Office and Environmental Health Center located in Washington DC. It also has three strategic alliances in Texas, Iowa, and western Pennsylvania, and a franchise in Portland, OR.

Industrial Division. Because recognizing industry's safety and health problems often requires specialized treatment, the council has arranged its industrial effort into sections guided by the Industrial Division. Each section is administered by its own executive committee, elected from its own membership.

The industrial membership of the Council is organized according to the following industries. These sections are designed to provide special help for all facets of the industrial section, as shown here:

- Aerospace. Missile and aircraft manufacturer, related components
- Automotive, Tooling, Metalworking, and Associated Industries. Machining fabrication, assembly manufacturing
- Cement, Quarry and Mineral Aggregates. Quarrying, processing and marketing
- Chemical. Manufacturing chemical compounds and substances
- Coal Mining. Underground and open pit coal operations
- Food and Beverage. Dairies, brewers, confectioners, distillers, canners and freezers, grain handling and processing, meat processing, restaurants, and related operation
- Forest Industries. Logging, pulp and paper, plywood, and related products
- Health Care. Hospital, patient, employee, visitor safety
- International Air Transport. Ground terminal safety affecting personnel and equipment
- Marine. Crew, passenger, vessel safety, stevedoring, shipbuilding, repair, cargo safety
- Metals. Foundries, ferrous and nonferrous manufacturing, fabricating, other metal products
- Mining. Metals and minerals extraction and processing

- Petroleum. Exploration, drilling, production, pipeline, marketing, retail
- Power Press and Forging. Metal stamping and forming, and forging
- Printing and Publishing. Newspaper and commercial publishing
- Public Employee. Employees and governments from local to national levels and associated activities
- Public Utilities. Communications, electric, gas, water, cable TV
- Railroad. Employee, passenger freight, and public safety
- Research and Development. Laboratory safety
- Rubber and Plastics. Manufacturers of rubber plastic products
- Textile and Apparel. Manufacturing and fabrication, natural and synthetic fibers, ginning and finished goods
- Trades and Services. Food service, hotels, motels, mercantile automotive, leather, financial institutions, warehouses, offices, recreational facilities.

Assignment of Council Industrial Division members to sections provides them greater interest-specific involvement with similar organizations and professionals. Many members volunteer their time and expertise to assist in accomplishing the council's mission by serving on committees and in other leadership capacities related to their respective personal safety and health specialties and interests.

The Industrial Division meets two times a year to review current occupational safety and health problems and to determine, on a national scale, the best procedures to follow in providing increasingly beneficial programs to industry. The division is made up primarily of industrial members of the council. It composes section general chairmen, vice-general chairment, and other members-at-large, drawn from business and industry member organizations, professional and trade associations, and other groups.

The Council also has divisions concerned with construction, labor, motor transportation, traffic, and agriculture. Each is concerned with programs, policies, and training.

Improvement of Council services through new technical materials and visual aids takes a major portion of each committee's meeting time. Membership solicitation is an ongoing program. Planning of National Safety Council programs is given careful consideration. Special committees are often assigned to work on problems unique to an industry and on which there are no ready program materials.

Program materials. The following Council publications have proven to be particularly useful for industrial and off-the job safety programs (unless otherwise stated, they are published monthly):

- *Family Safety & Health* (quarterly)
- *Today's Supervisor*
- *Safety & Health*
- *OSHA Updatge*
- *Product Safety Up-to-Date* (bimonthly)
- Section Newsletter (one for each of the 26 sections—six issues a year)
- *Safe Driver* (issued in three editions—Car, Truck, and Bus
- *Safe Worker*
- *Traffic Safety* (six issues a year).
- *Healthcare Supervisor*

In addition, the following statistical materials are also available from the council:

- *Accident Facts* (annually)
- Section Contest Bulletins (bimonthly)
- Work Injury and Illness Rates (annually)

Technical materials (see current Council General Materials Catalog for a complete listing):

- *Accident Prevention Manual for Business & Industry,* 3 vols.
  — *Administration & Programs* volume
  — *Engineering & Technology* volume
  — *Environmental Management* volume
- *Aviation Ground Operations Handbook*
- *Fundamentals of Industrial Hygiene*
- Industrial Data Sheets (a series; complete listing available)
- *Industrial Noise and Hearing Conservation*
- *Motor Fleet Safety Manual*
- *Occupational Health and Safety*
- *Power Press Safety Manual*
- *Safeguarding Concepts Illustrated*
- *Safety Guide for Health Care Institutions*
- *Safety Handbook for Office Supervisors.*

Training and motivational materials (see Poster Catalog for a complete listing):

- Banners
- Booklets
- Calendars
- Films
- Posters
- Supervisory training pamphlets
- Audiovisuals.

Meetings. The National Safety Council sponsors the annual Congress and Exposition, one of the largest conventions held anywhere. Nearly 200 general or specialized sessions, workshops, and clinics are held covering the full range of safety and health topics.

The annual meetings or special business meetings of a dozen allied organizaitons and associations are also held concurrent with the Congress, greatly enhancing the exchange of views and information in safety and health.

Special services. Special services available throughout the Council to support industrial safety programs include:

- *Library.* With a collection of more than 140,000 documents, of which 105,000 are indexed on an in-house computerized database, the National Safety Council's library is one of the most comprehensive safety and occupational health libraries in the world. The library is available to the general public, and it networks with other data bases for safety and health information.
- *Statistics.* The Council's statisticians provide a highly refined statistical capability. They are a recognized source of reliable, accurate, and authoritative data within the safety community. Equipped with data processing and research tools, they study various types of accident data in the continuing search for clues on the causes of accidents.

Safety Training Institute. Among the strengths of the National Safety Council's wealth of safety and health advocacy programs is its renowned Safety Training Institute (STI), which was founded in 1946. STI offers training of all types for industry managers, safety professionals, and others who supervise workplace activities, including every management level from corporate senior officers to shop foremen. More than 24,000 students have completed courses. Classes are held at STI's Chicago facilities, in the Council's regional offices, and on-site at company locations provided by organizations.

Course offerings cover such general subjects as principles of occupational safety and health, safety training methods, and safety management techniques, as well as specific topics that include industrial hygiene, hazard communication, ergonomics, product safety, chemical hazards, laboratory, hospital, advanced hospital safety, and advanced safety concepts.

The National Safety Council awards continuing education units (CEUs) for its STI courses. In addition to CEUs, STI students may earn the Council's

Advanced Safety Certificate by completing certain course tracks.

Further information about the STI is available from the annual *Safety Training Institute Course Catalog & Schedule*, wherein are listed dates of presentation, course content, and other courses.

The institute also offers two home study courses: Supervising for Safety and Protecting Workers' Lives. These courses are recognized by the National Home Study Council.

Two supervisor training courses are also available for purchase from STI and have been designed for presentation by in-house training personnel. These programs include a Management Development Program, which is oriented to the new supervisor, and Supervisors' Development Program for individuals with six or more months' experience. (Complete information on these other courses can be found in the National Safety Council's *General Materials Catalog*.)

The Forklift Truck Operators Training Course is an eight-hour program designed to help a company comply with training regulations. The course can be presented in the standard form or can be tailored to fit an individual company's facilities and needs.

A number of these courses offered by STI are also presented by local safety councils.

The STI can tailor courses and seminars to customer needs and present them at customer locations. Information on costs, scheduling, and specifics is available from the STI.

Consulting. The Council consulting services are available for occupational safety and health, motor fleet, traffic, industrial hygiene, environmental, community, product, motivation, agricultural, and research and statistical services.

The safety management consulting services can provide an analysis, recommendation, and structure for a safe and healthful working environment. Follow-up audits, training, and support services are also available. Information available upon request.

Speakers Bureau. The Speakers Bureau is an ideal resource for nationally known speakers who can cover a wide range of safety, health, environmental, and motivational subjects. Long or short presentations are available. Speakers are ideal for seminars, quarterly meetings, and conferences.

First Aid Insitute. The Council's First Aid Institute offers programs, certification of instructors, and meterials for teaching first aid and cardiopulmonary resuscitation (CPR) to employees. The courses of several levels meet requirements of governmental agencies and other organizations.

National Solid Waste Management Association
1730 Rhode Island Avenue, NW, Suite 1000
Washington DC 20036
202/659-4613
This association represents persons with responsibilities for the management of solid waste in public agencies. It publishes *Waste Age*.

Organisation for Economic Co-operation and Development (OECD)
(Organisation de Cooperation et de Developpment Economique)
2, rue Andre Pascal
F-75775 Paris Cedex 16, France
1 45248200
Representatives of West European governments, Australia, Canada, Japan, New Zealand, Turkey, and the United States. Seeks to stimulate economic cooperation between member countries, expand world trade, and coordinate aid for less developed areas of the world. Publishes *Energy Prices and Taxes*, quarterly, *Financial Market Trends*, 3/year, *Financial Statistics*, biweekly, *Indicators of Industrial Activity*, quarterly, *Main Economic Indicators*, monthly, etc.

Spill Control Association of America
400 Renaissance Center, Suite 1900
Detroit, MI 48243
313/567-0500
This organization consists of companies and individuals concerned with cleaning up spills of hazardous materials. It also represents manufacturers of products for spill control, cleanup, and the protection of individuals.

Product Manufacturers

Adhesive and Sealant Council
1627 K Street, NW, Suite 1000
Washington DC 20006-1707
202/452-1500

Asbestos Information Association
1745 Jefferson Davis Highway, Room 509
Arlington, VA 22202
703-979-1150

Business Council on Indoor Air Quality
1225 19th Street, Suite 300
Washington DC 20036
202/775-5887

Carpet and Rug Institute
310 Holiday Avenue
Dalton, GA 30720
404/278-3176

Chemical Specialties Manufacturers Association
1913 I Street, NW
Washington DC 20006
202/872-8110

Electric Power Research Institute
P.O. Box 10412
Palo Alto, CA 94303
415/855-2902

Formaldehyde Institute, Inc.
1330 Connecticut Avenue, NW
Washington DC 20036
202/822-6757

Foundation of Wall and Ceiling Industries
1600 Cameron Street
Alexandria, VA 22314-2705
703/548-0374

Gas Research Institute
8600 West Bryn Mawr Avenue
Chicago, IL 60631
312/399-8304

National Paint and Coatings Association
1500 Rhode Island Avenue, NW
Washington DC 20005
202/462-6272

Thermal Insulation Manufacturers Association
Technical Services
Air Handling Committee
1420 King Street
Alexandria, VA 22314
703/684-0474

Building Service Associations

Air-Conditioning and Refrigeration Institute
1501 Wilson Boulevard, 6th Floor
Arlington, VA 22209
703/524-8800

Air Conditioning Contractors of America
16th Street, NW
Washington DC 20036
202/483-9370

American Consulting Engineers Council
1015 15th Street, NW, Suite 802
Washington DC 20005
202/347-7474

Associated Air Balance Council
1518 K Street, NW
Washington DC 20005
202/737-0202

Association of Energy Engineers
4025 Pleasantdale Road, Suite 420
Atlanta, GA 30340
404/447-5083

Association of Specialists in Cleaning and
Restoration International
10830 Annapolis Junction Road, Suite 312
Annapolis Junction, MD 20701
301/604-4411

National Air Duct Cleaners Association
1518 K Street, NW, Suite 503
Washington DC 20005
202/737-2926

National Association of Power Engineers
3436 Haines Way, Suite 101
Falls Church, VA 22041
703/845-7055

National Energy Management Institute
601 North Fairfax Street, Suite 160
Alexandria, VA 22314
703/739-7100

National Environmental Balancing Bureau
1385 Piccard Drive
Rockville, MD 20850
301/977-3698

National Pest Control Association
8100 Oak Street
Dunn Loring, VA 22027
703/573-8330

Sheet Metal and Air Conditioning Contractors
National Association
4201 LaFayette Center Drive
Chantilly, VA 22021
703/803-2980

Unions

AFL-CIO
Department of Occupational Safety and Health
815 16th Street, NW
Washington DC 20006
202/637-5000

American Federation of Government Employees
80 F Street, NW
Washington DC 20001
202/737-8700

American Federation of State, County, and Municipal Employees
1625 L Street, NW
Washington DC 20036
202/429-1215

American Federation of Teachers
555 New Jersey Avenue, NW
Washington DC 20001
202/879-4400

Communication Workers of America
501 3rd Street, NW
Washington DC 20001
202/434-1160

International Union of Operating Engineers
1125 17th Street, NW
Washington DC 20036
202/429-9100

Service Employees International Union
1313 L Street, NW
Washington DC 20005

Environmental, Health, and Consumer Organizations

American Academy of Allergy and Immunology
611 East Wells Street
Milwaukee, WI 53202
414/272-6071

American Lung Association
1740 Broadway
New York, NY 10019

Consumer Federation of America
1424 16th Street, NW, Suite 604
Washington DC 20036

National Center for Environmental Health Strategies
1100 Rural Avenue
Voorhees, NJ 08043
609/429-5358

National Environmental Health Association
720 South Colorado Boulevard
South Tower, Suite 970
Denver, CO 80222
303/756-9090

National Foundation for the Chemically Hypersensitive
P.O. Box 9
Wrightsville Beach, NC 28480
517/697-3989

Occupational Health Foundation
1126 16th Street, NW
Washington DC 20036
202/842-7840

ELECTRONIC INFORMATION SYSTEMS

EPA-OLS (Online Library System)
U.S. Environmental Protection Agency
401 M Street, SW
Washington DC 20460
This on-line system is composed of the following seven data bases:

1. National Catalog Data Base: contains bibliographic citations of EPA reports that have been submitted to the National Technical Information Service, as well as the holdings of the 28 libraries in the EPA Library Network.
2. Hazardous Waste Data Base: contains references to key materials on hazardous waste in the EPA Library Network. Citations include books, EPA reports, and Office of Solid Waste and Emergency Response (OSWER) policy and guidance directives.
3. Clean Lakes Data Base: information on lake management protection and restoration.
4. Region I Data Base: holdings of the EPA Library in Boston.

5. Region V Data Base: holdings of the EPA Library in Chicago.
6. Region IX Data Base: holdings of the EPA Library in San Francisco.
7. Chemical Collection System: citations for items relating to chemicals.

Information on accessing the Online Library System is available through EPA headquarters or the regional offices.

ENVIROLINE

R.R. Bowler Electronic Publishing
121 Chanlon Road
New Providence, NJ 07974
908/404-6800
800/521-8110
This data base covers the world's environmental literature. It is comprehensive, indexing and abstracting more than 5,000 international publications dealing with the environment. The data base has more than 165,000 citations.

Environmental Bibliography

Environmental Studies Institute
International Academy at Santa Barbara
800 Garden Street, Suite D
Santa Barbara, CA 93101
805/965-5010
Environmental Bibliography has been on-line since 1973, with more than 426,000 records. More than 300 periodicals are indexed in this data base, providing a good access to articles on environmental research.

HAZARDLINE

Occupational Health Services, Inc.
11 West 42nd Street
New York, NY 10036
212/789-3535
800/445-6737
This data base has full-text information on more than 4,000 dangerous materials. It is constantly updated in order to provide a complete record of each substance listed.

National Safety Council

Library
1121 Spring Lake Drive
Itasca, IL 60143-3201
708/285-1121

The National Safety Council Library is a comprehensive source of safety and health information. It has 140,000 documents, of which 105,000 are on an in-house data base. The environmental documents relate to worker safety and health, and the public's concern with major environmental issues (e.g., safe drinking water, air pollution, radon.)

National Technical Information Services (NTIS)

U.S. Department of Commerce
5285 Port Royal Road
Springfield, VA 22161
703/487-4929
A bibliographic citation data base that consists of government-sponsored technical reports. Through this data base, unclassified reports are available for sale from federal agencies. Subjects include all areas of science, engineering, and technology. Coverage starts at 1964; there are more than 2 million entries.

PAPERCHEM

Institute of Paper Science and Technology
575 14th Street, NW
Atlanta, GA 30318
404/853-9500
This data base covers the literature on pulp and paper in all aspects, including environmental concerns.

SciSearch®

Institute for Scientific Information (ISI)
3501 Market Street
Philadelphia, PA 19104
215/386-0100
800/386-2911
SciSearch® is a multidisciplinary on-line index to the journal literature published in more than 3,000 leading scientific and technical journals.

TOXLINE & TOXLIT

National Library of Medicine
Toxicology Information Program
8600 Rockville Pike
Bethesda, MD 20894
301/496-1131
800/638-8480
The TOXLINE (Toxicology Information Online) and TOXLIT (Toxicology Literature from Special Studies) data bases are compilations of on-line bibliographic information covering the toxicological effects of drugs and chemicals.

Transportation Research Information Service (TRIS)
Transportation Research Board
2101 Constitution Avenue, NW
Washington DC 20418
202/334-5250
TRIS is a bibliographic data base consisting of information on air, rail, water, and road transport. It covers regulations, legislation, the environment, safety, construction, traffic control, and other areas related to transportation.

PUBLICATIONS

The books listed in this section supplement the specific references listed at the end of each chapter in this manual. The material cited in this bibliography was selected to provide safety and health professionals with information useful in dealing with the problems of worker health protection and hazard assessment in relation to environmental issues. This compilation should not be viewed as comprehensive coverage of the abundant literature on the environment, nor is any endorsement implied.

General Information

Cone JE, Hodgson MJ. *Problem Buildings: Building-Associated Illness and the Sick Building Syndrome.* From the series: *Occupational Medicine: State of the Art Reviews.* Hanley & Belfus, Inc., 210 South 13th Street, Philadelphia, PA 19107.

Godish T. *Indoor Air Pollution Control.* 1989. Lewis Publishers, 121 South Main Street, Chelsea, MI 48118.

Rajhans GS. *Report of the Interminsterial Committee on Indoor Air Quality.* 1988. Contact: G Rajhans, Health and Safety Support Services Branch Ministry of Labour, 400 University Avenue, 7th Floor, Toronto, Ontario, Canada M7A 1T7.

Sheet Metal and Air Conditioning Contractor's National Association, Inc. (SMACNA). *Indoor Air Quality.* 1988. 8224 Old Courthouse Road, Vienna, VA 22180.

U.S. Environmental Protection Agency and the Public Health Foundation. *Directory of State Indoor Air Contacts.* Updated 1991. (EPA Public Information Center (PM-211B), 401 M Street, SW, Washington DC 20460. 202/260-2080.)

U.S. Environmental Protection Agency. *Design for Good Indoor Air Quality: An Introduction for Design Professionals.* (In progress.)

U.S. Environmental Protection Agency. *Project Summaries: Indoor Air Quality in Public Buildings.* 1988. Contains findings of research on IAQ in 10 new public and commercial buildings and on building material emissions. (EPA Public Information Center (PM-211B), 401 M Street, SW, Washington DC 20460. 202/260-2080.)

U.S. Environmental Protection Agency and the U.S. Consumer Product Safety Commission. *The Inside Story: A Guide to Indoor Air Quality.* 1988. Addresses residential indoor air quality primarily but contains a section on offices. (EPA Public Information Center (PM-211B), 401 M Street, SW, Washington DC 20460. 202/260-2080.)

U.S. Environmental Protection Agency. *Sick Building Syndrome.* Indoor Air Quality Fact Sheet #4. Revised, 1991. (EPA Public Information Center (PM-211B), 401 M Street, SW, Washington DC 20460. 202/260-2080.)

U.S. Environmental Protection Agency. *Ventilation and Air Quality in Offices.* Indoor Air Quality Fact Sheet #3. Revised, 1990. (EPA Public Information Center (PM-211B), 401 M Street, SW, Washington DC 20460. 202/260-2080.)

World Health Organization. *Air Quality Guidelines for Europe.* 1987. WHO Regional Publications, European Series No. 23. Available for WHO Publications Center USA, 49 Sheridan Avenue, Albany, NY 12210.

Air (Atmospheric and Indoor)

Air & Waste Management Association, *Air Pollution Engineering Manual*, AP-40, 4th edition. New York: Van Nostrand Reinhold Co., 1992.

Quarles J Lewis WH. *The New Clean Air Act: A Guide to the Clean Air Program as Amended in 1990.* Washington DC: Morgan, Lewis & Bochuis, 1991.

U.S. Environmental Protection Agency, Office of Air Quality Planning & Standards. *Compilation of Air Pollutant Emission Factors,* AP-42, 4th edition (2 volumes), V.1; Stationary Point and Area Sources; V.2: Mobile Sources. Springfield, VA: National Technical Information Service, 1991.

U.S. Environmental Protection Agency, Office of Air Quality Planning & Standards. *National Air Quality & Emission Trends Report,* October 1992, 19th Annual Report. Washington DC: Environmental Protection Agency.

Asbestos

Keyes DL, Chesson J. *A Guide to Monitoring Airborne Asbestos in Buildings.* 1989. Environmental

Sciences, Inc., 105 East Speedway Boulevard, Tucson, AZ 85705.

U.S. Department of Health and Human Services, Public Health Service, U.S. Centers for Disease Control, National Institute of Occupational Safety and Health. *Testimony of NIOSH on the Occupational Safety and Health Administration's Proposed Rule on Occupational Exposure to Asbestos, Tremolite, Anthophyllite, and Actinolite.* June 1984, May 1990, and January 1991. NIOSH Docket Office, C-34, 4676 Columbia Parkway, Cincinnati, OH 45226.

U.S. Environmental Protection Agency. *A Guide to Respiratory Protection for the Asbestos Abatement Industry.* 1986. EPA 560/OTS 86-001. (TSCA Assistance Hotline (TS-799), 401 M Street, SW, Washington DC 20460. 202/554-1404.)

U.S. Environmental Protection Agency. *Abatement of Asbestos-Containing Pipe Insulation.* 1986. Technical Bulletin No. 1986-2.(TSCA Assistance Hotline (TS-799), 401 M Street, SW, Washington DC 20460. 202/554-1404.)

U.S. Environmental Protection Agency. *Asbestos Abatement Projects: Worker Protection.* Final Rule 40 CFR 763. February 1987. (TSCA Assistance Hotline (TS-799), 401 M Street, SW, Washington DC 20460. 202/554-1404.)

U.S. Environmental Protection Agency. *Asbestos Ban and Phaseout Rule.* 40 CFR 63.160 to 763.179. Federal Register, July 12, 1989. (TSCA Assistance Hotline (TS-799), 401 M Street, SW, Washington DC 20460. 202/554-1404.)

U.S. Environmental Protection Agency. *Asbestos in Buildings: Guidance for Service and Maintenance Personnel* (in English and Spanish). 1985. EPA 560/5-85-018. ("Custodial Pamphlet.") (TSCA Assistance Hotline (TS-799), 401 M Street, SW, Washington DC 20460. 202/554-1404.)

U.S. Environmental Protection Agency. *Asbestos in Buildings: Simplified Sampling Scheme for Surfacing Materials.* 1985. 560/5-85-030A. ("Pink Book"). (TSCA Assistance Hotline (TS-799), 401 M Street, SW, Washington DC 20460. 202/554-1404.)

U.S. Environmental Protection Agency. *Guidance for Controlling Asbestos-Containing Materials in Buildings.* 1985. EPA 560/5-89-024. (AKA "Purple Book.") (TSCA Assistance Hotline (TS-799), 401 M Street, SW, Washington DC 20460. 202/554-1404.)

U.S. Environmental Protection Agency. *Guidelines for Conducting the AHERA TEM Clearance Test to Determine Completion of an Asbestos Abatement Project.* EPA 560/5-89-001. (TSCA Assistance Hotline (TS-799), 401 M Street, SW, Washington DC 20460. 202/554-1404.)

U.S. Environmental Protection Agency. *Managing Asbestos in Place: A Building Owner's Guide to Operations and Maintenance Programs for Asbestos-Containing Materials.* 1990. (AKA "Green Book.") (TSCA Assistance Hotline (TS-799), 401 M Street, SW, Washington DC 20460. 202/554-1404.)

U.S. Environmental Protection Agency. *Measuring Airborne Asbestos Following an Abatement Action.* 1985. EPA 600/4-85-049. (AKA "Silver Book.") (TSCA Assistance Hotline (TS-799), 401 M Street, SW, Washington DC 20460. 202/554-1404.)

U.S. Environmental Protection Agency. *National Emission Standards for Hazardous Air Pollutants.* 40 CFR 61. April 1984. (TSCA Assistance Hotline (TS-799), 401 M Street, SW, Washington DC 20460. 202/554-1404.)

U.S. Environmental Protection Agency. *Transmission Electron Microscopy Asbestos Laboratories: Quality Assurance Guidelines.* 1989. EPA 560/5-90-002. (TSCA Assistance Hotline (TS-799), 401 M Street, SW, Washington DC 20460. 202/554-1404.)

U.S. Department of Labor, OSHA Regulations. 29 CFR 1910.1001. *General Industry Asbestos Standard.* 29 CFR 1926.58. *Construction Industry Asbestos Standard.* June 1986; amended September 1988. DOL-OSHA Docket, 200 Constitution Avenue, NW, Room N 2625, Washington DC 20210.

U.S. Department of Labor, OSHA Regulations. 29 CFR 1910.134. *Respiratory Protection Standard.* June 1974. DOL-OSHA Docket, 200 Constitution Avenue, NW, Room N 2625, Washington DC 20210.

Hill K, Piccirelli K. *Gale Environmental Sourcebook.* Detroit: Gale Research, Inc., 1992.

Sullivan TF (editor). *Directory of Environmental Information Sources,* 4th edition. Rockville, MD: Government Institutes, Inc., 1992.

U.S. Environmental Protection Agency, Office of Information Resource Management. *INFOTERRA/USA Directory of Environmental Sources.* Springfield, VA: National Technical Information Service, 1991.

World Environmental Directory, 6th edition. Silver Springs, MD: Business Publications, Inc., 1991.

Biological Topics

American Council of Governmental Industrial Hygienists. *Guidelines for the Assessment of Bioaerosols in the Indoor Environment*. 1989. 6500 Glenway Avenue, Building D-7, Cincinnati, OH 45211.

Morey P, Feeley J, Otten J. *Biological Contaminants in Indoor Environments*. 1990. American Society for Testing and Materials Publications, 1916 Race Street, Philadelphia, PA 19103.

Building Management, Investigation, and Remediation

Bazerghi H, Arnoult C. *Practical Manual for Good Indoor Air Quality*. 1989. Quebec Association for Energy Management. 1259 Berri Street, Suite 150, Montreal, Quebec, Canada H2L 4C7.

Hansen J. *Managing Indoor Air Quality*. 1991. Fairmont Press, 700 Indian Trail, Lilburn, GA 30247.

U.S. Department of Health and Human Services. Public Health Service. Centers for Disease Control. National Institute for Occupational Safety and Health. *Indoor Air Quality: Selected References*. 1989. (NIOSH Publications Dissemination, 4676 Columbia Parkway, Cincinnati, OH 45202. 513/533-8287.)

U.S. Department of Health and Human Services. Public Health Service. Centers for Disease Control. National Institute for Occupational Safety and Health. *Guidance for Indoor Air Quality Investigations*. 1987. (NIOSH Publications Dissemination, 4676 Columbia Parkway, Cincinnati, OH 45202. 513/533-8287.)

Weekes DM, Gammage RB. *The Practitioner's Approach to Indoor Air Quality Investigations*. Proceedings of the Indoor Air Quality International Symposium. 1989. American Industrial Hygiene Association, P.O. Box 8390, Akron, OH 44320.

Environmental Health

Archives of Enrivonmental Health. Heldref Publications. 1319 18th Street, NW, Washington DC 20036, 202/296-6267. This publication provides objective documentation on the effects of environmental agents on human health. Its articles describe research from epidemiology, toxicology, biostatistics, and biochemistry. Frequency: Bimonthly; Price: $90 per year.

Environmental Almanac. Boston: Houghton, Mifflin Co., 1992.

National Archives and Records Administration, 40 CFR, *Protection of the Environment;* 49 CFR, *Transportation;* 29 CFR, *Labor*. Washington DC: Government Printing Office.

National Archives and Records Administration, *Federal Register*. Washington DC: Government Printing Office.

Environmental Tobacco Smoke

National Research Council. *Environmental Tobacco Smoke: Measuring Exposures and Assessing Health Effects*. 1986. National Academy Press, 2001 Wisconsin Avenue, NW, Washington DC 20418.

U.S. Department of Health and Human Services. Public Health Service. *The Health Consequences of Involuntary Smoking, A Report of the Surgeon General*. 1986. Office on Smoking and Health, 1600 Clifton Road, NE (Mail Stop K50), Atlanta, GA 30333.

U.S. Department of Health and Human Services. Public Health Service. Centers for Disease Control. National Institute for Occupational Safety and Health. *Current Intelligence Bulletin 54: Environmental Tobacco Smoke in the Workplace— Lung Cancer and Other Health Effects*. DHHS (NIOSH) Publication No. 91-108. 1991. (NIOSH Publications Dissemination, 4676 Columbia Parkway, Cincinnati, OH 45202. 513/533-8287.)

U.S. Department of Health and Human Services. National Cancer Institute. Office of Cancer Communications. A series of one-page information sheets on all aspects of smoking in the workplace. For copies, call 1-800-4-CANCER.

U.S. Environmental Protection Agency. *Environmental Tobacco Smoke*. Indoor Air Quality Fact Sheet #5. 1989. (EPA Public Information Center (PM-211B), 401 M Street, SW, Washington DC 20460. 202/260-2080.)

U.S. Environmental Protection Agency. *Environmental Tobacco Smoke: A Guide to the Development of Effective Smoking Policies*. (In progress.) (EPA Public Information Center (PM-211B), 401 M Street, SW, Washington DC 20460. 202/260-2080.)

U.S. Environmental Protection Agency. *Health Effects of Passive Smoking: Assessment of Lung Cancer in Adults and Respiratory Disorders in Children*. (In progress.)

Hazardous Wastes or Training

NIOSH (National Institute for Occupational Safety and Health), OSHA (Occupational Safety and Health Administration), USCG (U.S. Coast

Guard), and EPA (Environmental Protection Agency). *Occupational Safety and Health Guidance Manual for Hazardous Waste Site Activities.* Cincinnati, OH: NIOSH, 1985.

U.S. Department of Transportation. *Emergency Response Guidebook.* Washington DC: U.S. DOT, 1987.

U.S. Environmental Protection Agency, Office of Emergency and Remedial Response, Hazardous Response Support Division. Environmental Response Team, *Standard Operating Safety Guides.* Washington DC: U.S. EPA, 1992.

Tokle G (ed). *Hazardous Materials Response Handbook,* 2nd ed. Quincy, MA: National Fire Protection Association, 1992.

U.S. Environmental Protection Agency. *Hazardous Materials Emergency Planning Guide, NRT-1.* Washington, DC: U.S. EPA, 1987.

Laboratory Practices

Clessere LS et al. *Standard Methods for the Examination of Water and Wastewater,* 18th edition. Washington DC: American Public Health Association, 1992.

U.S. Environmental Protection Agency, Office of Solid Waste and Emergency Response. *Test Methods for Evaluating Solid Waste Physical/Chemical Methods.* SW-86. Springfield, VA: National Technical Information Service, 1987 and updates.

Law

Arbuchle JG. *Environmental Law Handbook,* 11th edition. Rockville, MD: Government Institutes Inc. 1991.

Management

International Organization for Standardization (ISO). *ISO 9000 Standards.* 1987. ISO, 1, rue de Verembe, Case Postale 56, CH-1121 Geneva 20, Switzerland.

PCBs

U.S. Department of Health and Human Services. Public Health Service. Centers for Disease Control. National Institute for Occupational Safety and Health. *Current Intelligence Bulletin 45: Polychlorinated Biphenyls—Potential Health Hazards from Electrical Equipment Fires or Failures.* DHHS (NIOSH) Publication No. 86-111. 1977. Available from the National Technical Information Service, 5285 Port Royal Road, Springfield, VA 22161.

U.S. Department of Health and Human Services. Public Health Service. Centers for Disease Control. National Institute for Occupational Safety and Health. *A Recommended Standard for Occupational Exposure to Polychlorinated Biphenyls.* DHHS (NIOSH) Publication No. 77-225. 1977. Available from the National Technical Information Service, 5285 Port Royal Road, Springfield, VA 22161.

U.S. Environmental Protection Agency. *Transformers and the Risk of Fire: A Guide for Building Owners.* 1986. OPA/86-001.

Pesticides

Hayes WJ, Jr., Laws ER, Jr. *Handbook of Pesticide Toxicology* (3 volumes): v.1: General Principles; v.2 and v.3: Classes of Pesticides. Orlando, FL: Academic Press, 1991.

Radon

U.S. Environmental Protection Agency. *State Proficiency Report.* 1991. EPA 520/1-91-014. Available from state radon offices. List of laboratories that have demonstrated competence in radon measurement analysis.

Research and Development

Clayton GD, Clayton FE (eds). *Patty's Industrial Hygiene & Toxicology* (vols 2A, 2B, 2C): *Toxicology.* New York: Wiley 1981, 1982.

Lewis, RJ. *Sax's Dangerous Properties of Industrial Materials,* 8th edition (3 volumes). New York: Van Nostrand Reinhold, 1992.

Sittig M. *Handbook of Toxic and Hazardous Chemicals and Carcinogens,* 3rd edition. Park Ridge, NJ: Noyes Publications, 1991.

U.S. Department of Health and Human Services for Occupational Safety and Health. *NIOSH Pocket Guide to Chemical Hazards.* Washington D.C.: Government Printing Office, 1990.

Standards and Guidelines

American Conference of Government Industrial Hygienists. *Threshold Limit Values and Biological Exposure Indices.* 1992, 1993. 6500 Glenway Avenue, Building D-7, Cincinnati, OH 45211.

U.S. Department of Health and Human Services. Public Health Service. Centers for Disease Control. National Institute for Occupational Safety and Health. *NIOSH Recommendations for Occupational Safety and Health. Compendium of Policy Documents and Statements.* DHHS

(NIOSH) Publications 91-109. 1991. (NIOSH Publications Dissemination, 4676 Columbia Parkway, Cincinnati, OH 45202. 513/533-8287.)

U.S. Department of Labor. OSHA Regulations. 29 CFR 1910.1000. *OSHA Standards for Air Contaminants*. Available from the U.S. Government Printing Office, Washington DC 20402. 202/783-3238. Additional health standards for some specific air contaminants are also available in Subpart Z.

Training

American Industrial Hygiene Association (AIHA). P.O. Box 8390, 345 White Pond Drive, Akron, OH 44320. 216/873-2442. Sponsors indoor air quality courses in conjunction with meetings for AIHA members only.

American Society of Heating, Refrigerating, and Air-Conditioning Engineers (ASHRAE). 1791 Tullie Circle NE, Atlanta, GA 30329. Sponsors professional development seminars on indoor air quality.

Andrews LP (ed.). *Worker Protection During Hazardous Waste Remediation—Center for Labor Education and Research*. New York: Van Nostrand Reinhold, 1990.

Brunacini AV, Beagcron JD. *Workbook for Fire Command*. Quincy, MA: National Fire Protection Association, 1985.

Brunacini AB. *Fire Command*. Quincy, MA: National Fire Protection Association, 1985.

Chemical Manufacturers Association. *Site Emergency Response Planning*. Washington DC: Chemical Manufacturers Association, 1986.

FEMA (Federal Emergency Management Agency). *Fire Service Emergency Management Handbook*. Washington DC: FEMA, 1985.

Martin WF, Lippitt JM, Prothero TG. *Hazardous Waste Handbook for Health and Safety*. Boston: Butterworth Publishers, 1987.

Maslansky CJ, Maslansky SP. *Monitoring Instrumentation: A Manual for Emergency, Investigatory, and Remedial Responders*. New York: Van Nostrand Reinhold, 1993.

NIOSH Division of Training and Manpower Development and NIOSH-funded Educational Resource Centers. 4676 Columbia Parkway, Cincinnati, OH 45226. Provide training to occupational safety and health professionals and paraprofessionals.

OSHA. Accreditation of Training Programs for Hazardous Waste Operations; Notice of Proposed Rulemaking. 55 Federal Register 2776, January 26, 1990.

OSHA. Hazardous Waste Operations and Emergency Response. Interim Final Rule. 51 Federal Register 45654, December 19, 1986.

OSHA Training Institute. 155 Times Drive, Des Plaines, IL 60018. Provides courses to assist health and safety professionals in evaluating indoor air quality.

Ventilation and Thermal Comfort

ASHRAE materials are available from their Publication Sales Department, 1791 Tullie Circle, NE, Atlanta, GA 30329. 404/636-8400.

ASHRAE Guideline 1-1989. *Guideline for the Commissioning of HVAC Systems*. 1989.

ASHRAE Journal. October 1989 issue. Several articles describing ASHRAE Standard 62-1989.

ASHRAE Standard 52-76. *Method of Testing Air-Cleaning Devices Used in General Ventilation for Removing Particulate Matter*. 1976.

ASHRAE Standard 55-1981. *Thermal Environmental Conditions for Human Occupancy*. 1981.

ASHRAE Standard 62-1989. *Ventilation for Acceptable Indoor Air Quality*. 1989.

National Conference of States on Building Codes and Standards, Inc. *The Ventilation Directory*. 505 Huntmar Park Drive, Suite 210, Herndon, VA 22070. 703/481-2020. Summarizes natural, mechanical, and exhaust ventilation requirements of the model codes, ASHRAE standards, and unique state codes.

Water

Montgomery JH, Welhom LM. *Ground Water Chemicals Desk Reference*. Chelsea, MI: Lewis Publishers, 1989.

Appendix 3

Glossary of Acronyms

AABC Associated Air Balance Council

ACBM asbestos-containing building materials

ACM asbestos-containing materials

AEA Atomic Energy Act

AHS air-handling system

AHERA Asbestos Hazard Emergency Response Act

ASHRAE American Society of Heating, Refrigeration, and Air Conditioning Engineers

ALARA as low as reasonably achievable

ALLMW alpha low-level mixed waste

ALLW alpha low-level waste

ANC acid-neutralizing capacity

APC air pollution control

APEG alkaline metal hydroxide/polyethylene glycol

AR attributable risk

ASEAN Association of South East Asian Nations

ASTM American Society of Testing and Materials

ATEG sodium hydroxide/tetra-ethylene glycol

ATSDR Agency for Toxic Substance Disease Registry

BACT best available control technology

BAPEDAL Environmental Management Agency (Indonesia)

BAT best available technology; best available technology economically achievable

BCD base-catalyzed decomposition

BCSD Business Council for Sustainable Development

BCT best conventional pollution control technology

BEJ best engineering judgment

BOD biological oxygen demand

BPT best practice technology

BRI building-related illness

Bt British thermal unit, a measure of heat content

CAA Clean Air Act

CAS Number Chemical Abstract Service number

487

CCME Canadian Council of Ministers of the Environment

CCRIS Chemical Carcinogenesis Research Information System

CDC Centers for Disease Control and Prevention

CEPA Canadian Environmental Protection Act

CEQ Council of Environmental Quality

CERCLA Comprehensive Environmental Response, Compensation, and Liability Act of 1990 (Superfund)

CFC chlorofluorocarbon

cfm cubic feet per minute

CFR Code of Federal Regulations

CGS centimeter-gram-second

CH₄ methane

CIAP Climate Impact Assessment Program

CIH certified industrial hygienists

CITIES Convention on International Trade in Endangered Species

Cl-ClO_x chlorine-chlorine oxide

ClO_x oxides of chlorine

CO carbon monoxide

CO₂ carbon dioxide

COC chemical of concern; contaminant of concern

COD chemical oxygen demand

CRTK Community Right-to-Know Act (CRTK Law)

CSP certified safety professional

CWA Clean Water Act

DART Developmental and Reproductive Toxicology Database

DES diethylstilbestrol

DMBA dimethylbenzanthracene

DNA genetic material

DOD Department of Defense

DOE Department of Energy; Department of the Environment; design of experiments

DOJ Department of Justice

DOT Department of Transportation

DRE destruction and removal efficiency

DSL Domestic Substances List

EA Environmental Assessment; exhaust air

EC European Community; Environmental Coordinators

ECRA Environmental Responsibility and Cleanup Act

EDE effective dose equivalent

EDTA ethylene diamine tetra-acetic acid

EEC European Economic Community

EHS Extremely Hazardous Substance List (SARA section 302)

EIA environmental impact assessments

EIS environmental impact statement

ELF extremely low frequency

EMF electromagnetic field

EPA Environmental Protection Administration

EPCRA Emergency Planning and Community Right-to-Know Act

EPT emergency preparedness teams

ER emergency response

ERA ecological risk assessment

ETA event tree analysis

ETS environmental tobacco smoke

EU European Union

Fannie Mae Federal National Mortgage Association

FAO Food and Agriculture Organization

FFCA Federal Facilities Compliance Act

FHO Federal Health Office

FIFRA Federal Insecticide, Fungicide, and Rodenticide Act

FMEA Failure Modes and Effects Analysis

FNMA Federal National Mortgage Association

foc fraction of organic carbon

Form R Toxic Chemical Release Inventory Reporting Form

FRG former West Germany

FTA fault tree analysis

FWPCA Federal Water Pollution Control Act

G gauss

GAC granular activated carbon

GATT General Agreements on Trade and Tariffs

GCMs general circulation models

GC/MS gas chromatography-mass spectrometry

GEF global environment facility

GFDL Geophysical Fluid Dynamics Laboratory

GI gastrointestinal

GISS Goddard Institute for Space Sciences

GNP gross national product

GREEN general equilibrium environmental model

H hydrogen atom

HAPs hazardous air pollutants

HAZMAT hazardous materials

HCC high current configuration

HMIRO Hazardous Materials Incident Response Organization

HMR Hazardous Materials Regulations

HMTA Hazardous Materials Transportation Act

HO hydroxyl

HO$_2$ hydroperoxyl

HQ hazard quotient

HRA health risk assessment

HSDB Hazardous Substances Databank

HSWA Hazardous and Solid Waste Amendments of 1984

HVAC heating, ventilation, and air conditioning systems

Hz hertz

IAGA International Association of Geomagnetism and Aeronomy

IAQ indoor air quality

IATA International Air Transport Association

ICA International Civil Aviation Organization

ICAM integrated computer-aided manufacturing

ICC International Chamber of Commerce

IDEF integrated computer-aided manufacturing definition methodology

IEA Institute for Environmental Auditing

ILO International Labor Organization

IMDG International Maritime Dangerous Goods Regulations

IME integrated manufacturing and environment

IMF International Monetary Fund

IMO International Maritime Organization

INE National Ecology Institute (Instituto Nacional de la Ecologia; Mexico)

IPCC Intergovernmental Panel on Climate Change

IRIS Integrated Risk Information System data base

ISC Industrial Source Complex

ISO International Standards Organization

ISRA Industrial Site Recovery Act

ISV In-situ vitrification

ITC International Trade Commission

IUCN International Union for the Conservation of Nature

JEA Japan Environmental Agency

km kilometers

Km mixing factor

KPEG potassium hydroxide/polethylene glycol

LCC low current configuration

LAER lowest achievable emission rate

LDR land disposal restriction

LEPC Local Emergency Planning Committees

LOAEL lowest observable adverse effect level

LOEC lowest observed effect concentration

LUST leaking underground storage tank

MA mixed air

MACT maximum achievable control technology

MATC maximum acceptable toxicant concentration

mb millibars

MCLs maximum contaminant levels

MDI Bailey Mental Development Index

MFO mixed function oxidases

mG milligauss

MITI Ministry of International Trade and Industry

MOE Ministry of the Environment

MPRSA Marine Protection, Research, and Sanctuaries Act

MSDS Material Safety Data Sheet

MTD maximum tolerated dose

MWTA Medical Waste Tracking Act

NADCA National Air Duct Cleaners Association

NAFTA North American Free Trade Agreement

NAPL nonaqueous phase liquids

NARM naturally occurring and accelerator-produced radioactive material

NAS National Academy of Sciences

NCAR National Center for Atmospheric Research

nCi/g nanocuries per gram

DMNA dimethylnitrosamine

NEBB National Environmental Balancing Bureau

NEPA National Environmental Policy Act

NESHAPs National Emission Standards for Hazardous Air Pollutants

NGOs nongovernmental organizations

NICs newly industrializing countries

NIEH National Institutes for Environmental Health

NIMBY not-in-my-backyard

NIOSH National Institute for Occupational Safety and Health

NIS Newly Independent States

NL The Netherlands

nm nanometer

N$_2$O nitrous oxide

NO nitric oxide

NO$_2$ nitrogen dioxide

NO$_x$ nitrogen oxides

NOAA National Oceanic and Atmospheric Administration

NOAEL no observable adverse effect level

NOEC no observed effects concentration

NPDES National Pollutant Discharge Elimination System

NPL National Priorities List

NRC Nuclear Regulatory Commission; National Response Center

NRDC Natural Resources Defence Council

NREL National Renewable Energy Laboratory

NSPS new source performance standards

O oxygen

O$_2$ oxygen molecules

O$_3$ ozone

OA outdoor air

OAU Organization of African Unity

ODC ornithine decarboxylase

OECD Organization for Economic Cooperation and Development

OH hydroxl ions

OHMT Office of Hazardous Materials Transportation

OJT On-the-job training

OME Ontario Ministry of the Environment

O&M Operation and Maintenance

OSHA Occupational Safety and Health Administration

OSTP Office of Science and Technology Policy

OSWER Office of Solid Waste and Emergency Response

OUST Office of Underground Storage Tanks

Pb lead

PBPK physiologically based pharmacokinetic

PCB polychlorinated biphenyl

PCDDs polychlorinated dibenzo-p-dioxins

PCDFs polychlorinated dibenzofurans

PCM phase contrast microscopy

PCP pentachlorophenol

PEL permissible exposure level

POHC principal organic hazardous constituent

POTW publicly owned treatment works

PPA Pollution Prevention Act

ppb parts per billion

PPE personal protective equipment

PRA probabilistic risk assessment

PRP potentially responsible party

PSD prevention of significant deterioration

PSL Priority Substances List

Q fever Coxiella burnetti

r risk

RA return air

rad wastes radioactive wastes

RBD reliability block diagrams

RCRA Resource Conservation and Recovery Act

RF radiofrequency

RfCs reference concentrations

RfDs reference doses

RFI RCRA facility investigation

RMW regulated medical waste

ROC receptors of concern

RQ reportable quantity

RR relative risk

RSAC radiological safety analysis computer

RSPA Research and Special Programs Administration

RTECS Registry of Toxic Effects of Chemical Substances

SA supply air

SAMIS Standard Assembly Line Manufacturing Industrial Simulation

SARA Superfund Amendments and Reauthorization Act of 1986

SAR safety analysis review

SBS sick building syndrome

SDWA Safe Drinking Water Act

SEDESOL Secretariat for Social Development (Mexico)

SEDUE Secretariat of Urban Development and Ecology (Mexico)

SEGH Society for Environmental Geochemistry and Health

SERC State Emergency Response Commission

SERI Solar Energy Research Institute

SIC standard industrial clarification

SIO Scripps Institution of Oceanography

SIP state implementation plan

SIR standardized incidence ratios

SMR standard mortality ratio

SNA System of National Accounts

SNV Swedish Environmental Protection Agency

SO$_2$ sulfur dioxide

SPCC spill prevention control and countermeasure

SQC sediment quality criteria

SQCoc organic carbon-normalized SQC

SST supersonic transport aircraft

STEL short-term excursion limit

Superfund Comprehensive Environmental Response, Compensation, and Liability Act of 1980

T tesla

μT microtesla

TAB or T&B testing and balancing

TBS tight building syndrome

TCA trichloroethane

TCDD tetrachlorodibenzo-p-dioxin

TCE trichloroethylene

TCLP toxicity characteristic leaching procedure

TDG Transportation of Dangerous Goods

TDGR Transportation of Dangerous Goods Regulations

TEM transmission electron microscopy

TOMS total ozone mapping spectrometer

TPQ threshold planning quantity

TSD treatment, storage, and disposal

TSDF treatment, storage, and disposal facilities

TQ toxicity quotient

TRI Toxic Release Inventory

TSCA Toxic Substance Control Act

TQM total quality management

µg/dL micrograms of lead per decaliter of blood

UK United Kingdom

UN United Nations

UNCED United Nations Conference on Environment and Development

UNDP United Nations Development Programme

UNEP United Nations Environment Programme

UNESCO United Nations Educational, Scientific, and Cultural Organization

U.S. AID United States Agency for International Development

UST underground storage tanks

UV ultraviolet

VAV variable air volume

VOCs volatile organic compounds

V/m volts/meter

WCC wiring code configuration

WCED World Commission on Environment and Development

WCS World Conservation Strategy

WHO World Health Organization

WIPP Waste Isolation Pilot Plant

WMO World Meteorological Organization

WQA Water Quality Act of 1987

WWF World Wildlife Fund

Appendix 4

Glossary of Terms

A horizon: The layer in soil of maximum biological activity and of removal of dissolved or suspended materials in water.

Absorbed dose: The amount of material absorbed into the systemic circulation.

Accountability: The extent to which the environmental management function staff, first-line supervisors, and midlevel managers are expected to be answerable to senior-level executives for proper execution of preplanned strategies and tactics.

Achievement levels: A hierarchy of objective statements that describe the intended focus and benchmarks for the environmental management function.

Acid-neutralizing capacity: The content of neutralizing ions in a body of water.

Acutely hazardous wastes: Those wastes that, in small doses, are capable of causing death or significantly contributing to irreversible and/or incapacitating illness.

Adaptation options: Options defined by the National Academy of Sciences as those whose "effects come into play after climate has changed."

Adaptive control: Control that focuses on determining the most appropriate way in which to respond to changes.

Additive chemical interaction: An effect that occurs when the combined effect of two or more chemicals is equal to the sum of the effect of each chemical administered alone.

Adsorption: The transfer of contaminants from mobile (in groundwater) to immobile (attached to soil) state.

Advection: For surface water transport processes, the transport of water flowing in a particular direction (more or less horizontally), such as water flowing because of the current in a stream or river.

The process of dissolved chemicals within an aquifer unit moving at the average velocity of the groundwater. For atmospheric fate and transport, the movement of contaminants with an air mass in a predominantly horizontal direction. The process is dependent on wind speed and direction.

Aerobic respiration: Respiration in which oxygen is used to accept electrons.

Agenda 21: The broad new mandate for environmental action resulting from the Rio Conference on Environment and Development held in June 1992.

Air strippers: Devices that are principally used for light petroleum products such as those associated with gasoline. The process consists of discharging water at the top of a tower while simultaneously blowing air upward through the water stream. The passage of the air removes the volatile contaminants within the wastewater stream.

Allergic rhinitis: Nasal discharge, congestion, and interior swelling associated with many causes.

Anaerobic respiration: Respiration conditions in which oxygen is unavailable and sulfate or nitrate serves as an electron receptor.

Antagonistic chemicals: Chemicals that interfere with one another's toxic effects or one reduces the toxicity of another.

Anthropocentric: Interpreting reality in terms of human values and experience.

Aquatic ecosystems: Ecosystems that may be primarily freshwater (lakes, ponds, streams, rivers) or marine (ocean, sea coast, estuaries).

Aquatic sediment: A collection of debris—coarse to fine particles of mineral and organic matter deposited at the bottom of water bodies. Fifty to 90% of the volume is "interstitial" or "pore" water, depending on depth and degree of compaction.

Aquifer: A saturated, permeable geologic unit that can transmit significant or usable quantities of water under ordinary hydraulic gradients. There are many alternative definitions for aquifer, including that of a water-bearing rock. It is imperative for an industrial manager concerned with groundwater issues to understand how local and federal authorities define the term.

Aquitard: Geologic unit that inhibits groundwater flow due to its low permeability. Aquitards frequently separate aquifer units, but flow across an acquitard is possible. If little flow occurs, the unit is termed an *aquiclude.*

Area sources: Several sources distributed over a homogeneous surface area. Numerous, usually small, sources of contaminant emissions that are distributed across a specified surface area.

Assessment endpoint: "An explicit expression of the actual environmental value that is to be protected" (Chapter 12, U.S. EPA, 1992a).

Asthma: The illness involving reversible narrowing of airways typically triggered by a single source, such as chemicals like formaldehyde or isocyanates.

Atmospheric stability: A meteorological condition that affects dispersion of airborne contaminants. The atmosphere is said to be stable when there is little or no vertical movement of air masses. With little or no vertical mixing of the contaminant with the air, contaminant concentrations accumulate at ground level. Unstable atmospheric conditions result in vertical mixing of the air masses.

Attributable risk: The proportion of events that can be attributed to the exposure under study.

B horizon: In soils, layers of suspended materials, including residual oxides, silicate clays, or other transformed materials.

Base-catalyzed decomposition: A technology for removing chlorine molecules from contaminants such as PCBs, dioxins, and pentachlorophenols.

Bioaccumulation: The extent to which chemicals accumulate in living organisms.

Bioavailability: A measure of the amount of toxicant available to cause an effect.

Biocentric ethic: Ethical system that avoids all hierarchical postures and acknowledges the tight ecological relationships, the web of life that connects all living and nonliving entities.

Biochemical oxygen demand: A measure of the potential impact of wastewater organic matter on a stream or river.

Biological plausibility: A determination of whether there are reasonable or known physiologic mechanisms that can be evoked to explain the association of a suspected carcinogen and cancer.

Bioremediation technologies: Approaches that involve enhancing biodegradation of contaminants through the stimulation of indigenous soil and groundwater microbial populations or the addition of proprietary, natural microbial species.

Box model: A model in which the entire modeling region is contained in a single cell (box); it is used to obtain estimates of area source concentrations.

Bridging encapsulation: Abatement technique that involves coating an ACBM with a film similar to latex paint that effectively seals the material.

C horizon: Weathered mineral material that is unaffected by soil-forming processes except for the accumulation of salts or oxides of varying solubilities.

Cancer cluster investigations: Studies that investigate apparent excesses of incidence of cancer.

Capillary fringe: Transition zone from the partially saturated vadose zone to the water table (phreatic) surface.

Case control studies: Studies similar to prevalence studies because they examine the relationship between existing disease (e.g., childhood leukemia) and other potential exposure variables. Case control studies are extremely common in the environmental epidemiology literature and are frequently the first type of study performed in order to establish an associative link between a disease endpoint and an environmental exposure.

Cation exchange capacity (CEC): A measured factor in soil that has significant impact on the uptake of chemicals into plants. CEC varies as a function of the amount and distribution of clay and organic matter present.

Chelating agents: Agents that can be effective for immobilizing metals. Stable metal chelates may be strongly bound to the soil to prevent movement.

Chemical: Any natural or synthetic compound or single element that is undergoing movement or transformation in the environment.

Chemical oxygen demand: A measure of the oxygen that would be removed from a water body due to chemical reactions.

Chemical precipitation: Process that must be used when contaminants remain suspended or dissolved in water even when the water is completely still. Chemical precipitation removes contaminants through creation of molecules that are heavier than water, elimination of electrical charges that keep particles in suspension, or absorption of contaminants onto the surface of a treatment chemical added to the water. Commonly used chemicals for chemical precipitation are alum, ferric chloride, lime, and powdered activated carbon.

Chemical speciation/precipitation and complex formation: The trace metal cations in a soil solution form complexes and precipitates as a function of the content and type of organic matter, the presence of other oxide minerals, soil pH (measure of acidity or alkalinity), soil Eh (oxidation-reduction potential), and the CEC. The bioavailability of trace metal and organic (synthetic) chemical uptake is dramatically affected by the factors that influence complex formation. Due to chemical contamination, these effects have serious impacts when soil treatment options are considered.

Chemical treatment in-situ: Process that consists of immobilizing or detoxifying the contaminants, or flushing them from the soil. Immobilization methods are designed to prevent the contaminants from dissolving into water that flows through the soil. This prevents removal of the contaminants from the soil matrix that would allow it to ultimately reach a human receptor.

Chlorofluorocarbons: Synthetic hydrocarbon compounds containing chlorine and fluorine. They are inert stable compounds with other chemical properties that make them well suited for propellants, refrigerants, solvents, and other industrial uses.

Clean Air Act (CAA): A federal law that sets definite goals for emission reductions and ambient air quality improvements.

Clean Water Act (CWA): A federal law intended to limit the discharge of pollutants into navigable waters and to make the nation's waters fishable and swimmable again.

Closure: The period after which hazardous wastes are no longer accepted by a facility, and the operator completes treatment, storage, or disposal.

Climate variability: The variations of climatic parameters within a typical averaging period (Chapter 14, Gates & MacCracken, 1985).

Clustering: The discrete space-time aggregation of cancers or other adverse health endpoint (e.g., birth defects) in relatively small areas (Chapter 11, Schulte et al, 1987).

Cogeneration: An energy-generating process that takes advantage of the fact that electricity can be produced using steam heated by the combusted wastes.

Coherence: A criterion for judging epidemiologic studies in which the body of positive related facts must be consistent with the new evidence or associations uncovered by a study.

Cohort: A group of individuals under study who are initially disease free.

Coliform bacteria counts: An indicator of the disease potential of a wastewater stream.

Commissioning: A process in which a new HVAC system's performance is identified, verified, and documented to ensure proper operation and compliance with codes, standards, and design intentions.

Community-level endpoints: Endpoints that can be used to evaluate whether the composition (diversity and/or richness) and abundance of species have been altered as a result of exposure to the stressors.

Competitive performance structure: An approach that focuses on a strategic intent of building distinctive organizational competencies and capabilities and on a mission of preparing, protecting, and preserving the firm's environmental resources.

Complete carcinogens: Direct-acting carcinogens that may interact with the genetic material and stimulate cell proliferation through one or more mechanisms.

Compliance: Acting in agreement with a request or command.

Compliance schedules: Specific timetables by which an organization must bring itself into compliance with an applicable requirement. These schedules may be set forth in regulations, special judicial orders issued by a court, administrative orders issued by the U.S. Environmental Protection Agency (EPA), or less formal correspondence from a regulatory agency at any level.

Compliance structure: An approach that focuses on a strategic intent and a mission of maintaining compliance with mandates from environmental regulatory sources.

Compliance tracking: The process used to verify data collection and maintenance.

Compliance violations: Official citations by federal or state

Composting: The conversion of organic matter to a form that is useful as a soil additive and/or fertilizer.

Conceptual model: The final product of the problem formulation process.

Conditionally exempt small-quantity generator: A generator that does not generate more than about 220 lb or 25 gal (100 kg) of hazardous waste and no more than 2.2 lb (1 kg) of acute waste in a month. A conditionally exempt small-quantity generator is not subject to the full spectrum of hazardous waste management and permitting requirements.

Confined aquifer: An aquifer that has an overlying layer that does not allow direct contact of the aquifer with the atmosphere. Water in a confined aquifer is under pressure and wells penetrating into the aquifer will have a water level that reflects the pressure in the aquifer at the point of penetration. Also called an *artesian aquifer.* Compare with *Perched aquifer* and *Unconfined aquifer.*

Confounder: An agent that may act as another explanatory or associative variable as significant as the original study variable. Alcohol, smoking, and diet are frequent confounders in many epidemiologic studies.

Congruency evaluation: A means of evaluation that shows the accord and harmony between environmental management strategy and structure and the competitive performance strategy of the organization.

Container: Any portable device in which a material is stored, transported, treated, disposed of, or otherwise handled.

Contaminate: To make a substance unfit for its intended use or to pollute (*see* Pollution).

Contaminants of concern: Those chemicals most likely to contribute significantly to risk.

Contingent valuation techniques: A sample population is surveyed in order to determine the monetary valuation individuals will pay to avoid an untoward effect or how much compensation would be required if an adverse outcome occurs.

Contingency plan: A plan, readily available at each district, which contains provisions for environmental response in case hazardous materials spill on-site.

Convection: The transport of water because of density gradients. In this form of transport, the driving forces of the currents are density gradients resulting from temperature differences in deep lakes, and temperature and salinity differences in estuaries.

Corrosive: A characteristic hazardous pH less than or equal to 2, or greater than or equal to 12.5. pH must be determined by a pH meter using approved EPA test methods or equivalent.

Cyanide wastes: Wastes generated from cyanide plating solutions and simple cyanide solutions.

Darcy's law: A measure of the average or bulk velocity through a given cross section of a porous medium and is valid for steady flow with constant flux. The volumetric flow rate (Q) through a porous medium is directly proportional to the hydraulic conductivity (K), the hydraulic gradient (i), and the cross-sectional area of flow (A): $Q = -KiA$. The minus signifies that flow is from areas of high hydraulic head to areas of low hydraulic head.

Decommissioning: The process of retiring a facility from service in a manner that does not adversely impact human health or the environment.

Decontamination: The removal of radioactive contamination that has been deposited on surfaces or spread throughout a given work space.

Degradation: The transformation of a contaminant to other forms through decay, biodegradation, or other process.

Delivered dose: The amount of material fed to or inhaled by an experimental animal.

Deposition: The process by which particulates and reactive gases are deposited on the earth's surface from the atmosphere. Both wet and dry deposition occurs. Wet deposition occurs sporadically during specific rain or snow events; dry deposition occurs continuously under dry atmospheric conditions.

Dermatitis: Skin rashes attributable to irritants, light, and allergies. Various types of fibers (e.g., fiberglass) may be released from ventilation dust linings and produce strong irritant reactions. Rashes have also been attributed to relative humidities below 30%; however, these rashes may not be associated with visible medical findings.

Deterministic models: Models that produce single-value estimates of risk.

Detoxification: Neutralization or oxidation/reduction of a contaminant.

Diffusion: The dispersion of a contaminant relative to its advective movement. Diffusion reduces the central concentrations in a contaminant mass (plume), increases the concentration at the periphery of the mass, and expands the periphery.

Direct dumping: Object of pollution control efforts that includes sludge, radioactive and toxic wastes, and dredge spoils (i.e., silt from harbors and rivers).

Direct market-based values: Direct multiplication of the quantity of goods lost or damaged by the current market value can work for goods such as timber and fish, for which an existing market is in place.

Discharge: The quantity or process of water being lost from the saturated zone. Compare with *Recharge*.

Dispersion: Spreading of a contaminant source as it flows through an aquifer. For surface water transport processes, the scattering of particles because of the combined effect of shear and diffusion (molecular and turbulent). Usually, the combined effect of shear and transverse diffusion, represented as an effective dispersion, is orders of magnitude greater than other diffusive mechanisms acting in the direction of flow in rivers and estuaries.

Disposal: To discharge, deposit, inject, dump, spill, leak, or place wastes into or on land or waters.

Disposal facility: A facility or part of a facility where hazardous waste is intentionally placed and where it will remain after closure.

Dominant social paradigm: A set of standards, beliefs, values, and habits.

Dose: The mass of toxicant per mass receptor (e.g., mg toxicant/kg animal).

Dose-response relationship: Quantitative relationship between increasing dose of a chemical and any measured endpoint.

Dry deposition: Both direct settling of sulfate particles and to capture of sulfur dioxide gas by vegetation.

Drying beds: Drying beds are concrete-lined or clay-lined areas where solids are placed to dry prior to storing in drums for disposal. Solids placed in drying beds are collected from the solids settling basin and the oil/water/solids separator.

Ecological component: A part of the ecological system that might be individuals, populations, communities, or ecosystems.

Ecology: Scientific natural history.

Ecosystem-level endpoints: Endpoints that are used to evaluate the impacts of stressors associated with various environmental media on ecosystem function.

Ecosystems: Functional systems of complementary relationships between communities of organisms and their physical environment.

Ecotoxicology: The study of the fate and effects of toxic agents in ecosystems.

Effective plume height: The physical height of the stack adjusted by factors that raise the plume (as a result of buoyancy or momentum) or lower it (as a result of downwash or deflection).

Effluent: A discharge to waters.

Electric fields: Represent the forces or strength that electric charges exert on charges at a distance.

Elementary neutralization unit: A container used for neutralizing certain hazardous wastes.

Emission: A discharge to the air.

Endpoints: Characteristics of an ecological component that may be affected by exposure to a stressor.

Environment: All of the surrounding conditions and influences (physical and biological) affecting the development of living things; often refers to natural resources like air, land, or water.

Environmental audit: A process that seeks to verify documented expectations, typically regulations and policies, by conducting interviews, reviewing records, and making first-hand observations; also called a *compliance audit:*

Environmental chemodynamics: The process of describing the movement of hazardous materials in the environment

Environmental ethics: Disciplined reflection about moral relationships to the natural world.

Environmental laws: Acts or statutes passed by U.S. Congress and state legislatures to govern activities that pollute air, land, and water.

Environmental liability: The legal and financial responsibility of property owners to ensure that activities conducted on their property particularly with respect to hazardous substances are in compliance with applicable laws. Liability means that the owner must pay severe fines and is subject to prison sentences for violations.

Environmental management evaluation: The management's practice of determining the efficacy of formulated strategy compared against the objectives and expected effects (benchmarks) it set out to achieve.

Environmental programs: The group of laws, regulations, and other policy guidance aimed at protecting a specific resource or managing a particular environmental problem; for example, surface water quality programs and hazardous waste programs.

Environmental site assessment: A process that seeks to characterize a physical property or operation from an environmental view with an overall objective of understanding site-specific conditions. Information is collected through interviews, record reviews, and first-hand observations, and may also involve testing environmental media (e.g., air, water, soil) and facility characteristics (e.g., wastewater, insulation, paint, airflow).

Environmental toxicology: The science of poisons as they relate to earth's biosphere.

EP toxicity: A characteristic of hazardous wastes, whereby the extract from a waste contains certain constituents in excess of a listed concentration. Currently applies to certain metals and pesticides.

EPA identification number: The number assigned by EPA to each hazardous waste generator, transporter, and treatment, storage, or disposal facility.

Epigenetic carcinogens: Agents associated with the progression of preneoplastic cells to cancer by causing cell division and proliferation without direct interaction with the genetic material.

Equalization: Process used by industries that have widely varying waste streams. By combining the waste streams, the toxicity can be reduced through dilution. An example of this technique is textile mill wastewater. This waste stream tends to have high fluctuations in pH and biological oxygen demanding material. Using holding tanks, the waste streams can be combined to provide a more consistent wastewater quality. This higher consistency is essential for subsequent treatment processes.

Ethics: The disciplined reflection about moral problems.

Evapotranspiration: Water that is returned to the atmosphere by evaporation from the surface and by transpiration from plants.

Ex-situ technologies: Treatment technologies that require removal and material handling of the contaminated materials.

Exosphere: One of the outer regions of the atmosphere.

Exposure: Contact with a chemical or contaminated media.

Exposure profile: Final product of the exposure characterization that quantifies the magnitude of exposure through space and time.

Externalities: Indirect costs associated with environmental valuation.

Extraction procedure (EP): This process requires testing of representative samples of solid wastes subjected to an acetic acid leaching medium (5.0 pH).

Extremely low frequency fields: EMFs in the 3–3,000 hertz (Hz) range.

Fate and transport characterization: The process of describing the movement of hazardous materials in the environment.

Fault: Fracture or fracture zone along which there has been displacement of the opposite sides relative to one another.

Filter racks: Used filters are accumulated on a special filter rack and allowed to drain for at least 24 hours prior to disposal. The drained waste oil is then combined with other waste oil generated at the facility.

Filtration: Process is used to screen out contaminants from a wastewater stream. Filtration ranges from coarse screens with mesh sizes that can be seen by the naked eye to ultrafine screens that filter molecular constituents.

Foc: The fraction of organic carbon.

Fold: Bending or crumpling produced in stratified rocks (and sediments) by compressive stress.

Fracture: Break in rocks.

Gaussian plume model: A model most commonly used to represent plume dispersion in the near-field range. Simple expressions (Gaussian functions) are assumed to represent the dependence of pollutant concentration on lateral or vertical distance from the plume centerline (i.e., the advective path).

Generator: Any person, by site, whose act or process produces hazardous waste.

Genotoxic carcinogens: Compounds that interact with and alter the structural/functional characteristics of the genetic material (DNA) resulting in the conversion of the cell to the preneoplastic stage (cancer precursor cells).

Geologic structures: Features produced by the deformation of rocks. Geologic structures may occur at scales ranging from microscopic to continental (or oceanic). Significant types of geologic structures include folds, fractures, and faults.

Glasnost: Relaxation of control in the former USSR.

Good housekeeping: Routine facility maintenance to prevent emergencies and environmental degradation as well as to avoid environmental problems.

Granular activated carbon: Substance generally used to remove suspended and/or dissolved matter as well as tastes and odors in water. GAC, in many instances, can provide both filtration and absorption of contaminants.

Gravitational settling: An intermedia transfer mechanism in which particulate contaminants or contaminants adsorbed onto suspended particulates settle to surface media via gravitational attraction. The particulates are normally more than 20 μm in diameter (Chapter 5, U.S. EPA, 1988a).

Greenhouse gases: Gases that are highly transparent to incoming solar radiation but are relatively opaque to longwave radiation, similar (for descriptive purpose) to glass in a greenhouse (Chapter 14, Luther & Ellingson, 1985).

Groundwater: Water below the land surface in a zone of saturation.

Hazardous chemicals: Broad term used to include all the regulatory phrases that describe chemicals, such as hazardous substances (CERCLA), regulated substances (USTs), hazardous wastes (RCRA), hazardous materials (DOT), and toxic chemicals (SARA).

Hazardous Materials Transportation Act: A federal law that provides for the regulation of hazardous materials that are transported by air, water, rail, or highway.

Hazardous and Solid Waste Amendments (HSWA): A law that revised the RCRA statutes to include notification and technical provisions for underground storage tanks and established land disposal bans for certain hazardous wastes.

Hazardous materials: Broad term used to include all the regulatory phrases that describe chemicals, like hazardous substances (CERCLA), (UST), hazardous wastes (RCRA), hazardous materials (DOT), carcinogens (OSHA), and extremely hazardous substances (SARA).

Hazardous substance: Every hazardous substance or material designated under the following sections of environmental statutes: 311 of the Clean Water Act; 102 of CERCLA; or that is listed as hazardous characteristic under RCRA (unless regulation of that particular waste under RCRA was suspended by Congress).

Hazardous waste: Defined by the U.S. EPA under RCRA as a material that by definition is a *solid waste*—a solid, liquid, or gas that is a discarded material and is abandoned, recycled, or otherwise "inherently waste-like."

Under U.S. RCRA regulations, two distinct categories: (1) listed hazardous waste (i.e., listed in the regulations) and (2) characteristic hazardous waste (i.e., waste that is tested using agency testing methods for certain characteristics or properties). The Hazardous waste is (A) solid waste, or a combination of solid wastes which because of its quantity, concentration, or physical, chemical or infectious characteristics may (B) pose a substantial present or potential hazard ... when improperly treated, stored, transported, or disposed of, or otherwise managed [RCRA 1004(5), 42 U.S.C. 6903(5) (1976)].

Classified under Canadian TDG regulations as any of the following: any discarded goods listed in a schedule of dangerous goods; any industrial waste listed in a schedule; any waste mixtures or solutions, not fully specified, that have hazardous properties described by specified criteria.

Hazardous waste management units: Also known as solid waste management units, facilities such as surface impoundments, waste piles, land treatment areas, landfill cells, incinerators, tanks, and containment storage areas.

Healthy ecosystem: An ecosystem "having a high level of biodiversity, productivity, and habitability which lead naturally to the endpoint of diversity, productivity, and habitat preservation" (Chapter 12, Chapman 1991).

Heavy metals: Wastes generated from a variety of operations. Contaminants include arsenic, barium, chromium, cadmium, lead, mercury, silver, or selenium.

Hedonic pricing techniques: The price the public is willing to pay to either obtain or avoid a given environmental impact and how these are reflected in property values. For example, would electric rate payers absorb the large cost increases associated with underground line burial in order to avoid electromagnetic field exposures?

Hierarchical: A ladder of complexity with the higher ones more deserving than the lower ones.

High-level waste: Primary waste produced from the chemical reprocessing of spent fuel.

Hydraulic conductivity: A coefficient of proportionality between specific discharge and hydraulic gradient. The hydraulic conductivity is related to the permeability of an earth material for a given fluid at a given degree of saturation. Unless otherwise stated, it is assumed that the hydraulic conductivity is stated for a geologic material that is fully saturated with water. Determined from well-pumping tests. Compare with *Hydraulic gradient*.

Hydrologic cycle: The endless circulation of water among the ocean, atmosphere, and land.

Hydraulic gradient: The change in hydraulic head between two points divided by the distance between the points. Determined from a water level map. Compare with *Hydraulic conductivity*.

Hydraulic head: The product of fluid potential and the acceleration due to gravity of groundwater in an aquifer. Head is the sum of two components—the elevation head (elevation of the measuring point) and the pressure head (the incremental elevation due to the pressure exerted by the water at the point of measurement).

Hydrolytic dissipation: A process that uses nutrients, organics, and sunlight to accelerate the decomposition and evaporation of organics from the soil.

Hypersensitivity pneumonitis: Illness involving fever, cough, chest tightness, fatigue, and pulmonary infiltrates on a chest X-ray. These findings are also associated with changes in lung function related to a specific workplace exposure. Exposure sources include organic dusts; endotoxins; and aerosols from cooling towers that are contaminated with bacteria, molds, and fungi, such as *Alternaria aspergillus, Cladosporium, Fusarium,* and *Penicillium:*

Ignitable (I): A characteristic hazardous waste that exhibits qualities listed in 40 CFR 261.21, including:

- liquid with less than 24 percent alcohol by volume and a flashpoint less than 140 F (60 C)
- non liquid but is capable of causing fire through friction under standard pressure and tempeature; burns vigorously
- a compressed gas
- an oxidizer.
  EPA-approved test methods apply.

Ignitable wastes: Wastes including any flammable or combustible liquid or combustible solid.

Igneous rock: Rock formed by cooling of a magna or melt. Common igneous rocks include granite, basalt, and tuff.

Incompatibility: Refers to the mingling of chemicals and/or wastes which cause:

- corrosion or decay of containment materials
- production of heat, pressure, fire, explosion, violent reaction, toxic or flammable dusts, mists, fumes, or gases.

In-situ technologies: Treatment technologies that can be applied without large-scale removal of the contaminated material from its existing location.

Incidence rate: The number of new cases divided by the population at risk.

Individual-level endpoints: Endpoints that refer to stressor effects on single organisms, they are not usually considered as assessment endpoints.

Infections: Conditions resulting from myriad viruses, bacteria, fungi, and rickettsial organisms that cause human disease and can be transmitted within the indoor environment. These diseases include *Legionella,* tuberculosis, Pontiac fever, and *Coxiella burnetti* (Q fever).

Infiltration: Percolation of water through earth materials to the saturated zone of the groundwater system.

Injection wells: A well into which fluids are injected, primarily for disposal purposes.

Innovative environmental learning: An approach that emphasizes the error detection and correction process by searching for advantages that are inherently sustainable and that build distinctive environmental capabilities.

Instrumental value: Value attached to preserving plants and animals and other environmental entities because of their usefulness to human beings.

Interstitial monitoring: A leak-detection process in which instruments monitor the space between the UST and a double wall, partial barrier, or tank liner. With this approach it is possible to detect tank leaks before the contamination has spread to soil and/or groundwater.

Intrinsic value: Value attached to preserving plants and animals and other natural entities because they are believed to be valuable in and of themselves and not because of what they may add to human experience.

Ion exchange: An ion exchange process occurs when a dissolved chemical in the soil column substitutes itself for another chemical that is already adsorbed onto a mineral or soil surface. The primary ions affected by this process are cations (+ charge) and the measured factor in soil is known as the *cation exchange capacity* or *CEC.* The CEC of a soil will vary directly as a function of the amount and distribution of clay and organic matter present. The CEC has a significant impact on the uptake of chemicals into plants and can have potentially serious consequences for the home vegetable garden impacted by contaminated soils.

Judicial or administrative orders: Orders that result from litigation and typically direct that certain actions be taken.

Land-based point sources: An area toward which pollution control efforts are directed. Major pollutant sources include (1) logging, (2) agriculture, (3) dam construction and irrigation, (4) urbanization adjacent to coastal areas, i.e., partially treated or untreated sewage discharges, and (5) deposition of airborne pollutants such as nitrogen from power plants, industrial facilities, and agriculture.

Land disposal ban: The federal government has banned land disposal (placement of waste materials in a concrete vault, injection well, landfill, surface impoundment, or waste pile) for certain hazardous wastes that potentially migrate through soil and pollute groundwater.

Landfill: A disposal facility where waste is placed in or on land.

Landfilling: The final disposal of material that cannot be otherwise beneficially used, either because of the nature of the material or the local limitations to alternative approaches such as combustion.

Leachate: Any liquid filtered through or drained from waste.

Leach fields: A specifically designed series of perforated pipes placed in appropriate soil conditions for drainage onto which liquid wastes are applied.

Leadership control: Control based on past practices, precedence, and intuition and perception of the strategy or tactic leader.

Liabilities: Vulnerability to sources of harm.

Linearized multistage model: A model based on the theory that multiple events may be needed to yield tumor induction.

Liner: A continuous layer of materials lining a surface impoundment, landfill, landfill cell, or container.

Listed hazardous substances: The elements, compounds, and hazardous wastes appearing in Table 302.4 of the CERCLA reulations (40 CFR 302) are designated as hazardous substances. These substances have a specificed reportable quantity (see specific definition for RQ and tables contained in Section 3.0).

Losses: Actual harm sustained.

Low-temperature thermal desorption: The process of heating the soil to a temperature that boils the contaminants out of the soil.

Lowest observed adverse effect level: The lowest dose of a chemical in a study that produces statistically or biologically significant increases in the frequency or severity of adverse effects observed in the study between the exposed population and its appropriate control.

Lowest observed effect concentration: The lowest test concentration at which any adverse effect was observed. The geometric mean of these two values is an estimate of the threshold for chronic toxicity.

Low-level waste: Waste that is not spent fuel, high-level, or uranium or thorium mill tailings. Transuranic waste is definitely excluded from this category.

Magnetic fields: Fields produced by the motion of the electrical charge.

Major federal actions: A term defined by the Council on Environmental Quality (CEQ) as "actions with effects that may be major and which are potentially subject to federal control and responsibility." Includes adoption of rules, regulations, and policy; adoption of formal plans; and approval of specific projects.

Manifest: Shipping paper; to transport a shipment with an accompanying special shipping paper.

Matched pair: The original manifest plus the shipper's copy.

Maximum acceptable toxicant concentration: The chronic test results presented as two test concentrations: the no-observed-effects-concentration (NOEC) and the lowest-observed-effect-concentration (LOEC).

Measurement endpoints: Quantitative expressions of a measurable effect or characteristic designed to provide insight into potential impacts on the corresponding assessment endpoints.

Membrane processes: Reverse osmosis, ultrafiltration, electrolysis.

Mesosphere: The region of the atmosphere that lies above the stratopause.

Metamorphic rock: Rock formed from a preexisting rock or sediment by subjection to heat and/or pressure without complete melting. Metamorphic rocks include gneiss, schist, and marble.

Mineral: A naturally occurring solid with a defined crystal structure and a limited range of composition.

Mission: The highest level of achievement that is expected by the environmental management function that describes the ambitious, long-term strategic purpose of the function and is intended to guide decision-making and operating actions. The basic question that must be answered in order to determine an environmental management statement of mission is, "What is our business purpose?"

Mission statement: The ambitious long-term strategic purpose of the function and is intended to guide decision making and operating actions.

Mitigation options: Options defined by the National Academy of Sciences as those that "can take effect prior to the onset of climate change and slow its pace."

Mixed function oxidases: Oxidases that may alter xenobiotics (foreign compounds) to speed their excretion from the body.

Mixing efficiency: Factors such as the placement or movement of bookshelves, furniture, windows, and walls that change the pattern of air movement.

Mixing factor: The ratio of the amount of air required to dilute a contaminant to the ideal amount of air that should reduce it.

Molecular diffusion: The scattering of particles by random molecular motion, commonly characterized by Fick's law of diffusion.

Monitoring well: A petroleum UST leak-detection method in which a well is drilled in the vicinity of the UST and checked regularly to determine whether any product has leaked and is accumulating on or in the water table.

Moral: Concerning right and wrong.

MSDS: Material Safety Data Sheets required for certain industries by OSHA to warn workers about potential physical and health hazards of chemical products.

Multiple chemical sensitivity: A nonspecific malady that some individuals now state they suffer from because of years of exposure to a variety of synthetic chemicals.

Multiple comparisons procedures: Statistical procedures that control the overall false positive rate when several independent data sets are compared.

Multiple source: The combination of point and nonpoint sources, such as ones typically found in urban or industrial settings.

Negligence: Conduct that falls below the standard established by law for the protection of others against unreasonable risk of harm.

Neutralization: Process that is necessary in waste streams that have a wide range in pH that cannot be controlled through equalization. Waste streams that are either highly alkaline or acidic usually must be neutralized before other contaminants can be removed. This is particularly true of metal contaminants that remain dissolved in water only within limited pH ranges.

No observed adverse effect level: The dose at which there are no statistically or biologically significant increases in the frequency or severity of adverse effects observed in the study between the exposed population control populations. Effects are produced at this dose, but they are not considered adverse.

No observed effects concentration: The highest test concentration that caused no significant toxic effects.

No observed effect level: The dose at which there are no statistically or biologically significant increases in the frequency or severity of the effects observed in the study between exposed and control populations (i.e., effects of concern in the study are not produced at this dose).

Nonpoint Source: A discharge from an area rather than a specific point. An example of a nonpoint source discharge is contaminated stormwater leaving a facility via overland flow rather than through a specific pipe or point. *See also* Area Sources.

Odds ratio: The comparison of the number of exposed cases (a) × the number of nonexposed controls (d) ÷ by the number of exposed controls (b) × the number of nonexposed cases (c).

Offgassing: A well-known phenomenon of many materials used in building construction, interior spaces, furniture, and finishes. Chemicals used in bonding, covering, stabilizing, and insulating elements of the spaces we live and work in give off formaldehyde and VOCs.

On-site: The same property where the waste generator occurs; the property may be divided by public or private rights-of-way.

Opportunistic control: Control that focuses on the long-term strategy and structure of the environmental management function while remaining flexible enough to solve day-to-day problems and recognize new opportunities.

Ozone: A toxic and highly reactive gas that is used in industry as a disinfectant.

Ozone column: Total ozone in the stratosphere (that above a unit area of the earth's surface).

Paint wastes: Wastes generated by paint booths used in the sign and advertising display industry as well as in various wood finishing industries (including furniture) and sports firearm manufacture.

Pallets: Portable platforms used for handling, storing, and moving materials. These are used for drum storage.

Particle deposition: The settling of particles from the water body to the underlying bed.

Particle entrainment: The picking up or lifting or particles from the underlying bed of a water body by turbulent motion over the bed.

Particle settling: The sinking of particles having densities greater than the fluid of the water body, such as sediments or suspended solids.

Pasquill-Gifford dispersion coefficients: Numerical values of the standard deviations of atmospheric displacements about any point moving with the mean wind. The values are defined as a continuous, empirical function of downwind travel distance (or pollutant from a source) for each discrete stability class.

Pasquill-Turner atmospheric stability classification: The most often used inferential method of estimating the turbulent state of the atmosphere, in discrete classes, using solar intensity and wind speed as surrogate indicators.

Perched aquifer: Beds of clay, silt, or other materials of limited areal extent that present a restriction to flow of downward-moving water in the vadose zone may cause local areas of saturation above the regional water table. An unsaturated zone is present between the bottom of the perching bed and the water table. Compare with *Confined aquifer* and *Unconfined aquifer*.

Permeability: The ability of an earth material to transmit a fluid (usually taken to be water). Compare with *Porosity*.

Personal protective equipment: The set of equipment and clothing designed to protect workers from exposure to chemical, physical, or health hazards during specified activities; for example, respirator, neoprene gloves and boots, hard hat, and Tyvek suit.

pH: A measure of the acidity or alkalinity of any solution; pH 7 is neutral in a scale from 1 (acid) to 14 (alkaline)..

Phosphate phosphorus: A nutrient that will stimulate plant growth such as weeds and algae in receiving water bodies.

Physical in-situ methods: Methods that involve physical manipulation of the soils in place to immobilize or detoxify waste constituents. These technologies include in-situ heating, vitrification, and ground freezing (U.S. EPA, 1985).

Plating wastes: Wastes generated from electroplating operations that may be acidic or alkaline.

Plume height: The height of the release source. The physical height of the stack adjusted by factors that can either raise (buoyancy or momentum) or lower (downwash or deflection) the plume.

Point source: A specific emission point, such as an industrial stack associated with a refinery or other industrial process. This source can be defined in terms of height above ground, stack diameter, gas velocity, and temperature at the stack opening. A discharge from a specific point or conveyance such as a pipe, drain, conduit, or ditch that directs or funnels discharge water to surface water.

Pollutant: *See* Pollution.

Pollution: Contamination of the environment with synthetic material, heat, or other substances and conditions not native to that environment.

Pollution prevention: The reduction in volume and/ or toxicity of a waste prior to discharge.

Polymerization: The injection of a compound into the groundwater that transforms the groundwater into a gel-like nonmobile mass.

Population-level endpoints: Endpoints that can be used to determine risks to an entire population of a species.

Porosity: The percentage of an earth material that is open space. Compare with *Permeability*.

Potentiation chemical interaction: An effect that occurs when one chemical is nontoxic alone, but increases the toxicity of another chemical.

Potentiometric surface: The surface to which water in an aquifer would rise under hydrostatic pressure. For an unconfined aquifer, the potentiometric surface is the water table surface. For a confined aquifer, the potentiometric surface lies above the elevation of the aquifer in which the water is found.

Precipitation: An intermedia transfer mechanism that removes particulate and aerosol matter from the atmosphere. Particulate and aerosol matter serve as a nucleus for the condensation of raindrops. Raindrops generally remove particulates and aerosols greater than 1.0 μm in diameter from the air (Chapter 5, U.S. EPA, 1988).

Prevalence: The number of existing cases at a specified time divided by the size of the population under consideration.

Private nuisance: An unreasonable interference with another's use and enjoyment of his or her land or related personal or property interests.

Probabilistic risk assessment techniques: Techniques used to address a wide variety of process safety and reliability issues.

Problem formulation: The first phase of ecological risk assessment.

Product change: Modification of the end product to eliminate a waste.

Programs: Strategic operations.

Protocol: A typical procedure for completing an environmental audit.

Public nuisance: A nuisance that involves interference with a general public right.

Public water supplies: Supplies defined by SDWA as those that serve at least 25 people for at least 60 days per year.

Publicly owned treatment works (POTW): A system used to treat liquid industrial wastes owned by a state or municipality, including sewers if they convey wastewater to the system providing treatment.

RCRA: Resource Conservation and Recovery Act; a federal law that governs solid and hazardous waste management.

Reactive (R): A characteristic hazardous waste that exhibits various qualities listed in 40 CFR 261.23, including:
- reacts violently with water
- normally unstable and readily undergoes violent change without detonating
- forms potentially explosive mixtures with water
- is capable of explosion if subject to a strong initiating source
- EPS-approved test methods apply.

Reactive wastes: Wastes generated primarily by the photographic equipment and supplies industry. Specific wastes may include strong oxidizing agents, such as chromic acid, perchlorates, and permanganates used in metal finishing, and other reactive compounds, such as hypochlorites, peroxides, sulfides, nitrates, cyanides, and sodium hydroxide.

Real-time control: Control that deals with information systems that provide current information about the status of strategies and tactics.

Receptors of concern: Species that play key roles in ecosystem structure and/or function).

Recharge: The quantity or process of water being added to the saturated zone. Compare with *Discharge*.

Reclamation: Separating a particular component of a waste stream for reuse.

Recovery: The process of separating a particular component of a waste stream for reuse. The most common example is solvent recovery through distillation.

Recycling: The on-site or off-site use, with or without treatment, of a waste.

Reference concentrations: Values derived for inhalation exposure are called *reference concentrations (RfCs)* and are presented in units of milligrams per cubic meter (mg/m^3) (Chapter 5, Jarabek et al, 1989). The concentrations of chemical in air (in mg/m^3) that are not expected to cause adverse health effects in exposed populations assuming continuous chronic exposure.

Reference doses: Exposure limits for noncancer endpoints through the oral route of exposure. An estimate (with uncertainty spanning perhaps and order of magnitude or greater) of a daily exposure level for the human population, including sensitive subpopulations, that is likely to be without an appreciable risk of deleterious effects during a lifetime (Chapter 5, Barnes & Dourson, 1988, 480).

Refractory attack: A process in which sodium and potassium in the wastes can attack the brick lining of the thermal unit.

Regulate: The responsibility of federal and state agencies to govern the use and management of chemical products, and solid and hazardous wastes.

Regulated medical waste (RMW): Waste that is capable of causing disease in humans and that may pose a risk to individual and/or community health if not treated properly.

Regulated substance: The underground storage tank requirements define these as petroleum products, CERCLA hazardous substances, or RCRA hazardous wastes contained in USTs.

Regulations: Rules written by federal or state agencies that define procedures for carrying out environmental laws.

Relative risk: The ratio between the risk in exposed subjects and the risk in nonexposed subjects.

Regulatory framework: The set of federal, state, and local environmental programs that apply to the Western district location.

Release: Any spill, leaking, pumping, pouring, emitting, emptying, discharging, injecting, escaping, leaching, dumping, or disposing into the environment; this term has been expanded to include the gradual seepage or leaching of hazardous substances from storage areas or spills.

Reportable quantity: For any CERCLA hazardous substance, the reportable quantity established and identified in the CERCLA list. For extremely hazardous substances (EHSs) the reportable quantity is 1 lb (0.45 kg).

Retardation factor: A measure of the lack of movement of a chemical in the soil due to ion exchange and adsorptive processes. The retardation factor, Rf, is a commonly used term and is related to the distribution coefficient Kd: Rf = 1 + Kd × the bulk density of the solid (Ps) ÷ the total porosity of the soil. This equation is formally: Rf = 1 + Kd Ps/N.

Reuse: The use, reuse, or reclaiming of a waste. This includes using a waste as an ingredient or processing it to recover a useable product, for example, lead values from spent batteries and renovation of spent solvents.

Revealed preferences: The ability to infer value by observing the behavioral preferences exhibited by consumers. For example, what costs are the public willing to pay for different types of recreational facilities (swimming, fishing, hiking) located at similar distances?

Risk: Potential harm.

Risk characterization: The final phase of both HRA and ERA, in which the stressor-response and exposure profiles developed in the analysis phase are integrated to evaluate the likelihood of adverse ecological effects associated with exposure to a stressor(s).

Risk ratio: The ratio between the risk in exposed subjects and the risk in nonexposed subjects.

Rock: A natural aggregate of one or more minerals. Three major classes of rocks are recognized: sedimentary, igneous, and metamorphic.

Run-off: Any liquid that drains over land from a facility.

Run-on: Any liquid that drains over land onto a facility.

Safe Drinking Water ACT (SDWA): A federal law intended to provide for the safety of drinking water supplies throughout the nation by establishing and enforcing national drinking water quality standards.

Safety Kleen Parts Washer: Parts are washed in the maintenance shop in a Safety-Kleen parts washer. Spent shop solvents that accumulate in the parts washer are recycled by Safety Kleen.

SARA: Superfund Amendments and Reauthorization Act; strengthens Superfund and includes provisions for emergency planning by communities to respond to releases of certain hazardous substances.

Satori: A vivid esthetic experience of nature in Zen Buddhist spiritual beliefs.

Sanitary landfill: A landfill where municipal garbage and other nonhazardous solid wastes are disposed.

Saturated zone: The zone in earth materials in which the pore space is completely filled with liquid water. In the saturated zone, the pressure head is greater than atmospheric. Sometimes called the *phreatic zone*. Compare with *Soil zone* and *Vadose zone*.

Scale effects: The result of increases in the quantity of goods and services moving within countries and across borders. To the extent that increases in trade following liberalization lead to greater transportation needs, higher levels of manufacturing output, and general increases in the demand for raw and processed products, they can also impose greater wear and tear on natural ecosystems.

Sediment: Solid material—mineral or organic—that has been moved from its site of origin by air, water, or ice and has come to rest on the earth's surface. Sediments may be classified by grain size as well as by composition. Grain sizes for sediments, from coarse to fine, are gravel, sand, silt, and clay.

Sedimentary rock: Rock formed by accumulating sediments. Sedimentary rocks are characteristically stratified. They include chemical sedimentary rocks—such as limestone, gypsum, and salt deposits—as well as physical sedimentary rocks, such as sandstone and shale.

Sedimentation: Process is used to treat industrial wastewater when the contaminants are heavier than water and will drop to the bottom if the velocity of the waste stream is sufficiently slowed. Examples include waste streams that contain metal filings or large volumes of grit.

Sensitivity analysis: A type of analysis that defines the magnitude of the effect a given parameter has on model output and ranks each parameter according to its influence on model output (e.g., cancer risks).

Shear: Mixing due to variations in the fluid velocity at different positions in the water body. One example of this could occur in a lake where a significant decrease in temperature occurs with depth, thereby causing a thermal resistance (resistance of colder and, therefore, denser and lower-lying water to be displaced by warmer, lighter, and higher-lying water). A shear plane divides the surface current that follows the wind from the return currents that run counter to the wind (Chapter 5, Fair et al, 1968).

Site reconnaissance: Visual and physical inspection of a property.

Site safety and health plan: Required by OSHA in 29 CRF 1910.120 for hazardous waste sites; descirbes all potential hazards on-site and means for preventing employee exposure.

Small quantity generator: A generator that generates more than 220 lb (100 kg) and less than 2,205 lb (1,000 kg) (or about 25 to under 300 gal) of hazardous waste in a month and must comply with the hazardous waste management regulations. Under this category, hazardous waste may be accumulated on-site for up to 180 days (or up to 270 days if the waste has to be transported across a distance of 200 miles or more) without a permit, provided the waste accumulated on-site never exceeds 13,230 lb (6,000 kg).

Soil: Earth material that has been so acted upon by natural physical, chemical, and biological agents that it can support rooted plants. Whether a soil does, in fact, support plants depends on factors such as moisture content and chemical composition, as well as on its degree of weathering.

Soil column: The entire distance between the surface and bedrock.

Soil flushing: Any of a variety of techniques ranging from solvent flushing, ground leaching, and solution mining. The technique consists of injecting water or an aqueous solution into an area of contamination. Contaminants are mobilized into solution either because of their inherent solubility or the increased solubility associated with a chemical additive placed into the injected water.

Soil zone: Area in which evaporation and transpiration of water occurs. Compare with *Saturated zone* and *Vadose zone.*

Soil washing: An aqueous-based technology that uses mechanical processes to separate particles that contain contaminants.

Solid waste: Defined under U.S. RCRA as a solid, liquid, semisolid, or gas that is a discarded material and is abandoned, recycled, or otherwise "inherently waste-like." [U.S. RCRA, Section 1004(27)].

Solvent extraction: A process that uses an organic solvent to separate hazardous organic contaminants from oily wastes, soils, sludges and sediments, thereby reducing the volume of hazardous waste that must be treated.

Sorption: Transfer process in which dissolved chemicals in the groundwater become attached to sedimentary materials and/or organic matter. This process is described using the concept of partitioning.

Source: Equipment or activities that emit pollutants to the environment.

Source control: Modifications to the production process itself.

Source reduction: The handling of materials to reduce the volume of discarded materials.

Source segregation: The separation of listed hazardous wastes from nonhazardous wastes.

Specific discharge: The flow rate divided by the cross-sectional area across which the flow occurs.

Spent fuel: Nonreprocessed, irradiated nuclear fuel.

Spill assessment: The steps taken to identify and plan an appropriate response to a spill, and report spills to the regulatory authority.

Standard mortality ratio: The observed mortality rate divided by the expected rate. This relationship is frequently presented as SMR = obs ÷ exp.

STAR (stability array) data: Summaries of meteorological data, including seasonal or annual joint frequencies for each stability class, wind direction, and wind speed category. STAR data is available from the National Climatic Center (NCC), Asheville, North Carolina, for all National Weather Service (NWS) locations in the United States. STAR data from the NWS station most representative of the site should be used for modeling purposes.

Statutory and regulatory requirements: Directly imposed obligations upon facilities apart form permits, orders, and schedules.

Stock pile: A waste management unit; a noncontainerized accumulation of solid, nonflowing waste that is used for treatment or storage.

Storage: The quantity of water that is held in the pore space of an aquifer but that may be released (and begin to flow) when a hydraulic stress is applied to the aquifer.

Storage coefficient: For an unconfined aquifer, the specific yield, that is, the amount of water that drains from a soil due to the force of gravity.

Strategy: The process of determining and describing the long-term manner in which the environmental management function plans to contribute to the competitive performance of the firm, while determining ways to comply with regulatory demands.

Strategy formulation: The process of determining and describing the long-term manner in which the environmental management function plans to contribute to the regulatory compliance, environmental sustainability, and competitive performance standards of the firm.

Strategic and tactical control: Control that involves anticipating and minimizing potential deviations from intended outcomes.

Strategic intent: The envisioning of building and sustaining distinctive organizational competencies and capabilities for becoming recognized as world class (i.e., being better than every other company in the industry in at least one important aspect of environmental performance).

Strategic operations: The major efforts necessary to fulfill major purposes or resolve major problems, thus improving the potential for accomplishing the environmental management mission. Strategic operations (sometimes called *programs*) are countermeasure efforts that are designed and intended to prepare for risk(s), protect against danger(s), and preserve against environmental loss(es).

Strategies: The statements that represent the major portions of the strategic operations intended to guide decision-making and operating actions necessary to achieve the strategic operation goal. Strategies can also mean the actions needed to obtain and match available resources, under advantageous arrangements, and to improve the potential for achievement.

Stratosphere: The region of the atmosphere that is above the tropopause.

Stressor: Any chemical, physical, or biological entity or condition that has the potential to induce adverse effects on any *ecological component*, which might be individuals, populations, communities, or ecosystems (Chapter 12, Norton et al, 1992).

Strict liability: A type of liability that arises because some courts have decided that the handling of hazardous wastes is similar to the law of strict product liability, which applies to the manufacturers and distributors of defective products. Thus, there may be strict liability on those who generate or handle hazardous wastes for injuries or damages caused as a result of such activity.

Strong acid wastes: Wastes generated wherever any type of metal is formed or processed.

Strong alkaline wastes: Wastes generated from the manufacture of products with a pH >7.

Structural fit: A means of ensuring congruency between the strategic intent of the environmental management function, mission, strategy, and structure.

Subsurface flow system: The portion of the hydrologic cycle below the land surface.

Superfund: Comprehensive Environmental Response, Compensation, and Liability Act of 1980.

Surface impoundments: A facility formed primarily of earthen materials designed to hold liquid wastes that is not an injection well (holding, settling and aaeration pits, ponds, lagoons).

Surface water: Any body of water with its surface exposed to the atmosphere that flows into waters of the United States. U.S. waters are broadly defined to include lakes, rivers, estuaries, washes, bayous, creeks, conduits, irrigation ditches, and—most recently—wetlands.

Sustainable development: The kind of human activity that nourishes and fulfills the whole community of life on earth. The World Commission on Environment and Development defined the term as development that "meets the needs of the present without compromising the ability of future generations to meet their own needs" (WCED, 1987).

Synergistic chemical interaction: An effect that occurs when the combined effect of two or more chemicals is much greater than their individual effects when given alone.

Tactics: The statements that represent the most basic level of achievement and that are intended for use in implementing strategies. Tactics relate to small-scale actions that serve the strategy and include information that emphasizes the degree of attainment expected in terms of measurable quality, quantity, order, time, money, sequence, object, machine device, human factor, or behavior demanded for successful completion. Tactics can also mean the skillful and effective use of resources to solve a portion of some particular problem and contribute toward achievement of strategy.

Tank: A stationary device made primarily of no-nearthen materials, such as wood, concrete, steel, and plastic that is designed to hold hazardous waste.

Terrestrial ecosystems: Ecosystems classified according to the dominant vegetation types (primary producers). Examples are various kinds of forests, grasslands, and deserts.

Thermosphere: An outer region of the atmosphere.

Totally enclosed treatment facility: A facility treating hazardous waste, which is connected to the waste source to prevent release into the environment during treatment (an example is a pipe in which waste acid is neutralized).

Total quality management: A concept that can play a significant role in pollution prevention by optimizing the productivity and efficiency of an industry.

Total solids: The residue left in a drying dish after evaporation of a sample of wastewater and subsequent drying in an oven.

Total suspended solids: The residue retained on an ultrafine glass-fiber filter after a sample of wastewater has been passed through the filter.

Threshold dose: Some dose below which no adverse effect is observable.

Transporter: Any person engaged in the offsite transportation of hazardous waste by air, rail, highway, or water.

Transuranic waste: Waste that contains more than 100 nCi/g (nanocuries per gram) of a long-lived alpha-emitting transuranium radionuclides. High-level waste is not in this category and is separately defined. One curie (ci) is equal to the activity of 1 g of radium-226. A nanocurie is 10^{-9} Ci.

Treatment: The process(es) by which the characteristics of a waste are changed to reduce the material's volume and/or toxicity.

Trespass: Wrongful interference with another person's property rights. One whose pollution causes physical damage to another's property is liable for resulting damages in a trespass action.

Tropopause: The minimum temperature in the atmosphere, whose height varies with latitude and occurs at about 5 mi (8 km) near the poles and about 10 mi (16 km) at the tropics.

Troposphere: The region of the atmosphere below the tropopause.

TSCA: Toxic Substances Control Act; a federal law that applies primarily to releases of PCBs.

Turbulent diffusion: Scattering of particles by random turbulent motion (advective transport via turbulent motion in the form of eddies).

Ultrafiltration: A low-pressure membrane separation process.

Unconfined aquifer: An aquifer in which the top of the saturated zone (water table) is in direct contact with the atmosphere through the open pores of the earth material above. Also called a *water table aquifer.* Compare with *Confined aquifer* and *Perched aquifer.*

Underground injection: The disposal of fluids in a permitted well.

Underground storage tank (UST): Any tank or combination of tanks (including underground pipes connected thereto) that is used to contain an accumulation of regulated substances, and the volume of which is 10% or more beneath the surface of the ground.

Unlisted hazardous substances: RCA characteristic wastes that exhibit one or more of the I, C, R, and

E characteristics. These wastes are identified as "unlisted hazardous substances" in the CERCLA Table 302.4. I, C, and R characteristic wastes have an RQ of 100 pounds. E characteristic wastes have specific RQs depending on the constituent. The E characteristic waste RQ applies to the total quantity of waste not merely the toxic constituent.

Uptake: The movement of the contaminant across a biological boundary such as the intestinal mucosa, alveoli of the lung, or epidermis of the skin.

Vadose zone: The zone in earth materials in which the pore space is not completely filled with liquid water, although there ordinarily is moisture present. Sometimes called the *unsaturated zone:* The pressure head is less in the vadose zone than in the atmosphere. Compare with *Saturated zone* and *Soil zone.*

Vapor pressure: For a chemical, the dominant predictor of air space chemical concentration.

Verification: The validation of the process, as well as confirmation of specific predictions made on the basis of the risk assessment.

Vitrification: The process of converting contaminated soil to a durable glass-like material in which the wastes are crystallized.

Waldsterben: The German term for "forest death," explaining the adverse impacts of acid rain on forest ecosystems.

Waste: Defined under TDG regulations as "a product or substance intended for disposal."

Waste exchange: An administrative technique based on the premise that "one person's trash is another person's treasure." It is essentially a brokerage that handles wastes that another company could use as raw material. It then becomes identical to recovery/reuse.

Waste management: The use of a carefully planned program for handling wastes.

Wastestreams: Categories of wastes typically generated by a common source; for example, field wastes and acid wastewater.

Waste management units: A specific area or facility in which wastes are treated or stored, including surface impoundments, underground storage tanks and stock piles.

Wastewater: Water that has been changed chemically, biologically, and/or physically by human activity so that it is harmful to humans or natural systems.

Wastewater treatment unit: A tank that is part of a wastewater treatment facility that treats or stores hazardous wastewater.

Water table: The surface at which the water pressure in the pores of the porous geologic materials is exactly atmospheric. The elevation of the water table is identified by determining the elevation of water in a well that penetrates the vadose zone and is open to the top of the saturated zone.

Wetland ecosystems: Areas of transition between terrestrial and aquatic environments, e.g., bogs and salt marshes.

Worldview: A set of assumptions about what is real, what is true, and who can be trusted to know the facts about reality and truth.

Xenobiotics: Foreign compounds.

Zoning: For HVAC engineers, a process that defines which areas are to be served by an AHS.

Index